KEYS TO THE TREMATODA
VOLUME 1

KEYS TO THE TREMATODA VOLUME 1

Edited by

David I. Gibson, Arlene Jones and Rodney A. Bray

The Natural History Museum, London, UK

CABI *Publishing* and The Natural History Museum

CABI is a trading name of CAB International

CABI Head Office
Nosworthy Way
Wallingford
Oxfordshire OX10 8DE
UK

Tel: +44 (0)1491 832111
Fax: +44 (0)1491 833508
Email: cabi@cabi.org
Web site: www.cabi.org

CABI
745 Atlantic Avenue
8th Floor
Boston, MA 02111
USA

Tel: +1 617 682-9015
Email: cabi-nao@cabi.org

A catalogue record for this book is available from the British Library, London, UK

Library of Congress Cataloging-in-Publication Data
Keys to the Trematoda / edited by David I. Gibson, Arlene Jones, and Rodney Bray.
p. cm.
Includes bibliographical references (p.).
ISBN: 0-85199-547-0 (alk. paper)
1. Trematoda--Identification. 2. Trematoda--Classification. I. Gibson, David Ian. II. Jones, Arlene. III. Bray, Rodney Alan.

QL391.P7 K453 2001
592'.48—dc21

2001035389

ISBN-13: 978-0-85199-547-2

First published 2002
Transferred to print on demand 2017

Typeset by AMA DataSet Ltd, UK.
Printed and bound by CPI Group (UK) Ltd, Croydon, CR0 4YY

Contents

Contributors

Dr Rodney A. Bray, *Department of Zoology, The Natural History Museum, Cromwell Road, London SW7 5BD, UK*

Dr Tom H. Cribb, *Department of Microbiology and Parasitology, The University of Queensland, Brisbane 4072, Australia*

Dr Steve S. Curran, *Gulf Coast Research Laboratory, The University of Southern Mississippi, PO Box 7000, Ocean Springs, MS, 39566-7000, USA*

Professor Bernard Fried, *Department of Biology, Lafayette College, Easton, PA 18042-1778, USA*

Dr David I. Gibson, *Department of Zoology, The Natural History Museum, Cromwell Road, London SW7 5BD, UK*

Dr Arlene Jones, *Department of Zoology, The Natural History Museum, Cromwell Road, London SW7 5BD, UK*

Professor I. Kanev, *Institute of Experimental Pathology and Parasitology, Bulgarian Academy of Sciences, Bl. 25, Sofia 1113, Bulgaria.* Temporary address: *H.W. Manter Laboratory of Parasitology, University of Nebraska, Lincoln, NB 68588-0514, USA*

Dr Lotfi F. Khalil, *16 Marshalswick Lane, St Albans, Hertfordshire, AL1 4XG, UK*

Dr Katarzyna Niewiadomska, *W. Stefański Institute of Parasitology, Polish Academy of Sciences, Twarda 51/55, 00-818 Warsaw, Poland*

Professor Robin M. Overstreet, *Gulf Coast Research Laboratory, The University of Southern Mississippi, PO Box 7000, Ocean Springs, MS 39566-7000, USA*

Dr Thomas R. Platt, *Department of Biology, Saint Mary's College, Notre Dame, IN 46556, USA*

Dr Teresa Pojmańska, *W. Stefański Institute of Parasitology, Polish Academy of Sciences, Twarda 51/55, 00-818 Warsaw, Poland*

Dr Valentin Radev, *Institute of Experimental Pathology and Parasitology, Bulgarian Academy of Sciences, Bl. 25, Sofia 1113, Bulgaria*

Professor Klaus Rohde, *School of Biological Sciences, Division of Zoology, University of New England, Armidale, NSW 2351, Australia*
Dr Tomas Scholz, *Institute of Parasitology, Academy of Sciences of the Czech Republic, Branišovská 31, 370 05 České Budějovice, Czech Republic*
Dr John W. Smith, *29 Donmouth Road, Bridge of Don, Aberdeen AB23 8DR, UK*

Acknowledgements

Numerous individuals and organizations have helped generously with the production of this volume.

The editors are especially grateful to the contributors for the time and care which they have taken in the production of their chapters and for completing them on time.

Staff of the various museums and other institutes who have lent material to many of the contributors or have accommodated visiting contributors are gratefully acknowledged. These include: Mrs E.A. Harris (The Natural History Museum, London); Drs Ralph Lichtenfels, Eric Hoberg and Pat Pilitt from the United States National Parasite Collection (Beltsville, Maryland); Drs Jun Araki and the late Shunya Kamegai from the Meguro Parasitologica Museum (Tokyo, Japan); Drs Anna Kohn and Dely Noronha of the Instituto Oswaldo Cruz (Rio de Janeiro, Brazil); Prof. Jean-Lou Justine from the Muséum National d'Histoire Naturelle (Paris); and Dr Skip Sterner from the Harold W. Manter Laboratory of University of Nebraska State Museum (Lincoln, Nebraska).

K. Rohde also acknowledges Louise Streeting for help with the drawings, Maureen Heap for computing assistance and Dr David Gibson for some valuable advice.

R.M. Overstreet and S.S. Curran also thank: Dr Endre Willassen and Prof. Björn Berland, University of Bergen Museum of Zoology, and Li Qingkui, now of Tianjin Natural History Museum, People's Republic of China, for the loan of specimens, and Dr David Gibson for providing literature that was difficult to obtain. They also thank Dr I.M. Podvyaznaya and Prof. Oleg N. Bauer from the Zoological Institute, St Petersburg, Russia; Qui Zhaozhi of Nankai University, Prof. Tang Chongti of Xiamen University, Prof. Li Huizhu, Capital University of Medical Sciences, Beijing, and Dr Nie Pin of the Institute of Hydrology, Academia Sinica, People's Republic of China, for trying to help them to obtain specimens. They are grateful to Prof. Zhang Jian-ying and Ms Liu Lin of South China Normal University, Guangzhou, People's Republic of China, for their help with the Chinese literature and making useful translations. Maurice Dale Little and Antonio D'Alessandro of Tulane Medical

School, New Orleans, Louisiana, made the helminthological literature of the Faust Library and Paul Beaver Collection available to them. Kirsten Jensen from the University of Connecticut, Susan Plourde, Ervin Otvos, and Kim Overstreet from Ocean Springs, Mississippi translated literature. Their study was funded in part by the US Department of Commerce, National Marine Fisheries Service, US Gulf of Mexico Marine Stock Enhancement Program (US-GMSEP): Award Nos NA96FL0358 and NA06FL0501.

I. Kanev, V. Radev and B. Fried also thank library sources, which include the H.W. Manter Laboratory of Parasitology, at the University of Nebraska, Lincoln, Nebraska, and the Natural History Museums at Vienna, Austria, and Berlin, Germany. Specimens were also borrowed from collections at: the Natural History Museum, Vienna, Austria; the Natural History Museum, Berlin, Germany; the Natural History Museum, Geneva, Switzerland; the Museum for Central Africa, Tervuren, Belgium; the National Helminthological Collection, University of Mexico City, Mexico; the-Institute of Parasitology, Russian Academy of Sciences, Moscow, Russia; and the Institute of Experimental Pathology and Parasitology, Bulgarian Academy of Sciences, Sofia, Bulgaria. They would also like to thank Mrs Irina Petkova for her technical assistance and Mr Dimitar Vlaev for the drawings. Their work was supported in part by funds from the William C. Campbell Endowment Fund at the H.W. Manter Laboratory, University of Nebraska, and from the Kreider Professorship to B. Fried, Department of Biology, Lafayette College, Easton, Pennsylvania.

General Introduction

The Trematoda is the third group of animal parasitic helminths to be covered in a systematic series published by CAB *International*, following on from the *CIH Keys to the Nematode Parasites of Vertebrates*, edited by Anderson, Chabaud & Willmott (1974–1983) and *Keys to the Cestode Parasites of Vertebrates*, edited by Khalil, Jones & Bray (1994). The aim of this present work is to provide keys for the identification of the sexual adults of trematodes (normally found in vertebrate hosts) to help both specialist and non-specialist alike to identify material to the generic level. The work will appear in three volumes.

The first volume will comprise the Aspidogastrea and digenean groups belonging to the order Strigeida (including the blood flukes, brachylaemids, bucephalids, clinostomids, diplostomids, fellodistomids, gymnophallids, hemiuroids and strigeids: the Didymozoidae are excluded from this volume, as the chapter was not completed at the time of going to press, and will be included in a subsequent volume). The second volume will include the order Echinostomida (including the echinostomatids, fasciolids, notocotylids, paramphistomoids, pronocephalids and psilostomids) and part of the order Plagiorchiida (including the acanthocolpids, allocreadiids, lepocreadiids and opecoelids). The third volume will complete the Plagiorchiida (including the cryptogonimids, dicrocoeliids, heterophyids, lecithodendriids, microphallids, monorchiids, opisthorchiids, plagiorchiids and zoogonids).

Individual groups were tackled by international specialists selected by the editors. These contributors were provided with the basic classification of the Trematoda to be used in these volumes and a list of nominal genera in their groups taken from the Host–Parasite Catalogue and Database maintained in the Parasitic Worms Division, Department of Zoology, The Natural History Museum, London. Authors were also given general instructions for the preparation of their chapters but were not restricted in the way they approached their group unless it impinged on the work of other authors or upon the basic classification that we have used. Consequently, contributions do vary in, for example, the detail which authors consider necessary to diagnose taxa within their group and the style of the illustrations.

In the case of the Digenea (see Chapter 3), the basic unit to which we initially key down and the fundamental unit of the classification will be the superfamily. Each superfamily and family group is generally treated as a separate chapter. Preceding the keys, each chapter has an introduction where a brief history of the group and details of recent and/or present changes within it are given. Diagnoses are provided for genera and suprageneric groups; these are abbreviated in that features common to the members of a family group tend not to be repeated at lower levels in the classification. At least one figure for a representative of each genus considered valid is included.

As with any series of keys where the major groups are not usually readily recognized, the first key to the major subunits - here the superfamilies - is likely to be the weakest link in the chain. In the Digenea, in particular, homoplasy is rife and within-group variation is often considerable; users should recognize this point. Similarly, evidence for intergroup relationships based on the morphology of the sexual adult may be difficult to interpret, as progenesis has occurred in some groups and especially in those inhabiting the tissues. For example, blood fluke adults are probably homologous with the metacercarial stage of the related diplostomids, which also has a blood phase, and it is likely that transversotrematids have taken this one step further. The priority of these keys, however, is identification. Current knowledge, which in most groups is based on adult morphology, limits the extent to which phylogenetic relationships can be inferred.

One nomenclatural difference that readers may notice between this and some other works is that we have tried to follow the principle of coordination, as laid down in the International Code of Zoological Nomenclature (1999, Article 36), and have, therefore, treated family-group names as coordinate with the same authority. Hence, the authorities cited here will in some cases differ from those given in the existing literature.

The first two volumes will include general keys to the superfamilies included within them. The final volume will include a linking key to all groups. A cumulative bibliography will appear at the end of each volume and include only those references cited in the text; those used only as authorities for taxa are omitted. An index to family-group and generic names will be included at the end of Volumes 1 and 2, with a cumulative index completing Volume 3. A glossary of relevant specialized terminology will appear at the end of the third volume in order to avoid repetition, but readers should note that some variations in terminology do occur between groups.

The editors are grateful to the authors for their various contributions and to the people and organizations that have made available material and publications.

David I. Gibson, Arlene Jones and Rodney A. Bray
The Natural History Museum, London, UK

1 Class Trematoda Rudolphi, 1808

D.I. Gibson

Department of Zoology, The Natural History Museum, London, UK

Introduction

The Trematoda are a class of the phylum Platyhelminthes (flatworms), which are commonly referred to as flukes. The group comprises two subclasses, the Aspidogastrea and the Digenea. The former are a small but disparate group, composed of only a score of nominal genera parasitic in molluscs, fishes and chelonians. The members of the group are readily recognized by the nature of their attachment organ and the most recent classification was presented by Gibson & Chinabut (1984). The Digenea are a much larger group, comprising more than 2500 nominal genera, the vast majority of which use molluscs as primary hosts and vertebrates as final hosts. Since the middle of the 19th century, various attempts have been made to find useful criteria upon which an acceptable classification of this subclass could be based. Early attempts relying on sucker arrangements, i.e. *Monostoma* Zeder, 1800, *Distoma* Retzius, 1786, *Amphistoma* Rudolphi, 1801 and *Gasterostomum* von Siebold, 1848, initially at generic level and later at much higher levels, were unsatisfactory, as 'distomes' formed the majority of the group and the others, e.g. the 'monostomes' and 'amphistomes', were clearly not monophyletic. More modern versions of the classification, based on a greater variety of characters, have been presented by workers such as Poche (1926). Rather than use adult (i.e. sexual adult) morphology, some authors used features of cercarial morphology (e.g. La Rue, 1957; Cable, 1974) or asexual adult (daughter sporocyst/redia) morphology (Odening, 1961a), and others (e.g. Pearson, 1972) have suggested the use of life-history patterns. Consequently, there has been a wealth of, often conflicting, data from different life-history stages, which have complicated any speculation on the classification of and evolutionary relationships within the group. This has resulted in recent years in major works on the group (e.g. Yamaguti, 1971) essentially limiting their classification to the lower levels of the group. A cladistic analysis of the Digenea presented by Brooks *et al.* (1985, 1989), based on data from all life-history stages, included a classification at higher levels, but this has been heavily criticized by Gibson (1987) and Pearson (1992). Gibson (1987) discussed aspects of the

 Keys to the Trematoda, Vol. 1 (eds D.I. Gibson, A. Jones and R.A. Bray)

evolution of the Trematoda and, more recently, a study of the relationships within the group as indicated by combined morphological and molecular data has been presented by Cribb *et al.* (2001).

A simple classification of the Digenea, dividing the group into three orders, was suggested tentatively by Gibson & Bray (1994) and Gibson (1996). The three orders are: the Strigeida, with cercariae with a forked tail that tend to penetrate the next hosts; the Echinostomida, with cercariae that tend to encyst in the open; and the Plagiorchiida, with cercariae without a forked tail that tend to penetrate the next host. Although three orders are used in the organization of the present work, these characters, being both generalizations and also 'larval' features, cannot be used for keying the sexual adults from the definitive host. In this work, the Digenea are here keyed down directly to superfamily level using features of the sexual adult. The diagnoses of the Trematoda and Digenea presented below are essentially those of Gibson (1996).

CLASS TREMATODA RUDOLPHI, 1808

Diagnosis: Platyhelminthes: Neodermata. Permanent parasites generally of internal organs, primarily of molluscs, which harbour larval, asexual or occasionally sexual adult stages, and vertebrates, which harbour adult sexual stage. Molluscan host virtually always involved in life-history. Principally parasites of tissues in molluscs and gut of vertebrates, but may occur in wide variety of body-cavities, organs and tissues in vertebrates. Sexual adult usually with two organs of attachment (normally suckers or sucker-like structures), but occasionally only one and rarely none; one generally anterior and one ventral or posterior. Syncytial tegument armed with spines or not, bears tegumentary papillae or not. Gut virtually always present. Mouth usually within anterior attachment organ. Prepharynx linking mouth and pharynx present or absent. Muscular pharynx normally present. Intestine usually bifurcate, sometimes with single caecum, rarely absent; caeca sometimes branched or diverticulate, normally blind but occasionally open via anus(i) on body-wall or into excretory vesicle to form uroproct. Normally hermaphroditic, rarely dioecious. Testes one to many, commonly two. All or part of male terminal genitalia usually but not always enclosed in muscular sac; cirrus-sac if enclosing male duct only; sinus-sac if enclosing hermaphroditic duct (union of male and female ducts). Male terminal duct normally includes seminal vesicle, pars prostatica and ejaculatory duct. Genital atrium present or absent. Male and female terminal reproductive ducts usually open through common pore. Ovary normally single. Sperm storage apparatus (seminal receptacle) normally present; seminal disposal apparatus, commonly Laurer's canal, usually present. Mehlis' gland and oötype normally present. Eggs stored in uterus. Ciliated larva (miracidium) develops within egg. Vitellarium variable, exhibiting all forms between follicular and single compact mass. Excretory pore normally single in sexual adult, usually terminal; excretory vesicle normally saccular, I-, V- or Y-shaped; linked to system of tubules and flame-cells. Lymphatic system occasionally present.

Key to subclasses

1a. **Large ventral holdfast (attachment organ) covers most of ventral surface of body; holdfast subdivided by arrangement of loculi (called alveoli or rugae) or, rarely, into longitudinal row of separate suckers; in fishes and chelonians (rarely molluscs)** .... **Aspidogastrea Faust & Tang, 1936.** (Chapter 2)

1b. **Ventral attachment organ a sucker, occasionally absent, occasionally posteroventral; not normally covering more than half of ventral surface of body; not subdivided into alveoli; in all vertebrate groups (very rarely invertebrates)** .... **Digenea Carus, 1863.** (Chapter 3)

2 Subclass Aspidogastrea Faust & Tang, 1936

K. Rohde

School of Biological Sciences, Division of Zoology, University of New England, Armidale, New South Wales, Australia

Introduction

The class Aspidocotylea was established by Monticelli (1892a), replacing Aspidobothrii Burmeister, 1856 and Aspidobothria Monticelli, 1888. Faust & Tang (1936) introduced the name Aspidogastrea: this name is preferred here and in most recent works, on the basis of usage. Dollfus (1958) reviewed the Aspidogastrea as a subclass of the Trematoda, including redescriptions of some representative species. Yamaguti (1963) gave generic diagnoses and lists of species. Gibson & Chinabut (1984) distinguished two orders in the subclass Aspidogastrea, the Aspidogastrida with the single family Aspidogastridae and the subfamilies Aspidogastrinae, Cotylaspidinae and Rohdellinae, and the Stichocotylida with the families Stichocotylidae, Multicalycidae and Rugogastridae. Recent reviews of the group, including information on new taxa, host-specificity, geographical variation, life-cycles and phylogeny, were given by Rohde (1972, 1994). Relevant references can be found in these papers.

Aspidogastreans infect freshwater and marine molluscs and teleosts and freshwater turtles. Juveniles of one species, *Stichocotyle nephropis* Cunningham, 1884, have been found encapsulated in the intestinal wall of marine decapods, and worms resembling *Aspidogaster* Baer, 1827 have been found in the gill chamber of ascidians. The latter finding has yet to be verified (for references, see Rohde, 1972). Both the Aspidogastrea and the Digenea use molluscs as first hosts, which may be the original hosts of the trematodes (references in Rohde, 1972), although the nature of the invertebrate hosts of the Rugogastridae, Stichocotylidae and Multicalycidae is not known. In relation to the vertebrate hosts, the Digenea are almost entirely restricted to teleosts, amphibians, reptiles, birds and mammals, whereas the aspidogastreans are only known from chondrichthyans, teleosts and turtles. All of the species whose life-cycle is known use a mollusc as an obligate host and vertebrates as facultative or obligate final hosts (Rohde, 1972, 1994, and references therein). Nevertheless, Littlewood *et al.* (1999a), among others, have suggested that vertebrates are the original hosts of the trematodes, including the aspidogastreans. A postulated sequence of events in the evolution of the aspidogastrean groups was suggested by Gibson (1987).

Morphological (including ultrastructural) evidence and DNA studies have shown that the Aspidogastrea are the sister group of the Digenea (most recent papers, e.g. Littlewood *et al.*, 1999b; Litvaitis & Rohde, 1999). As such, they are a very ancient group. Although most species occur in teleosts, the greatest diversity at the family level and the most obvious differences between genera are found in species infecting chondrichthyans, suggesting that the original vertebrate hosts were chondrichthyans.

In works since Yamaguti's (1963) systematic treatment of the group there have been a number of changes. *Cotylogasteroides* Yamaguti, 1963 was synonymized with *Cotylogaster* Monticelli, 1892 by Fredericksen (1972), and Hendrix & Overstreet (1983) synonymized *Laterocotyle* Simpson & McGraw, 1979 and *Texanocotyle* Simpson & McGraw, 1979 with the same genus. Thoney & Burreson (1988) synonymized *Trigonostoma* Szidat, 1966, *Taeniocotyle* Stunkard, 1962 and *Macraspis* Olsson, 1868 with *Multicalyx* Faust & Tang, 1936 (*Macraspis* is pre-occupied by the beetle genus *Macraspis* MacLeay, 1819).

Pseudoaspidogaster and the subfamily Paraaspidogasterinae were established for a new taxon, *P. betwai*, from a freshwater fish in India by Agrawal & Sharma (1990) on the basis of a single row of alveoli on the ventral disc. The species is poorly described and the alveoli are not illustrated. It is likely to be a species of *Cotylaspis* Leidy, 1857 or *Aspidogaster*, as it has a cirrus-sac and a single testis characteristic of species of these two genera. The lack of several rows of alveoli may be due to the fact that the ventral disc was not yet fully developed; in *Aspidogaster conchicola* Baer, 1827, Williams (1942) observed that even relatively large specimens had discs not yet divided into alveoli.

Rugogaster hydrolagi Schell, 1973 resembles the Digenea in the presence of two (instead of one) caeca and a small ventral sucker-like structure, the anterior part of the larval posterior sucker that is retained by the adult (Rohde & Watson, 1992). However, studies using partial 28S ribosomal DNA (rDNA) have shown unambiguously that it must be included in the Aspidogastrea (see Litvaitis & Rohde, 1999).

SUBCLASS ASPIDOGASTREA FAUST & TANG, 1936

(Syns Aspidobothria Monticelli, 1888; Aspidocotylea Monticelli, 1892)

Diagnosis: Trematodes with single row of rugae, suckers or alveoli, or adhesive disc bearing three or four longitudinal rows of alveoli. Marginal bodies (ampullae storing secretion between marginal alveoli of adhesive disc) present or absent. Pharynx well developed; intestine with single caecum (two caeca in *Rugogaster* only), blind. Testes one, two or numerous, postovarian. Cirrus-sac present or absent; sinus-sac and hermaphroditic duct only in *Rohdella*. Common gonopore in forebody. Laurer's canal usually present, occasionally ending blindly. Uterus usually coiled in pre-ovarian field, sometimes also extending into postovarian field. Vitellarium follicular, in lateral fields, often connected posteriorly. Eggs operculate. Protonephridial system opens posteriorly. Adult parasitic in freshwater and marine molluscs, chondrichthyans, teleosts and freshwater turtles. Encapsulated juveniles of one species recorded from marine decapods. Known life-cycles involve molluscs and facultative or obligate vertebrate hosts; no parthenogenetic generations ('larval multiplication'). Larvae non-ciliated or with several ciliary patches, and with posterior sucker and excretory pores. Posterior appendage present or absent.

Key to families

1a. Rugae present on ventral attachment organ; intestine with two caeca; testes numerous .... Rugogastridae Schell, 1973.
1b. Rugae absent on ventral attachment organ; intestine with one caecum; testes one or two .... 2.

2a. Ventral attachment organ a single row of well-separated suckers .... Stichocotylidae Faust & Tang, 1936.
2b. Ventral attachment organ an adhesive disc consisting of alveoli separated by transverse septa .... 3.

3a. Adhesive disc a single row of deep alveoli .... Multicalycidae Gibson & Chinabut, 1984.
3b. Adhesive disc with three or four rows of alveoli .... Aspidogastridae Poche, 1907.

Family Aspidogastridae Poche, 1907

Diagnosis: Aspidogastrea with adhesive disc bearing three or four longitudinal rows of alveoli; marginal organs present. Intestine with single caecum. Hermaphroditic duct present or absent. One or two testes, postovarian, in mid- or more commonly in hindbody. Parasites as adults in molluscs, teleosts and turtles. Type-genus *Aspidogaster* Baer, 1827.

Key to subfamilies

1a. Hermaphroditic duct present .... Rohdellinae Gibson & Chinabut, 1984.
1b. Hermaphroditic duct absent .... 2.

2a. Adhesive disc bearing three longitudinal rows of alveoli .... Cotylaspidinae Chauhan, 1954.
2b. Adhesive disc bearing four longitudinal rows of alveoli .... Aspidogastrinae Poche, 1907.

Subfamily Aspidogastrinae Poche, 1907

Diagnosis: Adhesive disc bearing four longitudinal rows of alveoli. One or two testes. Cirrus-sac present or absent. Parasitic in molluscs, teleosts and occasionally freshwater turtles. Type-genus *Aspidogaster* Baer, 1827.

1a. Testes two .... ***Multicotyle*** **Dawes, 1941. (Fig. 2.3)**
Diagnosis: Testes two. Cirrus-sac present. Parasitic in molluscs and turtles. Type-species *M. purvisi* Dawes, 1941.
1b. Testis single .... **2.**

2a. Cirrus-sac present .... **3.**
2b. Cirrus-sac absent .... **4.**

3a. Head lobes present .... ***Lobatostoma*** **Eckmann, 1932. (Figs 2.4–2.6)**
Diagnosis: Head lobes present. Testis single. Cirrus-sac present. Parasitic in molluscs and teleosts. Type-species *L. ringens* (Linton, 1907).
3b. Head lobes absent .... ***Aspidogaster*** **Baer, 1827. (Figs 2.7, 2.8)**
(Syn. ?*Pseudoaspidogaster* Agrawal & Sharma, 1990)
Diagnosis: Head lobes absent. Testis single. Cirrus-sac present. Parasitic in molluscs, teleosts and occasionally freshwater turtles. Type-species *A. conchicola* Baer, 1827.

4a. Papillae on part of adhesive disc .... ***Lophotaspis*** **Looss, 1902. (Figs 2.9, 2.10)**
Diagnosis: Adhesive disc with numerous papillae. Testis single. Cirrus-sac absent. Parasitic in freshwater turtles. Type-species *L. vallei* (Stossich, 1899).
4b. Papillae on adhesive disc absent ....
.... ***Sychnocotyle*** **Ferguson, Cribb & Smales, 1999. (Figs 2.11, 2.12)**
Diagnosis: Papillae on adhesive disc absent. Testis single. Cirrus-sac absent. Parasitic in freshwater molluscs and turtles. Type-species *S. kholo* Ferguson, Cribb & Smales, 1999.

Subfamily Cotylaspidinae Chauhan, 1954

Diagnosis: Adhesive disc bearing three longitudinal rows of alveoli. Hermaphroditic duct absent. One or two testes. Cirrus-sac present or absent. Parasites of molluscs, teleosts and turtles. Type-genus *Cotylaspis* Leidy, 1857.

Key to genera

1a. Testes two .... ***Cotylogaster*** **Monticelli, 1892. (Figs 2.15, 2.17)**
(Syns *Cotylogasteroides* Yamaguti, 1968; *Texanocotyle* Simpson & McGraw, 1979; *Laterocotyle* Simpson & McGraw, 1979)
Diagnosis: Testes two. Cirrus-sac present. Parasitic in teleosts. Type-species *C. michaelis* Monticelli, 1892.
1b. Testis single .... **2.**

2a. Cirrus-sac present .... ***Cotylaspis*** **Leidy, 1857. (Fig. 2.16)**
(Syns *Platyaspis* Sinha, 1935; ?*Pseudoaspidogaster* Agrawal & Sharma, 1990)
Diagnosis: Testis single. Cirrus-sac present. Parasitic in molluscs and turtles. Type-species *C. insignis* Leidy, 1857.

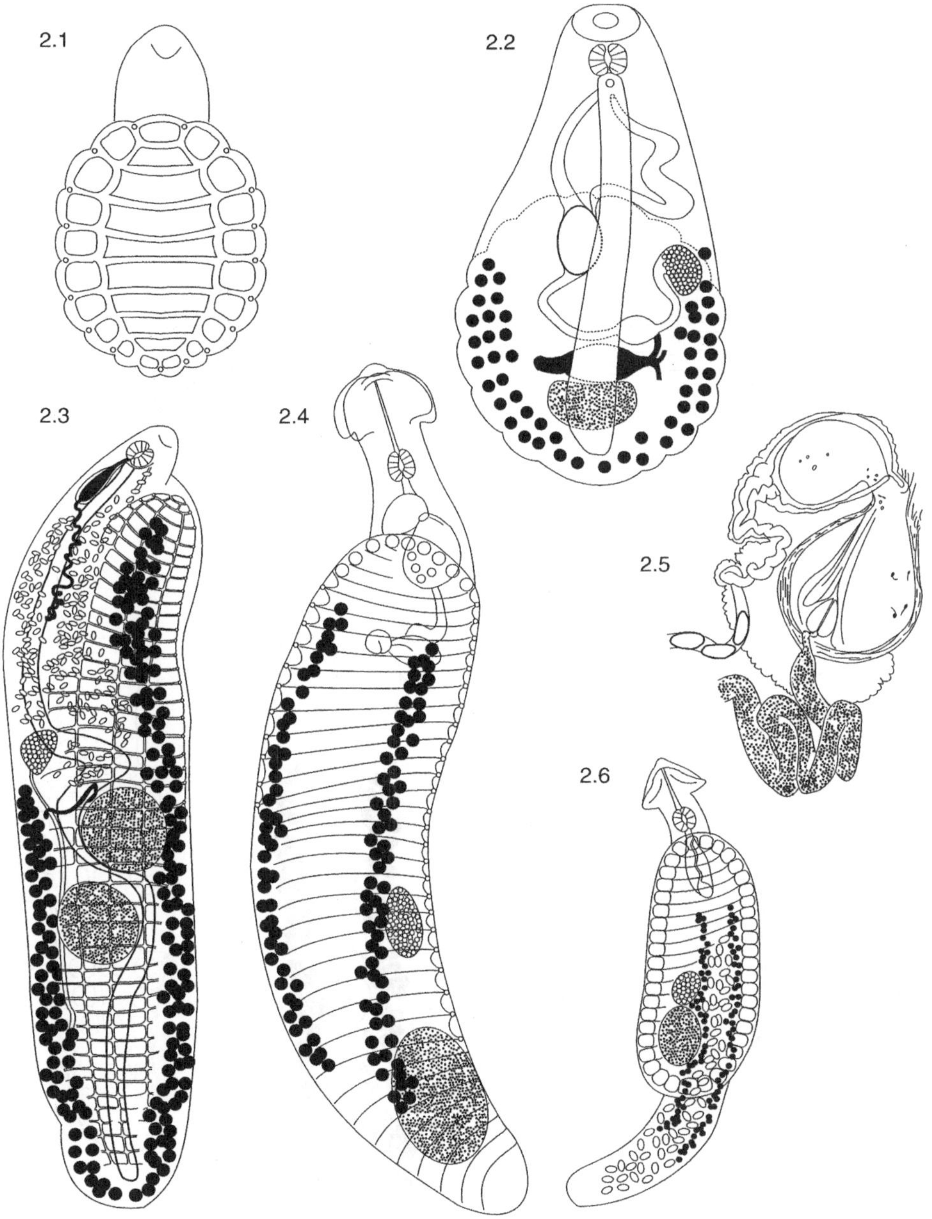

Figs 2.1–2.6.

2.1. *Lissemysia ocellata* Ramachandrula & Agarwal, 1984. (Modified after Ramachandrula & Agarwal, 1984.)

2.2. *Lissemysia ovata* Tandon, 1949. (Modified after Tandon, 1949.)

2.3. *Multicotyle purvisi* Dawes, 1941. (Modified after Rohde, 1971.)

2.4, 2.5. *Lobatostoma manteri* Rohde, 1973. 2.4. Whole-mount. 2.5. Terminal parts of reproductive ducts. (Modified after Rohde, 1973.)

2.6. *Lobatostoma platense* Mañe-Garzón & Spector, 1976. (Modified after Mañe-Garzón & Spector, 1976.)

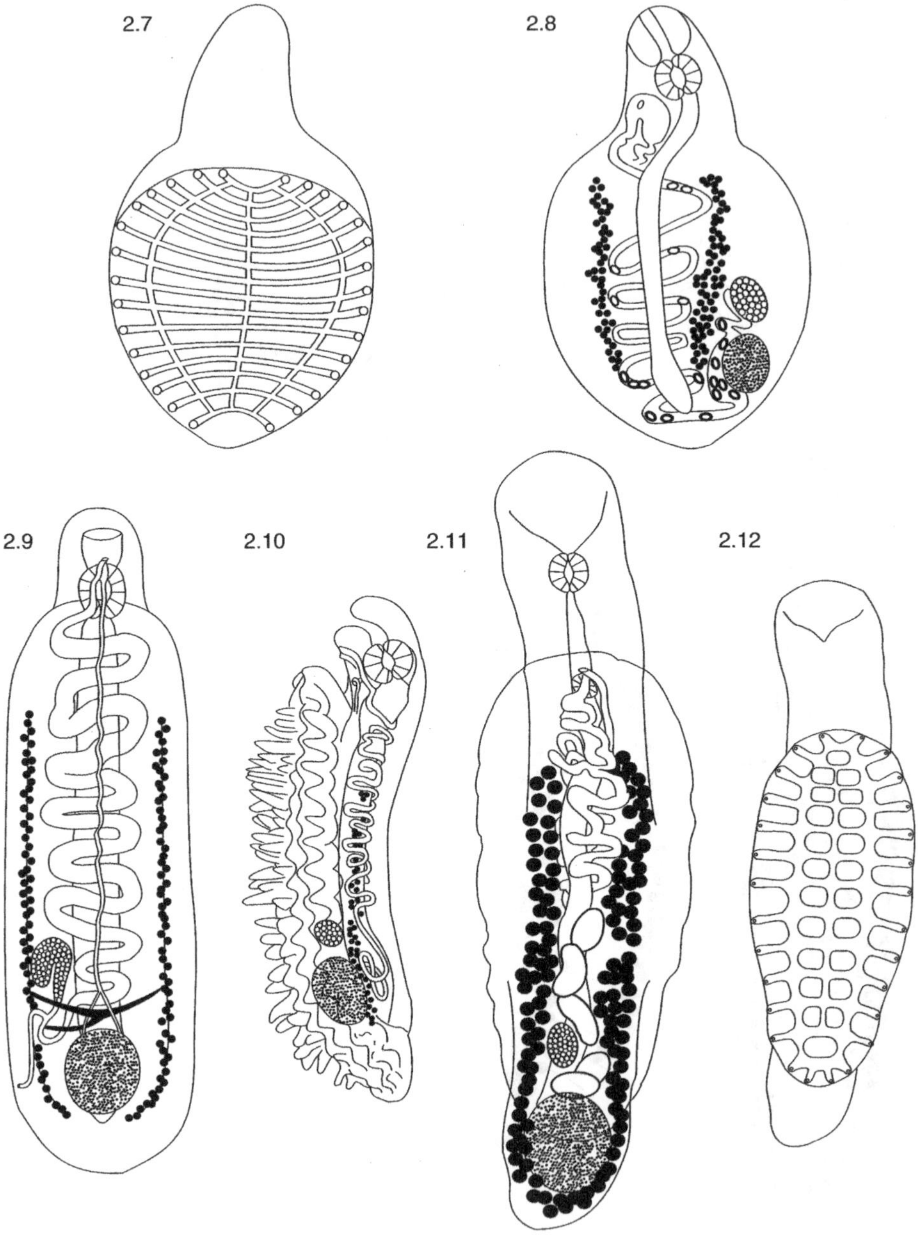

Figs 2.7–2.12.

2.7, 2.8. *Aspidogaster piscicola* Rawat, 1948. 2.7. Whole-mount showing ventral disc. 2.8. Whole-mount showing internal organs. (Modified after Yamaguti, 1963.)

2.9. *Lophotaspis vallei* (Stossich, 1899). Whole-mount showing internal organs. (Modified after Fuhrmann, 1928.)

2.10. *Lophotaspis interiora* Ward & Hopkins, 1931. Whole-mount, side-view showing details of ventral disc. (Modified after Yamaguti, 1963.)

2.11, 2.12. *Sychnocotyle kholo* Ferguson, Cribb & Smales, 1999. 2.11. Whole-mount showing internal organs. 2.12. Whole-mount showing ventral disc. (Modified after Ferguson *et al.*, 1999.)

2b. Cirrus-sac absent .... ***Lissemysia* Sinha, 1935. (Figs 2.1, 2.2)**
Diagnosis: Testis single. Cirrus-sac absent. Parasitic in molluscs and turtles. Type-species *L. indica* Sinha, 1935.

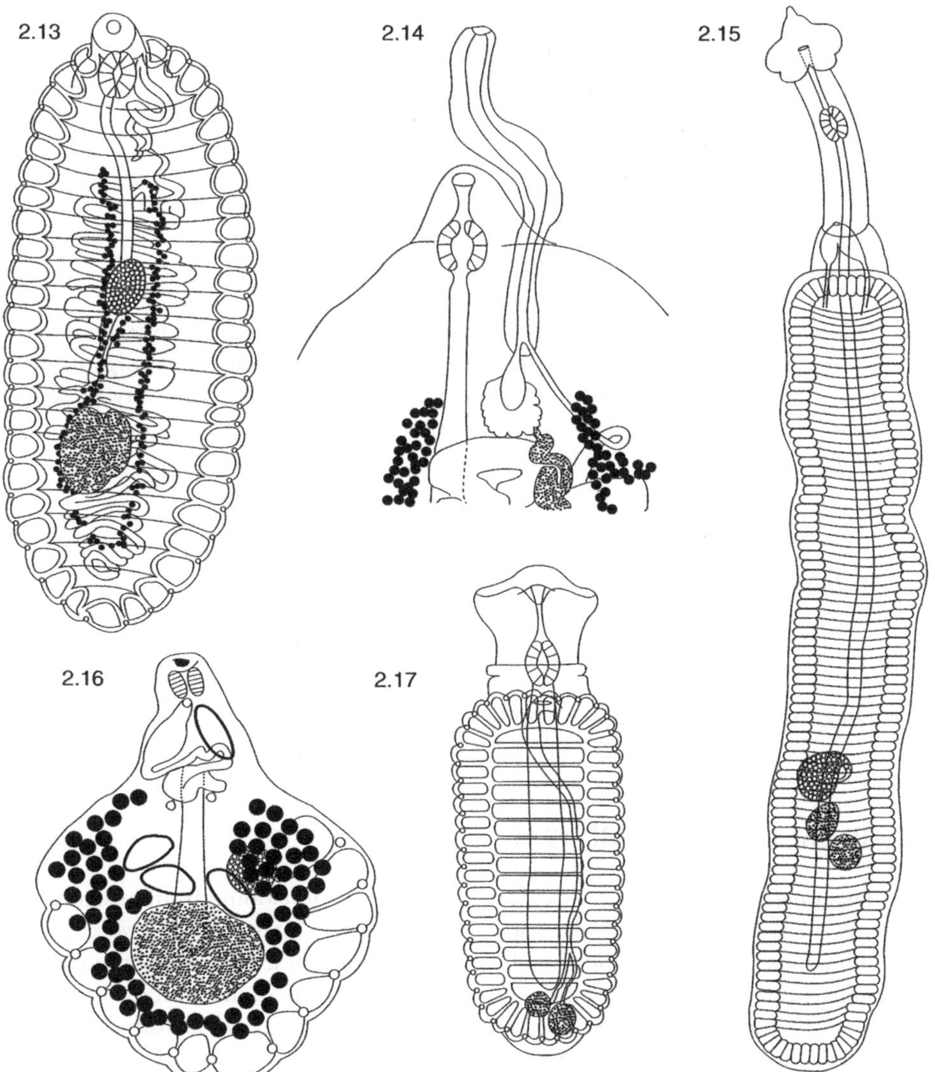

Figs 2.13–2.17.

2.13, 2.14. *Rohdella siamensis* Gibson & Chinabut, 1984. 2.13. Whole-mount. 2.14. Anterior end. (Modified after Gibson & Chinabut, 1984.)

2.15. *Cotylogaster occidentalis* Nickerson, 1902. (Modified after Fredericksen, 1972.)

2.16. *Cotylaspis sinensis* Faust & Tang, 1936. (Modified after Yamaguti, 1963.)

2.17. *Cotylogaster barrowi* (Huehner & Etges, 1972). (Modified after Huehner & Etges, 1972.)

Subfamily Rohdellinae Gibson & Chinabut, 1984

Diagnosis: Adhesive disc bearing three longitudinal rows of alveoli. Hermaphroditic duct present. Sinus-sac and sinus-organ present. Parasitic in freshwater teleosts. Type-genus *Rohdella* Gibson & Chinabut, 1984.

Genus *Rohdella* Gibson & Chinabut, 1984 (Figs 2.13, 2.14)

Diagnosis: With characters of subfamily. Type-species *R. siamensis* Gibson & Chinabut, 1954.

Family Multicalycidae Gibson & Chinabut, 1984

Diagnosis: Aspidogastrea with single row of deep alveoli separated by transverse septa; ventral foot-like appendages of septa (around openings of marginal organs) more or less well developed. Intestine with single caecum. Testis single. Ovary pretesticular, near posterior end of body. Parasites of holocephalans and elasmobranchs. Type-genus *Multicalyx* Faust & Tang, 1936.

Genus *Multicalyx* Faust & Tang, 1936 (Fig. 2.21)

(Syns *Macraspis* Olsson, 1868 (pre-occupied); *Taeniocotyle* Stunkard, 1962; *Trigonostoma* Szidat, 1966)

Diagnosis: With characters of family. Type-species *M. elegans* (Olsson, 1868).

Family Rugogastridae Schell, 1973

Diagnosis: Aspidogastrea with single row of rugae developing from posterior wall of anteriorly located, weakly developed ventral sucker. Intestine with two caeca. Testes numerous. Cirrus-sac and small genital atrium present. Common genital pore midventral near intestinal bifurcation. Ovary pretesticular, in forebody. Laurer's canal opens dorsally in forebody. Uterus fills much of body; eggs numerous. Vitellarium in lateral fields between level of ovary and ends of caeca. Parasitic in rectal glands of holocephalans. Type-genus *Rugogaster* Schell, 1973.

Genus *Rugogaster* Schell, 1973 (Figs 2.18, 2.19)

Diagnosis: With characters of family. Type-species *R. hydrolagi* Schell, 1973.

Family Stichocotylidae Faust & Tang, 1936

Diagnosis: Aspidogastrea with single row of well-separated suckers. Intestine with single caecum. Testes two. Ovary pretesticular. Vitellarium follicular, in lateral fields. Parasitic in elasmobranchs (encapsulated juveniles found in marine decapods). Type-genus *Stichocotyle* Cunningham, 1884.

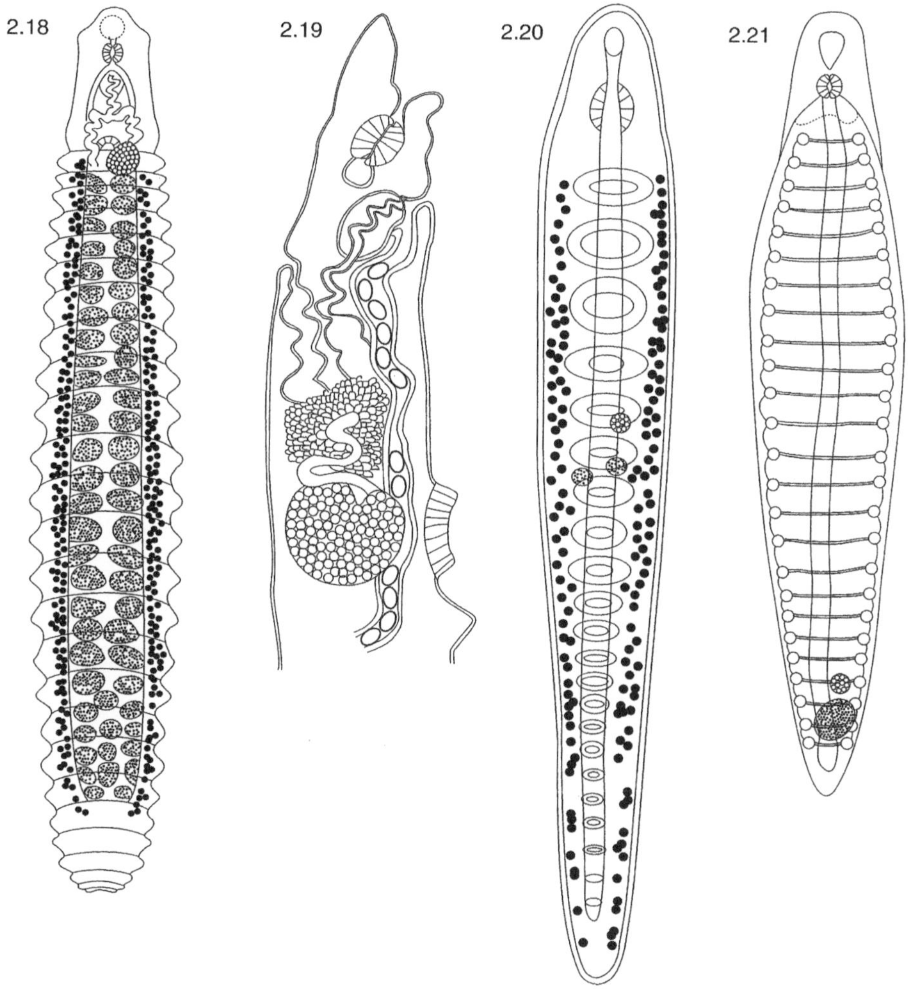

Figs 2.18–2.21.

2.18, 2.19. *Rugogaster hydrolagi* Schell, 1973. 2.18. Whole-mount. 2.19. Sagittal section through anterior end. (Modified after Schell, 1973.)

2.20. *Stichocotyle nephropis* Cunningham, 1884. (Modified after Nickerson, 1895.)

2.21. *Multicalyx elegans* (Olsson, 1868). Whole-mount. (Original.)

Genus *Stichocotyle* Cunningham, 1884 (Fig. 2.20)

Diagnosis: With characters of family. Type-species *S. nephropis* Cunningham, 1884.

3 Subclass Digenea Carus, 1863

D.I. GIBSON

Department of Zoology, The Natural History Museum, London, UK

Introduction

This group is discussed above under the Trematoda (Chapter 1).

SUBCLASS DIGENEA CARUS, 1863

Diagnosis: With alternation of generations; normally two asexual generations in mollusc and single sexual generation in vertebrate (rarely invertebrate). Normally hermaphroditic, occasionally partly or entirely dioecious. Generally small elongate-oval to tubular worms, but various other forms, including round and filamentous, occur. Tegument armed with spines or not. Usually with two muscular suckers, sometimes one, occasionally none. Mouth usually surrounded by oral sucker. Prepharynx present or absent. Muscular pharynx usually present. Oesophagus normally present. Intestine usually bifurcate, sometimes moniliform or saccular, rarely absent; caeca normally blind, occasionally unite, open to exterior through anus(i) or into excretory vesicle. Form of testes variable; one to many, often two. Cirrus-sac, occasionally sinus-sac, often envelops all or part of male terminal genitalia. Genital atrium present or absent. Common genital pore normally present; position variable, usually ventral on anterior body, occasionally in hindbody or even terminal. Ovary normally single, occasionally multiple; position and shape variable. Seminal receptacle present or absent; usually canalicular or uterine. Seminal disposal apparatus (e.g. Laurer's canal) present or absent. Uterus variable in size, form and distribution; terminal portion often forms muscular metraterm. Eggs normally oval and operculate, normally tanned, occasionally with spine or filament(s), occasionally fully embryonated. Vitellarium variable in shape and distribution, exhibiting all forms between follicular and single compact mass. Excretory pore single, usually terminal or subterminal, occasionally dorsal; vesicle saccular, I-, V- or Y-shaped. Parasitic as sexual adult in all vertebrate groups, rarely in invertebrates; usually present in gut or other body cavities, occasionally in blood-vessels or other tissues, rarely ectoparasitic under fish scales.

Key to the digenean superfamilies in this volume

This key is only pertinent to readers with adult specimens belonging to the order Strigeida. It is intended to direct them to the relevant superfamilies of this group only. Readers should be aware that specimens from other groups may key out here and that a key to all the superfamilies of the Digenea will appear in Volume 3 of this series. However, members of superfamilies appearing in Volumes 2 and 3 will tend to have a follicular vitellarium (usually), a cirrus-sac (usually), an I-shaped excretory vesicle (common) and a spiny tegument (common).

The following key will not work in 100% of cases, as there will always be one or two exceptions. However, we consider it stable enough to deal with the vast majority of relevant taxa.

1a. **Parasitic in alimentary canal and associated cavities and organs (body-cavity, gall-bladder, urinary system, mouth, nasal cavity, branchial cavity, lungs)** .... 4.
1b. **Not parasitic in alimentary canal and associated cavities and organs** .... 2.

2a. **Parasitic under scales of fishes in warmer waters** .... **Transversotrematoidea Witenberg, 1944.** (Chapter 53)
2b. **Parasitic in tissues of body** .... 3.

3a. **Parasitic in blood system (including heart, branchial, mesenteric and urinary bladder vessels, etc.)** .... **Schistosomatoidea Stiles & Hassall, 1898.** (Chapter 49)
3b. **Parasitic in body tissues (connective tissue, associated with muscles and other organs, eyes, ovary, branchial region, etc.) of marine (especially scombroid) fishes** .... **Hemiuroidea Looss, 1899 in part** (Didymozoidae Monticelli, 1888). (Not included in this volume)

4a. **Mouth well apart from anterior extremity; caecum single, elongate to saccular; ventral sucker absent; genital pore close to posterior extremity** .... **Bucephaloidea Poche, 1907.** (Chapter 15)
4b. **Mouth at or close to anterior extremity** .... 5.

5a. **Oral sucker present and ventral sucker usually present; mouth opens through oral sucker; parasitic in fishes and other vertebrate groups** .... 6.
5b. **Oral and ventral suckers absent; mouth opens through pharynx (though this does resemble oral sucker); parasitic in teleost fish (mainly tropical marine, occasionally freshwater)** .... **Bivesiculoidea Yamaguti, 1934.** (Chapter 5)

6a. Male and female terminal ducts unite to form hermaphroditic duct; prepharynx absent; testes normally pre-ovarian; excretory vesicle Y-shaped .... 7.

6b. Male and female ducts open separately into genital atrium or through common genital pore; prepharynx present or absent; testes pre- or postovarian; excretory vesicle I-, Y- or V- shaped .... 8.

7a. Hermaphroditic duct usually enclosed in sinus-sac; cirrus-sac absent; vitellarium variable, tubular to compact masses (usually two, occasionally seven or more, rarely one; follicular only in Ptychogonimidae); mainly in fishes, occasionally in amphibians and reptiles .... **Hemiuroidea Looss, 1899.** (Chapter 36)

7b. Hermaphroditic duct not enclosed in sinus-sac; cirrus-sac absent, but prostatic sac encloses seminal vesicle and pars prostatica; vitellarium follicular; testes postovarian (except in Leuceruthrinae); in freshwater and marine fishes .... **Azygioidea Lühe, 1909.** (Chapter 4)

8a. Cirrus-sac present or absent; prepharynx present or absent; vitellarium follicular; excretory vesicle I-, Y- or V-shaped; mainly in higher vertebrates .... 9.

8b. Cirrus-sac present; prepharynx absent; vitellarium tubular, branched; excretory vesicle Y-shaped; extremely large forms from marine teleosts .... **Hemiuroidea Looss, 1899 in part (Hirudinellidae Dollfus, 1932).** (Chapters 36, 42)

9a. Body with specialized holdfast organ situated posterior to ventral sucker; body often bipartite, with broad cup-shaped anterior region; genital pore at posterior extremity; body unspined .... **Diplostomoidea Poirier, 1886.** (Chapter 23)

9b. Body lacking specialized holdfast organ; body not normally distinctly bipartite; genital pore not usually at posterior extremity (but in posterior hindbody in Clinostomoidea and Brachylaimoidea – may be almost terminal); body may or may not be spined over at least part of its surface .... 10.

10a. Pharynx usually absent; genital pore posterior to middle of hindbody; adults in oesophagus or buccal cavity of piscivorous birds, occasionally reptiles, rarely amphibians and mammals .... **Clinostomoidea Lühe, 1901.** (Chapter 16)

10b. Pharynx present; genital pore in forebody (Gymnophalloidea), anterior to middle of body (Cyclocoeloidea) or posterior to middle of body (Brachylaimoidea); in various hosts and sites .... 11.

11a. Ventral sucker absent or at least poorly developed; oral sucker usually absent or at least poorly developed (occasionally present in Eucotylidae); body unarmed; caeca usually form cyclocoel; gonads close together near posterior extremity; usually in nasal sinuses, air sacs, trachea, abdominal cavities, orbit

and urinary tubules of birds (especially) and reptiles ....
.... **Cyclocoeloidea Stossich, 1902.** (Chapter 19)

11b. **Oral and ventral suckers present; body armed or not; caeca do not form cyclocoel; gonads may or may not be close together at posterior extremity; usually in gut of fishes, birds, mammals (occasionally reptiles and amphibians)** .... **12.**

12a. **Excretory vesicle with long stem and/or very long arms; genital pore anterior to ventral sucker; in fishes and birds (rarely mammals)** ....
.... **Gymnophalloidea Odhner, 1905.** (Chapter 30)

12b. **Excretory vesicle with short stem and short arms; genital pore posterior to ventral sucker; in birds and mammals (occasionally reptiles and amphibians)** .... **Brachylaimoidea Joyeux & Foley, 1930.** (Chapter 6)

4 Superfamily Azygioidea Lühe, 1909

D.I. Gibson

Department of Zoology, The Natural History Museum, London, UK

Introduction

The subfamily Azygiinae was erected by Lühe (1909) for *Azygia* Looss, 1899. It was first used at the family level by Odhner (1911b), who included most genera in the modern concept, i.e. *Azygia*, *Otodistomum* Stafford, 1904 and *Leuceruthrus* Marshall & Gilbert, 1905, but also the ptychogonimid *Ptychogonimus* Lühe, 1900. Earlier the same year, Goldberger had erected *Leuceruthrus* in a new subfamily, the Leuceruthrinae, and 'probably' a new family, the Leuceruthridae. The concept of the latter family was broadened by Travassos (1920) to include the Telorchiinae and the Azygiinae. Poche (1926) included the Azygiidae in a new superfamily, the 'Hemiurida'. More recently, Yamaguti (1953) recognized the Azygiidae to include all four genera listed below, but did not subdivide the family. Subsequently (Yamaguti 1958a, 1971), he split the group into three, the Azygiinae, the Leuceruthrinae and a new subfamily, the Proterometrinae. At the higher level, La Rue (1957) included the Azygiidae, along with the Bivesiculidae, in the superfamily Azygioidea and the suborder Azygiata, and Skrjabin & Guschanskaja (1958) included the Azygiidae, along with the Hirudinellidae, Liocercidae and Xenoperidae, in the Azygioidea and the suborder Azygiata, although they had previously (Skrjabin & Guschanskaja, 1956) included these groups in the suborder Hemiurata. In 1979, Gibson & Bray, in a detailed revision of the Hemiuroidea, included the Azygiidae within the latter superfamily, recognizing two subfamilies, the Azygiinae (syn. Proterometrinae) and Leuceruthrinae; numerous other families and subfamilies erected by various authors were considered synonyms (these are included below). In their cladistic analysis of the Digenea, Brooks *et al.* (1985) included the Azygiidae within their order Hemiuriformes, a group similar in conception to Gibson & Bray's (1979) Hemiuroidea, apart from their inclusion of the Bivesiculidae. Subsequently, Brooks *et al.* (1989), following criticism by Gibson (1987), hived off the Bivesiculidae into its own superfamily within the Hemiuriformes. Recently, the preliminary molecular studies of Blair *et al.* (1998) have indicated that the Azygiidae fall outside the hemiuroid clade. In the present work, therefore, the hemiuroid concept of Gibson & Bray (1979) is modified in that the Azygiidae, although still recognized as being closely related to the Hemiuroidea, are recognized at the superfamily level. Otherwise, the basic classification and terminology used here follows that of Gibson & Bray (1979).

The species of *Proterometra* described by Wu *et al.* (1997) from Chinese marine fishes are clearly not azygiids and are likely to be lepocreadiids.

Bundelatrema Agrawal, 1980, reported by Agarwal & Agrawal (1982) as an azygiid, is treated here as a *nomen nudum*. Since Agarwal & Agrawal (1982) refer to both species of the genus having a prepharynx, it is unlikely to be a hemiuroid or azygioid taxon.

Superfamily Azygioidea Lühe, 1909

Diagnosis: Body large or small, usually elongate. Ecsoma absent. Body surface smooth, without spines or plications. Oral and ventral suckers well developed; latter in middle or anterior half of body. Prepharynx absent. Pharynx well developed. Oesophagus usually short. Drüsenmagen absent. Gut-caeca terminate blindly close to posterior extremity. Testes two, in tandem, oblique or symmetrical, pre- or postovarian in hindbody. Seminal vesicle tubular, usually short, thin-walled, convoluted in forebody. Pars prostatica tubular. Prostatic sac present, surrounding pars prostatica and seminal vesicle only. Ejaculatory duct usually long and convoluted, but of variable length. Hermaphroditic duct short; at distal extremity of sinus-organ. Permanent sinus-organ variable in length, usually conical. Sinus-sac absent. Genital atrium usually well developed, variable in size. Genital pore mid-ventral in forebody. Ovary oval; pre- or post-testicular. Mehlis' gland pre-ovarian. Laurer's canal and uterine seminal receptacle present. Juel's organ and canalicular or blind seminal receptacle absent. Uterus entirely or almost entirely pre-ovarian, coiled mainly in hindbody. Eggs numerous, small, non-filamented. Vitellarium follicular, usually present laterally throughout much of hindbody, occasionally extending into forebody. Excretory vesicle Y-shaped; arms united in forebody or not. Parasitic in stomach or body-cavity of elasmobranchs and in stomach of freshwater teleosts and holosteans. Type-genus *Azygia* Lühe, 1900.

Family Azygiidae Lühe, 1909

(Syns Leuceruthridae Goldberger, 1911; Aphanhysteridae Guiart, 1938)

Diagnosis: With characters of superfamily. Type-genus *Azygia* Looss, 1899.

Key to subfamilies

1a. Testes postovarian .... Azygiinae Lühe, 1909.
1b. Testes pre-ovarian .... Leuceruthrinae Goldberger, 1911.

Subfamily Azygiinae Lühe, 1909

(Syns Aphanhysterinae Guiart, 1938; Gomtiotrematinae Gupta, 1955; Allogomtiotrematinae Yamaguti, 1958; Proterometrinae Yamaguti, 1958)

Diagnosis: Body normally large, occasionally small. Ventral sucker larger or smaller than oral sucker, in middle or anterior half of body. Testes in tandem, oblique or symmetrical, postovarian. Uterus entirely pre-testicular. Vitelline field may extend into forebody. Excretory arms may or may not unite in forebody. Parasitic in stomach or body-cavity of elasmobranchs and stomach or oesophagus of freshwater teleosts and holosteans. Type-genus *Azygia* Looss, 1899.

Key to genera

1a. **Testes symmetrical, near posterior extremity; vitelline follicles and uterine coils extending into forebody; in freshwater teleosts (North America) .... *Proterometra* Horsfall, 1933. (Fig. 4.1)**

Diagnosis: Body oval, small. Oral sucker large; ventral sucker small, situated at or just posterior to middle of body. Testes symmetrical, at posterior extremity. Sinus-organ small, conical. Uterine field extends from level of ovary into forebody. Vitellarium extends from level of testes or ovary anteriorly well into forebody, in lateral fields. Excretory arms united in forebody. Parasitic in gut of freshwater teleosts (in North America). Type-species *P. macrostoma* (Faust, 1918).

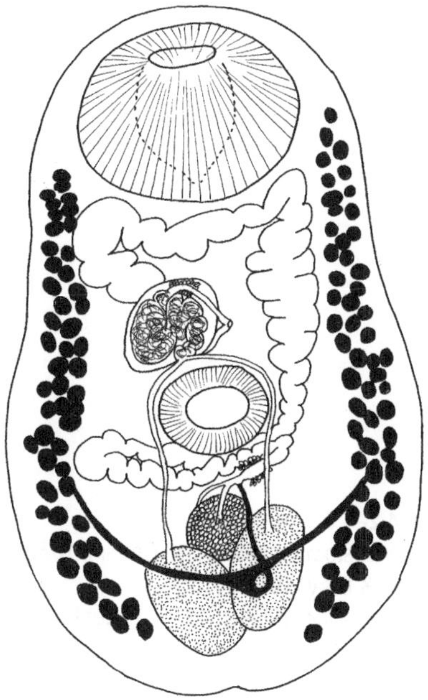

Fig. 4.1. *Proterometra macrostomum* (Faust, 1918). (After Horsfall, 1934.)

1b. **Testes tandem, oblique or occasionally symmetrical, usually well anterior to posterior extremity; vitelline follicles and uterine coils entirely or almost entirely confined to hindbody** .... **2.**

2a. **Vitelline follicles confluent posterior to testes; ventral sucker normally larger than oral sucker; parasitic in elasmobranchs** .... ***Otodistomum* Stafford, 1904. (Fig. 4.2)**

(Syns *Xenodistomum* Stafford, 1904; *Josstaffordia* Odhner, 1911[1]; *Aphanhystera* Guiart, 1938)

Diagnosis: Body large, spatulate to elongate. Ventral sucker larger than oral sucker, close to anterior extremity. Testes tandem or slightly oblique. Sinus-organ capable of considerable extension or contraction to form small papilla. Uterine field almost entirely between ovary and ventral sucker. Vitelline follicles extend in lateral fields posterior to ventral sucker, reaching back to post-testicular region, where fields are confluent. Excretory arms usually unite in forebody, but occasionally do not. Parasitic in stomach or body-cavity of elasmobranchs (sharks, rays and chimaeras). Type-species *O. veliporum* (Creplin, 1837).

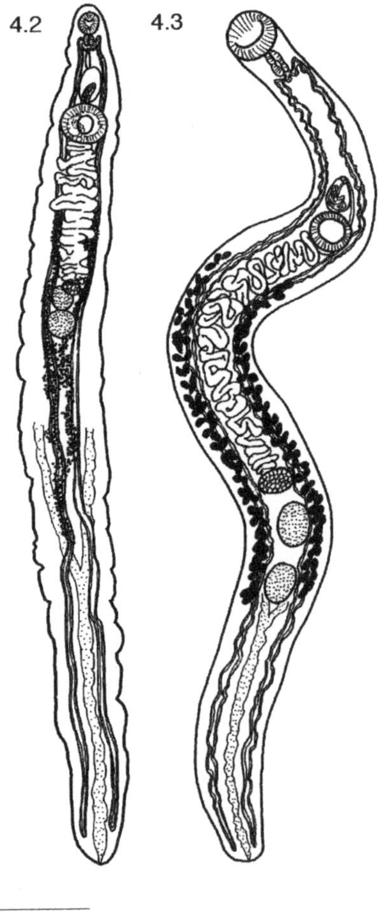

Figs 4.2–4.3.
4.2. *Otodistomum veliporum* (Creplin, 1837). (After Gibson & Bray, 1977.)
4.3. *Azygia lucii* (Müller, 1776). (After Looss, 1894.)

[1] Erected as a joke – see Gibson & Bray (1979).

2b. Vitelline follicles not confluent posterior to testes; oral sucker normally larger than ventral sucker; parasitic in freshwater teleosts and holosteans .... *Azygia* Looss, 1899. (Fig. 4.3)

(Syns *Megadistomum* Stafford, 1904; *Mimodistomum* Stafford, 1904; *Hassallius* Goldberger, 1911; *Eurostomum* MacCallum, 1921; *Gomtiotrema* Gupta, 1955, *nec* Sinha, 1934; *Allogomtiotrema* Yamaguti, 1958; *Pseudazygia* Yamaguti, 1971)

Diagnosis: Body medium to large, usually elongate, occasionally oval. Ventral sucker smaller than oral sucker, in anterior half of body. Testes tandem, occasionally somewhat symmetrical; anterior testis occasionally lateral to ovary. Sinus-organ small, papillate. Uterine field between ovary and ventral sucker. Vitelline follicles confined to hindbody, not confluent posterior to testes. Excretory arms apparently not united in forebody. Parasitic in stomach and intestine of freshwater teleosts and holosteans (mainly Holarctic). Type-species *A. lucii* (Müller, 1776).

Subfamily Leuceruthrinae Goldberger, 1911

Diagnosis: Body medium to large, elongate-oval. Ventral sucker smaller than oral sucker, near middle of body. Testes oblique, pre-ovarian, immediately posterior to ventral sucker. Prostatic sac small. Sinus-organ small, but well defined. Uterine field between ovary and ventral sucker, passing between testes. Vitelline follicles in lateral fields, extending almost throughout hindbody. Excretory arms unite in forebody. Parasitic in gut of freshwater teleosts and holosteans (in North America). Type-genus *Leuceruthrus* Marshall & Gilbert, 1905.

Genus *Leuceruthrus* Marshall & Gilbert, 1905 (Fig. 4.4)

Diagnosis: With characters of subfamily. Type-species *L. micropteri* Marshall & Gilbert, 1905.

4.4

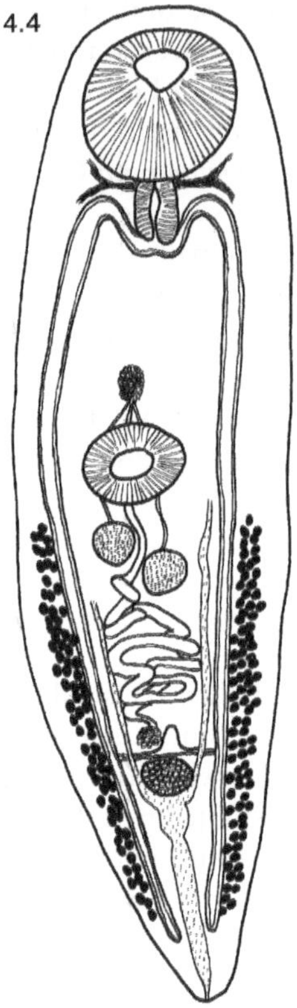

Fig. 4.4. *Leuceruthrus micropteri* Marshall & Gilbert, 1905. (After Marshall & Gilbert, 1905.)

5 Superfamily Bivesiculoidea Yamaguti, 1934

T.H. Cribb

Department of Microbiology and Parasitology, The University of Queensland, Brisbane, Australia

Introduction

The Bivesiculidae Yamaguti, 1934 are a small family of fewer than 30 species of trematodes found in the intestines of marine and freshwater fishes. The family group was originally proposed (Yamaguti, 1934) as a subfamily of the Monorchiidae Odhner, 1911 and later raised to full family status by Yamaguti (1939). The family is easily recognized by the complete lack of ventral and oral suckers and the V-shaped excretory vesicle. The digestive system has just one muscular structure, which is now usually (but not universally) interpreted as a pharynx (see Cribb *et al.*, 1994). Complete absence of oral and ventral suckers is possibly paralleled in the Angiodictyidae and the Notocotylidae (and related families). Among these families, the Bivesiculidae is distinctive in its V-shaped excretory vesicle and its life-cycle. Known life-cycles of bivesiculids have asexual stages in gastropods and distinctive furcocystocercous cercariae, which are eaten directly by the definitive host (e.g. Le Zotte, 1954; Cable & Nahhas, 1962; Pearson, 1968). The typical life-cycle thus has only two hosts. In some cases, it appears that a second intermediate host may be interpolated in the life-cycle (Cribb *et al.*, 1998); whether this is facultative or obligate is unknown. Despite being linked with the hemiuroids by Brooks *et al.* (1985), this family has no demonstrated close relationship to any other family, a view that is supported by recent molecular studies (Blair *et al.*, 1998). It thus appears to warrant its own superfamily, although the superfamily name does not seem to have been used previously.

Classification within the group is very simple. Only one subfamily other than the type-subfamily has been proposed, i.e. the Paucivitellosinae by Yamaguti (1971). Whereas the concept of this subfamily is perhaps supportable, it appears to have no real value as it is monotypic. The family Treptodemidae was proposed for *Treptodemus* Manter, 1961 by Yamaguti (1971), but this action has little merit and has received little support, although Shen (1995) continued to recognize the family for his new genus *Treptodemoides* Shen, 1995. These genera do not differ from other bivesiculid genera in

(eds D.I. Gibson, A. Jones and R.A. Bray)

any way that merits family-level separation. The view is taken here that all the genera are best included in Bivesiculidae without the recognition of subfamilies.

Five genera have been proposed. Of these, *Bivesicula* Yamaguti, 1934, the type-genus, is the largest. It is easily recognized and, in fact, encompasses sufficient variation for it to be eventually divided further. The other genera are all recognized here with differing levels of confidence. *Paucivitellosus* Coil, Reid & Kuntz, 1965 is easily distinguished because it has a highly reduced vitellarium and eggs that embryonate *in utero* (Coil *et al.*, 1965; Pearson, 1968). *Treptodemus* Manter, 1961 is easily recognized because it has a transversely elongate body and differences in the arrangement of the reproductive system as a result (Manter, 1961). *Bivesiculoides* Yamaguti, 1938 is distinguished from *Bivesicula* on the basis that the uterus does not pass posteriorly to the ovary. The type-species of *Bivesiculoides*, *B. atherinae* Yamaguti, 1938, has a strikingly pear-shaped pharynx, a condition shared with *B. fusiformis* Cribb, Bray & Barker, 1994 and *B. posterotestis* Durio & Manter, 1968 but not with *B. otagoensis* Manter, 1954 or *B. scari* Hafeezullah, 1971. This genus thus unites several species that may not strictly be closely related but appears a useful concept for the present. *Treptodemoides* Shen, 1995 was erected for *T. xinyingensis* Shen, 1995. It is concluded here that this species is a junior synonym of *Bivesiculoides fukenensis* Liu, 1995. Both species were described in 1995 from hemiramphid fishes from Chinese waters, but *B. fukenensis* was described in March, whereas *T. xinyingensis* was described in October (Liu, 1995; Shen, 1995). In erecting *Treptodemoides*, Shen (1995) made no mention of genera other than *Treptodemus*, which he continued to recognize within the Treptodemidae. The single species of *Treptodemoides*, here recombined as *Treptodemoides fukenensis* n. comb., is broadly similar to *Bivesicula*, but differs in that the uterus does not pass posterior to the testis. It differs from *Bivesiculoides* in being broader and having the ovary opposite the testis rather than anterior to it. The single species of *Treptodemus* and *Treptodemoides* are the only bivesiculids known from hemiramphid fishes, a distribution consistent with a relationship between these two genera.

In the light of the distinctions discussed above and the likelihood that, ultimately, more, not fewer, genera will be recognized, all five genera are recognized here.

Superfamily Bivesiculoidea Yamaguti, 1934

Diagnosis: Body flattened or plump, elongate to transversely elongate. Tegument armed with minute spines throughout. Eye-spots or eye-spot pigment usually visible near pharynx. Oral and ventral suckers absent. Prepharynx short or apparently absent. Pharynx present. Oesophagus distinct. Caeca paired, blind. Testis single, median or submedian, anywhere from middle of body to close to posterior extremity. Cirrus-sac present. External seminal vesicle present or absent. Genital pore on ventral mid-line, anywhere from anterior to posterior third of body. Ovary entire, usually anterior to testis, occasionally opposite. Laurer's canal present. Uterine seminal receptacle present. Uterus of varying length, may loop posterior to testis. Eggs large; capsules either thick-walled, tanned and containing unembryonated zygote or thin-walled, clear and containing embryonating eye-spotted miracidium. Vitellarium consisting of widely distributed extracaecal follicles or two small, lobed, largely intercaecal masses. Excretory pore at posterior margin; excretory vesicle V-shaped; arms may reach to pharynx or only to mid-body. Adults in intestine of marine or (rarely) freshwater teleosts; most common in tropical waters of the Pacific, Indian and Atlantic Oceans. Type-genus *Bivesicula* Yamaguti, 1934.

Family Bivesiculidae Yamaguti, 1934

(Syns Paucivitellosinae Yamaguti, 1971; Treptodemidae Yamaguti, 1971)

Diagnosis: With characters of superfamily. Type-genus *Bivesicula* Yamaguti, 1934.

Key to genera

1a. Vitellarium two small, symmetrical, lobed masses; eggs pale, thin-walled and embryonating to eye-spotted miracidium *in utero* .... *Paucivitellosus* Coil, Reid & Kuntz, 1965. (Fig. 5.1)

Diagnosis: Body small, squat. Eye-spots often remain prominent in adults. Ovary anterior to testis. Uterus extends to lateral margins of body and posterior to testis. Eggs thin-walled, clear, containing embryonating miracidium, which may have obvious eye-spots. Vitellarium two small, lobed masses. Excretory vesicle arms reach to eye-spots. In intestine of blenniid and mugilid teleosts; Indo-Pacific. Type-species *P. fragilis* Coil, Reid & Kuntz, 1965.

1b. Vitellarium follicular; follicles distributed widely; eggs yellow, thick-walled, not embryonating *in utero* .... 2.

2a. Body wider than long; testis and cirrus-sac opposite .... *Treptodemus* Manter, 1961. (Fig. 5.2)

Diagnosis: Body much wider than long. Caeca run parallel to edge of body, almost meeting in mid-line posteriorly. Testis and cirrus-sac intercaecal and symmetrically opposite. Ovary opposite testis. Uterus intercaecal and not passing posterior to testis. Eggs thick-walled, tanned, containing unembryonated zygote. Vitelline follicles extensive, inter- and extracaecal along caeca. Excretory vesicle arms intercaecal. In intestine of hemiramphid teleosts; Indo-Pacific. Type-species *T. latus* Manter, 1961.

2b. Body longer than wide; testis at least partly posterior to cirrus-sac .... 3.

3a. Uterus passing posterior to testis; arms of excretory vesicle extend well anterior to caeca .... *Bivesicula* Yamaguti, 1934. (Fig. 5.3)

Diagnosis: Body longer than wide. Caeca variable, may terminate anywhere between mid-body and posterior end of body. Testis in mid-line, in posterior half of body. Ovary anterior to testis. Uterus intercaecal, passing posterior to testis. Eggs thick-walled, tanned, containing unembryonated zygote. Vitelline follicles extensive, inter- and extracaecal, usually in fields along caeca. Excretory vesicle arms extend anterior to caeca. In intestine of many families of teleosts; Indo-Pacific and Atlantic. Type-species *B. claviformis* Yamaguti, 1934.

3b. Uterus not passing posterior to testis; arms of excretory vesicle extend no further anteriorly than to caeca .... 4.

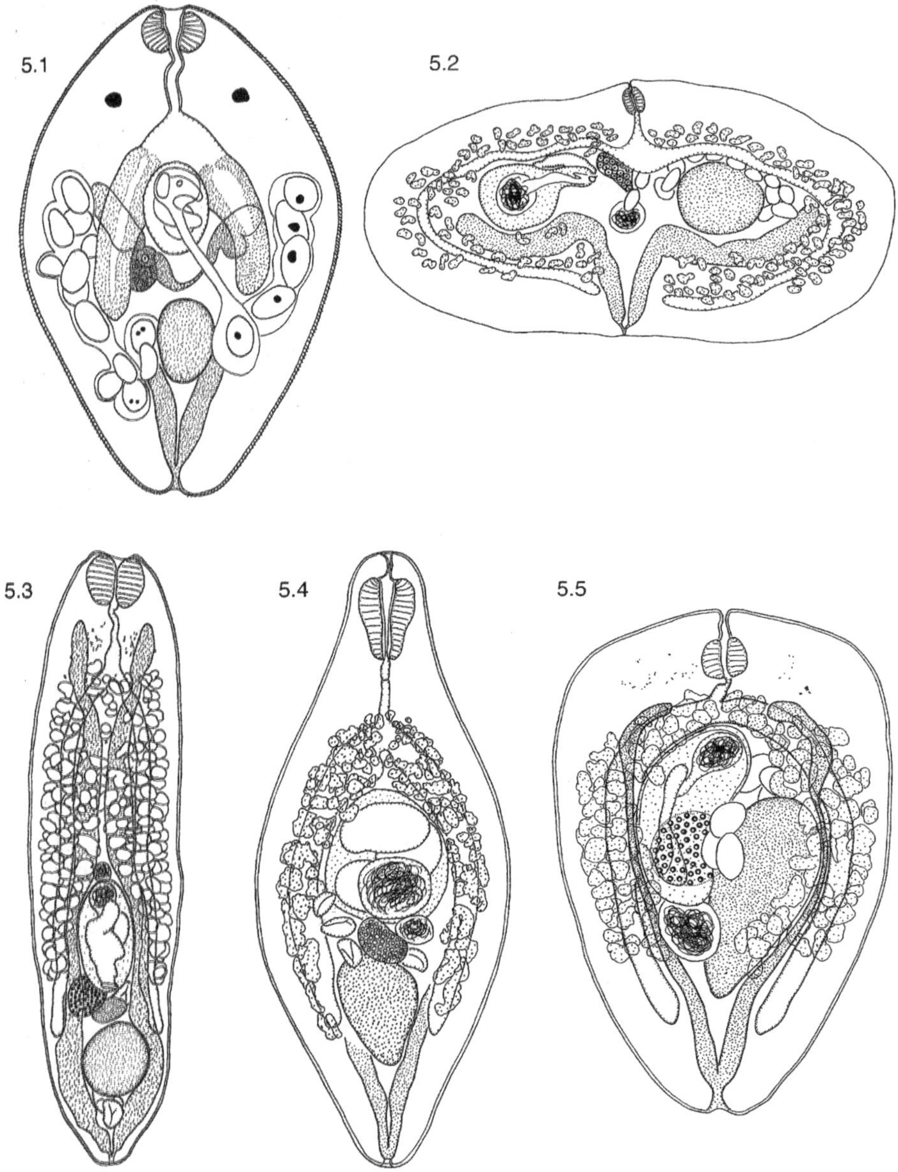

Figs 5.1–5.5.
5.1. *Paucivitellosus fragilis* Coil, Reid & Kuntz, 1965. (After Pearson, 1968.)
5.2. *Treptodemus latus* Manter, 1961. (After Manter, 1961.)
5.3. *Bivesicula claviformis* Yamaguti, 1934. (Original.)
5.4. *Bivesiculoides fusiformis* Cribb, Bray & Barker, 1994. (Original.)
5.5. *Treptodemoides fukenensis* (Liu, 1995) n. comb. (Original.)

4a. Body much longer than wide; length/width ratio 1 : ≥ 2; testis spherical to oval; ovary anterior to testis .... *Bivesiculoides* Yamaguti, 1938. (Fig. 5.4)

Diagnosis: Body longer than wide. Caeca terminate in posterior third of body. Testis in mid-line posterior to cirrus-sac. Ovary anterior to testis. Uterus intercaecal, never passing posterior to testis. Eggs thick-walled and tanned, containing unembryonated zygote. Vitelline follicles extensive, inter- and extracaecal, usually largely parallel caeca. Excretory vesicle arms terminate without crossing caeca. In intestine of several families of teleosts; Indo-Pacific and Atlantic. Type-species *B. atherinae* Yamaguti, 1938.

4b. Body only slightly longer than wide; length width ratio 1 : ≤ 1.7; testis much longer than wide; ovary opposite testis .... *Treptodemoides* Shen, 1995. (Fig. 5.5)

Diagnosis: Body only slightly longer than wide. Caeca terminate close to posterior end of body. Testis strikingly elongate. Ovary opposite testis. Uterus intercaecal, never passing posterior to testis. Eggs thick-walled, tanned, containing unembryonated zygote. Vitelline follicles extensive, inter- and extracaecal, largely along caeca. Excretory vesicle arms terminate ventral to caeca. In intestine of hemiramphid teleosts; Indo-Pacific. Type-species *T. fukenensis* (Liu, 1995) n. comb. (Syns *Bivesiculoides fukenensis* Liu, 1995; *T. xinyingensis* Shen, 1995, type by original designation.)

6 Superfamily Brachylaimoidea Joyeux & Foley, 1930

T. POJMAŃSKA

W. Stefański Institute of Parasitology, Polish Academy of Sciences, Twarda, Warsaw, Poland

Introduction

The superfamily Brachylaimoidea was first recognized by Allison (1943) (the original orthography being Brachylaemoidea) with Brachylaimidae Joyeux & Foley, 1930 as the type-family. It was the first, but not the only, erection of a higher taxon for this group of digeneans. La Rue (1957) established the suborder Brachylaimata within the order Strigeatoidea La Rue, 1926, and Travassos *et al.* (1969) the order Brachylaimiformes. The superfamily Brachylaimoidea was accepted by Mehra (1962a), Travassos & Kohn (1966) and Mas-Coma & Gallego (1975). The systems of Mehra (1962a) and Travassos & Kohn (1966) are very similar, differing only in the status of some taxa (at family or subfamily level), and comprised the Brachylaimidae, Harmotrematidae, Hasstilesiidae, Leucochloridiidae, Leucochloridiomorphidae, Liolopidae, Moreauiidae, Thapariellidae, Urorygmidae and Urotrematidae (or respective subfamilies). Yamaguti (1958a, 1971), who did not distinguish the taxon at superfamily level, mentioned all of these families (or subfamilies) as related to each other, together with the Ovariopteridae, which was proposed by Leonov *et al.* (1963).

The superfamily comprises a large group of digeneans, most of them the parasites of birds and mammals, characterized mainly by a very short or absent oesophagus, the position of the genital organs posterior to the ventral sucker and the structure of the excretory system, with a short excretory vesicle, two long, main lateral excretory canals and numerous flame-cells. In relation to the life-cycles, the common features are branched sporocysts and cercariae with a tail which is poorly developed or absent.

Three main patterns of life-cycle can be distinguished (data from many authors).

1. Life-cycle with two intermediate hosts, completed in an aquatic environment: mother-sporocyst → daughter-sporocyst → cercaria of furcocercous type, but with a poorly developed tail, leaving the first intermediate host → unencysted metacercaria in the second intermediate host → adult in water-birds (Leucochloridiomorphidae, perhaps Thapariellidae).

©CAB *International* and The Natural History Museum, London, 2002. *Keys to the Trematoda*, Vol. 1 (eds D.I. Gibson, A. Jones and R.A. Bray)

2. Life-cycle with two intermediate hosts, completed in a terrestrial environment: sporocyst (only one?) → cercariaeum with rudimentary, readily lost tail, leaving the first intermediate hosts → encysted or unencysted metacercaria in the second intermediate host → adult in birds and mammals, rarely in amphibians and reptiles (Brachylaimidae, Panopistidae).
3. Life-cycle with one intermediate host, completed in a terrestrial environment: sporocyst (only one) → cercariaeum with rudimentary tail or none, not leaving sporocyst → metacercaria unencysted or encysted inside sporocyst → adult in birds and mammals (Hasstilesiidae, Leucochloridiidae).

The classification adopted in this book generally follows the system of Travassos & Kohn (1966), but two families are rejected: the Urotrematidae Poche, 1925, which, according to La Rue (1957), belongs rather to the Plagiorchioidea; and the Liolopidae Dollfus, 1934, because of differences in the structure of the excretory system and life-cycle. Despite the aberrant morphology, the Moreauiidae Johnston, 1915 is retained within this superfamily, but its actual relationships and status remain uncertain. The family Ovariopteridae Leonov, Spasskii & Kulikov, 1963 is also retained on the basis of similarities in the morphology of the adult, although the life-cycle pattern in this family is not known.

The subfamily Panopistinae, included by many authors in the Brachylaimidae, is raised to full family status.

Three genera, *Urorygma*, *Michajlovia* and *Zeylanurotrema*, are considered as *incertae sedis*; the morphology of the adults corresponds with that of the Brachylaimoidea, but the combination of features of these particular genera does not permit their assignment to any particular family.

Urorygma Braun 1901 (Fig. 6.1), comprising only one species, was included by Odhner (1912) in the subfamily Harmostominae and family Harmostomidae. Kagan (1952) placed it in the subfamily Leucochloridiinae in the Brachylaimidae, although he stressed that some features (the extent of the vitellarium, the position of the gonads) do not fit the Leucochloridiinae. Yamaguti (1958a), who raised the subfamily Leucochloridiinae to family level, proposed within this family a new subfamily, the Urorygminae. Finally, Travassos & Kohn (1964) raised the subfamily Urorygminae to family level. This was not accepted by Yamaguti (1971). None of the proposed solutions appears to be satisfactory. The general appearance of *U. nanodes* Braun, 1901 (body shape, the proportions of the suckers in relation to body size, the lack of an oesophagus) resembles that of *Leucochloridium*, but the position of the gonads in the forebody does not fit with the brachylaimoids. Nothing is known of the life-cycle of this parasite. To establish the systematic position of the genus, more information on its morphology and biology is required.

Michajlovia Pojmańska, 1973 (Fig. 6.2) was included previously by Pojmańska (1973) in the Leucochloridiinae (family Brachylaimidae), because of the similarity in the general morphology of the adults to the representatives of this subfamily. Nevertheless, some characters (mainly the position of the genital pore on the ventral surface of the body) differentiate this genus from the genera in this family. The position of the genital pore and the well-developed pars prostatica resemble those of the Leucochloridiomorphidae, from which they differ in the position of the ovary, the course of the uterus and the extent of the vitellarium. Although there are descriptions

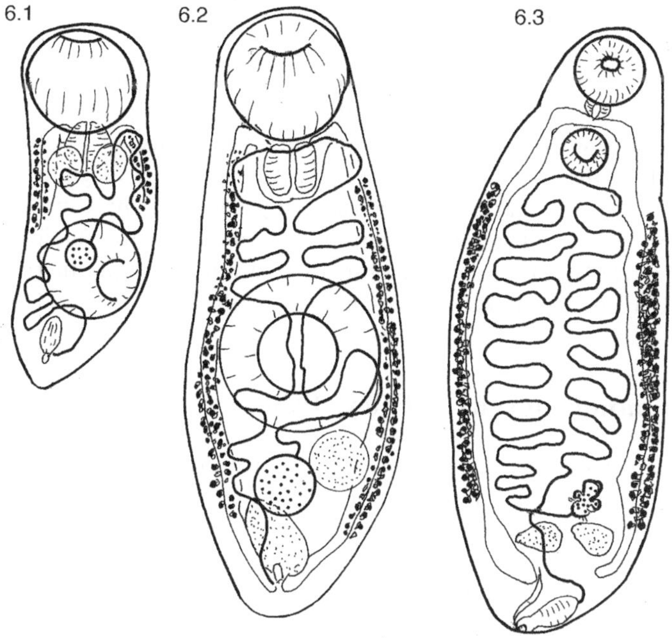

Figs 6.1–6.3. *Genera incertae sedis.*
6.1. *Urorygma nanodes* Braun, 1901. (After Travassos & Kohn, 1966.)
6.2. *Michajlovia migrata* Pojmańska, 1973. (After Pojmańska, 1973.)
6.3. *Zeylanurotrema lyriocephali* Crusz & Sanmugasunderam, 1973. (After Crusz & Sanmugasunderam, 1973.)

of several species in the literature which present the same features as the type-species (Pojmańska, 1973), nothing is known of their life-cycles, which make it impossible to establish the systematic position of the genus. It is likely that this is terrestrial, as the definitive hosts are mainly passerine birds.

Zeylanurotrema Crusz & Sanmugasunderam, 1973 (Fig. 6.3), proposed for a new species from the urinary bladder of a lizard, was placed by Crusz & Sanmugasunderam (1973) within the family Urotrematidae on the basis of such characters as the positions of the ventral sucker in the anterior region of body, of the gonads in the posterior region of the body and of the genital pore near the posterior extremity. But this genus differs from the Urotrematidae in the lack of an oesophagus, the position of the ovary, which in the Urotrematidae is situated far from the testes and close to the ventral sucker, in the absence of a cirrus-sac and in having ascending and descending uterine limbs. The position of the gonads close to the posterior extremity and close to each other, the lobation of the ovary, the lack of a cirrus-sac and the course of the uterus resemble the Ovariopteridae, from which this genus differs mainly in the position of the ovary anterior to rather than posterior to the testes. Cribb & Barton (1991) erected the subfamily Zeylanurotrematinae for this genus within the Brachylaimidae and stressed similarities with the panopistids, such as the body shape, the post-testicular position of the genital pore and the extent of the vitelline fields. Taking into account several important features that differentiate this genus from the Panopistidae and those in common with the Ovariopteridae, it seems difficult to accept such an affiliation on the basis of our current knowledge of this group. Nothing is known of the life-cycle; nevertheless, the presence of a rudimentary tail in juvenile forms (see Cribb & Barton, 1991) suggests that it may be similar to that of the brachylaimids or hasstilesiids. More information is required to resolve the problems concerning the systematic position of this genus.

In the present system the following features were taken into account for distinguishing taxa at the family level: the position of the genital pore, the vitelline extent, the arrangement of the gonads and the presence (or absence) of a cirrus-sac in adult forms, as well as the pattern of the life-cycle. All figures are diagrammatic in order to clarify the generic characters. This especially concerns the course of the uterus, which is an important generic feature, but in most brachylaimoids forms numerous coils that overlap each other and obscure the arrangement of other internal organs. Individual organs are marked in the same way in all figures.

Superfamily Brachylaimoidea Joyeux & Foley, 1930

Diagnosis: Body of variable shape, unarmed or armed with fine spines. Two, usually well-developed, suckers. Pharynx stout and muscular, often preceded by prepharynx; oesophagus very short or absent (sometimes detectable at cercarial and metacercarial stages only); caeca two, usually reaching close to posterior extremity, terminate blindly. Gonads posterior to ventral sucker, very often close to posterior extremity, in tandem or in triangle, with ovary usually between testes. Uterus and ejaculatory duct generally opening into common genital atrium. Common genital pore in hindbody. Only external seminal vesicle present. In some families pars prostatica very well developed. Cirrus-sac usually present. Mehlis' gland usually prominent; Laurer's canal present in most genera. Vitellarium follicular, in two lateral fields of variable extent. Uterus with at least one ascending and one descending limb, usually forming numerous coils; distal region generally forms distinct metraterm. Eggs small, operculate, numerous. Excretory system with short, Y-shaped vesicle and two long main canals running anteriorly as far as caecal bifurcation, then descending posteriorly and dividing on each side of body into two secondary canals. Flame-cell formula known for a few species only; it comprises numerous (> 100) flame-cells arranged in groups of three. Mainly parasitic in gut of birds and mammals, rarely reptiles and amphibians. Life-cycle with one or two intermediate hosts, both molluscs. Embryonic development usually in uterus. Miracidium hatches in intestine of snail. Sporocyst branched. Cercariae generally of cercariaeum type, with or without rudimentary tail (exception: Leucochloridiomorphidae with furcocercaria). Metacercaria with well-developed reproductive system, usually encysted in intermediate host. Type-genus *Brachylaima* Dujardin, 1843.

Key to families

1a. Body transversely elongate; genital pore lateral; gonads on antiporal side of body .... Moreauiidae Johnston 1915.

1b. Body longitudinally elongate; genital pore and gonads in longitudinal plane of body .... 2.

2a. Genital pore ventral, sometimes almost terminal .... 3.

2b. **Genital pore dorsal** .... **6.**

3a. **Ovary between testes; caeca terminating near posterior extremity; pars prostatica undetectable** .... **4.**

3b. **Ovary anterior to testes; caeca terminating anterior to region of gonads; pars prostatica prominent** .... **Leucochloridiomorphidae Yamaguti, 1958.**

4a. **Genital pore anterior to or between testes; vitellarium variable in extent** ..... **5.**

4b. **Genital pore posterior to testes; vitellarium in posterior part of body** .... **Panopistidae Yamaguti, 1958.**

5a. **Body elongate; gonad and cirrus-sac small; genital pore anterior to gonads or in region of anterior testis; vitellarium variable in extent but not reaching level of oral sucker** .... **Brachylaimidae Joyeux & Foley, 1930.**

5b. **Body oval; testes and cirrus-sac large; genital pore between testes or in region of anterior testis; vitelline fields short, in anterior part of body** .... **Hasstilesiidae Hall, 1916.**

6a. **Body elongate; suckers in anterior region of body; genital pore terminal; gonads near posterior end of body; ovary posterior to testes** .... **7.**

6b. **Body oval; ventral sucker in about middle of body; genital pore dorsal, sometimes almost terminal; gonads in middle of posterior region of body; ovary between testes or opposite anterior testis** .... **Leucochloridiidae Poche, 1907.**

7a. **Caeca not reaching posterior extremity; ventral sucker larger than oral; vitelline fields short, in region of ovary** .... **Thapariellidae Srivastava, 1955.**

7b. **Caeca reaching posterior extremity; ventral sucker smaller than oral; vitelline fields long, extending between ventral sucker and gonads** .... **Ovariopteridae Leonov, Spasskii & Kulikov, 1963.**

Genera incertae sedis

Genus *Urorygma* Braun, 1901 (Fig. 6.1)

Diagnosis: Body small, oval, unspined. Suckers large; ventral sucker postequatorial. Pharynx large, muscular; oesophagus absent; caeca reaching to posterior extremity. Gonads in forebody, partly dorsal to ventral sucker; testes symmetrical between caecal bifurcation and ventral sucker; ovary dorsal to ventral sucker, post-testicular. Genital pore at posterior extremity. Vitelline follicles in lateral fields in forebody. Uterine coils between caecal bifurcation and genital pore. Parasites of birds. Life-cycle unknown. Recorded from Europe. Type-species *U. nanodes* Braun, 1901.

Genus *Michajlovia* Pojmańska, 1973 (Fig. 6.2)

Diagnosis: Body relatively large, elongate, covered with fine spines. Suckers large; ventral distant from caecal bifurcation, subequatorial. Pharynx large, muscular; oesophagus absent; caeca reaching close to posterior extremity. Gonads between caecal extremities, arranged in triangle, with intertesticular ovary. Cirrus-sac small. Genital pore ventral, just posterior to gonads or in region of posterior testis. Vitelline fields not reaching caecal extremities, extend from caecal bifurcation (or region of oral sucker) to level of ovary or posterior testis. Uterus intercaecal, with characteristic medial ascending and descending arms, dorsal to ventral sucker. Parasites of birds. Life-cycle unknown. Recorded from Europe; possibly also occurs in North America. Type-species *M. migrata* Pojmańska, 1973.

Genus *Zeylanurotrema* Crusz & Sanmugasunderam, 1973 (Fig. 6.3)

Diagnosis: Body elongate to fusiform, covered with spines. Suckers well developed, in anterior part of body. Pharynx muscular; oesophagus absent; caeca not reaching posterior end of body. Gonads near posterior extremity, between or partly beyond caecal extremities, in triangle, with testes symmetrical and ovary pretesticular. Ovary strongly lobed. Cirrus-sac absent. Genital pore terminal or subterminal. Vitelline fields extending from level of ventral sucker to level of testes. Uterus with one ascending and one descending limb, occupies intercaecal field between ventral sucker and genital pore. Eggs small, numerous. Parasites of amphibians and reptiles. Life-cycle unknown. Recorded from Sri Lanka and Australia. Type-species *Z. lyriocephali* Crusz & Sanmugasunderam, 1973.

7 Family Brachylaimidae Joyeux & Foley, 1930

T. Pojmańska

W. Stefański Institute of Parasitology, Polish Academy of Sciences, Twarda, Warsaw, Poland

Introduction

The history of the family Brachylaimidae is complicated and has been confused from its very beginning. Even in the first publication dealing with these digeneans, Dujardin (1843) used two very similar generic names: *Brachylaime* and *Brachylaima.* When describing other new species of this group he later (Dujardin, 1845) changed the generic name to *Brachylaimus*. Then Blanchard (1847) stated that the generic name was orthographically incorrect and emended it to *Brachylaemus*. Although this orthographic correction was not accepted by Stiles & Hassall (1898), who returned to the name *Brachylaima*, Joyeux & Foley (1930) erected the family Brachylaemidae, based on Blanchard's orthography. In the meantime Braun (1899) had erected the genus *Harmostomum*, including in it the species described by Dujardin (1843) under the name *Brachylaime* (or *Brachylaima*) *advena*. Following this change in generic name, Braun (1900) erected the subfamily Harmostominae, and Odhner (1912) the family Harmostomidae. The latter was accepted by Witenberg (1925) and by Sinitsin (1931), who presented revisions of this group.

Since the papers by Dollfus (1934, 1935a), who accepted the interpretation of Joyeux & Foley (1930), the generic name *Brachylaemus* and consequently the family name Brachylaemidae have become fixed in the literature, although occasionally the names *Harmostomum* and Harmostomidae have also occurred. Brachylaemidae as the family name was accepted in major works by Skrjabin (1948), Yamaguti (1958a), Mehra (1962a) and Travassos & Kohn (1964, 1966).

In 1959 Kruidenier & Gallicchio discussed in detail the problem of priority of the generic name and the orthography of the family name. They stated that the first name used by Dujardin (1843) was *Brachylaime* (the name *Brachylaima* appeared on subsequent pages) and this name should be accepted as the generic name. They found no reason to correct the orthography proposed by Blanchard (1847), and consequently emended the family name from Brachylaemidae (first proposed by Miller, 1936) to Brachylaimidae. This view was accepted by Yamaguti, who used the name

Brachylaimidae in his two monographs (Yamaguti, 1971, 1975); subsequently, this family name has appeared more and more often in parasitological literature and is currently the one most commonly accepted. On the basis of this common usage and stability, this name is also retained here, although according to the law of priority the name Harmostomidae Braun, 1900 should theoretically be re-established.

Another question concerns the acceptance of the orthography of the generic name. Although Kruidenier & Gallicchio (1959) have proved the page priority of *Brachylaime* over *Brachylaima*, it is the second which is in common use in the current literature. Taking into account its common usage, nomenclatural stability and the fact that page priority no longer forms part of the International Code of Zoological Nomenclature, the generic name *Brachylaima* is retained here.

Rallitrema Travassos & Kohn, 1964, reduced by Yamaguti (1971) to subgeneric status, is considered a synonym of *Brachylaima*, because the two characters proposed by this author for separating subgenera (position of the genital pore and extent of the vitellarium) occur in all combinations in the representatives of the genus and are therefore not suitable for distinguishing two groups.

The subgenera *Postharmostomum* Witenberg, 1923 and *Serpentinotrema* Travassos & Kohn, 1964 are, however, maintained, as the anterior extent of the vitellarium readily enables us to classify the taxa to one or other subgenus.

The presence of two subfamilies within the family, proposed by Mehra (1962a) and Mas-Coma & Gallego (1975), is recognized on the basis of the position of the genital pore, the development of the suckers, the presence or absence of an oesophagus and body shape. However, Cribb & Barton's (1991) erection of the Zeylanurotrematinae is not accepted, as the genus *Zeylanurotrema* Crusz & Sammugasunderam, 1974 is considered *incertae sedis*.

Family Brachylaimidae Joyeux & Foley, 1930

(Syns Harmostomidae Braun, 1900; Brachylaemidae Joyeux & Foley, 1930; Brachylaimatidae Ulmer, 1952)

Diagnosis: Body elongate (rarely oval or subglobular), unarmed or armed with fine spines. Suckers usually well developed, but not large in relation to body size, generally situated in anterior part of body. Prepharynx usually present; pharynx muscular; oesophagus very short or absent; caeca long, usually terminating near posterior extremity. Gonads posterior to ventral sucker, near posterior extremity, arranged in tandem or in obtuse triangle, with ovary between testes. Cirrus-sac anterior to or near anterior testis, rarely between testes. Genital pore opening on ventral surface of body, median or submedian, anterior to or in region of anterior testis or between testes. Vitelline fields lateral, variable in extent. Uterus with ascending and descending limbs, coiled in intercaecal field, largely anterior to gonads, surpassing or not level of ventral sucker. Adults in alimentary tract of birds and mammals, rarely amphibians. Life-cycle with two intermediate hosts, both terrestrial molluscs. Daughter-sporocyst branched; cercaria with rudimentary, stumpy tail, leaves first intermediate host. Metacercaria encysted or not, in kidney or pericardium of second intermediate host. Type-genus *Brachylaima* Dujardin, 1843.

Key to subfamilies

1a. Body plump or elongate; suckers well developed; oesophagus practically absent .... Brachylaiminae Joyeux & Foley, 1930.
1b. Body filiform; suckers small; short oesophagus present .... Ityogoniminae Yamaguti, 1958.

Subfamily Brachylaiminae Joyeux & Foley, 1930

(Syns Harmostominae Braun, 1900; Brachylaeminae Joyeux & Foley, 1930; Brachylaimatinae Ulmer, 1952)

Diagnosis: Body usually armed with fine spines, at least anteriorly. Suckers well developed, in anterior part of body, sometimes ventral sucker almost equatorial. Oesophagus practically absent; caeca straight or winding. Gonads arranged in tandem or ovary displaced slightly to side, situated between caecal extremities. Genital pore pre- or intertesticular. Vitelline fields lateral, between anterior testis and ventral sucker or extending into forebody. Uterus forms close tranverse coils of ascending and descending limbs in posterior region of body or extending anterior to ventral sucker. Parasites of birds and mammals. Type-genus *Brachylaima* Dujardin, 1843.

Key to genera

1a. Genital pore between testes .... *Glaphyrostomum* Braun, 1901. (Fig. 7.1)
(Syn. *Cesartrema* Travassos & Kohn, 1964)
Diagnosis: Body linguiform or cylindrical. Suckers well developed, especially ventral sucker, which is situated almost equatorially. Oesophagus absent; caeca straight. Gonads in tandem, or with ovary displaced from median line. Cirrus-sac and genital pore intertesticular, in region of ovary. Vitelline fields extensive, reaching between caecal bifurcation and ovary. Uterine coils reaching to pharynx or even to level of oral sucker. Parasites of birds. North America and Brazil. Type-species *G. adherens* Braun, 1901.
1b. Genital pore anterior to or in region of anterior testis .... 2.

2a. Caeca symmetrical; two testes present .... 3.
2b. Caeca asymmetrical; one testis present .... *Parabrachylaima* Lotz & Corkum, 1975. (Fig. 7.2)
Diagnosis: Body elongate, armed with fine spines. Suckers in anterior part of body, close to each other. Pharynx muscular; oesophagus absent. Caeca unequal in length; one reaching posterior extremity, other terminating anterior to gonads. Gonads atypical; only one testis present, postovarian. Cirrus-sac and cirrus poorly developed; genital pore anterior to ovary. Vitelline fields rather short, extending from posterior to ventral sucker to near ovary. Uterus with coiled ascending and descending limbs, reaching forward to level of ventral sucker. Eggs not numerous, containing fully-developed miracidium. Parasites of kidney of terrestrial snail (*Euglandula rosea* – this is

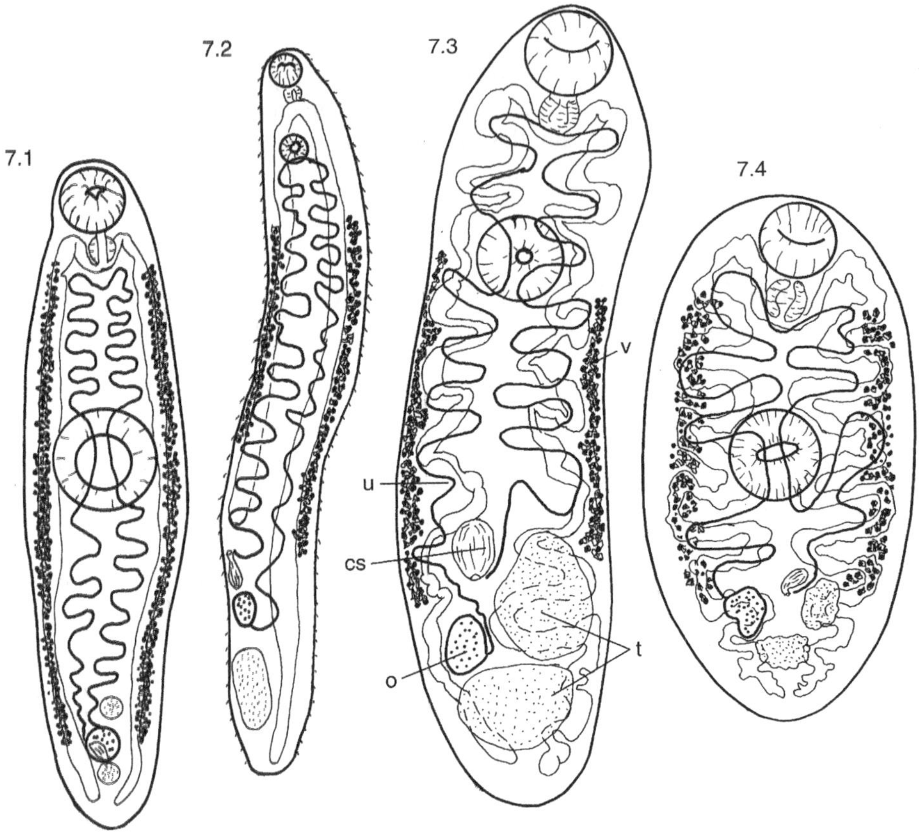

Figs 7.1–7.4.
7.1. *Glaphyrostomum adherens* Braun, 1901. (After Skrjabin, 1948.)
7.2. *Parabrachylaima euglandensis* Lotz & Corkum, 1975; note the presence of only one testis. (After Lotz & Corkum, 1975.)
7.3. *Postharmostomum* (*Postharmostomum*) *gallinum* Witenberg, 1923. cs, cirrus-sac; o, ovary; t, testis; u, uterus; v, vitellarium. (After Skrjabin, 1948.)
7.4. *Postharmostomum* (*Serpentinotrema*) *laruei* (McIntosh, 1934). (After McIntosh, 1934b.)

probably second intermediate host, in which progenetic metacercariae mature and produce eggs). North America. Type-species *P. euglandensis* Lotz & Corkum, 1975.

3a. Caeca straight or somewhat sinuous; gonads more or less tandem .... 4.
3b. Caeca very long, strongly winding; gonads in triangle .... *Postharmostomum* Witenberg, 1923. (Figs 7.3, 7.4)
(Syn. *Serpentinotrema* Travassos & Kohn, 1964)

Diagnosis: Body elongate or oval. Ventral sucker pre- or subequatorial. Oesophagus absent; caeca very long, longer than body, forming deep undulations throughout. Gonads smooth, or weakly lobed, arranged in triangle with ovary between testes or opposite anterior testis. Genital pore just anterior to or in region of anterior testis. Vitelline fields between ovary and ventral sucker or extending to caecal bifurcation. Uterus reaching anteriorly to caecal bifurcation, forms numerous coils occupying most of body. Parasites of caeca of birds and mammals. Life-cycle typical for family; metacercaria unencysted in pericardium of second intermediate host. Cosmopolitan. Type-species *P. gallinum* Witenberg, 1923.

Key to subgenera

1a. Vitellarium between ventral sucker and ovary .... *P.* (*Postharmostomum*) Witenberg, 1923. (Fig. 7.3)
Diagnosis: With characters of genus. Body elongate. Vitelline fields not reaching into forebody. Ovary between testes. Type-species *P.* (*Postharmostomum*) *gallinum* Witenberg, 1923.

1b. Vitellarium reaching level of caecal bifurcation .... *P.* (*Serpentinotrema*) Travassos & Kohn, 1964. (Fig. 7.4)
Diagnosis: With characters of genus. Body oval. Vitellarium extending into forebody and to level of caecal bifurcation. Ovary opposite anterior testis. Type-species *P.* (*Serpentinotrema*) *laruei* (McIntosh, 1934).

4a. Body elongate; vitellarium in middle third of body .... *Brachylaima* Dujardin, 1843. (Figs 7.5, 7.6)
(Syns *Brachylaime* Dujardin 1843; *Brachylaimus*, Dujardin, 1845; *Brachylaemus* Blanchard, 1847; *Harmostomum* Braun, 1899; *Heterolope* Looss, 1899; *Entosiphonus*, Sinitsin, 1931; *Centrodes* Travassos & Kohn, 1964; *Mazzantia* Travassos & Kohn, 1964; *Rallitrema* Travassos & Kohn, 1964)
Diagnosis: Body elongate to tongue-shaped, flattened or subcylindrical, usually covered with fine spines at least anteriorly. Suckers in anterior part of body. Oesophagus practically absent; caeca long, straight or slightly sinuous. Gonads in tandem, close to each other, near posterior extremity. Cirrus-sac elongate, small, anterior to or in region of anterior testis. Genital pore median or submedian, immediately anterior to or at level of anterior testis. Vitelline fields extending between anterior testis and ventral sucker or more anteriorly. Uterus reaches anteriorly as far as level of posterior border of ventral sucker or to caecal bifurcation. Parasites of birds and mammals. Life-cycle typical for family; metacercaria unencysted in second intermediate host. Cosmopolitan. Type-species *B. migrans* Dujardin, 1843.

4b. Body plump; vitellarium in posterior third of body .... *Ectosiphonus* Sinitsin, 1931. (Fig. 7.7)
Diagnosis: Body plump, elliptical to tongue-shaped. Ventral sucker almost equatorial. Oesophagus absent; caeca terminate near posterior extremity. Gonads in tandem or ovary slightly displaced laterally. Genital pore anterior to anterior testis. Vitellarium only in posterior part of body, extending from halfway between ventral sucker and anterior testis almost to posterior extremity. Uterus anterior to ventral sucker. Parasites of mammals. Life-cycle typical for family; metacercaria unencysted, in various organs of second intermediate host. North America. Type-species *E. rhomboideus* Sinitsin, 1931.

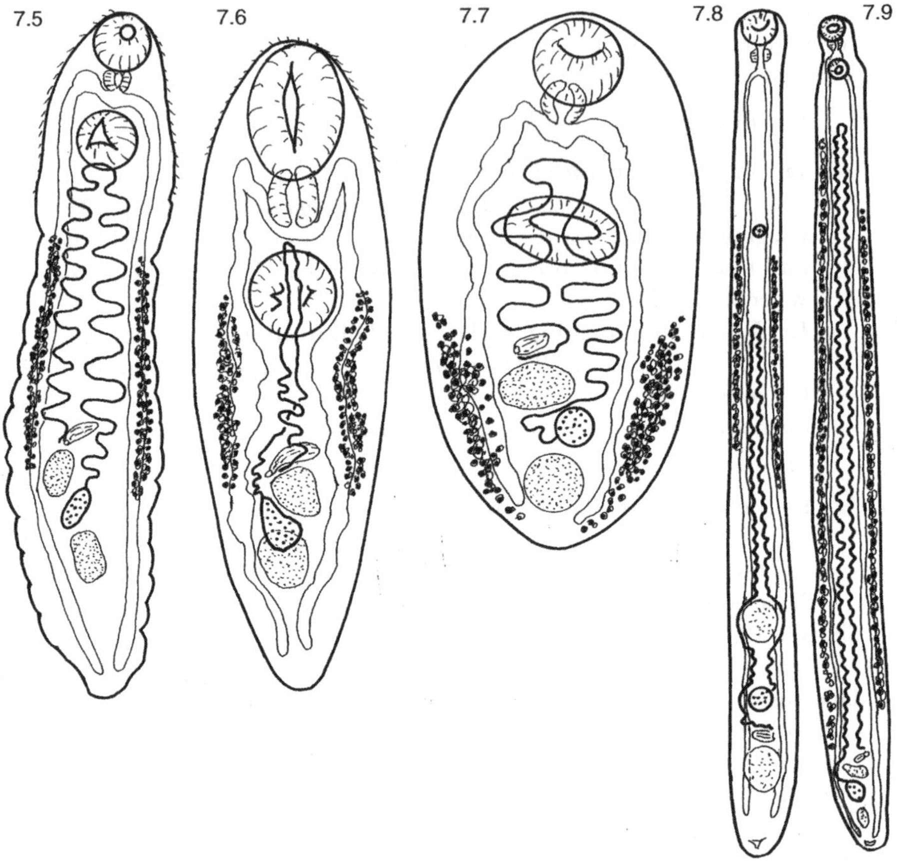

Figs 7.5–7.9.
7.5. *Brachylaima fulvum* Dujardin, 1843; note uterus not passing level of ventral sucker. (After Pojmańska, 1961.)
7.6. *Brachylaima spinosulum* Hofmann, 1899 – metacercaria; note uterus reaching to level of caecal bifurcation. (After Pojmańska, 1959.)
7.7. *Ectosiphonus ovatus* Sinitsin, 1931. (After Sinitsin, 1931.)
7.8. *Ityogonimus ocreatus* (Goeze, 1782). (After Yamaguti, 1971.)
7.9. *Scaphiostomum illatabile* Braun, 1901. (After Yamaguti, 1971.)

Subfamily Ityogoniminae Yamaguti, 1958

Diagnosis: Body very long, slender, filiform. Suckers small, in anterior part of body; ventral may be smaller than oral. Pharynx muscular, small; oesophagus very short; caeca slender, reaching close to posterior extremity. Gonads in tandem, near posterior extremity, between caecal extremities. Cirrus-sac small; genital pore anterior to anterior testis or between ovary and testes. Vitelline fields between ventral sucker and anterior testis. Uterus intercaecal; ascending limb does not reach ventral sucker. Parasites of birds and mammals. Development (known for *Scaphiostomum*) involves terrestrial molluscs as intermediate hosts. Type-genus *Ityogonimus* Lühe, 1899.

Key to genera

1a. Genital pore between testes; gonads well separated from each other; anterior limit of vitellarium reaching level of ventral sucker .... *Ityogonimus* Lühe, 1899. (Fig. 7.8)

(Syns *Dolichosomum* Looss, 1899; *Dolichodemas* Looss, 1900)

Diagnosis: Body filiform. Suckers small; ventral situated at distinct distance from oral and may be much smaller. Prepharynx present; pharynx small; oesophagus short. Gonads well separated. Cirrus-sac small, muscular. Genital pore between ovary and posterior testis. Vitellarium extending from level of ventral sucker no further than to level of anterior testis. Ascending and descending uterine limbs form loose coils. Parasites of mammals. Europe. Type-species *I. ocreatus* (Goeze, 1782).

1b. Genital pore pretesticular; gonads close to each other; vitellarium not reaching level of ventral sucker anteriorly .... *Scaphiostomum* Braun, 1901. (Fig. 7.9)

Diagnosis: Body filiform. Suckers small, close together; ventral may be smaller than oral. Prepharynx present; pharynx small; oesophagus very short. Gonads in tandem, close to each other, between caecal extremities. Cirrus-sac small. Genital pore just anterior to anterior testis. Vitellarium not reaching level of ventral sucker, terminating posteriorly at level of anterior testis. Uterus very long, forming numerous loosely arranged coils, anteriorly not reaching level of ventral sucker. Parasites of birds and mammals. North America and Brazil. Type-species *S. illatabile* Braun, 1901.

8 Family Hasstilesiidae Hall, 1916

T. POJMAŃSKA

W. Stefański Institute of Parasitology, Polish Academy of Sciences, Twarda, Warsaw, Poland

Introduction

This family has often been recognized at the subfamily level (Orlov *et al.*, 1934; Skrjabin, 1948; Mehra, 1962a) within the family Brachylaimidae, although its status at the family level has also been accepted (Yamaguti, 1958a, 1971; Travassos & Kohn, 1966; Cribb & Spratt, 1991). It resembles the Brachylaimidae in the position of the genital pore on the ventral side of the body and between or anterior to the testes, but differs from this family in the pattern of the life-cycle (resolved by Kasyanov, 1954; Rowan, 1955; Gvozdev & Soboleva, 1972, 1983). The most characteristic family features are: ovoid body shape, suckers rather small in relation to the body and anterior position of the vitellarium. The family comprises two genera, which differ in body shape, the extent of the caeca, the position of the genital pore and the arrangement of the gonads.

Family Hasstilesiidae Hall, 1916

Diagnosis: Body wide, at least in its anterior part, more or less ovoid, armed with fine spines. Both suckers rather small in relation to body size, close to each other, in anterior region of body. Gonads in triangle with ovary lateral or in tandem with ovary between testes. Testes larger than ovary. Cirrus-sac elongate, in median line of body. Vitelline fields short, between caecal bifurcation and gonads. Intestinal parasites of mammals. Life-cycle includes one intermediate host; cercariaeum with rudimentary tail; metacercaria remains unencysted inside ramified sporocyst. Type-genus *Hasstilesia* Hall, 1916.

Key to genera

1a. Body ovoid with rounded extremities; caeca extending posteriorly to posterior testis; gonads in triangle, with ovary lateral; testes much larger than ovary; genital pore between testes .... *Hasstilesia* Hall, 1916. (Fig. 8.1)

(Syn. *Skrjabinotrema* Orlov, Ershov & Badanin, 1934)

Diagnosis: Body ovoid, almost round. Caeca long, almost reaching posterior extremity. Testes very large, diagonal, separated by uterine coils. Ovary small, opposite or anterior to posterior testis, overlapping caecum. Cirrus-sac large; genital pore between testes. Vitellarium forming short fields between caecal bifurcation and anterior testis. Uterus in intercaecal field, forming several coils between caecal bifurcation and ovary. Parasites of rodents. North America. Type-species *H. tricolor* (Stiles & Hassall, 1894).

1b. Body fusiform with tapered posterior extremity; caeca extending to level of posterior testis; gonads in tandem in median line; genital pore in region of anterior testis .... *Strzeleckia* Cribb & Spratt, 1991. (Fig. 8.2)

Diagnosis: Body fusiform, with tapered posterior extremity. Caeca reaching only to level of posterior testis. Gonads arranged in tandem with ovary between testes, close to each other; testes larger than ovary. Cirrus-sac quite large. Vitelline fields rather narrow, between caecal bifurcation and anterior testis (sometimes not reaching this level). Uterus forms numerous coils occupying intercaecal field, enveloping ventral sucker laterally and anteriorly. Parasites of marsupials. Australia. Type-species *S. major* Cribb & Spratt, 1991.

8.1

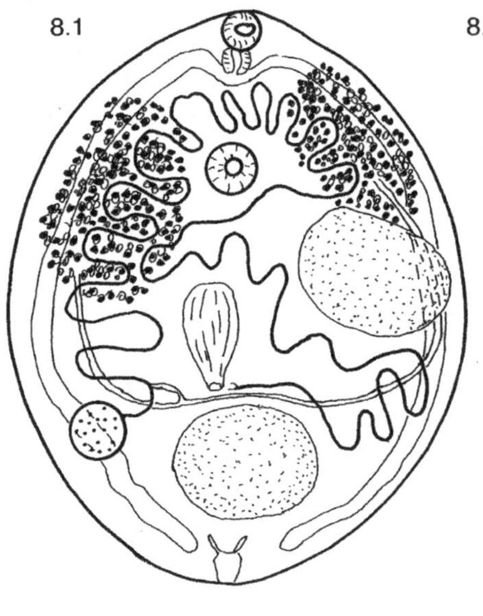

8.2

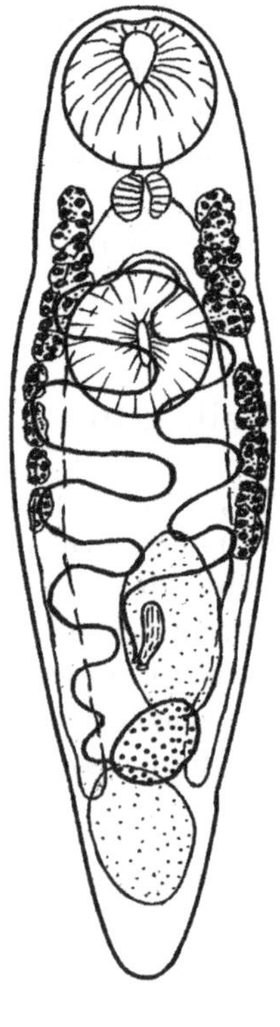

Figs 8.1–8.2.
8.1. *Hasstilesia tricolor* (Stiles & Hassall, 1894). (After Skrjabin, 1948.)
8.2. *Strzeleckia major* Cribb & Spratt, 1991. (After Cribb & Spratt, 1991.)

9 Family Leucochloridiidae Poche, 1907

T. Pojmańska

W. Stefański Institute of Parasitology, Polish Academy of Sciences, Twarda, Warsaw, Poland

Introduction

The history of the family Leucochloridiidae is very long and confused. The generic name *Leucochloridium* was introduced by Carus (1835) for the yellow-green-pigmented sporocysts found in the mollusc *Succinea putris*. Using metacercariae isolated from this sporocyst, Zeller (1874) and Heckert (1889) succeeded in experimentally obtaining adults in birds, which they believed to be identical with *Distoma macrostomum*, described earlier by Rudolphi (1803) from naturally infected birds. Subsequently, the generic name *Leucochloridium* has been established in the parasitological literature. However, at about the same time another generic name was introduced, *Urogonimus* Monticelli, 1888, with *U. macrostomum* (Rudolphi, 1803) as the type-species. Stiles & Hassall (1898) stated that, according to the law of priority, the name *Leucochloridium* is valid and *Urogonimus* is its synonym. Despite this statement, at the turn of the 19th century both generic names were in use, although generally treated as synonyms. Consequently, names for the higher taxa have been derived from both genera. Looss (1899) erected the subfamily Urogoniminae, and Poche (1907) the Leucochloridiinae. Only the latter name remained in use in the literature. Dollfus (1934) raised the subfamily Leucochloridiinae to the level of family. This name is retained here on the grounds of common usage and the maintenance of nomenclatural stability.

The validity of the two genera *Urogonimus* and *Leucochloridium* was questioned for a long time, as was the distinctness of *U. macrostomus* and *L. paradoxum*. In fact, the description of *Distoma macrostomum* given by Rudolphi (1803) was very poor. His original material was revised by Braun (1902), who united under the name of *U. macrostomus* several species of leucochloridiids characterized by the terminal position of the genital pore and gonads located in the hindbody. The original material of Rudolphi was revised once again and redescribed by Szidat (1936b). He had stated that Rudolphi's *Distoma macrostomum* from *Motacilla luscinia* differed from *Leucochloridium paradoxum* obtained experimentally by Zeller (1874) and by Heckert (1889). Kagan (1952) considered the differences as of generic value and established two genera:

Urogonimus Monticelli, 1888, with the type-species *U. macrostomus* (Rudolphi, 1803), based on Rudolphi's original specimens from *Motacilla luscinia* (as proposed by Monticelli, 1888), and *Leucochloridium* Carus, 1935, with the type-species *L. paradoxum* Carus, 1935, based on the specimens reared by Zeller (1874) in birds from metacercariae isolated from a yellow-green sporocyst. The publications by Schmidt (1965) and Lewis (1974) on the life-cycles of the representatives of *Urogonimus* and those of Pojmańska (1967, 1969, 1975, 1978) on the life-cycles of the representatives of *Leucochloridium* have revealed that these two genera also differ in the nature of the sporocyst (without and with brood sacs, respectively) and metacercariae (unencysted and encysted within the sporocyst, respectively). Subsequent to the revision of the family Leucochloridiidae by Kagan (1952), the validity of both genera was accepted by Bakke (1980), but not by Yamaguti (1971, 1975). Nevertheless, Yamaguti (1975) did accept the distinctness of *macrostomum* and *paradoxum*.

The division of the family by Yamaguti (1958a) into three subfamilies, the Urotocinae, Urorygminae and Leucochloridiinae, is not maintained, because *Urorygma* is here removed from the family, and the differences between *Urotocus* and *Leucochloridium*, which relate mainly to body shape and the development of the suckers, are not considered important at the subfamily level, and the structure of the sporocyst of *Urotocus* (described by Timon-David, 1957b) resembles that of *Urogonimus*.

Three genera, *Leucochloridium*, *Urogonimus* and *Urotocus* Looss, 1899, are distinguished in the family. *Neoleucochloridium* Kagan, 1952 is, according to the conception of Bakke (1980), reduced to subgeneric level.

Paraurotocus Travassoss & Kohn, 1966 is regarded as a synonym of *Urotocus*, because the only real difference in the generic diagnosis - a lack of the ventral sucker in the former - is doubtful; in *Urotocus* this sucker is rudimentary and hardly detectable.

Dollfusinus Biocca & Ferretti, 1958 for which Timon-David (1965) erected the subfamily Dollfusinusinae, is transferred to the family Panopistidae (which is in agreement with its inclusion by Mas-Coma & Gallego (1975) in the subfamily Panopistinae).

The genera *Urorygma* Braun, 1901 and *Michajlovia* Pojmańska, 1973 are considered *incertae sedis* (see Introduction to the superfamily Brachylaimoidea).

Genera of this family are distinguished on the basis of the proportions of the suckers, the position of the ventral sucker, the arrangement of the gonads, the position of the genital pore and the course of the uterus.

Family Leucochloridiidae Poche, 1907

Diagnosis: Body of medium size, oval to lanceolate, often covered with fine spines. Suckers usually well developed (except *Urotocus*); ventral sucker situated about middle of body. Pharynx muscular, well developed; oesophagus absent; caeca terminate near posterior extremity. Gonads in tandem or in triangle, with ovary between testes. Cirrus-sac present. Genital pore on dorsal surface of body, posterior to gonads, sometimes almost terminal. Laurer's canal present. Vitellarium forming two narrow lateral fields along caeca. Uterus with ascending and descending limbs, forming numerous coils, occupies entire intercaecal field. Parasites of alimentary tract (especially cloaca and bursa Fabricii) of birds. Life-cycle terrestrial, with one intermediate host; only one generation of sporocyst, strongly branched; cercariaeum without any trace of tail; metacercaria remaining unencysted or encysted inside sporocyst. Type-genus *Leucochloridium* Carus, 1835.

Key to genera

1a. Body elongate; suckers relatively small, in anterior region of body .... *Urotocus* Looss, 1899. (Fig. 9.1)
(Syn. *Paraurotocus* Travassos & Kohn, 1966)

Diagnosis: Body small, slender, resembling brachylaimids in shape, armed over entire surface with fine spines. Both suckers weakly developed, in anterior part of body; oral slightly larger than ventral, which is hardly detectable. Pharynx small, muscular. Oesophagus absent. Gonads in tandem, with ovary between testes. Genital pore almost terminal. Vitelline fields long, not reaching caecal bifurcation anteriorly, extending to level of ovary posteriorly. Uterine coils occupy entire intercaecal field between caecal bifurcation and posterior extremity. Parasites of cloaca or bursa Fabricii of birds. Life-cycle typical for family; metacercaria encysts inside sporocyst. Europe and North America. Type-species *U. rossitensis* (Mühling, 1898).

1b. Body oval; suckers well developed; ventral sucker in middle third of body .... 2.

2a. Genital pore subterminal; cirrus-sac small; uterus with two ascending and two descending limbs, crossing body horizontally posterior to ventral sucker .... *Urogonimus* Monticelli, 1888. (Fig. 9.2)

Diagnosis: Body more or less oval in shape. Both suckers well developed; pharynx large, muscular. Gonads between ventral sucker and caecal extremities, in tandem or more rarely in triangle. Cirrus-sac small in relation to size of gonads. Genital pore dorsal, often almost terminal. Vitelline fields long, extend along caeca from region of oral sucker to level of gonads, not reaching caecal extremities. Uterus forms numerous coils, ascending laterally to level of caecal bifurcation, then descending to level of posterior margin of ventral sucker, where it passes horizontally to opposite side of body and forms corresponding ascending and descending loops prior to reaching genital pore. Parasites of cloaca or bursa Fabricii of birds. Life-cycle typical for family; sporocyst strongly branched in hepatopancreas of snail; metacercariae unencysted, inside sporocyst. Cosmopolitan. Type-species *U. macrostomus* (Rudolphi, 1803).

2b. Genital pore distinctly dorsal; cirrus-sac well developed; uterus with one ascending and one descending limb, crosses body anterior to ventral sucker .. .. *Leucochloridium* Carus, 1835. (Figs 9.3–9.5)
(Syn. *Neoleucochloridium* Kagan, 1952)

Diagnosis: Body oval, armed with fine spines. Both suckers well developed. Pharynx muscular; caeca usually reach close to posterior extremity. Oral sucker, pharynx and ventral sucker close to each other, sometimes overlapping one another. Gonads between ventral sucker and caecal extremities, in triangle, sometimes almost in line. Cirrus-sac quite large in relation to size of gonads. Vitelline fields long, extending from region of oral sucker to caecal extremities, or more posteriorly. Uterus forming

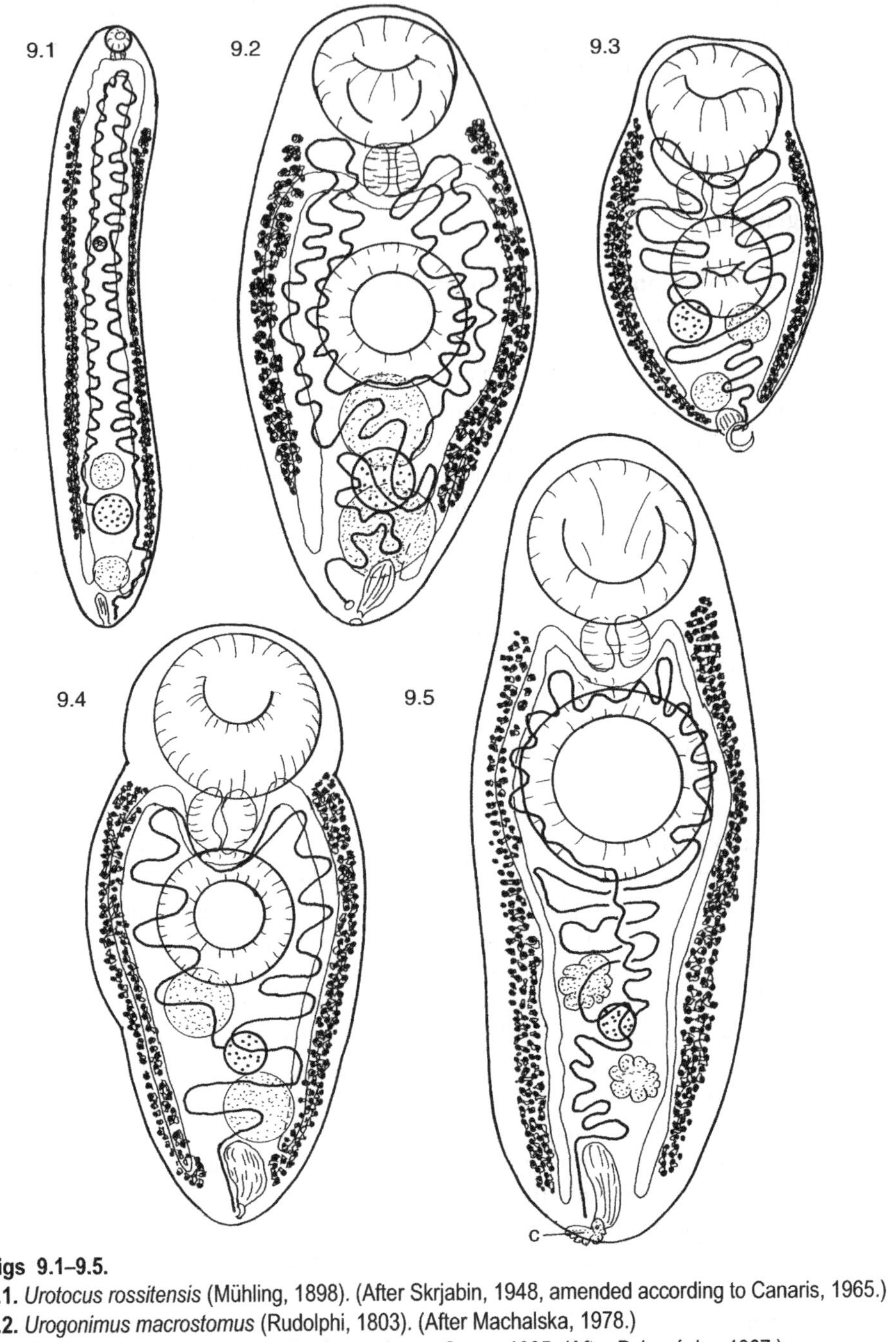

Figs 9.1–9.5.

9.1. *Urotocus rossitensis* (Mühling, 1898). (After Skrjabin, 1948, amended according to Canaris, 1965.)

9.2. *Urogonimus macrostomus* (Rudolphi, 1803). (After Machalska, 1978.)

9.3. *Leucochloridium* (*Leucochloridium*) *paradoxum* Carus, 1835. (After Pojmańska, 1967.)

9.4. *Leucochloridium* (*Papilloleucochloridium*) *vogtianum* (Baudon, 1881). (After Pojmańska, 1978.)

9.5. *Leucochloridium* (*Neoleucochloridium*) *holostomum* (Rudolphi, 1819). c, cirrus. (After Pojmańska, 1975.)

numerous coils, ascends to level of caecal bifurcation or region of oral sucker, crosses body horizontally between oral and ventral suckers, and descends to genital pore, occupying entire intercaecal area, sometimes extending beyond caeca both anteriorly and laterally. Parasites of cloaca or bursa Fabricii of birds (mainly Passeriformes). Life-cycle typical for family; sporocyst produces characteristic large sac-like ramifications (brood sacs), coloured and mobile, entering tentacles of snail; metacercariae encysted in these brood-sacs. Cosmopolitan. Type-species *L. paradoxum* Carus, 1835.

Key to subgenera

1a. Body rather small; ventral sucker equatorial; gonads in triangle; cirrus-sac relatively small; uterus passing caecal bifurcation anteriorly .... *L.* (*Leucochloridium*) Carus, 1835. (Fig. 9.3)

Diagnosis: Body oval or with tapered posterior extremity, armed with fine spines. Suckers large in relation to body size; oral usually slightly larger than ventral, situated in middle third of body. Gonads arranged in right-angled triangle, with ovary opposite anterior testis. Cirrus long, smooth. Genital pore dorsal, sometimes subterminal. Vitelline fields long, extending beyond caeca posteriorly, sometimes almost confluent. Uterus forming numerous transverse coils, reaches anteriorly to caecal bifurcation or to region of oral sucker. Brood-sacs of sporocysts coloured green, orange and brown. Type-species *L.* (*Leucochloridium*) *paradoxum* Carus, 1835.

1b. Body large; suckers in anterior part of body; gonads diagonal, almost in tandem; cirrus-sac large; uterus limited to intercaecal field .... 2.

2a. Body quite large; ventral sucker pre-equatorial; anterior testis just posterior to ventral sucker; cirrus smooth .... *L.* (*Papilloleucochloridium*) Bakke, 1980. (Fig. 9.4)

Diagnosis: Body elongate, tapered at posterior extremity, covered with very fine spines. Oral sucker distinctly larger than ventral; pharynx often overlaps both suckers. Gonads arranged in obtuse triangle, almost in tandem. Genital pore posterior to level of caecal extremities. Vitellarium extending from caecal bifurcation to about caecal extremities. Brood sac of sporocyst with brown cap on anterior tip and many wart-like protuberances on entire surface. Type-species *L.* (*Papilloleucochloridium*) *vogtianum* Baudon, 1881.

2b. Body very large; ventral sucker in anterior part of body; anterior testis far from ventral sucker; cirrus pustulated .... *L.* (*Neoleucochloridium*) Kagan, 1950. (Fig. 9.5)

Diagnosis: Body large in relation to other leucochloridiids, with rounded extremities, unarmed. Suckers well developed; ventral generally larger than oral, in anterior part of body. Gonads midway between ventral sucker and posterior extremity, arranged in triangle or diagonally almost in line, with ovary between testes; testes sometimes lobed. Genital pore distinctly dorsal. Cirrus with very characteristic pustules on its surface. Vitelline fields long, extending from region of oral sucker to about caecal extremities. Anterior tip of brood sac of sporocyst coloured brown. Type-species *L.* (*Neoleucochloridium*) *holostomum* (Rudolphi, 1819).

10 Family Leucochloridiomorphidae Yamaguti, 1958

T. Pojmańska

W. Stefański Institute of Parasitology, Polish Academy of Sciences, Twarda, Warsaw, Poland

Introduction

The genus *Leucochloridiomorpha*, erected by Gower (1938), has been placed in the subfamily Brachylaiminae (by Allison, 1943) or, more often, in the subfamily Leucochloridiomorphinae in the family Brachylaimidae (by Yamaguti, 1958a, 1971; Mehra, 1962a; Pojmańska, 1971, 1972a, b). Travassos & Kohn (1966) raised the latter subfamily to full family status. The resolution of the complete life-cycle of *L. constantiae* by Allison (1943) revealed that these digeneans differ from the other brachylaimids in developing in an aquatic environment and in having a free-living cercaria.

The main family characters are the length of the caeca, the position of the genital pore and ovary, the extent of the vitellarium, the course of the uterus and the presence of a prominent pars prostatica and of spines on the surface of the cirrus.

Following Mas-Coma & Gallego (1975), three genera are included in the family, although the life-cycles of representatives of *Ptyalincola* and *Amblosoma* are known only fragmentarily (Wootton & Murrell, 1967, and Pojmańska, 1972a, respectively).

Family Leucochloridiomorphidae Yamaguti, 1958

Diagnosis: Body elongate, unarmed. Suckers well developed. Pharynx muscular; oesophagus present. Caeca do not reach level of testes. Gonads at posterior end of body, arranged in triangle with ovary anterior to testes or overlapping region of testes. Genital pore ventral, posterior to or between testes. Cirrus-sac well developed; cirrus armed with spines. Pars prostatica very well developed. Vitellarium usually forming short lateral fields posterior to ventral sucker or in region of ventral sucker. Uterus with distinct metraterm, limited to region of gonads, or enveloping ventral sucker laterally and posteriorly. Parasites of aquatic birds, rarely of mammals. Life-cycle aquatic, with two intermediate hosts, both molluscs of family Viviparidae. Two generations of sporocysts; daughter-sporocyst branched; cercaria furcocercous, with short furcae, leaving first intermediate host; metacercaria in gonads or on surface of hepatopancreas of second intermediate host, unencysted. Type-genus *Leucochloridiomorpha* Gower, 1938.

Key to genera

1a. Body enlarged in middle; ventral sucker much larger than oral; genital pore between or posterior to testes; uterus with two ascending and two descending limbs, enveloping ventral sucker laterally and posteriorly .... 2.

1b. Body elongate; ventral sucker only slightly larger than oral; genital pore posterior to gonads; uterus confined to region of gonads .... *Amblosoma* Pojmańska, 1972. (Fig. 10.1)

Diagnosis: Body relatively large, elongate, rounded at both extremities. Distance between ventral sucker and gonads great. Oesophagus short; caeca reaching level of ovary. Gonads almost round; ovary distinctly anterior to testes; testes symmetrical. Cirrus-sac large, between testes. Genital pore posterior to gonads. Vitelline fields short, just anterior to testes, distant from ventral sucker. Uterus short, forming several coils in region of gonads, crosses body horizontally anteriorly to ovary and descends posteriorly at level of anterior margin of long pars prostatica. Metraterm large, oval. Parasites of cloaca of aquatic birds (*Aythya*). Sporocyst and cercaria not known. Metacercaria in viviparid snails. Europe, North America and Japan. Type-species *A. exile* Pojmańska, 1972.

2a. Testes symmetrical; genital pore between or posterior to testes; vitelline fields in region of ventral sucker, not reaching level of gonads .... *Leucochloridiomorpha* Gower, 1938. (Fig. 10.2)

(Syn. *Voelkeria* Travassos & Kohn, 1966)

Diagnosis: Body small, greatly enlarged in region of ventral sucker and tapered anteriorly. Ventral sucker enormous. Oesophagus short; caeca terminate just posterior to ventral sucker. Gonads occupy almost entire length of hindbody. Testes oval, symmetrical or slightly diagonal; ovary round, slightly anterior to or between testes, sometimes overlapping posterior margin of ventral sucker. Cirrus-sac in region of testes, or more posteriorly; pars prostatica very large, elongate, may overlap posterior margin of ventral sucker. Vitellarium forms short lateral fields, not reaching pharynx anteriorly and not passing middle of ventral sucker posteriorly. Uterus ascends on one side of body to level of anterior margin of ventral sucker, then descends to its posterior margin, crosses horizontally to opposite side, where it forms similar loops and descends to genital pore. Metraterm elongate. Parasites of cloaca of aquatic birds (*Anas* and *Aythya*). Life-cycle typical for family. Europe and North America. Type-species *L. constantiae* (Mueller, 1935).

2b. Testes diagonal; genital pore between ovary and testes; vitellarium posterior to ventral sucker, reaching level of ovary .... *Ptyalincola* Wootton & Murrell, 1967. (Fig. 10.3)

Diagnosis: Body small, enlarged in region of ventral sucker and tapered towards posterior extremity. Oesophagus short; caeca terminate posterior to ventral sucker, not

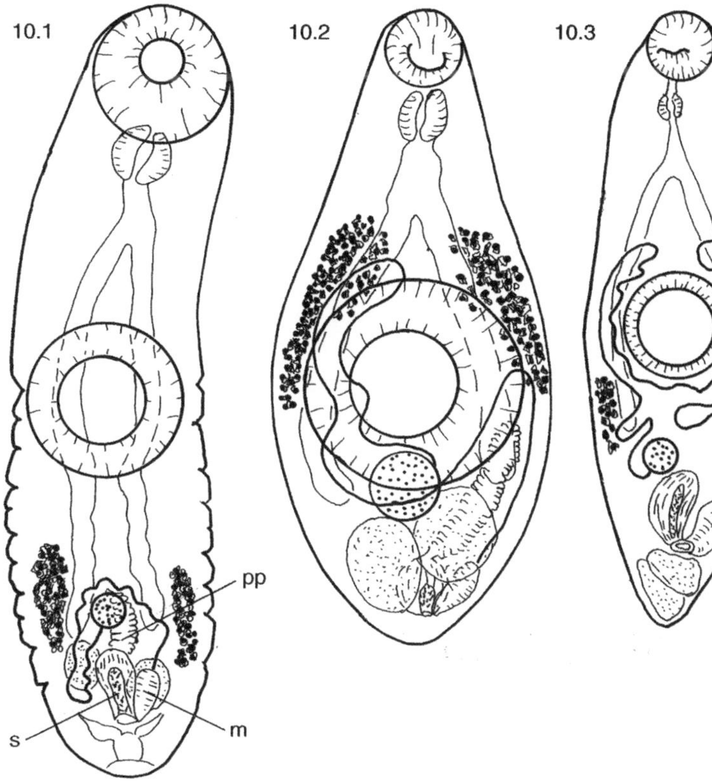

Figs 10.1–10.3.
10.1. *Amblosoma exile* Pojmańska, 1972. m, metraterm; pp, pars prostatica; s, spines on cirrus. (After Pojmańska, 1972a.)
10.2. *Leucochloridiomorpha lutea* (Baer, 1827). (After Pojmańska, 1971.)
10.3. *Ptyalincola ondatrae* Wootton & Murrell, 1967. (After Wootton & Murrell, 1967.)

reaching level of ovary. Ventral sucker larger than oral. Testes oval or transversely elongate, usually close to each other, at posterior extremity. Cirrus-sac large, just anterior to or between testes. Genital pore median, between or anterior to testes. Pars prostatica and metraterm ovoid. Ovary submedian, anterior and opposite to anterior testis. Vitelline fields short, extend from posterior margin of ventral sucker to region of gonads. Uterus with two ascending and two descending limbs, which envelop ventral sucker laterally and posteriorly, reaching just inside forebody. Parasites of rodents. Sporocyst and cercaria not known; metacercaria in *Anodonta*. North America. Type-species *P. ondatrae* Wootton & Murrell, 1967.

11 Family Moreauiidae Johnston, 1915

K. NIEWIADOMSKA

W. Stefański Institute of Parasitology, Polish Academy of Sciences, Twarda, Warsaw, Poland

Introduction

The monotypic subfamily Moreauiinae was established by Johnston (1915) for *Moreauia mirabilis* Johnston, 1915, a species having an aberrant morphology and unknown life-cycle. Johnston assigned it to the Harmostomidae Odhner, 1912 (later a synonym of the Brachylaimidae Joyeux & Foley, 1930). The same subfamily status within the Harmostomidae was accepted by Travassos (1928b). Its proper family relationship was, however, unknown. Dollfus (1934) placed it within the Liolopidae Odhner, 1912. Yamaguti (1958a) raised the group to full family status. This family, according to Travassos & Kohn (1966), is a member of the superfamily Brachylaimoidea Joyeux & Foley, 1930. Skrjabin (1948) reduced its rank to subfamily status among the Brachylaimidae.

The above-cited allocations for the Moreauiidae do not fit the aberrant morphology of *M. mirabilis*, especially the asymmetrical position of the gonads in relation to the cirrus-sac when the main axis of the body remains unchanged. This aberrant morphology rather suggests that the group should currently be treated within the Brachylaimoidea Joyeux & Foley, 1930, but of uncertain position and status.

Family Moreauiidae Johnston, 1915

Diagnosis: Body shape unusual, transversely elongate. Oral and ventral suckers small, of similar size; ventral sucker situated in middle of body. Prepharynx absent; pharynx globular; oesophagus very short; caeca long, narrow, reaching lateral extremities of body, reflexing and ending blindly at posterior end of body, near excretory vesicle. Testes lateral, on aporal side of body, separated from each other by ovary, transversely elongate, lobed, in loop of intestinal caecum. Large cirrus-sac on opposite side of body to testes, contains elongate seminal vesicle, long pars prostatica and spiny cirrus. Genital pore lateral. Ovary intertesticular. Seminal receptacle absent. Vitellarium follicular, extends in continuous lateral extracaecal fields, interrupted close to genital pore. Uterus runs transversely anterior to testis and posterior to ventral sucker, from region of ovary to genital pore. Eggs large, not numerous. Excretory vesicle Y-shaped, with arms giving off three main branches, which form anastomosing reticulum. In mammals (Monotremata). Australia. Type-genus *Moreauia* Johnston, 1915.

Genus *Moreauia* Johnston, 1915 (Fig. 11.1)

Diagnosis: With characters of family. Type-species *M. mirabilis* Johnston, 1915.

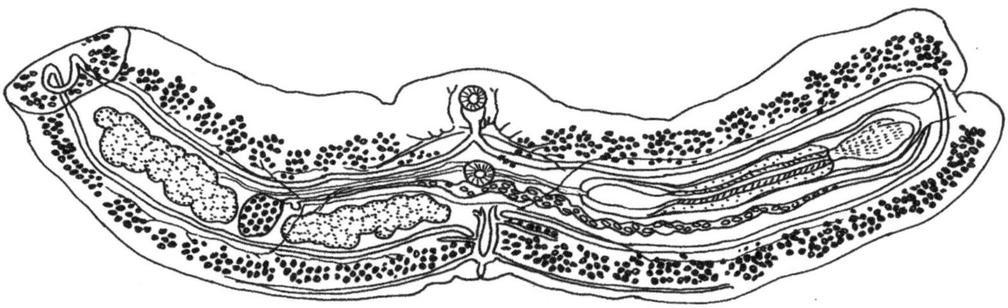

Fig. 11.1. *Moreauia mirabilis* Johnston, 1915. (Redrawn after Skrjabin, 1948.)

12 Family Ovariopteridae Leonov, Spasskii & Kulikov, 1963

T. POJMAŃSKA

W. Stefański Institute of Parasitology, Polish Academy of Sciences, Twarda, Warsaw, Poland

Introduction

The family Ovariopteridae was created by Leonov *et al.* (1963) for one species only and differs from the brachylaimids by the unusual shape of the ovary, the absence of a cirrus-sac and cirrus and the presence of a distinct genital sucker. The position of the ovary and the lack of a cirrus-sac bring this family close to the Thapariellidae, from which it differs in the extent of the vitellarium and uterus and in the absence of a cirrus. Nothing is known about the life-cycle of this species.

Family Ovariopteridae Leonov, Spasskii & Kulikov, 1963

Diagnosis: Body elongate, tongue-shaped, unarmed. Suckers in anterior part of body, relatively close to one another. Pharynx muscular, small; oesophagus absent; caeca terminating near posterior end of body. Gonads between caecal extremities; testes symmetrical; ovary post-testicular. Vitelline fields long and narrow, extending from level of ventral sucker to region of gonads. Uterus with one ascending and one descending limb, crosses body horizontally just anteriorly to ventral sucker. Genital pore terminal, surrounded by large genital sucker. Cirrus-sac and cirrus absent. Parasites of cloaca of birds. Life-cycle unknown. Type-genus *Ovarioptera* Leonov, Spasskii & Kulikov, 1963.

Genus *Ovarioptera* Leonov, Spasskii & Kulikov, 1963 (Fig. 12.1)

Diagnosis: With characters of family. Testes round. Ovary transversely elongate, deeply lobed, overlapping caeca. Parasites of Charadriiformes. Former USSR. Type-species *O. sobolevi* Leonov, Spasskii & Kulikov, 1963.

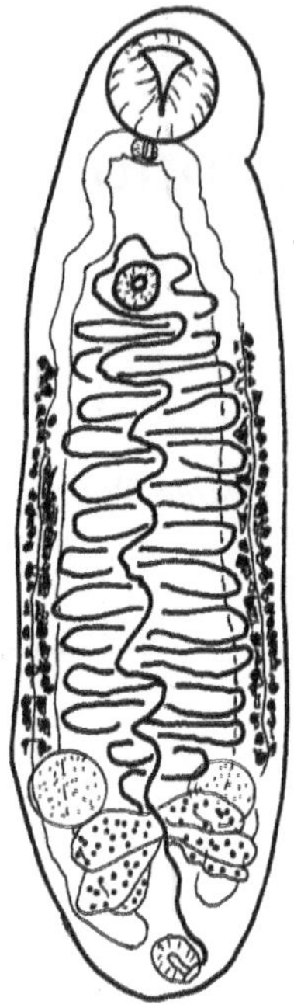

Fig. 12.1. *Ovarioptera sobolevi* Leonov, Spasskii & Kulikov, 1963. (After Leonov *et al.*, 1963.)

13 Family Panopistidae Yamaguti, 1958

T. Pojmańska

W. Stefański Institute of Parasitology, Polish Academy of Sciences, Twarda, Warsaw, Poland

Introduction

Panopistus Sinitsin, 1931, the type-genus of this family, has been either included in the subfamily Brachylaiminae Joyeux & Foley, 1930 (see Sinitsin, 1931; Kagan, 1952; Mehra, 1962a) or separated off in the subfamily Panopistinae Yamaguti, 1958 within the family Brachylaimidae (by Yamaguti, 1958a, 1971; Travassos & Kohn, 1966). The latter subfamily is here raised to full family status based on the general morphology and the type of life-cycle.

The panopistids differ from brachylaimids mainly in the general morphology of the adults, which more closely resembles that of the leuchochloridiids, while the position of the genital pore on the ventral surface of the body, as well as the type of life-cycle, with two intermediate hosts (Timon-David, 1965; Mas-Coma & Gallego, 1975), does more closely approximate that of the Brachylaimidae.

The generic status of *Pseudoleucochloridium* Pojmańska, 1959, reduced by Yamaguti (1971) to subgeneric level, is retained, as it differs from *Panopistus* in features considered here as of generic value.

Dollfusinus Biocca & Ferretti, 1958, included by Timon-David (1965) and Yamaguti (1971) in the family Leucochloridiidae, is included in the Panopistidae based on morphological similarities (such as the ventral position of the genital pore) and the type of life-cycle (with two intermediate hosts).

The family comprises four genera included previously in the subfamily Panopistinae: three by Mas-Coma & Gallego (1975) and one by Cribb (1992). These genera can be distinguished by the extent of the vitellarium, the course of the uterus and the position of the genital pore.

Family Panopistidae Yamaguti, 1958

Diagnosis: Body oval, armed or unarmed. Suckers well developed, in anterior part of body. Pharynx muscular; oesophagus absent; caeca reaching close to posterior extremity. Gonads in hindbody, with intertesticular ovary, arranged in tandem or somewhat in triangle. Genital pore opening on ventral surface, posterior to gonads. Vitellarium in posterior part of body, between ventral sucker and caecal extremities. Uterus occupying field posterior to ventral sucker or extending further anteriorly to level of caecal bifurcation. Eggs small, numerous. Intestinal parasites of mammals. Life-cycle terrestrial, with two intermediate hosts; metacercaria unencysted, in pericardium or kidney of second intermediate host. Type-genus *Panopistus* Sinitsin, 1931.

Key to genera

1a. Uterus not extending anteriorly to ventral sucker; anterior limits of vitelline fields do not reach level of ventral sucker .... *Panopistus* Sinitsin, 1931. (Fig. 13.1)

Diagnosis: Body elongate. Suckers in anterior part of body; ventral sucker at some distance from caecal bifurcation. Gonads arranged in tandem or ovary displaced slightly to one side. Genital pore posterior to caecal extremities. Vitellarium forming narrow, extracaecal fields in posterior part of body, between ventral sucker and caecal extremities. Uterus forming numerous coils of ascending and descending limbs, occupies field between ventral sucker and genital pore, not reaching posterior margin of ventral sucker. Life-cycle typical for family. North America. Type-species *P. pricei* Sinitsin, 1931.

1b. Uterus extending anteriorly to ventral sucker; vitelline fields reach level of ventral sucker .... 2.

2a. Ventral sucker close to caecal bifurcation; vitelline fields extend to caecal extremities; genital pore just posterior to caecal extremities .... *Pseudoleucochloridium* Pojmańska, 1959. (Fig. 13.2)

Diagnosis: Body oval, resembling Leucochloridiidae, unarmed. Suckers well developed; ventral near middle of body and close to caecal bifurcation. Gonads in tandem. Genital pore just posterior to gonads. Vitelline follicles forming narrow, lateral fields, between region of ventral sucker and caecal extremities. Uterus with two ascending and descending limbs, with anterior limit anterior to ventral sucker, crossing body horizontally posterior to ventral sucker. Sporocyst not known. Metacercaria in several terrestrial molluscs. Europe, Asia. Type-species *P. soricis* (Sołtys, 1952).

2b. Ventral sucker at some distance from caecal bifurcation; vitelline fields not reaching caecal extremities; genital pore posterior to caecal extremities .... 3.

3a. Genital pore almost terminal; uterus with two ascending and descending limbs .... *Dollfusinus* Biocca & Ferretti, 1958. (Fig. 13.3)

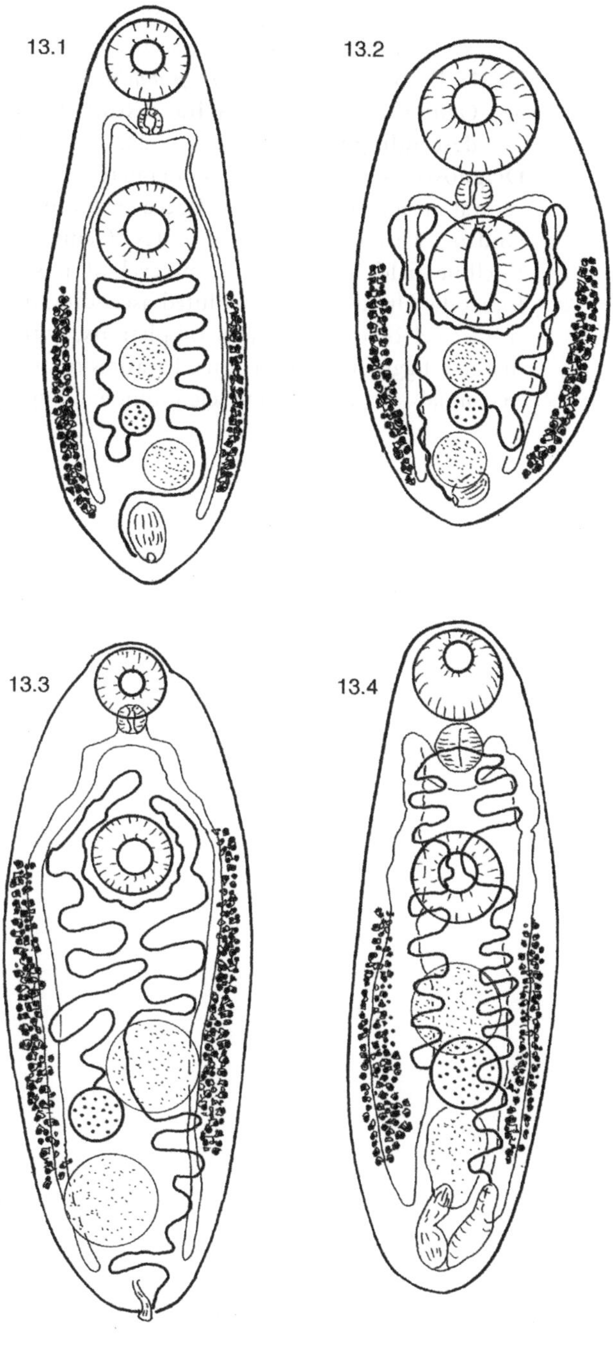

Figs 13.1–13.4.
13.1. *Panopistus pricei* Sinitsin, 1931. (After Sinitsin, 1931.)
13.2. *Pseudoleucochloridium soricis* (Sołtys, 1952) – metacercaria. (After Pojmańska, 1959.)
13.3. *Dollfusinus frontalis* Biocca & Ferretti, 1958. (After Biocca & Ferretti, 1958.)
13.4. *Dasyurotrema mascomai* Cribb, 1992. (After Cribb, 1992.)

Diagnosis: Body elongate, unarmed. Suckers in anterior part of body; ventral sucker at some distance from caecal bifurcation. Gonads arranged in triangle. Genital pore subterminal. Vitelline follicles form narrow lateral fields from region of ventral sucker to level of anterior margin of posterior testis. Uterus extends anteriorly into forebody, then descends and crosses body horizontally posterior to ventral sucker, forming corresponding ascending and descending coils on opposite side. Life-cycle typical for family. Europe. Type-species *D. frontalis* Biocca & Ferretti, 1958.

3b. Genital pore postcaecal, but distinctly ventral; uterus with one ascending and descending limb .... *Dasyurotrema* Cribb, 1992. (Fig. 13.4)

Diagnosis: Body plump to moderately elongate, armed with fine spines. Oral and ventral suckers of similar size. Caeca extending close to posterior extremity. Gonads in tandem, with ovary between testes. Genital pore posterior to caecal extremities. Vitelline fields extending from level of posterior testis to ventral sucker or into forebody. Uterus with convoluted ascending limb running dorsally to ventral sucker from level of ovary to level of pharynx, with similar descending limb; metraterm distinct. Eggs small, numerous. Parasites of alimentary tract and associated organs of marsupials. Australia. Life-cycle unknown. Type-species *D. mascomai* Cribb, 1992.

14 Family Thapariellidae Srivastava, 1953

T. Pojmańska

W. Stefański Institute of Parasitology, Polish Academy of Sciences, Twarda, Warsaw, Poland

Introduction

This family was erected by Srivastava (1953) and its status maintained by Travassos & Kohn (1966) and Yamaguti (1958a, 1971), while Mehra (1962a) treated it as a subfamily of the Brachylaimidae. The full family status is retained here, because of several features specific to these brachylaimids, i.e. the lack of a true cirrus-sac, the post-testicular position of the ovary and vitelline fields posterior to the testes. In relation to the life-cycle, only the definitive and second intermediate hosts are known.

The family comprises only one genus. Goodman's (1990) proposal to include *Amblosoma* Pojmańska, 1972 in this family is rejected. Although these two genera have some common features, especially in their general morphology, and probably have a similar life-cycle with two intermediate hosts (Viviparidae, according to Prudhoe, 1957, and Rai & Pande, 1967, cited from Goodman, 1990), there are, nevertheless, important differences, such as the position of the genital pore, the position of the ovary and the presence of a well-developed cirrus-sac, which align *Amblosoma* with the family Leucochloridiomorphidae. Perhaps these two families should be considered as closely related.

Family Thapariellidae Srivastava, 1953

Diagnosis: Body elongate, unarmed. Suckers well developed: oral smaller than ventral; pharynx well developed, muscular; oesophagus very short, practically absent; caeca do not reach posterior extremity. Gonads near posterior extremity; testes symmetrical; ovary post-testicular, submedian; true cirrus-sac absent, but prostatic glands well developed; cirrus armed with spines. Genital pore dorsoterminal. Vitelline fields very short, posterior to testes. Uterus short, forming several coils in region of gonads, crossing body horizontally between testes and ovary; eggs non-operculate. Parasites of birds. Sporocyst not known; metacercaria in viviparid snails. India. Type-genus *Thapariella* Srivastava, 1953.

Genus *Thapariella* Srivastava, 1953 (Fig. 14.1)

Diagnosis: With characters of family. Caeca do not extend posteriorly to level of ovary. Testes slightly diagonal, almost round in shape. Ovary smaller than testes. Type-species *T. anastomusa* Srivastava, 1953.

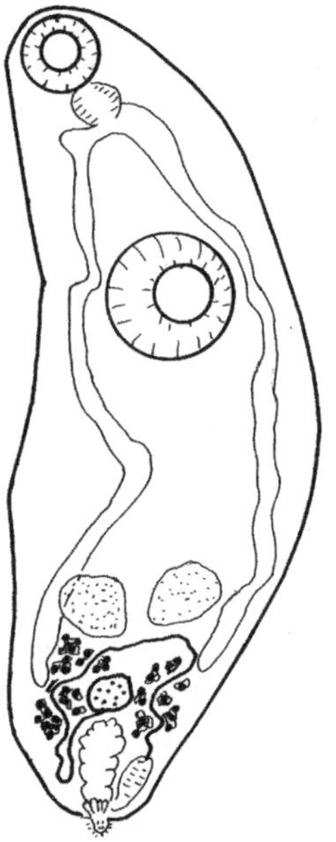

Fig. 14.1. *Thapariella anastomusa* Srivastava, 1953. (After Srivastava, 1953.)

15 Superfamily Bucephaloidea Poche, 1907

R.M. Overstreet & S.S. Curran

Gulf Coast Research Laboratory, The University of Southern Mississippi, USA

Introduction

For the purpose of this chapter, the superfamily Bucephaloidea Poche, 1907 can be considered to include only the Bucephalidae Poche, 1907. The Gasterostomatidae [*sic*] Braun, 1883 and the Sinicovothylacidae Yu, 1965 have at one time or another been considered separate families, but are herein considered synonyms and will be discussed below. Also, we tentatively do not accept the Nuitrematidae Kurochkin, 1975, but discuss it and provide the reported description.

The Bucephalidae is a cosmopolitan family parasitic in marine, brackish-water and freshwater fishes. Lamellibranch bivalves serve as molluscan intermediate hosts for the mother-sporocysts and cercariae, fishes serve as intermediate hosts for the metacercariae, and fishes and, at least in one case, an amphibian (*Bucephalopsis kweiyangensis* Chu, 1950 in the giant salamander *Megalobatrachus japonicus* in Kweichow Province, China; see Chu, 1950) serve as definitive hosts. Bucephalids differ from all other digeneans by the configuration of their digestive system, by the configuration of their terminal genitalia and by the presence of an anterior rhynchus for attachment that is disassociated from the digestive system. The rhynchus is an apical holdfast organ that differs in form, depending on the species of worm. The digestive system consists of a muscular pharynx that opens ventrally, an oesophagus and a simple caecum. The caecum superficially resembles that of a rhabdocoel turbellarian and may be sac-like or elongate but never bifurcate. Neither an oral nor ventral sucker is present in members of the family. The terminal genitalia are similar to those occurring in members of the Cyathocotylinae Mühling, 1898; however, unlike many cyathocotylines, bucephalids lack a true cirrus. Most members have an intromittent organ, typically termed a 'genital lobe' or 'genital tongue'. This structure, often with a papilla and partly surrounding a section of the ejaculatory duct, has been considered a cirrus by some workers. Yamaguti (1971) implied that it may also incorporate the terminal female duct, and we confirmed that arrangement in several species. It projects into the genital atrium and can protrude through the genital pore, as it may serve to deposit or guide sperm into the genital atrium of a mate. Even Yamaguti (1971), who defined a sac containing the undifferentiated 'cirrus portion' of the male duct as a 'false cirrus pouch', later called it a true cirrus-sac. At least some species appear to produce a spermatophore (Odhner, 1905; Ozaki, 1928; Crowcroft, 1947), and that structure will be discussed under the Prosorhynchinae Nicoll, 1914. The atrium is associated with gland-cells in many species.

The term cirrus-sac applied to bucephalids could be somewhat misleading, but the common practice of referring to the organ as such is retained in this chapter. The bucephalid terminal genitalia consist of the 'cirrus-sac' and is associated with a posterior sac, which functions as the genital atrium and opens externally by a genital pore. The complex occurs at the posterior end of the body, with the pore either terminal or subterminal on the ventral surface. The cirrus-sac contains a seminal vesicle at the proximal end, joining with the male duct. The male duct, also called the ejaculatory duct in the broad sense, consists of at least the pars prostatica, which may have several different types of associated prostatic cells, and a distal long or short, smooth-walled ejaculatory duct. The pars prostatica of many species has two or more well-differentiated portions. For purposes of this chapter, we refer to the distinct, proximal, usually narrow and often convoluted portion of the pars prostatica that joins with the seminal vesicle and occurs in at least the prosorhynchines as the 'seminal duct'.

Odhner (1905) noticed differences in the genital arrangement between species of *Prosorhynchus* Odhner, 1905 and forms then included in *Gasterostomum* von Siebold, 1848. Nicoll (1914a) separated the family into the Bucephalinae and Prosorhynchinae, based partly on Odhner's (1905) observations of the terminal genitalia. To our knowledge, subsequent classifications of the bucephalids relied mainly on rhynchal characters and the relative position of the ovary to the testes, and they did not stress the importance of the terminal genitalia for subfamily characters. Kniskern (1952) reviewed the characters important for taxonomic classification in the group and, along with Nagaty (1937), advocated that the structure of the rhynchus and the configuration of its accessory structures comprised the primary characters for separating subfamilies and genera. Both of these authors held that the location of the ovary in relation to the testes was of secondary importance for determining genera when adequate numbers of specimens were available. Kniskern (1952) considered peculiarities of the vitellarium as specific characters and considered the extent and direction of the caecum to be of doubtful taxonomic value, since intraspecific variation in this character had been observed in some. Yamaguti (1958a, 1971) created new subfamilies based on the relative position of the ovary to the testes in conjunction with rhynchal features and achieved a more accurate grouping of the bucephalid subfamilies than before, but he failed to recognize the relationship between *Paurorhynchus* Dickerman, 1954 and *Bellumcorpus* Kohn, 1962, and also he neglected to consider characters of the terminal genitalia.

Stunkard (1974) believed that rhynchal characters were poor characters at the generic and subfamily level and that they were better suited for classifying taxa at the specific level. Instead, he advocated the use of flame-cell patterns in cercariae for making generic comparisons. We interpret Stunkard's hypothesis to mean that mutations in genes underlying digenean attachment structures are more likely to be subjected to adaptive modification caused by ecological changes than are random mutations of the genetic components responsible for non-attachment structures, such as the excretory system or components of the reproductive system. His reasoning implied that, because the rhynchus was exposed to drastic host stimuli, its genetic component was not as conserved as that of the excretory system. In support of that argument, Stunkard (1974) compared the cercarial flame-cell pattern for *Prosorhynchus squamatus* Odhner, 1905 with that for *Prosorhynchus crucibulum* (Rudolphi, 1819) and proposed the new genus *Rudolphinus* to accommodate *P. crucibulum*. We agree that rhynchal characters are less informative for classifying some bucephalids at the generic level (e.g. *Alcicornis* MacCallum, 1917 and *Bucephalus* Baer, 1827, both members of Bucephalinae Poche, 1907). But Stunkard's emphasis on the excretory system as a source for generic characters is problematic because the excretory system has not been studied for many species. Furthermore, Stunkard (1974) seemed to be inconsistent as to what constituted a generic difference in flame-cell formulae. He seemed to have arbitrarily decided that the flame-cell pattern for *P. squamatus* {2[(3 + 4 + 4) + (3 + 4)]} was sufficiently different from that of *P. crucibulum* {2[(3 + 2 + 4) + (3 + 3 + 3)]} to warrant the creation of a new genus for the latter. Yet Stunkard (1975) then reported the pattern of flame-cells as {2[(2 + 2) + (2 + 2)]} for both *Rhipidocotyle transversale* Chandler, 1935 and *R. lintoni* Hopkins, 1954, but he also included *R. papillosa* (Woodhead, 1929), which had a flame-cell pattern of 2[(5 + 5 + 5 + 5) + (5 + 5 + 5 + 6)], *R. septpapillata* Krull, 1934, with a pattern of 2[(3 + 3 + 3) + (3 + 3 + 3)], and *R. lepisostei* Hopkins, 1954, with a pattern of 2[(4 + 4 + 4) + (4 + 4 + 4)]. We see no reason to subdivide *Prosorhynchus* or *Rhipidocotyle* Diesing, 1858 on the basis of these flame-cell patterns.

Rather than using components of the rhynchus to determine subfamilies, we advocate that the shape of the seminal vesicle, its length and, in some cases, the thickness of the prostatic duct should be used to group bucephalids into subfamilies. However, we need to emphasize that the terminal genitalia are both difficult to observe in some individuals and incorrectly illustrated or overly conventionalized in many described species. Consequently, all published illustrations and descriptions do not fit our keys, and there will be some exceptions in these basic features, especially in the shape of the seminal vesicle in immature and other specimens.

Characters derived from the rhynchus are best used for generic and specific determination. Also, the position of the ovary relative to the testes is a valuable generic character, as advocated by Kniskern (1952). We, however, advocate that specimens be heat-killed, because killing specimens under pressure, especially in a cool condition, can produce an abnormal alignment of gonads. Other reproductive traits, such as the structure of gland-cells within the cirrus-sac and the configuration of the female reproductive system, may prove to be valuable sources for generic characters in future work. We follow a modification of the classification of the Bucephalidae proposed by Skrjabin & Guschanskaja (1962), with some genera transferred into different subfamilies. We propose that the subfamilies Neidhartiinae Yamaguti, 1958 and Neoprosorhynchinae Yamaguti, 1958 not be recognized. We base our modified classification on the features of the terminal genitalia for subfamily designation and on the position of the ovary relative to the testes as well as intestinal and some rhynchal characters for generic-level distinctions. We consider the structure of the rhynchus to be informative in many but not all cases and therefore treat it as a less important generic character. Some gut features are useful at the subfamily level. For example, the gut is long and tubular in species of the Dolichoenterinae Yamaguti, 1958; long, tubular and recurving in species of the Paurorhynchinae Dickerman, 1954; and saccular in species of the Bucephalinae and Prosorhynchinae. The saccular gut of species in the Prosorhynchinae is usually directed anteriorly, but it may be directed either anteriorly or posteriorly in species of the Bucephalinae. For this reason, the gut orientation is best used as a specific-level character. At least one bucephaline species, *Prosorhynchoides ovatus* (Linton, 1900), has a long tubular gut that recurves.

Classification

The systematic placement of the Bucephalidae within Digenea has been debated since Odhner (1905) considered it a basal group on the basis of the presence of a ventral mouth and a simple sac-like or tubular gut. He presumed that the ventral gut of bucephalids was an ancestral condition homologous with that of rhabdocoel turbellarians. On this premise, he proposed that the Digenea be split into two suborders: the Gasterostomata, containing the bucephalids, and the Prosostomata, containing all other digenean groups. A single 'mother'-sporocyst generation in bucephalids has been considered an ancestral condition by some (e.g. Odening, 1974), but a secondary loss of the daughter-sporocyst is just as plausible (Gibson, 1987). Woodhead (1931) and Brooks *et al.* (1985) considered the bucephalids to have a daughter-sporocyst generation, but evidence for this is weak (see Gibson, 1987, for discussion of life-cycle). Numerous authors (e.g. Sewell, 1922; La Rue, 1926, 1957; Stunkard, 1946; Cable, 1974; Brooks *et al.*, 1985; Gibson, 1987) believed that there was evidence that the Bucephalidae was not the ancestral sister taxon to the rest of the digeneans. Sinitzin (1910) compared the morphological features of the cercaria of *Bucephalus polymorphus* Baer, 1827 with those of a furcocercous strigeoid cercaria of *Trichobilharzia ocellata* (La Valette, 1855) and came to the conclusion that the rhynchus was nothing more than a modified oral sucker and that the distome ventral sucker was modified to form the mouth and pharynx in bucephalids. Sewell (1922) considered, on the basis of the features of *Bucephalopsis haimeanus* Lacaze-Duthiers, 1854 as described by Tennent (1906), that the glands associated with the rhynchus appeared to be homologous to those of the cephalic glands of strigeoid cercariae, and he also believed that the similarity in form of the bucephalid gut and the rhabdocoel turbellarian gut was an example of convergence. Matthews (1973a) described the relationship between the nervous system and the rhynchus in *Prosorhynchus squamatus* as similar to that present in conventional digenean groups having an oral sucker, indicating to Gibson (1987) a homology between the bucephalid rhynchus and the digenean oral sucker. Matthews (1973a) also provided evidence that the rhynchus in the cercaria of *P. squamatus* was homologous to the anterior penetration organ present in the Strigeida (also referred to by some as the Strigeiformes). La Rue (1926) noted the similarity between the terminal genitalia of bucephalids and those of cyathocotylines, suggesting to him a phylogenetic link between the Bucephalidae and the superfamily 'Strigeoidea'. He proposed a new systematic classification of the Bucephalidae, erecting the order Strigeatoidea (Strigeiformes, after Travassos *et al.*, 1969) to contain the suborders Strigeata, Schistostomata and Bucephalata. This arrangement inherently adopted Sinitsin's (1910) contention that the bucephalid rhynchus was homologous to the oral sucker, and most taxonomists accept that relationship. La Rue (1957) augmented his original classification only slightly. Cable (1974) considered the bucephalids to be distinct from the 'Strigeiformes' but closely related to the Brachylaimidae Joyeux & Foley, 1930 and the Fellodistomidae Nicoll, 1909, based on the presence of the following common characters: eggs small, miracidia with cilia confined to patches or appendages, cercarial tail stem lacking flame-cells and cercaria shaped like an 'ox-head', even though only one non-bucephalid species, a fellodistomid, was noted to have a cercaria similar to that of bucephalids. None of these similarities is convincing. Brooks *et al.* (1985) determined that the bucephalids were closely related to the Brachylaemidae within the 'Strigeiformes', based primarily on cercarial tail features. We advocate the placement of the Bucephalidae within the conventional higher classification of trematodes as a derived family within the Strigeida, similar to that advocated by La Rue (1957). We follow the higher classification proposed for the Bucephalidae by Brooks *et al.* (1985) and consider the family to be in the order 'Strigeiformes' (as the Strigeida) and superfamily Bucephaloidea. Regardless, no clear agreement on the relationship between the Bucephalidae and the rest of the digeneans occurs in the literature. We present two strict consensus trees based on dissimilar outgroups, with the intention of providing some insight into the relationships among genera of the Bucephalidae. The first cladogram (Fig. 15.1) represents a polytomy based on an outgroup that consists of *Holostephanus anhingi* Mehra, 1943 and *Szidatia* sp., both cyathocotylids. We chose these outgroup species to test the hypothesis that the Bucephalidae possibly represents a derived group of digeneans within the Strigeida (see Brooks *et al.*, 1985). Moreover, we chose cyathocotylids because La Rue (1926) noted a striking similarity between the terminal genitalia of the cyathocotylids and the bucephalids. The tree represents a polytomy in which no relationship among the subfamilies is amplified. However, some relationships among prosorhynchine genera are supported, albeit weakly.

The second cladogram (Fig. 15.2) is derived using similar character data but with different outgroups. Instead of cyathocotylids, we used two outgroup species from separate genera in the Urotrematidae that infect freshwater fishes in China. These species have some adult features that superficially resemble those of the Bucephalidae, but their larval stages are unknown. The resulting tree supports monophyly of the Paurorhynchinae, the Prosorhynchinae and the Dolichoenterinae, but fails to support monophyly of the Bucephalinae. In addition, the tree indicates that the Paurorhynchinae is the sister group to the rest of the bucephalids and the Dolichoenterinae is the sister group to the paraphyletic branch that includes the Prosorhynchinae and the Bucephalinae. Neither tree is supported by a sufficient number of characters and both sets of outgroups are probably too distantly related to the Bucephalidae to provide a meaningful root for the bucephalids. Nevertheless, these trees provide a starting place for a future analysis based on more species and features.

In preparation for this chapter, we examined representatives from as many of the genera discussed below as we could obtain from lending museums or from our own collections; however, we were not able to procure specimens of species in many genera and were therefore forced to rely on the literature for our assessment of those genera. Our observations are presented below in paragraphs summarizing the history and characteristics of the above-named genera.

Family Bucephalidae Poche, 1907

The Bucephalidae was formerly known as the Gasterostomatidae [*sic*]. Poche (1907) recognized *Bucephalus polymorphus* as the earliest named 'gasterostomatid' and declared *Gasterostomum* a synonym of *Bucephalus*. As a result, the Bucephalidae, established for the oldest nominal gasterostomatid genus *Bucephalus*, replaced Gasterostomidae as the

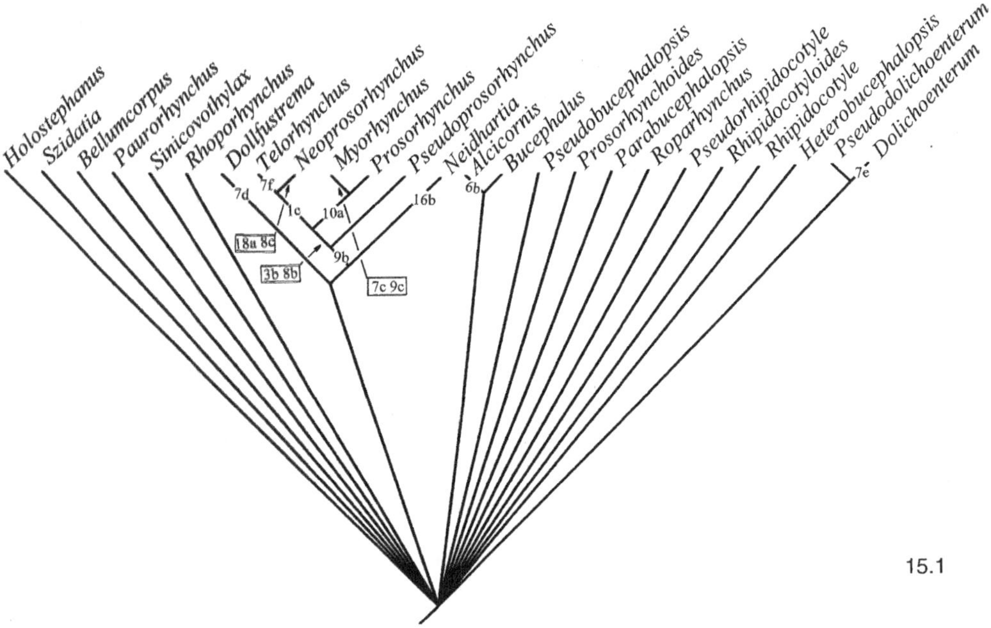

15.1

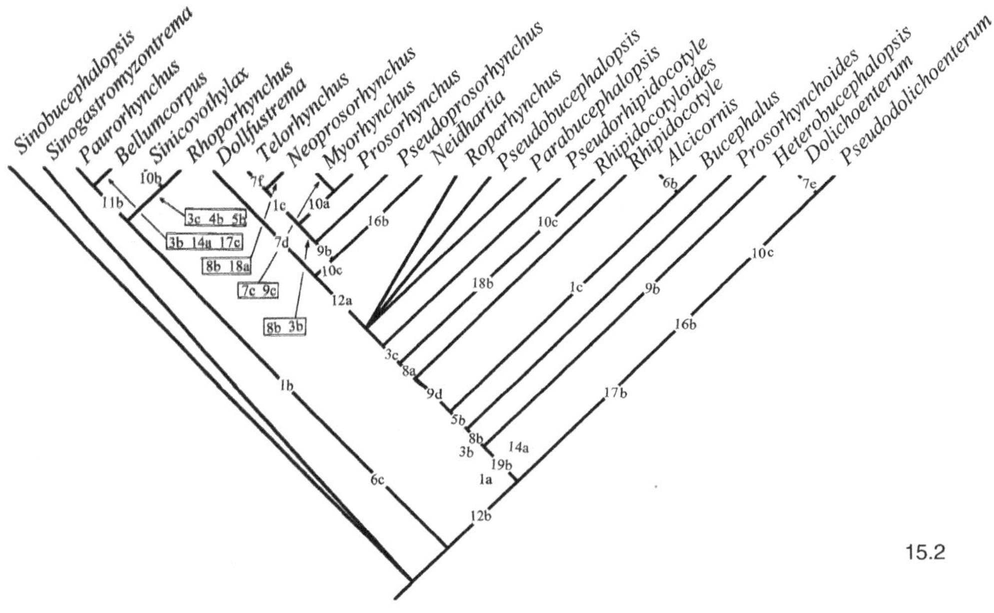

15.2

family name. Nicoll (1914a) declared *Gasterostomum* a *nomen nudum* and also considered it a junior synonym of *Bucephalus*. Recently (Anon., 1993), the International Commission on Zoological Nomenclature ruled indubitably that *Gasterostomum* was rejected as invalid and relegated it to a junior synonym of *Bucephalus*.

Subfamily Bucephalinae Poche, 1907

The Bucephalinae was erected to house all bucephalids consistent with Odhner's (1905) description of *Gasterostomum* von Siebold, 1848 (see Nicoll, 1914a). Odhner (1905) described *Gasterostomum* as having species with a simple sucker for a rhynchus, a vitellarium with two distinct groups of follicles, a genital lobe that

Figs 15.1–15.2. *(Opposite)* Phylogenetic trees for genera of the Bucephalidae using different outgroups. The proposed trees are each strict consensus trees derived from unrooted heuristic searches of maximum parsimony conducted using PAUP* Version 4.0b2. Morphological character states were derived from original generic descriptions and, when possible, from specimens. Numbers located on the branches indicate synapomorphic traits (shared character states that support monophyly of that branch). **15.1.** Outgroups are rooted by two species in the Cyathocotylinae, with the tree derived using 25 taxa and 18 unordered and equally weighted characters. Prosorhynchine genera form a clade. **15.2.** Outgroups are rooted by two species in the Urotrematidae, with the tree derived using 25 taxa and 19 unordered and equally weighted characters. The paurorhynchine genera form the basal bucephalid clade, and dolichoenterine genera (consisting of *Dolichoenterum* and *Pseudodolichoenterum*) form the sister clade to combined prosorhynchine and bucephaline genera. The prosorhynchine genera form the same clade in this tree and that in Fig 15.1, and the monophyly of the Bucephalinae is not supported in either tree.

Character states:

1. Body ellipsoid (1a)/ body elongate cylindrical (1b)/ body elongate relatively flat (1c).
2. Tegument spined (2a)/ tegument smooth (2b).
3. Pharynx opening anterior third of body (3a)/ around mid-body (3b)/ posterior to mid-body (3c).
4. Gut directed posteriorly, dorsally or recurving (4a)/ gut directed straight anteriorly (4b).
5. Gut shape tubular (5a)/ sac-like (5b).
6. Anterior holdfast organ with sucker (6a)/ with funnel (6b)/ with reduced sucker (6c).
7. Rhynchal modification none (7a)/ tentacles (7b)/ lobes (7c)/ spines (7d)/ papillae (7e)/ long spines (7f).
8. Ovary located between or at level of testes (8a)/ ovary pretesticular (8b)/ ovary post-testicular (8c).
9. Vitellarium in two bands entirely anterior to ovary (9a)/ in a confluent bunch entirely anterior to ovary (9b)/ in two bands not entirely anterior to ovary (9c)/ in two condensed bunches near ovary or not (9d).
10. Testes oblique (10a)/ testes symmetrical (10b)/ testes tandem (10c).
11. Testicular outline smooth (11a)/ dendritic or lobed (11b).
12. Seminal vesicle elongated (12a)/ spherical to ellipsoid (12b).
13. Pars prostatica straight (13a)/ bent slightly (13b)/ coiled once proximally as spermatic duct (13c).
14. Genital lobe well developed (14a)/ not well developed (14b).
15. Spermatophore formation absent (15a)/ present (15b).
16. Cirrus-sac wall up to 25 μm thick (16a)/ greater than 25 μm thick (16b).
17. Prostatic duct of uniform width (17a)/ proximal portion narrower than distal portion (17b)/ proximal portion wider than distal portion (17c).
18. Uterus entirely postovarian (18a)/ entirely pre-ovarian (18b)/ reaching anterior and posterior to ovary (18c).
19. Genital atrium poorly developed (19a)/ well developed (19b).

extends from the base of the genital atrium and may be extruded through the genital pore, a delicate-walled ejaculatory duct that extends into the genital atrium from the prostatic duct but does not evaginate, and a short uncoiled seminal vesicle. At the time, Nicoll (1914a) included *Bucephalus*, *Bucephalopsis* Diesing, 1855 and *Rhipidocotyle*, but he excluded *Gasterostomum* because he considered it a junior synonym of *Bucephalus*, which he named type-genus for the subfamily. *Nannoenterum* Ozaki, 1924, *Dolichoenterum* Ozaki, 1924, *Prosorhynchoides* Dollfus, 1929, *Bucephaloides* Hopkins, 1954, *Parabucephalopsis* Tang & Tang, 1963, *Pseudorhipidocotyle* Wang & Pan[1] in Long & Lee, 1964b, *Pseudobucephalopsis* Long & Lee, 1964, *Rhipidocotyloides* Long and Lee, 1964, *Pararhipidocotyle* Kohn, 1970, *Pseudodolichoenterum* Yamaguti, 1971, *Labratrema* Maillard, 1975, *Roparhynchus* Gupta & Kumari, 1978, *Heterobucephalopsis* Gu & Shen, 1983 and *Glandulorhynchus* Thatcher, 1999 have since been added. In this chapter, we consider the Bucephalinae to include all those genera with species that possess terminal genitalia consisting of a relatively thin- to moderately thick-walled (< 20 µm thick) cirrus-sac containing a spherical, ovate or ellipsoidal seminal vesicle that is usually less than twice as long as wide (except *Prosorhynchoides longicirrus* (Nagaty, 1937) n. comb. for *Bucephaloides longicirrus*) and seldom bipartite (e.g. bipartite in *Prosorhynchoides kweiyangensis* (Chu, 1950) n. comb. for *Bucephaloides kweiyangensis*). The pars prostatica connects to the posterior end of the seminal vesicle and runs straight to the posterior end of the cirrus-sac (the pars prostatica bends slightly near the junction of the seminal vesicle in *Rhipidocotyle*). The rhynchus of members of the Bucephalinae may be highly variable in form but generally consists of a simple sucker or a sucker with a dorsal modification. We include one genus (*Alcicornis*) in our key to the Bucephalinae that has a rhynchus radically different from those of other members of the subfamily. Members of *Alcicornis* have a long conical rhynchus that superficially resembles that found in members of the Prosorhynchinae; however, the tentacles associated with the rhynchus are structurally identical to the tentacles in species of *Bucephalus*. Also, two longitudinal bands of muscle that meet posteriorly, forming a V-shaped groove, are present on the ventral side of the rhynchal cone of *A. carangis* MacCallum, 1917. This muscular groove may be homologous to the rim of a sucker. We doubt that retractable tentacles evolved convergently in *Alcicornis* and *Bucephalus*, given their many other shared characters. Both marine and freshwater representatives are known from the Bucephalinae. We include the following genera in the key: *Bucephalus*, *Alcicornis*, *Rhipidocotyle*, *Prosorhynchoides* and *Heterobucephalopsis*, plus four genera, *Pseudorhipidocotyle*, *Parabucephalopsis*, *Pseudobucephalopsis* and *Roparhynchus*, that we consider *genera inquirenda*.

History and features of genera

Genus *Bucephalus* Baer, 1827

Bucephalus was initially established to represent the earliest known bucephalid, *B. polymorphus* Baer, 1827, on the basis of a cercaria from the European unionids (freshwater), *Unio pictorum* and *Anodonta* sp. von Siebold (1848) erected the genus

[1] Also transliterated as Pang.

Gasterostomum for all adult trematodes with a ventral mouth, believing that a European adult bucephalid that he named *Gasterostomum fimbriatum* von Siebold, 1848, and which Yamaguti (1958a) transferred to the genus *Bucephalus*, represented the adult stage of *B. polymorphus*, although this was not founded on experimental evidence. Even today, the adult stage of *B. polymorphus* has not been established convincingly. Despite this, most early European authors believed that *B. polymorphus* equated with *B. fimbriatum*. Diesing (1855) departed from the convention of his time and described the subgenus *Eubucephalus* to house *B. polymorphus*, but he rejected his own work 3 years later (Diesing, 1858). Nicoll (1914a) advocated the use of the name *Bucephalus* for bucephalids with tentacles, although he too had no direct evidence that cercariae designated as *B. polymorphus* developed into such adults. Various workers have shed light on some of the confusion surrounding *Bucephalus* by conducting life-history experiments that associated larval stages with adults. Woodhead (1930) discovered the life-cycle for *Bucephalus elegans* Woodhead, 1930, an American freshwater species with similar adult features to those of *G. fimbriatum*, and he consequently established adult and cercarial characters diagnostic for a species similar to both *B. polymorphus* and *G. fimbriatum*. A study by Taskinen *et al.* (1991) showed that another freshwater cercaria that resembled *B. polymorphus* developed to an adult belonging in *Rhipidocotyle*.

Results of some life-history studies have added questions to the problem of taxonomy within *Bucephalus*. For example, Maillard (1975) exposed unidentified bucephalid cercaria from *Cardium glaucum* in the Mediterranean Sea to several species of gobies and then fed those fishes to *Dicentrarchus labrax*. He obtained what he believed to be adult bucephalids, for which he erected *Labratrema* Maillard, 1975 and established the new combination *L. lamirandi* (Carrère, 1937). Based on the literature, that species strongly resembles both the adult form of *B. minimus* (Stossich, 1887) and the metacercarial stage of a species Matthews (1973b) experimentally obtained from the cercaria he believed to be *B. haimeanus*. The excretory system of Matthews' cercaria differed from that of the cercaria examined by Maillard, but we suspect the worms are very closely related forms that have both been considered synonymous with the cercaria referred to in the past as *Bucephalus* (*Bucephalopsis*) *haimeanus* (Lacaze-Duthiers, 1854). It is likely that the adult worm that Maillard described as *L. lamirandi* is actually *B. minimus* or a closely related species. In any event, we propose that the generic name *Labratrema* be considered a junior synonym of *Bucephalus*, producing *B. lamirandi* (Carrère, 1937) n. comb. Both *B. lamirandi* and *Bucephalus* sp. of Matthews (1973b) should be compared with *B. minimus* from *Dicentrarchus labrax* from off Trieste, Italy, in the Adriatic Sea.

To address the taxonomic problem of priority between *Bucephalus* and *Gasterostomum*, Baturo (1992) submitted an appeal to the Commission to review the problems associated with *Bucephalus*. She suggested, among other things, that *Bucephalus polymorphus* Baer, 1827 be considered the type-species for the genus *Bucephalus* and that *Gasterostomum* be considered a junior synonym of *Bucephalus*. Srivastava (1992) submitted a counter-appeal to the Commission, suggesting that *Gasterostomum* be conserved for adults currently referred to as *Bucephalus*, with *Gasterostomum fimbriatum* von Siebold, 1848 as type-species, and that *Rhipidocotyle galeatum* (Rudolphi, 1819) should take precedence over what is currently referred to as *B. polymorphus*, because life-cycle experiments suggested this to be the more likely classification of the species in question. The Commission (Anon., 1993) ruled that *Gasterostomum* be considered a junior objective synonym of *Bucephalus* and placed on the *Official Index of Rejected and*

Invalid Generic Names in Zoology, at least partly to avoid the repercussions of making drastic changes to the commonly accepted nomenclature within the group. By following the nomenclature established by the Commission's ruling, we treat *B. polymorphus* Baer, 1827 as the type-species, with the lectotype of *G. fimbriatum* von Siebold, 1848 (Zoological Museum, Humbolt University, Berlin, no. 1655b) designated as the neotype. The name '*fimbriatum* Siebold, 1848' was placed on the *Official Index of Rejected and Invalid Specific Names in Zoology*.

The primary generic characters for *Bucephalus* are a pretesticular ovary and features of the rhynchus. The rhynchus consists of a sucker that is topped by a ring of 6 to 21 retractable tentacles, depending on the species. The tentacles may have one or two lobes extending from the posterior surface, and retractable papillae may be present on the terminal end of tentacles. Most species have a relatively thin-walled cirrus-sac (< 15 µm thick). The male duct is straight and consists of a pars prostatica and a distal thin-walled ejaculatory region; the pars prostatica may have two or three distinct regions (but no seminal duct), with different types of prostatic cells. The ejaculatory duct opens into the genital atrium in close association with the protrusible genital lobe.

Genus *Alcicornis* MacCallum, 1917

Despite the superficially similar appearance of the funnel- or cone-shaped rhynchus with that of species in the Prosorhynchinae, we treat *Alcicornis* as a member of the Bucephalinae because the terminal genitalia are arranged in the same way as other members of the subfamily. The rhynchus in members of *Alcicornis* either has retained an ancestral state (cone-like structure) or has been modified from a sucker to a cone-like structure, in which case it should be in the Prosorhynchinae. There appears to be a muscular rim lining a ventral fold in the rhynchus of *A. carangis* MacCallum, 1917, which is similar in structure to the rim of a sucker in *Bucephalus* spp. A tentacle from either *Bucephalus* or *Alcicornis* may be variously equipped with no, one, or two posterior lobes, depending on the species, but a species with any particular arrangement may occur in either genus. For example, *A. carangis*, *A. cirrudiscoides* Velasquez, 1959 and *A. baylisi* Nagaty, 1937 have tentacles with two posterior lobes per tentacle, whereas *A. longicornutus* Manter, 1954 and *A. siddiqii* Nahhas & Cable, 1964 have tentacles that have only one posterior lobe. Similarly, *Bucephalus cynoscion* Hopkins, 1956 has tentacles with two posterior lobes and *B. scorpaenae* Manter, 1940 has tentacles with one posterior lobe. The tentacles from the two genera appear to be homologous, further indicating that the genera may be closely related. The number of posterior lobes is a specific character for both genera. Yamaguti (1971) considered *A. carangis* synonymous with *Bucephalus*, but we consider it a distinct genus on the basis of the cone-shaped rhynchus and lack of a typical sucker associated with the rhynchus. To consider *Alcicornis* in the Prosorhynchinae based on the funnel-shaped rhynchus would necessitate a less parsimonious evolutionary equation, in which both tentacles and the terminal genitalia would have to have evolved twice, rather than once if we placed the genus in the Bucephalinae.

Genus *Prosorhynchoides* Dollfus, 1929

The generic name for bucephalids having a simple sucker for a rhynchus has been debated for a long period of time. Diesing (1855) erected *Bucephalopsis* as a subgenus for the species *Bucephalus haimeanus* Lacaze-Duthiers, 1854, based on the cercaria from the oyster *Ostrea edulis* and clam *Cardium rusticum* in Mahon in the Balearic Isles, resulting in the combination *Bucephalus* (*Bucephalopsis*) *haimeanus* (Lacaze-Duthiers, 1854). Tennent (1906) believed that he had experimentally elucidated the life-cycle of the same cercaria but in a different oyster and a gar in North America. He considered the adult that he found in the gar to be conspecific with *Gasterostomum gracilescens*, which had been described from *Lophius piscatorius* from off Britain. Extrapolating from his North American work, he believed that *B. haimeanus* developed into *G. gracilescens*. His assumption has never been demonstrated experimentally. Nicoll (1914a) believed a cercaria from the British oyster that Lebour (1908) had experimentally shown to develop into a metacercaria similar to the adult named *G. gracilescens* was conspecific with *B. haimeanus* from the Spanish Mediterranean, and he raised *Bucephalopsis* to full generic status to house all marine bucephalids having a simple sucker for a rhynchus. He named *B. gracilescens* (Rudolphi, 1819) as the type-species. On the contrary, Matthews (1973b) believed that the cercaria he collected in Britain from the common cockle *Cerastoderma edule* (as *Cardium edule*), which resembled the cercaria described as *B. haimeanus* and developed into a metacercaria in gadid fish, resembled *Bucephalus minimus*, but even that association is suspect because the cercaria was not from the type-host or the type-locality of *B. haimeanus*. The convention of using *Bucephalopsis* to accommodate bucephalids having a simple sucker for a rhynchus was widespread in Europe and North America until relatively recently and is still currently recognized in parts of Asia, despite the fact that life-history studies have never confirmed that *B. haimeanus* develops to an adult form with a simple sucker for a rhynchus.

Linton (1898) misidentified a bucephalid having an unmodified sucker for a rhynchus that he collected from the tripletail, *Lobotes surinamensis*, caught off Massachusetts, USA, as *Monostomum orbiculare* (Rudolphi, 1819), but he later re-described it as *Gasterostomum ovatum* (Linton, 1900). Dollfus (1929) observed bucephalids from *L. surinamensis* caught off French Cameroon, and determined that they represented the same species as that reported by Linton (1900). Dollfus (1929) was aware that Nicoll (1914a) had elevated *Bucephalopsis* to generic status to contain species with a simple sucker for a rhynchus, but he believed that features of the digestive system of *G. ovatum* warranted the erection of the new genus and therefore established *Prosorhynchoides ovatus* (Linton, 1900) for *G. ovatum*. Linton (1940), apparently not pleased with or not aware of the decision of Dollfus, re-evaluated his specimens, did not accept *Prosorhynchoides* and instead created the new combination *Prosorhynchus ovatus* (Linton, 1900). In that redescription, the gut was described as first extending forward, then turning posteriorly and finally reaching the posterior end of the worm or a little beyond the posterior testis. This species lacks a conical rhynchus and the seminal vesicle is close to spherical and not elongated, indicating that it belongs in the Bucephalinae. We examined specimens used by both Linton (USNM no. 4872 (004872.01 and 004872.02)) and Dollfus (Muséum National d'Histoire Naturelle, Paris, seven specimens on three slides) and consider them conspecific as *Prosorhynchoides ovatus*, discounting Linton's (1940) assessment.

Nagaty (1937) did not consider features of the digestive system to be of importance at the generic level and considered *Prosorhynchoides* a junior synonym of *Bucephalopsis*. Hopkins (1954) pointed out that the cercariae of different species shed from many hosts, including *Cerastoderma edule*, have been implicated as the larval form for *B. haimeanus*, but never had any been shown to be identical, leaving the adult stage unidentifiable. Hopkins (1954) believed that the elevation of *Bucephalopsis* to generic status was unfounded and suggested that the name *Bucephalus haimeanus* be retained for the species with a cercaria consistent with the description by Lacaze-Duthiers (1854) based on specimens from *Ostrea edulis* and *Cardium rusticum* in the Spanish Mediterranean Sea. Hopkins (1954) erected *Bucephaloides* Hopkins, 1954 for all bucephalids with an unmodified sucker for a rhynchus and a pretesticular ovary and named *B. gracilescens* (Rudolphi, 1819) as the type-species. In doing so, he did not account for the available name *Prosorhynchoides* for *Bucephalopsis*. Srivastava & Chauhan (1973), like Hopkins (1954) and for the same reasons, did not accept *Bucephalopsis*, but instead proposed that all bucephalids with a simple sucker for a rhynchus and a pretesticular ovary should be placed with *Prosorhynchoides*, and *P. ovatus* should be the type-species. They proposed that *Bucephaloides* be considered a junior synonym of *Prosorhynchoides*, since the latter was the older name. The genus *Prosorhynchoides* requires a revision that will incorporate numerous new combinations.

Dayal (1948) based the monotypic genus *Neobucephalopsis* on the presence in *N. bagarius* Dayal, 1948 of a seminal receptacle, an organ that he contended no other bucephalid possessed. His specimens also had thin-shelled eggs. He did not incorporate sectioned or living material into his description to confirm the presence of the seminal receptacle. In any event, Srivastava & Chauhan (1973) reviewed the Indian gasterostomes and determined that the presence of a seminal receptacle was not important enough to warrant the erection of a new genus. They considered *Neobucephalopsis* a junior subjective synonym of *Prosorhynchoides*. Since no specimens are available, we consider some of the characters questionable and follow their proposal.

The conflict as to whether to accept *Prosorhynchoides* or *Bucephaloides* as the genus for all bucephalids with unmodified suckers is not easily solved. We generally agree with Nagaty (1937) in that characters associated with the digestive system are not strong generic features in the Bucephalidae; however, the tubular caecum of the species in question is very different in structure from that of other well-known members of the subfamily. The conflict is further promoted by the fact that Dollfus (1929) was fully aware of *Bucephalopsis* and considered *Prosorhynchoides ovatus* (Linton, 1900) to be in a separate genus from the bucephalids, having an unmodified anterior sucker and a pretesticular ovary. We, like many recent Canadian and British authors, choose to follow Srivastava & Chauhan (1973) and advocate that *Prosorhynchoides* be applied to all species of the Bucephalinae with a simple sucker for a rhynchus and a pretesticular ovary, since Hopkins (1954) did not propose any new generic features for *Bucephaloides* that were not already present in *Prosorhynchoides*. It follows that we consider *P. ovatus* the type-species. *Bucephalopsis* should be returned to subgeneric status and applied only for the cercaria consistent with that reported by Lacaze-Duthiers (1854), and *Bucephaloides* should be considered a junior synonym of *Prosorhynchoides*.

Genus *Heterobucephalopsis* Gu & Shen, 1983

Heterobucephalopsis gymnothoracis Gu & Shen, 1983 was described from *Gymnothorax undulatus* caught off the Xisha Islands, Guangdong Province, China (Gu & Shen,

1983). The authors established the monotypic genus resembling '*Bucephalopsis*' because the ovary was intertesticular and stated that 'it seems to be a new genus'. From their illustration, we observed that *H. gymnothoracis* closely resembles *Prosorhynchoides ovatus*, but it differs by having an ovary lying between oblique testes and by the distribution of the vitelline follicles being more lateral and more posterior in the body. Perhaps the genus belongs in the Dolichoenterinae, but we have not examined specimens and the illustration is not sufficient for determining fine points of the terminal genitalia. Based on similarities of the rhynchus and digestive caecum, we assume the species is related to *P. ovatus*, but we consider *Heterobucephalopsis* a *genus inquirendum*, with the type-species in need of redescription.

Genus *Rhipidocotyle* Diesing, 1858

Diesing (1858) erected *Rhipidocotyle* Diesing, 1858 (feminine gender) and suggested that *Gasterostomum gracilescens* and *G. minimum* (Wagener, 1852) belonged in the genus, but he did not name a type-species. Stiles & Hassall (1908) thought that *G. gracilescens* should serve as type, but Nicoll (1914a) placed *G. gracilescens* with *Bucephalopsis* because the name *Rhipidocotyle* implied that the rhynchus had a sucker in conjunction with a dorsally-attached hood. *Gasterostomum gracilescens* lacked such a structure (see Nicoll, 1914a; Kniskern, 1952). Nicoll (1914a) further demonstrated that *G. minimum* had such a structure, and he concluded that it belonged in *Rhipidocotyle* but that *G. gracilescens* did not. He proposed *Rhipidocotyle minima* (Wagener, 1852) as the type-species for *Rhipidocotyle*. Eckmann (1932) examined *G. galeatum* (Rudolphi, 1819) and concluded that it was synonymous with *Gasterostomum minimum* (Wagener, 1852). If Eckmann was correct in her assessment of Rudolphi's material, then *Rhipidocotyle galeata* (Rudolphi, 1819) Eckmann, 1932 should be accepted as the type-species (see Kniskern, 1952; Stunkard, 1976). Rudolphi (1819) originally reported this species from '*Centronoti glauci*' (presumably the carangid *Campogramma glaycos*) in the Mediterranean Sea. The Commission (Anon., 1993) has now ruled that *Rhipidocotyle* is an Official Name, with *Rhipidocotyle galeata* as the type-species, with *Monostoma galeatum* as the senior subjective synonym of *G. minimum*. The primary generic characters for *Rhipidocotyle* include a pretesticular ovary and a rhynchus consisting of a simple sucker having a dorsal hood or lobe, with some species additionally having lateral lobes. The prostatic duct is slightly bent.

As he did with *Dolichoenterum* Ozaki, 1924, Ozaki (1924, 1928) placed *Nannoenterum pentagonum* Ozaki, 1924 from *Leptocephalus myriaster*, a conger eel in Japan, in the Bucephalinae, on the basis of its having a rhynchus that is a sucker underlying a hood. Eckmann (1932), Nagaty (1937) and Yamaguti (1958a) considered *Nannoenterum* to be a synonym of *Rhipidocotyle*. The caecum is a short sac and the ovary is anterior to both testes, both characters common to *Rhipidocotyle*. The species has an ovate seminal vesicle and tubular pars prostatica, which also supports a strong resemblance to species of *Rhipidocotyle*. We concur with the synonymy also because the ovary is pretesticular and the rhynchus is consistent with those in some species of *Rhipidocotyle*.

Pararhipidocotyle jeffersoni Kohn, 1970 was described as the type-species and only species in *Pararhipidocotyle* Kohn, 1970 by Kohn (1970b). In that original description, the ovary was drawn as intertesticular (Kohn, 1970b). We examined paratypes (Collection of the Instituto Oswaldo Cruz nos 30.523 b–e) and determined that the ovary was anterior to the testes; the anterior testis in Kohn's illustration appears to be the ovary. Kohn re-examined the holotype and agrees with us (A. Kohn, personal communication). We therefore consider this genus a junior synonym of *Rhipidocotyle* and tentatively retain *R. jeffersoni* (Kohn, 1970) n. comb. until the species can be compared better with other species.

Genus *Pseudorhipidocotyle* Wang & Pan in Long & Lee, 1964

Long & Lee (1964b) made the genus *Pseudorhipidocotyle* available as *Pseudorhipidocotyle* Wang & Pan in Long & Lee 1964, as indicated in the footnote.[2] They considered the type-species *P. elopichthys* Long & Lee, 1964, even though, in their meeting presentation, they considered the name of the genus for that species as *Rhipidocotyloides* (*nomen nudum* in Long & Lee, 1964a) and also used (Long & Lee, 1964a) the same name, *Pseudorhipidocotyle*, for a species which they decided in their revision (Long & Lee, 1964b) to call *Rhipidocotyloides tsengi*. Unlike most of the article by Long & Lee (1964b), the description of *Pseudorhipidocotyle* was not translated into English. The description was also included in the compendium on Chinese parasites of fishes by Zhang *et al.* (1999).

Pseudorhipidocotyle elopichthys, according to Long & Lee (1964b), has a rhynchus with a scalloped dorsal hood over a simple sucker, similar or identical to that in *Rhipidocotyle septpapillata* Krull, 1934. The rhynchal hood appears scalloped, with seven sharp points in both species; terminal genitalia are typical of the Bucephalinae; and the gut and excretory vesicle are similar in placement and form in both species. The two differ in that the ovary is anterior to the testes and the vitelline follicles are in lateral bands in *R. septpapillata*, and the ovary is between the testes and vitelline follicles are in two clumps near the ovary in *P. elopichthys*. Both *R. septpapillata* and *P. elopichthys* have freshwater life-cycles, the former parasitic in *Lepomis gibbosus* (Centrarchidae) in the Potomac River, Virginia, USA, and the latter in *Elopichthys bambusa* (Cyprinidae) in the Taihu and Pearl Rivers in China. Since the placement of the ovary between the testes is the primary generic diagnostic feature of *Pseudorhipidocotyle* and some features are illustrated conventionally, we consider this taxon a *genus inquirendum*.

Genus *Rhipidocotyloides* Long & Lee, 1964

As indicated above, there is confusion regarding the name *Rhipidocotyloides*. Either by considering the name *Rhipidocotyloides* of Long & Lee (1964a) unavailable or

[2] A confusing chain of events transpired in regard to both the names and the genera *Pseudorhipidocotyle* and *Rhipidocotyloides*. The article by Long & Lee (1964b) treats these conflicting taxa, with their associated species, which had been reported earlier as new both by Long & Lee (1964a) and by Wang & Pan (1964) in presentations and abstracts of the same meeting of the Chinese Parasitology Society in Beijing during November 1963. Some of the conflict is obvious from the abstracts of the meeting published in 1964, and another insight was provided by Professor Zhang Jiang-ying and Liu Lin (personal communication). At the meeting, Long & Lee (1964a) reported *Rhipidocotyloides elopichthys* n. g. and n. sp. from the fish definitive host *Elopichthys bambusa*. They also reported *Pseudorhipidocotyle tsengi* n. g. and n. sp., a species with a rhynchus that is hood-shaped and does not have seven protrusions, from the fish definitive host *Ochelobius elongatus*. Wang & Pan (1964) simultaneously reported *Pseudorhipidocotyle* n. g., which included both the above species indicated plus one more. These were *P. macracetabulus* and *P. cheni* from *E. bambusa* and *P. micracetabulus* from *O. elongatus*. In their original published abstract, Long & Lee (1964a) provided a minimal generic diagnosis of *Rhipidocotyloides*, reporting a rhynchus with seven horn-like protrusions, an ovary between the testes, and the vitellarium as two small bunches nearly parallel to the ovary. Long & Lee (1964b) wished to respect the work by Wang & Pan and credited them as the authority of *Pseudorhipidocotyle*, as the genus was originally called *Rhipidocotyle* by Long & Lee (1964a), with the rhynchus having seven protrusions and with the type- and only species, *P. elopichthys* Long & Lee, 1964, infecting *E. bambusa*.

by considering Long & Lee (1964b) first revisers, we accept the availability of *Rhipidocotyloides* of Long & Lee (1964b). The monotypic *Rhipidocotyloides* is based on *R. tsengi* Long & Lee, 1964 from the intestine of *Ochelobius elongatus* (Cyprinidae) from the Taihu River in China. The English translation in the original description (Long & Lee, 1964b) incorrectly used the heading *Pseudorhipidocotyle tsengi* Long & Lee, 1964, which becomes a junior objective synonym of *R. tsengi*. This action probably occurred because the synonym was the binomen attributed to the species (as a *nomen nudum*) in the abstract by the authors (Long & Lee, 1964a). Wang & Pan (1964) referred to the species (also a *nomen nudum*) as *P. micracetabulus* in their abstract. The rhynchus of this species is similar to those present in members of *Rhipidocotyle*. We made an unsuccessful attempt to borrow material of this genus. Based on the illustration of the specimen, possibly somewhat distorted, the terminal genitalia are basically consistent with the Bucephalinae and the intertesticular ovary and the arrangement of the vitellarium indicate that the genus may be related to *Pseudorhipidocotyle*. We therefore consider *Rhipidocotyloides* as a *genus inquirendum*.

Genus *Pseudobucephalopsis* Long & Lee, 1964

The genus *Pseudobucephalopsis* was based on *P. spheroides* Long & Lee, 1964 by monotypy from the air-sac of the marine puffer *Spheroides oceculatus* in China. Long & Lee (1964b) recognized the new genus on the basis of a U-shaped muscular lip in the middle of the rhynchal sucker and an ovary that lies between the testes. We were unable to locate specimens. Other illustrated characters suggest a similarity to species of *Prosorhynchoides*. In the absence of specimens, we cannot be certain if the appearance of the rhynchus is an artefact of fixation or if in fact the U-shaped lip is a permanent feature. Also, a second species has been described (*P. belonnis* Gu & Shen, 1983). Another, *P. ganyuei* Tang & Tang, 1976, was considered to be *Pseudorhipidocotyle ganyuei* (Tang & Tang, 1976) by Wang & Wang (1998b).[3] We consider *Pseudobucephalopsis* a *genus inquirendum* until specimens can be evaluated.

Genus *Parabucephalopsis* Tang & Tang, 1963

We were unable to obtain the reference on *Parabucephalopsis prosthorchis* Tang & Tang, 1963 by Tang & Tang (1963), but we examined the description and illustration of the species by Long & Lee (1964b). The latter authors reported the parasite from the intestine of the fishes *Erythroculter ilishaeformis* and *E. darbryi* in the Taihu River, China, as final hosts and in the muscles of *Protosalanx hyalocranius* and *Neosalanx tangkahkeii* var. *taihuensis* from the same area as intermediate fish hosts. This species, also from the Yangtze, Ming Kiang and Pearl Rivers, has both of the testes anterior to the ovary and an anteriorly directed caecum (posteriorly in the related *Parabucephalopsis parasiluri* Wang & Wang, 1998). However, the direction of the gut does not appear to matter in members of the Bucephalinae as it does in those in the

[3] A thesis (Wang, 1995) concerning the Bucephalidae was printed with an English abstract and bound with the name 'Institute of Hydrobiology, Chinese Academy of Sciences, 1995' written where one would expect the publisher's name. Apparently a large number were printed, and it is probably a publication from a technical point of view, as defined by the International Commision on Zoological Nomenclature (1999). However, this work was not meant by the author to constitute a publication and was followed by a series of other publications (Wang & Wang, 1998a, b, c), which both repeat some aspects from the thesis and present different decisions about others. Some of the 1995 decisions will be indicated in this chapter. For example, the Provarinae Wang, 1995 was created for several bucephalines, but it was not mentioned by Wang & Wang (1998a, b).

Prosorhynchinae. The pars prostatica was illustrated as coiled. No specimen was available to us, and we consider this genus, now containing four species according to Wang & Wang (1998b), a *genus inquirendum* until specimens of *P. prosthorchis* and other species can be examined.

Genus *Roparhynchus* Gupta & Kumari, 1978

Gupta & Kumari (1978, indicated incorrectly as 1974 in the article) described *Roparhynchus nelsoni* Gupta & Kumari, 1978 from the freshwater fish *Xenentodon cancila* in the Satluj River at Ropar, India. The authors placed it in the Neoprosorhynchinae based on the presence of an ovary posterior to the right testis. They considered the presence of a sucker for a rhynchus and diagonal testes to be new generic characters for the subfamily. They illustrated a cirrus-sac with a spherical anterior seminal vesicle and tubular pars prostatica, consistent with the arrangement present in the Bucephalinae. The primary generic feature for *Roparhynchus* was the position of the ovary relative to the testes. The presence of a sucker for a rhynchus, the configuration of the male reproductive system and the arrangement of the vitellarium in two anterior fields are bucephaline distinctions. The uterus occupies a large portion of the body, being bounded anteriorly by only the rhynchus and vitellarium. It reaches posteriorly to the level of the genital pore. This is basically another freshwater species in the subfamily. Until heat-killed specimens can be examined, we consider this genus a *genus inquirendum* that shows similarities with other taxa requiring further investigation (i.e. *Parabucephalopsis* and *Pseudobucephalopsis*).

Genus *Glandulorhynchus* Thatcher, 1999

This genus was erected for specimens found in a new species of freshwater fish, *Hydrolycus* sp., from Brazil. Thatcher (1999) described *Glandulorhynchus turgidus* Thatcher, 1999 on the basis of glands associated with a sucker-like rhynchus, the presence of grape-like vitelline follicles in lateral bands and the presence of a swollen uterine reservoir that protrudes from the ventral portion of the body. One specimen is illustrated without the uterine reservoir, but the article was not clear if all specimens had this unusual structure or whether the structure could have been an artefact of fixation or preparation. The terminal genitalia are drawn as resembling those of the Bucephalinae, but the possibility of belonging to the Paurorhynchinae cannot be discounted. This taxon is in need of re-evaluation and is excluded from the key.

Subfamily Dolichoenterinae Yamaguti, 1958

Yamaguti (1958a) created the subfamily Dolichoenterinae to house *Dolichoenterum* Ozaki, 1924 and added *Paurorhynchus* based on the shape of the gut, the position of the opening to the pharynx and the presence of an intertesticular ovary or an ovary occurring at the level of the anterior testis. He later (Yamaguti, 1971) added *Bellumcorpus* Kohn, 1962 and *Pseudodolichoenterum* Yamaguti, 1971, and returned *Paurorhynchus* to the Paurorhynchinae. We regard the features used by Yamaguti (1958a, 1971) to unite the Dolichoenterinae to be generic features. We consider *Paurorhynchus* and *Bellumcorpus* to belong in the Paurorhynchinae, as explained above.

The Dolichoenterinae is a marine subfamily, superficially similar to the Bucephalinae and containing *Dolichoenterum* and *Pseudodolichoenterum*. We believe that Yamaguti (1958a) was correct in designating the new subfamily, but we add characteristics of the terminal genitalia to the subfamily diagnosis: cirrus-sac thick-walled, with length short relative to width when compared with most bucephalids; pars prostatica bending and narrow along the proximal portion, straight and relatively wide along the distal portion.

History and features of genera

Genus *Dolichoenterum* Ozaki, 1924

Ozaki (1924) described *Dolichoenterum longissimum* Ozaki, 1924 from *Leptocephalus myriaster* at Takamatsu, Onomichi and Shimonoseki, Japan. The original description was in Japanese, but Ozaki (1928) translated his earlier work into English and published it independently. In the translation, he placed *Dolichoenterum* in the Bucephalinae on the basis that the rhynchus had a sucker. Although similar to *Rhipidocotyle*, it was considered by Ozaki (1928) as belonging in a separate new genus because it had eight papillae on its rhynchal hood and because the caecum was long and tubular. Yamaguti (1958a) created the subfamily Dolichoenterinae to house bucephalids that had the ovary between the testis, a long tubular gut and a pharynx opening near the anterior end of the body. Eckmann (1932) considered *Dolichoenterum* synonymous with *Rhipidocotyle*, considering the eight dorsal papillae, gonad configuration and long tubular caecum as specific characters; however, Nagaty (1937) considered *Dolichoenterum* valid.

Genus *Pseudodolichoenterum* Yamaguti, 1971

Yamaguti (1971) erected *Pseudodolichoenterum* for *P. megagaster* (Corkum, 1968), which he transferred from *Rhipidocotyle*, recognizing that, with the exception of the rhynchus, most features of *Pseudodolichoenterum* were consistent with those of *Dolichoenterum*. He also recognized that the pars prostatica was narrow and bending proximally and wide and straight distally, but did not amend his diagnosis of the Dolichoenterinae to include this feature. Like Yamaguti (1971), we accept *Pseudodolichoenterum* as a member of the Dolichoenterinae.

Subfamily Paurorhynchinae Dickerman, 1954

Dickerman (1954) described *Paurorhynchus hiodontis* Dickerman, 1954 and erected the subfamily Paurorhynchinae to house it. Yamaguti (1958a) initially did not recognize this subfamily and placed *Paurorhynchus* in the Dolichoenterinae Yamaguti, 1958, presumably because *P. hiodontis* had a long tubular caecum like dolichoenterines. Later (Yamaguti, 1971), he recognized the Paurorhynchinae. Features of the rhynchus and terminal genitalia provide the primary characters that unite the subfamily. Members of the Paurorhynchinae have a weakly developed muscular apical pad or plug for a rhynchus. The terminal genitalia consist of a thin- to moderately thick-walled cirrus-sac (2.5–20.0 μm thick) and an elongate seminal vesicle (more than twice as long as wide). A seminal duct is absent in the subfamily and the pars prostatica is straight throughout its length. Secondary characters for the subfamily include an elongate and cylindrical body, usually devoid of spines, lobed or dendritic ovaries and testes, and all members have a freshwater life-cycle and have been found in the stomach or body-cavity, but never in the intestine of the definitive host. In addition to the two American genera, we tentatively include two Asian genera, as indicated below.

History and features of genera

Genus *Paurorhynchus* Dickerman, 1954

Paurorhynchus hiodontus Dickerman, 1954 was described from the body-cavity of the freshwater mooneye, *Hiodon tergisus*, in North America. Dickerman (1954) actually found immature specimens in the liver of some fish also infected with adults. He considered this species as remarkable among bucephalids because it lacked tegumental spines, the rhynchus was a weakly developed muscular apical pad and the adult stage occurred in the body-cavity of the host, with immature specimens in the liver. We find the genus distinct from all other genera in the subfamily because the pharynx opens between the anterior quarter and anterior half of the body, the oblique testes may be dendritic or not, and the ovary is lobed and lies dextrally at the level of the anterior testis. The seminal vesicle is extremely long and slender in *P. hiodontus*, and it appears sinuous proximally, prior to straightening out distally and connecting with a straight ejaculatory duct. The male duct is divided into a proximal pars prostatica and a thin-walled distal ejaculatory duct, which empties into a small genital atrium. The genital lobe is very small in *Paurorhynchus*.

Genus *Bellumcorpus* Kohn, 1962

Bellumcorpus major Kohn, 1962 was described from the stomach of *Salminus hilarii* caught in freshwater in Brazil. The spelling of the species was emended to *majus* by Yamaguti (1971). Kohn (1962) originally placed this species in the Neoprosorhynchinae and compared her species with *Neoprosorhynchus purius* Dayal, 1948. She distinguished *B. majus* from *N. purius* by the configuration of the testes, which are oblique in the former and tandem in the latter. Yamaguti (1971) included the genus in the Dolichoenterinae, probably because of similarities in features of the gut. We consider *B. majus* to be closely related to *Paurorhynchus* and place *Bellumcorpus* in the Paurorhynchinae on the basis that the tegument is smooth, the rhynchus is similar to that in *Paurorhynchus*, the ovary is dextral and lobed, and members of the two genera share some features of the terminal genitalia in common. *Bellumcorpus* differs from *Paurorhynchus* by having the pharynx opening near the mid-body rather than slightly anterior, and the ovary is more posterior relative to the anterior testis. Also, the genital lobe is larger in species of *Bellumcorpus*. Kohn (1970c) described a second species, *B. shubarti* Kohn, 1970, from *Salminus maxillosus* on the basis of it having lobed testes.

Genus *Sinicovothylax* Yu, 1965

Yu (1965) described *Sinicovothylax chujiangensis* Yu, 1965 and *Sinicovothylax mapaensis* Yu, 1965, from the swim-bladder of the freshwater fish *Hemibagrus elongatus* at Mapa, Chujiang Hsien, Kwangtung Province, China, with *S. chujiangensis* designated as the type-species. Yu (1965) erected the Sinicovothylacidae to house his new genus on the basis of a sac-like uterus. The terminal genitalia in the two species resemble those of members of the Paurorhynchinae by having an elongated seminal vesicle, a straight male duct, a small genital lobe and a rhynchus. Yu deposited type-material at the Department of Parasitology, Chungshan Medical College, Canton, but we have not examined the specimens. Wang & Wang (1998a) considered the Sinicovothylacidae a junior synonym for the Sinicovothylacinae, a subfamily which they retained, but Zhang *et al.* (1999) still accepted the family. We transfer *Sinicovothylax* to the Paurorhynchinae because the uterus in *Paurorhynchus* and a portion of the uterus in *Bellumcorpus* is sac-like and because the structure of the rhynchus in *Sinicovothylax* is similar to that in the Paurorhynchinae. Also, the body-shape, vitelline configuration, distribution of the gonads in the body and the fact that *Sinicovothylax* is reported from the swim-bladder of a freshwater fish all suggest that the genus has affiliation with the Paurorhynchinae.

Sinicovothylax is different from all other members of the subfamily because it is reported to have spines covering the tegument and the pharynx opens at the level of the posterior third of the body rather than at mid-body or anterior to mid-body; the caecum is a simple, sac-like structure directed anteriorly, never recurving posteriorly as in other genera; and the testes are symmetrical rather than tandem or oblique.

Genus *Rhoporhynchus* Wang, 1995 (also indicated as *Rhoporhynchus* Wang & Wang, 1998 by Wang & Wang (1998a, b, c)

Rhoporhynchus clarias Wang, 1995 was initially described by Wang (1995) in the monotypic genus *Rhoporhynchus* from the freshwater catfish *Clarias fuscus* in China; the genus was placed in the new subfamily Provarinae Wang, 1995. Later, Wang & Wang (1998c) described it as *R. lini* Wang & Wang, 1998 from the swim-bladder of the yellow catfish *Pseudobagrus fulvidraco* and considered it to be in the Bucephalinae but did not mention the Provarinae. Wang & Wang (1998a) considered *R. lini* as the only species in the genus. Presumably, *Rhoporhynchus lini* is a junior objective synonym of *R. clarias*. The illustration of *R. clarias* in Wang (1995) depicts a worm that appears ovoid, but it may have been flattened from an elongate-cylindrical state. The structure of the rhynchus is not clear, but it is probably consistent with those of the Paurorhynchinae, and the vitellarium and terminal genitalia are clearly similar to those in other members of the subfamily. We consider *Rhoporhynchus* to belong in the Paurorhynchinae and suspect that it is closely related to *Sinicovothylax*. Differing from other genera in the subfamily, it has a sinistral ovary and a uterus that coils in the posterior part of body, never reaching above the level of the pharynx. The gut seems to be cavernous and never recurves posteriorly and the pharynx opens in the posterior half of the body, as in *Sinicovothylax*. Species of both genera require additional investigation.

Subfamily Prosorhynchinae Nicoll, 1914

When Nicoll (1914a) split the bucephalids into the subfamilies Bucephalinae and Prosorhynchinae, he included all bucephalids that had a rhynchus lacking a sucker in the latter and stated: 'The definitions of these subfamilies are identical with the definitions given by Odhner (1905, pp. 296–297) for the genera *Gasterostomum* and *Prosorhynchus* Odhner, 1905 respectively.' Only three species of *Prosorhynchus* were known at the time. Odhner's (1905) diagnosis for *Prosorhynchus*, from which the subfamily diagnosis stems, emphasized the structure of the terminal genitalia. Despite this, no other authors have considered the structure of the terminal genitalia as an important generic or subfamily character since Nicoll (1914a). Instead, subsequent authors have treated the rhynchus as being the primary feature defining the Prosorhynchinae. Traditionally, all bucephalids having a disc-like or conical muscular rhynchus lacking a sucker have been included in the subfamily. We part from the traditional classification of the Prosorhynchinae and revert to the original intention of Nicoll (1914a) by emphasizing the importance of the structure of the terminal genitalia as the most important set of characters for distinguishing members of the subfamily. We include in the Prosorhynchinae all genera of bucephalids that have an elongate seminal vesicle that empties into a relatively narrow seminal duct, which typically coils once before connecting to the straight proximal portion of the pars prostatica. The position of the ovary relative to the testes and the position of the testes within the body should be considered generic features. The rhynchus varies among the genera but generally consists of a muscular disc or plug-shaped structure, often with a conical base. Almost all members of the subfamily are marine.

Spermatophores appear to be produced in some prosorhynchines. Crowcroft (1947) reported a spermatophore in *Telorhynchus arripidis* Crowcroft, 1947 with a yellowish sclerotized outer wall and a tapering stalk. He (Crowcroft, 1947) and Odhner (1905) considered spermatophores to be produced within the genital atrium by the surrounding gland-cells, with the genital lobe assisting in a spermatophore's formation rather than serving as a copulatory organ. The spermatophore is apparently deposited into a mate before another is produced. Spematophores are currently reported (e.g. Odhner, 1905; Ozaki, 1924, 1928; Crowcroft, 1947) only for species of *Telorhynchus*, *Prosorhynchus* and *Gotonius* Ozaki, 1924 (a synonym of *Prosorhynchus*). Odhner (1905) investigated the prosorhynchine *Prosorhynchus squamatus*, which produced a stalked spermatophore, and noted that contraction of the cirrus-sac caused the genital lobe to become erect and also a portion of the genital atrium to enlarge and protrude. The ejaculatory duct did not penetrate this portion of the genital lobe and did not turn inside-out like a typical digenean cirrus. He also studied the bucephaline *Prosorhynchoides gracilescens*, which did not produce a spermatophore. In it, the ejaculatory duct penetrated the lobe and the lobe erected when the cirrus-sac contracted, but the ejaculatory duct similarly did not turn inside-out. Erwin & Halton (1983), who studied spermatogenesis in the same species, referred to as *Bucephaloides gracilescens*, reported copulation to take place by means of an eversible cirrus. They discussed the importance of locomotory efficiency of the modified spermatozoa of that species to accomplish fertilization in the oötype. The 100 μm long spermatozoa have a head and tail of equal length that allows them to swim up the uterus. A comparison with the spermatozoa from the seminal vesicle and spermatophore of a prosorhynchine should provide valuable information in understanding variations in digenean reproduction. Moreover, in regard to the spermatophore, the prevalence, structure, function and species in which it occurs, as well as the function of both the genital atrium and genital lobe in species with and without spermatophores, require a critical evaluation.

A number of what we consider to be dubious subfamilies have been created to house genera that we believe belong in the Prosorhynchinae. Nagaty (1937) and Kniskern (1952) included *Neidhartia* Nagaty, 1937 and *Alcicornis* (which we transferred to the Bucephalinae) in the Prosorhynchinae. Yamaguti (1938a) erected *Pseudoprosorhynchus* and included it in the Prosorhynchinae. Later, he (Yamaguti, 1958a) reconsidered his earlier placement of *Pseudoprosorhynchus* in the Prosorhynchinae and created the Neidhartiinae, for both *Neidhartia* and *Pseudoprosorhynchus*. He contended that the plug-shaped rhynchus of *Neidhartia* and the disc-shaped rhynchus of *Pseudoprosorhynchus* were sufficiently different from the rhynchus of *Prosorhynchus* (and apparently sufficiently similar enough to each other) to necessitate the erection of the new subfamily. In addition to rhynchal characters, Yamaguti (1958a) considered the presence of an intertesticular ovary to unite the two genera and separate them from other genera in the Prosorhynchinae. We consider the terminal genitalia of both *Neidhartia* and *Pseudoprosorhynchus* to be consistent with other aspects of the subfamily diagnosis for the Prosorhynchinae, and we consequently consider these genera to belong in the Prosorhynchinae. We consider Neidhartiinae a junior synonym of Prosorhynchinae.

Neoprosorhynchus Dayal, 1948 was erected and added to the Prosorhynchinae. Yamaguti (1958a) created the Neoprosorhynchinae for *Neoprosorhynchus* on the basis that the pharynx was pre-equatorial, the caecum was short and the ovary was post-testicular. We consider all the features used by Yamaguti (1958a) to be of generic value and observed from the illustration of *Neoprosorhynchus purius* that the rhynchus and the terminal genitalia are consistent with those of other members of the Prosorhynchinae. Therefore, we consider *Neoprosorhynchus* to belong in the Prosorhynchinae, with the Neoprosorhynchinae a junior synonym of the Prosorhynchinae.

The Telorhynchinae Lebedev, 1968 was created for *Telorhynchus* Crowcroft, 1947 and *Telorhynchoides* Lebedev, 1968 (which we consider a synonym of *Telorhynchus* below). Crowcroft (1947) originally placed his new genus in the Prosorhynchinae. The terminal genitalia of *Telorhynchus* are described as resembling those in the Prosorhynchinae, but the rhynchus differs from others in the group by having a ring of long spines around the anterior rim. Lebedev (1968) apparently used features of the rhynchus to separate *Telorhynchus* from the Prosorhynchinae. We regard these features to be of generic importance only and consider the Telorhynchinae to be a junior synonym of the Prosorhynchinae.

Finally, the Folliculovarinae Wang, 1995 [*sic*: it should be Folliculovariinae] was created by Wang (1995) first in a thesis and later in an article (Wang & Wang, 1998a) to house *Folliculovarium* Gu & Shen, 1983, presumably based on species having a follicular ovary. Gu & Shen (1983) had placed their new genus in the Prosorhynchinae. We regard the presence of a follicular ovary to be a generic, if not suspect, feature and therefore consider the Folliculovarinae a junior synonym of the Prosorhynchinae.

The following genera have been added to the Prosorhynchinae in addition to those discussed above: *Gotonius* Ozaki, 1924, *Skrjabiniella* Issaitschikow, 1928, *Mordvilkovia* Pigulewsky, 1931, *Dollfustrema* Eckmann, 1934, *Parabucephalopsis* Tang & Tang, 1963, *Neodollfustrema* Long & Lee, 1964, *Paraprosorhynchus* Kohn, 1967, *Myorhynchus* Durio & Manter, 1968, *Chabaudtrema* Kohn, 1970, *Rudolphinus* Stunkard, 1974 and *Proopisthogyne* Koche, 1976, not all of which we accept.

History and features of genera

Genus *Prosorhynchus* Odhner, 1905

Odhner (1905) established the genus *Prosorhynchus* to house bucephalids that had a rhynchus lacking a sucker, and he named *P. squamatus* Odhner, 1905 as the type-species. He considered *P. squamatus* to be distinct from specimens identified as *Gasterostomum armatum* by van Beneden (1870) and used the name *P. aculeatus* (van Beneden, 1870) for those specimens. Odhner (1905) also transferred *G. crucibulum* (Rudolphi, 1819), which Rudolphi (1819) had described as *Monostomum crucibulum*, into his new genus, creating *P. crucibulum* (Rudolphi, 1819). Nagaty (1937) and Jones (1943) believed *P. squamatus* and *P. crucibulum* to be synonyms and therefore considered *P. crucibulum* (Rudolphi, 1819) the type; however, Brinkmann (1957) showed clear evidence for separating all three of Odhner's species. Based on our examination of material from the University of Bergen, Museum of Zoology, including specimens of *P. squamatus* (UBM 44848, one slide, and UBM 47987, two slides) and of *P. crucibulum* (UBM 44865, one slide, and UBM 44866, four slides) as well as material from The Natural History Museum, London, we follow the classification established by Brinkmann (1957) and consider *P. crucibulum* to be distinct from *P. squamatus* and therefore retain *P. squamatus* as the type.

Odhner (1905) made the combination *Prosorhynchus aculeatus* Odhner, 1905 for specimens described as *Gasterostomum crucibulum* Beneden, 1870. He considered these specimens to be conspecific with those reported as *G. armatum*, but Issaitschikow (1928) thought that the erection of his genus *Skrjabiniella* was warranted to accommodate *P. aculeatus*, as *S. aculeatus*, because the opening to the pharynx was so far posterior and because the testes were oblique rather than tandem. Jones (1943) later redescribed *S. aculeatus* and advocated that elongate species in *Prosorhynchus* having tandem testes and the vitellarium in two separate lateral groups be placed in *Prosorhynchus* (*sensu stricto*), while those species having an elliptical body with oblique testes and the vitellarium in one anterior arching field be placed in *Skrjabiniella*. Yamaguti (1958a, 1971) relegated Issaitschikow's (1928) genus to subgeneric status and included *P. magniovatus* Yamaguti, 1938 and *P. uniporus* (Ozaki, 1924) (as a synonym of *P. aculeatus*) in the subgenus *Skrjabiniella*. Manter (1953) believed that the generic-level features promoted by Issaitschikow (1928) were of specific-level value only and considered *Skrjabiniella* a junior synonym of *Prosorhynchus*. Brinkmann (1957), Kniskern (1952) and even Skrjabin & Guschanskaja (1962) also considered *Skrjabiniella* a junior synonym of *Prosorhynchus*. We agree.

Genus *Rudolphinus* Stunkard, 1974

Rudolphinus was erected to accommodate *P. crucibulum* (Rudolphi, 1819) on the basis of differences in cercarial flame-cell pattern between *P. crucibulum* and *P. squamatus*, as indicated above in the introduction. Stunkard (1974) argued that attachment organs of parasites were not well conserved genetically and were subject to more adaptive pressures than other organ systems. He also believed that the excretory system was the most conserved organ system in platyhelminths and therefore based his decisions in this case on the cercarial flame-cell pattern. Unless or until additional aspects of cercariae or other stages can be examined for all or several species in *Prosorhynchus* and related genera and the resulting observations demonstrate natural grouping of species, this approach does not seem practical. Because of that lack of information and the apparent wide variation in known flame-cell patterns of related bucephalids, we agree with Wang & Wang (1998a) and consider *Rudolphinus* a junior synonym of *Prosorhynchus*.

Genus *Gotonius* Ozaki, 1924

Gotonius facilis Ozaki, 1924 was described from the pyloric caeca of *Inimicus japonicus* along the coast of Takamatsu, Shikoku Island, Japan. Ozaki's (1928) subsequent English description is thorough and well illustrated, but he did not state the reasons why he decided to erect a new genus. We examined specimens from Ozaki's collection (Meguro Parasitological Museum Coll. No. 30049, two slides with four adults on one and 11 immature and early adults on the other) and saw no reason to resurrect the genus. We suspect it was created because the vitellarium is arranged in two anterior rows that were sometimes contiguous in an anterior arc (a dubious generic character) and because of tandem testes (which could be dependent on the elongated body). Eckmann (1932) considered *Gotonius* to be a synonym of *Prosorhynchus*, and Nagaty (1937) concurred. *Gotonius platycephali* (Yamaguti, 1934) was described from *Platycephalus indicus*, based on differing egg size, but Nagaty (1937) later considered it conspecific with *P. facilis*, the type-species. We follow Eckmann (1932) and consider *Gotonius* a junior subjective synonym of *Prosorhynchus*.

Genus *Paraprosorhynchus* Kohn, 1967

Paraprosorhynchus was originally spelled *Paraprosorphynchus* as an obvious typographical error (Kohn, 1967a). It is obvious because of its derivation and the expected spelling of the type-species, *Paraprosorhynchus jupe* Kohn, 1967, from *Promicrops guttatus* from the Atlantic Ocean off Brazil. The species was described in a new genus on the basis that the vitellarium lies in two separate bands in the anterior region of the body. Several species of *Prosorhynchus* have the vitellarium in two anterior bands, including *P. costai* Travassos, Artigas & Pereira, 1928, *P. promicropsi*, Manter, 1940, *P. gonoderus* Manter, 1940 and *P. chorinemi* Yamaguti, 1952. We do not consider the vitelline bands to constitute a generic character in this case and, like Wang & Wang (1998a), consider *Paraprosorhynchus* a junior synonym of *Prosorhynchus*. *Prosorhynchus jupe* should be compared closely with a large series of specimens of *P. promicropsi*.

Genus *Chabaudtrema* Kohn, 1970

Chabaudtrema rarus Kohn, 1970 was described from *Garrupa* sp. (currently considered a subgenus of *Epinephelus*) from the marine environment in Baía de Guanabara, Brazil. Kohn (1970a) included her new genus in the Prosorhynchinae, but she considered it closely related to *Bellumcorpus* (Paurorhynchinae). The description by Kohn (1970a) shows the terminal genitalia as consistent with the Prosorhynchinae; the vitellarium is in two lateral bands and the rhynchus is consistent with that of species in *Prosorhynchus*. The ovary appears post-testicular in the illustration in the description, and Kohn (1970a) named the new genus on that basis. After examining the holotype (Helminthological Collection of the Instituto Oswaldo Cruz No. 30.284), not an ideal preparation, we were unable to determine unquestionably whether the ovary was in the position described by Kohn (1970a) or in the position she described as the dextral testis. Kohn (1967b) had previously reported *P. ozakii* Manter, 1934, originally described from *Epinephelus niveatus* in Tortugas, Florida, from *Garrupa* sp. from Baía de Guanabara, Brazil, and that species differs from *C. rarus* only in the placement of the ovary relative to the testes. Since the quality of the single (?) specimen of *C. rarus* was poor, we consider *C. rarus* as *Prosorhynchus rarus* (Kohn, 1970) n. comb. until additional material shows otherwise. Consequently, we treat *Chabaudtrema* as a junior subjective synonym of *Prosorhynchus*.

Genus *Folliculovarium* Gu & Shen, 1983

Folliculovarium xishaense Gu & Shen, 1983 was described from *Epinephelus fasciatus* and *F. gymnothoracis* Gu & Shen, 1983 from *Gymnothorax undulatus*, with the former designated the type-species by the authors. As indicated above, Wang (1995) and later Wang & Wang (1998a) erected the Folliculovarinae [*sic*] for the genus, and presumably the subfamily was established on the basis of a follicular ovary. Although we have not examined specimens of either species, the figures in our copy of Gu & Shen (1983) do not clearly depict the ovary. The illustration of *F. gymnothoracis* is the 'follicular ovary', but it also shows two testes and a spherical structure just below the 'follicular ovary' that appears to us as possibly an ovary. Based on our examination of the ovary, testes and Mehlis' gland in numerous different bucephalids, we note that a testis, and perhaps Mehlis' gland, easily can be mistaken for the ovary. Until the female features can be corroborated, preferably with sectioned material, we treat the genus as a *genus inquirendum*. Even if the ovary is follicular, we would treat the genus as a member of the Prosorhynchinae.

Genus *Dollfustrema* Eckmann, 1934

Tseng (1930) described *Prosorhynchus vaneyi* Tseng, 1930 from the Asian freshwater snook *Siniperca shezeri* in Chung-King, Schezwan Province, China. Eckmann (1932) established *Dollfusina* Eckmann, 1932 and made the combination *D. vaneyi* (Tseng, 1930). Because *Dollfusina* was pre-occupied by a lamellibranch, Eckmann (1934) renamed it *Dollfustrema*, with *D. vaneyi* (Tseng, 1930) as the type-species. We did not examine the type of *P. vaneyi*, but the configuration of the terminal genitalia in the illustration of the original description (Tseng, 1930) was not clear. Nevertheless, we examined the type-specimen for *D. gravidum* Manter, 1940 (USNPC No. 16721), and its terminal genitalia resemble those of members of the Prosorhynchinae. Illustrations of *D. stromborhynchum* Manter &

Pritchard, 1961 and *D. bipapillosum* Manter & Pritchard, 1961 also show terminal genitalia consistent with members of the Prosorhynchinae.

Long & Lee (1964b) reported *D. vaneyi* from various freshwater hosts in China, but illustrated a specimen strongly resembling *Prosorhynchus thapari* Manter, 1953, a marine worm from the serranid *Plectropoma maculatum*. This specimen exhibited a pretesticular ovary, a condition not consistent with the original description of *D. vaneyi* (see Tseng, 1930). On the basis of observations of their own specimens, Long & Lee (1964b) created *Neodollfustrema* Long & Lee, 1964 for all members of *Dollfustrema* having an intertesticular ovary and designated *N. gravidum* (Manter, 1940) as the type-species. By their reckoning, this new genus included the species *D. echinatum* (Komiya & Tajimi, 1941), *D. californiae* Montgomery, 1957, *D. muraenae* Sogandares-Bernal, 1959, *D. bipapillosum* and *D. stromborhynchum*. They retained both *D. vaneyi*, the type-species, and *D. macracanthum* Hanson, 1950, which apparently has only one postovarian testis, in *Dollfustrema*. We have not examined either *D. vaneyi* or *D. macracanthum* but suspect that the ovary in a heat-killed specimen of *D. vaneyi* would be intertesticular or at the level of the anterior testis, as described by Tseng (1930). We also suspect that there is an anterior testis obscured by the vitellarium or uterus in *D. macracanthum*. A series of specimens should be examined to confirm the unique condition of possessing a single testis in more than aberrant material. In any event, we tentatively consider *Neodollfustrema* a junior synonym of *Dollfustrema* and restore all its species to their original combinations. The key generic features for *Dollfustrema* are the presence of circlets of short spines arranged in vertical bands of two or three spines per band around the rim of the rhynchus, which would otherwise resemble the rhynchus in species of *Prosorhynchus*, and the ovary at the level of the anterior testis or between the testes.

Genus *Mordvilkovia* Pigulewsky, 1931

Mordvilkovia elongata Pigulewsky, 1931 was described from the northern pike *Esox lucius* in the Dnieper Basin by Pigulewsky (1931). This species has been included in the Prosorhynchinae by subsequent authors. The original description was vague and the clarity of the illustration poor. Pigulewsky (1931) portrayed the terminal genitalia of his species as like those of members of the Bucephalinae, but the conical rhynchus, which perhaps is encircled by spines, was more reminiscent of *Dollfustrema*. *Dollfustrema* is the only other genus in the Prosorhynchinae that has freshwater representatives. Nagaty (1937) considered *M. elongata* synonymous with *Prosorhynchus crucibulum*, a marine parasite, but its freshwater habitat and the indistinctly drawn rhynchus preclude consideration of Nagaty's opinion. Jones (1943) and Kniskern (1952) considered the species distinct but transferred it to *Prosorhynchus*. Until specimens are discovered and examined, we believe that the freshwater life-history of *M. elongata* provides enough doubt to consider *Mordvilkovia* as an unlikely synonym of *Prosorhynchus*. Manter (1940) considered *Dollfustrema* as a possible synonym. Crowcroft (1947), without examining specimens, said he thought Manter was mistaken in assuming that spines of *Dollfustrema* corresponded to cuticular folds upon the rhynchus of *Mordvilkovia*, so he accepted *Mordvilkovia*. We consider *Mordvilkovia* a *genus inquirendum* and *M. elongata* a *species inquirenda* until specimens can be examined; we do not include the genus in the key because we cannot interpret key features from the description or illustration. In an attempt to obtain specimens, we enquired unsuccessfully about the whereabouts of any material from several museums in Europe. Dr I.M. Podvyaznaya, Department for Parasitic Worms, Zoological Institute, St Petersburg, Russia, could not find any slides of the material, even though that is where the type-material was apparently deposited (before a parasitological department was established in the mid-1930s). Prof. Oleg N. Bauer (retired from the same institute, personal communication) assumed they were lost. He added more useful and curious information. He knows of no specimen or subsequent report of *M. elongata*, despite numerous fish parasitologists studying the parasites of *E. lucius* and *E. reicherti* from many different Palaearctic areas from 1936 to the present. For this and other reasons, he also thinks *genus inquirendum* and *species inquirenda* are appropriate.

Genus *Myorhynchus* Durio & Manter, 1968

Myorhynchus pritchardae Durio & Manter, 1968 was described from an unidentified serranid purchased from a fish market in Noumea, New Caledonia. We examined the holotype (USNPC No. 63303) and only known specimen, and it resembled members of *Prosorhynchus* in general features. The two genera were considered synonyms by Wang & Wang (1998a). Durio & Manter (1968) observed that, even though the rhynchus is similar to that in species of *Prosorhynchus*, it has more extensive muscular lobes hanging from the rim of the conical portion. More importantly, the ovary lies in the extreme anterior end of the body and the vitellarium is in two lateral fields at the middle of the body. The egg is apparently unique in having the outer surface of the tanned layer of the shell with a granular texture encircled by a clear outer layer. This feature has not been described from any other species of bucephalid. This species probably most closely resembles *Prosorhynchus atlanticus* Manter, 1940, *P. pacificus* Manter, 1940 and *P. promicropsi* Manter, 1940, all also parasites from serranids. However, on the basis of the differences indicated above, we accept the genus and include it in the key.

Genus *Neoprosorhynchus* Dayal, 1948

Neoprosorhynchus purius Dayal, 1948 was described by Dayal (1948) from the intestine of *Epinephelus lanceolatus* caught at Puri, Orissa, India, and originally included in the Prosorhynchinae. It resembles species of *Prosorhynchus* in all respects except that the ovary is posterior to the testes (the primary generic distinction made by Dayal); also, it is said to have a Y-shaped excretory vesicle, but that structure was not figured. A second species, *N. dayali* Ahmad, 1984, was described from *Trichiurus muticus* from the Arabian Sea along the Bombay coast. Ahmad's (1982) illustration of *N. dayali* shows a post-testicular ovary, but he described a tubular (i.e. I-shaped) rather than Y-shaped excretory vesicle. This species appears to lack any similarity to *N. purius*, and the illustration shows terminal genitalia consistent with those of bucephalines. Both species need to be re-evaluated.

Genus *Proopisthogyne* Koche, 1976

Koche (1976) first reported and named *Proopisthogyne yamagutii* Koche, 1976 in a published abstract, but to our knowledge never published a formal description or an illustration of the species. In the abstract, *P. yamagutii* was described from *Platycephalus macrocanthus* in the Arabian Sea near Ratnagiri, Maharashtra, India, and placed in the Neoprosorhynchinae. It was reported to have a plug-shaped conical rhynchus, a Y-shaped excretory vesicle, a post-testicular ovary and a pharynx that opens anterior to the testes and ovary. We consider this taxon a *genus inquirendum* and do not include it in the key. If specimens of *P. yamagutii* become available, we assume that *Proopisthogyne* will be found to be synonymous with *Neoprosorhynchus*.

Genus *Neidhartia* Nagaty, 1937

Nagaty (1937) described *Neidhartia neidharti* Nagaty, 1937 and *N. ghardagae* Nagaty, 1937 from sea bass *Serranus* sp. in the Red Sea and named the former the type-species. He considered his species as members of the Prosorhynchinae. No clear distinction is apparent between the two species, and apparently no specimen has been deposited in a lending museum. Illustrations by Nagaty (1937) show specimens with a bulbous rhynchus that has a glandular base in the form of a cone that extends into the body proper. The anterior end of the rhynchus protrudes apically more than in other species in the Prosorhynchinae. The key diagnostic features for the genus are the shape of the rhynchus and an intertesticular ovary. In both species described by Nagaty, the pharynx opens between the posterior third and fourth of the body. There are several other nominal species, which are mentioned below. *N. microrhynchus* Chauhan, 1943 was described from *Psettodes erumei* off the west coast of India. This rather elongate species has a similar rhynchus, the excretory vesicle reaches the anterior region of the body and the pharynx opens between the middle and posterior thirds of the body. We examined the holotype of *N. polydactyli* Manter, 1953 (USNPC No. 587301), which was described from *Polydactylus plebeius* from Fiji (Manter, 1953). It was distinguished from previously described species of *Neidhartia* because the anterior region of the rhynchus is lobed and the excretory vesicle reaches only as far anteriorly as the mouth opening of the pharynx, which is located at the level of the posterior fourth of the body (Manter, 1953; personal observations). We also examined *N. macintoshi* Velasquez, 1959 (USNPC No. 68216, one slide with holotype and paratype), which was described from *Epinephelus bleekeri* in the Philippines. It differed from the other species by having the pharynx open just anterior to the mid-body. The extent of the excretory vesicle is not visible in the holotype. *N. coronata* Durio & Manter, 1968 was described from *Epinephelus* sp. off New Caledonia. This species has a single row of short spines circling the rhynchus. As with *N. macintoshi*, the excretory vesicle was not traced by Durio & Manter (1968). These authors suggested a relationship between *Neidhartia* and *Dollfustrema* based on the single circlet of spines. We believe that both the presence of rhynchal spines and the length of the excretory vesicle should be used as generic characters, but, since data have not been reported or are not evident for the length of the excretory vesicle in *N. macintoshi* and *N. coronata*, we hesitate to create new combinations. We suspect that *N. macintoshi* and *N. coronata* may belong in new genera. Future work involving new specimens or species is necessary to shed light on this predicament. We consider the most important generic features for *Neidhartia* to be the structure of the rhynchus and the position of the ovary relative to the testes. The presence of a coiling seminal duct links the genus to the Prosorhynchinae.

Genus *Pseudoprosorhynchus* Yamaguti, 1938

Yamaguti (1938a) originally placed the monotypic *Pseudoprosorhynchus synodi* Yamaguti, 1938 in the Prosorhynchinae, and no other species has been described in the genus. He (Yamaguti, 1958a) later created the Neidhartiinae and included *P. synodi* in that subfamily, differentiating *Pseudoprosorhynchus* from *Neidhartia* primarily on a short excretory vesicle. Manter (1940) considered *Pseudoprosorhynchus* a synonym of *Neidhartia* but did not explain why. We consider *Pseudoprosorhynchus* to be closely related to *Neidhartia*, since members have tandem testes and an intertesticular ovary, and, like Yamaguti (1958a), we think that the comparatively short excretory vesicle may be a good distinction between the two genera. However, as mentioned above, *N. polydactyli* has a short excretory vesicle and the excretory vesicle is not traceable in two other species in the genus. As a result, we think the generic status of *Pseudoprosorhynchus* needs to be re-examined, but we include it in the key.

Genus *Telorhynchus* Crowcroft, 1947

Telorhynchus arripidis Crowcroft, 1947 was described from the 'colonial salmon' *Arripis trutta* off southern Australia. Crowcroft (1947) established the new genus on the basis of specimens having a circlet of long spines (spikes) surrounding the rim of the rhynchus. The terminal genitalia are poorly illustrated, and their description indicates that they may be similar to those of other genera in the Prosorhynchinae, where the genus was placed by Crowcroft (1947). Skrjabin & Guschanskaja (1962) and Yamaguti (1958a, 1971) considered it to be in the Bucephalinae. Crowcroft (1947) described the cirrus-sac of *T. arripidis* as containing a tubular seminal vesicle that folded upon itself, i.e. probably what we call the seminal duct. This looped anteriorly before extending straight posteriorly and joining the 'primary' pars prostatica, apparently through a narrow aperture surrounded by a sphincter. The short and narrow ejaculatory duct passed through the genital lobe, which entered the genital atrium as in other bucephalids. The condition of having both a folded seminal vesicle and a coiling seminal duct is consistent with the Prosorhynchinae. The presence of a sphincter might be an apomorphic feature. Crowcroft (1947) described a saccular excretory vesicle, but Lebedev (1968) considered the group to contain a U-shaped vesicle. To our knowledge, Crowcroft's specimens were not deposited at a lending museum. Wang & Wang (1998a) transferred the species *T. cociellae* Gu & Chen, 1976 and *T. hippocampi* Shen, 1982 to *Dollfustrema*.

Lebedev (1968) erected *Telorhynchoides* Lebedev, 1968 for *T. longicollis* Lebedev, 1968, and the Telorhynchinae Lebedev, 1968 for both *Telorhynchoides* and *Telorhynchus*. He also reported *Telorhynchus arripidis*, described two new species of *Telorhynchus*, and described *Telorhynchoides longicollis*, all from *Arripis trutta* in the southern Great Australian Bight and some from the Tasman Sea. The species *T. longicollis* appears similar to those of *Telorhynchus* and was distinguished by Lebedev (1968) from the latter by having a rhynchus with 'antennae' rather than spines or spikes and the anterior neck portion of body that was dramatically elongated. All of the species are differentiated primarily by the number of rhynchal spikes or antennae present. We consider both of these traits to be specific characters at best, and therefore propose *Telorhynchoides* as a junior synonym of *Telorhynchus*. The status of the cirrus-sac and its contents in each of the species from *A. trutta* described by Lebedev (1968) is not clear, but *T. longicollis* appears to have a seminal duct, as described by Crowcroft (1947), and a spherical seminal vesicle, and the others, including *T. arripidis*, which should have a seminal duct, are illustrated with a spherical seminal vesicle but no seminal duct. Material of all of these species should be re-examined.

Family Nuitrematidae Kurochkin, 1975

Kurochkin (1975) described *Nuitrema strigeiformes* based on a very unusual digenean specimen among those collected by V.G. Leont'ev in 1969 during the 20th Parasitological Expedition of the Soviet Pacific Scientific Fisheries and Oceanography Research Institute (TINRO). The 1.8 mm long specimen was collected from the intestine of the sailfish reported as *Histiophorus orientalis* (a junior synonym of *Istiophorus platypterus*). No more specific information was given regarding the locality from which the sailfish was caught other than tropical Pacific Ocean (Kurochkin, 1975). *Nuitrema strigeiformes* shares features present in both the Strigeidae Railliet, 1919, and the Bucephalidae. The feature common to the Strigeidae is the presence of a bipartite body, the forebody being cup-shaped and the hindbody elongate. Bucephalid features include the presence of an apical attachment sucker (similar to a rhynchus), a posterior cirrus-sac (an internal seminal vesicle is illustrated but not discussed in the description), the presence of a simple sac-like intestine, relatively small eggs (16–18 × 12–14 μm) and a fish definitive host. Kurochkin (1975) created a new family within the Bucephalata for the single specimen.

Kurochkin (1975) observed that the overall appearance of his specimen superficially resembled that of many strigeoid species (i.e. species in *Apatemon* Szidat, 1928, *Cotylurus* Szidat, 1928 or *Strigea* Abildgaard, 1790); however, the forebody of *N. strigeiformes* was illustrated and described as having an apical sucker within a cup-like margin, disassociated from the digestive system. This feature is strikingly similar to the configuration of the forebody in species of *Apharyngostrigea* Ciurea, 1927, except that the apical sucker is an oral sucker in this strigeid. We believe that Kurochkin may have misinterpreted the features of the forebody in his specimen, an easy action to take, and advocate that sectioned material would clarify questions relating to the anatomy of the digestive system in his description. It is possible that, prior to being caught, the particular sailfish may have recently consumed a seabird infected with a strigeoid, which in turn may have attached to the fish's intestinal mucosa. This seems possible because of the wide range of dietary items in the sailfish (Evans & Wares, 1972) and the fact that 18 and 22 specimens of the alcid *Alle alle* were found in the stomach of two specimens of the related swordfish (Gibson, 1998). We consider it unusual for there to be no other record for this species in the sailfish, considering the number of sailfish examined for parasites. Until more specimens of this species become available for further examination, we propose that *Nuitrema strigeiformes* be considered a *species inquirenda*. Moreover, we tentatively do not accept the family Nuitrematidae until the record can be corroborated. Kurochkin (1975) provided the following overlapping diagnoses for the family and genus.

Family Nuitrematidae Kurochkin, 1975

Diagnosis: Body sharply subdivided into two parts; anterior portion thick-walled, bowl- or cup-shaped; posterior portion elongate with rounded posterior end. Tegument unarmed. Apical organ a sucker on elongate appendage extending from anterior body section. Pharynx opening ventrally on anterior body section. Intestine sac-like, extending posteriorly in forebody. Testes two, in hindbody. Ovary pre-testicular in hindbody. Vitelline follicles in two groups, one on either side of intestine in forebody. Cirrus-sac at posterior end of hindbody, opening ventrally at posterior extremity. Excretory pore subterminal, slightly posterior to genital pore. (No mention of tribocytic organ.) Type-genus *Nuitrema* Kurochkin, 1975.

Genus *Nuitrema* Kurochkin, 1975

Diagnosis: Body bipartite. Tegument lacking spines. Oral sucker absent. Pharynx opening ventrally on forebody. Intestine a single caecum extending posteriorly in forebody, sac-like with thick glandular walls. Testes tandem in hindbody. Ovary pretesticular, in anterior portion of hindbody, some distance from testes. Vitellarium two distinct groups of follicles arranged laterally in forebody. Uterus filled with eggs and looping throughout hindbody. Cirrus-sac in posterior portion of body. Genital pore subterminal on ventral surface. Excretory pore subterminal, opening slightly posterior to genital pore. In intestine of marine fish; tropical Pacific Ocean. Type-species *N. strigeiformis* Kurochkin, 1975.

Superfamily Bucephaloidea Poche, 1907

(Syns Gasterostomatiformes Skrjabin & Schulz, 1937; Gasterostomata Odhner, 1905; Alcicornata Skrjabin & Schulz, 1937)

Diagnosis: Body usually spinous. Anterior holdfast organ a rhynchus, with or without sucker. Mouth on ventral surface, associated with pharynx and oesophagus; oesophagus present in most species. Intestinal caecum simple, not bifurcating, elongate or sac-like, not associated with rhynchus. Oral and ventral suckers absent. Testes two. Cirrus-sac conspicuous, located posteriorly, containing internal seminal vesicle (lacking external seminal vesicle) and male duct; male duct consisting of extensive pars prostatica and distal ejaculatory duct leading into genital atrium, with seminal duct present or lacking; genital atrium a sac enclosing muscular genital lobe, with genital pore usually terminal. Ovary singular, spherical or rarely lobed. Vitellarium follicular. Uterus usually coiling throughout much of body, occasionally sac-like. Distal portion of uterus entering genital atrium. Excretory vesicle I-shaped, variable in length. Excretory pore terminal. Use lamellibranchs as first intermediate host, fishes as second intermediate host and fishes (one in salamander) as definitive host. Type-genus *Bucephalus* Baer, 1827.

Family Bucephalidae Poche, 1907

(Syns Gasterostomatidae Braun, 1883; Sinicovothylacidae Yu, 1965; Prosorhynchidae Kohn, 1970)

Diagnosis: With characteristics of the superfamily. Type-genus *Bucephalus* Baer, 1927.

Key to subfamilies (Figs 15.3–15.7)

1a. Seminal vesicle spherical to ovoid or ellipsoidal, usually less than twice as long as wide; seminal duct (distinct narrow coiling proximal portion of prostatic duct) absent; pars prostatica straight or curved .... 2.
1b. Seminal vesicle elongate, about twice as long as wide or longer; seminal duct present or absent; pars prostatica straight .... 3.

2a. Cirrus-sac thin- to moderately thick-walled (< 20 µm); pars prostatica straight (bending slightly in *Rhipidocotyle*); rhynchus almost always a simple sucker, unmodified or modified with lobes, tentacles or hood (except *Alcicornis*, with conical rhynchus that lacks sucker but has tentacles); in intestine of marine and freshwater fishes (rarely amphibians) .... Bucephalinae Poche, 1907.
2b. Cirrus-sac thick-walled (40–80 µm); pars prostatica curved, never straight; rhynchus a simple sucker with eight papillae or dorsal hood on anterior rim; in intestine of marine fish .... Dolichoenterinae Yamaguti, 1958.

3a. Pars prostatica relatively straight (seminal duct lacking or not apparent); rhynchus a small, weakly developed apical pad; in swim-bladder, body-cavity or possibly stomach of freshwater fishes ... Paurorhynchinae Dickerman, 1954.
3b. Pars prostatica not straight (seminal duct prominent and looping); rhynchus a muscular or glandular cone topped by muscular disc; disc unadorned or equipped with muscular lobes or circle of long or short spines; in intestine of marine (rarely freshwater) fishes .... Prosorhynchinae Nicoll, 1914.

Subfamily Bucephalinae Poche, 1907

(Syn. Provarinae Wang, 1995)

Diagnosis: Body more or less cylindrical, covered with minute spines. Rhynchus consisting of simple sucker with or without dorsal hood, lobe or retractable tentacles, rarely with muscular funnel topped by ring of retractable tentacles (*Alcicornis*). Intestinal caecum sac-like or tubular, running anteriorly, posteriorly or dorsally from pharynx. Testes smooth, spherical to ovoid, variably orientated. Cirrus-sac relatively thin- to moderately thick-walled (up to 20 µm thick). Seminal vesicle spherical or ovoid, rarely more than twice as long as wide (except *Prosorhynchoides longicirrus*), sometimes twisted (*Rhipidocotyloides tsengi*), rarely bipartite (e.g. *Prosorhynchoides kweiyangensis*). Seminal duct absent or extremely short. Pars prostatica proper usually straight (except some *Rhipidocotyle*). Ejaculatory duct thin-walled, without abundant glands, passing through posterior end of cirrus-sac, leading through muscular genital

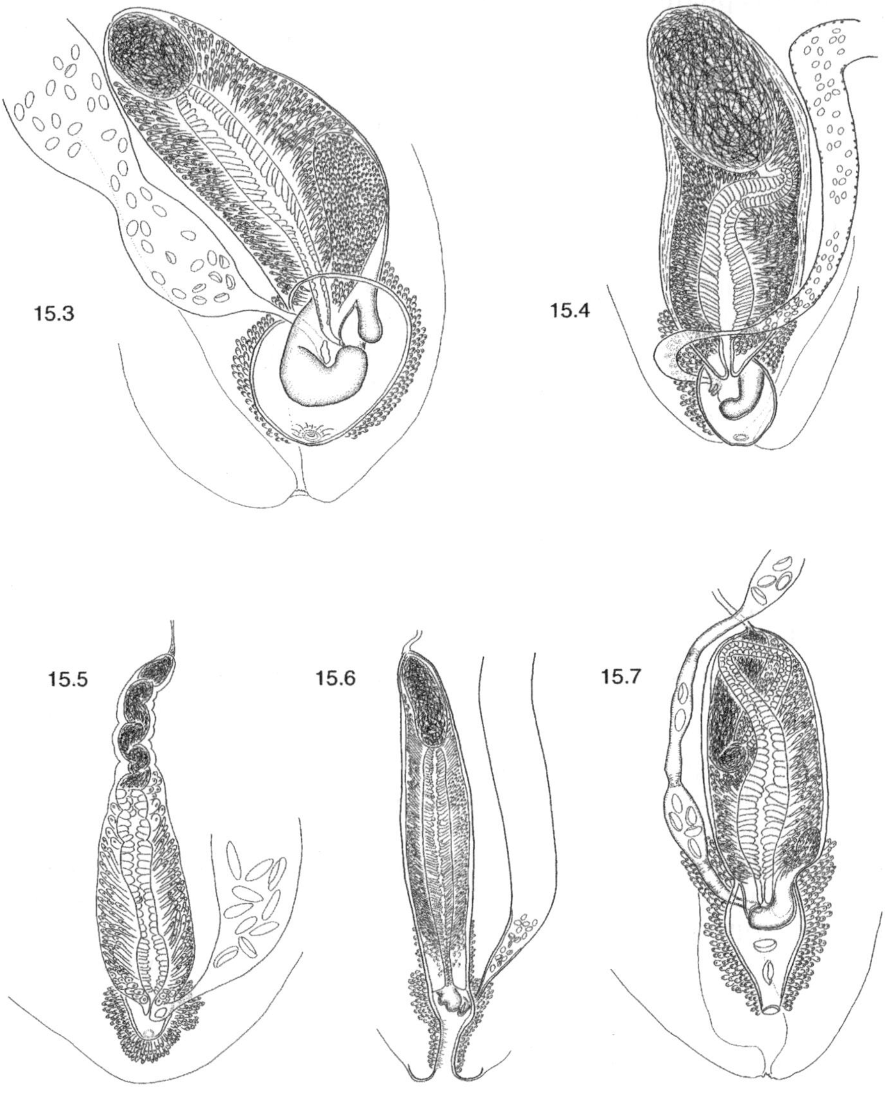

Figs 15.3–15.7. Subfamily characters.

15.3. Terminal genitalia of *Bucephalus scorpaenae* Manter, 1940.

15.4. Terminal genitalia of *Dolichoenterum longissimum* Ozaki, 1924. (Composite illustration from Meguro Parasitological Museum specimen 30078 and USNPC voucher 49155.)

15.5. Terminal genitalia of *Paurorhynchus hiodontis* Dickerman, 1954. (Original from USNPC paratype 37404.)

15.6. Terminal genitalia of *Bellumcorpus majus* Kohn, 1962. (Composite illustration from Instituto Oswaldo Cruz specimens 28746B–D.)

15.7. Terminal genitalia of *Prosorhynchus squamatus* Odhner, 1905. (Original from Bergen Zoological Museum Specimen 47987.)

lobe, opening into genital atrium. Genital atrium a thin-walled sac, enclosing genital lobe, with terminal genital pore. Ovary spherical to ovoid. Vitellarium two separated rows or bunches of follicles. Uterus tubular, coiling throughout body. In intestine of marine and freshwater fishes; also known from intestine of one amphibian. Type-genus *Bucephalus* Baer, 1827.

Key to genera

1a. Rhynchus with tentacles .... **2.**
1b. Rhynchus lacking tentacles, having simple sucker only or sucker with some dorsal modification .... **3.**

2a. Rhynchus a long funnel-shaped or short U-shaped glandular mass topped by ring of 7 to 21 fleshy retractable tentacles .... ***Alcicornis* MacCallum, 1917. (Fig. 15.8)**
Diagnosis: Body elongate. Rhynchus either long funnel-shaped or short U-shaped muscular structure, crowned by ring of 7 to 21 long fleshy retractable tentacles. Mouth near mid-body. Caecum sac-like, variably directed from pharynx. Testes oblique. Seminal vesicle spherical to ovoid. Pars prostatica straight. Ovary pretesticular. Vitelline follicles in anterior lateral rows. Excretory vesicle long, reaching anteriorly to near rhynchus. In marine fishes; cosmopolitan. Type-species *A. carangis* MacCallum, 1917.
2b. Rhynchus a simple sucker with seven to many retractable tentacles arranged around anterior rim of sucker .... ***Bucephalus* Baer, 1827. (Fig. 15.9)**
(Syns *Gasterostomum* von Siebold, 1848; *Eubucephalus* Diesing, 1855; *Labratrema* Maillard, 1975)
Diagnosis: Rhynchus a simple sucker, usually with seven tentacles around anterodorsal rim of sucker. Mouth near mid-body. Caecum sac-like, variably directed from pharynx. Testes oblique. Seminal vesicle spherical to ovoid. Pars prostatica straight. Ovary pretesticular. Vitelline follicles in two fields anterior to ovary. Excretory vesicle variable in length. In freshwater or marine fishes; cosmopolitan. Type-species *B. polymorphus* Baer, 1827.

3a. Rhynchus a simple sucker, with or without central muscular palp .... **4.**
3b. Rhynchus a simple sucker, covered by smooth hood; hood simple, muscular or with large muscular fleshy lobes or sharply pointed scalloped lobes ..... **8.**

4a. Rhynchus with central muscular palp .... ***Pseudobucephalopsis* Long & Lee, 1964. (Fig. 15.10)**
Diagnosis: Rhynchus a simple sucker (with reported central U-shaped palp structure, possibly fixation artefact). Mouth just posterior to mid-body. Caecum sac-like, directed dorsally from pharynx. Testes oblique. Ovary between testes. Vitellarium in two anterior clumps. Excretory vesicle short, reaching level of posterior testis. In marine fishes; Asia. Type-species *P. spheroides* Long & Lee, 1964.
4b. Rhynchus without central muscular palp .... **5.**

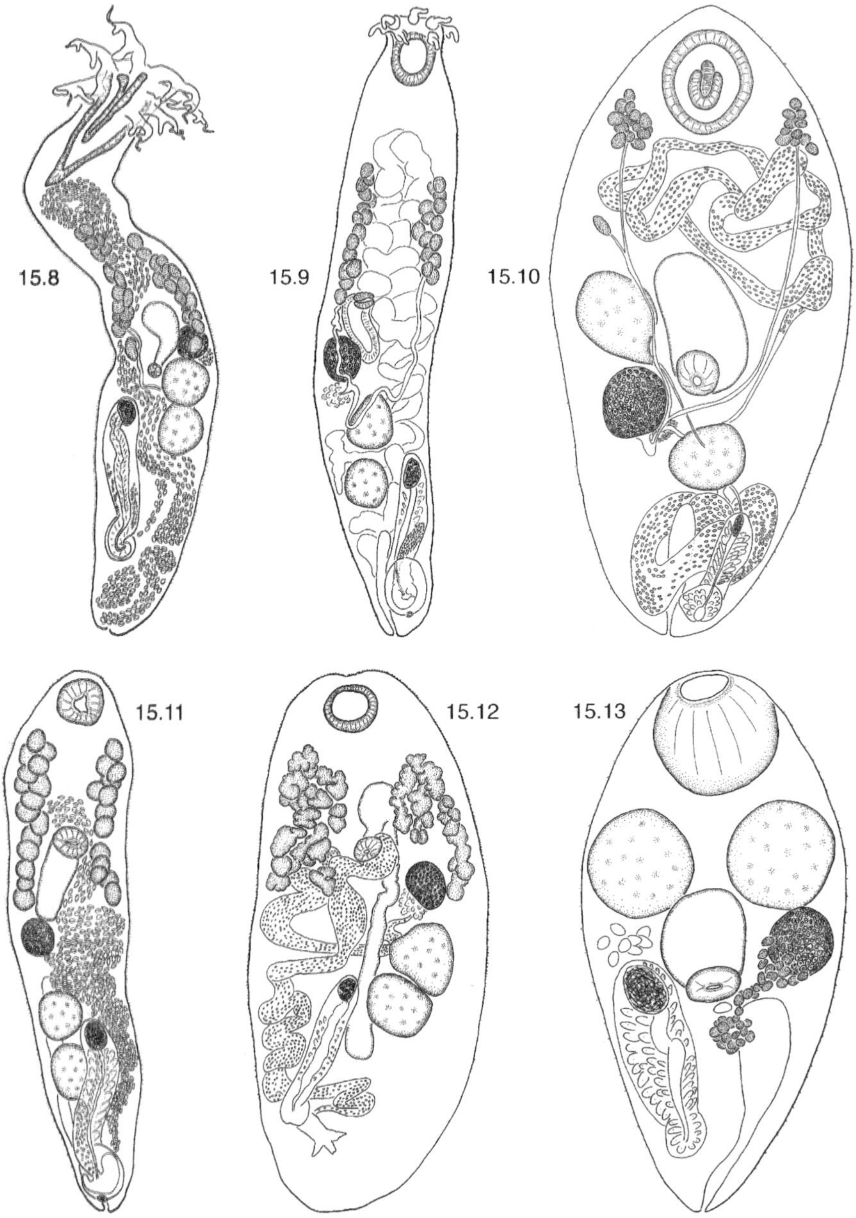

Figs 15.8–15.13. Bucephalinae.

15.8. *Alcicornis carangis* MacCallum, 1917. (Original from USNPC voucher 60249.)

15.9. *Bucephalus scorpaenae* Manter, 1940.

15.10. *Pseudobucephalopsis spheroides* Long & Lee, 1964. (Redrawn from Long & Lee, 1964b.)

15.11. *Prosorhynchoides gracilescens* (Rudolphi, 1819). (Original from British Museum No. 1987.9.22.26–125.)

15.12. *Prosorhynchoides ovatus* (Linton, 1900). (Original from Muséum National d'Histoire Naturelle specimen 30.11.1925.)

15.13. *Parabucephalopsis prosthorchis* Tang & Tang, 1963. (Redrawn from Long & Lee, 1964b.)

5a. Ovary pretesticular .... *Prosorhynchoides* Dollfus, 1929. (Figs 15.11, 15.12)
(Syns *Bucephalopsis* Diesing, 1855; *Neobucephalopsis* Dayal, 1948; *Bucephaloides* Hopkins, 1954)

Diagnosis: Rhynchus a simple sucker. Caecum sac-like, variably directed from pharynx, or an elongate tube, directed anteriorly then recurving posteriorly from pharynx, reaching posterior level of gonads. Testes oblique. Seminal vesicle spherical or ovoid. Pars prostatica straight. Genital atrium containing fleshy intromittent organ. Ovary pretesticular. Vitelline follicles in two anterior clumps or rows. Excretory vesicle variable in length. In marine and some freshwater fishes; cosmopolitan. Type-species *P. ovatus* (Linton, 1900).

5b. Ovary at level between testes or post-testicular .... 6.

6a. Ovary post-testicular .... *Parabucephalopsis* Tang & Tang, 1963. (Fig. 15.13)

Diagnosis: Body ellipsoidal. Rhynchus a simple sucker. Mouth just posterior to mid-body. Caecum sac-like, directed dorsally and anteriorly from pharynx. Testes symmetrical, anterior to caecum. Pars prostatica possibly coiling once. Ovary sinistral, post-testicular. Vitellarium two small clumps of follicles near ovary. Excretory vesicle reaching to level of pharynx. In freshwater fishes; Asia. Type-species *P. prosthorchis* Tang & Tang, 1963.

6b. Ovary between testes or at level of posterior testis .... 7.

**7a. Ovary at level of posterior testis; caecum sac-like ....
.... *Roparhynchus* Gupta & Kumari, 1978. (Fig. 15.14)**

Diagnosis: Body ellipsoidal. Rhynchus a simple sucker. Mouth in posterior half of body. Caecum sac-like, directed anteriorly from pharynx. Testes oblique. Ovary sinistral, posterior to anterior testis, at level of posterior testis. Vitelline follicles in two anterior clumps. Excretory vesicle short. In freshwater fishes; India. Type-species *R. nelsoni* Gupta & Kumari, 1978.

**7b. Ovary at level between testes; caecum elongate, tubular ....
.... *Heterobucephalopsis* Gu & Shen, 1983. (Fig. 15.15)**

Diagnosis: Rhynchus a simple sucker. Mouth in anterior third of body. Caecum elongate, tubular, directed posteriorly from pharynx, reaching to level of anterior testis. Testes oblique. Seminal vesicle convoluted. Ovary dextral, between testes. Vitelline follicles in two lateral bands in middle of body. In marine fishes; Asia. Type-species *H. gymnothoracis* Gu & Shen, 1983.

8a. Rhynchus a simple sucker covered by simple muscular hood or hood with three to five large fleshy lobes; pars prostatica slightly bent, not completely straight; ovary pretesticular .... *Rhipidocotyle* Diesing, 1858. (Figs 15.16–15.19)
(Syns *Nannoenterum* Ozaki, 1924; *Pararhipidocotyle* Kohn, 1970)

Diagnosis: Rhynchus a simple feeble sucker with dorsal hood; hood with or without associated lateral and/or frontal fleshy lobes. Mouth near mid-body. Caecum sac-like, variably directed from pharynx. Testes oblique. Seminal vesicle spherical to slightly ovoid. Pars prostatica curved, never straight. Ovary pretesticular. Vitellarium in two fields, anterior to ovary. Excretory vesicle variable in length. In freshwater and marine fishes; cosmopolitan. Type-species *R. galeata* (Rudolphi, 1819).

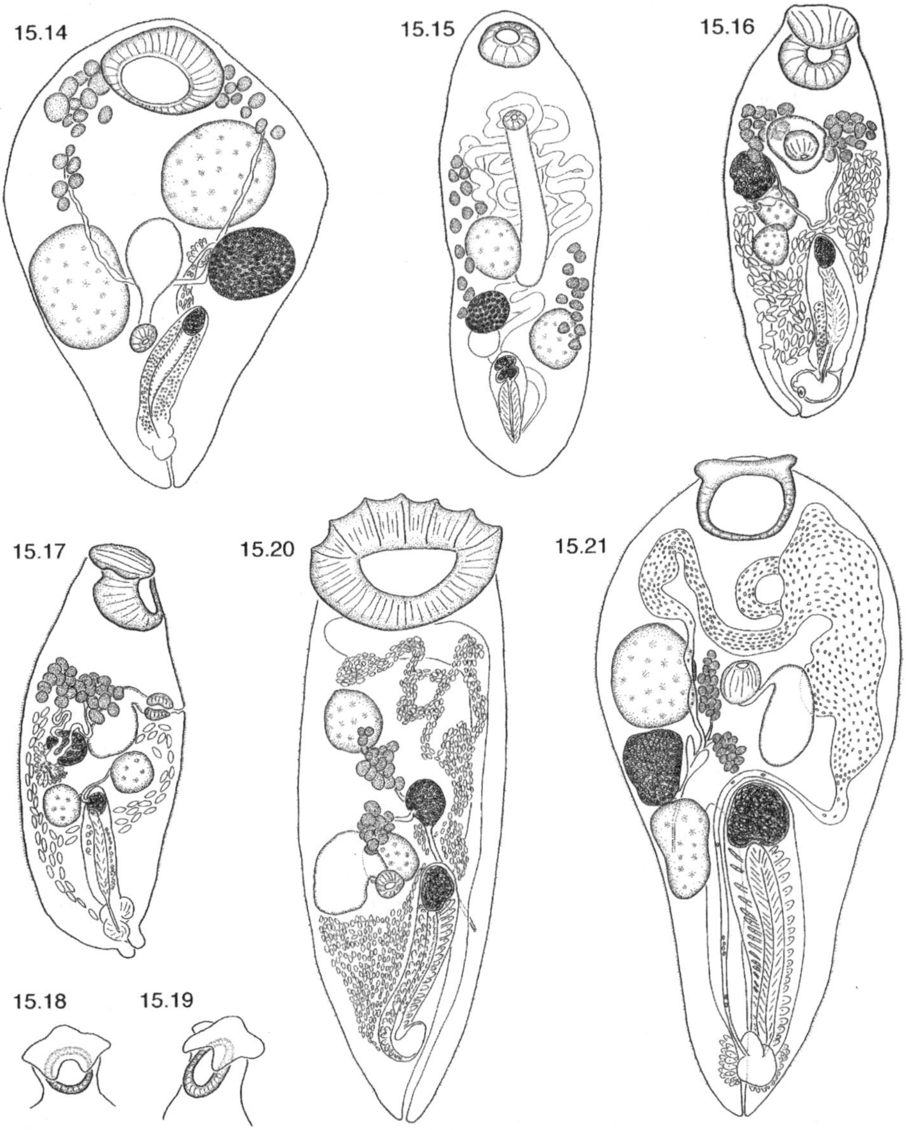

Figs 15.14–15.21. Bucephalinae.

15.14. *Roparhynchus nelsoni* Gupta & Kumari, 1978. (Redrawn from Gupta & Kumari, 1976.)

15.15. *Heterobucephalopsis gymnothoracis* Gu & Shen, 1983. (Redrawn from Gu & Shen, 1983.)

15.16. *Rhipidocotyle lintoni* Hopkins, 1954, ventral view.

15.17. *Rhipidocotyle lintoni* Hopkins, 1954, side-view.

15.18. Rhynchus of *Rhipidocotyle* sp.

15.19. Rhynchus of *Rhipidocotyle* sp.

15.20. *Pseudorhipidocotyle elopichthys* Long & Lee, 1964. (Redrawn from Long & Lee, 1964b.)

15.21. *Rhipidocotyloides tsengi* Long & Lee, 1964, although not apparent here, the seminal vesicle was illustrated by the original authors as twisted. (Redrawn from Long & Lee, 1964b.)

8b. Rhynchus a simple sucker, covered by smooth hood or hood with sharply pointed scalloped lobes, never with three to five fleshy muscular lobes; pars prostatica straight; ovary between testes or at level of either testis .... 9.

9a. Rhynchus a simple sucker with hood having sharply pointed scalloped lobes; pharynx opening near or below middle of body .... *Pseudorhipidocotyle* Wang & Pan, 1964. (Fig. 15.20)

(Syn. *Rhipidocotyloides* of Long & Lee (1964a) (*nomen nudum*))

Diagnosis: Rhynchus a sucker with dorsal hood having approximately seven scalloped lobes. Mouth in posterior half of body, posterior to the gonads. Caecum sac-like, directed dorsally, curving slightly posteriorly. Testes oblique to nearly tandem. Pars prostatica straight. Ovary between testes. Vitellarium in two clumps closely associated with ovary. Excretory vesicle long, reaching to level of rhynchus. In freshwater fishes; Asia. Type-species *P. elopichthys* Long & Lee, 1964.

9b. Rhynchus a simple sucker with smooth dorsal hood; ends of hood slightly protruding laterally; pharynx opening at level between anterior third and half of body .... *Rhipidocotyloides* Long & Lee, 1964. (Fig. 15.21)

(Syn. *Pseudorhipidocotyle* of Long & Lee (1964a) (*nomen nudum*))

Diagnosis: Body pyriform, wider anteriorly. Rhynchus a sucker with dorsal collar. Mouth between mid-body and anterior third of body. Caecum sac-like, directed posteriorly from pharynx. Testes tandem, dextral. Cirrus-sac large, almost half body length. Pars prostatica straight. Ovary dextral, between testes. Vitellarium in two clumps, closely associated with ovary. Uterus coiling anterior to cirrus-sac. Excretory vesicle I-shaped, reaching anteriorly as far as anterior extent of cirrus-sac. In freshwater fishes; Asia. Type-species *R. tsengi* Long & Lee, 1964.

Subfamily Dolichoenterinae Yamaguti, 1958

Diagnosis: Body greatly elongate, spined. Rhynchus a sucker with eight dorsal horn-like lobes or a dorsal reniform hood. Mouth near anterior quarter of body. Caecum tube-like, extending anteriorly, recurving posteriorly to posterior quarter of body or not. Testes tandem or oblique. Cirrus-sac thick-walled (40–80 vs. 1–20 μm thick in Bucephalinae), shorter and wider relative to body length than in other subfamilies. Seminal vesicle usually less than twice as long as wide. Seminal duct absent or not apparent. Pars prostatica curved and thin-walled proximally, straight and thick-walled distally. Ejaculatory duct thin-walled, passing through genital lobe. Ovary spherical, intertesticular. Vitelline follicles in two rows. Uterus tubular, coils generally confined between rows of vitelline follicles. Excretory vesicle long, reaching near rhynchus. In intestine of marine fishes. Type-genus *Dolichoenterum* Ozaki, 1924.

Key to genera

1a. Rhynchus a simple sucker with eight short fleshy lobes extending from dorsal aspect of anterior rim .... *Dolichoenterum* Ozaki, 1924. (Fig. 15.26)

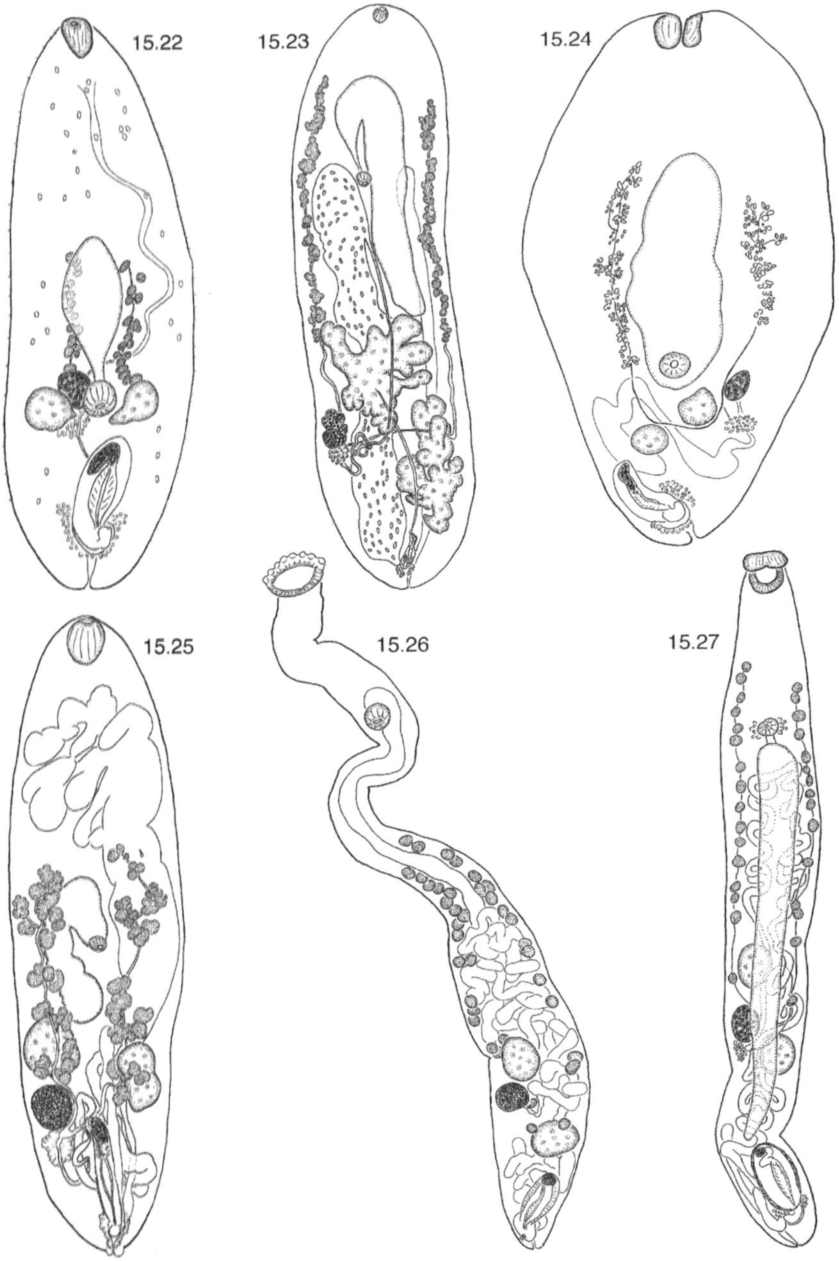

Figs 15.22–15.27. Paurorhynchinae and Dolichoenterinae.
15.22. *Sinicovothylax chujiangensis* Yu, 1965. (Redrawn from Yu, 1965.)
15.23. *Paurorhynchus hiodontis* Dickerman, 1954. (Dorsal view of USNPC paratype 374004.)
15.24. *Rhoporhynchus clarias* Wang, 1995. (Redrawn from Wang, 1995.)
15.25. *Bellumcorpus majus* Kohn, 1962. (Original from Instituto Oswaldo Cruz specimen 28746B.)
15.26. *Dolichoenterum longissimum* Ozaki, 1924. (Original from USNPC voucher 49155.)
15.27. *Pseudodolichoenterum megagaster* (Corkum, 1968). (Redrawn from Corkum, 1968.)

Diagnosis: Rhynchus a sucker with eight dorsal horn-like lobes. Caecum reaching posteriorly beyond level of cirrus-sac. Testes tandem. Ovary between testes. Vitelline follicles in two lateral bands between mid-body and posterior testis. In marine fishes; cosmopolitan. Type-species *D. longissimum* Ozaki, 1924.

1b. Rhynchus a simple sucker with a dorsal reniform hood .... *Pseudodolichoenterum* Yamaguti, 1971. (Fig. 15.27)

Diagnosis: Rhynchus a simple sucker with dorsal reniform hood. Pharynx surrounded by gland-cells. Testes oblique to nearly tandem. Ovary between testes. Vitelline follicles in two lateral rows entirely anterior to ovary. In caeca of marine fishes; Gulf of Mexico. Type-species *P. megagaster* (Corkum, 1968).

Subfamily Paurorhynchinae Dickerman, 1954

(Syn. Sinicovothylacinae Wang & Wang, 1998)

Diagnosis: Body elongate, cylindrical, usually unspined. Rhynchus a small, weakly-developed muscular apical pad, usually oval in frontal profile. Position of mouth variable. Caecum elongate, extending anteriorly from pharynx, recurving posteriorly in some genera. Testes smooth or lobed to dendritic. Cirrus-sac thin- to moderately thick-walled (2.5–20.0 μm). Seminal vesicle elongate, with length more than twice width; seminal duct lacking or not apparent. Pars prostatica and ejaculatory duct straight. Genital lobe present but small in some genera. Ovary spherical or lobed. Vitellarium in two bands of follicles, anterior to ovary. Uterus sac-like or with segments forming reservoirs for eggs. In swim-bladder, body-cavity or stomach of freshwater fishes. Type-genus *Paurorhynchus* Dickerman, 1954.

Key to genera

1a. Body spined; testes symmetrical .... *Sinicovothylax* Yu, 1965. (Fig. 15.22)

Diagnosis: Body spined. Mouth at level of posterior third of body. Caecum directed anteriorly, never recurving posteriorly. Testes symmetrical. Seminal vesicle elongate, ovoid. Genital lobe small. Ovary lobed, at level of testes. Uterus sac-like. Excretory vesicle short, reaching level of pars prostatica. In swim-bladder of freshwater fishes; China. Type-species *S. chujiangensis* Yu, 1965.

1b. Body smooth; testes oblique .... 2.

2a. Mouth opening in second quarter of body .... *Paurorhynchus* Dickerman, 1954. (Fig. 15.23)

Diagnosis: Body smooth. Caecum running anteriorly from pharynx then turning and running posteriorly. Mouth between anterior quarter and half of body. Testes oblique, dendritic or smooth. Seminal vesicle long, narrow, sinuous or spiralling. Pars prostatica narrow proximally, wider distally, without seminal duct. Genital lobe and genital atrium relatively small. Ovary dextral, lobed, at level of anterior testis. Uterus sac-like. Excretory vesicle long, reaching near rhynchus. In body-cavity of freshwater fishes; North America. Type-species *P. hiodontus* Dickerman, 1954.

2b. Mouth opening at middle or in posterior half of body .... 3.

3a. Ovary sinistral, posterolateral to opening of pharynx, anterior to testes or at level of anterior testis; uterus tubular, coiling in posterior part of body, not reaching anterior to pharynx .... *Rhoporhynchus* Wang, 1995. (Fig. 15.24)

Diagnosis: Body smooth. Mouth in posterior quarter of body. Caecum directed anteriorly from pharynx. Testes oblique. Seminal vesicle elongate. Ovary ovoid, sinistral. Vitelline follicles small. Uterus tubular, coiling entirely posterior to mouth. In freshwater fishes; China. Type-species *R. clarias* Wang, 1995.

3b. Ovary dextral, intertesticular or at level of posterior testis; uterus tubular but forming large sac-like reservoirs, extending throughout body .... *Bellumcorpus* Kohn, 1962. (Fig. 15.25)

Diagnosis: Body smooth. Mouth near middle of body. Caecum directed anteriorly from pharynx before recurving posteriorly. Testes oblique, rarely symmetrical. Seminal vesicle elongate, ovoid, rarely diminutive. Pars prostatica uniform in width. Genital lobe small but larger than in *Paurorhynchus*. Ovary dextral, intertesticular or at level of posterior testis, smooth or lobed. Uterus tubular but forming reservoirs, coiling throughout body. Excretory vesicle reaching middle of body. In stomach of freshwater fishes; South America. Type-species *B. majus* Kohn, 1962.

Subfamily Prosorhynchinae Nicoll, 1914

(Syns Neidhartiinae Yamaguti, 1958; Neoprosorhynchinae Yamaguti, 1958; Telorhynchinae Lebedev, 1968; Folliculovarinae Wang, 1995; Folliculovariinae Wang & Wang, 1998)

Diagnosis: Body elongate or ellipsoidal, covered with minute spines. Rhynchus a muscular and glandular funnel or plug, capped with muscular disc; disc with or without long or short spines or lobes. Caecum simple, sac-like, directed anteriorly from pharynx. Cirrus-sac relatively thin- to thick-walled (up to 45 µm thick in *Neidhartia*), containing seminal vesicle and male duct; seminal vesicle elongate; proximal portion of pars prostatica a seminal duct; seminal duct long, convoluted, relatively narrow; middle portion of male duct a pars prostatica proper, straight; distal portion of male duct an ejaculatory duct, thin-walled, slender, leaving posterior end of cirrus-sac and passing through muscular genital lobe. Spermatophore produced in genital atrium of some species. Genital atrium thin-walled, enclosing genital lobe, with terminal genital pore. Ovary variable in shape. Vitellarium usually anterior to ovary, either in two separated groups of follicles near or anterior to mid-body or in one confluent group in anterior region of body. Uterus tubular, coiling throughout body. Excretory vesicle variable in length. In intestine of marine (rarely freshwater) fishes. Type-genus *Prosorhynchus* Odhner, 1905.

Key to genera

1a. Rhynchus surmounted by flat disc surrounded by circlet of long spines or spine-like structures .... *Telorhynchus* Crowcroft, 1947. (Fig. 15.28)

(Syn. *Telorhynchoides* Lebedev, 1968)

Diagnosis: Body elongate. Rhynchus a long slender funnel surmounted by flat disc; disc circled by single row of long spines or spine-like structures. Mouth near

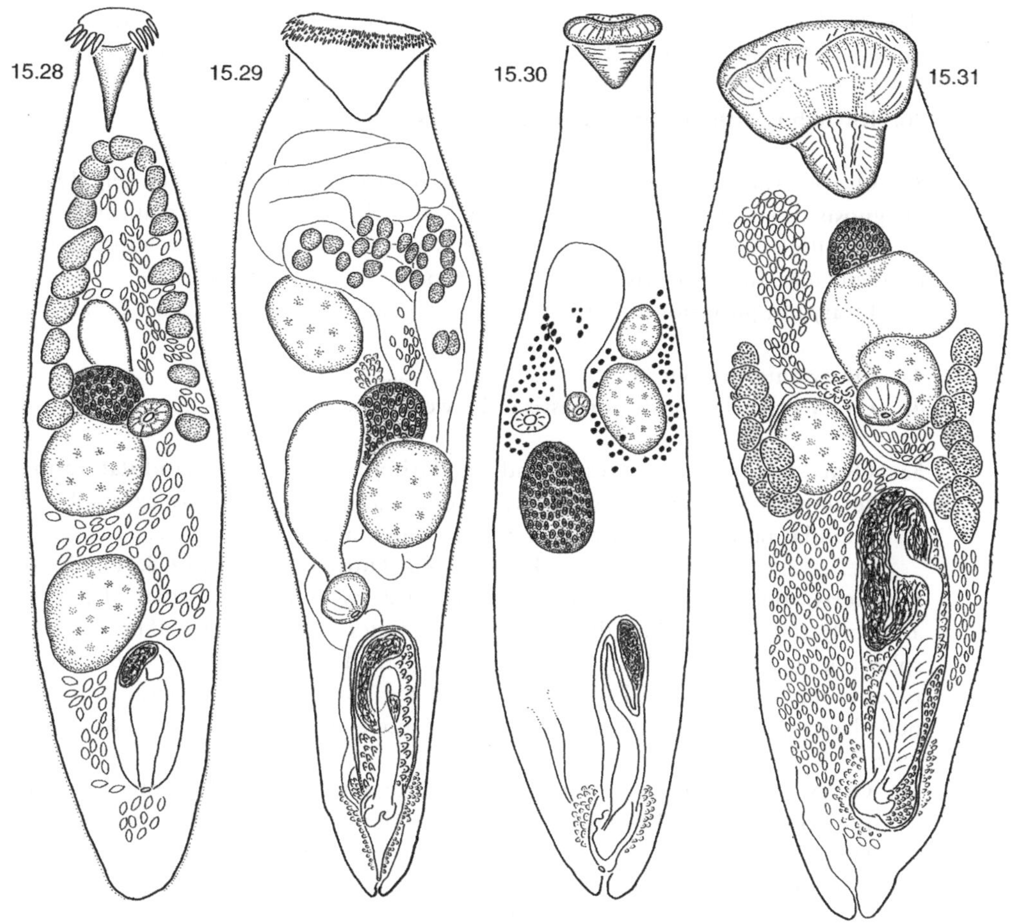

Figs 15.28–15.31. Prosorhynchinae.
15.28. *Telorhynchus arripidis* Crowcroft, 1947. (Redrawn from Crowcroft, 1947.)
15.29. *Dollfustrema stromborhynchum* Manter & Pritchard, 1961. (Redrawn from Manter & Pritchard, 1961.)
15.30. *Neoprosorhynchus purius* Dayal, 1948. (Redrawn from Skrjabin & Guschanskaja, 1962.)
15.31. *Myorhynchus pritchardae* Durio & Manter, 1968. (Original from USNPC holotype 63303.)

mid-body. Testes tandem. Spermatophores produced in genital atrium. Ovary spherical, pretesticular. Vitelline follicles arranged in confluent arc, anterior to other organs, not reaching into rhynchal region. In marine fishes; off New Zealand and Australia. Type-species *T. arripidis* Crowcroft, 1947.

1b. Rhynchus with or without apical disc, without long spines .... **2.**

2a. Rhynchus a shallow funnel surmounted by muscular disc encircled by three to four rows of small spines .... ***Dollfustrema* Eckmann, 1934. (Fig. 15.29)**

(Syns *Dollfusina* Eckmann, 1932, pre-occupied; *Neodollfustrema* Long & Lee, 1964)

Diagnosis: Body ellipsoidal, elongate in some species. Rhynchus a shallow funnel topped by muscular mass having three to four circlets of short spines. Mouth in posterior half of body. Caecum somewhat elongate. Testes oblique. Ovary smooth to globular, between testes or at level of anterior testis. Vitellarium in two distinct or indistinct clumps of follicles. In marine and freshwater fishes; cosmopolitan. Type-species *D. vaneyi* (Tseng, 1930).

2b. Rhynchus with one or no circlet of spines .... **3.**

3a. Ovary posterior to testes .... ***Neoprosorhynchus* Dayal, 1948. (Fig. 15.30)**

Diagnosis: Body elongate. Rhynchus funnel-shaped, surmounted by plain muscular disc. Mouth in anterior half of body. Testes tandem, sinistral. Ovary subspherical, post-testicular. In marine fishes; India. Type-species *N. purius* Dayal, 1948.

3b. Ovary not posterior to both testes .... **4.**

4a. Ovary anterior to testes .... **5.**

4b. Ovary at level of posterior testis or between testes .... **6.**

5a. Ovary far anterior to testes and vitellarium; rhynchus a glandular funnel surmounted by flat disc with large overhanging muscular flaps; eggshell with granular tanned layer encircled by clear outer layer .... ***Myorhynchus* Durio & Manter, 1968. (Fig. 15.31)**

Diagnosis: Body ellipsoidal. Rhynchus hooded by heavily muscular flap both anteriorly and dorsally. Mouth near mid-body. Testes oblique, near mid-body. Ovary spherical, far anterior in body. Vitellarium comprised of large follicles in two lateral fields at mid-body. Eggshell with granular tanned layer encircled by clear protective outer layer. In marine fish; South Pacific. Type-species *M. pritchardae* Durio & Manter, 1968.

5b. Ovary just anterior to testes, posterior to vitellarium; rhynchus a simple funnel surmounted by small disc with no flaps; eggshell with smooth tanned layer .... ***Prosorhynchus* Odhner, 1905. (Fig. 15.32)**

(Syns *Gotonius* Ozaki, 1924; *Skrjabiniella* Issaitschikow, 1928; *Paraprosorhynchus* Kohn, 1967; *Chabaudtrema* Kohn, 1970; *Rudolphinus* Stunkard, 1974)

Diagnosis: Body elongate or ellipsoid. Rhynchus funnel-shaped, covered by muscular disc or cap. Position of mouth variable. Testes oblique. Genital lobe mitten-shaped, with relatively large papilla. Ovary spherical, pretesticular. Vitellarium in one confluent anterior group of follicles or two lateral groups, always anterior to ovary. Excretory vesicle variable in length. In marine fishes; cosmopolitan. Type-species *P. squamatus* (Odhner, 1905).

6a. Ovary follicular (?); testes symmetrical to oblique .... ***Folliculovarium* Gu & Shen, 1983. (Fig. 15.33)**

Diagnosis: Body ellipsoidal. Rhynchus funnel-like, surmounted by muscular disc. Mouth at mid-body. Testes symmetrical to oblique. Ovary follicular (?). Vitellarium in one confluent group of follicles in anterior part of body. In marine fishes; China. Type-species *F. gymnothoracis* Gu & Shen, 1983.

6b. Ovary not follicular; testes tandem .... **7.**

7a. Ovary sinistral or medial; testes tandem on either left or right side of body; excretory vesicle long; rhynchus a muscular and glandular plug, with or without lobes .... ***Neidhartia* Nagaty, 1937. (Fig. 15.34)**

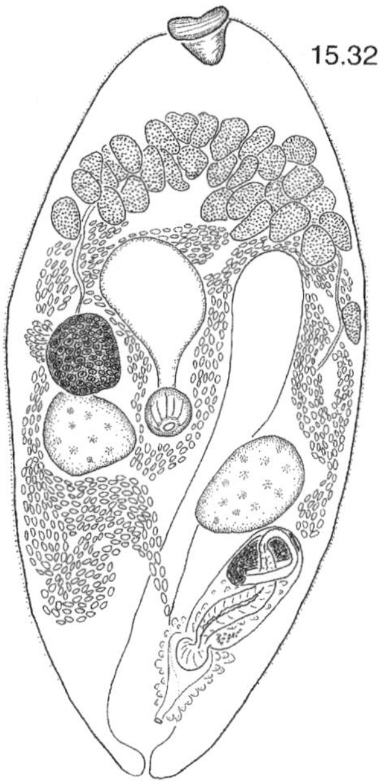

15.32

15.33

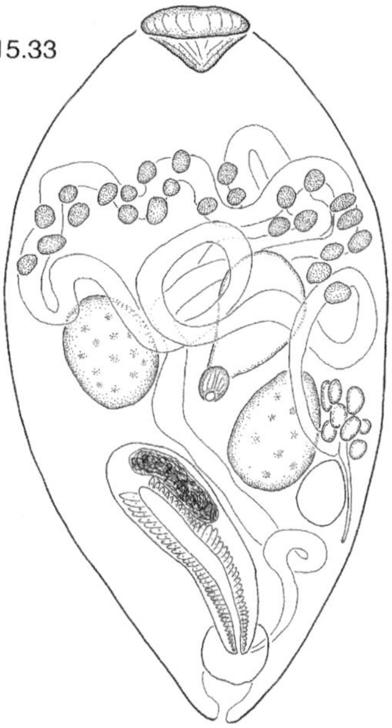

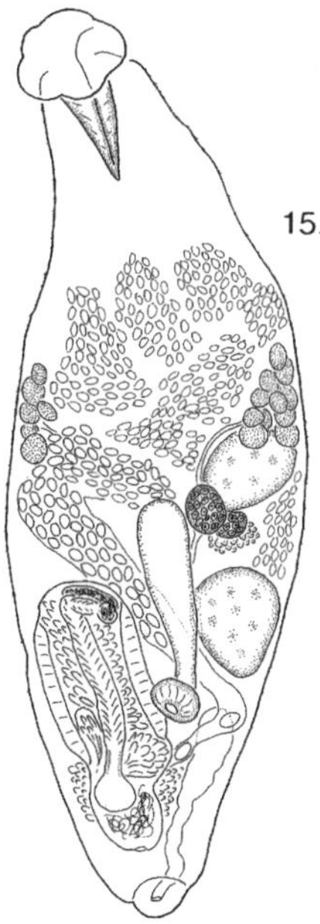

15.34

15.35

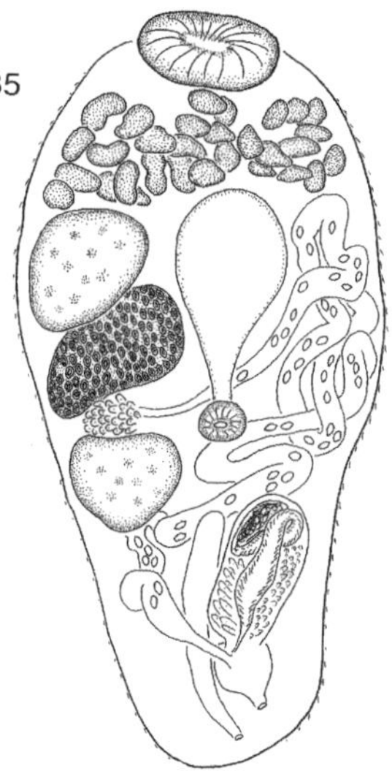

Figs 15.32–15.35. Prosorhynchinae. **15.32.** *Prosorhynchus squamatus* Odhner, 1905. (Original from BZM specimen 47987.) **15.33.** *Folliculovarium xishaense* Gu & Shen, 1983. (Redrawn from Gu & Shen, 1983.) **15.34.** *Neidhartia polydactyli* Manter, 1953. (Original from USNPC voucher 48730.) **15.35.** *Pseudoprosorhynchus synodi* Yamaguti, 1938. (Redrawn from Yamaguti, 1938a.)

Diagnosis: Body elongate or ellipsoidal. Rhynchus a blunt muscular, glandular mass (but *N. polydactyli* has lobes associated with rhynchus and *N. coronata* has single circlet of small spines around rhynchus). Mouth opening in posterior half of body. Testes tandem. Ovary at level between testes, sinistral or medial. Vitelline follicles in two indistinct groups. Excretory vesicle long, reaching almost to rhynchus (in species where extent is known). In marine fishes; cosmopolitan. Type-species *N. neidharti* Nagaty, 1937.

7b. Ovary dextral; testes tandem, dextral; excretory vesicle short, not reaching much past level of seminal vesicle; rhynchus a simple muscular disc .... *Pseudoprosorhynchus* Yamaguti, 1938. (Fig. 15.35)

Diagnosis: Body ellipsoidal. Rhynchus a simple muscular disc, non-glandular, with funnel base not apparent. Mouth at mid-body. Testes tandem, dextral. Ovary intertesticular, dextral. Vitellarium in one confluent mass, in anterior region of body. Excretory vesicle relatively short, reaching anteriorly to level of seminal vesicle. In marine fishes; off Japan. Type-species *P. synodi* Yamaguti, 1938.

16 Superfamily Clinostomoidea Lühe, 1901

I. Kanev,[1] V. Radev[1] & B. Fried[2]

[1]*Institute of Experimental Pathology and Parasitology, Bulgarian Academy of Sciences, Sofia, Bulgaria;* [2]*Department of Biology, Lafayette College, Easton, Pennsylvania, USA*

Introduction

Members of the superfamily Clinostomoidea are parasites of the buccal cavity, oesophagus and intestines of amphibians, reptiles, birds and occasionally mammals. The family-group name was first used by Lühe (1901) for the family Clinostomidae, which included *Clinostomum* Leidy, 1856. A year later, Pratt (1902) recognized the subfamily Clinostominae for the same genus. The group was first recognized at the superfamily level by Witenberg (1925). Dollfus (1931) erroneously applied the name Clinostomatidae to the family. The Liolopidae Odhner, 1912, Nephrocephalidae Travassos, 1929 (first used at the family level by Dollfus, 1931) and Harmotrematidae Yamaguti, 1933 (first used at the family level by Mehra, 1962a) have been listed within the superfamily Clinostomoidea by Mehra (1970). Nephrocephalidae is here considered a synonym of the Clinostomidae, but Liolopidae is considered valid with Harmotrematidae as a synonym.

For the past 20 years, the validity and the systematic position of many of these subfamilies and genera have been uncertain. For example, *Harmotrema* Nicoll, 1914 has been listed in four different family groups, including the Harmotrematidae, Brachylaimidae, Liolopidae and Clinostomidae. Consequently, the superfamily Clinostomoidea has been represented as comprising various different compositions of subfamily and family groups. The Liolopidae and Harmotrematidae have also been attributed to the Brachylaimidae. In our opinion, however, these parasites exhibit features that are more similar to those of the Clinostomoidea and for this reason we consider them as members of this superfamily.

Superfamily Clinostomoidea Lühe, 1901

Diagnosis: Body small to medium, spatulate, elongate, linguiform, pyriform or markedly tapered anteriorly, convex dorsally and concave ventrally. Tegumentary spines present or absent. Oral sucker small, rudimentary or well developed, muscular, may or may not be surrounded by collar-like flap. Prepharynx short to practically absent. Pharynx small, poorly developed. Oesophagus absent or short. Caeca two, blind, long, simple, smooth or with lateral diverticula. Ventral sucker small, flat and poorly developed or strongly muscular, close to oral sucker at anterior extremity or well separated from it in middle of body. Testes entire and oval, irregular in shape or deeply lobed, tandem or diagonal, in middle of hindbody or near posterior extremity. Cirrus-sac small or large, pretesticular, intertesticular, post-testicular or lateral to anterior testis. Seminal vesicle present or absent, voluminous inside or outside cirrus-sac. Pars prostatica present or absent. Genital pore(s) in hindbody, pretesticular, intertesticular, post-testicular or lateral to anterior testis; with common pore, median to submedian, or separate male and female pores, well apart, to left and right of median line. Ovary oval or irregular in shape, intercaecal, intertesticular, median or submedian. Vitellarium follicular; follicles in lateral fields in hindbody, reaching forward to level of cirrus-sac, ventral sucker or caecal bifurcation; fields may or may not unite posteriorly and/or anteriorly. Uterus intercaecal, in hindbody, mainly anterior or mainly posterior to terminal genitalia, saccular and short or tubular and long. Eggs small or large, may or may not be fully developed when released from uterus. Excretory system poorly or well developed; vesicle Y-shaped. Cercariae in freshwater and marine snails. Metacercariae in muscles and abdominal cavity of freshwater and marine fishes, snakes, frogs, salamanders or snails. Adults in buccal cavity, oesophagus or intestines of amphibians, reptiles, birds and occasionally in mammals, including humans. Type-genus *Clinostomum* Leidy, 1856.

Key to families

1a. Uterus mainly anterior to terminal genitalia; in buccal cavity and oesophagus of birds and reptiles .... Clinostomidae Lühe, 1901.

1b. Uterus mainly posterior to terminal genitalia; in intestine of amphibians and reptiles .... Liolopidae Dollfus, 1934.

17 Family Clinostomidae Lühe, 1901

I. KANEV,[1] V. RADEV[1] & B. FRIED[2]

[1]*Institute of Experimental Pathology and Parasitology, Bulgarian Academy of Sciences, Sofia, Bulgaria;* [2]*Department of Biology, Lafayette College, Easton, Pennsylvania, USA*

Introduction

Clinostomids are parasites of the buccal cavity, oesophagus and intestines of birds, reptiles and occasionally mammals, including humans (Kamo *et al.*, 1962). In reptiles, these parasites are found in three geographical regions: along the River Nile, Africa; Florida, North America; and Brazil, South America. Bird clinostomids are found in all continents. The family was first erected by Lühe (1901) for *Clinostomum* Leidy, 1856. This first system was further developed by Pratt (1902), who recognized the Clinostominae, and others, who added various taxa. These include: Odhner (1902, 1912) for *Nephrocephalus* and the Liolopinae, respectively; Nicoll (1914b) for *Harmotrema*; Baer (1923, 1930, 1959) for *Opisthophallus*, the Opisthophallinae and *Tremapoleipsis*, respectively; Witenberg (1925) for *Ithyoclinostomum*; Travassos (1928a, b) for *Euclinostomum*, *Odhneriotrema* and the Nephrocephalinae; Dollfus (1930, 1931, 1932a, 1950) for the Nephrocephalidae, *Clinostomatopsis* and *Clinostomoides*; Canavan (1933) for *Homoscaphis*; Yamaguti (1933) for the Harmotrematinae; van der Kuyp (1953) for *Tumaclinostomum*; Yamaguti (1958a) for the Euclinostominae and the Ithyoclinostominae; Mehra (1962a) for the Harmotrematidae; Travassos *et al.* (1969) for the Odhneriotrematinae; Pande & Baugh (1970) for *Metaclinostomum*; and Feizullaev & Mirzoeva (1983) for *Neutraclinostomum*. As a result of these works, clinostomes have been listed as comprising four families, six subfamilies and 15 genera. In the two most recent systems, the clinostomes were organized in two different ways. Skrjabin (1947a) gathered them all into one family, the Clinostomidae, with three subfamilies: the Clinostominae for *Clinostomum*, *Euclinostomum* and *Ithyoclinostomum*; the Opisthophallinae for *Opisthophallus* and *Odhneriotrema*; and the Harmotrematinae for *Harmotrema*. Yamaguti (1958a, 1971) organized the family into four subfamilies: the Clinostominae for *Clinostomum*, *Odhneriotrema*, *Clinostomatopsis* and *Clinostomoides*; the Ithyoclinostominae for *Ithyoclinostomum*; the Euclinostominae for *Euclinostomum*; and the Nephrocephalinae for *Nephrocephalus* and *Tremapoleipsis*.

The following are comments on some of the problem genera.

- *Clinostomum* was established by Leidy (1856) for adults from the buccal cavity and oesophagus of birds in the USA, referred to as *C. dubium*, and for metacercariae in fish, called *C. gracile*. The identity and the generic status of many species allocated to this genus are in doubt and many of the species have been transferred to *Clinostomatopsis*, *Euclinostomum*, *Ithyoclinostomum* and other genera mentioned below.
- *Mesogonimus* was established by Monticelli (1888) as a subgenus of *Distoma* for *D. reticulatum* Looss, 1885, a species transferred to *Clinostomum* by Braun (1899) and to *Clinostomatopsis* by Dollfus (1932a). However, because of features of the reproductive and digestive systems, this genus has been considered a synonym of the heterophyid genus *Heterophyes* Cobbold, 1886 (Skrjabin, 1947a; Yamaguti, 1958a, 1971).
- *Ithyoclinostomum* Witenberg, 1925 was established for adults with long, narrow bodies reaching 100 mm in length. These parasites were collected from the oesophagus of piscivorous birds from Brazil and had been named *Distoma dimorphum* by Diesing (1850). Braun (1899) split *D. dimorphum* into several separate species, one of which he (Braun, 1899, 1900) transferred to *Clinostomum*, as *C. dimorphus*. Witenberg (1925) established *Ithyoclinostomum* for these worms, for which Yamaguti (1958a) erected the subfamily Ithyoclinostominae. We consider these taxa valid.
- *Euclinostomum* Travassos, 1928 was erected for *Distoma heterostomum* Rudolphi, 1809, a worm with caeca having numerous long lateral branches in the hindbody. The same parasites had been placed in various genera, including *Clinostomum* by Braun (1899) and *Ithyoclinostomum* by Joyeux & Houdemer (1928). Skrjabin (1947a) and Yamaguti (1958a, 1971) both considered *Euclinostomum* a valid genus, but placed it in two different subfamilies: the Clinostominae (in Skrjabin, 1947a) and the Euclinostominae (in Yamaguti, 1958a, 1971). In our opinion, both *Euclinostomum* and the Euclinostominae are valid taxa.
- *Clinostomatopsis* Dollfus, 1932 was erected for worms collected in Brazil and described by Braun (1899) as *Clinostomum sorbens*. Braun (1900, 1902) revised *Clinostomum* and found that *C. sorbens* has a vitellarium, testes, cirrus-sac position and genital atrium which indicate that it does not fit well into *Clinostomum*. Based on Braun's (1902) suggestion, Dollfus (1932a) transferred *C. sorbens* to a new genus, *Clinostomatopsis*, but Skrjabin (1947a) did not recognize it, synonymizing it with *Clinostomum*. However, we agree with Yamaguti (1958a, 1971) in considering it valid.
- *Clinostomoides* Dollfus, 1950 was established for worms from the oesophagus of piscivorous birds in the Congo. The type-species, *C. brieni* Dollfus, 1950, was described as differing from *Clinostomum* in body size and in the genital pore being located ventrally to the posterior testis. The genus has subsequently been accepted as valid (e.g. Yamaguti, 1958a, 1971).
- *Tumaclinostomum* van der Kuyp, 1953 is a name given to metacercariae with long posterior caecal diverticula found in fishes in the Philippines and originally named *Euclinostomum multicaecum* by Tubangui & Masilungan (1935). van der Kuyp (1953) suggested that these metacercariae did not fit well into *Euclinostomum*, because of the long posterior caecal diverticula. However, Yamaguti (1971) synonymized *Tumaclinostomum* with *Euclinostomum*, an action with which we agree.

- *Metaclinostomum* Pande & Baugh, 1970 was established for metacercariae found in fishes and adults obtained from the buccal cavity of experimentally infected birds in India. In general morphology, it corresponds with *Euclinostomum*, except for the opening of the uterus at the anterior end of the uterine sac, the shape of the cirrus-sac and the number of caecal diverticula. Many authors, including Yamaguti (1971) and Feizullaev & Mirzoeva (1983), who revised the group, have considered *Metaclinostomum* a synonym of *Euclinostomum*.
- *Neutraclinostomum* Feizullaev & Mirzoeva, 1983 was established for worms from piscivorous birds of the genus *Phalacrocorax* in Venezuela and Angola. These worms had originally been described as *Clinostomum intermedialis* Lamont, 1920 and *C. phalacrocoracis* Dubois, 1932, respectively. Baer (1933) suggested that these two species are similar and perhaps identical. Feizullaev & Mirzoeva (1983) considered *C. phalacrocoracis* a synonym of *C. intermedialis*. *Neutraclinostomum* has a cirrus-sac, genital pores and other features that correspond with those of *Clinostomatopsis*. Although we agree with Feizullaev & Mirzoeva (1983) that the worms do not belong to *Clinostomum*, we can find no reason to recognize *Neutraclinostomum* as distinct from *Clinostomatopsis*.

In this work, the clinostomids are organized into four subfamilies.

Family Clinostomidae Lühe, 1901

(Syn. Nephrocephalidae Travassos, 1929)

Diagnosis: Body medium-sized to large, elongate or linguiform, stout or not, flat lanceolate or markedly tapered anteriorly, usually convex dorsally and concave ventrally. Tegument unarmed. Oral sucker small or well developed, may or may not be surrounded by collar-like fold when retracted. Ventral sucker stout, muscular, in anterior half of body. Prepharynx poorly developed or absent. Pharynx small, poorly developed. Oesophagus short or absent. Caeca long, simple; wall smooth or sinuous, may or may not possess lateral branches or diverticula. Testes smooth or slightly lobed, often irregular in shape, tandem, at about middle of hindbody or near posterior extremity. Cirrus-sac and genital pore(s) intertesticular, pretesticular, post-testicular or lateral to anterior testis; separate male and female genital pores occasionally present. Ovary intertesticular, median or submedian, oval, spherical or irregular in shape. Vitellarium follicular; follicles well developed, in lateral fields of hindbody, partly medial to caeca, usually reach to level of ventral sucker or into forebody, confluent posterior to hind testis. Uterus tubular, mainly anterior to terminal genitalia, mainly intercaecal, ascends on left from Mehlis' complex to level of ventral sucker, then descends on right towards genital pore. Excretory vesicle Y-shaped. Cercariae (oculate, pharyngeate, furcocercous, lophocercous, with elongate head organ) develop in rediae in freshwater pulmonate snails. Metacercariae in muscles and abdominal cavity of freshwater fishes, snakes, frogs, salamanders or snails. Adults mainly in buccal cavity or oesophagus of piscivorous birds, reptiles and occasionally mammals, including humans. Type-genus *Clinostomum* Leidy, 1856.

Key to subfamilies

1a. Adults in buccal cavity and oesophagus of reptiles .... Nephrocephalinae Travassos, 1929.
1b. Adults in buccal cavity and oesophagus of birds (occasionally mammals) . 2.

2a. Caeca with numerous lateral branches and diverticula .... Euclinostominae Yamaguti, 1958.
2b. Caeca simple, without lateral branches and diverticula .... 3.

3a. Body stout, linguiform, *c*. 5–30 mm long .... Clinostominae Lühe, 1901.
3b. Body slender, attenuated; up to 100 mm long .... Ithyoclinostominae Yamaguti, 1958.

Subfamily Clinostominae Lühe, 1901

Diagnosis: Body middle-sized to large, 5–30 mm long, stout, linguiform. Oral sucker and pharynx well developed. Ventral sucker muscular, well developed, in anterior half of body. Oesophagus practically absent. Caeca long, simple with more or less sinuous wall but without long lateral branches or diverticula. Testes smooth, irregular in shape or deeply lobed, in posterior half of body. Cirrus-sac and genital pore pretesticular, intertesticular, post-testicular or lateral to anterior testis. Ovary intertesticular, submedian. Vitelline fields well developed, may or may not extend into forebody. Uterus intercaecal, may or may not reach to ventral sucker. In buccal cavity and oesophagus of piscivorous birds. Type-genus *Clinostomum* Leidy, 1856.

Key to genera

1a. Genital pore post-testicular; vitelline fields extend into forebody .... *Clinostomoides* Dollfus, 1950. (Fig. 17.3)
Diagnosis: Body very large, stout. Oral sucker relatively small, may or may not be surrounded by rudimentary collar-like body fold. Ventral sucker muscular, well developed. Caeca long, sinuous, but without lateral branches, even though some sinuous folds may resemble diverticula. Testes oval, in posterior half of body. Cirrus-sac intertesticular, or overlaps posterior testis. Genital pore post-testicular. Vitelline fields extend from posterior end of body to close to level of caecal bifurcation. Uterus may or may not reach level of ventral sucker. Africa and Asia. Type-species *C. brieni* Dollfus, 1950.
1b. Genital pore pre- or intertesticular; vitelline follicles restricted to hindbody or extend into forebody .... 2.

2a. Cirrus-sac and genital pore pretesticular or lateral to anterior testis; vitelline fields extend to level of ventral sucker; uterus usually reaches level of ventral sucker .... *Clinostomum* Leidy, 1856. (Fig. 17.1)

Diagnosis: Body medium to large, linguiform, stout, convex dorsally and concave ventrally. Oral sucker surrounded by collar-like fold when retracted. Ventral sucker muscular, well developed. Caeca long, simple, with more or less sinuous wall, especially in anterior half of body, but without long lateral branches or diverticula. Testes smooth or irregular in shape, in posterior half of body. Cirrus-sac and genital pore pretesticular or lateral to anterior testis. Ovary intertesticular, submedian. Vitelline follicles in lateral fields between posterior extremity and level of ventral sucker. Uterus intercaecal, between Mehlis' gland and ventral sucker. Cosmopolitan. Type-species *C. complanatum* (Rudolphi, 1819).

2b. Cirrus-sac and genital pore intertesticular; vitelline fields extend into forebody; uterus may not reach level of ventral sucker .... ***Clinostomatopsis* Dollfus, 1932. (Fig. 17.2)**

(Syn. *Neutraclinostomum* Feizullaev & Mirzoeva, 1983)

Diagnosis: Body stout, linguiform, convex dorsally and concave ventrally. Oral sucker may or may not be surrounded by collar-like fold. Ventral sucker muscular, well

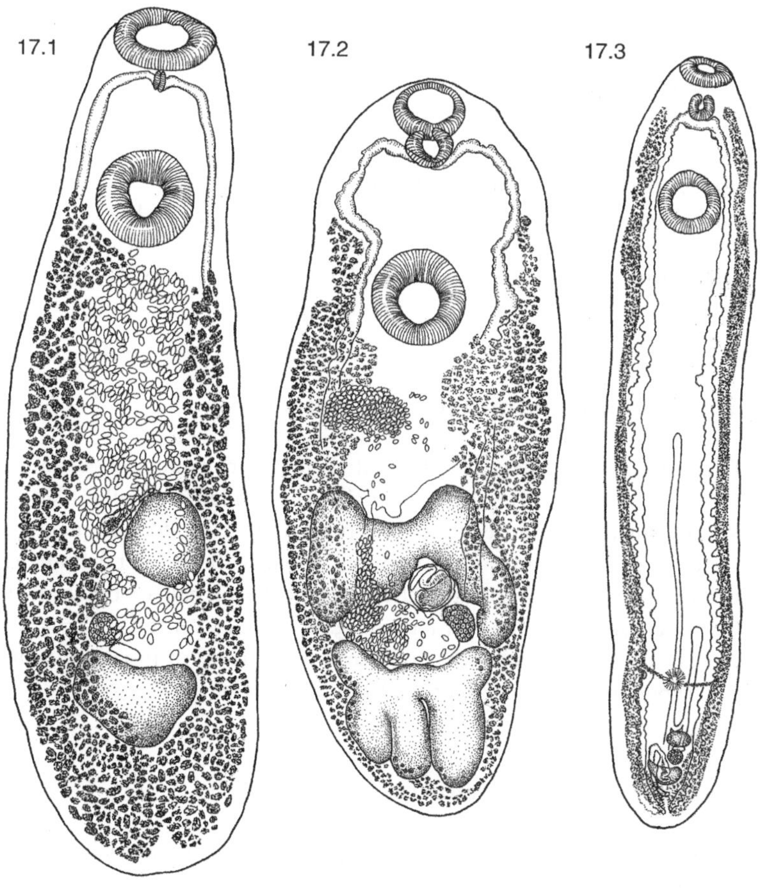

Figs 17.1–17.3.
17.1. *Clinostomum complanatum* (Rudolphi, 1819). (Original.)
17.2. *Clinostomatopsis sorbens* (Braun, 1899). (Original.)
17.3. *Clinostomoides brieni* Dollfus, 1950. (Original.)

developed. Caeca simple, long, sinuous, but without lateral branches or diverticula. Testes large, irregular in shape or deeply lobed, in posterior half of body. Cirrus-sac and genital pore intertesticular. Vitelline fields extend from posterior end of body into forebody. Uterus intercaecal, may or may not reach level of ventral sucker. Neotropical. Type-species *C. sorbens* (Braun, 1899).

Subfamily Euclinostominae Yamaguti, 1958

Diagnosis: Body stout, linguiform. Oral sucker may or may not be surrounded by collar-like fold. Ventral sucker well developed, relatively close to oral sucker. Caeca long, with numerous lateral diverticula branching from lateral walls between level of ventral sucker and posterior extremity. Testes irregular in shape, in posterior half of body. Cirrus-sac small, oval, pretesticular. Genital pore immediately pretesticular, median or submedian. Ovary intertesticular, submedian, oval. Vitelline fields well developed, extend from posterior end of body to ventral sucker; follicles surround caecal branches in hindbody. Uterus reaches forward to ventral sucker. In buccal cavity and oesophagus of piscivorous birds. Europe, Asia, Africa, America. Type-genus *Euclinostomum* Travassos, 1928.

Genus *Euclinostomum* Travassos, 1928 (Fig. 17.4)

(Syns *Tumaclinostomum* van der Kuyp, 1953; *Metaclinostomum* Pande & Baugh, 1970)

Diagnosis: With characters of subfamily. Type-species *E. heterostomum* (Rudolphi, 1809).

Subfamily Ithyoclinostominae Yamaguti, 1958

Diagnosis: Body slender, attenuated for a considerable distance anteriorly, up to 100 mm long. Oral sucker small, surrounded by collar-like rudimentary flap. Ventral sucker near anterior extremity of body, close to oral sucker. Pharynx present. Oesophagus very short. Caeca simple, long, without lateral branches or diverticula. Ovary, testes, vitellarium and uterus in posterior half of body; ovary intertesticular. Considerable space free of internal organs between ventral sucker and anterior limit of vitellarium. Adults in buccal cavity and oesophagus of piscivorous birds. South America. Type-genus *Ithyoclinostomum* Witenberg, 1925.

Genus *Ithyoclinostomum* Witenberg, 1925 (Fig. 17.5)

Diagnosis: With characters of subfamily. Type-species *I. dimorphum* (Diesing, 1850).

Subfamily Nephrocephalinae Travassos, 1928

(Syns Opisthophallinae Baer, 1933; Odhneriotrematinae Travassos, Freitas & Kohn, 1968)

Diagnosis: Body middle-sized to large. Oral sucker and pharynx poorly or well developed. Ventral sucker well developed, in middle of anterior half of body. Oesophagus present or absent. Caeca simple, long, undulating, with numerous sinuous branches, especially in anterior half of body. Testes smooth, oval or irregular, close together in posterior half of body. Cirrus-sac lateral to posterior testis. Genital pore(s) submedian, intertesticular or post-testicular; male and female genital pores united or well separated to left and right side of median line. Ovary intertesticular, median, oval or irregular in shape. Vitelline fields well developed, extend from posterior extremity to level of caecal bifurcation. In buccal cavity and oesophagus of reptiles. Type-genus *Nephrocephalus* Odhner, 1902.

Key to genera

1a. **Male and female genital pores well separated to left and right of median line, post-testicular; oral sucker well developed .... *Nephrocephalus* Odhner, 1902. (Fig. 17.6)**
(Syns *Opisthophallus* Baer, 1923; *Tremapoleipsis* Baer, 1959)

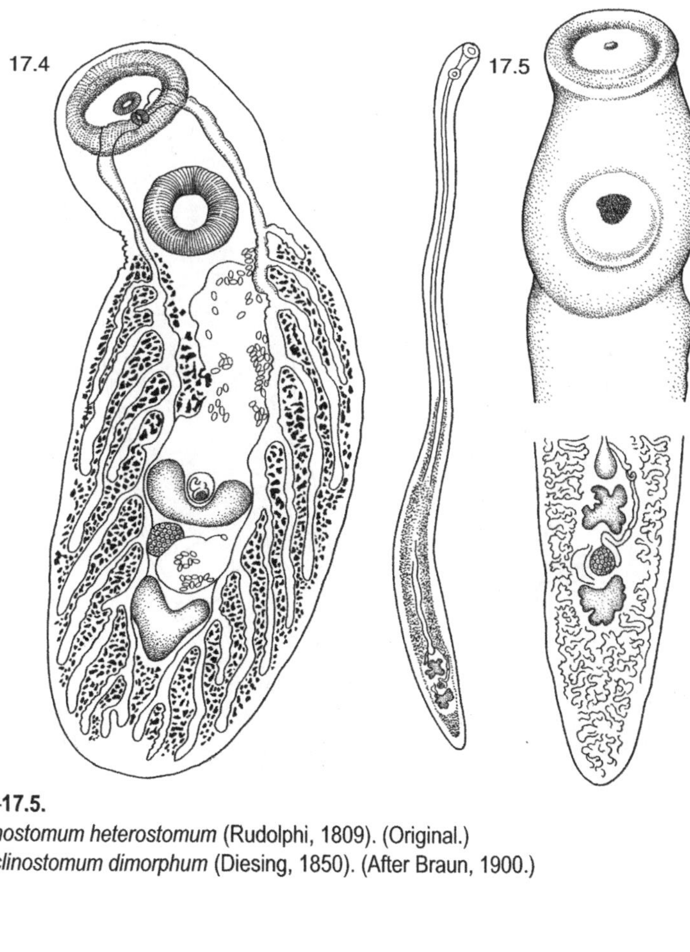

Figs 17.4–17.5.
17.4. *Euclinostomum heterostomum* (Rudolphi, 1809). (Original.)
17.5. *Ithyoclinostomum dimorphum* (Diesing, 1850). (After Braun, 1900.)

Diagnosis: Body medium-sized, elongate. Oral sucker well developed. Testes close to one another in posterior third of body. Cirrus-sac and genital pores post-testicular. Genital pores well separated, to left and right of median line. Vitelline fields present laterally between posterior extremity and level of caecal bifurcation. Africa. Type-species *N. sessilis* Odhner, 1902.

1b. Common genital pore present, intertesticular; oral sucker much reduced .... *Odhneriotrema* Travassos, 1928. (Fig. 17.7)
(Syn. *Homoscaphis* Canavan, 1933)

Diagnosis: Body large, elongate, with rounded extremities. Oral sucker very much reduced. Pharynx small. Oesophagus short or absent. Caeca sinuous, especially in forebody, terminating at posterior extremity. Testes in posterior third of body, well separated from each other. Cirrus-sac intertesticular. Male and female genital pores united, intertesticular, in median line. Vitelline fields lateral, extending from posterior extremity to level of ventral sucker. North and South America. Type-species *O. microcephala* (Travassos, 1922).

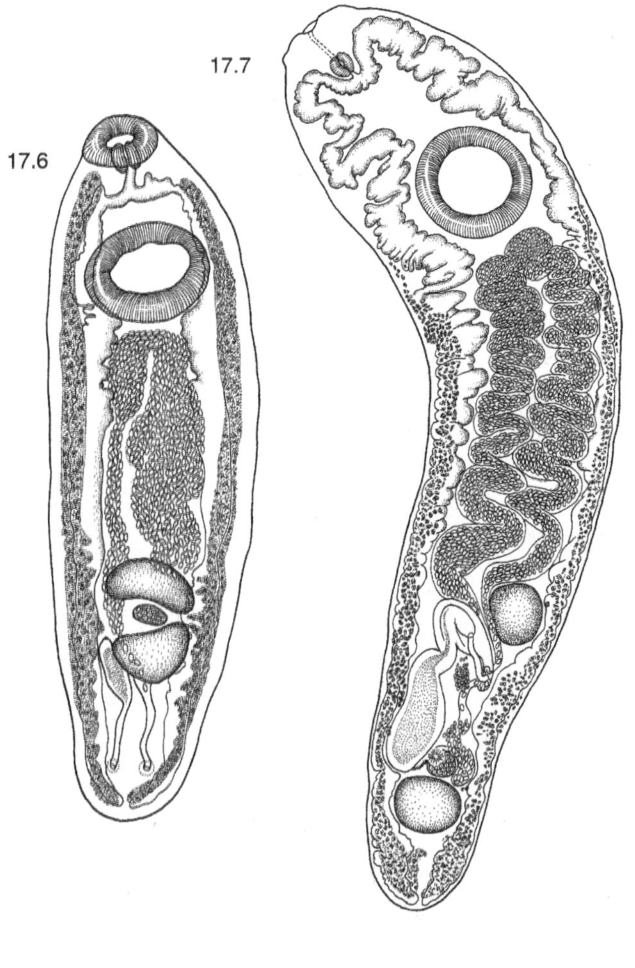

Figs 17.6–17.7.
17.6. *Nephrocephalus sessilis* Odhner, 1902. (Original.)
17.7. *Odhneriotrema microcephala* (Travassos, 1922). (Original.)

18 Family Liolopidae Odhner, 1912

K. Niewiadomska

W. Stefański Institute of Parasitology, Polish Academy of Sciences, Twarda, Warsaw, Poland

Introduction

The family Liolopidae Odhner, 1912 comprises only four genera; these are all parasites of amphibians and reptiles. The common characters of these genera are the intertesticular position of the ovary and the so-called 'strigeid-like' excretory system, which consists, in the anterior part of the body, of two lateral loops formed by intra- and extracaecal canals joined anteriorly and posteriorly (Fig. 18.1).

For the genera *Liolope* Cohn, 1902 (parasitic in amphibians), *Helicotrema* Odhner, 1912 and *Hapalotrema* Looss, 1899 (parasites of reptiles), Odhner (1912) originally created the subfamily Liolopinae as a part of the Harmostomidae Braun, 1900 (currently a synonym of the Brachylaimidae Joyeux & Foley, 1930). *Harmotrema* Nicoll, 1914 was assigned by Nicoll (1914b) to the Clinostomidae Lühe, 1901. Yamaguti (1933) formed the new subfamily, the Harmotrematinae, for this genus. Mehra (1962a) raised it to the rank of family as the Harmotrematidae, also encompassing *Helicotrema* and two monotypic subfamilies, the Harmotrematinae and a new subfamily, the Helicotrematinae. The separate status of the family Liolopidae with two subfamilies was accepted by Dollfus (1934). He excluded *Hapalotrema* from the Liolopidae, considering it to belong to the Spirorchiidae Stunkard, 1921, but retained the subfamily Liolopinae for the three genera discussed above. As a second subfamily, he included the Moreauiinae Johnston, 1915, created by Johnston (1915) for the monotypic genus *Moreauia* Johnston, 1915, a parasite of mammals (Monotremata). The latter genus, however, has a rather uncertain status, due to its aberrant morphology and host group, but it has been considered within the Brachylaimidae (see Chapter 11 on the Moreauiidae).

The systematic assignation of the liolopid family group varies according to different authors. Tubangui & Masilungan (1936) placed it in the Clinostomidae, and Skrjabin at first (Skrjabin, 1947a) in the Brachylaimidae and later (Skrjabin, 1962) in the Liolopidae. Yamaguti (1971) retained the Liolopidae for parasites of amphibians

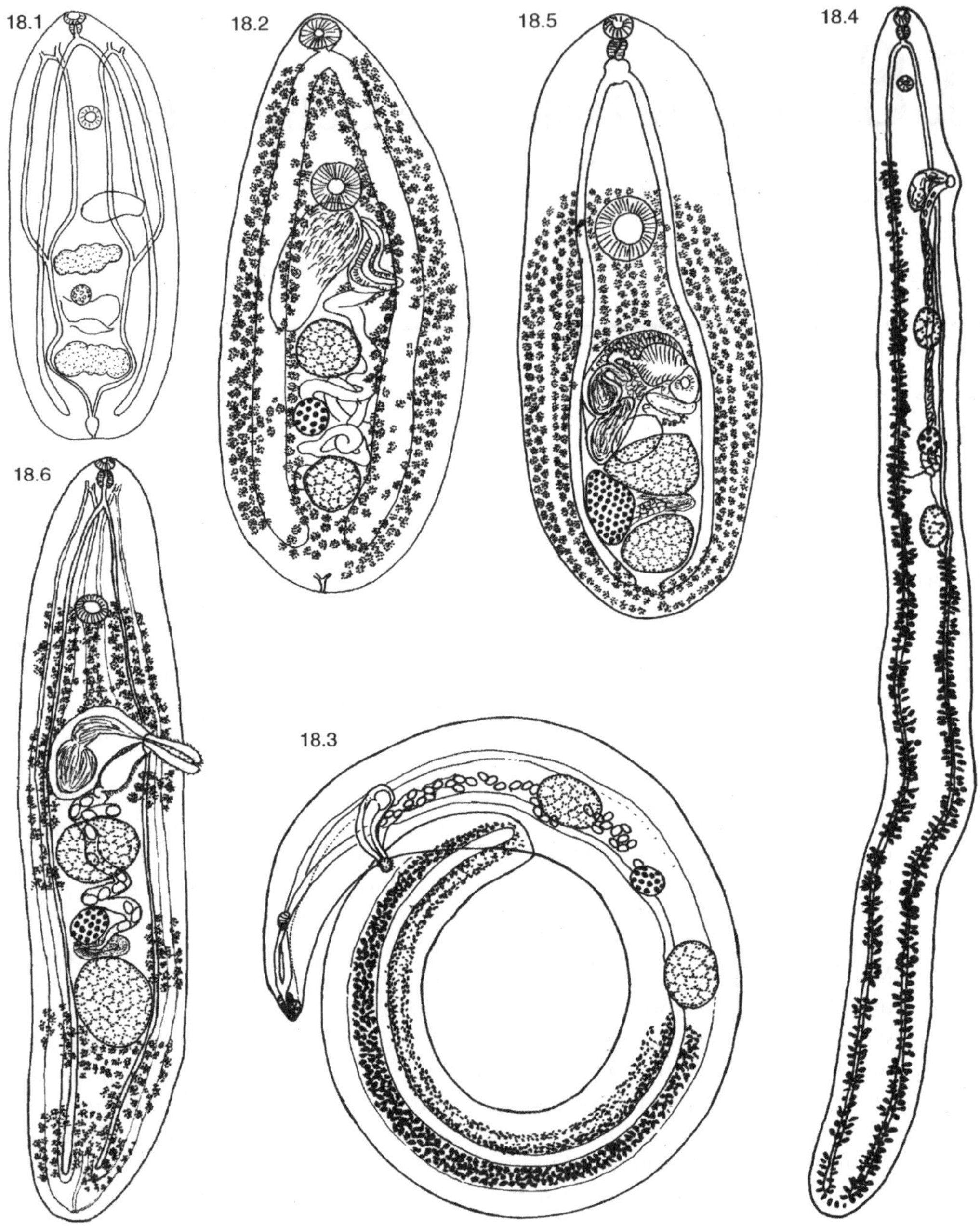

Figs 18.1–18.6.

18.1. Juvenile *Harmotrema infecundum* Nicoll, 1914 showing 'strigeid-like' excretory system. (Redrawn and simplified after Dollfus, 1950.)

18.2. *Liolope copulans* Cohn, 1902. (After Yamaguti, 1971.)

18.3. *Helicotrema magniovatum* Odhner, 1912. (After Skrjabin, 1962.)

18.4. *Helicotrema asymmetrica* Travassos, 1922. (After Skrjabin, 1962.)

18.5. *Dracovermis occidentalis* Brooks & Overstreet, 1978. (After Brooks & Overstreet, 1978.)

18.6. *Harmotrema laticaudae* Yamaguti, 1933. (After Yamaguti, 1933.)

and the Harmotrematidae for parasites of reptiles. In the present work, the Liolopidae are considered as a separate family within the superfamily Clinostomoidea Lühe, 1901.

Brooks & Overstreet (1978) accepted the family Liolopidae with three genera, but emphasized their host-specificity: amphibians (*Liolope*), turtles and lizards (*Helicotrema*) and crocodilians and snakes (traditionally *Harmotrema*). They demonstrated the morphological differences between *Harmotrema* from snakes and from crocodilians, and removed those from crocodilians into a separate genus, *Dracovermis* Brooks & Overstreet, 1978. This conception of the family is accepted here and used as the basis for the present work.

Family Liolopidae Odhner, 1912

(Syns Liolopinae Odhner, 1912; Harmotrematinae Yamaguti, 1933; Harmotrematidae Yamaguti, 1933; Helicotrematinae Mehra, 1962)

Diagnosis: Body small or moderately large, elliptical, linguiform or slender. Tegument unarmed. Oral sucker and pharynx present; prepharynx absent; oesophagus short or practically absent; caeca long, reaching posterior end of body. Ventral sucker at different distances from anterior end of body. Testes round, oval or irregular, tandem, separate, usually well separated, situated in posterior, middle or anterior third of body length. Cirrus-sac between ventral sucker and anterior testis, containing bipartite seminal vesicle, pars prostatica and spiny cirrus. External seminal vesicle absent or (rarely) present. Genital atrium present. Genital pore submedian or lateral, in hindbody. Ovary irregular, intertesticular. True seminal receptacle absent. Laurer's canal distinct or indistinct. Uterine seminal receptacle present (described for two species). Uterus rather short; in hindbody mainly posterior to cirrus-sac and genital pore; metraterm well differentiated. Vitellarium follicular, in lateral fields along caeca. Excretory vesicle Y-shaped. In intestine of amphibians and reptiles. Type-genus *Liolope* Cohn, 1902.

Key to genera

1a. Oesophagus absent; cirrus-sac abutting posterior margin of ventral sucker; in amphibians .... *Liolope* Cohn, 1902. (Fig. 18.2)

Diagnosis: Body elliptical; oral and ventral suckers well developed; oesophagus absent; caeca long, extending near end of body. Gonads in posterior half of body. Testes tandem, intercaecal. Cirrus-sac pretesticular, abutting posterior margin of ventral sucker, with seminal vesicle, prostatic duct and cells, ejaculatory duct and spined cirrus. Genital atrium present. Ovary intertesticular, intercaecal, submedian, rounded. Uterus short, winding, with prominent metraterm. Eggs large. Vitelline follicles extending from intestinal bifurcation to end of body, extra- and intercaecal. Excretory system with terminal pore, post-testicular bladder and paired inter- and extracaecal tubules joined anteriorly and posteriorly. In amphibians. Congo, Japan. Type-species *L. copulans* Cohn, 1902.

1b. Oesophagus short or practically absent; cirrus-sac pretesticular, not close to ventral sucker; in reptiles .... 2.

2a. Body significantly elongated, slender, may be spirally coiled; suckers close to each other; vitelline fields far from ventral sucker; in lizards .... *Helicotrema* Odhner, 1912. (Figs 18.3, 18.4)

Diagnosis: Body long, slender, may be spirally coiled; oral and ventral suckers small, similar in size; ventral sucker near intestinal bifurcation. Pharynx small; oesophagus short; caeca long and narrow, extending close to end of body. Gonads in anterior or middle part of body. Testes tandem, median or submedian. Cirrus-sac between ventral sucker and anterior testis, contains bipartite seminal vesicle, pars prostatica and spined cirrus. External seminal vesicle absent. Genital pore ventrolateral. Ovary intertesticular, median or submedian, on same side of body as testes. Uterus short, extending forward from oötype to genital pore, with well-differentiated metraterm. Eggs large. Vitelline follicles extend along greater part of intestine leaving different portions of anterior region of caeca free depending upon individual species (and even specimens). Excretory vesicle Y-shaped. In lizards. South America. Type-species *H. magniovatum* Odhner, 1912.

2b. Body linguiform or elongate; suckers well apart from one another; vitelline fields extend forward at least to level of ventral sucker; in crocodilians and snakes .... 3.

3a. Gonads near posterior third of body; vitelline follicles in most available space posterior to ventral sucker; in crocodilians .... *Dracovermis* Brooks & Overstreet, 1978. (Fig. 18.5)

Diagnosis: Body pyriform or elongate; oral and ventral suckers well developed; ventral sucker larger than oral. Oesophagus present; caeca long, extending to posterior end of body. Gonads in posterior third of body. Testes tandem, intercaecal. Cirrus-sac pretesticular, containing bipartite seminal vesicle, pars prostatica, ejaculatory duct and spined eversible cirrus. Genital pore ventral to left caecum or almost so. Ovary intertesticular, intercaecal, dextral. Uterus short, winding, with prominent metraterm. Eggs large. True seminal receptacle presumed lacking. Vitelline follicles occupy most of available space from slightly anterior to ventral sucker to end of body; fields tend to be continuous. Excretory system with terminal or dorsoterminal excretory pore, post-testicular vesicle and paired inter- and extracaecal tubules united anteriorly and posteriorly. In crocodilians. USA, India, Philippines, Congo. Type-species *D. occidentalis* Brooks & Overstreet, 1978.

3b. Gonads near middle of body; vitelline follicles along caeca or intercaecal, fields may be discontinuous; in snakes .... *Harmotrema* Nicoll, 1914. (Fig. 18.6)

Diagnosis: Body elongate, moderately large; oral and ventral suckers rather small; ventral sucker larger than oral. Pharynx small; oesophagus short; caeca long, reaching to posterior end of body. Gonads in posterior half of body. Testes tandem, intercaecal. Cirrus-sac pretesticular, containing bipartite seminal vesicle, pars prostatica and spined eversible cirrus. Genital pore ventral to left caecum. Ovary median or submedian, intertesticular, intercaecal, usually near posterior testis. Uterus short, winding, with prominent metraterm. Small vitelline follicles distributed along caeca, extending from slightly anterior to ventral sucker to end of body, forming two complete or interrupted bands, sometimes occupying intercaecal space. Excretory system with terminal or dorsoterminal excretory pore; post-testicular vesicle and paired inter-

and extracaecal tubules united anteriorly and posteriorly. In snakes. Congo, Japan, Philippines. Type-species *H. infecundum* Nicoll, 1914.

19 Superfamily Cyclocoeloidea Stossich, 1902

I. Kanev,[1] V. Radev[1] & B. Fried[2]

[1]*Institute of Experimental Pathology and Parasitology, Bulgarian Academy of Sciences, Sofia, Bulgaria;* [2]*Department of Biology, Lafayette College, Easton, Pennsylvania, USA*

Introduction

Cyclocoeloids are mainly parasites of birds and reptiles and usually inhabit sites other than the gut. They are treated here with the members of the Strigeida, but current indications from molecular biology suggest a relationship with the members of the Echinostomida (Cribb *et al.*, 2001). The group was first erected as a subfamily by Stossich (1902), but raised to family level by Kossack (1911) and superfamily level by Henry (1923). It originally included *Monostomum mutabile* Zeder, 1800 and several other species of the genus *Monostomum* Zeder, 1800. Monticelli (1888) included *Monostoma* Zeder, 1800 and several other genera in one family, the Monostomidae Cobbold, 1877, known also as the 'Monostomata group'. Brandes (1892) revised the family Monostomidae and erected a new genus, *Cyclocoelum*. Stossich (1902) completed a revision of the type-species of the latter genus, *M. mutabile*, and erected three new genera, *Haematotrephus*, *Ophthalmophagus* and *Typhlocoelum*. Following the work of Brandes (1892) and Stossich (1902), these species and genera were re-examined, redescribed and reorganized by numerous authors, of whom the most important are Fuhrmann (1904, 1928), Kossack (1911), Harrah (1922), Witenberg (1923, 1926, 1928), Morishita (1924b), Joyeux & Baer (1927), Stunkard (1929), Bhalerao (1935), Khan (1935), Lal (1939), Dollfus (1948), Bykhovskaya-Pavlovskaya (1949), Bashkirova (1950), La Rue (1957), Chatterji (1958), Yamaguti (1958a, 1971), Dubois (1959), Odening (1961b), Feizullaev (1980) and Feizullaev & Mirzoeva (1983). As a result of these studies on the superfamily Cyclocoeloidea, six families, nine subfamilies and seven tribes were recognized as shown below in the table.

Most of the subsequent publications tended to consider only the morphology and biology of the adult worms. However, several papers, including those of La Rue (1957), Odening (1961b) and Feizullaev & Mirzoeva (1983), included information on the morphology and life-history of the larval stages. Some of the most interesting works were those of Witenberg (1923, 1926, 1928), who created a 'periodical system' for the superfamily Cyclocoeloidea. His periodical system was similar to that created by

Families	Subfamilies	Tribes
Monostomidae		
Cyclocoelidae	Cyclocoelinae	'Cyclocoelea'
	Ophthalmophaginae	'Ophthalmophagea'
		'Contracoelea'
	Haematotrephinae	'Haematotrephea'
		'Wardianea'
	Hyptiasminae	'Hyptiasmea'
Typhlocoelidae	Typhlocoelinae	
	Polycyclorchiinae	
Eucotylidae	Eucotylinae	
	Tanaisiinae	
Bothrigastridae	Bothrigastrinae	'Bothrigastrea'
Neocyclocoelidae		

Mendeleev for the chemical elements. In his system, Witenberg (1923) established tables in which he listed the tribes vertically and the genera were arranged horizontally. Tribes were arranged on the basis of information on the internal morphology of the adults - mainly the shape, size and position of the ovary and testes. Generic groups were separated and arranged based mainly on the morphology and position of the uterus, vitellarium and genital pore. Organized in this way the system comprised a table of 35 cells, indicating an arrangement of seven tribes and five genera per tribe. Most of the tribes and genera were created by Witenberg (1923, 1926, 1928); however, he was able to justify only 15 genera.

Current systems for the arrangement of the Cyclocoeloidea are very complicated. *Bothrigaster* Dollfus, 1948 is considered a valid genus by some authors and belonging to its own family group, namely the Bothrigastridae Dollfus, 1948. Other authors have doubted the validity of *Bothrigaster* and have synonymized it with *Ophthalmophagus* Stossich, 1902 and *Spaniometra* Kossack, 1911. Similar conflicting and contradictory information has also been published for many of the families, subfamilies, tribes and genera of this superfamily, and it is not always easy to determine which of these opinions is correct.

Our concept of the Cyclocoeloidea, presented below, is based on literature studies and detailed examination of specimens.

Superfamily Cyclocoeloidea Stossich, 1902

Diagnosis: Body medium-sized to large, lanceolate, linguiform, with rounded extremities or markedly tapered anteriorly, flattened but convex dorsally and concave ventrally. Tegument unarmed. Oral sucker present, absent or poorly developed. Ventral sucker absent or poorly developed. Prepharynx short or apparently absent. Pharynx present, usually well developed. Oesophagus short, rarely long or winding. Caeca simple or with diverticula along medial walls, may or may not be united in cyclocoel posteriorly. Testes usually two, usually close to each other near posterior extremity, sometimes juxtaposed, equatorial or pre-equatorial, occasionally follicular.

Cirrus-sac present or absent. Genital pore extracaecal at level of pharynx or intercaecal just posterior to pharynx, occasionally more posterior. Uterus intercaecal or extracaecal, extending laterally beyond caeca and vitelline fields. Ovary intercaecal, pretesticular, intertesticular or post-testicular, forming triangle with or in line with testes. Vitelline fields ventral and lateral to caeca, reaching to or beyond level of caecal bifurcation anteriorly, may or may not be united posteriorly. Excretory vesicle simple or Y-shaped, at posterior extremity, with dorsoterminal pore. Eggs with fully-developed miracidium. Adults in nasal sinuses, air-sacs, abdominal cavity, buccal cavity, oesophagus or urinary tubules of kidneys of birds and reptiles, occasionally in mammals. Cercariae develop in rediae in freshwater or terrestrial snails. Metacercariae occur in rediae or in snails, fishes, snakes or frogs. Cosmopolitan. Type-genus *Cyclocoelum* Brandes, 1892.

Key to families

1a. Oral sucker well developed; anterior extremity may be delimited from rest of body by annular muscular thickening in form of cone or triangle; caeca may or may not unite posteriorly to form cyclocoel at posterior extremity; testes and ovary in middle of body; vitelline fields not reaching to posterior end of body; in urinary tubules of kidneys of birds .... Eucotylidae Cohn, 1904.

1b. Oral sucker absent or at least very poorly developed; anterior extremity not delimited from rest of body by annular muscular thickening; caeca unite posteriorly forming cyclocoel; testes and ovary usually in posterior half of body; vitelline fields reaching to posterior end of body, may or may not be united posteriorly; in trachea and air-sacs of birds .... 2.

2a. Testes two; caeca simple, without diverticula; uterine coils intercaecal or extending laterally beyond caeca and vitelline fields; (adults occasionally found in mammals, including humans) .... Cyclocoelidae Stossich, 1902.

2b. Testes two or numerous (follicular); caeca simple or with diverticula covering entire medial wall; uterine coils intercaecal or may overlap caeca laterally .... Typhlocoelidae Harrah, 1922.

20 Family Cyclocoelidae Stossich, 1902

I. Kanev,[1] V. Radev[1] & B. Fried[2]

[1]*Institute of Experimental Pathology and Parasitology, Bulgarian Academy of Sciences, Sofia, Bulgaria;* [2]*Department of Biology, Lafayette College, Easton, Pennsylvania, USA*

Introduction

The family Cyclocoelidae, originally erected as a subfamily by Stossich (1902), is the largest and most problematical family in the superfamily Cyclocoeloidea. Over 50 genera, tribes, subfamilies and families are included in the systematic organization of Cyclocoelidae, which was first raised to full family status by Kossack (1911).

Taxa that are associated with our systematic reorganization of the subfamily Cyclocoelinae Stossich, 1902 are commented upon below.

Cyclocoelum was established by Brandes (1892) for *Monostomum mutabile* Zeder, 1800 and its allied species, *M. flavum* Mehlis, 1831, *M. arcuatum* Brandes, 1892, *M. tringae* Brandes, 1892 and *M. ellipticum* Rudolphi, 1809. *M. flavum* was later separated from *Cyclocoelum* and placed in the genus *Typhlocoelum* Stossich, 1902, *M. arcuatum* in the genus *Hyptiasmus* Kossack, 1911, *M. tringae* in the genera *Haematotrephus* Stossich, 1902 and *Corpopyrum* Witenberg, 1923, and *M. ellipticum* in *Distoma* Retzius, 1786 by Looss (1899, 1901). Because of this, *Cyclocoelum* and its type-species, *C. mutabile*, were described with two main characters: testes and ovary located on opposite corners of a triangle; and uterine coils intercaecal, not extending laterally beyond the caeca and vitellarium. The validity of *Cyclocoelum* is not in doubt. However, its identity and synonyms are disputed and involve different generic groups, some of which are listed and discussed below.

Hyptiasmus was established by Kossack (1911) for adults from the nasal cavity and hypothalamus of birds in Europe. Originally, these parasites were named *Monostoma arcuatum* Brandes, 1892, *Cyclocoelum robustum* Stossich, 1902 and *C. adolphi* Stossich, 1902. The type-species, *H. arcuatus* (Brandes, 1892), was described with the ovary and testes forming a diagonal line, the uterus extending laterally beyond the caeca and the vitelline fields united posteriorly. The validity of *Hyptiasmus* has been disputed and its name synonymized with *Cyclocoelum* (by Joyeux & Baer, 1927, and Lal, 1939) and *Ophthalmophagus* Stossich, 1902 (by Gower, 1938, 1939). Conversely, Bashkirova (1950), Yamaguti (1958a, 1971) and Feizullaev (1980) considered *Hyptiasmus* a valid genus, an opinion with which we concur after the examination of specimens labelled with this generic name at the H.W. Manter Laboratory of Parasitology, University of Nebraska, Lincoln, USA (slides 30742, 30758, 30722, 30738, 30746, 30730, 30728, 30736, 30768, 30725, 31053, 30723) and the US National Parasite Collection, Beltsville, Maryland (slides 441 and 50178).

Allopyge was established by Johnston (1913) for digeneans from the air-sac of birds in Australia, after Kossack (1911) had found that *Hyptiasmus adolphi* (Stossich, 1902) Kossack, 1911 (originally described by Stossich (1902) as *Cyclocoelum adolphi*) and *H. ominosus* Kossack, 1911 have a genital pore in a postpharyngeal position, undulating caeca and absence of uterine coils posterior to the hind testis. Johnston (1913) considered that *A. antigones* was identical to the European specimens described by Kossack (1911) as *H. ominosus* and *H. adolphi*. Consequently, Dubois (1959) doubted the validity of *Allopyge* and synonymized *A. antigones* with *C.* (*Hyptiasmus*) *adolphi*. For this reason, Sprehn (1932) transferred *A. adolphi* to *Hyptiasmus*. Conversely, Bashkirova (1950) and Yamaguti (1971) considered *Allopyge* a valid genus and accepted *A. antigones* as its type-species. Feizullaev (1980) also recognized the validity of *Allopyge*, but considered *A. adolphi* as its only valid species. Our opinion is based on an examination of the original drawings and descriptions of *Cyclocoelum adolphi* and *Hyptiasmus ominosus* by Stossich (1902) and Kossack (1911) and on an examination of adults labelled as *Hyptiasmus* (see above).

Cycloprimum was established by Witenberg (1923) for adult worms from birds in Europe, which Stossich (1902) had described as *Cyclocoelum exile* and Lühe (1909) called *Monostomum exile* (Stossich, 1902). In his diagnosis, Witenberg (1923) noted that *Cycloprimum* did not differ from *Cyclocoelum*, but he separated them based on his idea of 'periodical rows' in the Cyclocoelidae. Although Yamaguti (1958a, 1971) listed *Cycloprimum* as a valid genus, numerous authors, including Bashkirova (1950), Joyeux & Baer (1927) and Feizullaev (1980), synonymized it with *Cyclocoelum*. We agree with the latter stance.

Mediopharyngeum is one of three subgenera of *Cyclocoelum* erected by Witenberg (1923) based on the location of the genital pore and pharynx. The name *Mediopharyngeum* was used for adults with genital pores in the middle of the pharyngeal region. As no type designation was made by Witenberg (1923), Yamaguti (1971) and other authors suppressed the name. Feizullaev (1980) synonymized *Mediopharyngeum* with *Cyclocoelum*; we agree with this.

Prohyptiasmus Witenberg, 1923 is a much discussed and controversial cyclocoelid genus. The type-species, *P. robustus*, a parasite of birds in Europe, was originally described and illustrated by Stossich (1902) as *Cyclocoelum robustum*. It was later attributed to several different genera, i.e. *Monostoma* by Lühe (1909), *Hyptiasmus* by Kossack (1911), Sprehn (1932) and Bashkirova (1950), *Prohyptiasmus* by Witenberg (1923), *Cyclocoelum* by Joyeux & Baer (1927) and Lal (1939), *Harrahium* by Witenberg (1928), *Ophthalmophagus* by Gower (1938) and *Cyclocoelum* (*Hyptiasmus*) by Dubois (1959). Witenberg (1923, 1928) used three different names for this digenean: *Prohyptiasmus* and *Harrahium*, as indicated above, plus *Stossichium*, a new genus, which is discussed below. Bashkirova (1950), Yamaguti (1971), Feizullaev (1980) and others recognized *Prohyptiasmus* as valid but listed different morphological characteristics. For example, according to Witenberg (1923), Bashkirova (1950) and Yamaguti (1971), the genital pore of the type-species, *P. robustus*, is prepharyngeal, while Feizullaev

(1980) described and illustrated the genital pore as postpharyngeal. This and other morphological differences described by Feizullaev (1980) for his material suggests to us that under the concept *P. robustus* there are adults belonging not only to two different species but also to two different genera. Based on this, a new species name and new genus, *Neohyptiasmus feizullaevi*, are hereby proposed for the *Prohyptiasmus robustus* of Feizullaev (1980).

Transcoelum was created by Witenberg (1923) for *Hyptiasmus oculeus* Kossack, 1911, a digenean from the hypothalamus of *Fulica atra* in Europe. *T. oculeus* was described as differing from *Hyptiasmus* in its different bird hosts and distribution of uterine coils. Yamaguti (1971) recognized *Transcoelum* as a valid genus, but other authors have doubted its validity and placed it in synonymy, i.e. with: *Ophthalmophagus* Stossich, 1902 by Gower (1938); *Cyclocoelum* Brandes, 1892 by Joyeux & Baer (1927); *Hyptiasmus* Kossack, 1911 by Sprehn (1932) and Bykhovskaya-Pavlovskaya (1949); and *Cyclocoelum* (*Hyptiasmus*) by Dubois (1959). Based on our investigation of specimens from the different parasite collections mentioned above, we found no significant differences between *Transcoelum* and *Hyptiasmus* and consider them identical.

Antepharyngeum Witenberg, 1923 was erected by Witenberg (1923) as a subgenus of *Cyclocoelum* for adults with genital pores located anteriorly to the pharynx. Its validity was doubted by many authors and it is now considered a synonym of *Cyclocoelum*.

Morishitium was erected by Witenberg (1928) for *Cyclocoelum vagum* described by Morishita (1924a) from the nasolacrimal ducts of birds. *M. vagum* differs from *Cyclocoelum* in the position of the ovary, testes and genital pore. However, Joyeux & Baer (1927), Dollfus (1948) and Bashkirova (1950) synonymized this genus with *Cyclocoelum* (*Pseudhyptiasmus*). Nevertheless, Yamaguti (1971) and Feizullaev (1980) recognized *Morishitium* as valid and listed *Pseudhyptiasmus* Dollfus, 1948 as its synonym. Based on our investigation of specimens from the different collections mentioned above, we agree with this decision.

Stossichium was erected by Witenberg (1928) for the same adults from *Fuligula cristata* in Europe which Stossich (1902) had originally named *Cyclocoelum robustus* and he (Witenberg, 1923) had renamed *Prohyptiasmus robustus*. For reasons given above for *Prohyptiasmus*, *Stossichium* is considered a synonym of *Prohyptiasmus*.

Receptocoelum was proposed by Lal (1939) for species of *Cyclocoelum* that possess a seminal receptacle. However, it was later found that all members of the genus possess this structure. Hence, *Receptocoelum* is considered a synonym of *Cyclocoelum*.

The subgenus *Pseudhyptiasmus* of Dollfus (1948) is based on *Pseudhyptiasmus vagum* (Morishita, 1924), which Morishita (1924a) originally described as *Cyclocoelum vagum* and Witenberg (1928) renamed *Morishitium vagum*. For the reasons described above for *Morishitium*, the subgenus *Pseudhyptiasmus* is considered a synonym of *Morishitium*.

Skrjabinocoelum Kurashvili, 1953 is monotypic and based on *S. petrowi* Kurashvili, 1953, a parasite of the body-cavity of *Lymnocryptes minima* in Georgia. It is described with the testes and ovary in a transverse line near the posterior extremity of the body and vitelline fields confluent both anteriorly, posterior to the intestinal bifurcation, and posteriorly. In these features it does not appear to fit well into the subfamily Cyclocoelinae or into other groups of the family Cyclocoelidae. However, based on the reports of Kurashvili (1957), Bykhovskaya-Pavlovskaya (1962), Kasimov *et al.* (1962), Vaidova (1978) and Feizullaev (1980), we consider *Skrjabinocoelum* a valid genus within the Cyclocoelinae.

Philoculum Savinov, 1958 possesses the same morphological characteristics as those described above for *Allopyge*. Feizullaev (1980) considered *Philoculum* identical with *Allopyge*. There is no reason to disagree with this opinion.

Neocyclocoelum Feizullaev, 1980 is morphologically similar to *Morishitium*. Both are parasites of terrestrial birds in South-East Asia (Japan and China, respectively). These morphological and biological similarities suggest that they may represent the same taxon. The type-species of *Neocyclocoelum*, *N. elongatum*, was a new name given by Feizullaev (1980) for the same parasite from *Cyanopolius cyanus* in Nankin, China, which Harrah (1922) had originally described as *Cyclocoelum elongatum*. Feizullaev (1980) suggested that this parasite does not belong to *Cyclocoelum* or to the family Cyclocoelidae, because its life-cycle involves terrestrial snails and terrestrial birds as hosts and adult worms contain eggs with miracidia that are not fully developed. Based on this, Feizullaev (1980) erected *Neocyclocoelum* and a new family, the Neocyclocoelidae. The same characters and life-cycle patterns have also been described for species of *Morishitium*. For example, Timon-David (1955, 1957a) described the life-history of a parasite known, following Yamaguti (1958a), as *M. dollfusi* (Timon-David, 1950). The adult worms of this species correspond in morphology to the adults which Feizullaev (1980) described and illustrated as *Neocyclocoelum*. The first intermediate hosts for *M. dollfusi* are land snails of the species *Helicella* (*Helicopsis*) *arenosa* and *Leucochroa candidissima*. The final hosts are domestic pigeons, *Corvus corone* and *Passer montanus*. We consider *Neocyclocoelum* a synonym of *Morishitium*.

Porphyriotrema was erected by Duggal & Toor (1986) for *Cyclocoelum odeningi* Dubois, 1956 and *C. bikanerensis* Duggal & Toor, 1985 from birds in the Punjab, India. We doubt the validity of this genus and consider it a synonym of *Cyclocoelum*.

All genera commented upon below from *Ophthalmophagus* Stossich, 1902 to *Szidatiella* Yamaguti, 1958 are associated with our systematic organization of the subfamily Ophthalmophaginae Harrah, 1922.

Ophthalmophagus singularis Stossich, 1902, from the orbit of birds in Europe, was described by Stossich (1902) and designated as the type-species of *Ophthalmophagus*. *O. singularis* was removed by Joyeux & Baer (1927) to *Spaniometra* Kossack, 1911 and by Sprehn (1932) to *Hyptiasmus* Kossack, 1911. Contrary to these opinions, *Ophthalmophagus* has been considered valid by many authors, including Bashkirova (1950), Yamaguti (1958a, 1971) and Feizullaev (1980). Dollfus (1948) recognized two subgenera, *Ophthalmophagus* and *Geowitenbergia* Dollfus, 1948. Harrah (1922) created a new tribe, the 'Ophthalmophagea', and a new subfamily, the Ophthalmophaginae. In all diagnoses *Ophthalmophagus* is described with a post-testicular ovary that is in line with the testes, uterine coils extending laterally beyond the caeca and vitelline fields, and vitelline fields that are confluent posteriorly. However, the positions of the cirrus-sac and genital pore were described with different characteristics by various authors, suggesting that under the name of *Ophthalmophagus* there may have been described parasites of another closely related generic group. Nevertheless, we consider *Ophthalmophagus* a valid genus and list *Geowitenbergia* as a synonym.

Bothriogaster Fuhrmann, 1904 was created for parasites found in the intestine of birds in Java. The type-species, *B. variolaris* Fuhrmann, 1904, differed from *Ophthalmophagus* mainly in the presence of a ventral sucker. Because of this, Fuhrmann (1904) considered *Bothriogaster* different from the Cyclocoelinae and placed it in the Fasciolidae. Kossack (1911) examined numerous parasite collections in Europe and found that a poorly-developed ventral sucker occurs in some adults of the subfamily Cyclocoelinae. Based on this, Kossack (1911) transferred *Bothriogaster* to the Cyclocoelinae and placed it in synonymy with *Spaniometra* Kossack, 1911. Contrary to this opinion, Witenberg (1923) considered both *Bothriogaster* and *Spaniometra* valid and established for them a new tribe, the Bothriogastrea. Dollfus (1948) agreed with Witenberg (1923) but found that the name *Bothriogaster* was pre-occupied and renamed it *Bothrigaster*, erecting for it the new family-group names Bothrigastrinae and Bothrigastridae. Yamaguti (1958a, 1971) considered both *Bothrigaster* and *Spaniometra* valid names, but Bashkirova (1950) and Feizullaev (1980) synonymized *Bothrigaster* with *Spaniometra* and *Ophthalmophagus*, respectively. Travassos *et al.* (1969) recognized as valid both *Bothrigaster* and the Bothrigastridae. Based on the examination of adults labelled as *Bothriogaster* (deposited in the Institute of Oswaldo Cruz, Rio de Janeiro, Brazil), we consider *Bothrigaster* as recognizable within the Ophthalmophaginae. However, *Spaniometra* Kossack, 1911 is based on *Monostomum oculobium*, which Cohn (1902) described from the orbit of *Vanellus melanogaster*. *S. oculobia* differs from *Ophthalmophagus* with regard to the pre-equatorial position of the testes and the postpharyngeal position of the genital pore. Harrah (1922) synonymized *Spaniometra* with *Cyclocoelum*, while Stunkard (1934) and Feizullaev (1980) synonymized it with *Ophthalmophagus*. Dubois (1959) listed *Spaniometra* as a subgenus of *Ophthalmophagus*, while Bashkirova (1950) and Yamaguti (1958a, 1971) recognized it as valid. We suggest that, if Kossack (1911) is correct and if *Bothrigaster* and *Spaniometra* are two different names used for one and the same parasite, then priority belongs to *Spaniometra*.

Promptenovum was erected by Witenberg (1923) for the same parasite which van Beneden (1858) described and illustrated as *Monostomum mutabile* Zeder, 1800 and Stossich (1902) renamed *Cyclocoelum ovopunctatus*. The type-species, named by Witenberg (1923) as *Promptenovum vanbenedeni*, a parasite of the infra-orbital sinus of birds, was described as differing from *Cyclocoelum* in that: the genital pore is immediately postpharyngeal; the ovary is post-testicular, inside the posterior caecal arch; the vitelline fields are united posteriorly; and the uterine coils extend laterally beyond the caeca and vitelline fields. Witenberg (1926) found that the systematic position of *Promptenovum* is complicated and problematical. Yamaguti (1958a, 1971) recognized *Promptenovum* and even erected a new subfamily, the Promptenovinae. However, the original description and illustration of *Monostomum mutabile* of van Beneden (1858) showed morphological characters, except for the position of the genital pore, very similar to those of *Ophthalmophagus*. These similarities indicate that *Promptenovum* is closely related to *Ophthalmophagus* and that both belong to the same subfamily. In support of this suggestion is the fact that, while working with parasites labelled and deposited in parasite collections as *Ophthalmophagus*, we found specimens that corresponded to *Promptenovum*. These specimens show that *Promptenovum* is not identical to *Szidatitrema* Yamaguti, 1971 (syn. *Szidatiella* Yamaguti, 1958), as suggested by Yamaguti (1958a, 1971) and Feizullaev (1980). Our opinion is that *Promptenovum* is a valid genus with its own distinct morphological characteristics.

Contracoelum and its type-species, *C. skrjabinianum*, a parasite from *Plegadis falcinellus* in Armenia, were described and illustrated by Witenberg (1926) with testes juxtaposed and pre-equatorial. Yamaguti (1958a) considered *Contracoelum* a valid genus. However, no subsequent specimens with such juxtaposed testes have been found in Armenia. Hence, it has been considered an invalid name by many authors.

Dubinina (1937), Bashkirova (1950), Dubois (1959) and Feizullaev (1980) listed *Contracoelum* as a synonym of *Ophthalmophagus*, while Joyeux & Baer (1927) synonymized it with *Spaniometra*. The pre-equatorial position of the testes and the postpharyngeal position of the genital pore suggest to us that Joyeux & Baer (1927) were correct.

Geowitenbergia is a subgenus of *Ophthalmophagus* established by Dollfus (1948) for adults possessing uterine coils that do not extend into the extracaecal field. The type-species, *O.* (*G.*) *nasicola* (Witenberg, 1923), a parasite from the nasal sinus of *Rallus aquaticus* in Russia, was placed in *Spaniometra* by Joyeux & Baer (1927), in *Hyptiasmus* by Sprehn (1932) and in *Ophthalmophagus* by Feizullaev (1980). Based on examined specimens, we consider *Geowitenbergia* a synonym of *Ophthalmophagus*.

Szidatiella Yamaguti, 1958 was erected for *Cyclocoelum vogeli*, described by Szidat (1932) from the trachea and air-sacs of *Francolinus ahantensis* in Liberia. Unfortunately, the generic name is pre-occupied and was replaced with *Szidatitrema* by Yamaguti (1971).

Haematotrephus Stossich, 1902 was erected for parasites from the air-sacs of charadriid birds in Europe with gonads arranged in a triangle close to the posterior caecal union and no uterine coils between the testes. The type-species, *H. lanceolatum*, from *Himantopus rubropterus* in Italy was originally described by Wedl (1858) as *Monostomum lanceolatum*. Later, the same parasite was synonymized with *M. mutabile* Zeder, 1800 by Diesing (1850) and *M. ventricosum* (Rudolphi, 1802) by Rivolta & Delprato (1881); it was later placed in *Haematotrephus* by Stossich (1902), in *Cyclocoelum* by Joyeux & Baer (1927) and Dubois (1959) and in *Haematotrephus* (*Haematotrephus*) by Dollfus (1948). Witenberg (1923) erected a new tribe, the 'Haematotrephea'. Dollfus (1948) recognized subgenera and erected a new subfamily, the Haematotrephinae. Conversely, Joyeux & Baer (1927) and Dubois (1959) doubted the validity of *Haematotrephus* and considered it a subgenus of *Cyclocoelum*. Our opinion (based on an examination of specimens deposited at the H.W. Manter Laboratory of Parasitology, University of Nebraska, Lincoln, USA (slides 34347, 34348) and the US National Parasite Collection, Beltsville, Maryland (slides 7915, 7916)) is that *Haematotrephus* is a distinct and valid genus.

Corpopyrum Witenberg, 1923 was erected to include adults of *Haematotrephus brasilianum* (Stossich, 1902) and *Cyclocoelum tringae* (Brandes, 1892). The type-species, *C. kossacki* Witenberg, 1923, a parasite from the abdominal cavity of *Tringa alpina*, differed from *Haematotrephus* and *Cyclocoelum* in the presence of a tapered anterior body, vitelline fields separated from the caeca and testes inside the caecal union and not separated from one another by the uterus. Dollfus (1948) and Yamaguti (1958a, 1971) considered *Corpopyrum* valid, but Joyeux & Baer (1927) and Dubois (1959) included it in *Cyclocoelum*, and Lal (1939), Bashkirova (1950) and Feizullaev (1980) synonymized it with *Haematotrephus*. The latter opinion is accepted here.

Haematoprimum was created by Witenberg (1923) for adults from *Numenius arquata* in Europe possessing testes that are not separated by the uterus and vitelline fields that are separated from the caeca and intercaecal uterus. The type-species, *H. fasciatum*, was originally described by Stossich (1902) as *Haematotrephus fasciatus*. Kossack (1911) transferred this species to *Cyclocoelum* and Witenberg (1923) erected *Haematoprimum* for it. Joyeux & Baer (1927) synonymized *Haematoprimum* with *Cyclocoelum*, but Lal (1939), Bykhovskaya-Pavlovskaya (1949), Bashkirova (1950) and

Feizullaev (1980) synonymized it with *Haematotrephus*. There is no reason to disagree with this opinion.

Uvitellina Witenberg, 1923 differs from *Haematotrephus* only in that the vitelline fields are united posteriorly. The type-species, *U. pseudocotylea*, and a second species, *U. magniembria*, were described by Witenberg (1923) from adults in the air-sacs of a bird in Turkestan. Bykhovskaya-Pavlovskaya (1949) and Bashkirova (1950) synonymized the two species, while Joyeux & Baer (1927) and both Lal (1939) and Macko & Feige (1960) synonymized *Uvitellina* with *Cyclocoelum* and *Haematotrephus*, respectively. Dollfus (1948) listed *Uvitellina* as a subgenus of *Haematotrephus*. Bykhovskaya-Pavlovskaya (1949), Bashkirova (1950), Yamaguti (1958a, 1971) and Feizullaev (1980) considered *Uvitellina* a valid genus; this is accepted here.

Wardianum and a new tribe, the Wardianea, were erected by Witenberg (1923) for adults with a pyriform body, testes juxtaposed inside the posterior caecal union and pretesticular ovary. The type-species, *W. triangulare*, a parasite of the air-sacs of *Tringa maculata* in the USA, was originally described by Harrah (1922) as *Cyclocoelum triangularum*. Dubois (1959) synonymized *W. triangulare* with *Cyclocoelum* (*Haematotrephus*) *tringae* (Brandes, 1892), while Macko & Feige (1960) synonymized it with *Haematotrephus lanceolatum* (Wedl, 1858). Joyeux & Baer (1927) synonymized *Wardianum* with *Cyclocoelum*, but Lal (1939) synonymized it with *Haematotrephus*, a decision with which we concur. Contrary to this opinion, Bashkirova (1950) and Yamaguti (1958a, 1971) considered *Wardianum* valid.

Harrahium Witenberg, 1926 was erected for adults possessing testes that are located close to one another and an ovary at the level of the anterior testis. It was based on a parasite originally found in *Totanus melanoleucus* in the USA and named *Cyclocoelum halli* by Harrah (1922). Yamaguti (1958a, 1971) treated *Harrahium* as a valid name, but Joyeux & Baer (1927), Dollfus (1948), Bykhovskaya-Pavlovskaya (1949), Bashkirova (1950) and Feizullaev (1980) synonymized it with either *Cyclocoelum* or *Haematotrephus*. We concur with the latter stance.

Family Cyclocoelidae Stossich, 1902

(Syn. Bothrigastridae Dollfus, 1948)

Diagnosis: Body medium-sized to large, flat, lanceolate or markedly tapered anteriorly. Oral sucker absent. Ventral sucker absent or poorly developed. Mouth subterminal; prepharynx short; pharynx strong, well developed. Oesophagus short, rarely long and winding. Caeca long, united posteriorly forming cyclocoel, simple without diverticula on inner wall, straight or undulating. Testes oval or irregular in shape, diagonal, close to each other at posterior extremity, or tandem and markedly separated either in middle of body or pre-equatorially. Cirrus-sac prebifurcal or postbifurcal. Genital pore prepharyngeal, pharyngeal or postpharyngeal. Ovary, oval, intercaecal and post-testicular, intertesticular or pretesticular, forming triangle with or in line with testes. Uterus intercaecal or extending laterally beyond caeca and vitellarium. Vitellarium follicular; fields ventral and lateral to caeca, reaching to or beyond level of caecal bifurcation anteriorly, may or may not be united posteriorly. Excretory vesicle simple or Y-shaped, at posterior extremity, with dorsoterminal pore. Tail-less cercaria (cercariaeum) develops in redia, in freshwater or terrestrial snails.

Metacercariae in redia with cercariae or in snail tissue. Adults in nasal sinus, air-sacs and abdominal cavities of birds. Type-genus *Cyclocoelum* Brandes, 1892.

Key to subfamilies

1a. Ovary between testes .... Cyclocoelinae Stossich, 1902.
1b. Ovary not intertesticular .... 2.

2a. Ovary post-testicular .... Ophthalmophaginae Harrah, 1922.
2b. Ovary pretesticular .... Haematotrephinae Dollfus, 1948.

Subfamily Cyclocoelinae Stossich, 1902

(Syns Cyclocoeliinae Witenberg, 1923; Hyptiasminae Dollfus, 1948)

Diagnosis: Posterior testis within posterior union of caeca; anterior testis post-equatorial, in line or forming triangle with ovary and posterior testis. Genital pore prepharyngeal or postpharyngeal. Ovary intertesticular. Vitelline fields ventral and lateral to caeca, may or may not be united posteriorly. Uterine coils intercaecal or extracaecal, extending laterally to caeca and vitelline fields. Type-genus *Cyclocoelum* Brandes, 1892.

Key to genera

1a. Testes in transverse row .... *Skrjabinocoelum* Kurashvili, 1953. (Fig. 20.7)
Diagnosis: Testes and ovary in transverse line near posterior extremity. Genital pore postpharyngeal. Vitelline fields united both anteriorly, just posterior to intestinal bifurcation, and posteriorly. Uterine coils intercaecal, not extending laterally beyond caeca and vitelline fields. Parasite of body-cavity of birds. Georgia and Azerbaidzhan. Type-species *S. petrowi* Kurashvili, 1953.
1b. Testes tandem or diagonal .... 2.

2a. Genital pore prepharyngeal .... 3.
2b. Genital pore postpharyngeal .... 5.

3a. Vitelline fields united posteriorly .... *Hyptiasmus* Kossack, 1911. (Fig. 20.4)
(Syn. *Transcoelum* Witenberg, 1923)
Diagnosis: Testes and ovary in a tandem line. Genital pore prepharyngeal. Vitelline fields ventral and lateral to caeca, not reaching caecal bifurcation anteriorly, united posteriorly. Uterine coils extended laterally beyond caeca and vitelline fields. Parasitic mainly in nasal cavity, hypothalamus and infra-orbital sinuses of birds. Cosmopolitan. Type-species *H. arcuatus* (Brandes, 1892) Kossack, 1911.
3b. Vitelline fields not united posteriorly .... 4.

4a. Testes oblique, form triangle with ovary; uterine coils intercaecal .... *Cyclocoelum* Brandes, 1892. (Fig. 20.1)

(Syns *Cycloprimum* Witenberg, 1923; *Mediopharyngeum* Witenberg, 1923; *Antepharyngeum* Witenberg, 1923; *Receptocoelum* Lal, 1939; *Porphyriotrema* Duggal & Toor, 1986)

Diagnosis: Testes oblique, forming triangle with ovary. Genital pore prepharyngeal. Vitelline fields ventral and immediately lateral to caeca, not united posteriorly. Uterine coils not extending beyond outer wall of caeca. Parasitic mainly in nasal and naso-lachrimal sinuses of birds; occasionally in mammals, including humans. Cosmopolitan. Type-species *C. mutabile* (Zeder, 1800) Brandes, 1892.

4b. Testes tandem, in straight line with ovary; uterine coils overlap caeca laterally .... *Prohyptiasmus* Witenberg, 1923. (Fig. 20.5)

(Syn. *Stossichium* Witenberg, 1928)

Diagnosis: Testes entire, diagonal and in line with ovary, in posterior half of body. Genital pore prepharyngeal. Vitelline fields ventral and lateral to caeca, not united posteriorly, not reaching caecal bifurcation anteriorly. Uterine coils extending laterally beyond caecal and vitelline fields. Parasitic mainly in nasal sinuses and air-sacs of birds. Cosmopolitan. Type-species *P. robustus* (Stossich, 1902) Witenberg, 1923.

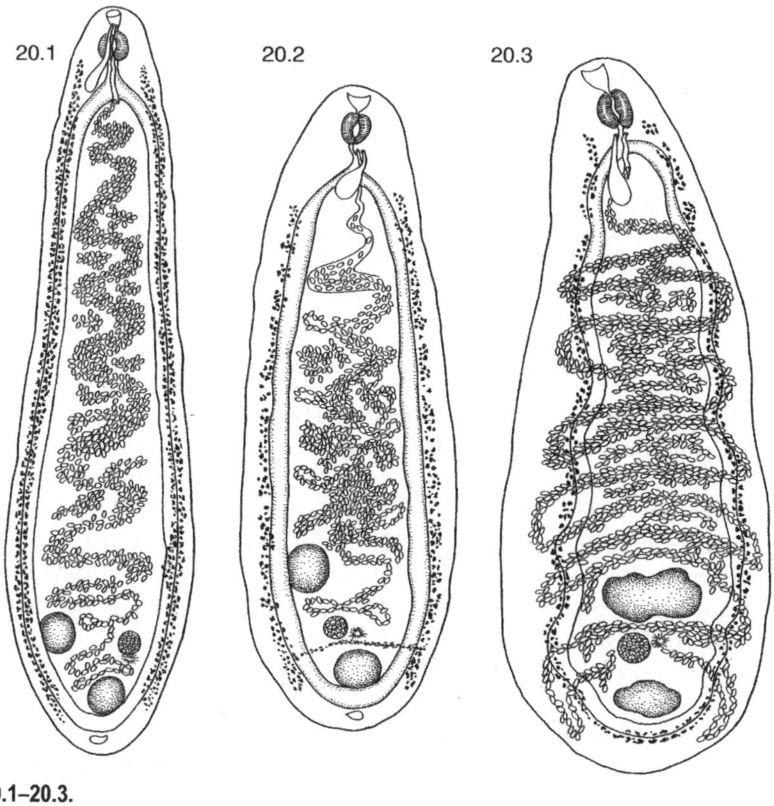

Figs 20.1–20.3.
20.1. *Cyclocoelum mutabile* (Zeder, 1800) Brandes, 1892. (Original.)
20.2. *Morishitium vagum* (Morishita, 1924) Witenberg, 1928. (Original.)
20.3. *Allopyge adolphi* (Stossich, 1902) Johnston, 1913. (Original.)

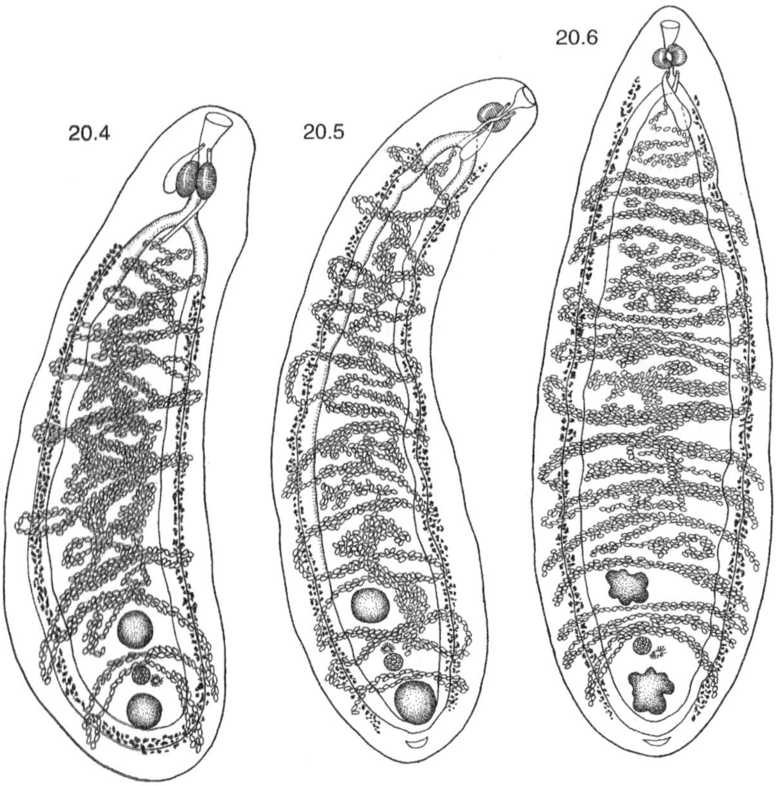

Figs 20.4–20.6.
20.4. *Hyptiasmus arcuatus* (Brandes, 1892) Kossack, 1911. (Original.)
20.5. *Prohyptiasmus robustus* (Stossich, 1902) Witenberg, 1923. (Original.)
20.6. *Neohyptiasmus feizullaevi* nom. nov. (Original.)

5a. Vitelline fields not united posteriorly .... **6.**
5b. Vitelline fields united posteriorly .... ***Allopyge* Johnston, 1913. (Fig. 20.3)**
(Syn. *Philoculum* Savinov, 1958)

Diagnosis: Testes and ovary forming a diagonal line in posterior half of body. Genital pore postpharyngeal. Vitelline fields ventral and lateral to caeca, reaching anteriorly to or beyond level of caecal bifurcation, united posteriorly. Uterine coils extend laterally beyond caeca and vitelline fields. Parasitic mainly in nasal cavity, hypothalamus and infra-orbital sinuses of birds. Cosmopolitan. Type-species *A. adolphi* (Stossich, 1902) Johnston, 1913.

6a. Testes rounded, entire; uterine coils not extending beyond caeca .... ***Morishitium* Witenberg, 1928. (Fig. 20.2)**
(Syns *Pseudhyptiasmus* Dollfus, 1948; *Neocyclocoelum* Feizullaev, 1980)

Diagnosis: Testes oblique, in line with ovary. Genital pore postpharyngeal. Vitelline fields ventral and immediately lateral to caeca, not united posteriorly. Uterus not

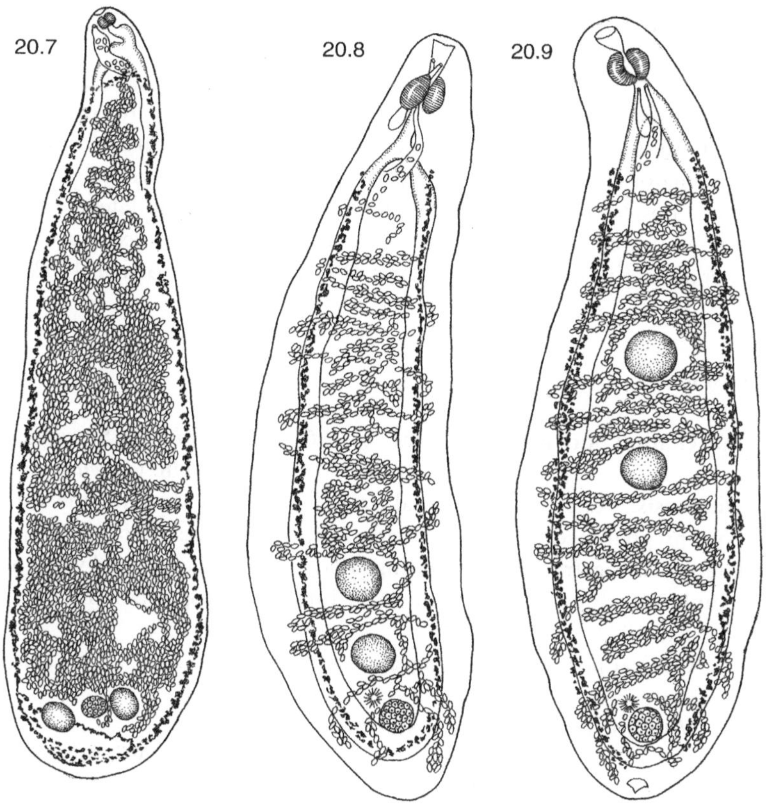

Figs 20.7–20.9.
20.7. *Skrjabinocoelum petrowi* Kurashvili, 1953. (After Kurashvili, 1953.)
20.8. *Ophthalmophagus singularis* Stossich, 1902. (Original.)
20.9. *Spaniometra variolaris* (Fuhrmann, 1904) Dollfus, 1948. (Original.)

extending beyond outer wall of caeca. Parasitic mainly in nasal and naso-lachrimal sinuses of birds. Cosmopolitan. Type-species *M. vagum* (Morishita, 1924) Witenberg, 1928.

6b. Testes lobed; uterine coils reach lateral margin of body .... *Neohyptiasmus* n. g. (Fig. 20.6)

Diagnosis: Testes lobed, diagonal and in line with ovary, in posterior half of body. Genital pore postpharyngeal. Vitelline field ventral and lateral to caeca, not united posteriorly, reaching to or beyond level of caecal bifurcation anteriorly. Uterine coils extracaecal, reaching to lateral margin of body. Parasitic mainly in nasal sinuses and air-sacs of birds. Cosmopolitan. Type-species *N. feizullaevi* nom. nov. pro *Prohyptiasmus robustus* of Feizullaev (1980).

Subfamily Ophthalmophaginae Harrah, 1922

(Syn. Bothrigasterinae Dollfus, 1948)

Diagnosis: Testes diagonal, close to each other, forming line or triangle with ovary at the posterior extremity or in tandem, well separated, with anterior testis equatorial or pre-equatorial. Genital pore prepharyngeal or postpharyngeal. Ovary post-testicular, inside posterior caecal union. Vitelline fields ventral and lateral to caeca, may or may not be united posteriorly. Uterine coils intercaecal or extracaecal and reaching to lateral margins of body. Type-genus *Ophthalmophagus* Stossich, 1902.

Key to genera

1a. Genital pore prepharyngeal ...... *Ophthalmophagus* Stossich, 1902. (Fig. 20.8)
(Syn. *Geowitenbergia* Dollfus, 1948)

Diagnosis: Testes postequatorial, oval, oblique, tandem in line with ovary. Genital pore prepharyngeal. Vitelline fields ventral and lateral to caeca, united posteriorly. Uterine coils extracaecal, extending laterally beyond caeca and vitelline fields. Parasitic mainly in orbit and nasal sinuses of birds. Cosmopolitan. Type-species *O. singularis* Stossich, 1902.

1b. Genital pore postpharyngeal .... 2.

2a. Vitelline fields united posteriorly ....
.. *Promptenovum* Witenberg, 1923. (Fig. 20.10)

Diagnosis: Testes oval, close to middle of posterior half of body, in line with ovary. Genital pore postpharyngeal. Vitelline fields ventral and lateral to caeca, united posteriorly. Uterine coils extending laterally to caeca and vitelline fields. Parasitic mainly in orbit and nasal sinuses of birds. Cosmopolitan. Type-species *P. vanbenedeni* Witenberg, 1923.

2b. Vitelline fields not united posteriorly .... 3.

3a. Testes tandem, in line with ovary ...... *Spaniometra* Kossack, 1911. (Fig. 20.9)
(Syns *Bothriogaster* Fuhrmann, 1904; *Contracoelum* Witenberg, 1926; *Bothrigaster* Dollfus, 1948)

Diagnosis: Testes oval, tandem, well apart from each other, in line with ovary; anterior testis equatorial or pre-equatorial. Genital pore postpharyngeal. Vitelline fields, ventral and lateral to caeca, not united posteriorly. Uterine coils extracaecal, reaching laterally to caeca and vitelline fields. Parasitic mainly in orbit and nasal sinuses of birds. Cosmopolitan. Type-species *S. oculobia* (Cohn 1902) Kossack, 1911.

3b. Testes diagonal, form triangle with ovary ....
.. *Szidatitrema* Yamaguti, 1971. (Fig. 20.11)
(Syn. *Szidatiella* Yamaguti, 1958)

Diagnosis: Testes oval, diagonal, close to posterior extremity, forming triangle with ovary. Genital pore postpharyngeal. Vitelline fields lateral to caeca, not united posteriorly. Uterine coils over-reaching caeca but not extended laterally beyond vitelline fields. Parasitic mainly in trachea and air-sacs of birds. West Africa. Type-species *S. vogeli* (Szidat, 1932) Yamaguti, 1971.

Subfamily Haematotrephinae Dollfus, 1948

Diagnosis: Testes oval or irregular in shape, diagonal, close to each other, inside posterior caecal union, forming triangle with ovary. Genital pore prepharyngeal or postpharyngeal. Ovary pretesticular, submedian, opposite to anterior testis. Vitelline fields ventral and lateral to caeca, may or may not be united posteriorly. Uterine coils well developed, extracaecal, reaching laterally beyond caeca and vitelline fields. Type-genus *Haematotrephus* Stossich, 1902.

Key to genera

1a. **Vitelline fields not united posteriorly; genital pore postpharyngeal .... *Haematotrephus* Stossich, 1902. (Fig. 20.12)**
(Syns *Corpopyrum* Witenberg, 1923; *Haematoprimum* Witenberg, 1923; *Harrahium* Witenberg, 1923; *Wardianum* Witenberg, 1923)

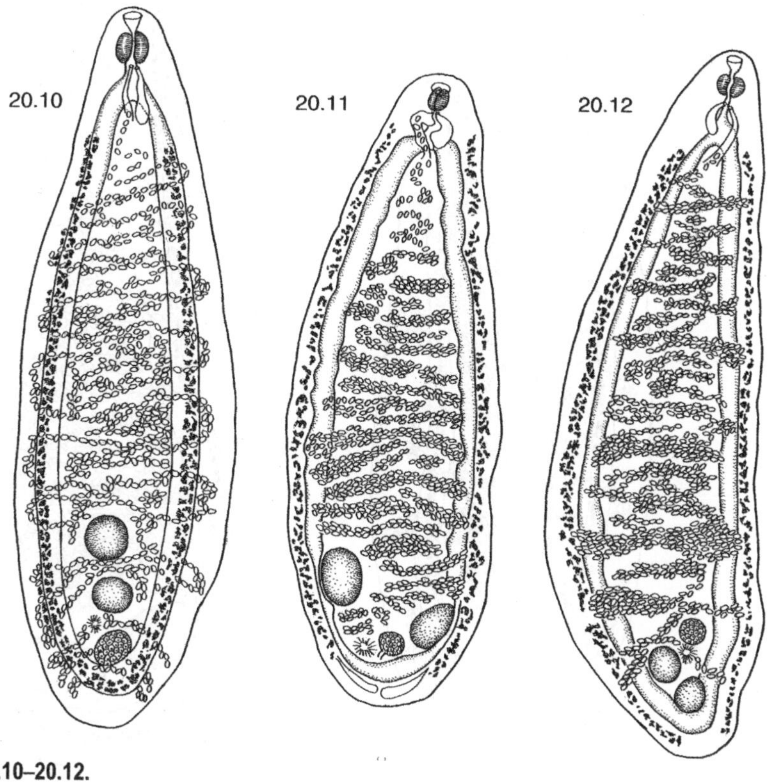

Figs 20.10–20.12.
20.10. *Promptenovum vanbenedeni* Witenberg, 1923. (Original.)
20.11. *Szidatitrema vogeli* (Szidat, 1932) Yamaguti, 1971. (After Szidat, 1932.)
20.12. *Haematotrephus lanceolatum* (Wedl, 1858) Stossich, 1902. (Original.)

Diagnosis: Testes oval, diagonal, close to each other, inside posterior caecal union, forming triangle with ovary. Genital pore postpharyngeal. Ovary pretesticular, submedian, opposite anterior testis. Vitelline fields ventral and lateral to caeca, not united posteriorly. Uterine coils reach to lateral margins of body. Parasitic mainly in air-sacs and body-cavities of birds. Cosmopolitan. Type-species *H. lanceolatum* (Wedl, 1858) Stossich, 1902.

1b. Vitelline fields united posteriorly; genital pore pre- or postpharyngeal .... 2.

2a. Genital pore postpharyngeal .... *Uvitellina* Witenberg, 1923. (Fig. 20.13)

Diagnosis: Testes oval or irregular in shape, diagonal, close to each other, inside posterior caecal arch, forming triangle with ovary. Genital pore postpharyngeal. Ovary pretesticular submedian, on opposite side of body to anterior testis. Vitelline fields ventral and lateral to the caeca, united posteriorly. Uterine coils well developed,

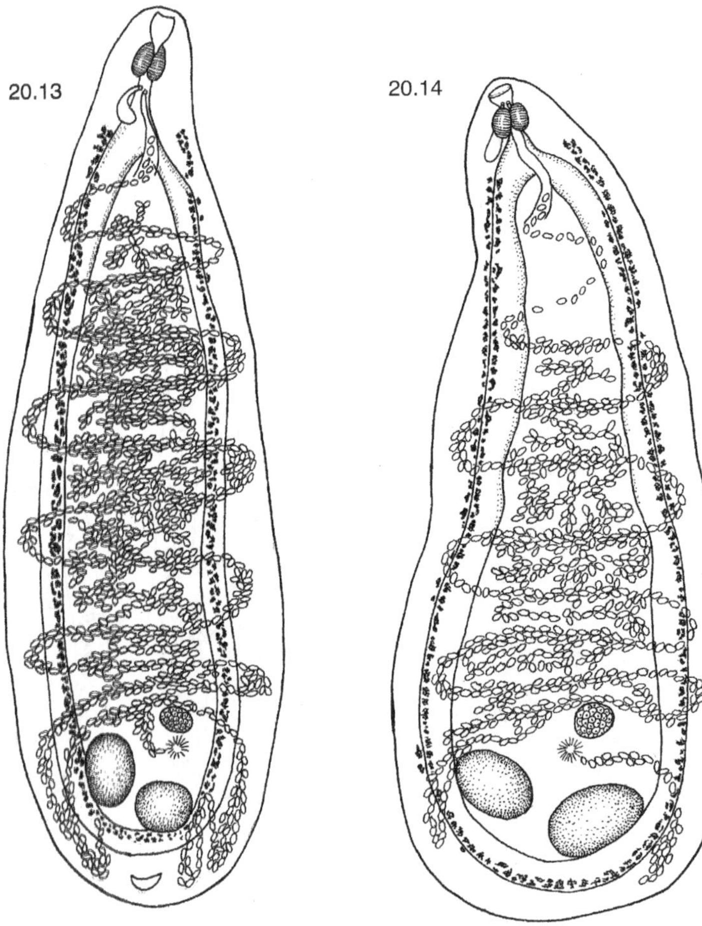

Figs 20.13–20.14.

20.13. *Uvitellina adelphus* (Johnson, 1917) Witenberg, 1923. (Original.)

20.14. *Neohematotrephus brasilianum* (Stossich, 1902) n. comb. (Original.)

extracaecal, reaching laterally beyond caeca and vitelline fields. Parasitic mainly in air-sacs and body-cavities of birds. Cosmopolitan. Type-species *U. adelphus* (Johnston, 1917) Witenberg, 1923.

2b. Genital pore prepharyngeal .... *Neohaematotrephus* n. g. (Fig. 20.14)

Diagnosis: Testes oval or irregular in shape, diagonal, close to each other and inside posterior caecal union, forming triangle with ovary. Genital pore prepharyngeal. Ovary pretesticular, submedian, on opposite side of body to anterior testis. Vitelline fields ventral and lateral to caeca, united posteriorly. Uterine coils reaching to lateral margin of body. Parasitic mainly in air-sacs and body-cavities of birds. South America. Type-species *N. brasilianum* (Stossich, 1902) n. comb.

21 Family Eucotylidae Cohn, 1904

I. Kanev,[1] V. Radev[1] & B. Fried[2]

[1]*Institute of Experimental Pathology and Parasitology, Bulgarian Academy of Sciences, Sofia, Bulgaria;* [2]*Department of Biology, Lafayette College, Easton, Pennsylvania, USA*

Introduction

The family Eucotylidae was established by Cohn (1904) for adult worms found in the urinary tubules of the kidneys of aquatic birds. The main familial characters were: the presence of an annular cervical thickening in the form of a cone or triangle at the anterior extremity; testes symmetrical and pre-equatorial; caeca simple and not united posteriorly; and genital pore median, between the ovary and caecal bifurcation. Skrjabin (1924, 1947b), Freitas (1951, 1959) and Yamaguti (1958a, 1971) presented several different classifications for the Eucotylidae. We include within the family Eucotylidae two subfamilies and seven genera.

Eucotyle Cohn, 1904 was erected for adult worms from the kidneys and urinary tubules of birds in Europe. The type-species, *E. nephritica* (Mehlis in Creplin, 1846), was originally found in *Colymbus arcuatus* in Germany and described by Mehlis (in Creplin, 1846) as *Monostomum nephriticum*. It was characterized by: the anterior extremity delimited from the rest of the body by an annular muscular thickening in the form of a cone or triangle; testes extracaecal; and vitelline fields pretesticular. Our studies suggest that adults with two different morphological forms have been described under the name *Eucotyle*. In one form, the testes are pre-equatorial and the vitelline fields are both pretesticular and post-testicular. Such specimens have tended to be identified as *E. nephritica*, *E. cohni* Skrjabin, 1924, *E. hassalli* Price, 1930 and *E. popovi* Skrjabin & Evranova, 1942. The second morphological form differs in the equatorial position of the testes and short vitelline fields that are restricted to the pretesticular region. These characteristics, found in adults from Europe, Asia and North America, have been identified and labelled as *E. zakharovi* Skrjabin, 1920, *E. wehri* Price, 1930 and *E. warreni* Schell, 1967. Because of these morphological differences, we consider that the latter specimens do not fit *Eucotyle* (*sensu stricto*) and we have transferred them to a new genus, *Neoeucotyle*, as *N. zakharovi* (Skrjabin, 1920) n. comb. (type-species), *N. wehri* (Price, 1930) n. comb. and *N. warreni* (Schell, 1967) n. comb.

Tanaisia was established by Skrjabin (1924) for adults from the urinary tubules of the kidneys of various birds in Turkestan, with *T. fedtschenkoi* Skrjabin, 1924 as the type-species. It was described with intercaecal, diagonal and deeply lobed testes, and the anterior extremity of the body not delimited from the rest of the body by an annular muscular thickening. Freitas (1951) erected the subfamily Tanaisiinae for this genus.

Paratanaisia was erected by Freitas (1959) for species of *Tanaisia* and *Tamerlania* Skrjabin, 1924 with long vitelline fields extending both anteriorly and posteriorly to the ovary. The worms were first found in Brazil, in domestic pigeons and chickens, and named *Tamerlania bragai* by Santos (1934). Also in Brazil, similar worms were found in the kidneys of *Cryptorellus tataupa* and *Ramphastos todo*, which Freitas (1951) described, illustrated and named as *Tanaisia robusta* and *T. confusa*, respectively. We consider *Paratanaisia* to be a valid genus, distinct from *Tamerlania*, based on a wider vitelline distribution. Adults from pigeons in Venezuela, described as *Tanaisia domestica* Nasir, 1972, and in Taiwan, described as *T. ectorchis* Fischthal, 1975, with vitelline fields extending both anteriorly and posteriorly to the ovary, are transferred to this genus as *P. domestica* (Nasir, 1972) n. comb. and *P. ectorchis* (Fischthal, 1975) n. comb.

Tamerlania was erected by Skrjabin (1924) for adults from the kidneys of birds in Turkestan. The type-species, *T. zarudnyi* Skrjabin, 1924, was described as differing from *Tanaisia* in the symmetrical, juxtaposed testes, which are entire pre-equatorial, and in the presence of a ventral sucker. Yamaguti (1958a, 1975) listed *Tamerlania* as a subgenus of *Tanaisia*. Freitas (1951, 1959) synonymized the two. We consider *Tamerlania* as a valid genus with morphological characteristics somewhat between *Tanaisia* and *Paratanaisia*.

Lepidopteria Nezlobinski, 1926 and its type-species, *L. atra* Nezlobinski, 1926, a parasite of the kidneys of *Fulica atra* and *F. a. americana* in Macedonia, were described with characters corresponding to those of *Tanaisia*. Although Yamaguti (1958a, 1971) recognized *Lepidopteria* as a subgenus of *Tanaisia*, we have followed Byrd & Denton (1950) in considering it a synonym of the latter.

Ohridia was a second genus erected by Nezlobinski (1926) for adults in the kidneys of birds in Macedonia. The type-species, *O. panuri* Nezlobinski, 1926, was described with features similar to those of the type-species of *Tanaisia*, *T. fedtschenkoi*, a parasite widespread and common in various waterfowl in Macedonia, Bulgaria and other countries in this region. Although Yamaguti (1958a, 1971) recognized *Ohridia* as a subgenus of *Tanaisia*, we have followed Byrd and Denton (1950) in considering it a synonym of the latter.

Prohystera was erected by Korkhaus (1930) for parasites of birds in Germany with characters typical of *Tanaisia*. Ejsmont (1932) synonymized the type-species, *P. rossittensis* Korkhaus, 1930, with the type-species of *Tanaisia*, *T. fedtschenkoi*, and *Prohystera* has been considered a synonym of *Tanaisia* by subsequent authors.

Family Eucotylidae Cohn, 1904

Diagnosis: Body middle-sized to large, elongate, flattened, subcylindrical or cylindrical; anterior extremity may or may not be delimited from rest of body by annular muscular thickening in form of cone or triangle. Oral sucker well developed, muscular, subterminal. Ventral sucker absent or poorly developed. Prepharynx very short, often not visible. Pharynx present. Oesophagus short, dilated or not. Caeca simple, somewhat undulating or smooth, may or may not be united posteriorly forming cyclocoel. Testes extracaecal or intercaecal, often overlapping caeca, smooth, irregular or strongly lobed, tandem, diagonal or symmetrical, equatorial or post-equatorial. Cirrus-sac present or absent. Genital pore median or submedian, between caecal bifurcation and ovary. Ovary intercaecal, pretesticular, median or submedian, smooth, irregular or lobed. Vitelline fields extracaecal or overlapping caeca, entirely pretesticular, entirely postovarian or in both regions, not reaching posterior extremity. Uterus long, usually intercaecal, may not extend anteriorly past caecal bifurcation, posteriorly beyond caecal union and laterally beyond caeca. Tail-less cercaria (cercariaeum) develops in sporocysts in freshwater and land snails; some stylet cercariae from freshwater snails are also considered as larvae of *Eucotyle*. Metacercaria in same sporocysts and snail where cercariae develop. Adults in kidney tubules of birds. Type-genus *Eucotyle* Cohn, 1904.

Key to subfamilies

1a. Body with annular cervical thickening in form of cone or triangle; caeca not forming cyclocoel; testes extracaecal or overlapping caeca; cirrus-sac present .... Eucotylinae Skrjabin, 1924.

1b. Body without annular cervical thickening; caeca form cyclocoel posteriorly; testes intercaecal; cirrus-sac absent .... Tanaisiinae Freitas, 1951.

Subfamily Eucotylinae Skrjabin, 1924

(Syns Eucotylinae Freitas, 1951; Eucotylinae Yamaguti, 1958)

Diagnosis: Body elongate, middle-sized, flattened, subcylindrical; anterior extremity delimited from rest of body by annular muscular thickening in form of cone or triangle. Oral sucker well developed, subterminal. Ventral sucker absent. Prepharynx short, often not visible. Pharynx present. Oesophagus short, may or may not be dilated. Caeca somewhat undulating or not, end blindly. Testes extracaecal or more or less overlapping caeca, elongate, irregular in shape or lobed, symmetrical, pre-equatorial or equatorial. Cirrus-sac pyriform, preovarian, containing seminal vesicle. Genital pore more or less median between caecal bifurcation and ovary. Ovary pretesticular, submedian, intercaecal, irregular in shape or lobed. Vitelline fields extracaecal, pretesticular, may or may not extend into post-testicular region. Uterus very long, intercaecal or overlapping caeca, entirely between caecal bifurcation and posterior extremity. Parasitic in kidney tubules of aquatic birds. Type-genus *Eucotyle* Cohn, 1904.

Key to genera

1a. Testes pre-equatorial; vitelline fields both pretesticular and post-testicular .... *Eucotyle* Cohn, 1904. (Fig. 21.1)

Diagnosis: Testes pre-equatorial. Vitelline fields both pretesticular and post-testicular. Cosmopolitan. Type-species *E. nephritica* (Creplin, 1846) Cohn, 1904.

1b. Testes equatorial; vitelline fields only pretesticular .... ***Neoeucotyle*** **n. g. (Fig. 21.2)**

Diagnosis: Testes equatorial. Vitelline fields only pretesticular. Holarctic. Type-species *N. zakharovi* (Skrjabin, 1920) n. comb.

Subfamily Tanaisiinae Freitas, 1951

(Syn. Tanaisiinae Yamaguti, 1958)

Diagnosis: Body elongate, linguiform, without annular cervical thickening. Oral sucker and pharynx moderately developed. Ventral sucker absent or poorly developed. Oesophagus short. Caeca simple, united posteriorly forming cyclocoel. Testes mainly

21.1

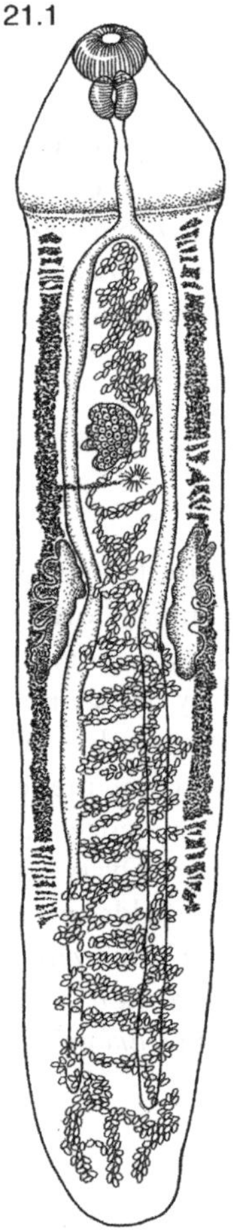

21.2

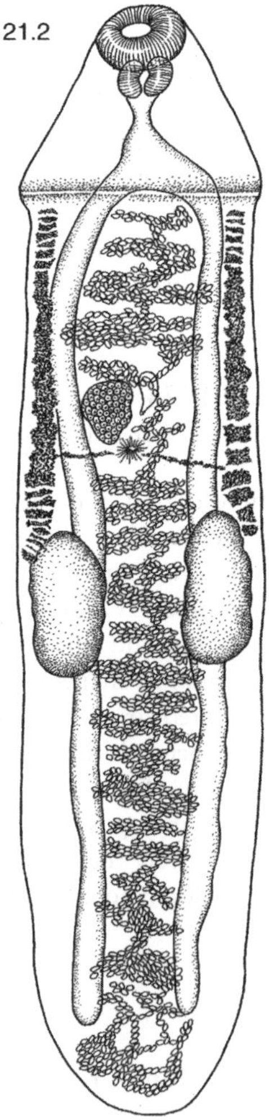

Figs 21.1–21.2.
21.1. *Eucotyle nephritica* (Mehlis in Creplin, 1846) Cohn, 1904. (Original.)
21.2. *Neoeucotyle zakharovi* (Skrjabin, 1920) n. comb. (Original.)

intercaecal, may or may not overlap caeca; oval, irregular or lobed, tandem, diagonal or symmetrical, pre-equatorial or equatorial. Cirrus-sac absent. Genital pore median between intestinal bifurcation and ovary. Ovary pretesticular, median or submedian, intercaecal, smooth, irregular or lobed. Vitelline fields extracaecal or overlapping caeca, postovarian, may or may not extend into pre-ovarian region. Uterus very long, intercaecal or overlapping caeca, reaching anteriorly just beyond caecal bifurcation and posteriorly to cyclocoel. Parasitic in the kidney tubules of birds. Type-genus *Tanaisia* Skrjabin, 1924.

Key to genera

1a. Testes tandem or diagonal, deeply lobed .... ***Tanaisia* Skrjabin, 1924. (Fig. 21.3)**
(Syns *Lepidopteria* Nezlobinski, 1926; *Ohridia* Nezlobinski, 1926; *Probystera* Korkhaus, 1930)
Diagnosis: Testes tandem or diagonal, deeply lobed. Vitelline fields only postovarian. Cosmopolitan. Type-species *T. fedtschenkoi* Skrjabin, 1924.
1b. Testes symmetrical .... **2.**

2a. Vitelline fields extend both pre- and post-testicular .... ***Paratanaisia* Freitas, 1959. (Fig. 21.4)**
Diagnosis: Testes symmetrical, equatorial, irregular in shape or slightly lobed. Vitelline fields extend in both pre-ovarian and postovarian fields. South America. Type-species *P. bragai* (Santos, 1934) Freitas, 1959.
2b. Vitelline fields restricted to post-testicular region .... ***Tamerlania* Skrjabin, 1924. (Fig. 21.5)**
Diagnosis: Testes symmetrical, pre-equatorial, smooth or irregular in shape. Vitelline field short, postovarian. Asia, Europe, North and South America, perhaps cosmopolitan. Type-species *T. zarudnyi* Skrjabin, 1924.

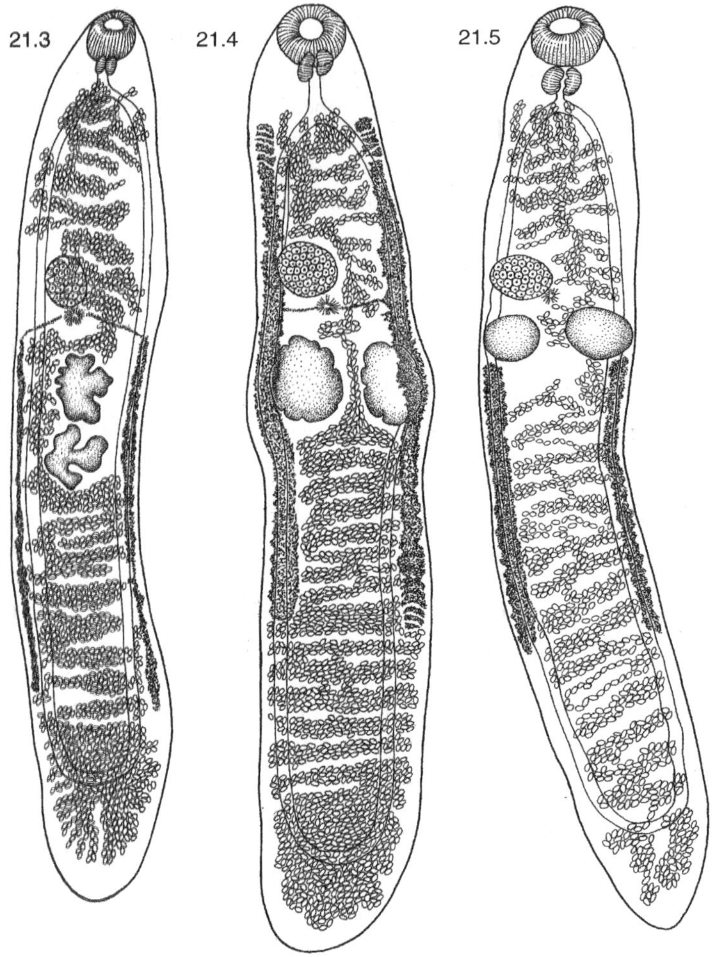

Figs 21.3–21.5.
21.3. *Tanaisia fedtschenkoi* Skrjabin, 1924. (Original.)
21.4. *Paratanaisia bragai* (Santos, 1934) Freitas, 1959. (Original.)
21.5. *Tamerlania zarudnyi* Skrjabin, 1924. (Original.)

22 Family Typhlocoelidae Harrah, 1922

I. KANEV,[1] V. RADEV[1] & B. FRIED[2]

[1]*Institute of Experimental Pathology and Parasitology, Bulgarian Academy of Sciences, Sofia, Bulgaria;* [2]*Department of Biology, Lafayette College, Easton, Pennsylvania, USA*

Introduction

Typhlocoelids are mainly parasites of the trachea of birds. The family group was originally established at the subfamily level by Harrah (1922) for *Typhlocoelum* Stossich, 1902 and *Tracheophilus* Skrjabin, 1913. Later, the group was revised and raised to full family status by Bittner & Sprehn (1928). Subsequent comments on the classification have been given by Travassos (1929), Lal (1936), Dollfus (1948), Bykhovskaya-Pavlovskaya (1949), Bashkirova (1950), Yamaguti (1958a, 1971), Dubois (1959) and Feizullaev (1980). In the present work, the family Typhlocoelidae is comprised of three subfamilies and six genera.

Typhlocoelum was established by Stossich (1902) for adults possessing caeca with numerous short diverticula, which were collected from various species of *Anas* in Europe. Adults attributed to this genus were first studied and named by Rudolphi (1809) as *Distoma cucumerinum* and by Mehlis (1831) as *Monostomum flavum*. *Typhlocoelum flavum* (Mehlis, 1831) was designated by Stossich (1902) as the type-species. Kossack (1911) synonymized *M. flavum* with *D. cucumerinum* and indicated the latter as the type-species of *Typhlocoelum*. Harrah (1922) erected for *Typhlocoelum* a new subfamily, the Typhlocoelinae, which Bittner & Sprehn (1928) raised to family level. Dollfus (1948) and Macko & Busa (1960) recognized the validity of *Typhlocoelum*, the Typhlocoelinae and the Typhlocoelidae. Bashkirova (1950) and Yamaguti (1958a, 1971) recognized *Typhlocoelum* and the Typhlocoelinae as valid but doubted the validity of the family Typhlocoelidae. The results of our comparative study suggest that, under the name of *Typhlocoelum*, digeneans belonging to two different morphological groups have been described. In one of these forms, described as *T. cucumerinum* (Rudolphi, 1809), the testes are lobed, comprising about 20 testicular bodies, which seem to form part of the lobed testes, and the vitelline fields are not united at the posterior end of the body. In the second form, named *T. americanum* Manter & Williams, 1928, the testes are two lobed bodies and the vitelline fields are united posteriorly. Based on these and several other differences, such as the position of the genital pore and presence of a ventral sucker, a new genus, *Manterocoelum*, is erected to accommodate specimens previously described as *T. americanum* and here called *M. americanum* (Manter & Williams, 1928) n. comb.

Tracheophilus, based on *T. sisowi* Skrjabin, 1913, was erected for parasites of the trachea of ducks in Europe by Skrjabin (1913). It differs from *Typhlocoelum cucumerinum* (Rudolphi, 1809) in the smooth shape of the testes, the prepharyngeal position of the genital pore and the arrangement of the vitelline fields, which are usually united posteriorly. Because of these differences, *Tracheophilus* was considered a valid genus by Bykhovskaya-Pavlovskaya (1949), Bashkirova (1950) and Yamaguti (1958a, 1971). The validity of *Tracheophilus* is doubtful according to some workers, and it has been synonymized with *Typhlocoelum* by Dubois (1959), Feizullaev (1980) and others. Dollfus (1948) listed *Tracheophilus* as a subgenus of *Typhlocoelum*. In our opinion, *Tracheophilus* is a valid genus.

Typhlultimum is one of 16 genera designated by Witenberg (1923, 1926) on the basis of his 'periodical system' for the Cyclocoelidae. The type-species, *T. sarcidiornicola* (Megnin, 1890) from *Sarcidiornis melanota* in Madagascar, has been listed under four different genera: *Monostoma* Zeder, 1800 in Megnin (1890); *Typhlocoelum* Stossich, 1902 in Stossich (1902); *Tracheophilus* Skrjabin, 1913 in Skrjabin (1913); and *Typhlultimum* Witenberg, 1923 in Witenberg (1923, 1926). Witenberg (1923, 1926) did not describe or illustrate *Typhlultimum*, except for a schematic figure showing the position of the uterus, ovary and testes. Because of this and Joyeux & Baer's (1927) observations that the figure given by Megnin (1890) was inaccurate, *Typhlultimum* has subsequently been considered invalid and a synonym of *Typhlocoelum*.

Neivaia was erected by Travassos (1929) for adults with simple caeca lacking diverticula. These adults were found in the trachea of anatid birds from Brazil and had been described by Travassos (1921) as *Typhlocoelum neivai*. However, they do not readily accord with *Typhlocoelum* because the caeca are simple and lack diverticula and the vitelline fields are united posteriorly. Based on these and several other characters, Travassos (1929) erected *Neivaia* to accommodate this species. Meanwhile, it became known that adults with the same morphology, found in the same bird hosts in Brazil, had been deposited in the Museum of Natural History, Vienna (the Natterer Collection). Originally, these worms had been described by Diesing (1850) and named *Monostomum cymbium*. The same specimens were later re-examined by Monticelli (1892b), Kossack (1911) and Dubois (1959). Although well studied, the systematics of these digeneans is in dispute. Monticelli (1892b) suggested that *M. cymbium* is identical with *M. flavum* Mehlis, 1831. Stossich (1902) considered *M. flavum* a member of *Typhlocoelum*, but did not include *M. cymbium* in the same genus. Kossack (1911) tentatively listed *M. cymbium* in *Typhlocoelum*, but doubted this systematic position. Dubois (1959) noted that *M. cymbium* and *Neivaia neivai* (Travassos, 1921) are identical in both morphology and biology, and transferred the former to *Neivaia*. Nevertheless, Bashkirova (1950) transferred *N. neivai* to *Cyclocoelum* and *N. cymbium* to *Tracheophilus*. Yamaguti (1958a, 1971) synonymized *Neivaia* with *Cyclocoelum*. Joyeux & Baer (1927) considered *M. cymbium* and *Typhlocoelum sisowi* identical and used the older name, *T. cymbium* (Diesing, 1850). Travassos *et al.* (1969) and Feizullaev (1980) considered *Neivaia* a valid genus and placed it in the subfamily Typhlocoelinae. Specimens of *Neivaia* suggest that in its general morphology, except for the caeca, it is more similar and perhaps more closely related to the Typhlocoelidae than to any other group. If this is the case, the taxonomic significance of the caeca is less than previously believed. Based on this assumption, we place *Neivaia* in the family Typhlocoelidae, but in a separate subfamily, the Neivaiinae n. subfam.

Typhlophilus was erected by Lal (1936) for adult worms from the intestine of birds in India. The type-species, *T. shovellus* Lal, 1936, was considered to differ from *Typhlocoelum* in the position of the ventral sucker and the presence of uterine coils between the testes. Although Dollfus (1948) and Yamaguti (1958a) considered *Typhlophilus* to be valid, Bashkirova (1950) treated *T. shovellus* as a species of *Typhlocoelum* and Bykhovskaya-Pavlovskaya (1949), Dubois (1959) and Yamaguti (1971) synonymized *T. shovellus* with *Typhlocoelum cucumerinum* (Rudolphi, 1809). Consequently, *Typhlophilus* is now considered a synonym of *Typhlocoelum*.

Polycyclorchis eudocimi was described by Pence & Bush (1973) from the trachea of *Eudocimus albus* in the USA as a new genus and species. It differs from other cyclocoelids in having numerous testes that are scattered along the inner margin of the cyclocoel. In its general morphology, *Polycyclorchis* is similar to adult worms of the family Typhlocoelidae, except that, like *Neivaia*, it lacks diverticula along the inner margin of the intestinal wall. For this reason, Pence & Bush (1973) erected *Polycyclorchis* in the Cyclocoelidae, but did not link it with either the Typhlocoelidae or the Typhlocoelinae. Additionally, Feizullaev (1984) considered that *Polycyclorchis* could not be included in either the Cyclocoelinae or the Typhlocoelinae and erected the Polycyclorchiinae. In our opinion, the multiple testes found in *Polycyclorchis* are typical for the family Typhlocoelidae, as they occur in adults of *Typhlocoelum*, and the simple intestine without diverticula occurs in *Neivaia*. We have recognized the Polycyclorchiinae on the basis of the following characters: the multiple testes differentiate it from the Neivaiinae n. subfam. and the simple intestine from the Typhlocoelinae.

Family Typhlocoelidae Harrah, 1922

Diagnosis: Body middle-sized to large, spatulate, flattened, elliptical, convex dorsally and concave ventrally, with rounded extremities, spined. Oral sucker absent. Mouth subterminal. Prepharynx short. Pharynx well developed. Oesophagus short, rarely long, curved. Caeca long, united posteriorly forming cyclocoel, simple or possessing diverticula along inner wall. Ventral sucker absent or poorly developed. Testes oval or deeply lobed, numerous, irregularly scattered along inner margin of cyclocoel, or two, with posterior testis located just inside posterior caecal union and anterior testis diagonal, close to inner intestinal wall. Cirrus-sac at level of oesophagus, may or may not extend posteriorly to intestinal bifurcation. Seminal vesicle present. Genital pore pre- or postpharyngeal. Ovary oval or round, opposite anterior testis. Vitellarium follicular; fields caecal and extracaecal, from level of pharynx anteriorly to posterior end of body, may or may not be united in post-testicular region. Uterus intercaecal, between posterior testis and pharynx. Excretory vesicle small, with dorsoterminal pore. Excretory plexus extending into peripheral parenchyma. Tail-less cercaria (cercariaeum) develops in redia in freshwater and terrestrial snails. Metacercaria in redia along with cercariae, or encysts in snail tissue. Adults in trachea or occasionally in air-sacs or intestines of birds. Type-genus *Typhlocoelum* Stossich, 1902.

Key to subfamilies

1a. Caeca with diverticula along inner wall .... Typhlocoelinae Harrah, 1922.
1b. Caeca simple without diverticula .... 2.

2a. Testes two .... Neivaiinae n. subfam.
2b. Testes multiple with about 100 follicles ....
.... Polycyclorchiinae Feizullaev, 1984.

Subfamily Typhlocoelinae Harrah, 1922

(Syn. Typhlocoeliinae Witenberg, 1923)

Diagnosis: Body, stout, wider at anterior end than at posterior end. Ventral sucker absent or poorly developed. Caeca possess diverticula along inner wall. Testes two, oval or slightly lobed, or divided into about 20 testicular masses. Genital pore may be pharyngeal, or pre- or postpharyngeal. Vitelline fields ventral and lateral to caeca, may or may not be united posteriorly. Type-genus *Typhlocoelum* Stossich, 1902.

Key to genera

1a. Testes strongly lobed, forming network of *c.* 20 testicular masses ....
.... *Typhlocoelum* Stossich, 1902. (Fig. 22.1)
(Syns *Typhlultimum* Witenberg, 1923; *Typhlophilus* Lal, 1936)
Diagnosis: Ventral sucker poorly developed. Testes strongly lobed, forming network of *c.* 20 testicular masses. Genital pore pharyngeal or postpharyngeal. Vitelline fields not united posteriorly. Cosmopolitan. Type-species *T. cucumerinum* (Rudolphi, 1809) Stossich, 1902.
1b. Testes two .... 2.

2a. Testes rounded, oblique; ventral sucker absent; genital pore prepharyngeal .... *Tracheophilus* Skrjabin, 1913. (Fig. 22.2)
Diagnosis: Ventral sucker absent. Testes two, rounded, oblique. Genital pore prepharyngeal. Vitelline fields united posteriorly. Cosmopolitan. Type-species *T. sisowi* Skrjabin, 1913.
2b. Testes strongly or moderately lobed; ventral sucker present; genital pore pharyngeal .... *Manterocoelum* n. g. (Fig. 22.3)
Diagnosis: Ventral sucker present. Testes two, strongly or moderately lobed. Genital pore pharyngeal. Vitelline fields united posteriorly. Cosmopolitan. Type-species *M. americanum* (Manter & Williams, 1928) n. comb.

Subfamily Neivaiinae n. subfam.

Diagnosis: Body similar in width at both anterior and posterior ends. Ventral sucker absent. Caeca simple, without diverticula on inner wall. Testes two, slightly lobed or

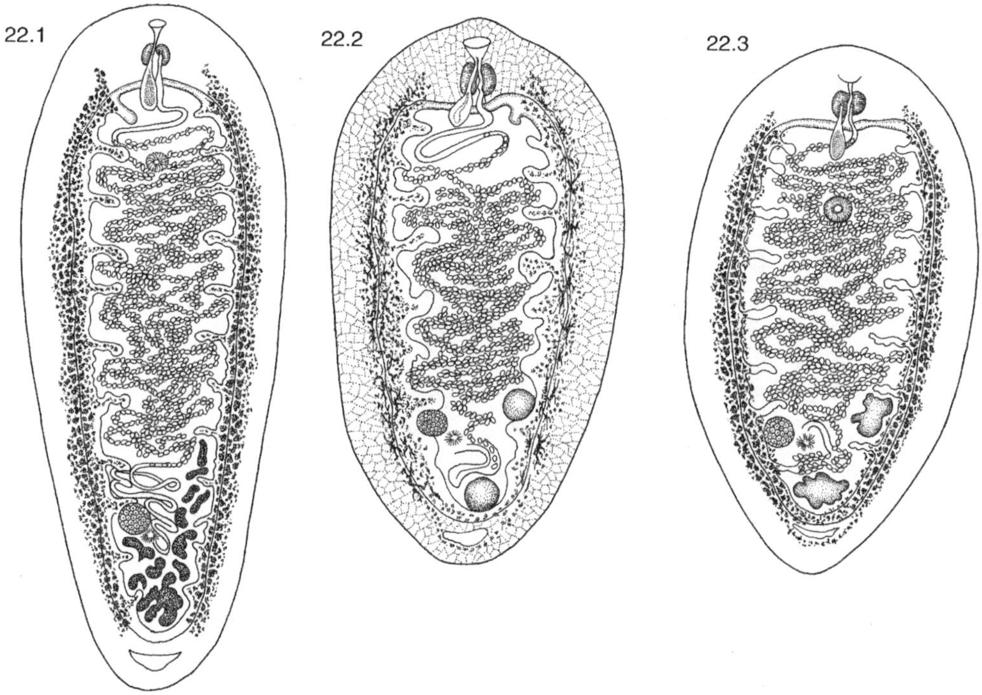

Figs 22.1–22.3.
22.1. *Typhlocoelum cucumerinum* (Rudolphi, 1809) Stossich, 1902. (Original.)
22.2. *Tracheophilus sisowi* Skrjabin, 1913. (Original.)
22.3. *Manterocoelum americanum* (Manter & Williams, 1928) n. comb. (Original.)

irregular in shape. Genital pore prepharyngeal. Vitelline fields lateral to caeca, united posteriorly. Type-genus *Neivaia* Travassos, 1929.

Genus *Neivaia* Travassos, 1929 (Fig. 22.4)

Diagnosis: With characters of subfamily. Neotropical. Type-species *N. cymbium* (Diesing, 1850) Dubois, 1959 (syn. *N. neivai* (Travassos, 1921)).

Subfamily Polycyclorchiinae Feizullaev, 1984

Diagnosis: Body narrower at anterior end than posterior end. Ventral sucker absent. Caeca simple, without diverticula on inner wall. Testes multiple, *c.* 100 in number, irregularly scattered along inner margin of cyclocoel from near posterior union to oesophageal region anteriorly. Genital pore prepharyngeal. Vitelline fields ventral and lateral to caeca, united posteriorly. Type-genus *Polycyclorchis* Pence & Bush, 1973.

Genus *Polycyclorchis* Pence & Bush, 1973 (Fig. 22.5)

Diagnosis: With characters of subfamily. Nearctic. Type-species *P. eudocimi* Pence & Bush, 1973.

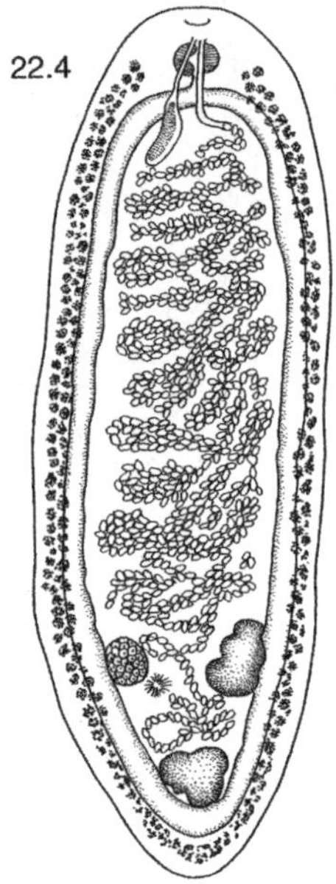

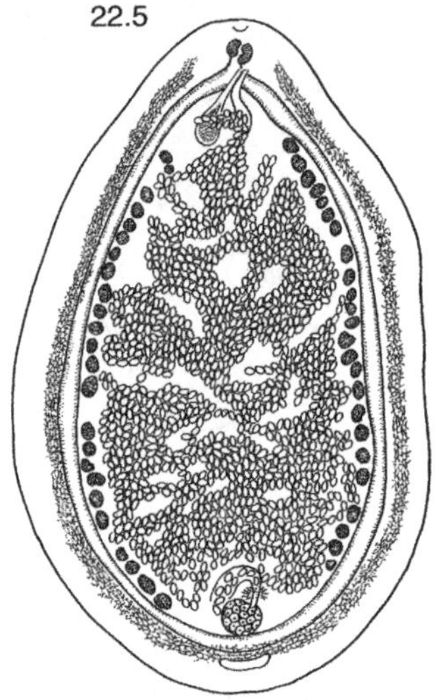

Figs 22.4–22.5.
22.4. *Neivaia cymbium* (Diesing, 1850) Dubois, 1959. (Original.)
22.5. *Polycyclorchis eudocimi* Pence & Bush, 1973. (Original.)

23 Superfamily Diplostomoidea Poirier, 1886

K. NIEWIADOMSKA

W. Stefański Institute of Parasitology, Polish Academy of Sciences, Twarda, Warsaw, Poland

Introduction

Members of the superfamily Diplostomoidea Poirier, 1886 are distinctly different from other groups of trematodes in possessing a unique holdfast organ. This sucker-like or bilobed structure, situated on the ventral surface of the body, posterior to the ventral sucker, plays both adhesive and digestive roles. Species belonging to the Diplostomoidea occur in mammals, birds and reptiles. This group was initially separated as the family Holostomidae by Blanchard (1847). Its rank was raised by Railliet (1919) to superfamily level, as the Strigeoidea, with the type-family Strigeidae (= Holostomidae) and the subfamilies Strigeinae, Polycotylinae, Cyathocotylinae, Alariinae and Braunininae. The basis for the systematics of the diplostomoids was the work of Poche (1926), who separated within his supersuperfamily Strigeida Poche, 1926 the family Strigeidae for all bisegmented species and Cyathocotylidae for non-segmented forms. The rank of the superfamily varied according to different authors: Dubois (1938, 1953) accepted the superfamily status, it was recognized as the suborder Strigeata La Rue, 1926 by La Rue (1926) and as the order Strigeidida La Rue, 1926 by Sudarikov (1959). La Rue (1957) reduced its rank to superfamily level as the Strigeoidea Railliet, 1819. Recently, Gibson (1996) has used the name Diplostomoidea Poirier, 1886 for the superfamily of this group, as this is the older family-group name (based on Diplostomidae Poirier, 1886). In contrast, Yamaguti (1958a, 1971) avoided the problem of the higher systematics of trematodes and, consequently, did not consider the rank of the diplostomatoids. Although it did not affect his systematic approach, for convenience he separated the digeneans according to their hosts (parasites of fish, amphibians, reptiles, birds and mammals). He also used both the morphology of adults and life-cycle patterns, when known, in his taxonomic arrangement of families and subfamilies.

At this time a fundamental series of works, those of Dubois (1938, 1953, 1968, 1970a, b, 1982, 1987, 1989), Yamaguti (1958a, 1971) and Sudarikov (1959, 1960a, b,

1961, 1997) presented versions of the systematics of the Diplostomoidea; these systems differ in both the number of taxa included and their synonymy.

The general division of the Diplostomoidea is based on the host-specificity of the adults and such characters of their morphology as the structure and shape of the forebody[1] and holdfast organ, the distribution of the vitellarium (in whole body, in forebody or hindbody), the presence or absence of bisegmentation of the body, the presence or absence of the cirrus-sac or paraprostate, and the structure of the copulatory apparatus (Figs 23.1–23.8). Life-cycle arrangement and morphological characters of the developmental stages have only occasionally been included.

The Diplostomoidea, except for the family Cyathocotylidae, have the cirrus-sac and cirrus replaced by an atypical copulatory apparatus and terminal genitalia. The posterior end of the body is occupied by the copulatory bursa with a large or small opening, with or without a genital cone or with a genital bulb. The seminal vesicle leads into the ejaculatory duct, which may or may not unite with the terminal portion of the uterus to form a hermaphroditic duct that opens through the wall of the copulatory bursa, at the tip of the genital cone or on the apex of a small papilla dorsally to the genital bulb. The uterus and ejaculatory duct may also have separate pores. Some other structures, such as a circular muscle ring in the copulatory bursal walls, sucker-like structures, a preputial fold around the genital cone, a paraprostate (a structure independent of the usual pars prostatica), an ejaculatory pouch (a thick-walled part of the ejaculatory duct) and others, may be present and modified in different taxonomic groups.

The Diplostomoidea contain six families. Two of them, the Brauninidae Bosma, 1931 and Bolbocephalodidae Strand, 1935, are monotypic, but their morphological characters warrant an independent status. Nothing is known of their life-cycles or of the morphology of the cercariae and metacercariae in these families. Four other families, the Cyathocotylidae Mühling, 1898, Strigeidae Railliet, 1919, Diplostomidae Poirier, 1886 and Proterodiplostomidae Dubois, 1936, are rich in species.

For many species belonging to the families Cyathocotylidae, Strigeidae and Diplostomidae, the life-cycles and the morphology of the cercariae and metacercariae are well known. Dubois (1968, 1970a) in his 'Synopsis des Strigeidae et des Diplostomatidae', included a lot of such information on the life-cycles and morphology of the cercariae and metacercariae (and mesocercariae when present). These data, however, did not influence the systematics of these families, being somewhat subordinated to the concept of polytypic genera and species characterized by wide geographical distribution. This concept does not reflect the proper relationship between the different morphological units at the specific and generic levels.

The morphology of the metacercariae is similar in species of related genera, and for this reason metacercarial forms were distinguished by generic names: *Tetracotyle* De Filippi, 1854, *Diplostomulum* Brandes, 1892, *Neascus* Hughes, 1927, *Prohemistomulum* Ciurea, 1933 and *Neodiplostomulum* Dubois, 1938. The main premise for such discrimination was the structure of the reserve bladder. The simplest structure is characteristic for *Diplostomulum* metacercariae: three longitudinal canals (two lateral with ramifications directed posteriorly and one median) connected anteriorly (posterior to the

[1] The fore- and hindbody in diplostomatoids differ from those generally accepted in Digenea. They are distinguished by a more or less distinct constriction; the folliform or cup-shaped forebody with oral and ventral suckers and holdfast organ plays an adhesive role, while the cylindrical or conical hindbody contains the reproductive system (see Figs 23.2–23.4).

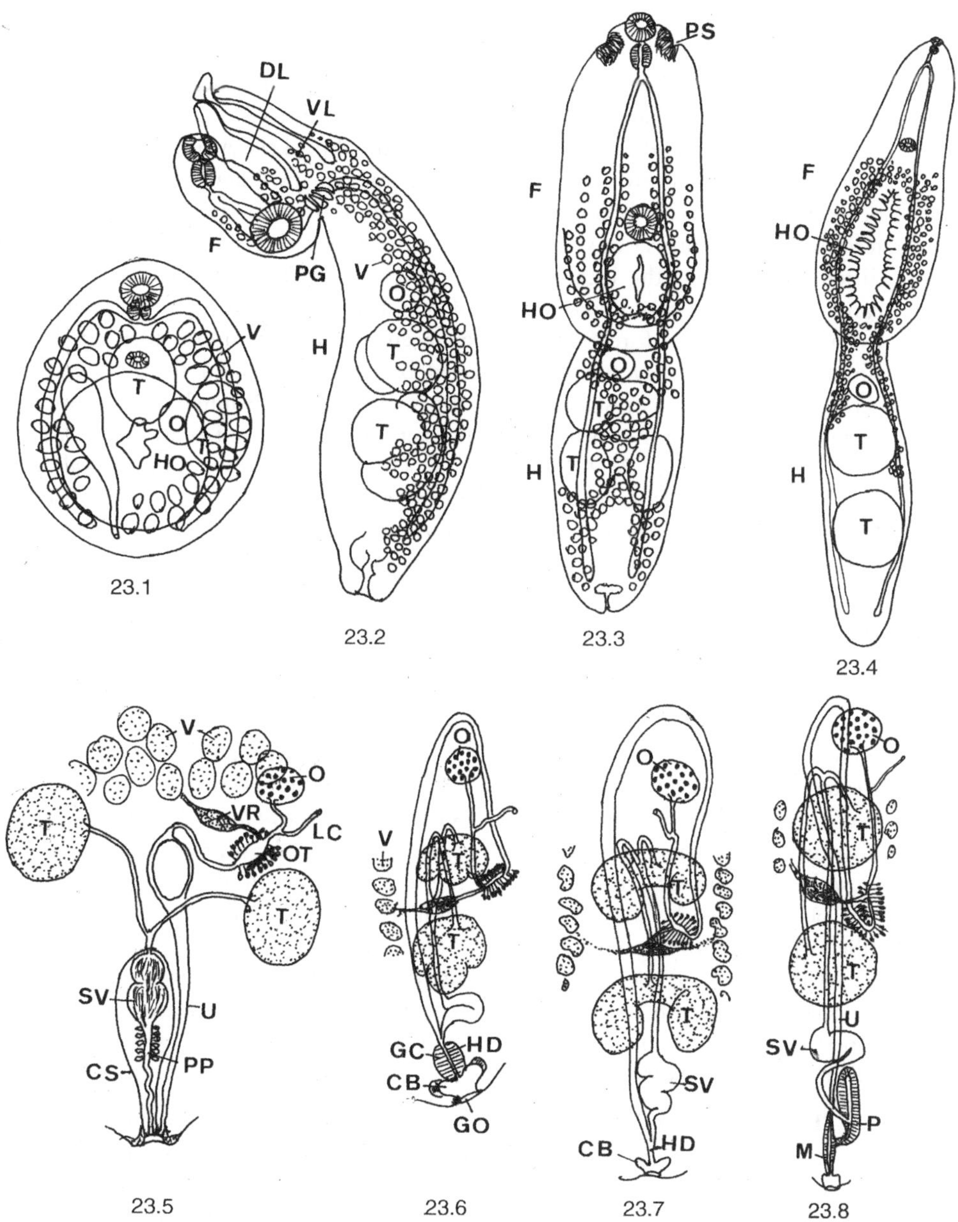

Figs 23.1–23.8. General morphology (23.1–23.4) and structure of genital system (23.5–23.8) of main groups of diplostomoids. 23.1, 23.5. Cyathocotylidae. 23.2, 23.6. Strigeidae. 23.3, 23.7. Diplostomidae. 23.4, 23.8. Proterodiplostomidae. CB, copulatory bursa; CS, cirrus-sac; DL, dorsal lobe of holdfast organ; F, forebody; GC, genital cone; GO, genital opening (pore); H, hindbody; HD, hermaphroditic duct; HO, holdfast organ; LC, Laurer's canal; M, metraterm; O, ovary; OT, oötype with Mehlis' gland; P, paraprostate; PG, proteolytic gland; PP, pars prostatica; PS, pseudosucker; SV, seminal vesicle; T, testis; U, uterus; V, vitellarium; VL, ventral lobe of holdfast organ; VR, vitelline reservoir. (Original.)

pharynx) and posteriorly (anterior to the ventral sucker) with a system of ramifications at the end of which are enlarged pockets containing round or oval excretory bodies. A more complicated structure of the reserve bladder indicates a *Neodiplostomulum* metacercaria, which has two or three additional commissures connecting the lateral and medial canals, and excretory bodies in terminal pockets. A similar reserve bladder occurs in the subfamily Alariinae, but in some genera there are also commissures between the posterior branches; excretory bodies are also situated in pockets that terminate the ramifications. The *Neascus* represents a different structure. The main difference concerns the localization of the excretory bodies, which are free in the canals of a reserve bladder composed of one median and two ramified lateral canals forming a net in the forebody of the metacercaria. The main excretory canals in the hindbody may or may not be connected by commissures. In the *Tetracotyle* metacercaria, which has a cup-shaped body, the reserve bladder is much more complicated, forming a net filling the entire body; the excretory bodies are free in the canals of the reserve bladder (Figs 23.9–23.13). Szidat (1969) added two more larval genera, *Tylodelphylus* and *Tetracotyloides*, basing them only on morphological differences between *Tylodelphys* Diesing, 1850 and *Tetracotyle* De Filippi, 1854, respectively. He did not take into account the structure of the reserve bladder.

The reserve bladder develops during the growth of the metacercaria, starting from commissures or small outgrowths of the main excretory tubes in the excretory system of the cercaria. The only exception is the *Prohemistomulum* (Fig. 23.14). The reserve bladder in cyathocotylid metacercariae is a part of the excretory system of the cercaria. Four main excretory tubes, arising from the excretory vesicle, form two loops connected anteriorly; the median tubes unite a small distance anteriorly to the ventral sucker. In some cyathocotylid species the excretory bodies are present in the main excretory tubes of the cercaria, as, for example, in *Cercaria laevissimus* Nasir, Acuna & Guevarra, 1966. There are also species (e.g. *C. monogasica* Nasir, Hamana & Diaz, 1969) in which the main excretory tubes give rise to a system of blindly ending ramifications just posterior to the intestinal bifurcation. Both main tubes and ramifications are filled with excretory bodies.

The forms of the metacercaria listed above vary in their morphology and especially in relation to their excretory system. They can be reduced to four main types: 'diplostomulum', 'neascus', 'prohemistomulum' and 'tetracotyle'. The diagnoses of these types are as follows.

- **'Diplostomulum':** Free and without cyst of parasite origin, or encapsulated with or without cyst of parasite origin. Forebody round, oval or elongate, slightly concave ventrally; hindbody small, conical. Pseudosuckers present or absent. Reserve bladder of three longitudinal canals (one median and two lateral with posterior branches) connected by anterior and posterior commissures only or with additional commissures, with system of ramifications with terminal enlarged pockets filled by round or oval excretory bodies. In fishes, amphibians, reptiles and mammals. Adult genera include *Diplostomum*, *Neodiplostomum*, *Alaria.*
- **'Neascus':** Encapsulated, with, or rarely without, cyst wall of parasite origin; cyst may be loosely fitting. Clusters of cysts (syncysts) may be formed. Forebody foliaceous or oval; hindbody well developed. Pseudosuckers present or absent. Reserve bladder composed of ramified median and two lateral canals forming net

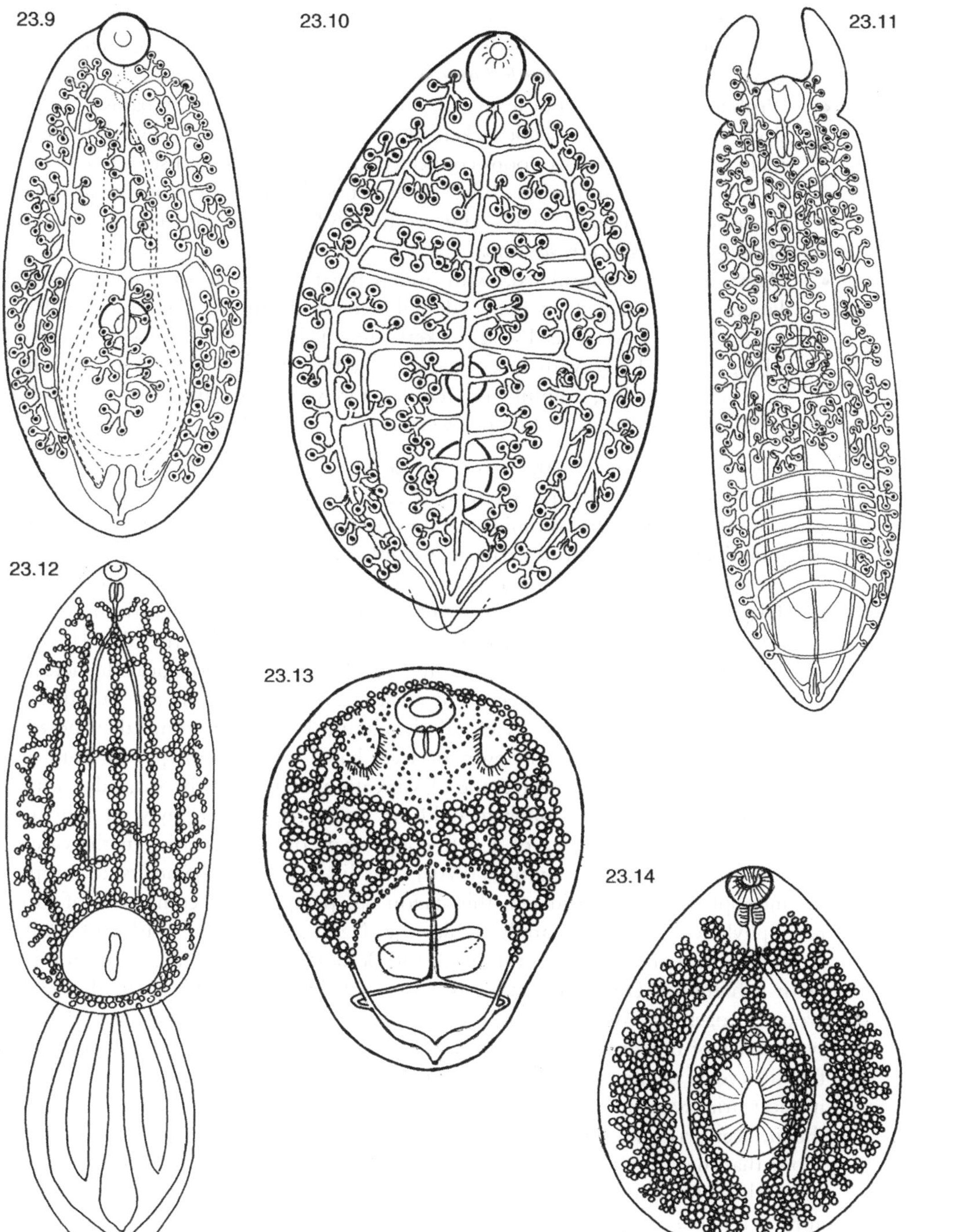

Figs 23.9–23.14. Metacercarial types in diplostomoids. 23.9–23.11. 'Diplostomulum' (23.9. *Diplostomulum*; 23.10. *Neodiplostomulum* (after Odening, 1965); 23.11. *Alaria* (after Pearson, 1956)). 23.12. 'Neascus'. 23.13. 'Tetracotyle'. 23.14. 'Prohemistomulum'. (Figs 23.9, 23.12–23.14: original.)

in forebody; in hindbody, median excretory canals may or may not be connected by commissures; excretory bodies free in canals. In fishes and oligochaetes. Adult genera include *Uvulifer*, *Posthodiplostomum*, *Bolbophorus*.

- **'Prohemistomulum':** Encapsulated, with thick-walled cysts of parasite origin. Body round or oval. Pseudosuckers absent. Reserve bladder composed of four main canals forming two loops connected anteriorly to ventral sucker; system of blindly-ending ramifications filled by excretory bodies arise from main canals. In fishes, amphibians, snails and leeches. Adult genera include *Cyathocotyle*, *Holostephanus*, *Paracoenogonimus*.
- **'Tetracotyle':** Encapsulated, with well-defined cyst wall of parasite origin. Forebody cup-shaped or flattened and concave ventrally; hindbody small or relatively well developed. Pseudosuckers present. Reserve bladder forms network filling entire body with free excretory bodies in canals. In snails, leeches, oligochaetes, fishes, amphibians, reptiles, rarely in birds and mammals. Adult genera include *Strigea*, *Cotylurus*, *Apatemon*.

One metacercarial type is characteristic for each of the two families, i.e. the Strigeidae and Cyathocotylidae with the 'tetracotyle' and 'prohemistomulum', respectively. In contrast, the family Diplostomidae embraces two types of metacercariae ('diplostomulum' and 'neascus'). Information available on the structure of the metacercariae of the family Proterodiplostomidae is very poor and lacks data on the reserve bladder.

Diagnoses of some taxa of the Diplostomoidea reflect the significance of the metacercarial type in relation to the systematics of this superfamily. This was confirmed by Shoop (1989), who took into consideration morphological characters of 27 species of adults and seven species of metacercariae representing two families, the Diplostomidae and Strigeidae. As a result, he obtained five sister groups that showed a remarkable concordance of phylogenetic trees based on morphotypes of adult and metacercarial stages. He accepted the classification of the family Strigeidae with a cup-shaped forebody and bilobed holdfast organ, the monophyly of which is supported by the occurrence of a 'tetracotyle' type of metacercaria. In the case of the Diplostomidae *sensu* Dubois (1970a), he did not find characters to support its monophyly. He considered this taxon as paraphyletic, composed of three monophyletic groups of family rank: the Neodiplostomidae (with subfamilies Crassiphialinae and Neodiplostominae), the Bolbophoridae and the Diplostomidae (with subfamilies Diplostominae and Alariinae). In this classification, the genera grouped together have a common metacercarial type. It seems, however, that the systematics of this group will be more readily discernible if more abundant data on the life-cycles of the various genera within the groups can be accumulated, the metacercarial types more precisely defined and the characters of the morphology of the metacercariae and cercariae also taken into consideration. Shoop's analysis clearly shows that the current systematics of the superfamily Diplostomoidea does not reflect the real evolutionary relationships and needs reconstructing according to a new approach based on an accumulation of data from various sources.

At higher taxonomic levels, the value of host-specificity to the main vertebrate groups appears to be overestimated. Such host groups, as distributed among the Diplostomidae, should be reflected at subfamily level (Diplostominae, Crassiphialinae and Alariinae).

At lower taxonomic levels, the genera are well discriminated based on the morphology of the cercaria, especially the number and position of the penetration glands and the flame-cell formula (Ginetsinskaya & Dobrovolskij, 1968; Niewiadomska, 1970, 1971, 1973). This corresponds to many of Dubois' subgenera, which should be raised to generic level. Consequently, the genus should be defined as a group of species having a similar morphology of the cercaria, metacercaria and adult (see Niewiadomska, 1973). Such a concept of the genus is applied in the present work.

Figures showing the morphology of representatives of genera have been redrawn and simplified, mainly after the monographs of Dubois (1938, 1968, 1970a) but also from other authors, as indicated.

Superfamily Diplostomoidea Poirier, 1886

(Syn. Strigeoidea Railliet, 1919)

Diagnosis: Characterized by presence of holdfast organ posterior to ventral sucker, terminal or subterminal position of genital opening at posterior end of body and absence of seminal receptacle. Oral and ventral sucker present; prepharynx present or absent. Pharynx and oesophagus present; caeca long, blind, reaching end of body or not. Life-cycle with generations of mother- and daughter-sporocysts; cercaria longifurcate furcocercous; metacercaria with reserve bladder filled with excretory bodies; mesocercaria may occur in life-cycle of some small groups. In intestine of reptiles, birds and mammals. Type-genus *Diplostomum* Nordmann, 1832.

Key to families

1a. Body massive with small caudal appendage, not divided into fore- and hindbody; pseudosuckers absent; genital complex in middle or in posterior part of body; cirrus-sac present .... 2.

1b. Body divided into fore- and hindbody; pseudosuckers may be present; genital complex situated in hindbody; cirrus-sac absent .... 3.

2a. Body pyriform, oval or linguiform; holdfast organ round or oval, large or small; gonads and uterus outside holdfast organ .... Cyathocotylidae Mühling, 1898.

2b. Body cordiform, with ventral concavity enclosing holdfast organ; gonads and loops of uterus inside large cordiform holdfast organ .... Brauninidae Wolffhügel, 1903.

3a. Forebody tubular, cup-shaped or bulbiform, enclosing bilobed holdfast organ; hindbody usually cylindrical; pseudosuckers present .... Strigeidae Railliet, 1919.

3b. Forebody flattened, foliate, calyciform or bulbous, with or without pseudosuckers; hindbody long or short; holdfast organ round or oval .... 4.

4a. Forebody bulbous with cup-shaped thickening at base; hindbody conical; holdfast organ in form of two transverse lips .... Bolbocephalodidae Strand, 1935.

4b. Forebody flattened or calyciform; hindbody cylindrical or conical; holdfast organ round or oval .... 5.

5a. Forebody foliate, spatulate or calyciform; holdfast organ round or oval with cavity opening via ventral slit; paraprostate absent .... Diplostomidae Poirier, 1886.

5b. Forebody flattened; holdfast organ round or elongate, with cavity opening by median slit, with or without papillae; paraprostate present .... Proterodiplostomidae Dubois, 1936.

24 Family Diplostomidae Poirier, 1886

K. NIEWIADOMSKA

W. Stefański Institute of Parasitology, Polish Academy of Sciences, Twarda, Warsaw, Poland

Introduction

The family Diplostomidae Poirier, 1886 comprises parasites of numerous orders of birds and mammals. Generally, it was accepted that genera belonging to this family were grouped into subfamilies according to host: parasites of mammals, the Alariinae; and parasites of birds, the Diplostominae. Then, among the Diplostominae, the tribes Diplostomini, Crassiphialini and Codonocephalini were distinguished based on morphological characters (Dubois, 1938, 1953, 1970a). A similar classification, but using higher ranks (superfamily, families and subfamilies), was proposed by Sudarikov (1960a). Dubois (1970a) drew attention to the 'neascus' type of metacercariae as characteristic of the Crassiphialini (or Crassiphialinae). In fact, the species having this type of metacercariae also have cercariae of a similar morphology, i.e. the number and position of the penetration glands and the flame-cell formula (when the life-cycles are known). But metacercariae of the 'neascus' type are also characteristic for *Ornithodiplostomum* Dubois, 1936, *Posthodiplostomum* Dubois, 1936, *Posthodiplostomoides* Williams, 1969, *Bolbophorus* Dubois, 1935 and *Conodiplostomum* Dubois, 1937. These genera have been classified (Sudarikov, 1960a; Dubois, 1970a; Yamaguti, 1971) in the Diplostominae or Diplostomini. Gibson (1996) recognized the subfamily status of the Crassiphialinae as distinct from the Diplostominae and not a subdivision of it, as was accepted by the above authors. He also transferred *Ornithodiplostomum* and *Posthodiplostomum* to the Crassiphialinae due to their 'neascus' type of metacercariae. Such an enlarged concept of the Crassiphialinae makes the classification of the genera at the adult stage very difficult. Following Sudarikov (1960a), Dubois (1970a) and Yamaguti (1971), the main character which distinguishes the Diplostominae and Crassiphialinae is the distribution of the vitellarium: in the fore- and hindbody, or within (or almost entirely within) the hindbody, respectively. Some details of the structure of the holdfast organ and shape of the testes were also utilized. However, the five genera included, with metacercariae of the 'neascus' type, have their vitellarium in the fore- and hindbody, and the structure of the holdfast organ is characteristic for the

Diplostominae. This makes separation of these subfamilies less distinct at the adult stage. In general, I accept the division of the Diplostomidae, according to host-specificity and metacercarial types, into four subfamilies: Diplostominae, Crassiphialinae, Alariinae and Codonocephalinae. The genera with known 'neascus' metacercariae have been added to the existing subfamily Crassiphialinae. The only exclusion is *Mesoophorodiplostomum,* with an unknown metacercarial type but which is morphologically very close to *Posthodiplostomum*. The main difference is the inter-testicular position of the ovary. An appropriate key for the subfamilies is proposed.

The monotypic subfamily Codonocephalinae Sudarikov, 1959 was separated off by Sudarikov (1959) from the Strigeidae. A life-cycle study (Niewiadomska, 1964) has shown, however, that the morphology of the early and progenetic metacercaria did not correspond with the characters of this family. Dubois (1970a) included this species in the subfamily Diplostominae with the rank of tribe, but I prefer to recognize this taxon as a subfamily, with equivalent rank to the Diplostominae, Crassiphialinae and Alariinae.

The status of *Neodiplostomum* Railliet, 1819, *Fibricola* Dubois, 1932 and *Conodiplostomum* Dubois, 1937 represents a special problem. Having a 'neascus' type of metacercaria (see *C. perlatum* Ciurea, 1911), *Conodiplostomum* should be raised to full generic status, as Sudarikov (1971) proposed in his work on strigeoid metacercariae.

Neodiplostomum and *Fibricola* were distinguished by Dubois (1970a) on the basis of host-specificity and distribution of the vitellarium and placed into two subfamilies, Diplostominae and Alariinae, respectively. However, the difference based on distribution of the vitellarium is not clear. *Neodiplostomum* has the vitellarium in both parts of the body, while *Fibricola* represents a series of species from those having the vitellarium accumulated in the forebody to those with the vitellarium in both parts. This was comprehensively analysed earlier by Pearson (1959), who united these genera and *Conodiplostomum* as subgenera of *Neodiplostomum.* Dubois & Angel (1972) regarded the mammalian species with the vitellarium distributed as in *Neodiplostomum* as belonging to the subgenus *Triloborchidiplostomum* Dubois & Angel, 1972. This controversial relationship between *Neodiplostomum* and *Fibricola* was reconsidered by Cribb & Pearson (1993). They compared their own material of diplostomids from Australian mammals with representatives from birds and showed that the distribution of the vitellarium represented various patterns from extensive (typical for bird parasites) to that confined to the forebody. On the other hand, adaptation to the host groups does not appear to warrant a generic distinction, as the infection of birds by three species of *Fibricola* has been proved experimentally by Leigh (1954), Ulmer (1955) and Seo (1989). In conclusion, Cribb & Pearson (1993) accepted a conservative approach and placed Australian species from mammals (previously included in *Fibricola*) into the genus *Neodiplostomum* without any subgenera. However, besides these experimental data, evidence from birds naturally infected by *Fibricola* is lacking. Resolution of the real relationship between these genera needs to be supported by both morphological and molecular evidence. Despite these difficulties, I retain the present generic and subfamily status of *Neodiplostomum* and *Fibricola*.

Procrassiphiala Verma, 1936, created for two Indian species, was originally defined by a very short diagnosis. Dubois (1938) considered it insufficient to justify its validity, but placed it among the Diplostomini. Later (Dubois, 1953), he recognized *Procrassiphiala* as a *genus inquirendum* and (Dubois, 1970a) included in it one more species, *Neodiplostomum fungiloides* Semenov 1927, but excluded *P. cuckooi* Verma, 1936, considering this a '*nomen delendum*'. Yamaguti (1971) left this genus in the Diplostomini. Lack of new data on this genus has inclined me to accept the status of *genus inquirendum*.

Yamaguti (1971) created the genus *Pseudocercocotyla* Yamaguti, 1971 for the type- and only species, *P. rudis* (R. Gupta, 1963). He separated this species from *Cercocotyla* Yamaguti, 1939, based mainly on the size and shape of the hindbody, which is stout in *Pseudocercocotyla* and long, slender and posteriorly enlarged in *Cercocotyla*. The other characters given in the diagnoses are the same. The illustrations of Yamaguti (1971) (figs 1431 and 1436, respectively) clearly showed that the specimen of *P. rudis* is more constricted. In my opinion, the differences between these genera (in fact between two species) are of significance at the specific level only and *Pseudocercocotyla* should be recognized as a synonym of *Cercocotyla*.

Yamaguti (1971) also mentioned the generic name *Prosthodiplostomum* Kozicka, 1958 (footnote on p. 648). This appears to be a *lapsus*, as no such name was mentioned in any paper by Kozicka either in 1958 or subsequently.

Paratylodelphus Saxena, 1952 was created by Saxena (1952), but the description is very superficial and inadequate. This generic name should be considered a *nomen nudum*.

The status of *Neoparadiplostomum* Bisseru, 1957 as a synonym of *Fibricola* Dubois, 1932 was finally established by Dubois (1981). He transferred the type-species, *N. magnitesticulatum* Bisseru, 1957, from *Neodiplostomum* Railliet, 1919 (Diplostominae) to *Fibricola* (Alariinae), based on morphological characters. He interpreted the occurrence of the aberrant host for this family (crocodiles) as a result of feeding on infected murids.

Proeputiodiplostomum Gupta, 1962 is an invalid name. Gupta (1962) erected this genus for an undescribed species, *P. indicum* Gupta, 1962, which is a *nomen nudum*.

Family Diplostomidae Poirier, 1886

(Syn. Alariidae Tubangui, 1922)

Diagnosis: Body composed of two parts: forebody foliate, spatulate or calyciform; hindbody cylindrical or coniform. Forebody with or without pseudosuckers (lappets); holdfast organ massive, round or elliptical, with compact gland at base. Oral sucker and ventral sucker present. Pharynx present; oesophagus short; caeca long, usually terminating blindly close to posterior extremity. Testes two, symmetrical, asymmetrical or oval, arranged tandem, occasionally diagonal or juxtaposed, in hindbody. Ovary pretesticular; Laurer's canal present. Vitellarium follicular, distributed in both parts of body or mainly in fore- or hindbody. No cirrus-sac. Seminal vesicle post-testicular. Copulatory bursa usually present. Genital pore dorsoterminal or dorso-subterminal. Parasites of birds and mammals. Cercaria furcocercous; metacercaria in fish and amphibians, occasionally in mammals. Mesocercaria may be included in some life-cycles. Type-genus *Diplostomum* Nordmann, 1832.

Key to subfamilies

1a. Parasites of mammals .... Alariinae Hall & Wigdor, 1918.
1b. Parasites of birds .... 2.

2a. Forebody infundibular, with large proteolytic gland and indistinct holdfast organ; hindbody much longer than forebody; metacercaria progenetic with fully-developed genital system .... Codonocephalinae Sudarikov, 1959.
2b. Forebody more or less spatulate; holdfast organ sucker-like with median slit; proteolytic gland in posterior part of holdfast organ; hindbody from shorter to longer than forebody; metacercaria with undeveloped genital system .... 3.

3a. Vitellarium in hindbody, occasionally also in forebody; copulatory bursa may be protrusible (metacercaria a 'neascus')[1] .... Crassiphialinae Sudarikov, 1960.
3b. Vitellarium in fore- and hindbody; copulatory bursa not protrusible (metacercaria a 'diplostomulum')[1] .... Diplostominae Poirier, 1886.

Subfamily Diplostominae Poirier, 1886

Diagnosis: Body more or less distinctly bipartite, hindbody from shorter or similar in length to much longer than forebody. Pseudosuckers present or absent. Vitellarium in fore- and hindbody. Genital cone present or absent. Copulatory bursa not protrusible. Parasites of birds. Metacercaria of 'diplostomulum' type. Type-genus *Diplostomum* Nordmann, 1832.

Key to genera

1a. Pseudosuckers or homologous structures present .... 2.
1b. Pseudosuckers or homologous structures absent .... 13.

2a. Body completely retroflexed .... *Harvardia* Baer, 1932. (Fig. 24.1)
Diagnosis: Body bipartite; forebody strongly concave; hindbody cylindrical or bluntly conical, reflexed, with its posterior end close to anterior extremity of forebody. Two mobile anterior lobes or pseudosuckers lateral to oral sucker. Holdfast organ eversible, fungiform, with central cavity. Oral sucker and pharynx well developed; ventral sucker large, in middle of forebody. Testes tandem, in inverted horseshoe; posterior testis larger than anterior. Ovary elliptical, anteroventral to anterior testis. Laurer's canal very long, opening at angle formed by dorsal flexion of forebody. Vitellarium reaches level of ventral sucker. Copulatory bursa very deep, cylindrical, with muscular wall. Uterus and ejaculatory ducts unite at right-angle to form hermaphroditic duct at base of bursa. In *Phalacrocorax*. Africa, India. Type-species *H. sandgroundi* Baer, 1932.

1 Data included as these larval features have been used as important distinguishing criteria.

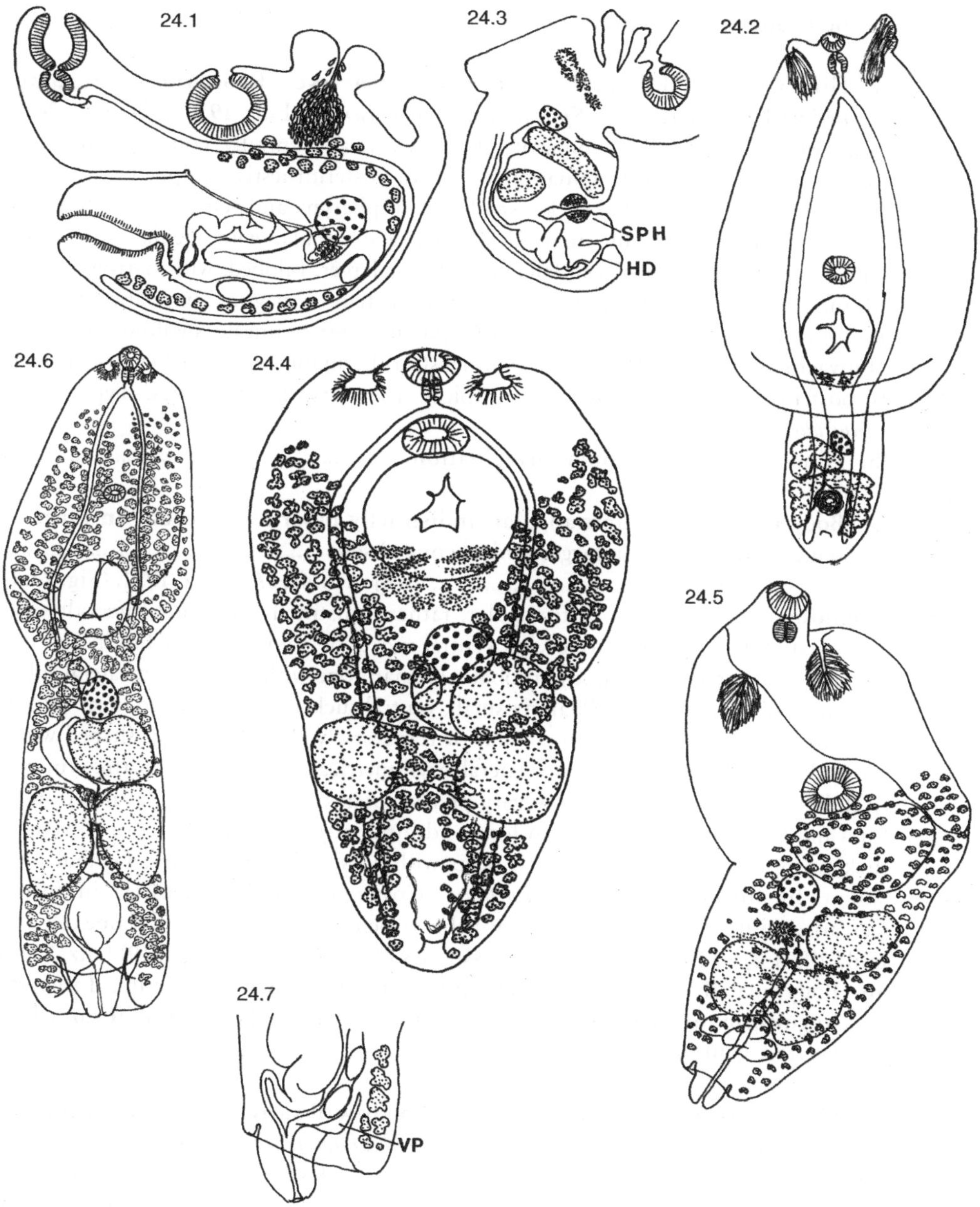

Figs 24.1–24.7.

24.1. *Harvardia sandgroundi* Baer, 1932. Scheme of sagittal section with incomplete genital system (without testes). (After Dubois, 1970a.)

24.2, 24.3. *Sphincterodiplostomum musculosum* Dubois, 1936. 24.2. Entire worm. 24.3. Scheme of sagittal section showing dorsal sphincter (SPH) and terminal genitalia. HD, hemaphroditic duct. (After Dubois, 1970a.)

24.4. *Hysteromorpha triloba* (Rudolphi, 1819). (Original.)

24.5. *Neoharvardia pandubi* Gupta, 1963. (After Dubois, 1970a.)

24.6, 24.7. *Adenodiplostomum triangulare* (Johnston, 1904). 24.6. Entire worm. 24.7. Terminal genitalia showing ventral pad (VP). (After Dubois, 1970a.)

2b. Body not completely retroflexed .... 3.

3a. Hindbody with dorsal tubular invagination at level of posterior testis, equipped with sphincter ..... ***Sphincterodiplostomum*** **Dubois, 1936. (Figs 24.2, 24.3)**
Diagnosis: Body distinctly bipartite; forebody oval, ventrally concave; hindbody ovoid, with deep tubular invagination at level of posterior testis, equipped with globular sphincter in middle. Oral sucker smaller than pharynx and ventral sucker. Pseudosuckers well developed, muscular. Holdfast organ circular with median slit. Testes tandem; anterior asymmetrical, lateral; posterior larger, bilobed, with narrow ventral isthmus. Ovary ellipsoidal, pretesticular, oval, median or diagonal in relation to anterior testis. Vitellarium ? (not examined). Copulatory bursa cuneate, shallower than dorsal invagination, with subterminal opening; short hermaphroditic duct enters through its posterior wall. In *Agamia*. Neotropical. Metacercariae in fishes. Type-species *S. musculosum* Dubois, 1936.
3b. Hindbody without tubular invagination .... 4.

4a. Body pyriform or subtriangular, indistinctly bipartite; forebody in form of pouch; holdfast organ large; genital cone absent .... ***Hysteromorpha*** **Lutz, 1931. (Fig. 24.4)**
Diagnosis: Body pyriform or triangular, rather indistinctly bipartite; forebody oval with greatest diameter close to level of ventral sucker or holdfast organ; hindbody conical. Anterior extremity of body trilobate, median lobe with oral sucker, lateral lobes separated by pseudosuckers. Pharynx small; ventral sucker well developed, near intestinal bifurcation. Holdfast organ circular or transversely elongate, funnel-shaped when protruded. Testes tandem; anterior lateral, asymmetrical; posterior claviform, symmetrical, bilobed, concave ventrally. Ovary pretesticular, oval, submedian, at junction of fore- and hindbody. Vitellarium in fore- and hindbody, beginning anterior to ventral sucker. Copulatory bursa moderately deep, with dorso-subterminal pore. Genital cone absent. In ardeids, cormorants, *Burhinus*. Cosmopolitan. Metacercariae in fishes. Cercariae with two pre- and two paracetabular penetration glands and flame-cell formula $2[(2+2)+(2+2+[2])]=20$. Type-species *H. triloba* (Rudolphi, 1819) Lutz, 1931.
4b. Body usually distinctly or exceptionally indistinctly bipartite; forebody linguiform or spatulate; holdfast organ small to large, fungiform when everted; genital cone present .... 5.

5a. Copulatory bursa with circular fold of body-wall surrounding genital cone in form of prepuce .... ***Neoharvardia*** **Gupta, 1963. (Fig. 24.5)**
Diagnosis: Body distinctly bipartite, strongly retroflexed. Pseudosuckers muscular, cup-shaped. Fore- and hindbody of equal length. Forebody with deep concavity enclosing holdfast organ; hindbody conical. Holdfast organ massive, without gland-cells in its wall. Testes tandem; anterior asymmetrical, elliptical, submedian; posterior bilobed, transversely enlarged. Ovary oval, pretesticular, situated at level of anterior margin of anterior testis, near anteriorly margin of hindbody. Vitellarium in fore- and hindbody, not extending anteriorly to ventral sucker. Copulatory bursa with terminal pore and protrusible genital cone, surrounded at base by circular fold of body-wall (prepuce). Hermaphroditic duct passes through genital cone and opens terminally.

Uterus unites with ejaculatory duct at base of genital cone. In cormorants. India. Type-species *N. pandubi* Gupta, 1963.

5b. Bursa without circular fold of body-wall around base of genital cone .... 6.

6a. Large unicellular gland-cells with ducts opening on ventral surface grouped around ventral sucker; genital cone with semicircular ventral pad ....
.. *Adenodiplostomum* Dubois, 1937. (Figs 24.6, 24.7)

Diagnosis: Body distinctly bipartite; ventral sucker surrounded by large unicellular gland-cells opening on to ventral surface. Holdfast organ rather small, with central cavity. Ovary ellipsoidal, pretesticular, near anterior margin of hindbody. Testes tandem, large; anterior asymmetrical, lateral; posterior bilobed. Vitellarium extending from close to level of intestinal bifurcation to extremity of hindbody, partly in strand ventral to testes. Copulatory bursa large, dorso-subterminal; genital cone with semicircular ventral pad projecting from dorsal region of anterior wall of bursa, encloses hermaphroditic duct. In alcedines. Australia. Type-species *A. triangulare* (Johnston, 1904) Hughes, 1929.

6b. No large unicellular glands around ventral sucker, body distinctly or indistinctly bipartite; genital cone present or absent .... 7.

7a. Ventral sucker reduced or absent; vitellarium concentrated around holdfast organ .... 8.

7b. Ventral sucker present; vitellarium in both parts of body .... 10.

8a. Vitellarium follicular; copulatory bursa very small ....
... *Austrodiplostomum* Szidat & Nani, 1951. (Fig. 24.8)

Diagnosis: Body linguiform, indistinctly bipartite; hindbody small, conical. Pseudosuckers present. Oral sucker and pharynx well developed; ventral sucker rudimentary or absent. Holdfast organ oval with median slit. Testes tandem; anterior testis asymmetrical in shape, smaller than symmetrical posterior testis. Ovary round, small, pretesticular, anterior to or opposite anterior testis. Vitellarium extending into both body regions, concentrated around holdfast organ. Copulatory bursa very small, with terminal pore; genital cone absent. In cormorants. South America. Metacercariae of 'diplostomulum' type, in fishes. Cercariae with two pairs of pre- and paracetabular penetration glands, flame-cell formula $2[(1 + 1) + (1 + 1 + [1])] = 10$. Type-species *A. mordax* Szidat & Nani, 1951.

8b. Vitellarium digitiform; copulatory bursa moderate or large .... 9.

9a. Sucker-like copulatory bursa surrounded by large bell-shaped skirt ....
.... *Bursatintinnabulus* Tehrany, Dronen & Wardle, 1999. (Figs 24.9, 24.10)

Diagnosis: Body indistinctly bipartite; forebody slightly spatulate, longer than hindbody. Pseudosuckers present. Holdfast organ simple and inconspicuous or well developed, conspicuous sac, round or oval. Oral sucker present; ventral sucker and prepharynx absent; pharynx present; oesophagus short or absent; caeca terminating some distance from posterior extremity. Testes smooth, tandem. Ovary median to dextral, immediately pretesticular. Vitellarium distributed mainly in forebody, extending ventrolaterally into hindbody to level of posterior testis; vitelline follicles form digital processes. Copulatory bursa sucker-like, surrounded by large bell-shaped skirt occupying 25–30% of whole body. Genital cone absent. Uterus unites with ejaculatory duct just anterior to anterior margin of copulatory bursa. Eggs large. In *Pelecanus* and *Morus*. North America. Type-species *B. macrobursus* (Dronen, Tehrany & Wardle, 1998) Tehrany, Dronen & Wardle, 1999.

9b. Sucker-like copulatory bursa without bell-shaped skirt .... *Bursacetabulus* Dronen, Tehrany & Wardle, 1999. (Figs 24.11, 24,12)

Diagnosis: Body indistinctly bipartite; forebody slightly spatulate, longer than conical hindbody. Pseudosuckers present. Holdfast organ inconspicuous and pouch-like or conspicuous and well developed, circular to elliptical. Oral sucker present; ventral sucker and prepharynx absent; pharynx present; caeca terminating some distance from posterior extremity. Testes smooth to slightly lobed, tandem. Ovary median to slightly dextral, immediately pretesticular. Vitellarium with digitiform processes, distributed mostly in forebody with processes extending into holdfast organ and ventrolaterally into hindbody to level of posterior testis. Copulatory bursa sucker-like. Genital cone absent. Uterus unites with ejaculatory duct just anterior to anterior margin of copulatory bursa. Eggs large. In *Pelecanus* and *Morus*. North America. Type-species *B. pelecanus* Dronen, Tehrany & Wardle, 1999.

10a. Body distinctly, rarely indistinctly, bipartite; ovary in anterior part of hindbody, variable in position; genital cone absent .... *Diplostomum* von Nordmann, 1832. (Fig. 24.13)

(Syns *Hemistomum* Diesing, 1850; *Proalaria* La Rue, 1926)

Diagnosis: Body usually bipartite; hindbody shorter or longer than forebody, rarely conical. Anterior extremity trilobate; pseudosuckers present. Oral sucker, ventral sucker and pharynx small; caeca reaching close to posterior extremity. Holdfast organ round or elliptical, with median slit. Testes tandem; anterior asymmetrical; posterior symmetrical, bilobed, ventrally concave, with lobes more or less longitudinally elongate. Ovary ellipsoidal or ovoid, pretesticular, lateral to and contiguous with anterior testis. Vitellarium in fore- and hindbody, extending forward beyond ventral sucker. Copulatory bursa in form of cavity of variable depth with opening of short hermaphroditic duct at its base. Small genital papilla very occasionally present. In intestine of various birds. Cosmopolitan. Metacercariae of 'diplostomulum' type, in fishes. Cercariae with four large penetration glands posterior to ventral sucker; flame-cell formula $2[(1 + 1 + 1) + (1 + 1 + 1 + [2])] = 16$. Type-species *D. spathaceum* (Rudolphi, 1819) Olsson, 1876.

10b. Body distinctly or indistinctly bipartite; ovary near beginning of hindbody; genital cone present .... 11.

11a. Body distinctly bipartite; anterior testis asymmetrical .... *Dolichorchis* Dubois, 1961. (Figs 24.14, 24.15)

Diagnosis: Body bipartite; forebody linguiform or spatulate; hindbody conical or sacciform. Pseudosuckers present; holdfast organ round or oval, with median slit. Oral and ventral suckers and pharynx feebly developed. Ventral sucker normally close to holdfast organ. Testes tandem; anterior asymmetrical in shape; posterior larger, symmetrical, bilobed, with elongate lobes. Ovary pretesticular, round or oval, median or lateral, near borderline between fore- and hindbody. Vitellarium extending from close to level of ventral sucker or intestinal bifurcation to copulatory bursa.

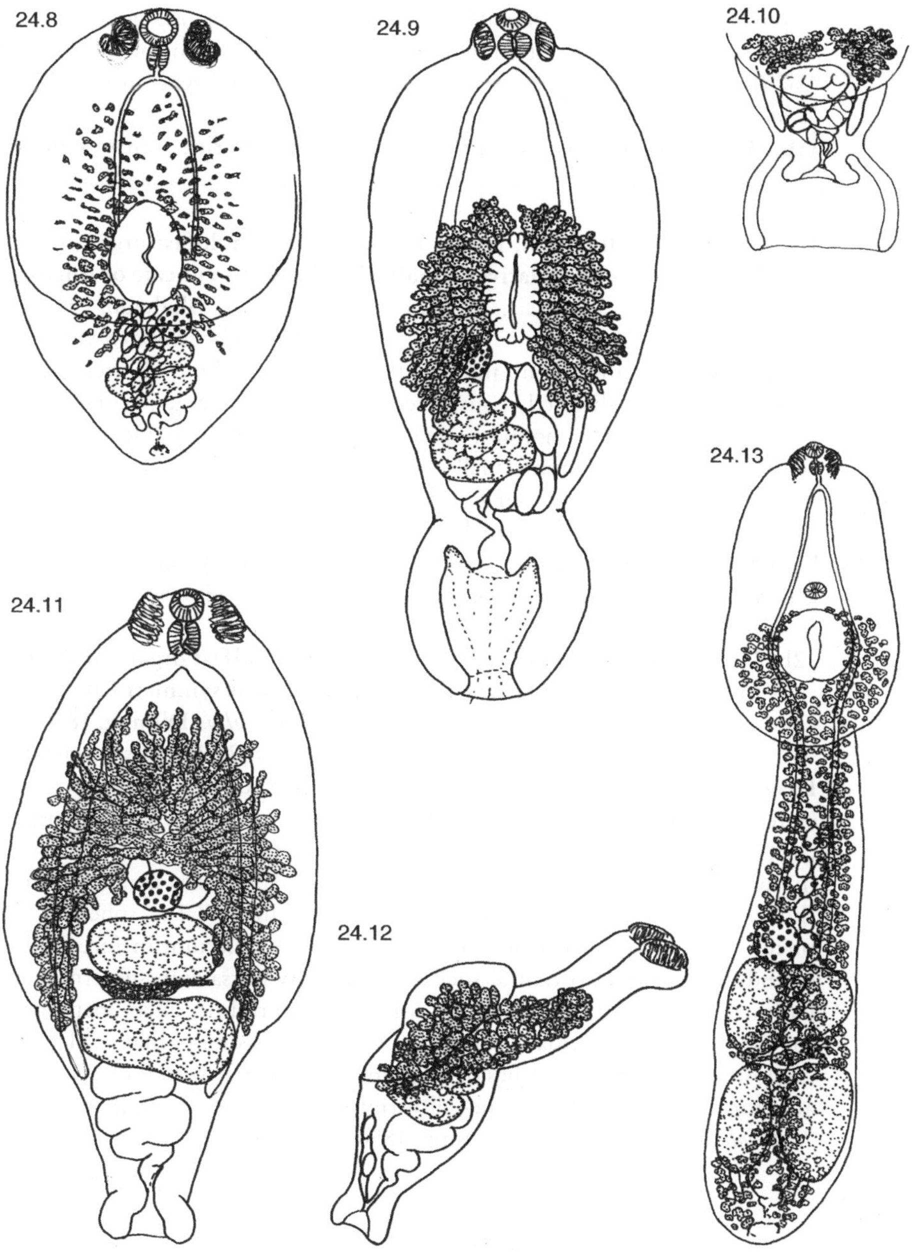

Figs 24.8–24.13.

24.8. *Austrodiplostomum compactum* (Lutz, 1928). (After Dubois, 1970a.)

24.9. *Bursatintinnabulus bassanus* Tehrany, Dronen & Wardle, 1999. (After Tehrany *et al.*, 1999.)

24.10. Posterior end of *B. macrobursus* (Dronen, Tehrany & Wardle, 1998), with bell-shaped skirt surrounding copulatory bursa. (After Dronen *et al.*, 1999.)

24.11, 24.12. *Bursacetabulus pelecani* Dronen, Tehrany & Wardle, 1999. 24.11. Entire worm. 24.12. Lateral view showing holdfast organ. (After Dronen *et al.*, 1999.)

24.13. *Diplostomum spathaceum* (Rudolphi, 1819). (Original.)

Copulatory bursa with terminal or subterminal pore, encloses genital cone with hermaphroditic duct opening terminally. In numerous fish-eating birds. Africa, Australia, Asia (India, Vietnam). Metacercariae of 'diplostomulum' type, in fishes. Type-species *D. marahoueense* Baer, 1957.

11b. Body linguiform, indistinctly bipartite; anterior testis symmetrical or asymmetrical .... 12.

12a. Hindbody cylindrical, longer than forebody; holdfast organ near intestinal bifurcation; testes relatively small, anterior testis cuneate or linguiform .... *Glossodiplostomoides* Bhalerao, 1942. (Fig. 24.16)

(Syn. *Pseudoglossodiplostomum* Dubois, 1944)

Diagnosis: Body linguiform, indistinctly bipartite; forebody shorter than hindbody, cochleariform, not distinctly constricted off from much larger, cylindrical, hindbody. Pseudosuckers present; holdfast organ large, circular, with median slit. Oral and ventral suckers and pharynx small. Testes relatively small, tandem or slightly oblique, near middle of body; anterior testis cuneate or linguiform; posterior testis reniform or linguiform. Ovary ellipsoidal, median, pretesticular. Vitellarium extending into both body regions, reaching ventral sucker. Copulatory bursa spacious, enclosing genital cone, with terminal or subterminal pore. In Accipitridae. India. Type-species *G. hieraetii* (Vidyarthi, 1938) Bhalerao, 1942.

12b. Hindbody conical, shorter than forebody; holdfast organ far from intestinal bifurcation; testes relatively large, anterior testis symmetrical .... *Tylodelphys* Diesing, 1850. (Fig. 24.17)

(Syns *Glossodiplostomum* Dubois, 1932; *Prodiplostomum* Ciurea, 1933)

Diagnosis: Body linguiform, indistinctly bipartite; hindbody conical. Anterior extremity not distinctly trilobate; pseudosuckers present. Holdfast organ oval; its cavity opens by median slit. Oral and ventral suckers and pharynx rather small; ventral in middle of forebody. Testes tandem, symmetrical, ventrally concave, forming horseshoe; anterior testis wider than posterior. Ovary ellipsoidal or spherical, submedian, pretesticular, near anterior margin of hindbody. Vitellarium in fore- and hindbody, extending anteriorly to ventral sucker. Copulatory bursa with subterminal or (rarely) terminal pore; small genital cone with hermaphroditic duct opening terminally. In Accipitridae, Ardeidae and Podicipedidae. Cosmopolitan. Metacercariae of 'diplostomulum' type, in fishes or amphibians. Cercariae with four penetration glands anterior to ventral sucker; flame-cell formula $2[(2 + 2) + (2 + [2])] = 16$. Type-species *T. clavata* (Nordmann, 1832) Diesing, 1850.

Figs 24.14–24.20. (*Opposite*)
24.14. *Dolichorchis marahoueense* Baer, 1957.
24.15. Terminal genitalia of *D. tregenna* showing copulatory bursa with genital cone. (After Dubois, 1970a.)
24.16. *Glossodiplostomoides hieraetii* (Vidyarthi, 1938). (After Dubois, 1970a.)
24.17. *Tylodelphys clavata* (Nordmann, 1832). (Original.)
24.18, 24.19. *Lophosicyadiplostomum nephrocystis* (Lutz, 1928). 24.18. Entire worm. 24.19. Oral sucker and pharynx. (After Dubois, 1970a.)
24.20. *Neodiplostomum attenuatum* (Linstow, 1906). (After Dubois, 1970a.)

24.14

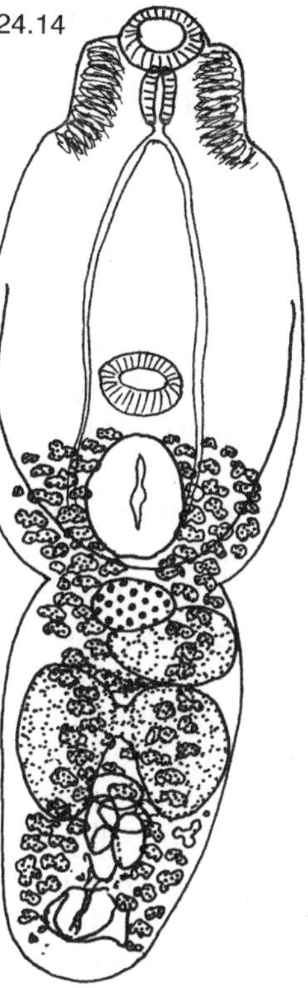

24.15

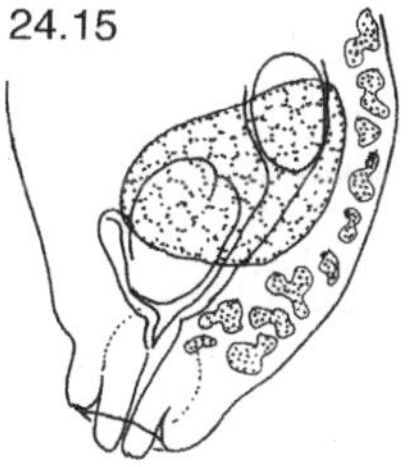

24.16

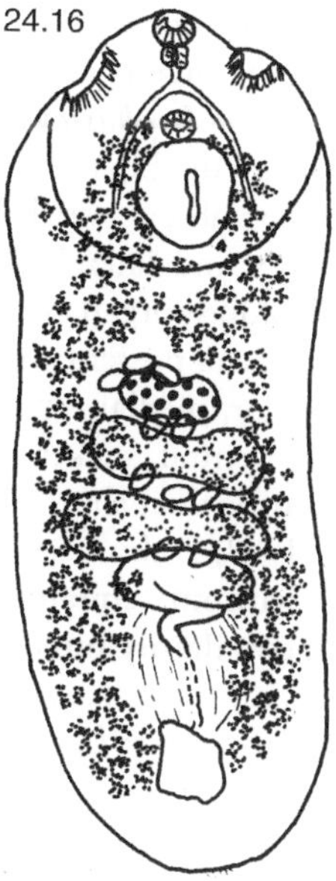

24.17

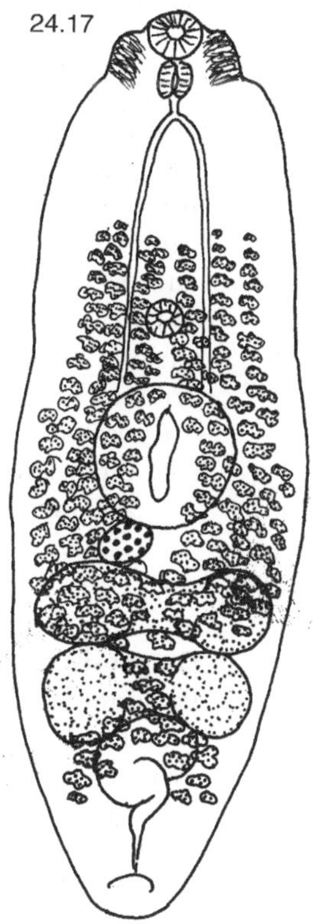

24.19

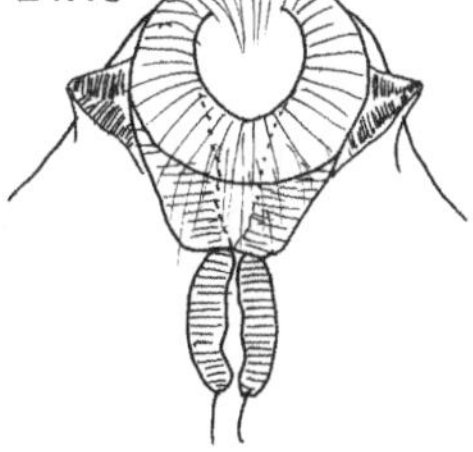

24.20

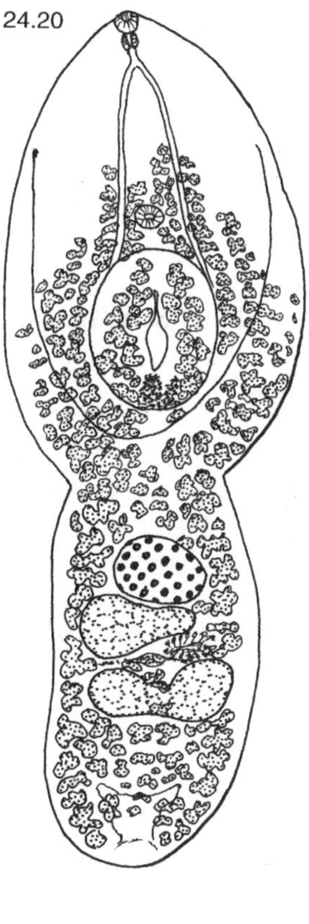

24.18

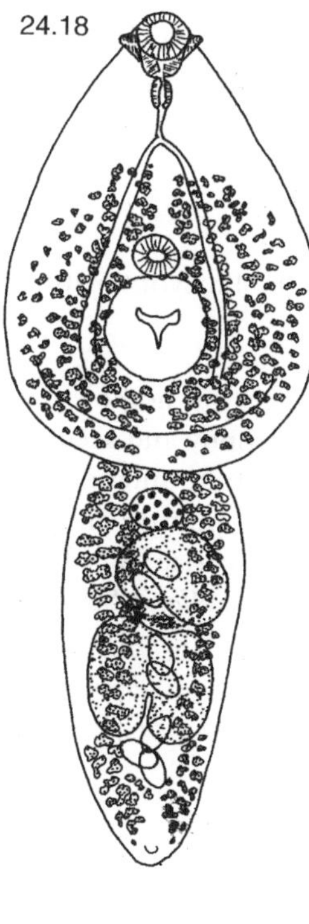

13a. Oral sucker with equatorial muscular ring; anterior testis reniform; copulatory bursa reduced .... *Lophosicyadiplostomum* Dubois, 1936. (Figs 24.18, 24.19)

Diagnosis: Body distinctly bipartite; forebody oval, ventrally concave; hindbody fusiform; pseudosuckers absent. Oral sucker elliptical with equatorial muscular ring surrounding it dorsally and laterally; pharynx small. Ventral sucker small, in mid-region of forebody; holdfast organ circular with central cavity of variable shape. Testes tandem; anterior reniform; posterior bilobed, symmetrical, with deep median groove. Ovary oval, pretesticular, median or submedian. Vitellarium in fore- and hindbody, reaching anterior to ventral sucker. Copulatory bursa reduced; genital pore subterminal; hermaphroditic duct short. In Passeriformes. Brazil, Venezuela. Type-species *L. nephrocystis* (Lutz, 1928) Dubois, 1937.

13b. Oral sucker without equatorial muscular ring; anterior testis usually asymmetrical; copulatory bursa small or large .... *Neodiplostomum* Railliet, 1919. (Fig. 24.20)

(Syns *Conchogaster* Lutz, 1928; *Triplostomum* Lutz, 1928; *Neodiplostomoides* Vidyarthi, 1938)

Diagnosis: Body distinctly bipartite; forebody spatulate; hindbody cylindrical or oval. Pseudosuckers absent. Oral and ventral suckers and pharynx present. Holdfast organ round or oval, with median slit. Testes of similar size, tandem; anterior in general asymmetrical; posterior symmetrical, bilobed. Ovary reniform or ellipsoidal, pretesticular, median or submedian, situated close to borderline between fore- and hindbody, rarely near middle of hindbody. Vitellarium may extend almost to intestinal bifurcation. Copulatory bursa small or large; genital cone absent; hermaphroditic duct opens directly into bursa. In Falconiformes and Strigiformes. Cosmopolitan. Metacercariae of 'neodiplostomulum' type, in amphibians; paratenic hosts reptilians and mammals. Cercariae with two pairs of pre- and paracetabular penetration glands; flame-cell formula $2[(1 + 1 + 1) + (1 + 1 + [1])] = 12$. Type-species *N. spathulaeforme* (Brandes, 1888) Railliet, 1919.

Subfamily Alariinae Hall & Wigdor, 1918

(Syns Pharyngostominae Szidat, 1936; Fibricolinae Sudarikov, 1960; Procyotrematinae Yamaguti, 1971)

Diagnosis: Body more or less distinctly bipartite. Pseudosuckers or homologous structures usually present. Holdfast organ with tendency to hypertrophy and accumulation of vitelline follicles inside. Vitellarium in forebody, may penetrate anterior part of hindbody. Testes tandem or opposite. Ovary pretesticular or opposite anterior testis. In mammals (Carnivora, Rodentia, Marsupialia). Metacercariae of 'diplostomulum' type. Mesocercaria occurs in most life-cycles. Type-genus *Alaria* Schrank, 1788.

Key to genera

1a. Posterior part of body subdivided into ventral and dorsal conical lobes .... *Podospathalium* Dubois, 1932. (Figs 24.21, 24.22)

Diagnosis: Body divided into anterior and posterior part, with anatomy not reflecting distinct fore- and hindbody. Forebody flat, with lateral folds curved ventrally and covering long, oval holdfast organ. Pseudosuckers absent. Oral and ventral suckers and pharynx small, similar in size. Hindbody morphologically indistinct, narrow; its anterior region contains gonads; posterior region forms dorsal and ventral conical lobes. Anterior testis asymmetrical, lateral, opposite oötype; posterior testis oval or reniform, median. Ovary ovoid, anterior to vitelline reservoir and posterior testis. Vitellarium in fore- and hindbody but concentrated in forebody. Well-developed periprostate, surrounding ejaculatory duct, lies between intestinal caeca and dorsal to posterior testis. Uterus and ejaculatory duct open into small cavity preceding tubular genital atrium; latter with muscular anterior wall giving rise to long blind ventral diverticulum. Genital opening slightly ventral on dorsal lobe of foot. Excretory pore terminal on ventral lobe. In Didelphiidae. Brazil. Type-species *P. pedatum* (Diesing, 1850) Dubois, 1932.

1b. Posterior part of body not divided into two conical lobes .... 2.

2a. Body ascidiform, indistinctly bipartite; copulatory bursa very large and retractile .... *Bursotrema* Szidat, 1960. (Fig. 24.23)

Diagnosis: Body indistinctly bipartite; forebody saccular with oblique aperture, constricted at level of anterior limit of vitellarium. Pseudosuckers absent; holdfast organ stalk-like, with median canal. Oral and ventral suckers small; pharynx well developed. Testes tandem, in middle of hindbody; anterior testis small; posterior large, bilobed. Ovary transversely oval, dorsolateral, adjacent to anterior testis. Vitellarium confined to middle region of body, extends to posterior testis. Vitelline reservoir intertesticular. Seminal vesicle large, sinuous. Ejaculatory duct unites with metraterm to form hermaphroditic duct. Copulatory bursa very large, thick-walled and retractile, with genital cone and large dorsal aperture. In marsupials. Argentina. Metacercariae in amphibians. Type-species *B. tetracotyloides* Szidat, 1960.

2b. Typical diplostomid shape; body indistinctly or distinctly bipartite; copulatory bursa moderate or small .... 3.

3a. Holdfast organ elongate, rarely round, with vitelline follicles inside, usually covering ventral sucker; vitellarium restricted to forebody; testes symmetrical .... 4.

3b. Holdfast organ round or oval; vitellarium concentrated in forebody but also penetrates to hindbody; testes tandem .... 6.

4a. Pseudosuckers absent; testes lobed; holdfast organ large, cordiform .... *Pharyngostomum* Ciurea, 1922. (Fig. 24.24)

Diagnosis: Body indistinctly bipartite; forebody slightly longer than hindbody or of similar size; hindbody stout, ovoid. Pseudosuckers absent. Oral sucker small; ventral sucker present or rudimentary; pharynx large, muscular. Holdfast organ huge, cordiform, occupying almost entire concavity of forebody. Testes large, lobate, juxtaposed.

Ovary median or submedian, pretesticular, at junction between fore- and hindbody. Vitellarium in forebody, may reach pharyngeal level; follicles also accumulated in holdfast organ. Oötype and vitelline reservoir between ovary and testes. Seminal vesicle followed by long ejaculatory duct. Hermaphroditic duct opens on dorsal wall of copulatory bursa; genital cone absent. Genital pore terminal. In carnivorous mammals. Europe, Asia, Africa. Metacercariae of 'diplostomulum' type, in amphibians. Cercariae with two pairs of pre- and paracetabular penetration glands; flame-cell formula 2[(1 + 1) + (1 + 1 + [1])] = 10. Type-species *P. cordatum* (Diesing, 1850) Ciurea, 1922.

4b. Pseudosuckers present; testes entire, generally ovoid; holdfast organ linguiform or oval .... **5.**

5a. Forebody spatulate; holdfast organ slightly longer than broad or round .... ***Pharyngostomoides* Harkema, 1942. (Fig. 24.25)**

(Syn. *Parallelorchis* Harkema & Miller, 1961)

Diagnosis: Body indistinctly bipartite, oval; forebody spatulate, shorter than or similar in size to conical or subcylindrical hindbody. Pseudosuckers present; oral and ventral suckers of similar size; pharynx present. Holdfast organ large oval or round with median slit or anteriorly bilobed; its anterior margin extends beyond middle of forebody and may overlie ventral sucker and reach pharynx. Testes oval, symmetrical, in anterior part of hindbody. Ovary reniform, median, near posterior end of forebody. Oötype and vitelline reservoir between ovary and testes, at junction of fore- and hindbody. Vitellarium mainly in forebody, extending from level of ventral sucker to near anterior margin of testes, penetrates holdfast organ. Seminal vesicle posterior to testes; muscular ejaculatory pouch may be present. Copulatory bursa small or deep, with small genital papilla. Ejaculatory duct and uterus unite at base of papilla. Genital pore dorsal, subterminal. In procyonid mammals. USA (Florida). Mesocercariae in branchiobdellid annelids associated with crayfish. Cercariae with two pairs of pre-acetabular or pre- and postacetabular penetration glands; flame-cell formula 2[(2 + 2 + 2) + (2 + 2 + [2])] = 24. Metacercariae of 'diplostomulum' type, develop during trans-entero-pulmonary migration in definitive host. Trans-mammary transmission of mesocercariae was also demonstrated. Type-species *P. procyonis* Harkema, 1942.

5b. Forebody marsupiform; holdfast organ elongate, linguiform .... ***Procyotrema* Harkema & Miller, 1959. (Fig. 24.26)**

Diagnosis: Body large, elongate, subcylindrical; forebody marsupiform, longer and slightly narrower than cylindrical hindbody. Pseudosuckers present; oral sucker larger than ventral sucker; pharynx larger than oral sucker. Holdfast organ linguiform, large, extending posteriorly from region of ventral sucker to posterior end of forebody, contained inside forebody pouch. Testes approximately equal, smooth or irregular, symmetrical, close to anterior end of hindbody. Ovary transversely elongate, at anterior end of hindbody. Oötype just posterior to ovary, anterior to or between anterior margins of testes. Vitellarium confined to posterior two-thirds of forebody, extending into lateral regions of holdfast organ. Seminal vesicle followed by long ejaculatory duct which unites with uterus near genital pore and opens on elevated ventral wall of well-developed copulatory bursa. Ejaculatory pouch absent. In racoons. USA. Mesocercariae in anurans. Cercariae with two pairs of pre-acetabular penetration

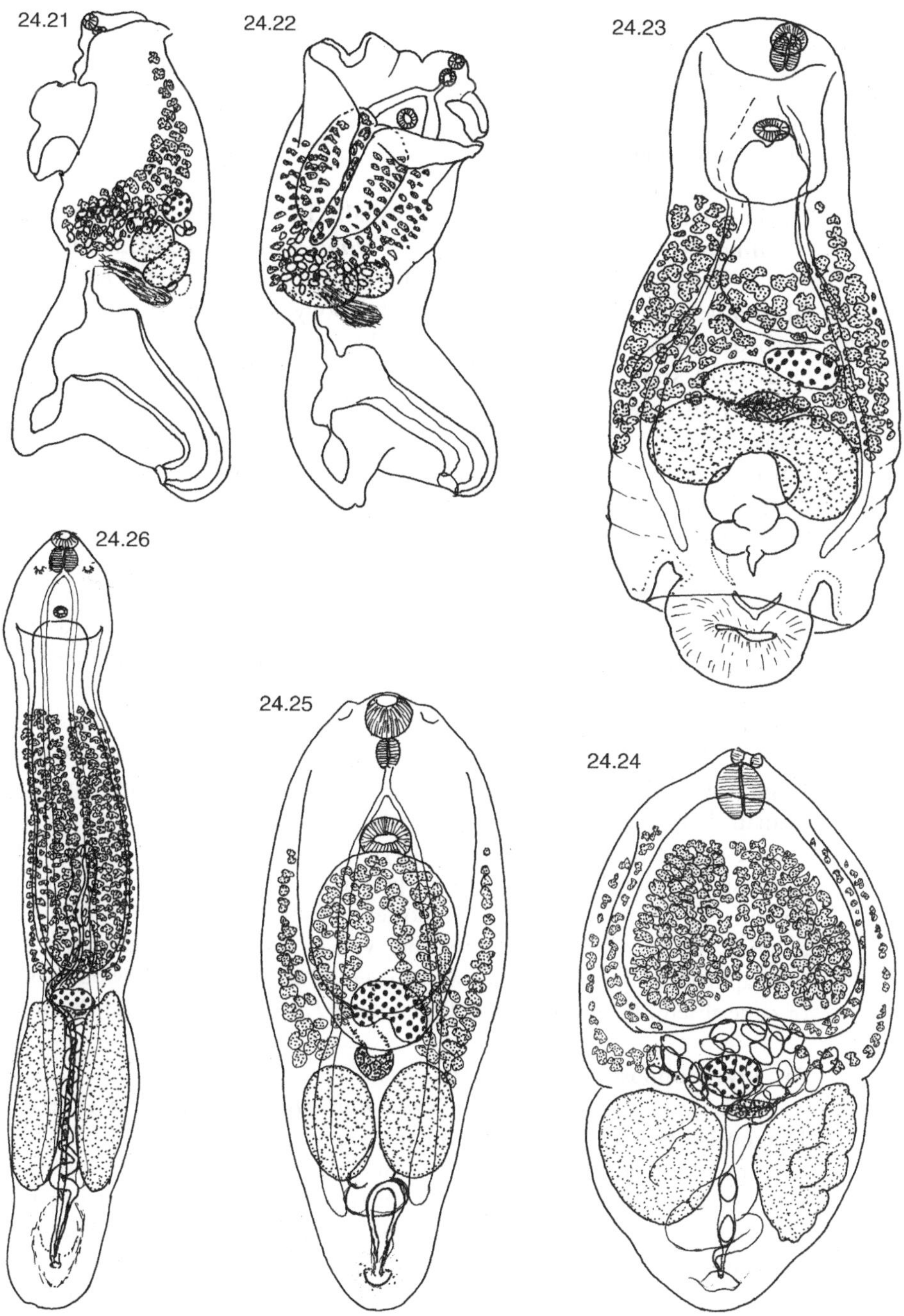

Figs 24.21–24.26.
24.21, 24.22. *Podospathalium pedatum* (Diesing, 1850). (After Dubois, 1970a.)
24.23. *Bursotrema tetracotyloides* Szidat, 1960. (After Dubois, 1976.)
24.24. *Pharyngostomum cordatum* (Diesing, 1850). (After Dubois, 1970a.)
24.25. *Pharyngostomoides procyonis* Harkema, 1942. (After Dubois, 1970a.)
24.26. *Procyotrema marsupiformis* Harkema & Miller, 1959. (After Dubois, 1970a.)

glands; flame-cell formula 2[(2 + 2) + (2 + 2 + [2])] = 20. Type-species *P. marsupiformis* Harkema & Miller, 1959.

6a. Pseudosuckers absent .... **7.**
6b. Pseudosuckers present .... **8.**

7a. Forebody spatulate; vitellarium usually in forebody, occasionally penetrates hindbody .... ***Fibricola* Dubois, 1932. (Figs 24.27, 24.28)**
(Syns *Theriodiplostomum* Dubois, 1944; *Neoparadiplostomum* Bisseru, 1957)
Diagnosis: Body distinctly bipartite; forebody spatulate; hindbody shorter than forebody, occasionally longer. Pseudosuckers absent. Oral sucker, ventral sucker and pharynx of similar size. Holdfast organ round or oval, one-sixth to one-third of forebody in length. Testes tandem; anterior ovoid or cuneiform, asymmetrical in shape; posterior larger, almost symmetrical, occasionally trilobed. Ovary ovoid or cuneiform, submedian or sublateral, at junction of fore- and hindbody, pretesticular. Oötype opposite anterior testis. Vitellarium in forebody, may extend anteriorly to ventral sucker, occasionally penetrates hindbody reaching copulatory bursa. Genital cone absent; copulatory bursa with dorso-subterminal pore. In various mammals. Australia, North America, Asia. Metacercariae of 'diplostomulum' type, in amphibians. Cercariae with two pairs of pre-acetabular penetration glands; flame-cell formula 2[(1 + 1 + 1) + (1 + 1 + [1])] = 12. Type-species *F. cratera* (Barker & Noll, 1915) Dubois, 1932.
7b. Forebody cup-shaped; vitellarium in fore- and hindbody ....
.... ***Prudhoella* Beverley-Burton, 1960. (Fig. 24.29)**
Diagnosis: Body not distinctly bipartite, cylindrical; forebody cup-shaped with ventral median cleft; hindbody longer than forebody. Pseudosuckers absent; oral sucker small; ventral sucker poorly developed; pharynx present. Holdfast organ linguiform and protrusible. Testes lobed, tandem, in posterior part of hindbody. Ovary and oötype anterior to testes. Vitellarium in posterior half of forebody, holdfast organ and anterior region of hindbody. Seminal vesicle followed by ejaculatory duct; short hermaphroditic duct opens medially on ventral wall of copulatory bursa. Genital bulb or cone absent. In otters. Africa. Type-species *P. rhodesiensis* Beverley-Burton, 1960.

8a. Ovary opposite anterior testis ....
.... ***Cynodiplostomum* Dubois, 1936. (Figs 24.30, 24.31)**
Diagnosis: Body distinctly bipartite; forebody spatulate; hindbody ovoid, of similar length to forebody or shorter. Pseudosuckers present; oral and ventral suckers and pharynx well developed. Holdfast organ round, rather small, posterior to middle of forebody at some distance from ventral sucker. Anterior testis oval or reniform, asymmetrical, lateral; posterior much larger, symmetrically bilobed. Ovary oval or reniform, lateral, opposite anterior testis, at junction of fore- and hindbody. Oötype between ovary and testes. Vitellarium occupies posterior region of forebody, commencing at level of ventral sucker. Seminal vesicle voluminous. Copulatory bursa relatively large, with dorso-subterminal pore. In carnivorous mammals. Africa. Type-species *C. azimi* (Nazmi Gohar, 1933) Dubois, 1936.
8b. Ovary pretesticular .... **9.**

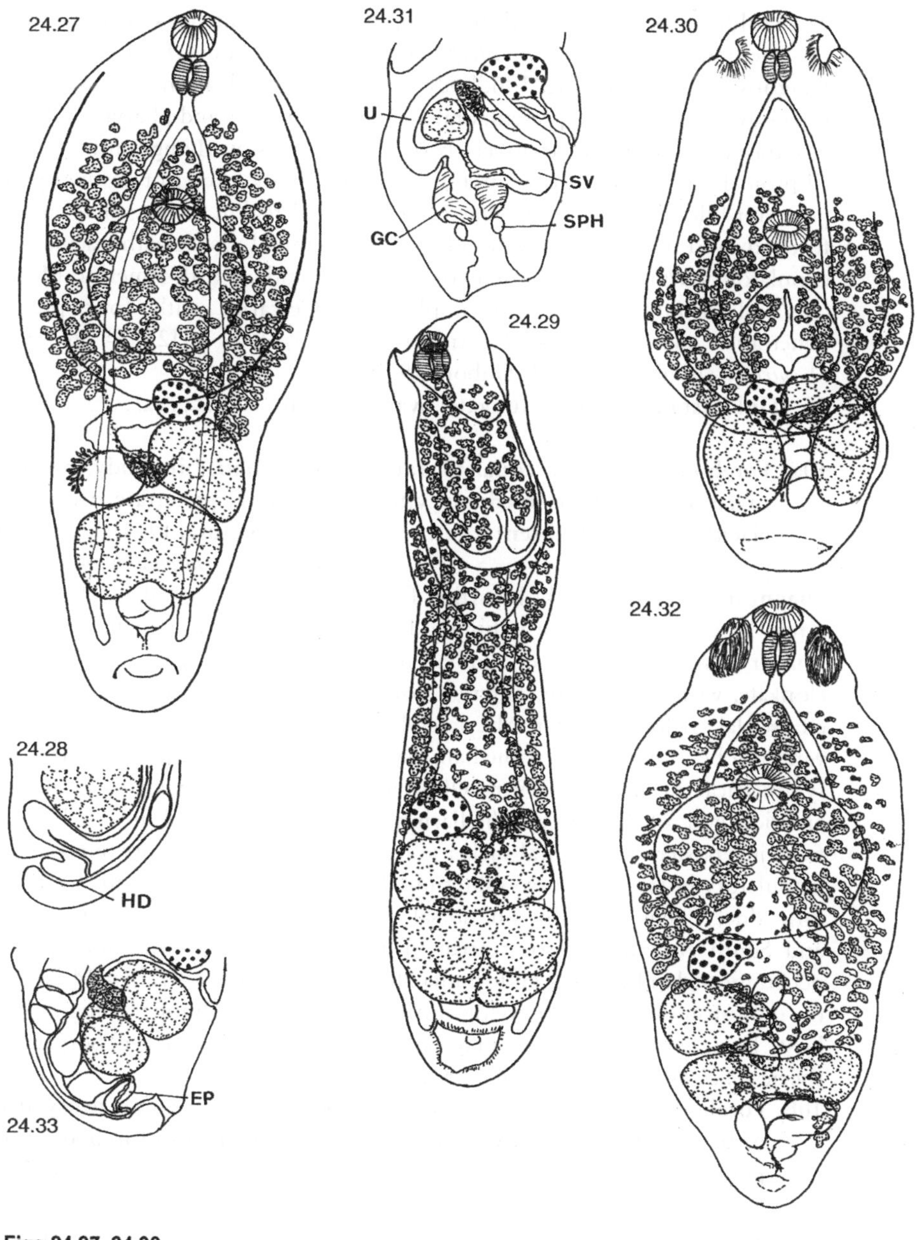

Figs 24.27–24.33.

24.27, 24.28. *Fibricola cratera* (Barker & Noll, 1915). 24.27. Entire worm. 24.28. Terminal genitalia showing long hermaphroditic duct (HD). (After Dubois, 1970a.)

24.29. *Prudhoella rhodesiensis* Beverley-Burton, 1960. (After Dubois, 1970a.)

24.30, 24.31. *Cynodiplostomum azimi* (Nazmi Gohar, 1933). 24.30. Entire worm. 24.31. Terminal genitalia showing genital cone (GC) and copulatory bursa with sphincter (SPH) at base. U, uterus; SV, seminal vesicle. (After Dubois, 1970a.)

24.32, 24.33. *Didelphodiplostomum variabile* (Chandler, 1932). 24.32. Entire worm. 24.33. Terminal genitalia showing ejaculatory pouch (EP). (After Dubois, 1970a.)

9a. Holdfast organ of medium size, round or transversely oval; ovary submedian .... *Didelphodiplostomum* Dubois, 1944. (Figs 24.32, 24.33)

Diagnosis: Body bipartite; hindbody ovoid to conical, narrower and shorter than forebody. Pseudosuckers present; oral and ventral suckers and pharynx well developed. Holdfast organ round or transversely oval, sometimes overlying ventral sucker. Ovary ellipsoidal, submedian, between fore- and hindbody. Anterior testis asymmetrical; posterior bilobed. Vitellarium may extend beyond region of intestinal bifurcation in forebody and posterior to testes in hindbody. Seminal vesicle with ejaculatory pouch; copulatory bursa without genital cone. In didelphids. North America. Metacercariae of 'diplostomulum' type, in amphibians. Cercariae with four pre-acetabular penetration glands; flame-cell formula 2[(2 + 2) + (2 + [2])] = 16. Type-species *D. variabile* (Chandler, 1932) Dubois, 1944.

9b. Holdfast organ large, elongate; ovary median .... 10.

10a. Forebody shorter to equal to, rarely longer than, hindbody; pseudosuckers invaginated; ovary median; vitellarium may penetrate into hindbody; ejaculatory pouch absent .... *Paralaria* Krause, 1914. (Figs 24.34, 24.35)

(Syn. *Enhydridiplostomum* Dubois, 1944)

Diagnosis: Body distinctly bipartite; forebody elongate, spatulate, shorter to equal, rarely longer than claviform hindbody. Pseudosuckers invaginated. Ventral sucker smaller than oral sucker or similar in size; pharynx large. Holdfast organ oval, elongate, with median slit; anterior border over-reaches middle of forebody. Testes ellipsoidal, trilobed posteriorly; lateral lobes may be subdivided into dorsal and ventral lobes; anterior testis median, asymmetrical, cuneate; posterior larger, may be symmetrical or massive. Ovary ellipsoidal or reniform, median, in middle or at anterior end of hindbody. Vitellarium densely distributed in forebody, extends from just posterior to ventral sucker to level of ovary or copulatory bursa in hindbody. Copulatory bursa with dorso-subterminal opening. Genital cone absent. Hermaphroditic duct opens ventrally on dorsal wall of copulatory bursa. In otters. North and South America. Mesocercariae in anurans. Cercariae with four paracetabular penetration glands; flame-cell formula 2[(1 + 1 + 1) + (1 + 1 + [2])] = 14. Metacercariae of 'diplostomulum' type, develop during trans-entero-pulmonary migration in final host. Encysted metacercariae may be present in muscles of accessory hosts. Type-species *P. clathrata* (Diesing, 1850) Krause, 1914.

10b. Forebody longer than hindbody; pseudosuckers auricular; ovary at junction of fore- and hindbody; vitellarium mainly confined to forebody; ejaculatory pouch present .... *Alaria* Schrank, 1788. (Figs 24.36, 24.37)

(Syn. *Conchosomum* Railliet, 1896)

Diagnosis: Body indistinctly bipartite; forebody linguiform, concave; hindbody cylindrical, usually shorter than forebody. Pseudosuckers present, forming ear-like projections. Oral and ventral suckers rather small; pharynx relatively large. Holdfast organ round to elongate, variable in length; anterior margin may reach pharynx. Testes of different size and form, multi- or bilobed, tandem; anterior asymmetrical, opposite oötype; posterior symmetrical, larger. Ovary oval, median, pretesticular, at junction of fore- and hindbody. Vitellarium mainly in forebody, may penetrate holdfast organ and extend into hindbody. Seminal vesicle with either ejaculatory pouch or ejaculatory duct with muscular region. Copulatory bursa small, with subterminal dorsal pore. Hermaphroditic duct opens at tip of small papilla. In Carnivora. Eurasia and tropical North America. Mesocercariae in anurans. Paratenic hosts of mesocercariae are included in life-cycle. Cercariae with two pairs of pre-acetabular penetration glands; flame-cell formula $2[(2 + 2 + 2) + (2 + 2 + (2))] = 24$. Metacercariae of 'diplostomulum' type, develop during trans-entero-pulmonary migration in final host. Type-species *A. alata* (Goeze, 1782) Krause, 1914.

Subfamily Codonocephalinae Sudarikov, 1959

Diagnosis: Body distinctly bipartite; hindbody much longer than forebody. Holdfast organ of aberrant diplostomid structure, with large proteolytic gland near posterior end of forebody. Pseudosuckers absent. Type- and only genus *Codonocephalus* Diesing, 1850.

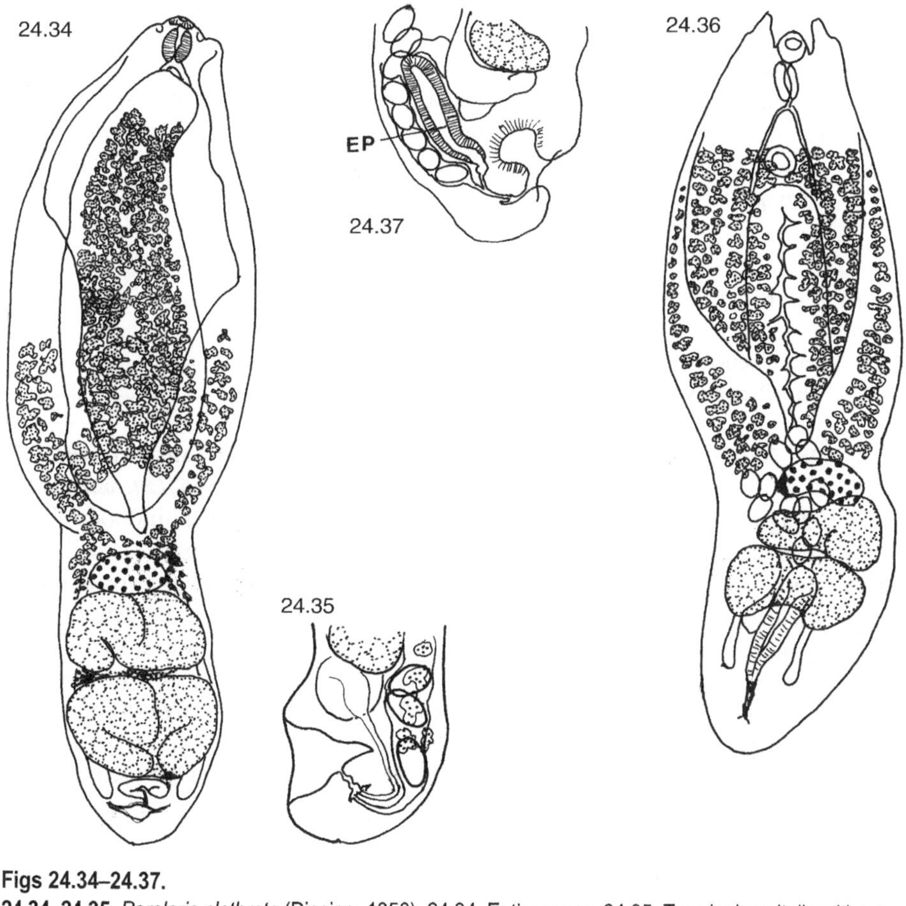

Figs 24.34–24.37.
24.34, 24.35. *Paralaria clathrata* (Diesing, 1850). 24.34. Entire worm. 24.35. Terminal genitalia without ejaculatory pouch. (After Dubois, 1970a.)
24.36, 24.37. *Alaria marcianae* (La Rue, 1917). 24.36. Entire worm. 24.37. Terminal genitalia with ejaculatory pouch (EP). (After Dubois, 1970a.)

Genus *Codonocephalus* Diesing, 1850 (Figs 24.38–24.40)

Diagnosis: Body bipartite; forebody short, infundibular or cup-shaped, with crenulated borders; hindbody long, cylindrical. Pseudosuckers absent; ventral sucker and pharynx of approximately same size, larger than oral sucker. Holdfast organ small with large proteolytic gland at junction of fore- and hindbody. Testes tandem, lobed, with lobes directed posteriorly, in posterior part of hindbody. Ovary round, median, pretesticular. Vitellarium in hindbody. Copulatory bursa relatively small, with terminal pore provided with muscular ring. Genital cone protrusible, surrounded by circular fold (in form of prepuce), with hermaphroditic duct opening at apex. In ardeids. Palaearctic. Progenetic metacercariae in frogs. Cercariae with four large penetration glands anterior to ventral sucker; flame-cell formula 2[(2 + 2 + 2) + (2 + 2 + [2])] = 24. Type- and only species *C. urniger* (Rudolphi, 1819).

Subfamily Crassiphialinae Sudarikov, 1960

Diagnosis: Body more or less distinctly bipartite; pseudosuckers or homologous structures present or absent. Vitellarium in hindbody or almost entirely so, rarely extending into forebody. Prepuce-like fold around genital cone present or absent. Copulatory bursa may be protrusible. Metacercariae of 'neascus' type. Type-genus *Crassiphiala* Van Haitsma, 1925.

Key to genera

1a. Pseudosuckers or homologous structures present .... **2.**
1b. Pseudosuckers or homologous structures absent .... **7.**

2a. Vitellarium restricted to hindbody .... ***Pulvinifer* Yamaguti, 1933. (Fig. 24.41)**
(Syn. *Laterostrigea* Yang, 1962)

Diagnosis: Body bipartite; forebody wide, concave ventrally; hindbody elongate, cylindrical, 1.25–4 times longer than forebody. Pseudosuckers present; suckers well developed, with ventral larger than oral; pharynx small; holdfast organ pulvinate with median slit. Gonads in posterior half of hindbody. Testes tandem, subspherical or elongate. Ovary pretesticular or opposite anterior testis, round or oval. Vitellarium confined to hindbody. Seminal vesicle without ejaculatory pouch. Copulatory bursa small; genital cone absent; hermaphroditic duct opens directly into bursa; genital pore subterminal. In Charadriiformes. Metacercariae of 'neascus' type, in Oligochaeta. Holarctic. Type-species *P. macrostomum* (Jägerskiöld, 1900) Dubois, 1938.
2b. Vitellarium in fore- and hindbody .... **3.**

3a. Testes transversely elongate, posterior testis bilobed; copulatory bursa evaginable; genital cone with prepuce-like structure ....
.... ***Posthodiplostomoides* Williams, 1969. (Fig. 24.42)**

Diagnosis: Body distinctly bipartite; forebody oval or elliptical, ventrally concave; hindbody oval or conical. Pseudosuckers present; oral and ventral suckers well

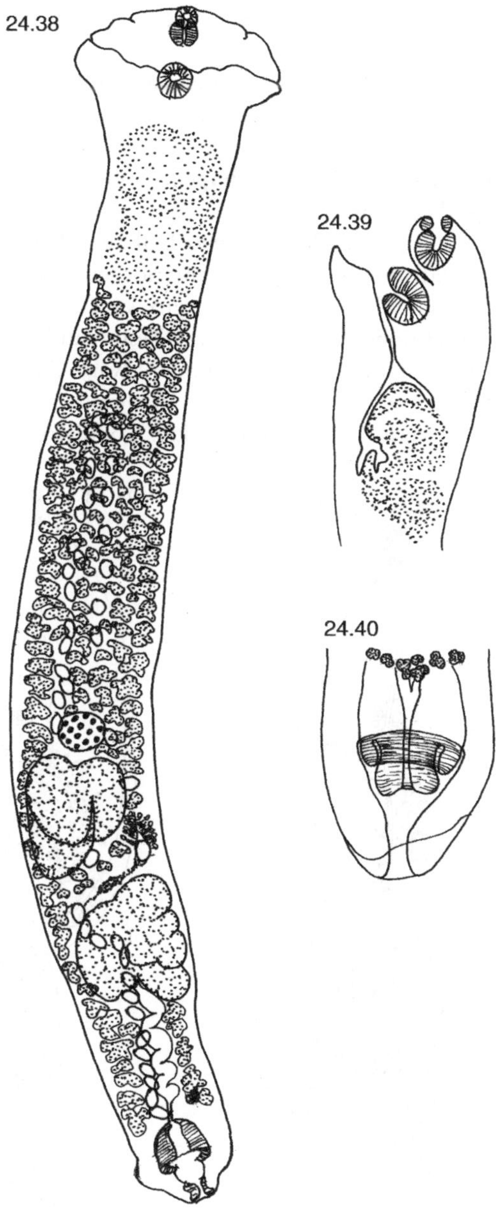

Figs 24.38–24.40. *Codonocephalus urniger* (Rudolphi, 1819). 24.38. Entire worm. 24.39. Section through forebody showing holdfast organ with proteolytic gland. (After Dubois, 1970a.) 24.40. Terminal genitalia. (After Sudarikov, 1959.)

developed; pharynx small; holdfast organ almost round, with cavity opening by median slit. Testes tandem, of similar size; anterior symmetrical; posterior bilobed with lobes directed anteriorly. Ovary pretesticular, median, oval, near junction of fore- and hindbody. Vitellarium in both parts of body, reaching anteriorly beyond ventral sucker. Copulatory bursa evaginable, with circular fold surrounding genital cone in form of prepuce. Hermaphroditic duct formed by union of uterus and ejaculatory duct, passes through genital cone and opens at its apex. In ardeids. Africa.

Metacercariae of 'neascus' type in cyprinid fishes. Type-species *P. leonensis* (Williams, 1967) Williams, 1969.

3b. Testes of different shape, copulatory bursa not evaginable; genital cone present or absent .... 4.

4a. Testes lobate, sinuous, twisted, V-, H- or N-shaped; copulatory bursa with muscular bulb at base; genital cone present .... *Bolbophorus* Dubois, 1935. (Fig. 24.43)

Diagnosis: Body distinctly or indistinctly bipartite; anterior extremity trilobate; pseudosuckers large; oral sucker, ventral sucker and pharynx small. Holdfast organ circular or oval, with median slit. Hindbody subcylindrical. Testes tandem, lobate, in form of V, H or N, occupying most of middle hindbody. Ovary ellipsoidal, pretesticular, submedian, near anterior end of hindbody. Vitellarium in fore- and hindbody, usually not extending anteriorly to ventral sucker. Copulatory bursa with

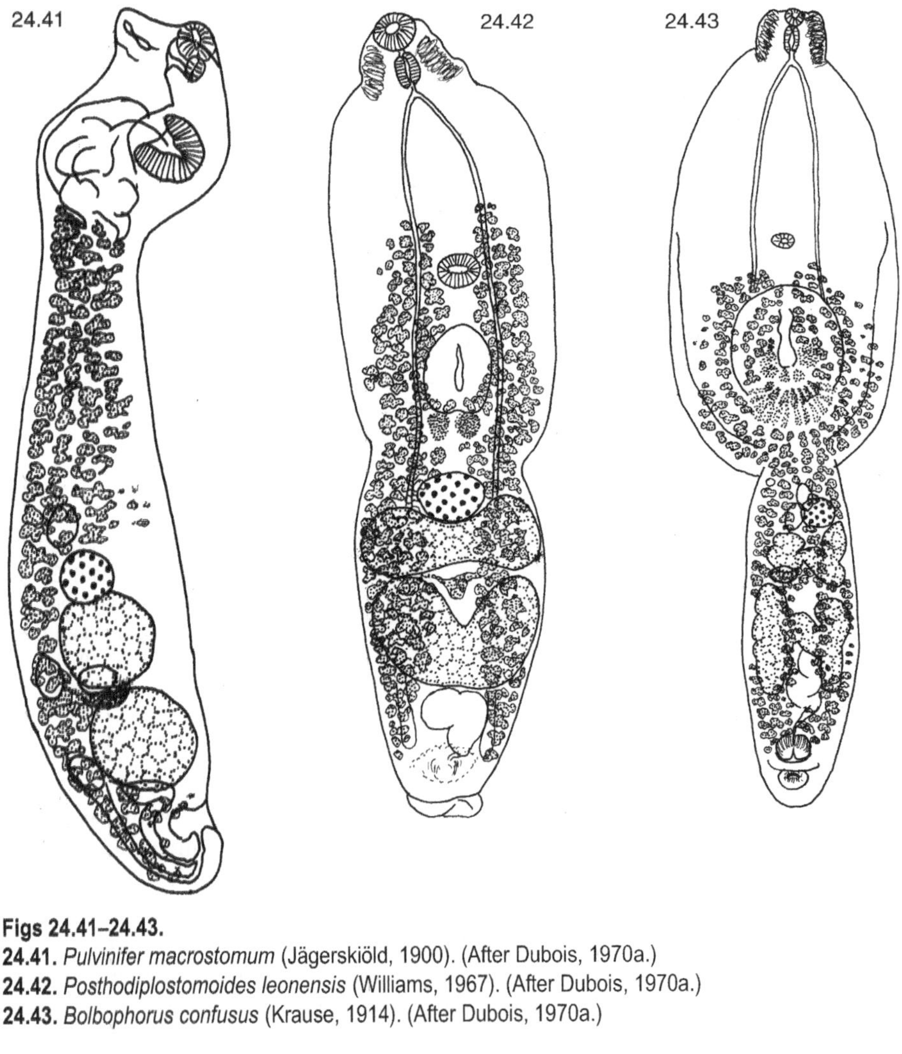

Figs 24.41–24.43.
24.41. *Pulvinifer macrostomum* (Jägerskiöld, 1900). (After Dubois, 1970a.)
24.42. *Posthodiplostomoides leonensis* (Williams, 1967). (After Dubois, 1970a.)
24.43. *Bolbophorus confusus* (Krause, 1914). (After Dubois, 1970a.)

subterminal pore; muscular genital bulb at base of bursa, dorsal to genital cone. In Pelecaniformes and Ardeae. Holarctic, Ethiopian and Australian regions. Metacercariae of 'neascus' type in fishes. Cercariae with rudimentary digestive system, six penetration glands in two rows lateral to rudimentary ventral sucker when developed; flame-cell formula 2[(2 + 2) + (2 + 2 + [2])] = 20. Type-species *B. confusus* (Krause, 1914) Dubois, 1935.

4b. Testes massively lobed, tranversely elongate or bilobed; muscular bulb at base of copulatory bursa absent; genital cone absent .... 5.

5a. Holdfast organ with two wing-like lobes; vitellarium extends from level of oral sucker to ovary; muscular vaginal sphincter present .... *Pseudoscolopacitrema* Palmieri, Krishnasamy & Sullivan, 1979.
(Figs 24.44, 24.45)

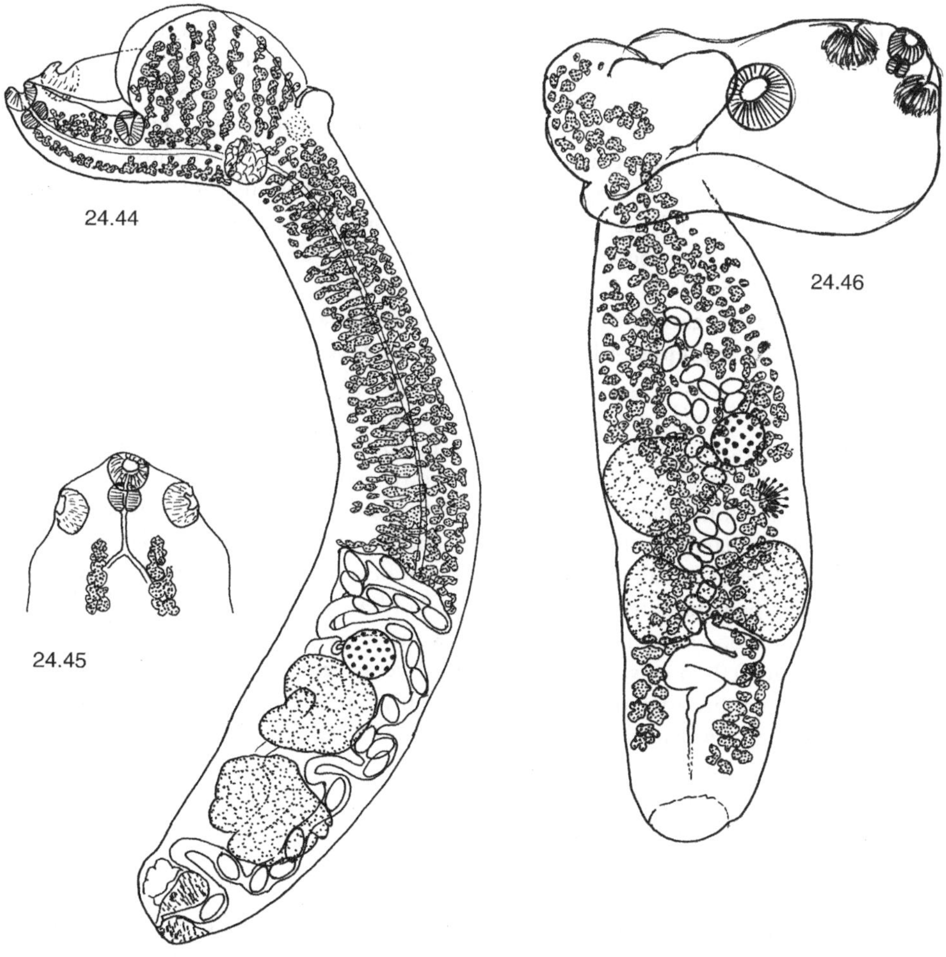

Figs 24.44–24.46.
24.44, 24.45. *Pseudoscolopacitrema otteri* Palmieri, Krishnasamy & Sullivan, 1979. 24.44. Entire worm. 24.45. Anterior end with pseudosuckers. (After Palmieri *et al.*, 1979.)
24.46. *Allodiplostomum scolopacis* Yamaguti, 1935. (After Dubois, 1970a.)

Diagnosis: Body distinctly bipartite; forebody concave, shorter than long, slightly conical hindbody. Holdfast organ with two fan- or wing-like lobes, occupying major portion of concavity of forebody. Pseudosuckers well developed. Oral and ventral suckers well developed; pharynx smaller than oral sucker. Testes tandem, large, lobed. Ovary pretesticular, round or oval, in posterior part of hindbody. Vitelline follicles abundant in fore- and hindbody between base of oral sucker and level of ovary. Seminal vesicle intertesticular. Muscular vaginal sphincter present. In otters. India, Europe. Type-species *P. otteri* Palmieri, Krishnasamy & Sullivan, 1979.

5b. Holdfast organ without two wing-like lobes; vitellarium not reaching beyond ventral sucker; muscular vaginal sphincter absent .... 6.

6a. Holdfast organ trilobed; vitellarium extends into median lobe of holdfast organ; prepuce-like structure around genital cone absent .... *Allodiplostomum* Yamaguti, 1935. (Fig. 24.46)

Diagnosis: Body bipartite; forebody linguiform, with ventral concavity; hindbody about twice as long as forebody, subcylindrical. Pseudosuckers present; ventral sucker larger than oral sucker; pharynx small. Holdfast organ massive, trilobed. Testes tandem; anterior asymmetrical; posterior symmetrical, bilobed. Ovary round, close beside anterior testis, situated near middle of hindbody. Vitellarium extends into median lobe of holdfast organ. Seminal vesicle without ejaculatory pouch. Copulatory bursa with moderately large, almost terminal pore. Hermaphroditic duct opens on tip of papilliform protuberance. In Charadriiformes. Japan. Type-species *A. scolopacis* Yamaguti, 1935.

6b. Holdfast organ not divided into lobes; vitellarium not extending into holdfast organ; prepuce-like fold around genital cone present .... *Subuvulifer* Dubois, 1952. (Figs 24.47, 24.48)

(Syns *Choanochenia* Yang, 1959; *Neochoanochenia* Yang, 1965; ?*Cotylostoma* Yang, 1965)

Diagnosis: Body bipartite; forebody oval, cochleariform; hindbody subcylindrical, about two to three times longer than forebody; pseudosuckers present. Holdfast organ round, small, one-quarter to one-third of forebody length. Ventral sucker larger than oral sucker; pharynx rather small. Testes tandem, massive, lobed, similar in size. Ovary spherical or ellipsoidal, median, pretesticular, near middle of hindbody. Vitellarium in hindbody, penetrates forebody to level of holdfast organ, but less densely distributed than in hindbody. Seminal vesicle large, without ejaculatory pouch. Copulatory bursa encloses genital cone, which is surrounded ventrolaterally at its base by prepuce-like fold. Hermaphroditic duct opens terminally on genital cone. In alcedines. Asia. Type-species *S. halcyonae* (Gogate, 1940) Dubois, 1952.

7a. Vitellarium restricted to hindbody .... 8.
7b. Vitellarium in fore- and hindbody .... 11.

8a. Ventral sucker absent; copulatory bursa with large sucker at base; genital cone absent .... *Cercocotyla* Yamaguti, 1939. (Fig. 24.49)

(Syn. *Pseudocercocotyla* Yamaguti, 1971)

Diagnosis: Body distinctly bipartite; forebody oval, spoon-shaped; hindbody long, slender, enlarged posteriorly. Pseudosuckers and ventral sucker absent; oral sucker and pharynx very small. Holdfast organ small, close to posterior end of forebody. Testes tandem, ovoid or transversely elongate, unlobed. Ovary round, pretesticular, median, in middle or in posterior region of hindbody. Vitellarium confined to posterior region of hindbody or extends to its anterior border. Seminal vesicle without ejaculatory pouch. Medium or large copulatory bursa with wide terminal aperture, without genital cone, with large sucker at base ventral to opening of hermaphroditic duct. In alcedines. India, Japan. Type-species *C. cerylis* Yamaguti, 1939.

8b. Ventral sucker present, rarely absent; copulatory bursa without large sucker at base; retractile genital cone and prepuce-like structure present .... 9.

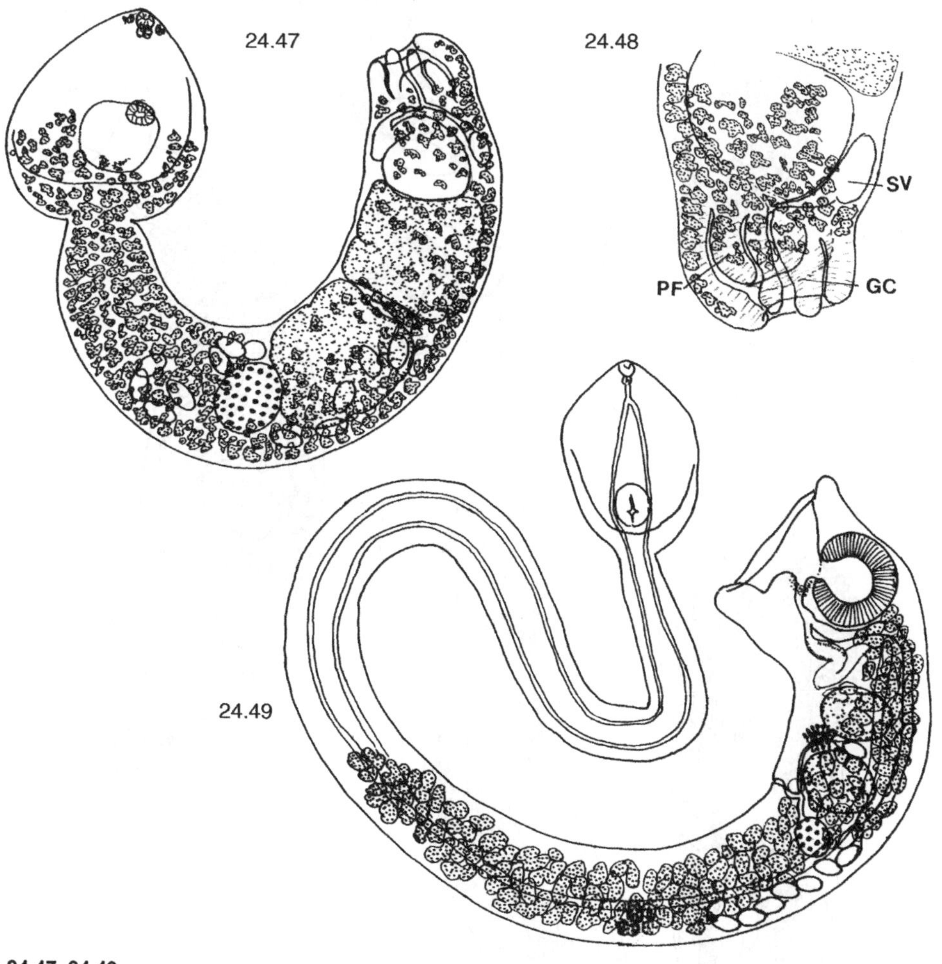

Figs 24.47–24.49.
24.47, 24.48. *Subuvulifer halcyonae* (Gogate, 1940). 24.47. Entire worm. 24.48. Terminal genitalia showing prepuce-like fold (PF) at base of genital cone (GC). SV, seminal vesicle. (After Dubois, 1970a.)
24.49. *Cercocotyla cerylis* Yamaguti, 1939. (After Dubois, 1970a.)

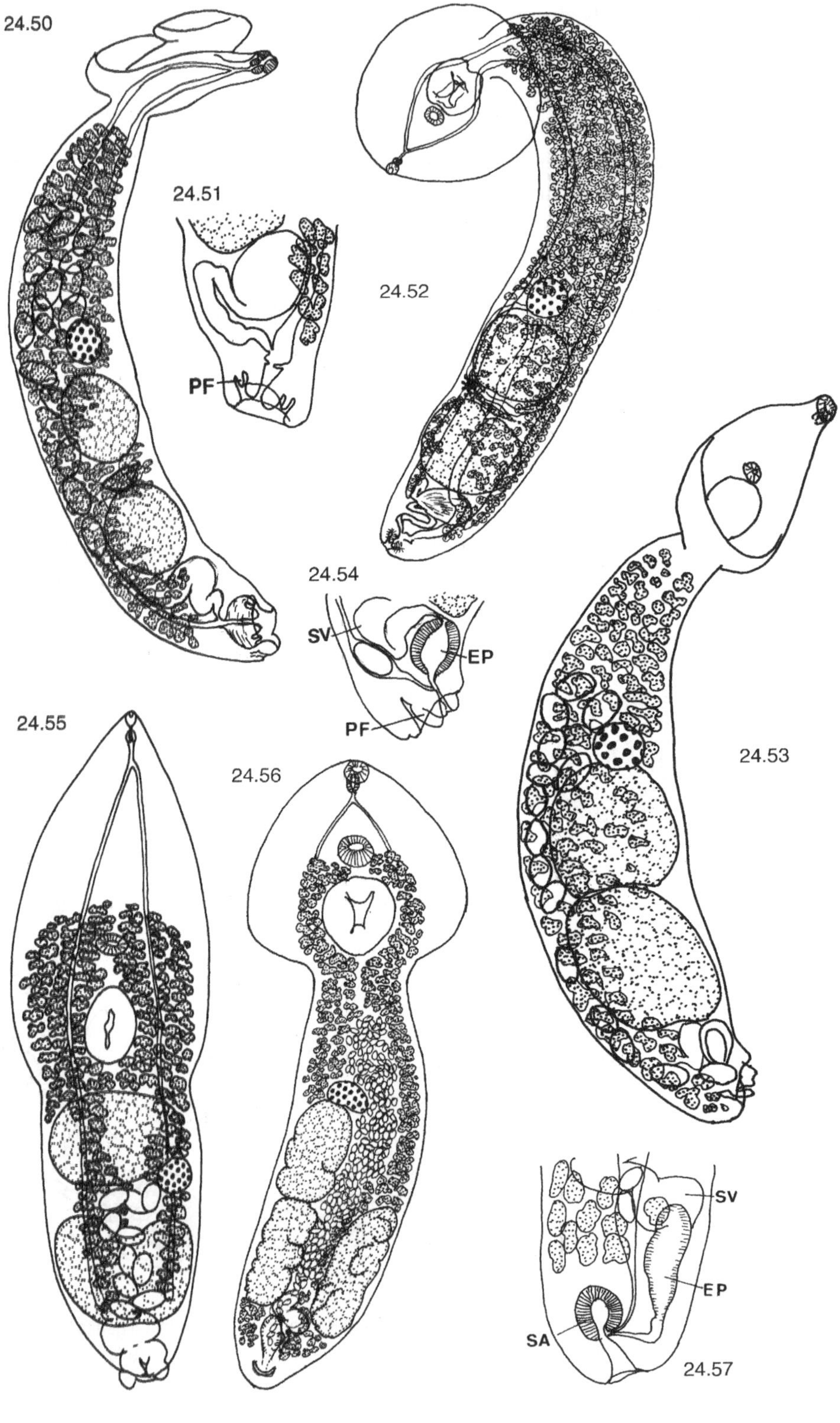
24.50
24.51
PF
24.52
24.54
SV
EP
PF
24.55
24.56
24.53
SV
EP
SA
24.57

9a. Holdfast organ large, occupying entire concavity of forebody; testes large unlobed; ejaculatory pouch absent .... *Crassiphiala* Van Haitsma, 1925. (Figs 24.50, 24.51)

Diagnosis: Body bipartite; forebody much shorter than hindbody, oval, with ventral concavity; pseudosuckers absent. Holdfast organ large, bulbous, occupying entire concavity of forebody. Oral sucker and pharynx well developed; ventral sucker rudimentary or absent. Testes large, tandem, rounded or subcylindrical, unlobed. Ovary ellipsoidal, pretesticular, situated in anterior part of hindbody. Vitellarium restricted to hindbody. Seminal vesicle without ejaculatory pouch. Copulatory bursa with terminal pore, contains retractile genital cone. Hermaphroditic duct opens terminally on genital cone. In alcedines. USA. Metacercariae of 'neascus' type, in fishes. Type-species *C. bulboglossa* Van Haitsma, 1925.

9b. Holdfast organ small, in posterior part of forebody; ejaculatory pouch present; genital cone protrusible .... 10.

10a. Testes round or transversely elongate; ejaculatory pouch small; copulatory bursa very small; genital cone without preputial fold .... *Pseudodiplostomum* Yamaguti, 1934. (Fig. 24.52)

Diagnosis: Body distinctly bipartite; forebody oval, cochleariform; hindbody long, subcylindrical, two to three times longer than forebody. Pseudosuckers absent; ventral sucker larger than oral sucker; pharynx small. Holdfast organ small, round, with median slit. Testes tandem, round or transversely elongate; posterior symmetrical. Ovary ellipsoidal or spherical, median, pretesticular, in posterior half of hindbody. Vitellarium in hindbody. Seminal vesicle followed by small ejaculatory pouch. Copulatory bursa small, with terminal opening, contains protrusible genital cone. Hermaphroditic duct opens at apex of cone. In alcedines. Japan, India. Type-species *P. cochleariforme* Yamaguti, 1934.

10b. Testes round or oval, dorsoventrally elongated; ejaculatory pouch small to moderate; copulatory bursa moderate or large; genital cone half-enclosed in prepuce-like fold .... *Uvulifer* Yamaguti, 1934. (Figs 24.53, 24.54)

(Syn. *Prochoanochenia* Yang, 1965)

Diagnosis: Body distinctly bipartite; forebody oval, cochleariform; hindbody claviform, much longer than forebody. Pseudosuckers absent; ventral sucker smaller than oral; pharynx small. Holdfast organ round, small, 15–30% of forebody length. Testes round, tandem, in posterior region of hindbody. Ovary ellipsoidal or spherical,

Figs 24.50–24.57. (*Opposite*)

24.50, 24.51. *Crassiphiala bulboglossa* Van Haitsma, 1925. 24.50. Entire worm. 24.51. Terminal genitalia showing prepuce-like fold (PF) surrounding genital cone. (After Dubois, 1970a.)

24.52. *Pseudodiplostomum major* Yamaguti, 1939. (After Dubois, 1970a.)

24.53, 24.54. *Uvulifer semicircumcisus* Dubois & Rausch, 1950. 24.53. Entire worm. 24.54. Terminal genitalia showing ejaculatory pouch (EP) following seminal vesicle (SV) and prepuce-like fold (PF) at base of genital cone. (After Dubois, 1970a.)

24.55. *Mesoophorodiplostomum pricei* (Krull, 1934). (After Dubois, 1970a.)

24.56, 24.57. *Scolopacitrema cubrensis* Sudarikov & Rykovsky, 1958. 24.56. Entire worm. 24.57. Terminal genitalia showing seminal vesicle (SV) with ejaculatory pouch (EP) and sucker-like genital atrium (SA). (After Dubois, 1970a.)

pretesticular. Vitellarium in hindbody, often confined to its posterior region. Seminal vesicle followed by muscular ejaculatory pouch situated dorsally. Copulatory bursa moderate or large, with protrusible genital cone half enclosed by ventrolateral preputial fold. Hermaphroditic duct opens at apex of cone. In alcedines. Holarctic, Neotropical and Oriental regions. Metacercariae of 'neascus' type, in fishes. Cercariae with rudimentary caeca, three pairs of penetration glands arranged in rows and flame-cell formula 2[(1 + 1 + 1) + (1 + 1 + 1 + [2])] = 16. Type-species *U. gracilis* Yamaguti, 1934.

11a. Ovary intertesticular; copulatory bursa evaginable .... *Mesoophorodiplostomum* Dubois, 1936. (Fig. 24.55)

Diagnosis: Body more or less distinctly bipartite; forebody oval; hindbody cylindrical. Pseudosuckers absent; oral sucker and pharynx small; ventral sucker larger than oral; holdfast organ small, oval, with median slit. Testes tandem, different in size and shape; anterior transversely oval, large; posterior reniform, larger. Ovary round, lateral, intertesticular, contiguous with anterior testis. Vitellarium extends from beyond ventral sucker in forebody to posterior margin of hind testis. Copulatory bursa evaginable, campaniform when expanded, opens dorsally. Hermaphroditic duct opens on conical eminence at its base. In larids. USA. Metacercariae in fishes. Type-species *M. pricei* (Krull, 1934) Dubois, 1936.

11b. Ovary pretesticular; copulatory bursa not evaginable .... 12.

12a. Vitellarium extending into greater part of holdfast organ; copulatory bursa without genital cone, with large sucker at base .... *Scolopacitrema* Sudarikov & Rykovsky, 1958. (Figs 24.56, 24.57)

Diagnosis: Body distinctly bipartite; forebody flat, oval, with ventral concavity; hindbody subcylindrical to claviform, longer than forebody. Pseudosuckers absent. Oral and ventral suckers and pharynx well developed. Holdfast organ massive, mushroom-like, with longitudinal groove. Testes tandem, relatively large, occupy more than half of hindbody. Ovary ellipsoidal, pretesticular, distant from junction of fore- and hindbody. Vitellarium extends throughout length of hindbody, and into both forebody and holdfast organ. Seminal vesicle with ejaculatory pouch, situated dorsally. Copulatory bursa with terminal opening. Short hermaphroditic duct opens at tip of small genital papilla close to opening of sucker-like genital atrium. In Scolopacidae. Europe. Type-species *S. cubrensis* Sudarikov & Rykovsky, 1958.

12b. Vitellarium not extending into holdfast organ; copulatory bursa with genital cone, without sucker at base .... 13.

13a. Copulatory bursa small, not evaginable; genital cone without circular fold (prepuce); ejaculatory pouch absent .... *Conodiplostomum* Dubois, 1937. (Figs 24.58, 24.59)

Diagnosis: Body bipartite; forebody spatulate or linguiform; hindbody cylindrical or conical. Pseudosuckers absent; oral and ventral suckers and pharynx relatively small; holdfast organ oval or round, with median slit. Testes of similar size, tandem, symmetrical. Ovary pretesticular, round or transversely oval, near junction of fore- and hindbody. Vitellarium in fore- and hindbody, reaching forward from level of ventral sucker to intestinal bifurcation. Copulatory bursa small or voluminous.

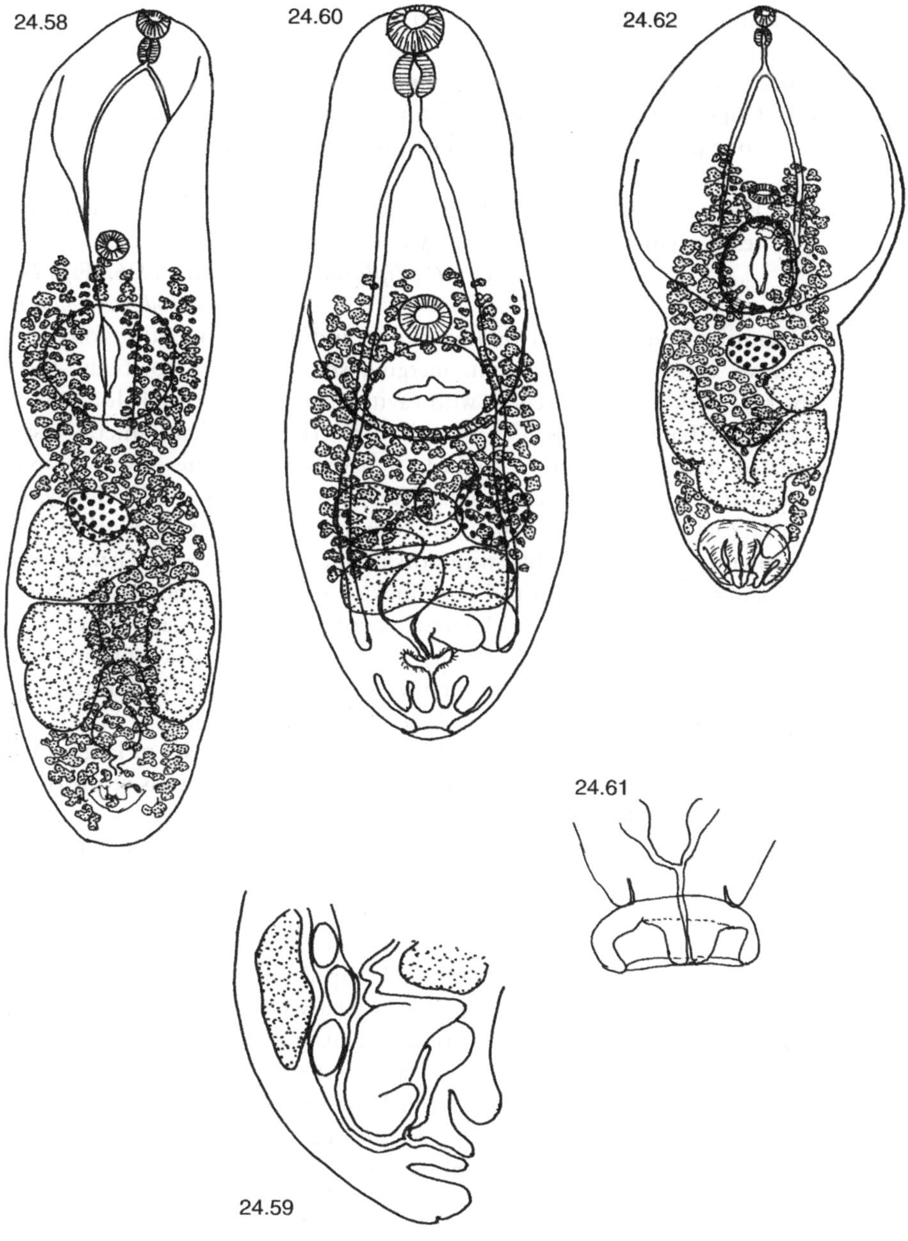

Figs 24.58–24.62.

24.58, 24.59. *Conodiplostomum perlatum* (Ciurea, 1911). 24.58. Entire worm. 24.59. Terminal genitalia with genital cone. (After Dubois, 1970a.)

24.60, 24.61. *Ornithodiplostomum scardinii* (Schulman, 1952). 24.60. Entire worm. 24.61. Protruded terminal genitalia. (After Sudarikov & Kurochkin, 1968.)

24.62. *Posthodiplostomum cuticola* (Nordmann, 1832). (After Dubois, 1970a.)

Hermaphroditic duct opens at tip of small genital cone. In Falconiformes, Strigiformes and Passeriformes. Cosmopolitan. Metacercariae of 'neascus' type, in fishes. Type-species *C. spathula* (Creplin, 1829) Dubois, 1937.

13b. Copulatory bursa evaginable; prepuce-like structure surrounding genital cone present; ejaculatory pouch present or absent .... 14.

14a. Body elliptical or linguiform, indistinctly bipartite; testes massive; anterior testis asymmetrical; ejaculatory pouch present .... *Ornithodiplostomum* Dubois, 1936. (Figs 24.60, 24.61)

(Syn. *Prolobodiplostomum* Baer, 1959)

Diagnosis: Body elliptical or linguiform, indistinctly bipartite; forebody ventrally concave; hindbody short, stout, merging smoothly with forebody. Pseudosuckers absent; holdfast organ circular, with cavity opening via ventral slit. Oral and ventral suckers and pharynx small, similar in size. Testes tandem; anterior asymmetrical, situated laterally; posterior bilobed, transversely elongate. Ovary ellipsoidal, pretesticular or opposite anterior testis. Vitellarium in fore- and hindbody, reaching to ventral sucker region. Seminal vesicle voluminous; ejaculatory pouch may be present. Copulatory bursa evaginable, with dorso-subterminal opening. Genital cone encloses hermaphroditic duct formed by fusion of uterus and ejaculatory duct. In Podicipedidae and Anatidae. Holarctic region, Africa. Metacercariae of 'neascus' type, in fishes. Cercariae with rudimentary alimentary system, lacking ventral sucker, three pairs of penetration glands in posterior part of body, flame-cell formula $2[(2 + 2) + (2 + [2])] = 16$. Type-species *O. ptychocheilus* (Faust, 1917) Dubois, 1936.

14b. Body more or less distinctly bipartite; testes moderate or large; anterior testis asymmetrical; ejaculatory pouch absent; posterior testis reniform, bilobed to twisted, with large indentation anteriorly .... *Posthodiplostomum* Dubois, 1936. (Fig. 24.62)

(Syn. *Choanouvulifer* Lung, 1966)

Diagnosis: Body distinctly bipartite; forebody flat or concave, oval, sometimes elongate or lanceolate; hindbody oval or claviform to subcylindrical. Pseudosuckers absent; holdfast organ round or almond-shaped, with cavity opening via median slit. Oral and ventral sucker feebly developed; pharynx small. Testes tandem, different in size and shape; anterior asymmetrical, lateral; posterior larger, bilobed, reniform or cordiform, sometimes twisted, with large indentation anteriorly. Ovary ellipsoidal, pretesticular, median, lateral or diagonal and tangential to anterior testis. Vitellarium in fore- and hindbody, extending to various levels in relation to ventral sucker. Copulatory bursa evaginable, with terminal or subterminal opening. Genital cone surrounded by prepuce, encloses hermaphroditic duct, which is formed at its base by union of uterus and ejaculatory duct; ejaculatory pouch absent. In Ciconiiformes. Cosmopolitan. Metacercariae of 'neascus' type, in fishes. Cercariae with rudimentary alimentary system; ventral sucker rudimentary or absent; three pairs of penetration glands in posterior part of body; flame-cell formula $2[(2 + 2) + (2 + 2 + [2])] = 20$. Type-species *P. cuticola* (Nordmann, 1832) Dubois, 1936.

25 Family Bolbocephalodidae Strand, 1935

K. NIEWIADOMSKA

W. Stefański Institute of Parasitology, Polish Academy of Sciences, Twarda, Warsaw, Poland

Introduction

The family, as the Bolbocephalidae, was created by Dubois (1934) for *Bolbocephalus intestiniforax* Dubois, 1934. Later, Strand (1935) proved that this generic name was pre-occupied and renamed it *Bolbocephalodes* and the family the Bolbocephalodidae. The aberrant morphology of the type-species of the genus and family showed unclear relationships with other diplostomatoids. Dubois (1936, 1938, 1953) considered it as a distinct subsuperfamily, the Bolbocephalodines, of the same rank as the Strigeines and Diplostomines within the superfamily 'Strigeides', or (Dubois, 1970b) the Bolbocephalodoinea in the superfamily Strigeoidea. Sudarikov (1960a, b) aligned it at the family level within the superfamily Diplostomatoidea Nicoll, 1937, while Yamaguti (1971) placed the subfamily Bolbocephalodinae within the family Strigeidae Railliet, 1919.

Family Bolbocephalodidae Strand, 1935

(Syn. Bolbocephalidae Dubois, 1934)

Diagnosis: Body distinctly bipartite; forebody massive, bulbous with cup-shaped thickening at base; hindbody small, conical or subovate. Pseudosuckers absent. Oral sucker absent; pharynx poorly developed; digestive tract reduced. Ventral sucker present. Holdfast organ in form of two transverse lips. Anterior testis asymmetrical, lateral; posterior larger, bilobed, in shape of dorsally-inverted horseshoe. Seminal vesicle winding ventral and posterior to posterior testis. Ovary opposite anterior testis. Uterus coiling in medial field of hindbody, ventral to ovary and testes. Vitellarium enclosed in cup-shaped thickening. Short hermaphroditic duct opens terminally on massive genital cone. Copulatory bursa occupied in most part by genital cone, may be evaginated through wide terminal pore. In Ciconiiformes. Italy, Syria. Type- and only genus *Bolbocephalodes* Strand, 1935.

Genus *Bolbocephalodes* Strand, 1935 (Figs 25.1, 25.2)

(Syn. *Bolbocephalus* Dubois, 1934 pre-occ.)

Diagnosis: With characters of family. Type-species *B. intestiniforax* (Dubois, 1934) Strand, 1935.

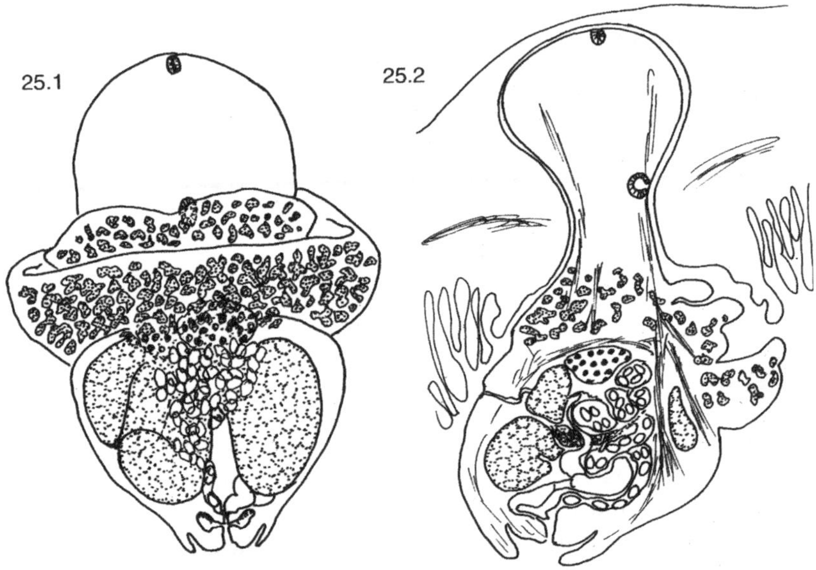

Figs 25.1–25.2. *Bolbocephalodes intestiniforax* (Dubois, 1934). 25.1. Entire worm. 25.2. Longitudinal section. (After Dubois, 1938.)

26 Family Brauninidae Wolf, 1903

K. Niewiadomska

W. Stefański Institute of Parasitology, Polish Academy of Sciences, Twarda, Warsaw, Poland

Introduction

Subfamily status was established by Wolf (1903) for the genus *Braunina* Heider, 1900. This genus was described by Heider (1900) without a specific designation. Wolf (1903) re-examined Heider's material, compared it with newly collected specimens and described *B. cordiformis* Wolf, 1903 as the type-species. The subfamily status was not accepted by Lutz (1935), but was restored by Szidat (1936a) and raised to family level by Dubois (1938, 1953). Dubois (1936) allocated it to the superfamily 'Cyathocotylides'. The family Brauninidae was mentioned by La Rue (1957) and Yamaguti (1971) as equivalent to the Cyathocotylidae Mühling, 1898.

Morphological characters justifying the familial status of this group include the structure of the holdfast organ and the presence of gonads and part of the uterus and caeca within it; these are features not apparent in the Cyathocotylidae. In addition, the host-group, marine mammals (cetaceans), is quite different.

Family Brauninidae Wolf, 1903

Diagnosis: Body cordiform, ventrally convex, with short conical or truncate caudal appendage. Holdfast organ large, overlaid by ventral fold forming deep cavity; on dorsal surface of body are two lateral oval protuberances caused by holdfast organ, plus small median one corresponding to pharynx. Oral and ventral suckers absent. Subterminal oral opening leads to pharynx. Oesophagus short; caeca sinuous, terminate in holdfast organ. Testes multilobed, symmetrical, in ventral part of holdfast organ. Cirrus-sac long, extending to base of holdfast organ; ejaculatory duct opens into copulatory bursa. Ovary small, round, close to right or left testis. Vitelline follicles in dorsal part of holdfast organ, distributed symmetrically. Uterus in ventral part of holdfast organ, with pore ventral to ejaculatory duct. In Cetacea. Coast of Europe, subtropical North America. Type- and only genus *Braunina* Heider, 1900.

Genus *Braunina* Heider, 1900 (Figs 26.1-26.3)

Diagnosis: With characters of family. Type-species *B. cordiformis* Wolf, 1903.

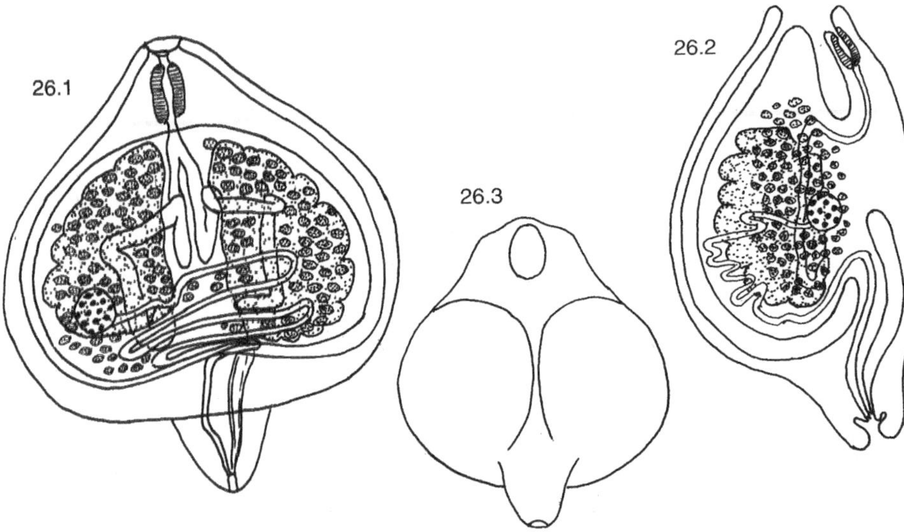

Figs 26.1–26.3. *Braunina cordiformis* Wolf, 1903, scheme of anatomy in ventral (26.1), lateral (26.2) and dorsal view (26.3). (All after Sudarikov, 1961.)

27 Family Cyathocotylidae Mühling, 1898

K. Niewiadomska

W. Stefański Institute of Parasitology, Polish Academy of Sciences, Twarda, Warsaw, Poland

Introduction

The family Cyathocotylidae is one of two families (the second is the family Brauninidae) having the characters both of the diplostomatoids and those of other digeneans. It is characterized, on the one hand, by the lack of a division of fore- and hindbody and by the presence of a cirrus-sac that encloses the seminal vesicle, pars prostatica and an eversible ejaculatory duct that forms a cirrus, and, on the other, by a sucker-like holdfast organ extending over all or part of the ventral surface of the body. The testes and ovary are round or oval and variable in position. Another characteristic feature is the broad host spectrum of the family (reptiles, birds and mammals) and even of particular genera, e.g. *Cyathocotyle* Mühling, 1896, which has species in reptiles and birds, and *Mesostephanus* Lutz, 1933, with species in birds and mammals. Generally, cyathocotylids demonstrate great morphological variability at the adult and also the metacercarial stage. On the contrary, the morphology of the cercaria is much more homogeneous. All cyathocotylid cercariae have the same structure of the excretory system, which is composed of four main excretory ducts united anteriorly; two lateral and two median ducts are also united anteriorly to the ventral sucker. The differences between the genera and subfamilies depend on the number and arrangement of the flame-cells in groups (single, two or three), the length of the furca and the presence or absence of fin-folds. Also, prosobranch snails are the specific host group of mother- and daughter-sporocysts. The cercariae are quite well known and many species have been described, but many of them have unknown life-cycles. According to the number and arrangement of flame-cells in the excretory system, four groups were established by Dubois (1951): *Vivax* Sewell, 1922 (36 flame-cells, in groups of three), *Novena* Dubois, 1951 (24 flame-cells, in groups of three), *Tetis* Sewell, 1922 (14 flame-cells; formula not clear) and *Tauiana* Szidat, 1933 (12 flame-cells, in groups of two). However, the relationships of these groups of cercariae in relation to the systematics of the family is not clear.

The subfamily status of cyathocotylids within the Holostomidae was recognized by Mühling (1898) after he had described *Cyathocotyle prussica* Mühling, 1896, the first representative of this group. Later, Poche (1926) created a separate family for them. At present, the family Cyathocotylidae comprises five subfamilies: Cyathocotylinae Mühling, 1898, Prosostephaninae Szidat, 1936, Prohemistominae Lutz, 1935, Szidatiinae Dubois, 1938 and Muhlinginae Mehra, 1950.

Szidat (1936a) created one more subfamily, the Pseudhemistominae, for the monotypic genus *Pseudhemistomum* Szidat, 1936. This genus, however, exhibits characters of the Cyathocotylinae and was rightly included in this subfamily by Dubois (1953) and Sudarikov (1961). The retention of the Pseudhemistominae by Yamaguti (1971) was unjustified.

The genera assigned to particular subfamilies differ according to various authors. The most satisfactory I find to be that of Dubois (1987), but with some modifications concerning the genera *Serpentostephanus* Sudarikov, 1961, *Neogogatea* Chandler & Rausch, 1947 and *Mesostephanoides* Dubois, 1951. Taking into account the characters such as the vitellarium entirely or largely confined to the holdfast organ, testes tandem, dorsal and overlaid by holdfast organ, in the type-genus *Prosostephanus* Lutz, 1935, the genus *Serpentostephanus* should be included in the subfamily Prosostephaninae, together with *Prosostephanus* and *Duboisia* Szidat, 1936. *Neogogatea*, which exhibits a cylindrical posterior part of the body with the whole or part of the gonads and vitellarium confined to the holdfast organ, should be considered a member of the subfamily Szidatiinae Dubois, 1938, together with *Szidatia* Dubois, 1938 and *Gogatea* Lutz, 1935, as proposed by Yamaguti (1971). *Mesostephanoides* was placed in the Szidatiinae by Dubois (1987), based on its specificity to reptiles. The incorporation of this genus within the subfamily Prohemistominae Lutz, 1935 by Yamaguti (1971) appears to me more appropriate.

Linstowiella Szidat, 1933 was unjustifiably synonymized with *Paracoenogonimus* Katsurada, 1914 by Sudarikov (1961) and Dubois (1983b). Mishchenko (1974) re-examined the life-cycle of *L. viviparae* (Linstow, 1877) and the morphology of particular developmental stages that justified the validity of *Linstowiella*.

Fengcotyle was erected by Tang & Tang (1989) for a Chinese bird parasite with the cirrus-sac situated horizontally and posteriorly to the testes, vitelline follicles surrounding the large holdfast organ and reaching to the posterior end of the body, testes lying obliquely and the ovary anterior to the left testis. These characters correspond with *Holostephanoides* Dubois, 1983, but the authors did not compare these genera. In spite of differences in host-specificity (*Holostephanoides* was found in fish), these genera can be considered synonymous.

The subfamily Muhlinginae was created by Mehra (1950) for the monotypic representative *Muhlingina lutrai* Mehra, 1950. This species, more precisely described by Mehra (1956), is characterized by the structure of the holdfast organ, which is unique for a diplostomoid. This is a parenchymatous and glandular rather than muscular organ, with a cavity reduced to a narrow slit close to the end of the body. It opens into a ventral cavity via a fairly large opening posterior to the pharynx. Other characters fit the family Cyathocotylidae.

Family Cyathocotylidae Mühling, 1898

Diagnosis: Diplostomoidea, with body generally undivided, oval, pyriform or linguiform, with small or sometimes elongate caudal appendix. Holdfast organ round or oval; oral sucker present; ventral sucker present or absent; pseudosuckers absent. Oesophagus short; caeca usually reaching close to posterior end of body. Position of ovary and testes variable. Cirrus-sac present, occasionally rudimentary, enclosing seminal vesicle, pars prostatica and cirrus. Genital pore terminal. Eggs large, not numerous. Vitellarium follicular, variable in extent. Life-cycle with longifurcate furcocercous cercaria having excretory system composed of four stems united anteriorly, two lateral and two median joined anterior to ventral sucker; ventral sucker present or absent. Metacercaria of 'prohemistomulum' type, with sucker-like holdfast organ and crown-like reserve bladder (Fig. 27.14). Parasites of reptiles, birds and mammals. Mother- and daughter-sporocysts in gastropods (Prosobranchia); metacercariae in fishes, amphibians and aquatic invertebrates. Type-genus *Cyathocotyle* Mühling, 1898.

Key to subfamilies

1a. Holdfast organ more or less strongly developed; ventral sucker near intestinal bifurcation; vitellarium extensive, surrounding, but not extending into, holdfast organ; gonads in various positions .... Cyathocotylinae Mühling, 1898.

1b. Holdfast organ variable in shape and size; vitellarium otherwise distributed; ventral sucker present or absent; gonads mainly in posterior part of body .... 2.

2a. Body with small cup-shaped anterior part and massive posterior region; vitellarium extensive, situated posterior to long holdfast organ; holdfast organ long, with cavity reduced to narrow slit, prolonged deep into posterior part of body; ventral sucker absent .... Muhlinginae Mehra, 1950.

2b. Body linguiform, pyriform or ovoid, with or without caudal appendage; vitellarium confined or not to holdfast organ; holdfast organ round or oval; ventral sucker present or absent .... 3.

3a. Body linguiform, pyriform or ovoid, with small caudal appendage devoid of gonads; holdfast organ small; ventral sucker (when present) in middle of body; vitellarium surrounding holdfast organ .... Prohemistominae Lutz, 1935.

3b. Body flattened or massive, with or without ventral concavity and caudal appendage; holdfast organ round, oval or elongate; ventral sucker variable in position; vitellarium confined to holdfast organ .... 4.

4a. Body with ventral concavity, with or without caudal appendage; holdfast organ large, occupying whole or part of ventral concavity; vitellarium entirely or largely confined to holdfast organ .... Prosostephaninae Szidat, 1936.

4b. Body linguiform; posterior portion elongate, well developed, enclosing some or all of gonads, with caudal appendage absent; holdfast organ moderate, in middle part of body; vitellarium lateral to holdfast organ or inside it .... Szidatiinae Dubois, 1938.

Subfamily Cyathocotylinae Mühling, 1898

(Syn. Pseudhemistominae Szidat, 1936)

Diagnosis: Body massive, round, oval or pyriform, with or without ventral concavity. Holdfast organ large, round or oval, with or without central cavity. Oral sucker terminal; ventral sucker near intestinal bifurcation, usually covered by holdfast organ. Pharynx well developed; oesophagus very short; caeca not reaching end of body. Testes diagonal, in posterior part of body. Cirrus-sac well developed, at posterior end of body. Genital pore terminal. Ovary variable in position in relation to testis, intertesticular or opposite anterior testis. Oötype postovarian. Vitellarium composed of comparatively large follicles, may or may not surround gonads, does not extend into holdfast organ. Parasites of reptiles, birds and mammals. Type-genus *Cyathocotyle* Mühling, 1896.

Key to genera

1a. Body massive, without ventral concavity; holdfast organ large, with central cavity, extends over ventral surface of body .... *Cyathocotyle* Mühling, 1896. (Fig. 27.1)

(Syns *Paracyathocotyle* Szidat, 1936; *Neocyathocotyle* Mehra, 1943)

Diagnosis: Body massive, oval, pyriform or fusiform, without ventral concavity. Holdfast organ large, round, with aperture of different shape, elevated above ventral surface. Oral sucker and pharynx well developed; ventral sucker small, near intestinal bifurcation, in some species absent or not visible, covered by holdfast organ. Oesophagus very short. Testes round or elongate, diagonal or symmetrical, varying in position. Cirrus-sac well developed, claviform, with large seminal vesicle at its base. Genital pore subterminal. Ovary round, small, variable in position and ventral to testes. Vitellarium in form of coarse follicles surrounding holdfast organ in peripheral part of body and overlying caeca, usually does not extend into organ. In different groups of birds, rarely in reptiles. Europe, Asia. Metacercariae of 'prohemistomulum' type, in fishes or leeches; one species in *Bithynia*. Cercariae, with flame-cell formula $2[(3+3)+(3+[3])]=24$, developing in Prosobranchia (*Bithynia*, *Bellamya*) or Pulmonata (*Bulinus*). Type-species *C. prussica* Mühling, 1896.

1b. Body massive with ventral concavity occupied by holdfast organ .... 2.

2a. Body round or oval; ventral concavity filled by large holdfast organ .... *Holostephanoides* Dubois, 1983. (Fig. 27.2)

(Syn. *Fengcotyle* Tang & Tang, 1989)

Diagnosis: Body round or oval; hypertrophied holdfast organ with large central cavity filling entire ventral concavity. Oral and ventral suckers well developed; ventral

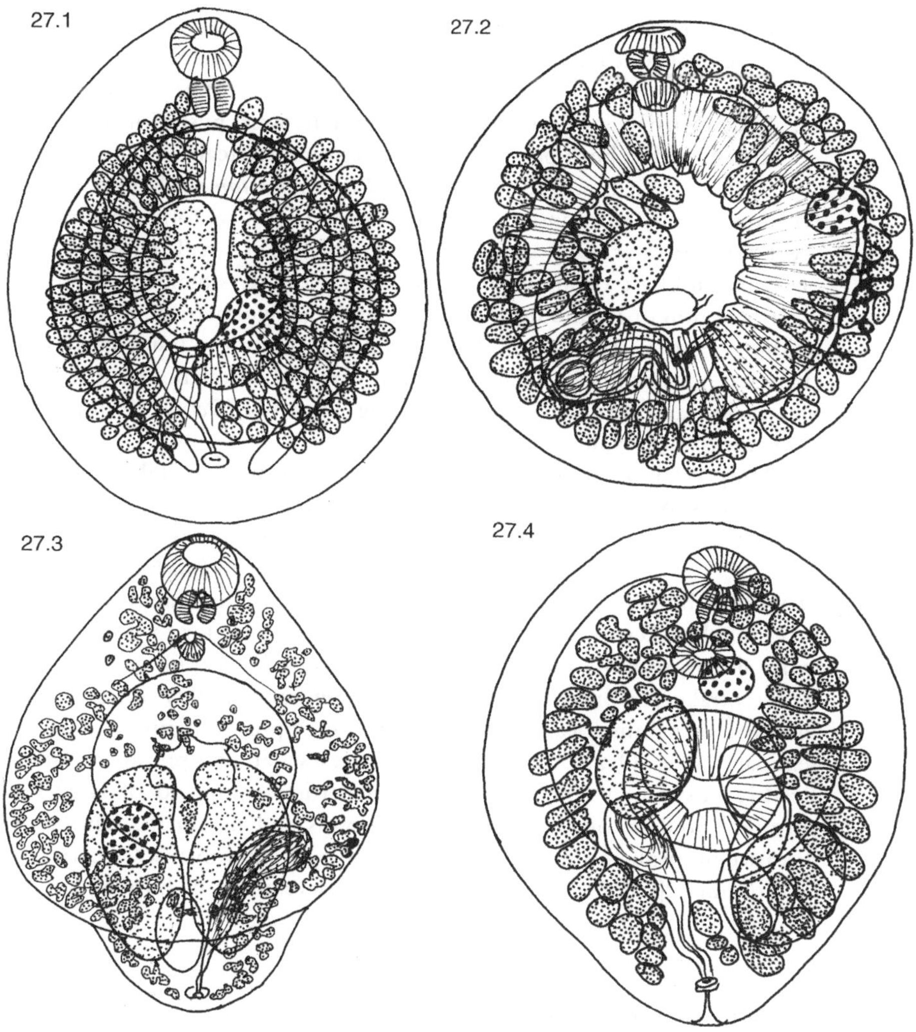

Figs 27.1–27.4.
27.1. *Cyathocotyle bithyniae* Sudarikov, 1973. (After Sudarikov, 1974.)
27.2. *Holostephanoides ictaluri* (Vernberg, 1952). (After Dubois, 1983a.)
27.3. *Pseudhemistomum unicum* Szidat, 1936. (After Dubois, 1938.)
27.4. *Holostephanus luehei* Szidat, 1936. (After Dubois, 1938.)

smaller, anterior to holdfast organ; pharynx near size of ventral sucker. Testes diagonal in posterior part of body. Cirrus-sac transverse, near posterior end of body, contains bipartite seminal vesicle. Ovary subequatorial, submedian, opposite anterior testis. Vitellarium composed of large follicles surrounding holdfast organ. In birds. North America, China. Metacercariae of 'prohemistomulum' type, in fishes. Cercariae, with flame-cell formula $2[(3 + 3 + 3) + (3 + 3 + [3])] = 36$, developing in Viviparidae (Prosobranchia). Type-species *H. ictaluri* (Vernberg, 1952) Dubois, 1983.

2b. Body round, oval or pyriform; holdfast organ rather small, occupies only part of ventral concavity .... 3.

3a. Body consists of large anterior and narrower posterior parts; vitellarium distributed in both parts of body; testes symmetrical .... *Pseudhemistomum* Szidat, 1936. (Fig. 27.3)

Diagnosis: Body with anterior region flattened, oval or foliiform; posterior region short and narrower. Oral sucker large; ventral sucker small, poorly developed, near pharynx. Holdfast organ round with central cavity, situated in about middle of anterior region or in middle of body. Testes large, symmetrical, posterior to cavity of holdfast organ. Cirrus-sac well developed. Ovary ventral to testes, on opposite side to cirrus-sac. Vitellarium of small follicles distributed throughout entire body between oral sucker and terminal genital opening. In birds (*Sterna*). Europe. Type-species *P. unicum* Szidat, 1936.

3b. Body oval or pyriform, not divided into parts; vitellarium surrounding holdfast organ; testes symmetrical or diagonal .... *Holostephanus* Szidat, 1936. (Fig. 27.4)

(Syn. *Cyathocotyloides* Szidat, 1936)

Diagnosis: Body round, oval or pyriform, with ventral concavity variable in development. Oral sucker well developed; ventral smaller, anterior to holdfast organ, sometimes absent; pharynx rather small. Holdfast organ sucker-like, of variable size, with large central cavity or hollow, situated inside ventral concavity. Testes oval or elongate, diagonal or symmetrical. Cirrus-sac well developed, containing bipartite internal seminal vesicle. Ovary round or oval, pretesticular or opposite to anterior testis. Relative position of gonads variable. Vitellarium composed of large follicles that surround holdfast organ and gonads. In birds and mammals. Europe, Asia. Metacercariae of 'prohemistomulum' type, in fishes or amphibians. Cercariae, with flame-cell formula $2[(2+2+2)+(2+2+[2])]=24$, developing in Bithyniidae. Type-species *H. luehei* Szidat, 1936.

Subfamily Muhlinginae Mehra, 1950

Diagnosis: Body composed of small cup-shaped anterior part and massive posterior region. Oral sucker well developed, terminal; ventral sucker absent. Pharynx well developed; oesophagus short; intestinal caeca outside holdfast organ, terminate at small distance from posterior end of body. Ventral cavity in anterior part of body encloses round, sucker-like part of holdfast organ with funnel-shaped opening, whose disc-shaped margin usually projects as two eversible flaps or lobes. Holdfast organ prolonged deep into posterior part of body almost to extremity, terminates as glandular, feebly muscular parenchymatous structure, with cavity reduced to narrow slit. Gonads outside holdfast organ; testes entire, almost symmetrical; ovary dorsal to anterior testis. Cirrus-sac large, stout, muscular. Uterus outside holdfast organ; metraterm well developed; eggs few, large. In mammals (mustelids). India. Type- and only genus *Muhlingina* Mehra, 1950.

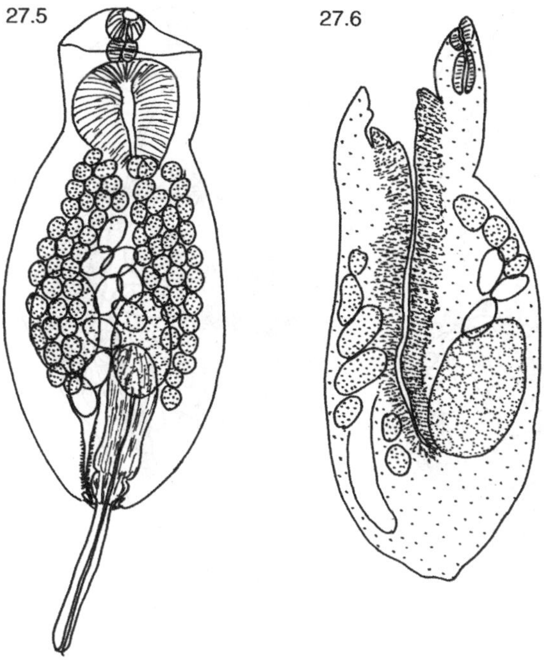

Figs 27.5–27.6. *Muhlingina lutrai* Mehra, 1950. 27.5. General morphology. 27.6. Sagittal section showing narrow slit of glandular parenchymatous holdfast organ. (After Mehra, 1956.)

Genus *Muhlingina* Mehra, 1950 (Figs 27.5, 27.6)

Diagnosis: With characters of subfamily. Type-species *M. lutrai* Mehra, 1950.

Subfamily Prohemistominae Lutz, 1935

Diagnosis: Body oval, linguiform or pyriform, concave ventrally, with small caudal appendage. Oral sucker, ventral sucker and pharynx quite well developed; ventral sucker near middle of body. Oesophagus short or quite long; caeca long or terminating well short of posterior extremity. Holdfast organ round, with median slit or central cavity, rather small, posterior to ventral sucker. Testes round or oval, tandem. Cirrus-sac long, lateral to testes. Genital opening terminal. Ovary variable in position, submedian, posterolateral to anterior testis or intertesticular. Vitellarium surrounding holdfast organ, in posterior part of body. In reptiles, birds and mammals. Type-genus *Prohemistomum* Odhner, 1913.

Key to genera

1a. Holdfast organ small, round, overlies anterior testis; vitellarium horseshoe-shaped, encircles holdfast organ; cirrus-sac reaching anterior testis .... *Mesostephanoides* Dubois, 1951. (Fig. 27.7)

Diagnosis: Body oval, flattened, ventrally concave, with small conical posterior appendage. Oral sucker and pharynx moderately developed; ventral sucker smaller than oral sucker, situated in middle of body. Oesophagus rather long; caeca

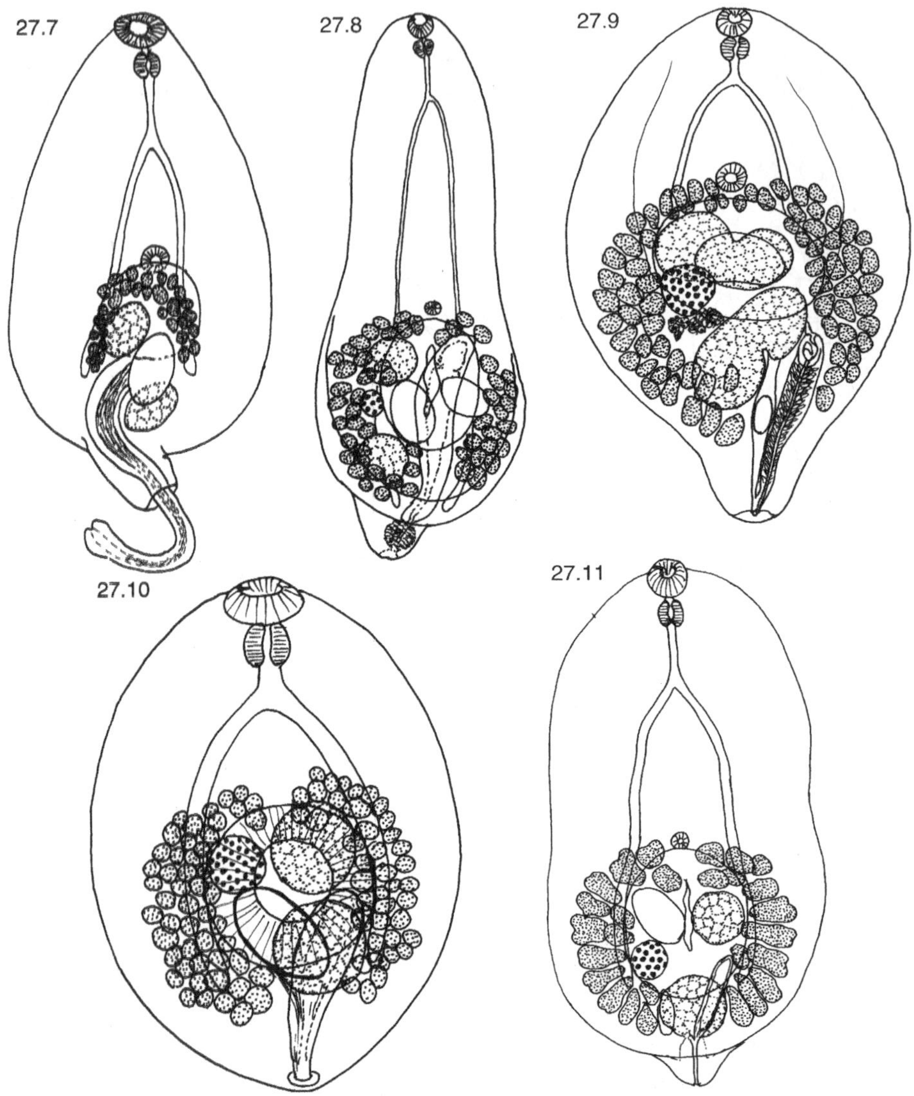

Figs 27.7–27.11.
27.7. *Mesostephanoides burmanicus* (Chatterji, 1940). (After Dubois, 1951.)
27.8. *Mesostephanus appendiculatoides* (Price, 1934). (After Dubois, 1938.)
27.9. *Prohemistomum odhneri* (Travassos, 1924). (After Sudarikov, 1961.)
27.10. *Linstowiella viviparae* (Linstow, 1877). (After Mishchenko, 1974.)
27.11. *Paracoenogonimus ovatus* Katsurada, 1914. (Original.)

terminating well short of posterior extremity. Holdfast organ small, round or oval, with median cavity, situated just posterior to ventral sucker. Testes ovoid, tandem or diagonal; anterior testis intercaecal. Cirrus-sac large, claviform, containing seminal vesicle, prostatic complex and spinulate protrusible cirrus. Ovary round, postero-lateral to anterior testis or intertesticular. Vitellarium horseshoe-shaped, encircles holdfast organ anteriorly, close to ventral sucker. Uterus with few large eggs; metraterm with muscular wall but without sphincter. Genital pore terminal. In reptiles. Myanmar. Type-species *M. burmanicus* (Chatterji, 1940) Dubois, 1951.

1b. Holdfast organ round, oval or transversely oval, situated in posterior part of body; vitellarium well developed, forming two lateral fields of large follicles, which are united anteriorly or not; cirrus-sac small or large, may reach anterior testis .... 2.

2a. Vaginal sphincter present .... *Mesostephanus* Lutz, 1935. (Fig. 27.8)
(Syn. *Gelanocotyle* Sudarikov, 1961)

Diagnosis: Body oval or linguiform, foliiform or cochleariform, with ventral concavity in posterior part of body and small dorsoterminal appendage at end of body. Oral sucker, ventral sucker and pharynx moderately developed; oesophagus rather short; caeca reaching close to posterior extremity. Ventral sucker at about middle of body. Holdfast organ protrusible, small or medium in size, round, with median slit or central cavity. Testes subspherical or oval, tandem. Cirrus-sac well developed. Genital pore terminal. Ovary round, posterior or posterolateral to anterior testis, opposite cirrus-sac. Vitellarium of large follicles, posterior to ventral sucker, surrounding holdfast organ and gonads but not confluent posteriorly. Vaginal sphincter ellipsoidal, sometimes indistinct but always present. In birds and mammals. Cosmopolitan. Metacercariae of 'prohemistomulum' type, in fishes or amphibians. Cercariae with flame-cell formula $2[(3 + 3 + 3) + (3 + 3 + [3])] = 36$. Type-species *M. fajardensis* (Price, 1934) Lutz, 1935.

2b. Vaginal sphincter absent .... 3.

3a. Body elongate posteriorly but without caudal appendage; ventral concavity in form of shallow pocket which may overlie holdfast organ; cirrus-sac well developed .... *Prohemistomum* Odhner, 1913. (Fig. 27.9)

Diagnosis: Body small, oval or pyriform, flattened, with deep ventral concavity, without caudal appendage. Oral sucker, ventral sucker and pharynx of similar size. Holdfast organ relatively small, round or oval, with median slit, in middle of body, just posterior to ventral sucker, sometimes overlain by posterior margin of ventral concavity. Testes transversely elongate, tandem, in posterior part of body. Cirrus-sac well developed. Genital pore terminal or subterminal. Ovary round, submedian or lateral near anterior testis. Vitellarium of large follicles in two lateral fields, posterior to holdfast organ or united anterior to it. Vaginal sphincter absent. In predatory birds and frigate-birds, occasionally in mammals. Subtropical North America. Metacercariae of 'prohemistomulum' type, in fishes. Cercariae with flame-cell formula $2[(3 + 3 + 3) + (3 + 3 + [3])] = 36$. Type-species *P. vivax* (Sonsino, 1892) Odhner, 1913.

3b. Body linguiform or oval; caudal appendage present or absent; body without pocket-like ventral concavity; cirrus-sac variable in size .... 4.

4a. Oral sucker and pharynx well developed; ventral sucker absent; cirrus-sac well developed; vitellarium in two separate lateral fields .... *Linstowiella* Szidat, 1933. (Fig. 27.10)

Diagnosis: Body small, oval, with or without very short caudal appendage. Oral sucker and pharynx well developed; ventral sucker absent. Oesophagus short; caeca extending close to end of body. Holdfast organ inconspicuous, with median slit. Testes tandem or slightly diagonal. Cirrus-sac well developed, containing seminal vesicle, pars prostatica and cirrus. Genital pore terminal. Ovary opposite to anterior testis. Vitellarium of large follicles, in posterior part of body, enveloping holdfast organ laterally. In birds (Anatidae, accidentally Laridae). Europe. Metacercariae of 'prohemistomulum' type, in molluscs and fishes. Cercariae with flame-cell formula 2[(3 + 3) + (3 + [3])] = 24, developing in Viviparidae. Type-species *L. viviparae* (Linstow, 1877) Szidat, 1933.

4b. **Oral sucker and pharynx relatively small; ventral sucker present; cirrus-sac rather small; vitellarium may unite closely anterior or posterior to holdfast organ .... *Paracoenogonimus* Katsurada, 1914. (Fig. 27.11)**

Diagnosis: Body small, oval, flattened, with or without conspicuous ventral concavity; caudal appendage absent or very small. Oral sucker and pharynx rather well developed; ventral sucker present, just anterior to holdfast organ. Oesophagus relatively long; caeca extend almost to posterior end of body. Holdfast organ rather small, with median slit, situated in posterior part of body. Testes nearly oval, oblique or tandem, at various distances apart, in posterior half of body. Cirrus-sac moderately developed, club-shaped. Genital opening terminal. Ovary round, small, submedian or lateral, near anterior testis. Vitellarium surrounding holdfast organ and gonads. Uterus with large eggs. Vaginal sphincter absent. In birds, rarely in mammals. Europe. Metacercariae of 'prohemistomulum' type, in fishes. Cercariae with flame-cell formula 2[(3 + 3 + 3) + (3 + 3 + [3])] = 36, developing in Viviparidae. Type-species *P. ovatus* Katsurada, 1914.

Subfamily Prosostephaninae Szidat, 1936

Diagnosis: Body massive, with ventral concavity, with or without conical caudal process. Oral sucker rather small; ventral sucker present or absent. Pharynx well developed; oesophagus short or almost absent; caeca long, extending to posterior end of body or not. Holdfast organ round, large, occupying whole or part of ventral concavity. Testes large, tandem. Cirrus-sac well developed, long, may reach anterior testis. Genital opening terminal. Ovary laterally intertesticular. Vitellarium of coarse follicles, entirely or partly confined to holdfast organ, surrounding gonads or only overlying testes. Uterus with few large eggs. In reptiles and mammals. Type-genus *Prosostephanus* Lutz, 1935.

Key to genera

1a. **Holdfast organ large, occupying whole ventral concavity; testes tandem, large, in central part of body .... *Prosostephanus* Lutz, 1935. (Fig. 27.12)**
(Syns *Travassosella* Faust & Tang, 1938; *Tangiella* Sudarikov, 1961)

Diagnosis: Body oval or pyriform, with ventral concavity; posterior end in form of blunt conical process. Oral sucker and pharynx well developed; ventral sucker present or absent. Oesophagus very short; caeca ending at posterior margin of testis. Holdfast organ large, occupying whole ventral concavity, extends forwards to level of pharynx or oral sucker and overlaps ventral sucker. Testes oval, large or very large, tandem, dorsal in mid-region of body. Cirrus-sac well developed, ventral to testes, reaching anterior testis. Ovary round, comparatively small, median or submedian, ventral to anterior testis; oötype intertesticular. Vitellarium of coarse follicles, entirely or largely confined to holdfast organ, forming two lateral groups. Uterus extends past anterior testis anteriorly. Eggs not numerous, large. In mammals. Asia. Metacercaria of 'prohemistomulum' type, in fishes. Cercariae with flame-cell formula $2[(3 + 3 + 3) + (3 + 3 + [3])] = 36$. Type-species *P. industrius* (Tubangui, 1922) Lutz, 1935.

1b. Body pouch-like with holdfast organ inside; testes large or relatively small, variable in position, in posterior part of body .... 2.

2a. Holdfast organ large, reaching to pharynx, enclosed in ventral concavity; vitellarium forms two lateral masses, in middle part of body; ventral sucker close to pharynx .... *Duboisia* Szidat, 1936. (Fig. 27.13)

Diagnosis: Body oval or pyriform with pocket-like ventral concavity. Holdfast organ large, oval, enclosed in concavity, its anterior border reaching to pharynx. Oral sucker and pharynx well developed; ventral sucker small, overlain by holdfast organ, or absent. Testes oval or rounded, diagonal or symmetrical, in posterior part of body. Cirrus-sac long, strongly developed, reaches anterior margin of ovary. Genital pore terminal. Ovary round or oval, anterior to posterior testis. Vitellarium of large follicles that form two lateral masses and extend into holdfast organ and reach level of intestinal bifurcation. Vaginal sphincter present. In birds (*Ciconia* and *Alcedo*). Europe, Asia. Type-species *D. syriaca* (Dubois, 1934) Szidat, 1936.

2b. Holdfast organ in posterior part of body; ventral sucker far from intestinal bifurcation; body without deep ventral concavity; vitellarium horseshoe-shaped, posterior to ventral sucker .... *Serpentostephanus* Sudarikov, 1961. (Figs 27.14, 27.15)

Diagnosis: Body small, oval, with small ventral concavity. Caudal appendage poorly developed or absent. Oral sucker, ventral sucker and pharynx well developed. Holdfast organ large, round, without cavity, extending over posterior part of ventral surface; encloses gonads, uterus and vitellarium; its anterior border reaching or covering ventral sucker. Testes laterally enlarged or round, tandem, in posterior part of body. Cirrus-sac moderately developed, reaching level of ovary. Genital pore terminal. Ovary oval, laterally intertesticular. Vitelline follicles large, surrounding gonads, mainly inside holdfast organ. Uterus with few large eggs. In snakes. Tunisia. Type-species *S. natricis* (Dubois, 1958) Sudarikov, 1961.

Subfamily Szidatiinae Dubois, 1938

(Syn. Gogatinae Mehra, 1947)

Diagnosis: Body composed of flattened, elongate anterior part and shorter, cylindrical posterior part. Holdfast organ oval or elongate, with median slit, variable in size.

27.12

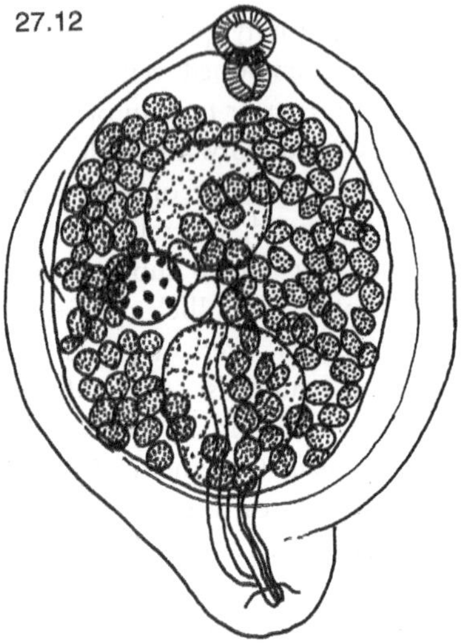

27.13

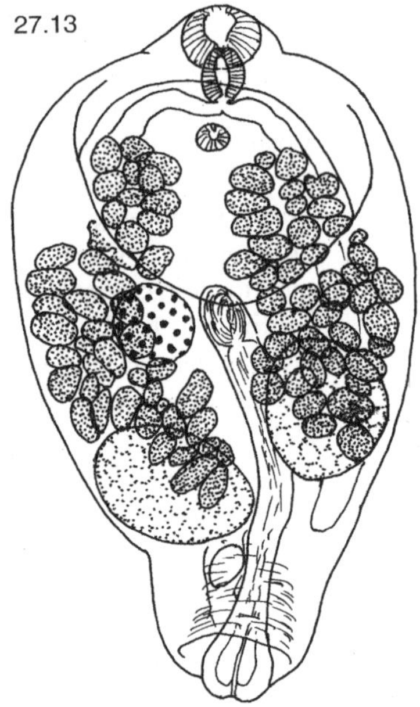

27.14

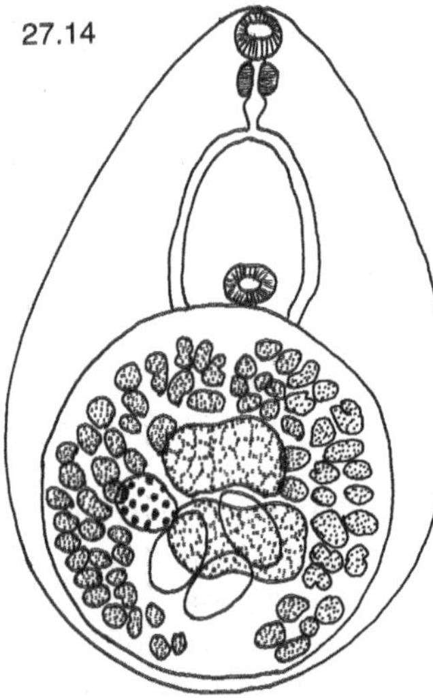

27.15

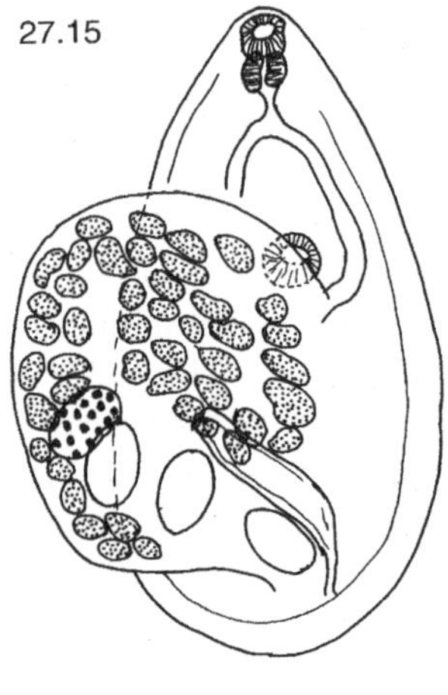

Figs 27.12–27.15.
27.12. *Prosostephanus industrius* (Tubangui, 1922). (After Dubois, 1938.)
27.13. *Duboisia syriaca* (Dubois, 1934). (After Dubois, 1938.)
27.14, 27.15. *Serpentostephanus natricis* (Dubois, 1958). (After Dubois, 1958.)

Ventral sucker present or absent. Testes tandem, partly or entirely situated in posterior part of body. Genital pore terminal. Ovary intertesticular. Vitellarium lateral to holdfast organ or inside it, reaching back to posterior testis. In snakes and birds. Type-genus *Szidatia* Dubois, 1938.

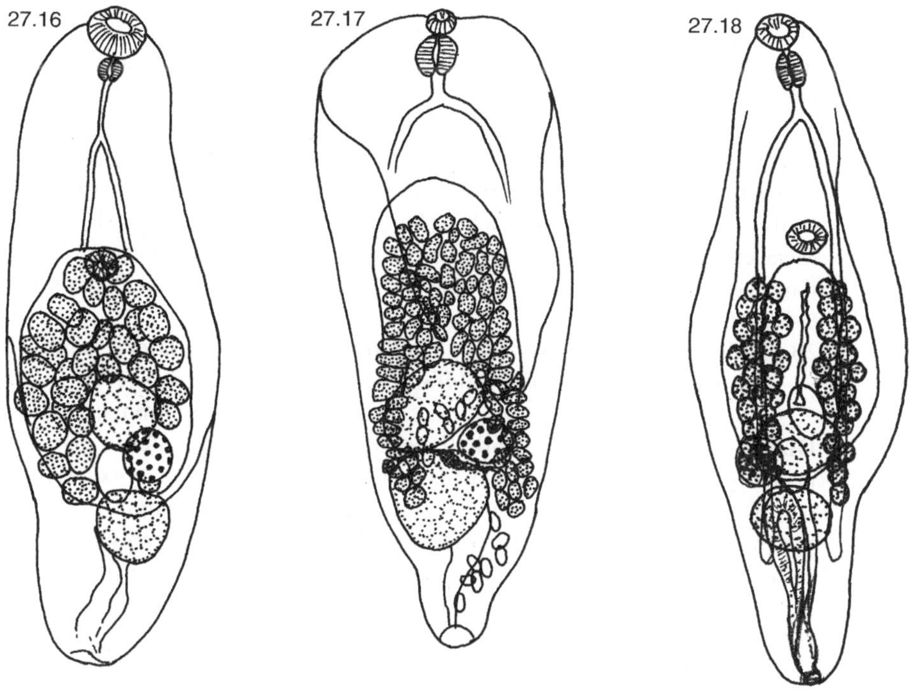

Figs 27.16–27.18.
27.16. *Gogatea serpentium* (Gogate, 1932). (After Dubois, 1938.)
27.17. *Neogogatea bubonis* Chandler & Rausch, 1947. (After Sudarikov, 1961.)
27.18. *Szidatia joyeuxi* (Hughes, 1929). (After Dubois, 1951.)

Key to genera

1a. Body elongate, linguiform, with distinct elongate posterior part containing posterior testis; holdfast organ oval, large, with vitellarium confined within it; testes round, in posterior half of body .... *Gogatea* Lutz, 1935. (Fig. 27.16)

Diagnosis: Body elongate, linguiform; anterior part flattened with shallow ventral concavity; posterior part cylindrical. Oral sucker well developed; ventral sucker smaller, situated in middle of ventral concavity; pharynx small. Holdfast organ oval, well developed, situated posterior to ventral sucker or overlying it. Testes round, tandem; posterior testis in cylindrical part of body. Cirrus-sac well developed. Genital pore terminal. Ovary round, submedian, intertesticular. Vitellarium of large follicles confined to holdfast organ. Eggs large. In snakes. Asia. Type-species *G. serpentium* (Gogate, 1932) Lutz, 1935.

1b. Body spatulate or concave anteriorly; posterior part of body narrower, conical; holdfast organ small or large, oval or elongate; vitellarium confined or partly confined to holdfast organ, in form of lateral bands confluent anteriorly or not; testes oval or transversely elongate, in posterior part ...... 2.

2a. Holdfast organ elongate, linguiform; vitellarium in form of horseshoe, mainly confined to holdfast organ, reaching back to posterior testis .... *Neogogatea* Chandler & Rausch, 1947. (Fig. 24.17)

Diagnosis: Body elongate, spatulate; anterior part flattened; posterior narrower, conical. Oral sucker present; ventral sucker absent. Pharynx large; oesophagus short. Holdfast organ large, oval, in middle or in posterior part of body, occupying about half of body length. Testes oval, tandem, in posterior part of body. Cirrus-sac elongate, feebly developed. Genital pore terminal. Ovary oval, laterally intertesticular or opposite anterior testis. Vitellarium of large follicles, in form of elongate horseshoe, confined to holdfast organ, reaching back to posterior testis. In birds. North America. Metacercariae of 'prohemistomulum' type, in fishes and anurans. Cercariae with flame-cell formula 2[(3 + 3 + 3) + (3 + 3 + [3])] = 36. Type-species *N. bubonis* Chandler & Rausch, 1947.

2b. Holdfast organ small, oval; vitellarium forms two lateral bands extending from anterior margin of holdfast organ to intertesticular level .... *Szidatia* Dubois, 1938. (Fig. 27.18)

Diagnosis: Body elongate, with flattened, wide, concave anterior part and narrow posterior part containing gonads. Oral sucker and pharynx well developed; oesophagus short; caeca terminating at level of posterior margin of posterior testis. Ventral sucker in middle of anterior part of body. Holdfast organ small, elongate, with median slit. Testes small, transversely elongate, tandem, in posterior part of body. Cirrus-sac well developed, reaching posterior testis. Genital pore terminal. Ovary round, small, median or submedian, intertesticular. Uterus with few large eggs. Vitellarium in two lateral fields between anterior margin of holdfast organ and intertesticular level. In snakes. Africa. Metacercariae of 'prohemistomulum' type, in anurans and fishes. Type-species *S. joyeuxi* (Hughes, 1929) Dubois, 1938.

28 Family Proterodiplostomidae Dubois, 1936

K. Niewiadomska

W. Stefański Institute of Parasitology, Polish Academy of Sciences, Twarda, Warsaw, Poland

Introduction

Dubois (1936) separated the diplostomoids of reptiles from those of birds and mammals into a distinct family, which he called the 'Proterodiplostomatidae'. This not very rich group differs from other diplostomoids not only by its host group but also by the presence of a so-called paraprostatic gland or paraprostate. This organ is not comparable with any part of the distal genitalia of distomes and is not well developed in this group of diplostomatoids. The paraprostate is an independent organ with the shape of a thin- or thick-walled tubule or pouch, surrounded by gland-cells and containing their secretion. The paraprostate efferent duct is independent of the ejaculatory duct, and consequently proterodiplostomids have three genital ducts: uterus, ejaculatory duct and paraprostate. The topography of the genital ducts in the vicinity of the copulatory bursa includes: a ventrally located uterus in the form of a tube, which may or may not be modified, opening into the copulatory bursa at the base of the genital cone; an ejaculatory duct descending dorsally to the uterus and uniting with the latter to form the hermaphroditic duct; and the paraprostate duct descending into the genital cone and opening at its tip. This arrangement may sometimes be greatly modified. The topography of the genital ducts, the possible development of the paraprostate in the Proterodiplostomidae and its origin and function have been analysed by Byrd & Reiber (1942).

The systematics of the Proterodiplostomidae, first elaborated by Dubois (1938, 1953), was based on host-specificity, holdfast organ shape and size, the presence or absence of papillae situated at the borders of the median slit of the holdfast organ and the distribution of the vitellarium. Dubois considered that the family comprised two 'supersubfamilies' - parasites of crocodiles and chelonians and parasites of snakes: the first with two subfamilies, the 'Proterodiplostomatinae' (holdfast organ small or of medium size, papillate, vitellarium in fore- and hindbody) and the Polycotylinae (holdfast organ small or of medium size, papillate, vitellarium in forebody); and the second with one, the 'Ophiodiplostomatinae' (holdfast organ massive, devoid of papillae, vitellarium in fore- and hindbody). Similar systematics based on the same characters but adjusted to the rank of superfamily and comprising the families 'Proterodiplostomatidae' and 'Ophiodiplostomatidae' were adopted by Sudarikov (1960b). He distinguished the subfamilies 'Proterodiplostomatinae', Polycotylinae and Massoprostatinae in the first family, and 'Ophiodiplostomatinae' and Proalarioidinae in the second. This division was based on the distribution of the vitelline follicles, which in the Polycotylinae are concentrated in the forebody, but in small numbers, and reach back to the posterior testis in the hindbody. He created separate families for the genera *Massoprostatum* Caballero, 1948 and *Proalarioides* Yamaguti, 1933. This systematic arrangement was also accepted by Yamaguti (1971), who added the new subfamily Crocodilicolinae.

Dubois (1970c, 1979) revised the family Proterodiplostomidae and presented (Dubois, 1979) a new key to the genera of diplostomids of reptiles. The main division, as in the case of his previous key, was based on host-specificity and remained unchanged, but some new synonyms were added. Among the latter, the subfamily Crocodilicolinae was synonymized with the Polycotylinae.

A new approach to the systematics of Proterodiplostomidae was proposed by Brooks *et al.* (1992). This cladistic classification is based on the arrangement of the terminal genitalia. These authors considered that there were four conditions: (i) paraprostate fused with the ejaculatory duct and uterus opening separately; (ii) uterus fused with the paraprostate and the ejaculatory duct; (iii) paraprostate fused first with the uterus and then with the ejaculatory duct; and (iv) paraprostate opening separately. As a result of their cladistic approach, they confirmed the separateness of species specific to crocodilians and chelonians and those species specific to snakes. Inside these groups, however, the various subdivisions comprise other subsets of species. Although an interesting step in the systematics of the group, this work is too preliminary to be of use in the present key.

The genera *Neelydiplostomum* Gupta, 1958 and *Petalodiplostomum* Dubois, 1936 were recognized by Dubois (1979) as synonyms of *Herpetodiplostomum* Dubois, 1936 and *Ophiodiplostomum* Dubois, 1936, respectively, although Sudarikov (1960b) had synonymized *Neelydiplostomum* with *Pseudoneodiplostomum* Dubois, 1936. Dubois (1970b) also recognized as superfluous the subgenus *Pseudoneodiplostomoides* created by Yamaguti (1954) for *Pseudoneodiplostomum crocodili* Yamaguti, 1954 on the basis of some structures which appear not to be unique.

Data on the life-cycles and morphology of the metacercariae are very scarce and cannot be used as a basis for the verification of the taxonomic status of the family. At present, the classification of Dubois (1979) is accepted as a basis for further elaboration.

Family Proterodiplostomidae Dubois, 1936

Diagnosis: Body more or less distinctly bipartite; forebody flattened, with or without pseudosuckers; hindbody cylindrical, claviform or conical, sometimes with thick-walled capsule or series of suckers. Holdfast organ variable in size, may be provided with papillae. Oral and ventral suckers present; oesophagus slender; caeca may or may not reach posterior end of body. Testes round or oval, tandem or slightly diagonal.

Ovary pretesticular; oötype intertesticular, occasionally pretesticular. Vitellarium in fore- and hindbody or restricted to either fore- or hindbody. Seminal vesicle posterior to hind testis; paraprostate commonly present. Genital cone usually present. Genital atrium wide or reduced, sometimes with sucker-like form. Uterus may or may not extend into forebody, opens together with ejaculatory duct at base of genital atrium or at tip of genital cone either in form of hermaphroditic duct or by separate pores. Parasitic exclusively in reptiles. Type-genus *Proterodiplostomum* Dubois, 1936.

Key to subfamilies

1a. Parasites of snakes; holdfast organ large without papillae .... 2.
1b. Parasites of crocodiles and chelonians; holdfast organ rather small, with or without papillae (lacking in parasites of chelonians) .... 3.

2a. Pseudosuckers absent; vitellarium in fore- and hindbody .... Ophiodiplostominae Dubois, 1936.
2b. Pseudosuckers present; vitellarium essentially in forebody, but several follicles extend into anterior hindbody ..... Proalarioidinae Sudarikov, 1960.

3a. Vitellarium confined to hindbody .... Massoprostatinae Yamaguti, 1958.
3b. Vitellarium in fore- and hindbody .... 4.

4a. Vitellarium mainly in forebody, with posterior limit not extending beyond posterior testis .... Polycotylinae Monticelli, 1888.
4b. Vitellarium in fore- and hindbody, extending beyond posterior testis .... Proterodiplostominae Dubois, 1936.

Subfamily Massoprostatinae Yamaguti, 1958

Diagnosis: Body bipartite; forebody small, spoon-shaped; hindbody cylindrical, very long. Pseudosuckers absent; oral and ventral suckers small; holdfast organ small, round, with papillate margins of aperture. Pharynx small; oesophagus short; caeca reaching close to posterior end of body. Testes tandem but slightly diagonal, near posterior extremity. Ovary pretesticular. Vitellarium extends full length of hindbody. Paraprostate long, sinuous, opening beside ejaculatory duct at tip of small genital cone. Uterus opens separately. Copulatory bursa small, with terminal aperture. In crocodilians. Neotropical. Type- and only genus *Massoprostatum* Caballero, 1948.

Genus *Massoprostatum* Caballero, 1948 (Figs 28.1, 28.2)

Diagnosis: With characters of subfamily. Type-species *M. longum* Caballero, 1948.

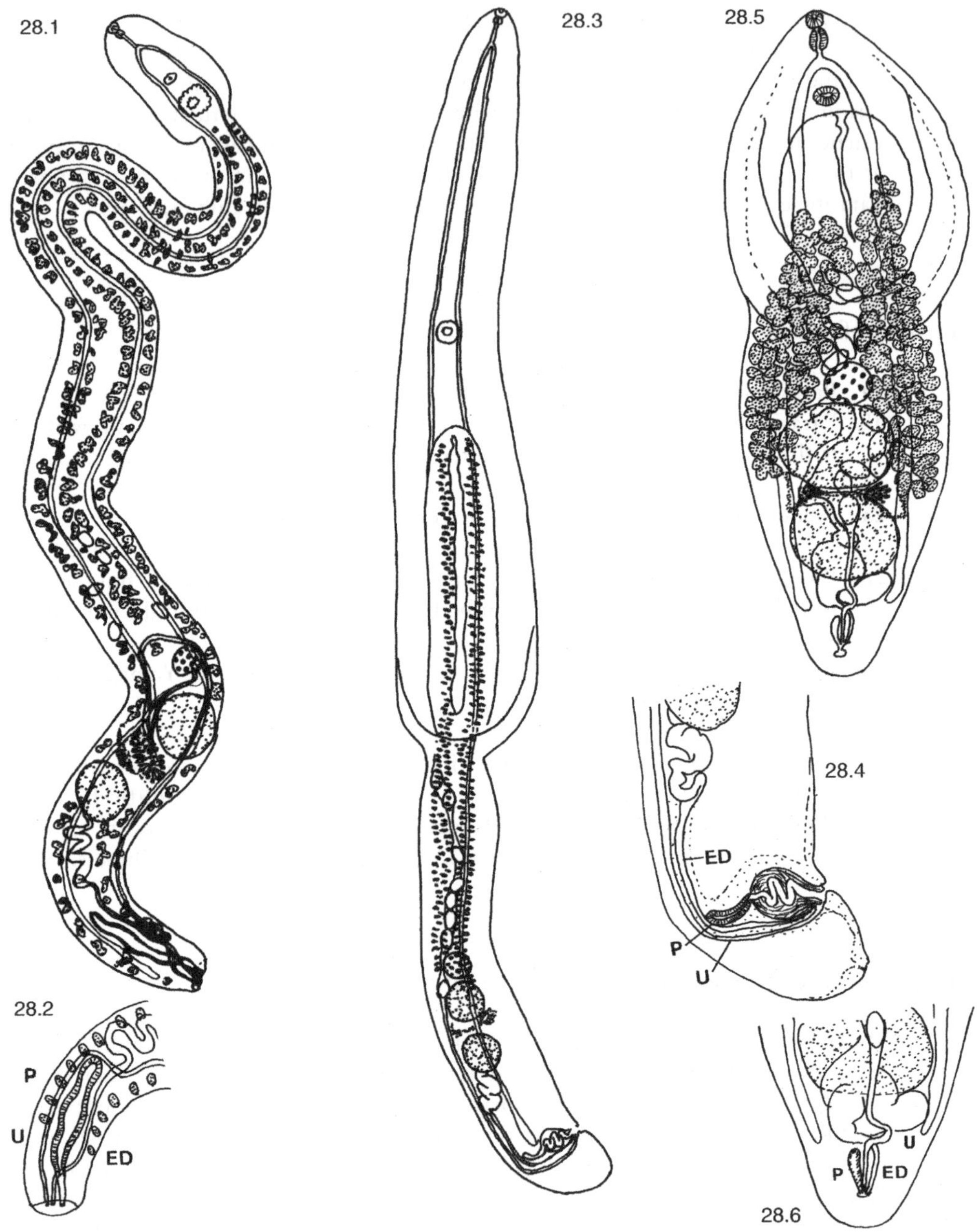

Figs 28.1–28.6.
28.1, 28.2. *Massoprostatum longum* Caballero, 1948. 28.1. Entire worm. 28.2. Terminal genitalia showing separate openings of paraprostate (P), ejaculatory duct (ED) and uterus (U). (After Sudarikov, 1960b.)
28.3, 28.4. *Heterodiplostomum lanceolatum* Dubois, 1936. 28.3. Entire worm. 28.4. Terminal genitalia showing paraprostate (P) with paraprostate muscular sac and separate openings of hermaphroditic duct and uterus (U). ED, ejaculatory duct. (After Dubois, 1938.)
28.5, 28.6. *Ophiodiplostomum spectabile* Dubois, 1936. 28.5. Entire worm. 28.6. Terminal genitalia showing paraprostate (P), ejaculatory duct (ED) and uterus (U) opening side by side. (After Dubois, 1938.)

Subfamily Ophiodiplostominae Dubois, 1936

Diagnosis: Body distinctly bipartite; pseudosuckers absent; holdfast organ large, without papillae. Testes tandem; ovary pretesticular. Vitellarium in fore- and hindbody, not extending anteriorly beyond anterior margin of holdfast organ; oötype and vitelline reservoir intertesticular. Paraprostate small, clavate; hermaphroditic duct absent. Parasites of snakes. Type-genus *Ophiodiplostomum* Dubois, 1936.

Key to genera

1a. Paraprostate small, tubular; eversible paraprostate duct enclosed in muscular paraprostate sac; uterus and ejaculatory duct opening side by side into greatly reduced copulatory bursa; vitellarium entering holdfast organ and reaching anterior testis posteriorly .... *Heterodiplostomum* Dubois, 1936. (Figs 28.3, 28.4)

Diagnosis: Body distinctly bipartite; forebody flattened, elongate, lanceolate; hindbody cylindrical, elongate. Pseudosuckers absent; oral sucker and pharynx very small; ventral sucker larger; holdfast organ longitudinally elongate, about 40% of forebody length, with non-papillate median slit. Oesophagus very short; caeca reaching close to posterior extremity. Testes tandem, just posterior to middle of hindbody. Ovary pretesticular; oötype intertesticular. Vitellarium entering holdfast organ; follicles in two parallel bands running from anterior end of holdfast organ to anterior testis. Tubular paraprostate enters the muscular sac with eversible paraprostate duct, which resembles cirrus-sac. Ejaculatory duct and uterus open side by side into small copulatory bursa; latter with dorso-subterminal opening. In snakes. Neotropical. Type-species *H. lanceolatum* Dubois, 1936.

1b. Paraprostate elongate, small; paraprostate sac absent; uterus and ejaculatory duct open at tip of minute genital cone; vitellarium more extensive in hindbody than forebody .... *Ophiodiplostomum* Dubois, 1936. (Figs 28.5, 28.6)

(Syn. *Petalodiplostomum* Dubois, 1936)

Diagnosis: Body bipartite; forebody oval to pyriform, ventrally concave; hindbody cylindrical or conical, similar in size to forebody. Pseudosuckers absent; oral sucker smaller than ventral sucker; holdfast organ massive, oval, about half of forebody length, with non-papillate median slit. Pharynx larger than oral sucker; oesophagus very short; caeca not reaching posterior end of body. Testes tandem, situated in posterior region of hindbody. Ovary pretesticular. Oötype intertesticular. Vitellarium forming lateral clusters of follicles between posterior region of holdfast organ and anterior margin of posterior testis. Paraprostate very small. Ejaculatory duct and uterus open at tip of minute genital cone; latter enclosed in greatly reduced copulatory bursa with dorsosubterminal opening. In snakes. Neotropical. Type-species *O. spectabile* Dubois, 1936.

Subfamily Polycotylinae Monticelli, 1888

(Syn. Crocodilicolinae Yamaguti, 1971)

Diagnosis: Body bipartite; hindbody may be provided dorsally with numerous suckers or sucker-like structures; pseudosuckers absent; holdfast organ with papillate median slit. Testes tandem. Ovary pretesticular; oötype and vitelline reservoir intertesticular or between ovary and anterior testis; vitellarium confined to forebody or extends into hindbody to level of ovary or anterior testis. In crocodilians and chelonians. Type-genus *Polycotyle* Willemoes-Suhm, 1870.

Key to genera

1a. Dorsal surface of hindbody with 14–15 small suckers; oötype between ovary and anterior testis .... *Polycotyle* Willemoes-Suhm, 1870. (Fig. 28.7)

Diagnosis: Body distinctly bipartite; forebody spatulate; hindbody subcylindrical, with median longitudinal row of suckers on dorsal side; these suckers increase in size posteriorly. Pseudosuckers absent; oral and ventral suckers rather small; pharynx very small; holdfast organ elliptical, with papillate median slit, about quarter as long as forebody. Testes tandem, in anterior part of hindbody. Ovary pretesticular; oötype and vitelline reservoir between ovary and anterior testis. Vitellarium confined to posterior region of forebody. Paraprostate strongly developed, occupying posterior half of hindbody together with seminal vesicle and copulatory bursa. Ejaculatory duct joined by efferent duct of paraprostate and then uterus. Common opening situated at tip of genital cone. Copulatory bursa spacious, with wide dorso-subterminal aperture and muscular, sucker-like structure on bursal wall. In crocodilians. Nearctic. Type-species *P. ornata* Willemoes-Suhm, 1870.

1b. Suckers on dorsal surface of hindbody absent; oötype between testes .... 2.

2a. Hindbody with thick-walled, sucker-like dorsal 'capsule' opening mid-dorsally near posterior end of body .... *Cystodiplostomum* Dubois, 1936. (Figs 28.8, 28.9)

Diagnosis: Body distinctly bipartite; forebody elongate; hindbody subcylindrical, shorter than forebody, with thick-walled, sucker-like dorsal 'capsule' which opens dorsally near posterior end of body. Pseudosuckers absent; oral and ventral suckers moderately developed, ventral larger than oral; holdfast organ elliptical, with papillate median slit, about 20% of forebody length. Pharynx small; oesophagus short; caeca reaching posterior extremity. Testes tandem, in anterior region of hindbody. Ovary pretesticular, close to junction between fore- and hindbody. Vitellarium confined to forebody, reaching level of ventral sucker. Paraprostate long, tubular, reaching posterior testis; opens independently of ejaculatory duct and uterus into subterminal copulatory bursa. In crocodilians. Neotropical. Type-species *C. hollyi* Dubois, 1936.

2b. Dorsal 'capsule' absent .... 3.

3a. Copulatory bursa enormous, occupying about half of hindbody length, with voluminous genital cone; testes nearly symmetrical .... *Paradiplostomum* La Rue, 1926. (Figs 28.10, 28.11)

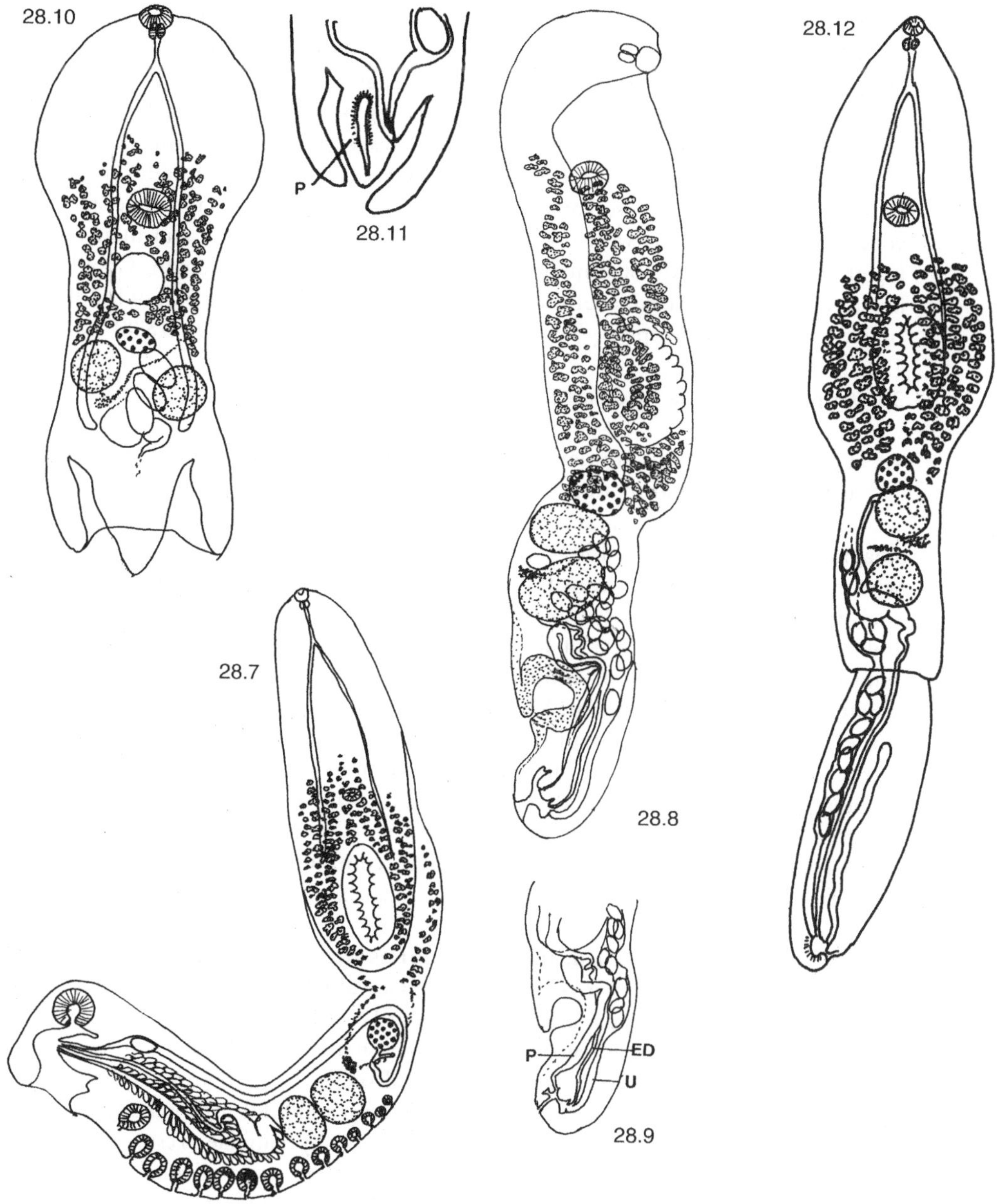

Figs 28.7–28.12.

28.7. *Polycotyle ornata* Willemoes-Suhm, 1870. (After Dubois, 1938.)

28.8, 28.9. *Cystodiplostomum hollyi* Dubois, 1936. 28.8. Entire worm. 28.9. Terminal genitalia showing separate opening of paraprostate (P) and common aperture of ejaculatory duct (ED) and uterus (U). (After Dubois, 1938.)

28.10, 28.11. *Paradiplostomum abbreviatum* (Brandes, 1888). 28.10. Entire worm. 28.11. Terminal genitalia with common opening of ejaculatory duct and uterus on ventral side of genital cone. P, paraprostate. (After Dubois, 1938.)

28.12. *Prolecithodiplostomum constrictum* Dubois, 1936. (After Dubois, 1938.)

Diagnosis: Body indistinctly bipartite; forebody oval or pyriform; hindbody subcylindrical. Pseudosuckers absent; oral and ventral suckers well developed; holdfast organ circular, small, 10–12% of forebody length, with papillae in its cavity. Pharynx small; oesophagus short; caeca terminating posterior to testes. Testes almost symmetrical, in anterior half of hindbody. Ovary pretesticular, almost median. Vitellarium mainly in posterior region of forebody, extending from anterior to ventral sucker to testes. Vitelline reservoir posteromedial to right testis. Paraprostate small, in voluminous genital cone with opening at its tip. Ejaculatory duct and uterus open together on ventral side of genital cone about halfway along its length. Copulatory bursa voluminous, with wide terminal opening, occupying posterior part of hindbody. In crocodilians. Neotropical. Type-species *P. abbreviatum* (Brandes, 1888) La Rue, 1926.

3b. Copulatory bursa inconspicuous; testes tandem .... 4.

4a. Paraprostate long, tubular, with separate opening into copulatory bursa; uterus and ejaculatory duct have common opening .... *Prolecithodiplostomum* Dubois, 1936. (Fig. 28.12)

Diagnosis: Body bipartite; forebody lanceolate; hindbody cylindrical or claviform, with more or less marked constriction near middle. Pseudosuckers absent; ventral sucker larger than oral, situated slightly anterior to middle of forebody; pharynx small; holdfast organ oval, 15–25% of forebody length, with papillate median slit. Testes tandem, in anterior half of hindbody. Ovary directly pretesticular. Oötype and vitelline reservoir intertesticular. Vitellarium confined to posterior half of forebody; posterior limit near ovary. Paraprostate long, tubular, opens separately into copulatory bursa. Ejaculatory duct and uterus have common opening at base of bursa. Copulatory bursa small or large, with dorso-subterminal opening. In crocodilians. Neotropical. Type-species *P. constrictum* Dubois, 1936.

4b. Paraprostate thick-walled, opens through common pore with uterus and/or ejaculatory duct .... 5.

5a. Paraprostate thick-walled, muscular, unites with ejaculatory duct anteriorly or laterally, receives uterus and forms thick-walled hermaphroditic duct .... *Pseudocrocodilicola* Byrd & Reiber, 1942. (Figs 28.13, 28.14)

Diagnosis: Body distinctly bipartite; forebody flattened, lanceolate, longer than cylindrical hindbody. Pseudosuckers absent; oral sucker larger than pharynx; ventral sucker larger than oral, situated in middle or anterior to middle of forebody; holdfast organ rather small, oval, with median slit bordered by papillae. Testes tandem in anterior part of hindbody. Ovary pretesticular; oötype intertesticular. Vitellarium confined to posterior part of forebody; its anterior border may extend forward beyond ventral sucker. Paraprostate small, muscular, ellipsoidal, surrounded by relatively few, large gland-cells. Ejaculatory duct discharges into paraprostatic canal. Uterus unites with paraprostate to form dilated hermaphroditic duct surrounded by thick-walled secondary pouch, which opens into small copulatory bursa. In crocodilians. Nearctic. Type-species *P. americaniense* Byrd & Reiber, 1942.

5b. Thick-walled hermaphroditic duct absent .... 6.

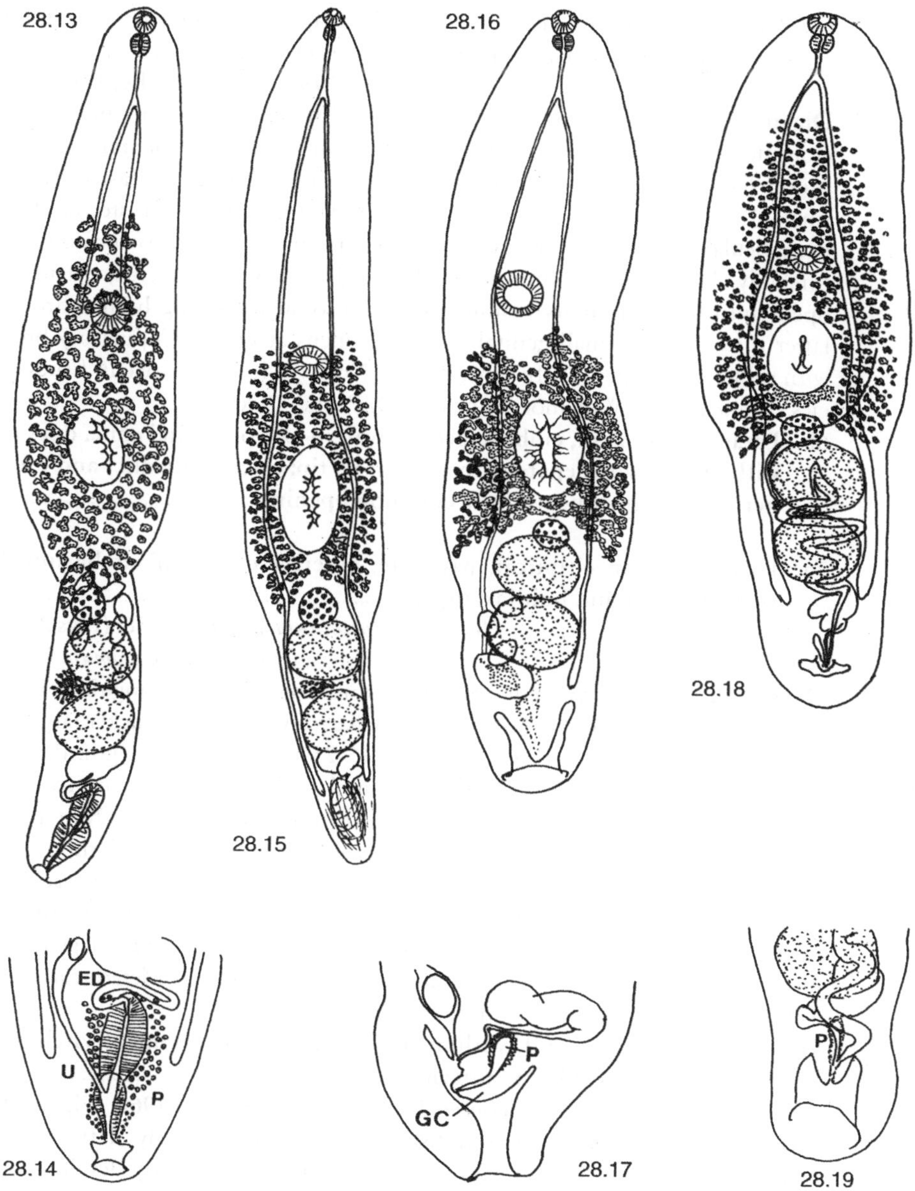

Figs 28.13–28.19.

28.13, 28.14. *Pseudocrocodilicola georgiana* Byrd & Reiber, 1942. 28.13. Entire worm. 28.14. Terminal genitalia with ejaculatory duct (ED) discharging into paraprostatic canal which unites with uterus (U) to form short hermaphroditic canal surrounded by secondary sac. P, paraprostate. (After Byrd & Reiber, 1942.)

28.15. *Crocodilicola pseudostoma* (Willemoes-Suhm, 1870). (After Dubois, 1938.)

28.16, 28.17. *Herpetodiplostomum caimancola* (Dollfus, 1935). 28.16. Entire worm. 28.17. Terminal genitalia showing common opening of ejaculatory duct and uterus at mid-length of genital cone (GC) and pore of paraprostate (P) at its apex. (After Dubois, 1938.)

28.18, 28.19. *Cheloniodiplostomum testudinis* (Dubois, 1936). 28.18. Entire worm. 28.19. Terminal genitalia showing separate opening of paraprostate (P) at apex of genital cone and hermaphroditic duct opening into copulatory bursa. (After Dubois, 1938.)

6a. Paraprostate elliptical, thick-walled, unites with uterus; ejaculatory duct opens into uterus near its fusion with paraprostate .... *Crocodilicola* Poche, 1926. (Fig. 28.15)

Diagnosis: Body rather indistinctly bipartite; forebody linguiform; hindbody shorter, elongate or conical. Pseudosuckers absent; oral sucker and pharynx small; ventral sucker larger than oral, situated near middle of forebody; holdfast organ elliptical, small, 12-20% of forebody length, with papillate median slit. Testes tandem. Ovary pretesticular; oötype intertesticular. Vitellarium confined to posterior part of forebody. Paraprostate elliptical, thick-walled, unites with uterus; ejaculatory duct opens into uterus near its fusion with paraprostate, forming hermaphroditic duct; latter runs through small genital cone and opens at its tip. Copulatory bursa small, with subterminal opening. In crocodilians. Nearctic. Metacercariae in fishes. Type-species *C. pseudostoma* (Willemoes-Suhm, 1870) Poche, 1926.

6b. Paraprostate small, club-shaped or oval, in axis of genital cone, does not unite with uterus, with separate opening at its tip; ejaculatory duct and uterus form hermaphroditic duct or have common opening .... 7.

7a. Holdfast organ elliptical, papillate; anterior extent of vitellarium between ventral sucker and holdfast organ .... *Herpetodiplostomum* Dubois, 1936. (Figs 28.16, 28.17)

(Syn. *Neelydiplostomum* Gupta, 1958)

Diagnosis: Body indistinctly bipartite; forebody elongate, linguiform; hindbody cylindrical, shorter than forebody. Pseudosuckers absent; ventral sucker larger than oral; holdfast organ elliptical or oval, papillate, 15-25% of forebody length. Pharynx small; intestinal bifurcation well anterior to ventral sucker; caeca do not reach posterior extremity. Testes tandem, may occupy more than half of hindbody. Ovary pretesticular, close to forebody. Oötype and vitelline reservoir intertesticular. Vitellarium in forebody, mainly in posterior region posterior to ventral sucker; posterior limit reaches ovary or anterior testis. Paraprostate small, in axis of genital cone, opening at its tip. Common opening of ejaculatory duct and uterus in middle of ventral side of genital cone. In crocodilians and turtles. Neotropical and Oriental regions. Metacercariae in fishes. Type-species *H. caimancola* (Dollfus, 1935) Dubois, 1936.

7b. Holdfast organ round, without papillae; anterior extent of vitellarium near intestinal bifurcation or anterior to ventral sucker .... *Cheloniodiplostomum* Sudarikov, 1960. (Figs 28.18, 28.19)

Diagnosis: Body more or less indistinctly bipartite; forebody flattened; hindbody cylindrical. Oral and ventral suckers and pharynx rather poorly developed. Holdfast organ round, with aperture devoid of papillae. Testes tandem; ovary pretesticular. Oötype and vitelline reservoir intertesticular. Vitellarium mainly confined to forebody, but small number of follicles penetrate hindbody to level of anterior testis. Anterior limit of vitellarium reaches intestinal bifurcation. Paraprostate small, occupies longitudinal axis of genital cone and opens independently at its tip. Uterus and ejaculatory duct form hermaphroditic duct which opens into copulatory bursa. Parasites of chelonians. Neotropical. Type-species *C. testudinis* (Dubois, 1936) Sudarikov, 1960.

Subfamily Proterodiplostominae Dubois, 1936

Diagnosis: Body distinctly bipartite; pseudosuckers absent; holdfast organ of medium size or small, with papillate median slit. Testes tandem. Ovary pretesticular; oötype and vitelline vesicle intertesticular. Vitellarium in fore- and hindbody, with anterior limit at level of ventral sucker or sometimes anterior to it and posterior border extending beyond posterior testis. Parasites of crocodiles. Type-genus *Proterodiplostomum* Dubois, 1936.

Key to genera

1a. Paraprostate small, occupying central part of pointed genital cone; vitellarium in hindbody, reaching posterior testis; copulatory bursa with wide subterminal aperture .... *Pseudoneodiplostomum* Dubois, 1936. (Figs 28.20, 28.21)
(Syn. *Pseudoneodiplostomoides* Yamaguti, 1954)

Diagnosis: Body distinctly bipartite; forebody scoop-shaped; pseudosuckers absent; holdfast organ elongate, with papillate median slit; hindbody cylindrical or subcylindrical posteriorly. Oral sucker and pharynx small; ventral sucker larger than oral. Caeca narrow, terminate near end of body or posterior to hind testis. Testes tandem, in middle of hindbody. Ovary pretesticular. Vitellarium mainly in forebody, extending to region of ventral sucker but also occurs in hindbody reaching intertesticular level. Paraprostate small, close to or within base of genital cone; in middle of cone it unites with ejaculatory duct and opens at its tip. Uterus opens ventrally, at base of genital cone, on ventral wall of copulatory bursa; bursa with wide subterminal aperture. In crocodilians. Africa, Australia, Oriental and Neotropical regions. Metacercariae in fishes. Type-species *P. thomasi* (Dollfus, 1935) Dubois, 1936.

1b. Paraprostate well differentiated or not; genital cone absent; copulatory bursa reduced; vitellarium may extend beyond posterior testis .... 2.

2a. Paraprostate, ejaculatory duct and uterus enclosed in capsule of circular muscle fibres .... *Capsulodiplostomum* Dwivedi, 1966. (Fig. 28.22)

Diagnosis: Body bipartite; forebody foliate; hindbody long, truncate. Pseudosuckers absent; oral and ventral suckers small; holdfast organ elliptical, papillate, about 40% of forebody length. Pharynx small; oesophagus short; caeca terminating at level of posterior margin of posterior testis. Testes tandem, close to each other; posterior testis in middle of hindbody. Ovary median, immediately pretesticular. Vitellarium extending from ventral sucker level to middle of posterior testis. Paraprostate, ejaculatory duct and terminal part of uterus enclosed in long elliptical capsule of circular fibres. Copulatory bursa reduced; genital cone absent. Ejaculatory duct and uterus open separately. In crocodilians. India. Type-species *C. crocodilinum* Dwivedi, 1966.

2b. Muscular capsule enclosing muscular paraprostate, ejaculatory duct and uterus absent .... 3.

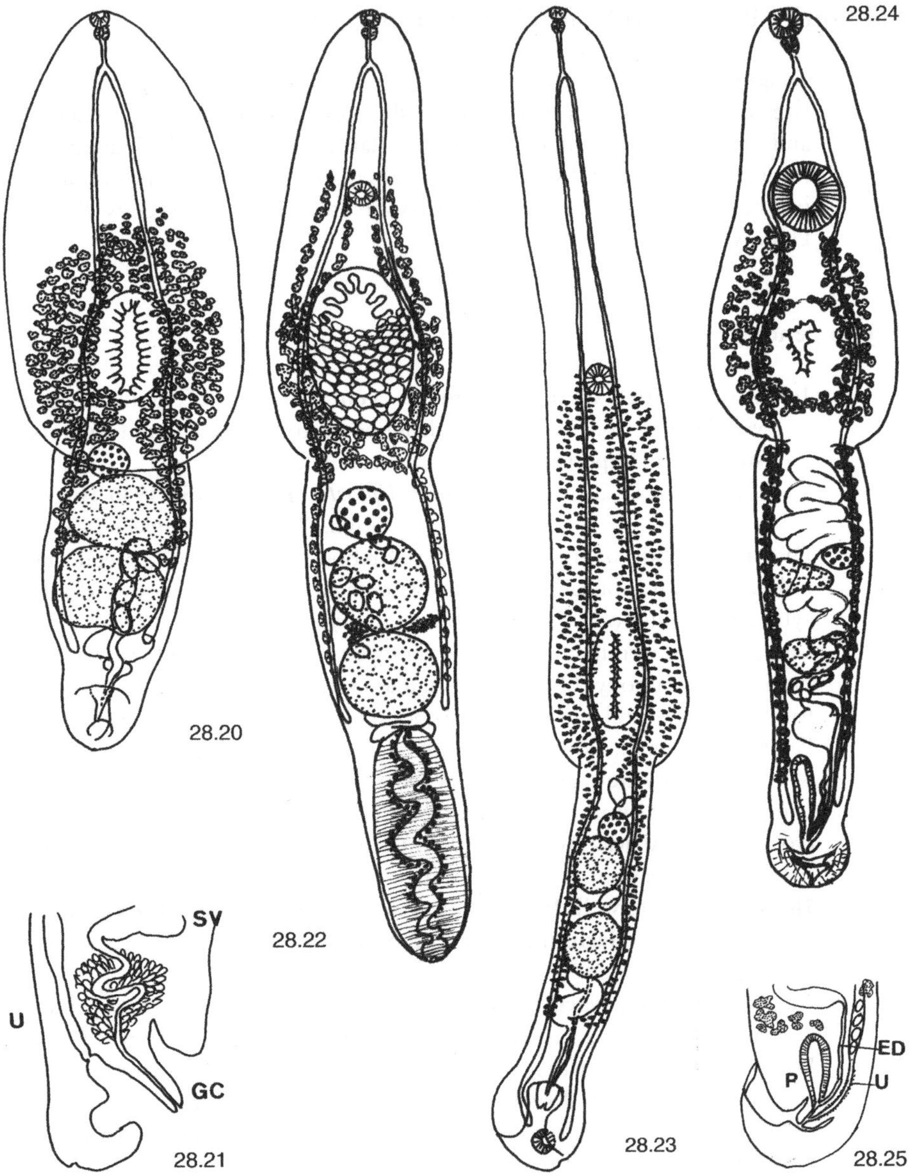

Figs 28.20–28.25.

28.20. *Pseudoneodiplostomum thomasi* (Dollfus, 1935). (After Dubois, 1938.)

28.21. Terminal genitalia of *P. siamense* (Poirier, 1886) showing ejaculatory duct surrounded by prostatic cells and uterus (U) opening separately. The junction of paraprostate and ejaculatory duct was not delineated. GC, genital cone; SV seminal vesicle. (After Dubois, 1938.)

28.22. *Capsulodiplostomum crocodilinum* Dwivedi, 1966. (After Dwivedi, 1966.)

28.23. *Mesodiplostomum gladiolum* Dubois, 1936. (After Dubois, 1938.)

28.24, 28.25. *Archaeodiplostomum acetabulatum* (Byrd & Reiber, 1942). 28.24. Entire worm. 28.25. Terminal genitalia showing ejaculatory duct (ED) joining paraprostate (P) and forming short hermaphroditic duct, which opens at apex of genital cone. U, uterus. (After Byrd & Reiber, 1942.)

3a. Paraprostate not differentiated; uterus and ejaculatory duct open together at apex of retractile cone; copulatory bursa bulbous with lobed margins and small posteroventral diverticulum .... *Mesodiplostomum* Dubois, 1936. (Fig. 28.23)

Diagnosis: Body distinctly bipartite; forebody elongate, linguiform; hindbody shorter, cylindrical. Pseudosuckers absent; oral sucker and pharynx small; ventral sucker larger, near middle of forebody; holdfast organ elliptical, with papillate median slit, small, about 15% of forebody length, situated far posteriorly. Testes tandem; posterior near middle of hindbody. Ovary pretesticular. Vitellarium extending from region of ventral sucker to short distance posterior to testes. Paraprostate not differentiated. Uterus and ejaculatory duct form short hermaphroditic duct which opens at apex of retractile genital cone. Copulatory bursa bulbous with small postero-ventral diverticulum and lobed margin of mid-dorsal aperture. In crocodilians. Neotropical. Type-species *M. gladiolum* Dubois, 1936.

3b. Paraprostate well differentiated; copulatory bursa muscular or spacious with large opening; uterus unites with ejaculatory duct or opens separately .... 4.

4a. Paraprostate claviform, well developed, joining ejaculatory duct at base of genital cone and then uterus; genital cone lacks sucker-like structure; copulatory bursa muscular .... *Archaeodiplostomum* Dubois, 1944. (Figs 28.24, 28.25)

Diagnosis: Body distinctly bipartite; forebody flattened; hindbody elongate. Pseudo-suckers absent; oral sucker small; ventral sucker large; holdfast organ oval, with opening bordered by papillae, in posterior region of forebody occupying quarter of forebody length. Pharynx small; caeca almost reaching posterior end of body. Testes tandem, in middle third of hindbody. Ovary pretesticular, lateral. Vitellarium extending from ventral sucker to near end of body. Paraprostate claviform, well developed, uniting with ejaculatory duct at base of genital cone and subsequently with extremity of uterus. Short hermaphroditic duct at apex of genital cone. Copulatory bursa muscular, with subterminal opening. In crocodilians. USA (Florida). Type-species *A. acetabulatum* (Byrd & Reiber, 1942) Dubois, 1944.

4b. Paraprostate tubular, well developed; ejaculatory duct opens with para-prostate efferent duct at apex of small genital cone; latter attached to sucker-like muscular structure at posterior extremity; uterus opens between cone and this structure; copulatory bursa spacious with large opening .... *Proterodiplostomum* Dubois, 1936. (Figs 28.26, 28.27)

Diagnosis: Body bipartite; forebody spatulate; hindbody elongate, cylindrical, direct continuation of forebody. Pseudosuckers absent; oral and ventral suckers moderately developed; pharynx small or moderately developed; holdfast organ elliptical, elongate (about 30% of forebody length), with papillae on margin of median slit. Testes tandem; anterior near middle of hindbody. Ovary pretesticular. Vitellarium extends from level of ventral sucker to posterior to hind testis. Paraprostate well developed, tubular, reaching close to posterior testis; fine efferent duct of paraprostate unites with ejaculatory duct and opens at tip of small genital cone; latter has sucker-like muscular structure attached to it; uterus opens separately

28.26

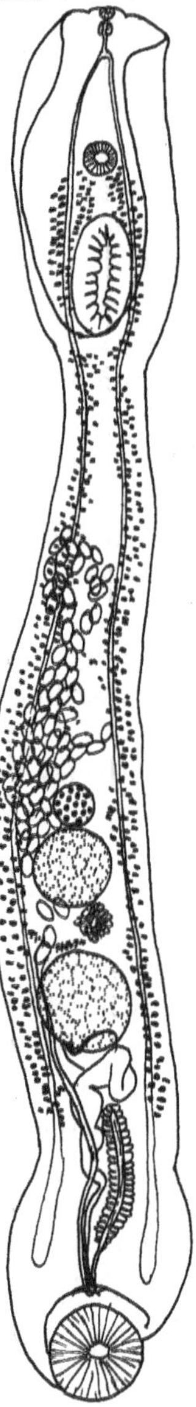

28.27

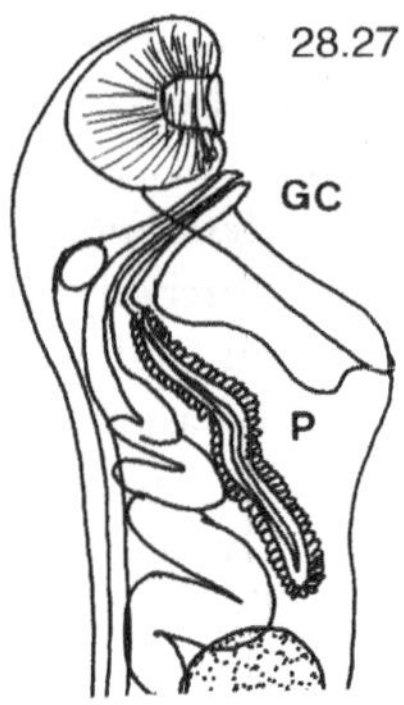

28.28

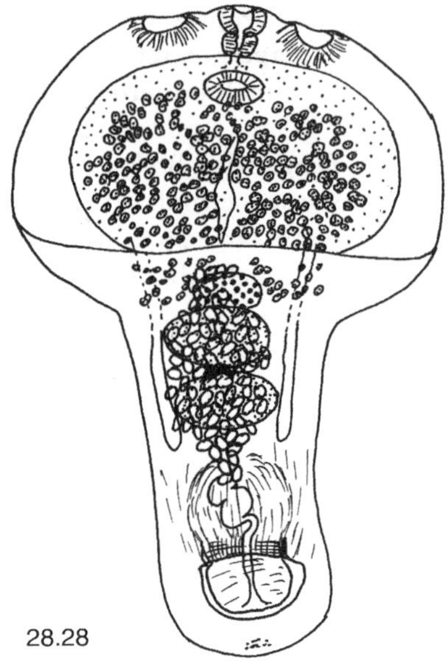

Figs 28.26–28.28.

28.26–28.27. *Proterodiplostomum longum* (Brandes, 1888). 28.26. Entire worm. 28.27. Terminal genitalia showing common opening of ejaculatory duct and paraprostate (P) and separate opening of uterus. GC, genital cone. (After Dubois, 1938.)

28.28. *Proalarioides serpentis* Yamaguti, 1933. (After Dubois, 1938.)

between this structure and genital cone. Copulatory bursa spacious, with large subterminal opening. In crocodilians, occasionally in snakes. Neotropical. Type-species *P. longum* (Brandes, 1888) Dubois, 1936.

Subfamily Proalarioidinae Sudarikov, 1960
(Syn. Travassosstominae Bhalerao, 1938)

Diagnosis: Body more or less distinctly bipartite. Forebody circular with pronounced posteroventral fold and ventral concavity; hindbody cylindrical, rounded terminally. Pseudosuckers present; oral and ventral suckers well developed; ventral near intestinal bifurcation; holdfast organ large, round, half to two-thirds length of forebody, with vitelline follicles around median longitudinal slit. Pharynx small; oesophagus short; caeca reaching to about middle of hindbody. Testes and ovary tandem, in anterior half of hindbody. Vitelline follicles around and in holdfast organ, occasionally penetrate hindbody. Ejaculatory duct and distal part of uterus form wide, straight, hermaphroditic duct enclosed in muscular bulb, which opens at tip of conical prominence in copulatory bursa. Copulatory bursa opens dorsally close to posterior extremity. In snakes. Palaearctic and Oriental. Type- and only genus *Proalarioides* Yamaguti, 1933.

Genus *Proalarioides* Yamaguti, 1933 (Fig. 28.28)
(Syn. *Travassosstomum* Bhalerao, 1938)

Diagnosis: With characters of subfamily. Metacercariae in frogs. Type-species *P. serpentis* Yamaguti, 1933.

29 Family Strigeidae Railliet, 1919

K. NIEWIADOMSKA

W. Stefański Institute of Parasitology, Polish Academy of Sciences, Twarda, Warsaw, Poland

Introduction

The family Strigeidae embraces a distinct group of species with a characteristic cup-shaped forebody and a holdfast organ in the form of two (ventral and dorsal) lobes. First, it was separated off by Nitzsch (1819), as the genus *Holostomum*, from *Amphistoma* Rudolphi, 1801 (later considered a synonym of *Strigea* Abildgaard, 1790). Blanchard (1847) erected the family Holostomidae, but the basis for the systematics of this group was the work of Railliet (1919), who created the family Strigeidae for all bisegmented species and Cyathocotylidae for non-segmented ones. However, it was Dubois (1938) who restricted the family Strigeidae to genera with a cup-shaped forebody and bilobed holdfast organ. He also distinguished the subfamilies Strigeinae and Duboisiellinae according to their host group (birds and mammals, respectively), and the tribes Strigeini and Cotylurini within the Strigeinae. In the opinion of Sudarikov (1959), these tribes have subfamilial status. He also distinguished the subfamily Pseudapatemoninae, considering the stopper-like structure of the holdfast organ characteristic of *Pseudapatemon* Dubois, 1936, and reconfirmed this opinion in 1984 (Sudarikov, 1984). The separation of *Pseudapatemon* at the rank of subfamily was also supported by Zazornova & Sysoev (1993) in their phylogenetic analysis of the Strigeidae. However, the supposed uniqueness of the structure of the holdfast organ of *Pseudapatemon* is not readily proved; in some specimens (see, for example, Dubois, 1968, fig. 252) it is not visible, and consequently it seems an unsuitable feature to be used at the subfamily level.

The Strigeidae are mainly specific to birds. Only one genus, *Duboisiella* Baer, 1938, occurs in mammals, and for this reason it is placed in a separate subfamily, the Duboisiellinae Baer, 1938. This separate status is also supported by morphological characters that are not present in any other diplostomoid genera. The holdfast organ of *Duboisiella*, composed of short dorsal and ventral lobes, is situated at the proximal end of a very long forebody cavity reaching back to the anterior testis. The uterus enters the forebody and extends to the first quarter of its length. Of all described strigeids, this is the only one with such a uterine distribution; the uterus does not usually extend into the forebody.

The criterion of host-specificity was also the basis for the creation of a distinct family, the Neostrigeidae Bisseru, 1956, for three species described from the intestine of *Crocodilus niloticus*: *Neostrigea africana* Bisseru, 1956, *N. leiperi* Bisseru, 1956 and *Prostrigea arcuata* Bisseru, 1956. Despite the importance of specificity in the systematics of Diplostomoidea, Dubois (1968) considered the crocodile an accidental host for these species and the establishment of a separate family as unnecessary. Based on their morphology, he proved that *N. africana* is a synonym of the cosmopolitan species *Strigea falconis* Szidat, 1928. He transferred *N. leiperi* to the genus *Strigea* Abildgaard, 1790 and *Prostrigea arcuata* to the genus *Parastrigea* Szidat, 1928. Sudarikov (1959), although considering the correspondence between these species and strigeids, preserved the separate status of the Neostrigeidae but as an addendum to the Strigeidae (a superfamily in Sudarikov's system). The results of both Sudarikov (1959) and Dubois (1968) justify the incorporation of the species in question within the family Strigeidae and the synonymizing of the Neostrigeidae with this family.

The genus *Chabaustrigea*, created by Sudarikov (1959) for *Strigea geoduboisi* Chabaud, Golvan & Rousselot, 1956, was based mainly on holdfast organ structure and the presence of a seminal receptacle, which is unique for strigeids. Yamaguti (1971) regarded these characters as not having generic value and, in the case of the seminal receptacle, not certain. For these reasons this genus is considered a synonym of *Strigea*.

Agamostrigea Lutz, 1933 is a *lapsus calami* for *Apharyngostrigea* Ciurea, 1927 (see Lutz, 1933, p. 400, German translation).

Strigeids are also uniform when the morphology of life-cycle stages is analysed. In all known life-cycles, the metacercariae are of the 'tetracotyle' type. In species of the genus *Strigea*, a mesocercarial stage is also included in the life-cycle.

Family Strigeidae Railliet, 1919

(Syn. Neostrigeidae Bisseru, 1956)

Diagnosis: Body composed of two parts: forebody cup-shaped to tubular or bulbiform; hindbody usually cylindrical. Forebody with or without pseudosuckers. Holdfast organ inside forebody, composed of ventral and dorsal lobes usually separated by deep slit, with proteolytic gland almost always distinctly separated, posterior to ventral sucker and near border between fore- and hindbody. Oral sucker and pharynx usually present; oesophagus short; caeca long, terminating blindly close to posterior end of body. Reproductive organs in hindbody. Testes tandem, bi-, tri- or multilobed, rarely reniform or cordiform, more or less horseshoe-shaped. Cirrus-sac and paraprostate absent. Copulatory bursa more or less developed, with conus or bulbus genitalis inside, and with terminal or dorsoterminal aperture. Ovary pretesticular; Laurer's canal present. Vitellarium follicular, in both parts of body or confined to hindbody or forebody. In birds, rarely mammals. Cercariae furcocercous; metacercariae of 'tetracotyle' type, in vertebrates and invertebrates; mesocercariae may be present. Type-genus *Strigea* Abildgaard, 1790.

Key to subfamilies

1a. **Parasites of birds; vitellarium in both parts of body or confined to hindbody .... Strigeinae Railliet, 1919.**
1b. **Parasites of mammals; vitellarium in forebody .... Duboisiellinae Baer, 1938.**

Subfamily Strigeinae Railliet, 1919

Diagnosis: Cup-shaped forebody separated from hindbody by more or less pronounced constriction; constricted 'neck' region of hindbody may be present. Pharynx occasionally absent. Vitellarium in fore- and hindbody or confined to hindbody. Uterus extends anteriorly but does not extend into forebody. Copulatory bursa well or weakly developed, with genital cone or papilla inside; genital bulb may be present. Hermaphroditic duct opens at tip of genital cone or papilla, or lateral to genital cone. In birds, accidentally in reptiles. Metacercariae of 'tetracotyle' type, in invertebrates (leeches, gastropods), fishes, amphibians and birds. Mesocercariae may be present in life-cycle (in amphibians). Type-genus *Strigea* Abildgaard, 1790.

Key to genera

1a. **Vitellarium in fore- and hindbody .... 2.**
1b. **Vitellarium in, or mainly in, hindbody .... 4.**

2a. **Forebody with vitelline follicles accumulated into two symmetrical masses localized in lateral expansions of dorsal lobe of holdfast organ .... *Parastrigea* Szidat, 1928. (Fig. 29.1)**
(Syns *Prostrigea* Bisseru, 1956; *Chaseostrigea* Ukoli, 1967; *Brasiliana* Ukoli, 1967)
Diagnosis: Body distinctly bipartite; forebody ovoid or pyriform, with narrow opening and two symmetrical bulbous expansions in which vitelline follicles are concentrated; hindbody sac-like, claviform or cylindrical, curved dorsally. Oral sucker close to ventral sucker, feebly developed; pharynx present (occasionally absent). Testes tandem, large, lobed, in posterior half of body. Ovary oval, sometime lobed, near middle of hindbody or more anterior. Vitellarium extends into lateral expansions of forebody, confined in hindbody mainly to pretesticular region but extends further posteriorly, reaching posterior extremity. Copulatory bursa with muscular ring, weakly delimited from body parenchyma or not; genital cone with hermaphroditic duct opening at its apex. In various birds. Cosmopolitan. Metacercariae of 'tetracotyle' type, in anurans. Cercariae with flame-cell formula $2[(1 + 1) + (1 + 1 + [1])] = 10$, anterior excretory commissure, which may not be closed, two groups of seven penetration glands lateral and posterior to ventral sucker. Type-species *P. cincta* (Brandes, 1888) Szidat, 1928.
2b. **Vitellarium uniformly distributed in both parts of body .... 3.**

3a. **Pharynx absent .... *Apharyngostrigea* Ciurea, 1927. (Fig. 29.2)**
(Syn. *Ridgeworthia* Verma, 1936)

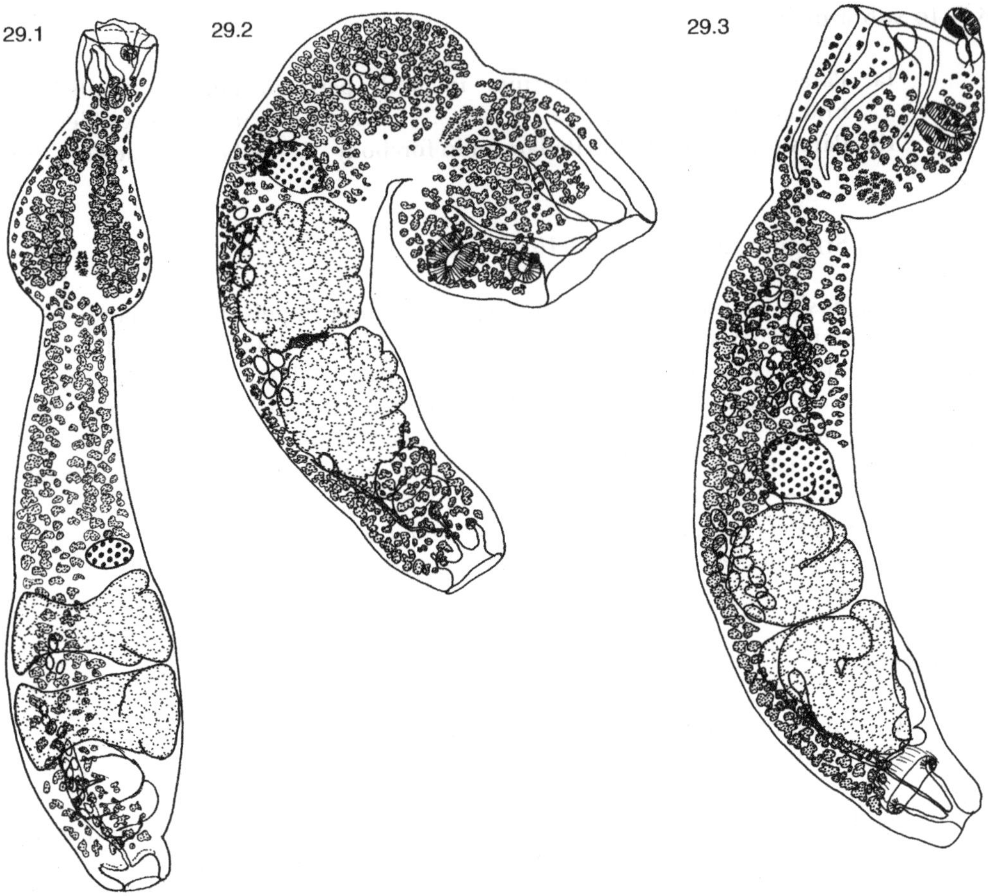

Figs 29.1–29.3.
29.1. *Parastrigea intermedia* Tubangui, 1932. (After Dubois, 1958.)
29.2. *Apharyngostrigea cornu* (Zeder, 1800). (After Dubois, 1968, modified.)
29.3. *Strigea falconis* Szidat, 1928. (After Dubois, 1968.)

Diagnosis: Body bipartite; forebody oval, pyriform or calyciform, lacking lateral expansions; hindbody cylindrical to claviform, curved dorsally, with constriction near region of copulatory bursa. Oral and ventral suckers feebly developed. Pharynx absent. Testes multilobed. Ovary reniform, near middle of hindbody. In forebody, vitellarium extends into dorsal and lateral body wall and into holdfast organ (mainly dorsal lobe); in hindbody, it reaches copulatory bursa. Genital cone more or less distinctly delimited from body parenchyma. Muscular ring surrounding genital cone feebly developed or indistinct. Hermaphroditic duct opens at apex of genital cone. In ardeids. Cosmopolitan. Metacercariae of 'tetracotyle' type in fishes. Cercariae with flame-cell formula $2[(2 + 2) + (2 + 2 + [2])] = 20$, two excretory commissures, one anterior and one posterior to ventral sucker, two groups of seven or four penetration glands anterior and lateral to ventral sucker. Type-species *A. cornu* (Zeder, 1800) Ciurea, 1927.

3b. Pharynx present .... ***Strigea* Abildgaard, 1790. (Fig. 29.3)**
(Syns *Holostomum* Nitzsch, 1819; *Gongylura* Lutz, 1933; *Neoalaria* Lal, 1939; *Neostrigea* Bisseru, 1956; *Chabaustrigea* Sudarikov, 1959)

Diagnosis: Forebody cup- or funnel-shaped, clearly separated from hindbody; hindbody cylindrical, curved dorsally, without long, slender neck region. Pseudosuckers present; oral and ventral sucker large, similar in size or with ventral larger; pharynx well developed. Holdfast organ of dorsal and ventral lobes reaching to anterior margin of forebody. Testes multilobed or rounded, tandem, in median or posterior part of hindbody. Ovary round or oval. In forebody, vitellarium extends into body-wall and holdfast organ lobes; in hindbody, mostly in anterior region and ventral to copulatory bursa. Large copulatory bursa with genital cone delimited from body parenchyma and surrounded by muscular ring. Short ejaculatory duct opens at apex of genital cone. In various birds. Cosmopolitan. Life-cycle with mesocercariae in anurans; metacercariae of 'tetracotyle' type, in amphibians, snakes, birds or mammals. Cercariae with flame-cell formula $2[(1 + 1) + (1 + 1 + [1])] = 10$, without excretory commissures, with two groups of four penetration glands distributed in line posterior to or lateral to ventral sucker. Type-species *S. strigis* (Schrank, 1788) Abildgaard, 1790.

4a. Body very long; hindbody 5–25 times longer than forebody; long, slender neck region present .... **5.**

4b. Body of moderate length; hindbody < 6 times longer than forebody; neck region small or absent .... **7.**

5a. Forebody small with large opening; hindbody 8.5–25 times longer than forebody; vitellarium confined to hindbody .... ***Nematostrigea* Sandground, 1934. (Fig. 29.4)**

Diagnosis: Body bipartite; forebody small, oblong, round or subinfundibular, similar to diplostomids but with holdfast organ of two lobes and well-developed proteolytic gland. Oral sucker smaller than ventral; pharynx small. Hindbody 8.5–25 times longer than forebody, with long slender neck region and broader posterior region containing gonads. Testes tandem, lobed, near posterior extremity. Ovary subglobular. Vitellarium confined to hindbody. Copulatory bursa relatively large, with terminal opening; genital cone protrusible, traversed by uterus; ejaculatory duct unites with uterus at different levels within cone. Mainly in predatory birds. Europe, North America, Africa. Type-species *N. serpens* (Nitzsch, 1819) Sandground, 1934.

5b. Hindbody 5–17 times longer than forebody; vitellarium may penetrate forebody .... **6.**

6a. Forebody small, calyciform; hindbody, with neck region, 6–17 times longer than forebody; copulatory bursa not protrusible; vitellarium predominantly in hindbody, in forebody confined to base of holdfast organ .... ***Ophiosoma* Szidat, 1928. (Fig. 29.5)**

Diagnosis: Body bipartite; forebody small, calyciform, equatorially constricted; hindbody 6–17 times longer than forebody, cylindrical or broadened distally. Pseudosuckers absent; oral and ventral sucker and pharynx poorly developed. Testes multilobed or round, tandem, near posterior extremity. Ovary round, median. Vitellarium in forebody confined to base of holdfast organ, distributed mainly in

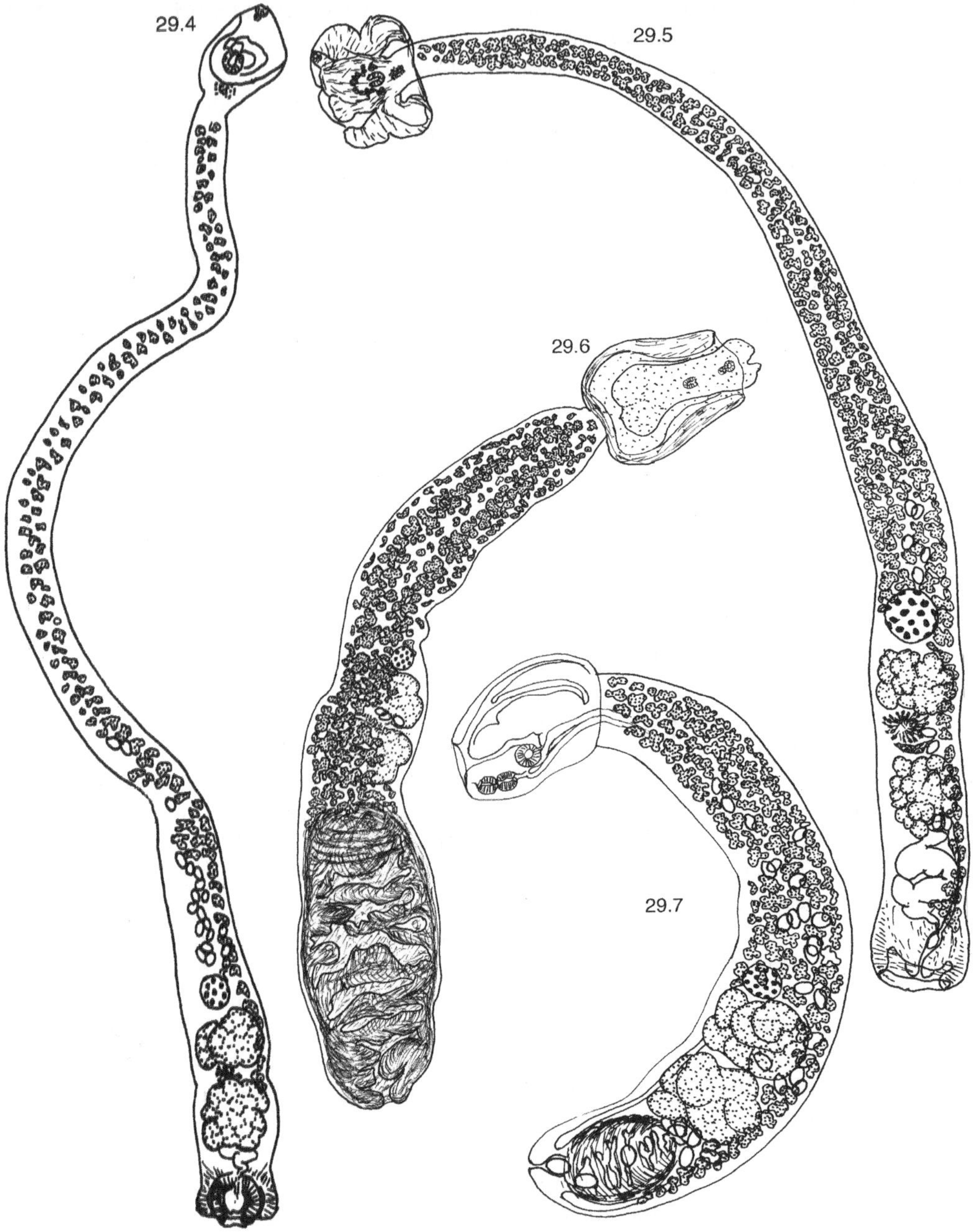

Figs 29.4–29.7.
29.4. *Nematostrigea serpens* (Nitzsch, 1819). (After Dubois, 1968.)
29.5. *Ophiosoma patagiatum* (Creplin, 1846). (After Dubois, 1968.)
29.6. *Cardiocephaloides longicollis* (Rudolphi, 1819) with voluminous copulatory bursa.
29.7. *C. brandesi* (Szidat, 1928) with small copulatory bursa. (After Dubois, 1968.)

hindbody, reaches copulatory bursa. Copulatory bursa with terminal opening, feebly delimited, with muscular ring. Genital cone not delimited from body parenchyma. Short hermaphroditic duct opens at apex of genital cone. In various birds. Holarctic, Neotropical, Oriental regions. Type-species *O. patagiatum* (Creplin, 1846) Dubois, 1937.

6b. Forebody relatively small, cordiform, pear-shaped; hindbody with neck region, 4–8.5 times longer than forebody; copulatory bursa protrusible; vitellarium confined to hindbody .... *Cardiocephaloides* Sudarikov, 1959. (Figs 29.6, 29.7)

(Syn. *Cardiocephalus* Szidat, 1928 *nec* Broili, 1904)

Diagnosis: Body distinctly bipartite. Forebody relatively small, cordiform, pear-shaped, rarely bulbiform. Hindbody long, 4–8.5 times longer than forebody, with clavate or cylindrical neck region, enlarged near posterior end. Oral sucker, ventral sucker and pharynx feebly developed. Holdfast organ well developed. Gonads situated in posterior part of hindbody. Testes tandem, relatively small, lobate, rarely unlobed. Ovary oval or reniform, rather small. Vitellarium confined to hindbody; vitelline follicles extend mainly between ovary and anterior end of hindbody, reaching copulatory bursa ventrally. Copulatory bursa evaginable, voluminous, moderate or small, with terminal opening provided with sphincter or not, sometimes delimited by transverse constriction. Genital cone at base of bursa, encloses short hermaphroditic duct, formed by junction of uterus and long ejaculatory duct. Eggs numerous. In larids and penguins. Europe, Africa, Neotropical, subtropical North America, Australia. Metacercariae in marine fishes (Sparidae, Scombresocidae). Cercariae with two pairs of pre- and postacetabular penetration glands, with flame-cell formula $2[(2+2)+(2+2+[2])]=20$, developing in Nassidae (Prosobranchia). Type-species *C. longicollis* (Rudolphi, 1819) Dubois, 1982.

7a. Forebody pouch-like; pseudosuckers conspicuous, roll-shaped with auricular expansions; vitellarium may extend into base of forebody .... *Schwartzitrema* Pérez Vigueras, 1941. (Figs 29.8, 29.9)

(Syn. *Schwartziella* Pérez Vigueras, 1940)

Diagnosis: Body bipartite; forebody rhomboid, ovoid or pyriform; holdfast organ with dorsal and two ventrolateral lobes; hindbody claviform or ovoid to saccular, with or without neck region. Pseudosuckers well developed, in form of longitudinal muscular rolls, with proximal (larger) and distal auricular expansions. Oral sucker, ventral sucker and pharynx well developed. Testes bilobed, tandem; anterior asymmetrical, posterior symmetrical; both in posterior part of hindbody. Ovary ovoid. Vitellarium confined to hindbody or may extend into base of forebody. Copulatory bursa small or relatively large, with terminal opening. Ejaculatory duct and uterus unite near base or apex of genital cone. In Pelecaniformes. USA, Cuba. Type-species *S. schwartzi* (Pérez Vigueras, 1940) Pérez Vigueras, 1941.

7b. Forebody cup-shaped or subglobular; pseudosuckers not conspicuously roll-shaped; vitellarium tends to be confined to hindbody .... 8.

8a. Holdfast organ massive; genital cone or genital bulb absent .... *Pseudapatemon* Dubois, 1936. (Figs 29.10, 29.11)

(Syn. *Eroliostrigea* Yamaguti, 1971)

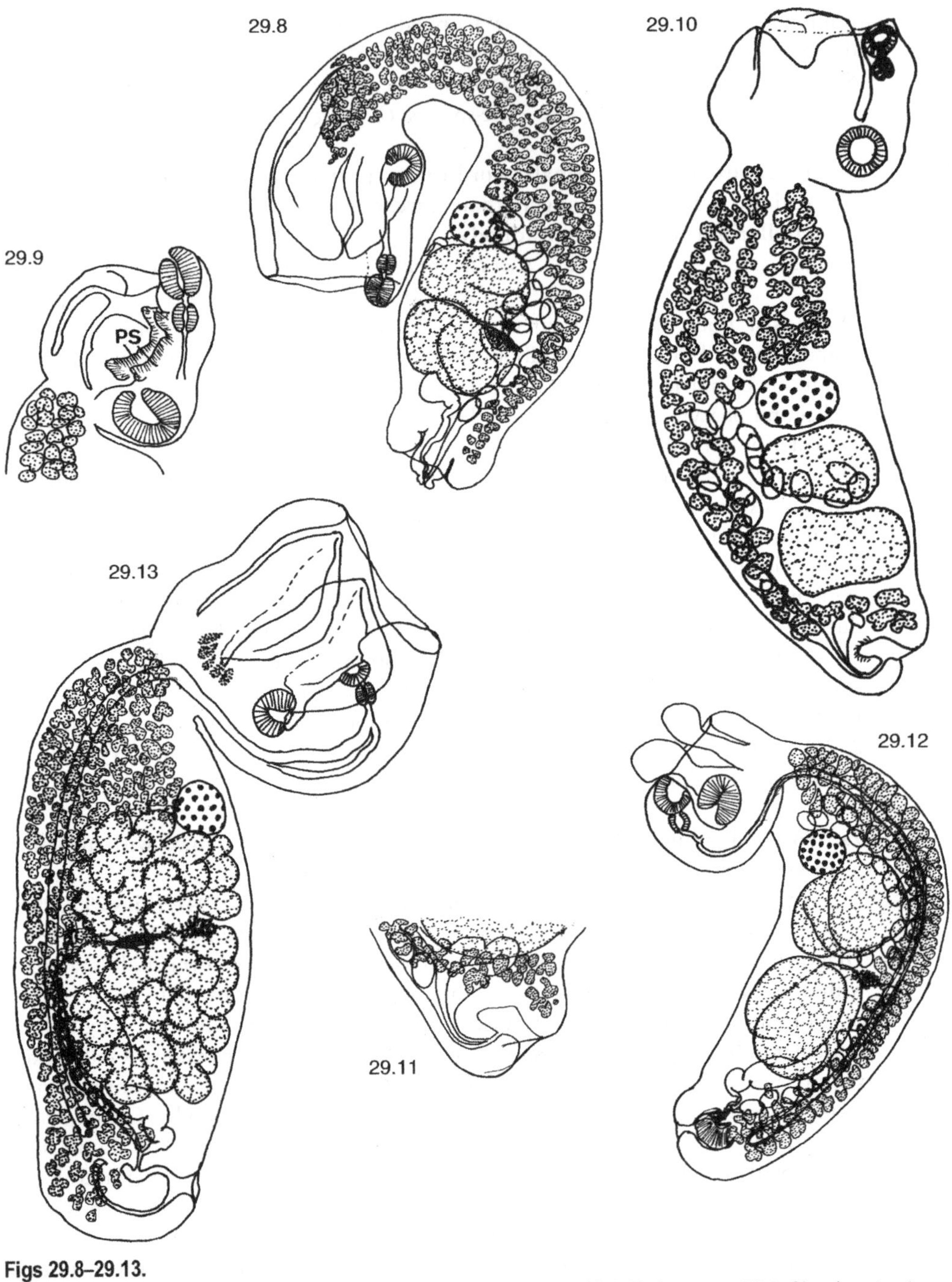

Figs 29.8–29.13.

29.8, 29.9. *Schwarzitrema novaehollandiae* Dubois & Pearson, 1967. 29.8. Entire worm. 29.9. Showing structure of pseudosuckers (PS). (After Dubois, 1968.)

29.10, 29.11. *Pseudapatemon mamilliformis* (Tubangui, 1932). 29.10. Entire worm. 29.11. Showing terminal genitalia without conus or bulbus genitalis. (After Dubois, 1968.)

29.12. *Cotylurus cornutus* (Rudolphi, 1808). (After Dubois, 1968.)

29.13. *Ichthyocotylurus platycephalus* (Creplin, 1825). (Original.)

Diagnosis: Body distinctly bipartite; forebody in shape of flat cup or subglobular, wider than long, without prominent pseudosuckers, with massive holdfast organ. Hindbody cylindrical, claviform, saccular or subreniform, without neck region. Oral and ventral suckers well developed; pharynx small. Testes tandem; anterior asymmetrical, posterior symmetrical, bilobed, situated in posterior or middle part of hindbody. Ovary reniform or oval. Vitellarium tends to be confined to hindbody, may enter holdfast organ but not forebody wall. Copulatory bursa with subterminal dorsal opening, without genital cone; distal part of uterus recurved, joining ejaculatory duct and opening into bursa on small protuberance on its posterior wall. Ejaculatory pouch present. In Scolopacidae. Europe, North America, Philippines. Type-species *P. elassocotylus* (Dubois, 1934) Dubois, 1936.

8b. Holdfast organ lobed; genital cone or genital bulb present .... **9.**

9a. Genital bulb present .... **10.**

9b. Genital cone present .... **11.**

10a. Forebody spherical; proteolytic gland dispersed; testes trilobed with smooth lobes .... ***Cotylurus* Szidat, 1928. (Fig. 29.12)**

(Syns *Choanodiplostomum* Vigueras, 1944; *Cotylurostrigea* Sudarikov, 1961)

Diagnosis: Body bipartite; forebody globular, subglobular or cup-shaped, with lobed holdfast organ and dispersed proteolytic gland; hindbody saccular, reniform or fusiform, more or less arched, without neck region. Pseudosuckers, oral sucker, ventral sucker and pharynx present. Testes tandem, trilobed with smooth lobes, situated in middle of hindbody. Ovary round or reniform. Vitellarium tends to be confined to ventral and lateral parts of hindbody, may penetrate into forebody wall or holdfast organ. Copulatory bursa pocket-shaped with dorso-subterminal opening and protrusible genital bulb inside. Genital cone absent. Genital pore at apex of small protuberance at base of genital bulb. In Anatidae, Charadriidae and Scolopacidae. Holarctic, Neotropical and Oriental regions. Metacercaria of 'tetracotyle' type, in snails and leeches. Cercariae with flame-cell formula $2[(2+2)+(2+2+[2])]=20$, excretory commissure anterior to ventral sucker, four penetration glands anterior to ventral sucker and intercaecal. Type-species *C. cornutus* (Rudolphi, 1808) Szidat, 1928.

10b. Forebody cup-shaped; proteolytic gland compact; testes trilobed with lobulate lobes .... ***Ichthyocotylurus* Odening, 1969. (Fig. 29.13)**

Diagnosis: Body bipartite; forebody cup-shaped, with lobed holdfast organ and well developed proteolytic gland; hindbody cylindrical, more or less curved dorsally. Pseudosuckers, oral sucker, ventral sucker and pharynx present. Testes tandem, trilobed with lobulate lobes, in middle or posterior part of hindbody. Ovary transversely oval. Vitellarium confined to hindbody or may penetrate into base of forebody. Copulatory bursa pocket-shaped. Genital cone absent. Genital bulb relatively large; genital pore at tip of small protuberance near base of bulb. In piscivorous birds. Holarctic. Metacercariae of 'tetracotyle' type, in fishes. Cercariae with two pairs of postacetabular penetration glands, flame-cell formula $2[(2+2)+(2+2+[2])]=20$, excretory commissure anterior to ventral sucker. Type-species *I. platycephalus* (Creplin, 1825) Odening, 1969.

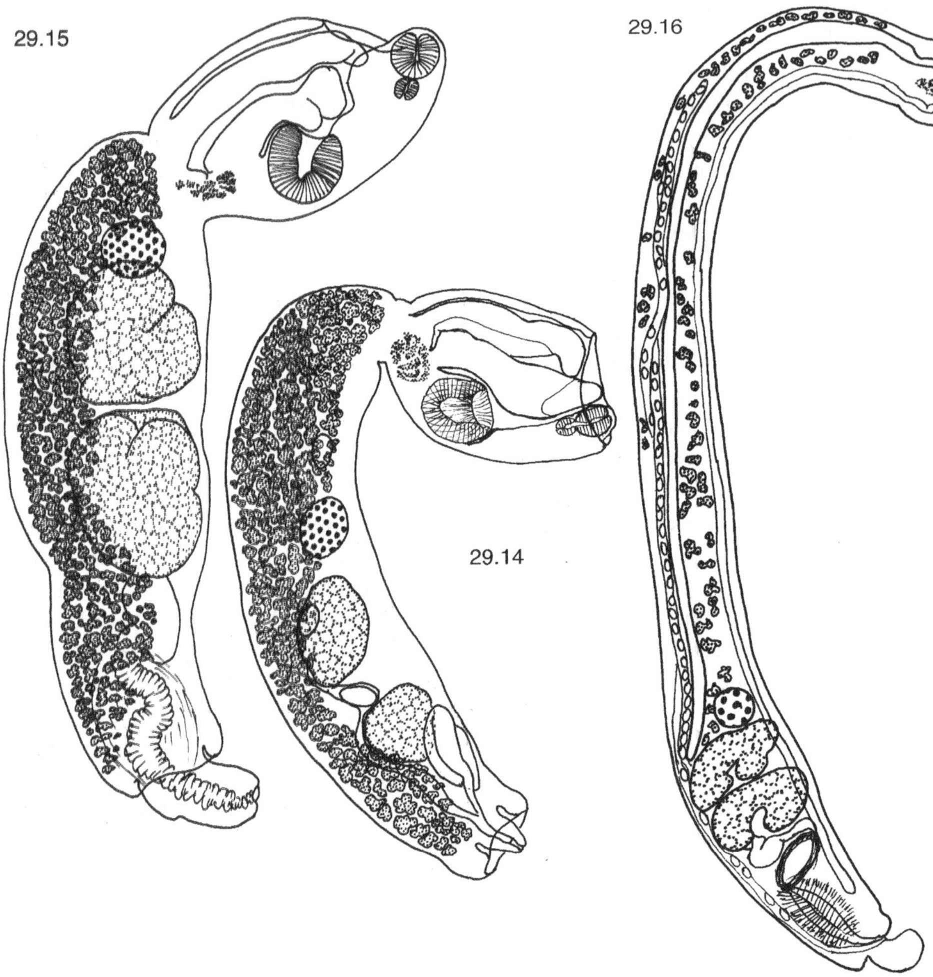

Figs 29.14–29.16.
29.14. *Apatemon gracilis* (Rudolphi, 1819). (After Dubois, 1968.)
29.15. *Australapatemon anseris* (Dubois, 1967). (After Dubois, 1968.)
29.16. *Duboisiella proloba* Baer, 1938. (After Dubois, 1968.)

11a. Genital cone small, not or poorly delimited from parenchyma; ejaculatory duct without internal rugae .... *Apatemon* Szidat, 1928. (Fig. 29.14)
(Syn. *Pseudostrigea* Yamaguti, 1933)

Diagnosis: Body bipartite; forebody infundibular, cup-shaped or spherical; hindbody fusiform, reniform or subcylindrical, more or less arched, with neck region absent. Oral and ventral suckers well developed; ventral larger than oral; pharynx present. Testes tandem in posterior half of body. Ovary reniform or ovoid. Vitellarium in hindbody, confined to ventral and subventral fields. Seminal vesicle situated dorsally to hind testis. Uterus extends anterior to ovary. Copulatory bursa terminal, rather small; muscular ring absent; genital cone not or poorly delimited from surrounding parenchyma. Hermaphroditic duct, formed by union of distal part of uterus and ejaculatory duct, opens close to apex of cone. Mostly in piscivorous birds. Holarctic region, Brazil, Congo. Metacercariae of 'tetracotyle' type in fishes. Cercariae with two groups of four small postacetabular penetration glands and flame-cell formula 2[(1 + 1) + (1 + 1 + [1])] = 10; excretory commissure posterior to ventral sucker; alimentary system rudimentary. Type-species *A. gracilis* (Rudolphi, 1819) Szidat, 1928.

11b. Genital cone large, well delimited from parenchyma; ejaculatory duct with internal rugae .... *Australapatemon* Sudarikov, 1959. (Fig. 29.15)

Diagnosis: Body bipartite; forebody globular, cup- or bell-shaped, with holdfast organ composed of two lobes; hindbody elongate saccular, with neck region absent. Oral and ventral suckers well developed; ventral sucker larger than oral. Testes tandem, irregular in shape, in middle or anterior half of hindbody. Ovary round, oval or reniform. Vitellarium confined to hindbody. Copulatory bursa large with terminal opening. Genital cone large (one-third to one-fifth of hindbody length), encloses long ejaculatory duct with internal rugae. In anseriform birds. Cosmopolitan. Metacercariae of 'tetracotyle' type, in leeches. Cercariae with flame-cell formula 2[(1 + 1) + (2 + 2 + [1])] = 14; excretory commissures anterior and posterior to ventral sucker, may be incomplete; penetration glands in two groups of four, posterior to ventral sucker; alimentary system well developed. Type-species *A. intermedius* (Johnston, 1904) Sudarikov, 1959.

Subfamily Duboisiellinae Baer, 1938

Diagnosis: Body long, cylindrical, not clearly divided into fore- and hindbody. Forebody tubular, with long cavity extending to anterior testis; end of cavity delimits short hindbody. Holdfast organ small, situated at proximal end of forebody cavity, composed of dorsal and ventral lobes. Proteolytic gland diffuse. Oral sucker small; pharynx large; ventral sucker rudimentary, visible only on sections. Testes tandem, lobed, in anterior part of hindbody. Seminal vesicle followed by ejaculatory pouch and 'periprostate'; latter with muscular wall surrounded by glandular cells. Ovary pretesticular, round, dorsal. Oötype with Mehlis' gland and vitelline reservoir, posterior to ovary, close to anterior testis. Vitellarium confined to lateral walls of forebody and within holdfast organ. Uterus extending longitudinally into ventral part of forebody, reaching three-quarters of its length. Eggs numerous, large. Genital pore dorso-subterminal. In mammals. Venezuela, Panama. Type- and only genus *Duboisiella* Baer, 1938.

Genus *Duboisiella* Baer, 1938 (Fig. 29.16)

Diagnosis: With characters of subfamily. Type-species *D. proloba* Baer, 1938.

30 Superfamily Gymnophalloidea Odhner, 1905

R.A. Bray

Department of Zoology, The Natural History Museum, London, UK

Introduction

In this volume the superfamily Gymnophalloidea is used in a more restricted sense than has recently been adopted by Gibson (1996), in that the Brachylaimoidea and Bucephaloidea are recognized as separate superfamilies, but more inclusively than the system suggested by Hall *et al.* (1999), in that the Fellodistomidae are included. These decisions are made for convenience, rather than new phylogenetic insights, although recent molecular evidence (Cribb *et al.*, 2001) does not indicate a close relationship between the Bucephalidae, on the one hand, and the Fellodistomidae and Tandanicolidae, on the other.

The earliest use of the superfamily name I have detected is that of Brooks *et al.* (1985), who included in it only the nominal family. Later, Brooks *et al.* (1989) added the Fellodistomidae. When Cable (1953) drew attention to the affinities of the Gymnophallinae and the Fellodistomidae, he based several of his points on characteristics of the genus *Bacciger* Nicoll, 1924, which is now considered to belong in the family Faustulidae Poche, 1926. Molecular evidence has indicated that this family is distant from the true fellodistomes and close to the zoogonids (Hall *et al.*, 1999; Cribb *et al.*, 2001). Similar evidence (Hall *et al.*, 1999) has confirmed the close relationship of the gymnophallids and the true fellodistomes and tandanicolids.

The families covered in this superfamily include, therefore, the Gymnophallidae, Fellodistomidae and Tandanicolidae. In addition, two enigmatic families, the Callodistomidae and Botulisaccidae, are included for convenience of identification.

Superfamily Gymnophalloidea Odhner, 1905

Diagnosis: Body small to medium-sized, rarely large; tegument armed or not. Oral and ventral suckers present. Prepharynx absent or short. Pharynx well developed. Oesophagus long to almost absent. Caeca blind, two, short to long, or one. Testes two, tandem, oblique or symmetrical. Cirrus-sac present or absent. Seminal vesicle and pars prostatica internal when cirrus-sac present; pars prostatica often well developed. Genital atrium small, apparently absent or well developed. Genital pore usually in forebody, occasionally in anterior hindbody, ventromedial to ventrolateral. Ovary usually pretesticular, sometimes post- or intertesticular. Laurer's canal present. Seminal receptacle uterine or canalicular. Uterus usually extensive, filling much of hindbody, occasionally restricted to pregonadal region. Eggs operculate, small, numerous. Vitellarium usually follicular in two lateral fields, but follicles may be compacted into two (rarely one) masses. Excretory vesicle Y-, V- or I-shaped; pore terminal or almost so. Parasitic as adults in gut, occasionally other body organs, of birds, mammals and fishes. Type-genus *Gymnophallus* Odhner, 1900.

Key to families

1a. Cirrus-sac absent; in birds, rarely mammals .... Gymnophallidae Odhner, 1905.
1b. Cirrus-sac present (rarely absent); in fishes .... 2.

2a. Oesophagus rectilinear; cirrus-sac usually and genital pore always in fore-body .... 3.
2b. Oesophagus with lateral outgrowths; cirrus-sac and genital pore in hind-body .... Botulisaccidae Yamaguti, 1971.

3a. Canalicular seminal receptacle present; excretory vesicle I-shaped; in gall-bladder of fishes .... Callodistomidae Odhner, 1910.
3b. Uterine seminal receptacle present; excretory vesicle V- or Y-shaped; in intestine or other organs, of mainly marine fishes .... 4.

4a. Accessory copulatory organ present in genital atrium .... Tandanicolidae Johnston, 1927.
4b. Accessory copulatory organ absent .... Fellodistomidae Nicoll, 1909.

31 Family Gymnophallidae Odhner, 1905

T. SCHOLZ

Institute of Parasitology, Academy of Sciences of the Czech Republic, Branišovská, České Budějovice, Czech Republic

Introduction

This is a small group of digeneans occurring in the intestine, gall-bladder and bursa Fabricii of a narrow spectrum of marine birds of the orders Charadriiformes (shore birds), including gulls, and Anseriformes (diving ducks); two taxa, *Gymnophalloides seoi* Lee, Chai & Hong, 1993 and *G. heardi* Ching, 1995, have been described from mammals. The type-genus, *Gymnophallus*, was established by Odhner (1900b) with *G. deliciosus* (Olsson, 1893) as its type-species. The same author (Odhner, 1905) proposed the subfamily Gymnophallinae (within the family Heterophyidae) to accommodate this genus. According to the Principle of Coordination, *International Code of Zoological Nomenclature*, Article 36(b), Odhner (1905) is the authority for the family, not Dollfus (1939), as reported by Yamaguti (1971), or Morozov (1955), as generally used in the literature (Morozov, 1955; James, 1964; Bartoli, 1974; Bowers *et al.*, 1990; Ching, 1995a). Yamaguti (1958a) introduced another subfamily, the Parvatrematinae, to accommodate *Parvatrema* Cable, 1953; this subfamily is not accepted here because the differential criteria used to separate it from the Gymnophallinae are variable within genera of both subfamilies and thus unsuitable (see Yamaguti, 1958a, 1971; James, 1964). To date, seven gymnophallid genera have been established (Hoberg, 1981; Ching, 1995a).

Typical gymnophallid life-cycles may involve bivalves as first and second intermediate hosts, bivalves and polychaetes as second intermediate hosts or bivalves as the only hosts; gastropods or brachiopods are exceptionally second intermediate hosts (Bartoli, 1974; Ching, 1995a). Cercariae are furcocercous. Metacercariae are never encysted (previous reports on encysted metacercariae referred to a capsule formed by the host) and are quite common in marine bivalves (Stunkard & Uzmann, 1958; Loos-Frank, 1971; Bartoli, 1974; Lauckner, 1983). In the life-cycles of some species, a developmental stage unique among trematodes, i.e. a germinal sac, may be present (Szidat, 1962; James, 1964; Ching, 1982; Galaktionov, 1996).

The taxonomy of the family, inclusive of the generic classification, is unsatisfactory, mainly due to the considerable homogeneity of its members, their small body size and difficulties in the observation of internal structures because of the presence of numerous eggs (Bartoli, 1974). Permanent preparations are usually difficult to study and many internal features are readily visible only in live specimens. Moreover, existing descriptions, including those of the type-species of several genera, are often too simple or even erroneous and based on a limited number of specimens, sometimes only one worm (see James, 1964). Species diagnoses of some taxa do not always conform with those of the relevant genus (see Ching, 1995b), and there have been numerous transfers of individual species between genera (Bowers & James, 1967; Loos-Frank, 1970), including the synonymy of the type-species of both *Lacunovermis* Ching, 1965 and *Meiogymnophallus* Ching, 1965, with taxa previously placed in other genera (Pekkarinen & Ching, 1994). Several gymnophallids, including the type-species of *Gymnophalloides* Fujita, 1925, *G. tokiensis* Fujita, 1925, were described from metacercariae or cercariae and their adults are unknown. It is probable that some cercariae, metacercariae and adults, established as distinct species under different names, in fact represent developmental stages of the same taxa (Ching, 1995a). The existence of sibling species (Bartoli, 1972; Bowers *et al.*, 1990, 1996) makes it difficult or even impossible to distinguish some taxa solely on the basis of the morphology of the adults, and knowledge of the complete life-cycles is required for species differentiation (Bartoli, 1974; Ching, 1995a).

All the above-mentioned problems limit the possibility of solving taxonomic problems within the group, including those related to the generic classification, unless the life-cycles of most taxa are elucidated and the detailed morphology of their developmental stages is known. Despite considerable progress in the taxonomy, biology and ecology of the Gymnophallidae, James' (1964) statement, 'the present system is, therefore, simply an attempt to bring together, organize and interpret present knowledge and cannot, of course, be considered as final but may facilitate future research', is still valid.

Ching (1995a) listed 11 morphological features considered to be key characters for differentiating gymnophallid genera. However, there are numerous inaccuracies and mistakes in these characters (see Table 1 of Ching, 1995a), e.g. in the description of the tegumental spines of *Pseudogymnophallus* Hoberg, 1981 (see Hoberg, 1981), the shape of the seminal vesicle (both bipartite and undivided in *Gymnophalloides*) (see Ching, 1995b) or the extent of uterine loops in several genera. Examination of type- and voucher specimens (see below) and a review of the literary data by the present author cast doubts on the reliability of some characters used for differentiating gymnophallid genera. Therefore, remarks on three taxa of dubious validity are included below.

Meiogymnophallus Ching, 1965 was proposed as a new name for *Gymnophalloides* sensu James (1964) *nec* Fujita, 1925, which was distinguished from *Gymnophalloides* Fujita, 1925 by the absence of a ventral pit anterior to the ventral sucker (Ching, 1965, 1972). The validity of *Meiogymnophallus* was questioned by Bartoli (1965, 1974). Yamaguti (1971), apparently misunderstanding the reasons for its erection (see Ching, 1972), synonymized it with *Gymnophalloides*. Bowers *et al.* (1990) differentiated it from the related genera *Gymnophallus* and *Parvatrema* by body-size, the nature of the genitalia, the flame-cell formula and the position of the parasite in the host. However, species of these genera exhibit variation in all the above-mentioned characters. As to the site of infection, the criterion considered to be the most important for the separation of *Meiogymnophallus*, *Gymnophallus* and *Parvatrema* by Bowers *et al.* (1990), species of all three genera may occur in the intestine of the definitive hosts (Stunkard & Uzmann, 1958; Bartoli, 1974; Hoberg, 1981; Ching, 1995a), which nullifies this as a discriminative character. *Meiogymnophallus* closely resembles *Parvatrema* in almost all key characters (Ching, 1995a), the only marked differences being the structure of the genital pore, which is wide in *Parvatrema* but small in *Meiogymnophallus*, and the shape of the genital atrium (wide, oval in *Parvatrema* versus tubular in *Meiogymnophallus*). However, the size and position of the genital pore vary to a great extent within individual specimens of the same *Parvatrema* species (James, 1964). For this reason, James (1964) concluded that 'this character can no longer be considered so important taxonomically'. Moreover, the size of the genital pore and its position depend on contraction of the worms and may easily be misinterpreted (Pekkarinen & Ching, 1994). The unsuitability of the morphology of the genital pore and atrium for generic differentiation has been demonstrated by Pekkarinen & Ching (1994), who admitted that a transverse fold on the ventral sucker of *Parvatrema affinis* (Jameson & Nicoll, 1913) may have been mistaken for the wide genital pore in their previous studies. As a consequence, the authors synonymized the type-species of *Meiogymnophallus*, *M. multigemmulus* Ching, 1965, with *P. affinis*, originally described as *Gymnophallus affinis* and then placed in *Parvatrema* by Cable (1953). Considering the doubtful reliability of the characters used to separate these genera, *Meiogymnophallus* is here considered to be a synonym of *Parvatrema*.

Lacunovermis Ching, 1965 was established by Ching (1965) to accommodate the species *L. conspicuus* Ching, 1965, which has recently been synonymized with *L. macomae* (Lebour, 1908) (syn. *Cercaria macomae* Lebour, 1908) by Pekkarinen & Ching (1994). *Lacunovermis* resembles *Gymnophalloides* in possessing the ventral pit, compact vitellarium, lateral projections on the oral sucker, V-shaped excretory vesicle and small body-size. The main difference between them is the structure of the genital pore, which is small and lacking papillae in *Gymnophalloides* but wide and papillate in *Lacunovermis* (see Ching, 1965, 1972, 1995a). This feature, however, varies within members of the same genus, as is the case in *Gymnophalloides heardi* and *G. seoi* (see Lee *et al.*, 1993; Ching, 1995b; Figs 31.3 and 31.4). Conversely, the type-species of *Lacunovermis*, *L. macomae*, has a relatively small genital pore, much smaller than the ventral pit (see Loos-Frank, 1970, fig. 2A, B). *Lacunovermis* and *Gymnophalloides* differ in several other key characters (see table 1 of Ching, 1995a), but these are questionable or apparently erroneous: (i) the seminal vesicle in *Gymnophalloides* can be undivided (*G. heardi*) (see Ching, 1995b) as in *Lacunovermis*; in addition, this feature may vary within a species (see Bowers *et al.*, 1990); (ii) the genital atrium, reported to be tubular in *Gymnophalloides*, can be shallow in this genus (*G. seoi*) (see Lee *et al.*, 1993), which corresponds to *Lacunovermis*; and (iii) the extent of the uterus in species of both genera is identical (see Ching, 1965, 1973; Fig. 31.3). Consequently, the invalidity of *Lacunovermis* and its synonymy with *Gymnophalloides*, first proposed by Yamaguti (1971), is accepted here: hence, the type-species, *L. macomae*, becomes *Gymnophalloides macomae* n. comb.

Paragymnophallus Ching, 1973 was established to accommodate specimens identified by Odhner (1900b, 1905) as *Gymnophallus somateriae* (Levinsen, 1889), but apparently differing from Levinsen's material. Ching (1973) differentiated *Paragymnophallus* from the most closely related *Gymnophallus* mainly by the presence of a wide genital pore situated at a distance anterior to the ventral sucker. As discussed above, this feature should not be used as a basis for the separation of genera because of its variability and the possibility of the misinterpretation of its structure and position (see James, 1964; Pekkarinen & Ching, 1994).

Other differences between *Paragymnophallus* and *Gymnophallus* listed by Ching (1995a) are negligible or questionable, as in the cases of: the shape of the tegumental spines, reported to be broad in the former genus (but without SEM evidence) versus sharp (spines of *Gymnophallus* may be broad and serrate (Bartoli, 1983; unpublished observations)); the shape of the excretory vesicle, which is not V-shaped in *Paragymnophallus* but Y-shaped as in *Gymnophallus* (see the original description of Ching, 1973); and the extent of the uterus, which in *Paragymnophallus* is not limited to the hindbody but also forms loops in the forebody (see Ching, 1973), as in *Gymnophallus*. Therefore, it is considered appropriate to suppress *Paragymnophallus* as a synonym of *Gymnophallus*.

Family Gymnophallidae Odhner, 1905

Diagnosis: Small to minute worms; body oval to pyriform. Tegument spined, with sharp, simple or broad, serrate spines, covering most of body-surface except for small areas, usually around ventral sucker and genital pore. Oral sucker subterminal, larger than ventral sucker and sometimes up to twice as large, with or without muscular lateral projections. Ventral sucker equatorial to slightly postequatorial or even in posterior third of body. Prepharynx lacking; pharynx well developed; oesophagus absent or short. Intestinal caeca short, terminate blindly, not reaching mid-body, widely divergent, often wide, with (*Pseudogymnophallus* only) or without dorsal diverticula. Testes two, symmetrical or slightly oblique, posterolateral to ventral sucker. Seminal vesicle present, divided (bi- or rarely tripartite) or undivided. Pars prostatica present or absent; if present, tubular or oval; if absent, prostatic glands surround genital atrium. Cirrus-sac and cirrus absent. Short ejaculatory duct uniting with metraterm to form hermaphroditic duct; genital atrium tubular or transversely oval. Genital pore medial, postbifurcal in forebody, inconspicuous at anterior margin of ventral sucker or wide, pit-like, at some distance anterior to ventral sucker; pore may be surrounded by muscle fibres. Ovary usually anterior to testes, rarely intertesticular. Laurer's canal present; seminal receptacle present or absent. Vitellarium paired or single, compact, lobed or follicular, situated anteriorly, laterally or closely posteriorly to ventral sucker. Uterus forms loops in forebody, hindbody or both. Eggs small, operculate, embryonated when laid. Excretory vesicle V- or Y-shaped, with long lateral arms reaching to pharyngeal level, with or without diverticula. Excretory collecting duct short, ciliated; flame-cells in doublets, 16 or 24 in total. Excretory pore terminal. Type-genus *Gymnophallus* Odhner, 1900.

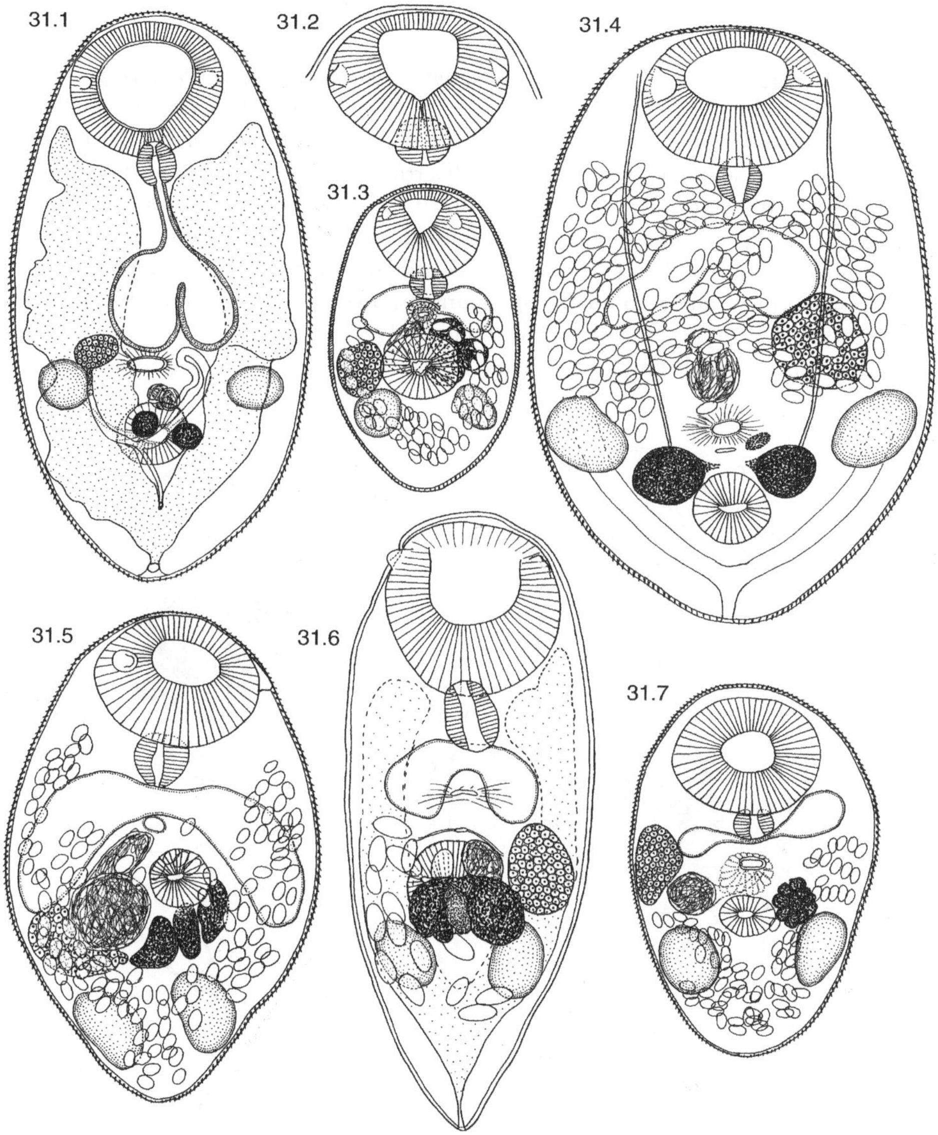

Figs 31.1–31.7.

31.1. *Gymnophalloides tokiensis* Fujita, 1925. Dorsal view. (Original (neotype; USNPC 60899).)

31.2, 31.3. *Gymnophalloides heardi* Ching, 1995. 31.2. Oral sucker with lateral projections. (Original (paratype; USNPC 83872).) 31.3. Ventral view; note presence of large genital pore some distance anterior to ventral sucker. (Original (paratype; USNPC 83872).)

31.4. *Gymnophalloides seoi* Lee, Chai & Hong, 1993. Ventral view; note position and size of genital pore. (Original (paratype; USNPC 82478).)

31.5. *Gymnophalloides macomae* (Lebour, 1908) n. comb. Dorsal view. (Original (paratype; USNPC 72252).)

31.6. *Parvatrema borealis* Stunkard & Uzmann, 1958. Ventral view. (Original ('type or cotype' (specimen not designated); USNPC 56235).)

31.7. *Parvatrema borinquenae* Cable, 1953. Ventral view. (Original (holotype; USNPC 47875).)

Key to genera

1a. Ventral pit present .... ***Gymnophalloides*** **Fujita, 1925. (Figs 31.1–31.5)**
(Syn. *Lacunovermis* Ching, 1965)

Diagnosis: Minute worms. Oral sucker with lateral projections. Ventral pit present. Caeca short, enlarged, without dorsal diverticula. Seminal vesicle bipartite or undivided. Genital pore variable in size, from inconspicuous close to anterior margin of ventral sucker to wide at distance anterior to sucker. Ovary pretesticular. Vitellarium compact or lobed. Excretory bladder V-shaped. In intestine of Charadriiformes and Anseriformes, rarely in mammals (humans, rice rat). North and South America, Europe, Asia. Type-species *G. tokiensis* Fujita, 1925.

1b. Ventral pit absent .... **2.**

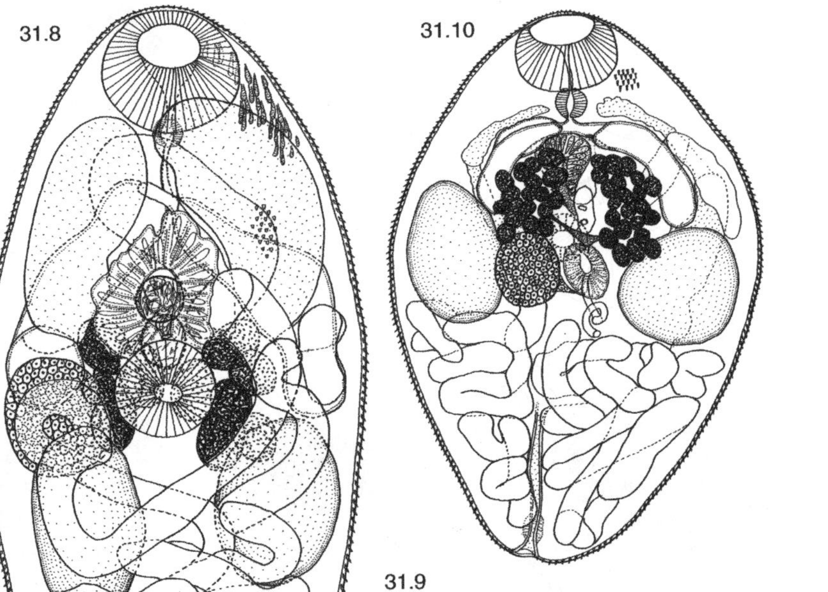

Figs 31.8–31.10.
31.8. *Gymnophallus deliciosus* (Olsson, 1893). Ventral view. (Original (BMNH 1984.6.7.2).)
31.9. *Gymnophallus* sp. from gall-bladder of *Larus argentatus*, Cross Mullet, C. Mayo, Eire. Lateral view of ventral sucker and male terminal genitalia. (Original (BMNH 1991.7.2.64–67).)
31.10. *Pseudogymnophallus alcae* Hoberg, 1981. Dorsal view. (Original (paratype; USNPC 75936).)

2a. Oral sucker with lateral projections; excretory vesicle V-shaped ....
.. *Parvatrema* Cable, 1953. (Figs 31.6–31.7)
(Syn. *Meiogymnophallus* Ching, 1965)

Diagnosis: Minute worms. Oral sucker with lateral projections. Ventral pit absent. Caeca enlarged, without dorsal diverticula. Seminal vesicle undivided. Prostatic duct oval. Genital pore usually wide, situated at distance anterior to ventral sucker. Ovary pretesticular. Vitellarium compact. Excretory bladder V-shaped. In intestine of Charadriiformes and Anseriformes. North and South America, Europe, Asia. Type-species *P. borinquenae* Cable, 1953.

2b. Oral sucker without lateral projections; excretory vesicle Y-shaped .... 3.

3a. Ovary pretesticular; intestinal caeca without dorsal diverticula ....
.. *Gymnophallus* Odhner, 1900. (Figs 31.8, 31.9)
(Syn. *Paragymnophallus* Ching, 1973)

Diagnosis: Small worms. Oral sucker without lateral projections. Ventral pit absent. Caeca not enlarged, without dorsal diverticula. Seminal vesicle bipartite, rarely tripartite. Prostatic duct elongate, curved posteroventrally, surrounded by numerous prostatic cells. Genital pore inconspicuous, close to anterior margin of ventral sucker. Ovary pretesticular. Vitellarium usually follicular, arranged in two groups. Excretory bladder Y-shaped. In gall-bladder, bursa Fabricii and intestine of Charadriiformes and Anseriformes. North and South America, Europe, North Africa, Asia. Type-species *G. deliciosus* (Olsson, 1893).

3b. Ovary intertesticular; intestinal caeca with dorsal diverticula..............................
.. *Pseudogymnophallus* Hoberg, 1981. (Fig. 31.10)

Diagnosis: Small worms. Oral sucker without lateral papillae. Ventral pit absent. Caeca not enlarged, with dorsal diverticula. Seminal vesicle undivided. Prostatic duct elongate, curved posteroventrally, surrounded by numerous prostatic cells. Genital pore inconspicuous, close to anterior margin of ventral sucker. Ovary between testes. Vitellarium follicular, arranged in two groups. Excretory bladder Y-shaped. In gall-bladder and intestine of Charadriiformes (Alcidae). North America (Alaska, Canada). Type-species *P. alcae* Hoberg, 1981.

Note added in proofs

Cremonte (2001) described a new genus, *Bartolius*, to accommodate *Bartolius pierrei* n. g., n. sp., described from metacercariae from *Darina solenoides* (King) and adults from *Larus dominicanus* Lichtenstein from Argentina. The new genus is distinguished from other genera of the Gymnophallidae by the post-testicular position of the ovary, the oral sucker of *B. pierri* is without lateral projections, a ventral pit is absent and the genital pore is inconspicuous (Cremonte, 2001).

32 Family Botulisaccidae Yamaguti, 1971

R.A. Bray

Department of Zoology, The Natural History Museum, London, UK

Introduction

Overstreet (1969) corrected the description by Caballero y Caballero *et al.* (1955) of *Botulisaccus pisceus* Caballero, Bravo Hollis & Grocott, 1955 and considered the worm a fellodistomid. With the dismemberment of the older concept of the Fellodistomidae (as espoused, for example, by Bray, 1988) by Hall *et al.* (1999), it is not clear to which family this genus should now belong. Examination of worms kindly lent and donated by Dr Overstreet has confirmed the morphology, but not clarified the placement of this form. The spinous tegument suggests faustulid affinities, but the uterine seminal receptacle, the anteriorly placed Laurer's canal opening and the cirrus-sac with the complex, strongly muscular ejaculatory duct are not usual faustulid features. Yamaguti (1971) recognized a new family for this genus and, considering the combination of unusual characters exhibited by this worm, including the genital pore and cirrus-sac in the hindbody and the oesophagus with distinct lateral outgrowths, it appears to be a reasonable action and is followed here. The family is included in the superfamily Gymnophalloidea due to its sharing major characters with the Fellodistomidae (see above).

Family Botulisaccidae Yamaguti, 1971

Diagnosis: Body pyriform. Tegument spinous. Oral sucker subterminal, spherical to infundibuliform. Ventral sucker in anterior half of body. Prepharynx short. Pharynx large, elongate. Oesophagus long, with large lateral outpocketings, which lie dorsal or anterodorsal to ventral sucker. Intestinal bifurcation at about level of posterior edge of ventral sucker. Caeca reach just past testes, blind. Testes two, symmetrical, in mid-hindbody, largely extracaecal; entire. Cirrus-sac elongate-oval, straight to arcuate, in anterior hindbody. Internal seminal vesicle bipartite. Pars prostatica wide, vesicular. Ejaculatory duct thick-walled muscular, deeply folded. Genital pore in anterior hindbody, dextral. Ovary between anterior regions of testes, oval, entire. Uterine seminal receptacle present. Laurer's canal opening posteriorly to ovary. Uterus mainly post-testicular. Eggs thick-shelled, operculate. Metraterm short. Vitellarium in two lateral fields of irregular follicles, from immediately anterior to testes, extending almost to level of ventral sucker. Excretory pore terminal. Excretory vesicle V-shaped; arms extending into forebody. In marine fishes (*Albula*). Pacific and Atlantic Oceans. Type-genus *Botulisaccus* Caballero, Bravo Hollis & Grocott, 1955.

Genus *Botulisaccus* Caballero, Bravo Hollis & Grocott, 1955 (Figs 32.1, 32.2)

Diagnosis: With characters of family. Type-species *B. pisceus* Caballero, Bravo Hollis & Grocott, 1955.

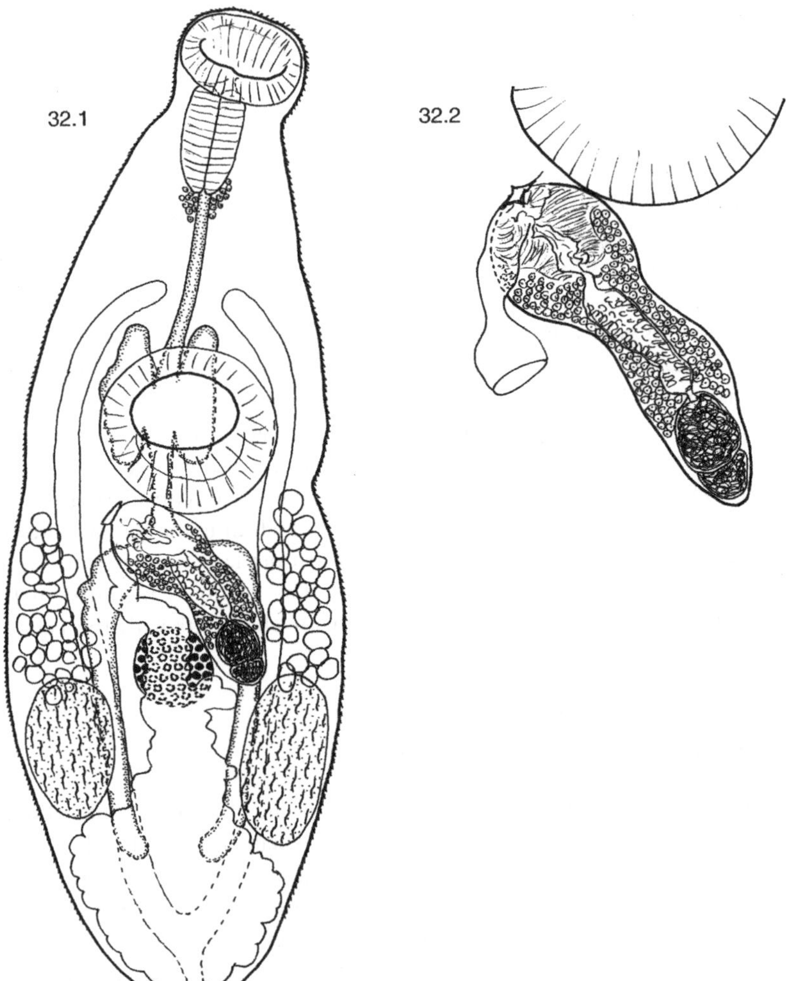

Figs 32.1–32.2. *Botulisaccus pisceus* Caballero, Bravo Hollis & Grocott, 1955. 32.1. Ventral view, uterus in outline. (Original.) 32.2. Terminal genitalia. (Original.)

33 Family Callodistomidae Odhner, 1910

R.A. Bray

Department of Zoology, The Natural History Museum, London, UK

Introduction

Odhner (1910a) erected this taxon as a subfamily to include the African genera *Callodistomum* Odhner, 1902, *Cholepotes* Odhner, 1910 and *Thaumatocotyle* Odhner, 1910 *nec* Scott, 1904. The latter, being pre-occupied, was renamed *Braunotrema* by Price (1930) and placed in its own family by Yamaguti (1971). All forms recognized herein are reported from the gall-bladder of freshwater fishes. The family has been enlarged by the addition of South American freshwater and also marine forms. The marine forms can probably best be placed elsewhere (see below), but the South American forms are retained in the family. The family is very similar to the Fellodistomidae, which is almost exclusively marine, but apparently differs in having an I-shaped excretory vesicle, a poorly defined pars prostatica, a canalicular seminal receptacle, eggs containing an embryo with eye-spots and by inhabiting the gall-bladder.

Yamaguti (1971) described the testes of *Teratotrema* Travassos, Artigas & Pereira, 1928 as 'close together behind cirrus-pouch'. Travassos *et al.* (1928) had suggested that a single testis was present, but its lobation could cause it to be interpreted as two superimposed testes. Kohn & Fernandes (1987) confirmed that only a single testis is present and my examination of a specimen of *T. dubium* Travassos, Artigas & Pereira, 1928 (FIOCRUZ 32.252) has corroborated this. Other details have been derived from this specimen. The male duct within the cirrus-sac is unusual in that the only evidence of sperm is in the distal portion, but as only a single specimen is available it may well be an anomalous observation. The monotypic subfamily Teratotrematinae Yamaguti, 1958 was distinguished from the Callodistominae by the testis in the forebody. It is not recognized in this key as it adds nothing to our understanding of the relationships in the family.

As pointed out by Bray & Gibson (1980), Yamaguti (1971, p. 118) based *Guptatrema* Yamaguti, 1971 on *Steringophorus lethrini* Gupta, 1955, which he also placed in *Fellodistomum* Stafford, 1904 in the same work (p. 70). Bray & Gibson (1980) also suggested that the host and the overall morphology indicated that the species might be a cryptogonimid, probably belonging in *Paracryptogonimus* Yamaguti, 1934. A second species assigned to this genus, *G. papillonae* Toman, 1992, from the pomacanthid *Pomacanthodes imperator* from the Seychelles (Toman, 1992), appears to be rather distinct in several ways. The much more extensive follicular vitellarium and the uterine distribution indicate that the species is probably an opecoelid.

Pirupalkia Puga & Figueroa, 1989 was placed in the family Callodistomidae by Puga & Figueroa (1989). I have examined specimens apparently identical to *P. queulensis* Puga & Figueroa, 1989 (BNMH 1996.2.9.7-11), but which were identified as *Villarrealina peruanus* Bolaños & Salas, 1982. Bolaños & Salas (1982) allocated their species to the Allocreadiidae, subfamily Urorchiinae, but I believe that it is probably an opecoelid and that *Villarrealina* and *Pirupalkia* are synonymous.

Callodistomoides Yamaguti, 1970 was originally placed in this family by Yamaguti (1970), but it was recognized as a synonym of the fellodistomid genus *Steringophorus* Odhner, 1905 by Bray & Gibson (1980).

Travassos *et al.* (1969) considered *Pseudoprosthenhystera* Kloss, 1966 synonymous with *Prosthenhystera* Travassos, 1920, without comment. Yamaguti (1971) pointed out that the vitelline distribution as described by Kloss (1966) precluded inclusion in the Callodistomidae. The action by Travassos *et al.* implies that Kloss's description is in error and is followed here.

Family Callodistomidae Odhner, 1910

(Syn. Teratotrematinae Yamaguti, 1958)

Diagnosis: Body flattened to cylindrical. Tegument unarmed. Oral sucker subterminal. Ventral sucker in mid- to anterior body. Prepharynx short or apparently absent. Pharynx oval. Oesophagus distinct. Intestinal bifurcation in forebody. Caeca blind, reach close or fairly close to posterior extremity. Testes present, one in forebody or two in hindbody; latter symmetrical to oblique. Cirrus-sac present. Internal seminal vesicle coiled, tubular. Pars prostatica weakly developed. Ejaculatory duct distinct, coiled. Genital pore median, in forebody. Ovary pre- or post-testicular, in hindbody. Canalicular seminal receptacle present. Laurer's canal present. Uterus extensive, mainly in hindbody, extends post-testicularly. Eggs thin-shelled, containing embryos with distinct eye-spots. Vitellarium in two lateral fields of follicles, in forebody or anterior hindbody. Excretory pore terminal. Excretory vesicle I-shaped. In gall-bladder of freshwater chondrosteans and teleosts. Africa and South America. Type-genus *Callodistomum* Odhner, 1902.

Key to genera

1a. Testis single, in forebody .... *Teratotrema* **Travassos, Artigas & Pereira, 1928. (Figs 33.1, 33.2)**

Diagnosis: Body flattened, oval. Prepharynx short. Oesophagus short, but distinct. Intestinal bifurcation in anterior forebody. Testis single, in forebody, just posterior to cirrus-sac, subglobular, entire. Cirrus-sac postbifurcal. Internal seminal vesicle sinuous, narrow. Pars prostatica and ejaculatory duct not distinctly differentiated; distal male

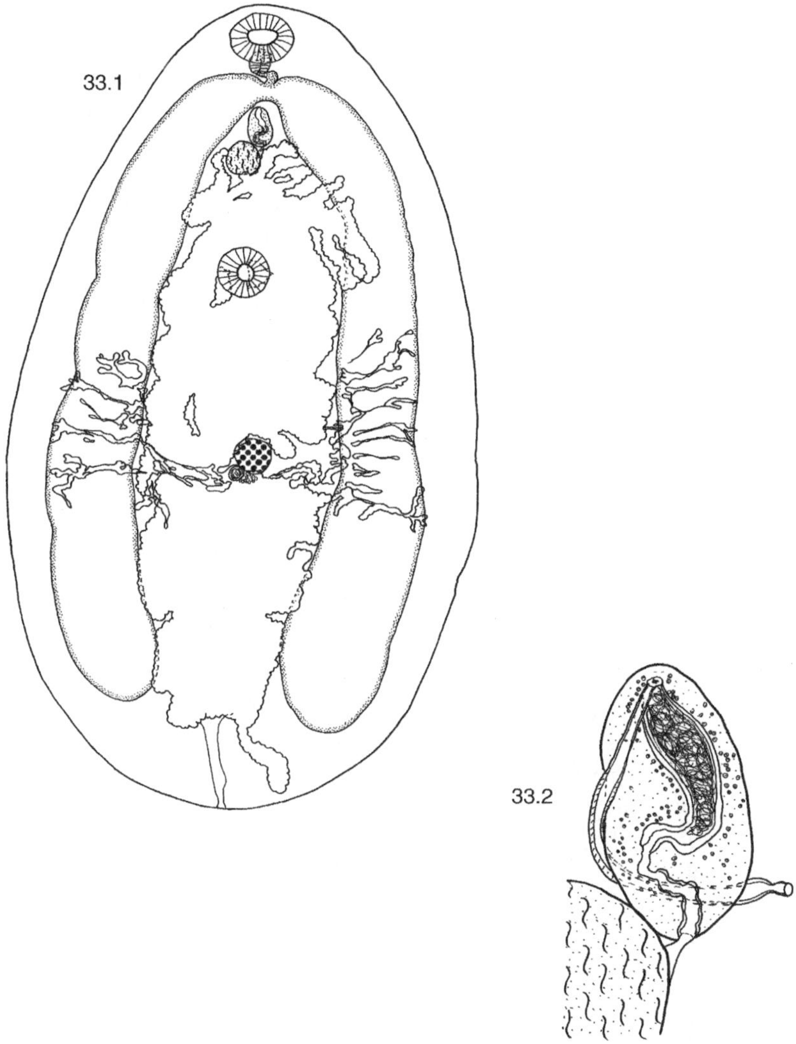

Figs 33.1–33.2. *Teratotrema dubium* Travassos, Artigas & Pereira, 1928. 33.1. Ventral view, uterus in outline (FIOCRUZ 32.252). (Original.) 33.2. Terminal genitalia and testis, note sperm in distal internal male duct. (Original.)

duct muscular. Genital pore median, just postbifurcal. Ovary subglobular, median just in anterior half of hindbody. Laurer's canal opens at level of seminal receptacle. Uterus intercaecal, between about level of caecal extremities and testis. Metraterm fairly muscular. Vitellarium two symmetrical lateral fields of elongate, branching follicles, between just posterior to level of ventral sucker and just posterior to level of ovary. Excretory vesicle reaches posterior margin of uterus. In freshwater teleosts. South America. Type-species *T. dubium* Travassos, Artigas & Pereira, 1928.

1b. Testes two, symmetrical in hindbody or at ventral sucker level .... **2.**

2a. Uterus restricted to hindbody; African .... **3.**

2b. Uterus reaching into lateral parts of forebody; Neotropical .... ***Prosthenhystera*** **Travassos, 1922. (Figs 33.3, 33.4)**
(Syn. *Pseudoprosthenhystera* Kloss, 1966)

Diagnosis: Body flattened, oval. Ventral sucker in anterior half of body. Prepharynx short. Oesophagus distinct. Intestinal bifurcation in anterior to mid-forebody. Testes two, subsymmetrical, in anterior hindbody, overlapping caeca. Cirrus-sac postbifurcal. Internal seminal vesicle oval, thick-walled. Pars prostatica and ejaculatory duct not distinctly differentiated; distal male duct muscular, sinuous. Genital pore just postbifurcal, median. Ovary pretesticular, sinistrally submedian. Laurer's canal opens dorsally just posterior to ventral sucker. Uterus fills most of hindbody and encroaches into lateral forebody up to level of oral sucker. Metraterm muscular. Vitellarium in two symmetrical lateral fields between testes and cirrus-sac. Excretory vesicle extends

33.3

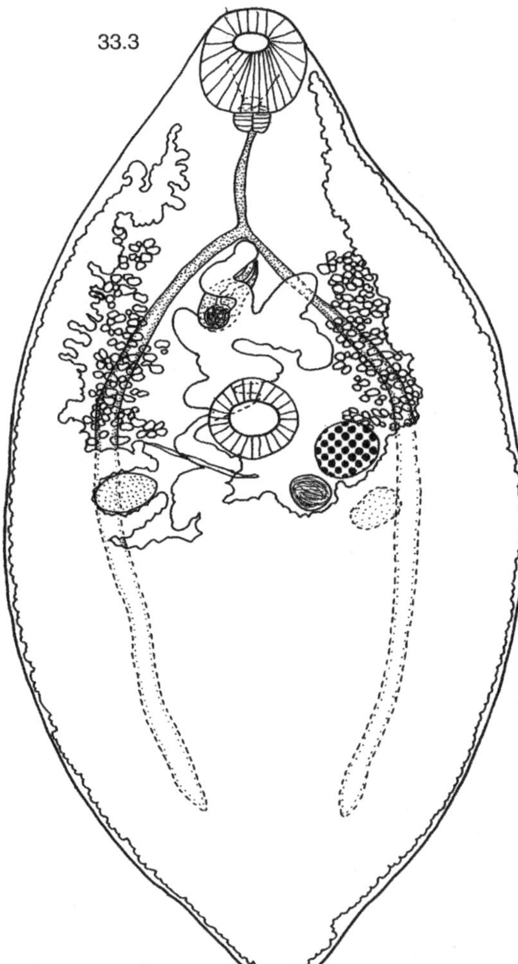

33.4

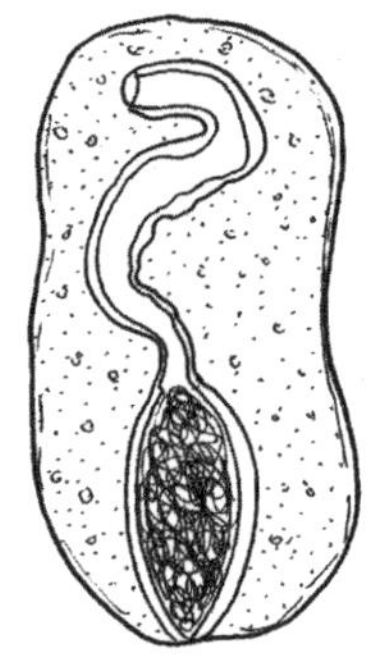

Figs 33.3–33.4. *Prosthenhystera obesa* (Diesing, 1850). 33.3. Ventral view, uterus in outline (FIOCRUZ 31.818a). (Original.) 33.4. Cirrus-sac. (Original.)

almost to testes. In freshwater teleosts. South America. Type-species *P. obesa* (Diesing, 1850).

3a. Body flattened; testes extracaecal; uterus extending extracaecally ....
.... ***Callodistomum* Odhner, 1902. (Figs 33.5, 33.6)**

Diagnosis: Body flattened, pyriform in outline. Oral sucker large. Ventral sucker large, in anterior half of body. Prepharynx absent. Pharynx small, oval. Oesophagus distinct. Intestinal bifurcation in mid-forebody. Testes two, symmetrical, at about level of ventral sucker, oval, smooth or slightly indented. Cirrus-sac pyriform, mainly in forebody, contains many gland-cells. Internal seminal vesicle wide, coiled, tubular. Pars prostatica not clearly differentiated. Ejaculatory duct long, thick-walled. Genital pore median, bifurcal. Ovary between testes, irregularly oval. Uterus fills most of hindbody, runs dorsal to ventral sucker, extends into extracaecal fields. Eggs non-operculate. Metraterm weakly developed. Vitellarium in two symmetrical fields,

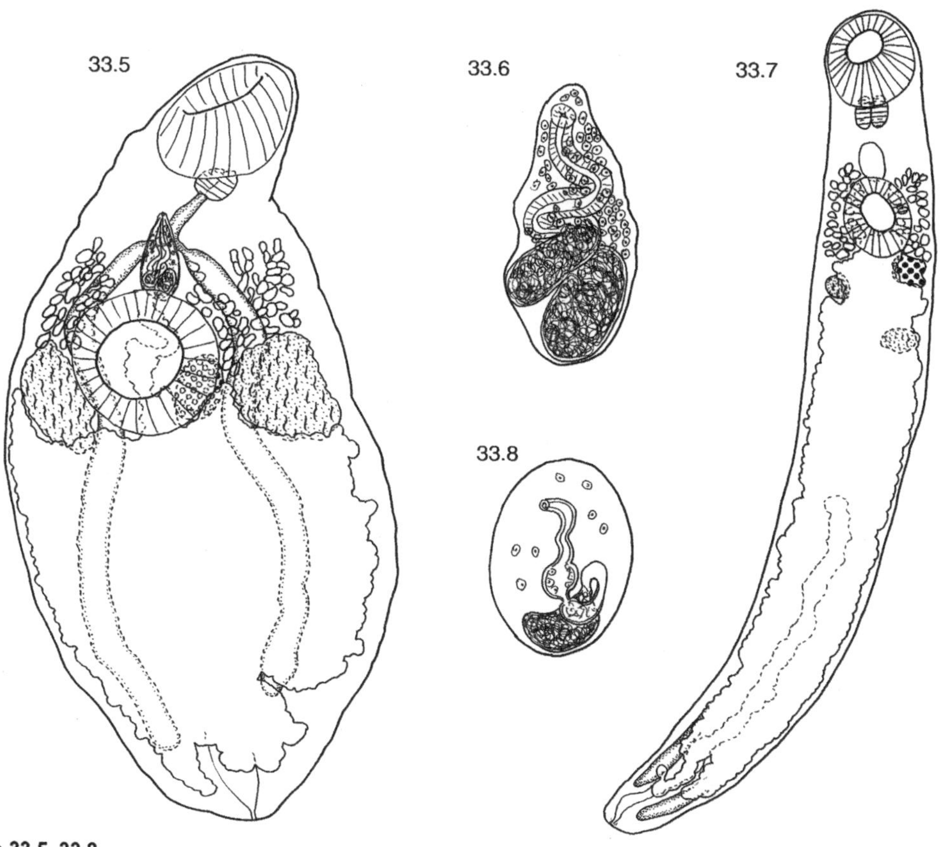

Figs 33.5, 33.8.
33.5–33.6. *Callodistomum diaphanum* Odhner, 1902. 33.5. Ventral view of specimen from *Ctenopoma kingsleyae*, Senegal (CIH 3870). (Original.) 33.6. Terminal genitalia. (Original.)
33.7, 33.8. *Cholepotes ovofarctus* (Odhner, 1902). 33.7. Ventral view, uterus in outline, specimen from *Synodontis schall*, Sudan, anterior part of gut and details of cirrus-sac not seen in this specimen (BMNH 1973.12.11.57–60). (Original.) 33.8. Cirrus-sac reconstructed from sections. (Original.)

immediately anterior to testes; follicles elongate and sometimes branched. Excretory vesicle reaches almost to ovary. In freshwater chondrosteans and teleosts. Africa. Type-species *C. diaphanum* Odhner, 1902.

3b. Body cylindrical; testes overlying caeca; uterus intercaecal .... *Cholepotes* Odhner, 1910. (Figs 33.7, 33.8)

Diagnosis: Body cylindrical. Ventral sucker in anterior third of body. Prepharynx absent. Oesophagus distinct. Intestinal bifurcation in mid-forebody. Testes oblique in anterior hindbody, lateral, overlying caeca. Cirrus-sac in forebody. Internal seminal vesicle coiled tubular. Pars prostatica weakly developed. Ejaculatory duct long, coiled. Genital pore median, postbifurcal. Ovary oval or irregular, pretesticular, lateral, on same side as posterior testis. Uterus intercaecal, fills most of hindbody. Eggs non-operculate. Vitellarium in two lateral fields, in region between cirrus-sac and anterior testis on one side and cirrus-sac and ovary on other. Excretory vesicle traced to mid-uterine field. In freshwater teleosts. Africa. Type-species *C. ovofarctus* (Odhner, 1902).

34 Family Fellodistomidae Nicoll, 1909

R.A. Bray

Department of Zoology, The Natural History Museum, London, UK

Introduction

The Fellodistomidae Nicoll, 1909 is a large family of marine fish digeneans, characterized by restricted fields of vitelline follicles. The history of the family since its erection as a subfamily of the order Prosostomata Odhner, 1905 by Nicoll (1909) has been somewhat chequered. Numerous subfamilies and genera have been added and removed from it. Detailed reviews of these changes can be read in Bray & Gibson (1980) and Bray (1988). Bray (1988) recognized seven subfamilies, namely the Fellodistominae Nicoll, 1909, Baccigerinae Yamaguti, 1958, Lintoniinae Yamaguti, 1970, Monascinae Dollfus, 1947, Monodhelminthinae Dollfus, 1937, Proctoecinae Skrjabin & Koval, 1957 and Tergestiinae Skrjabin & Koval, 1957. Two recent studies have presented important data that alter the conception we have of the family and its constituent subfamilies. Dove & Cribb (1995) published the description of a new genus, *Coomera* Dove & Cribb, 1995, which has characters of both the subfamilies Fellodistominae and Monascinae. They assigned it to the subfamily Fellodistominae, but acknowledged 'the fact that the subfamilial classification of the family is far from resolved'. Hall *et al.* (1999) inferred the phylogeny of the family using partial 18S ribosomal RNA (rRNA) and morphology and came to the conclusion that the family should be dismembered thus. The Tandanicolinae Johnston, 1927 (senior synonym of the Monodhelminthinae) should be given family status and is the sister group to the Gymnophallidae Odhner, 1905, and together they form the sister group to the Fellodistomidae *sensu stricto*. The Baccigerinae should also be given family status, and is, in fact, closer to the Zoogonidae Odhner, 1902 than to the Fellodistomidae *sensu stricto*. I have followed both these suggestions, but readers should note that Baccigerinae is now considered a junior synonym of the Faustulidae Poche, 1926 (Cribb *et al.*, 1999; Hall *et al.*, 1999). In the molecular study by Hall *et al.* (1999), *Coomera* is the sister group to *Proctoeces* Odhner, 1911, i.e. it is not closest to members of either of the subfamilies considered related by Dove & Cribb (1995). In fact, the genera *Fellodistomum* Stafford, 1904, *Complexobursa* Oshmarin & Mamaev, 1963,

Proctoeces and *Coomera* form a monophyletic group in the analysis of Hall *et al.* (1999). There now seems to be no good morphological reason for sustaining the separation of the subfamilies Proctoecinae, Monascinae and Lintoniinae from the Fellodistominae. Differences in life-cycle stages, particularly that in *Lintonium* Stunkard & Nigrelli, 1930 (see discussion in Bray, 1988), have been used to differentiate subfamilies, but this is not considered useful here for the identification of adults (and there is some doubt as to the reliability of the life-cycle studies of *Lintonium*).

Yamaguti (1938a) had doubts as to the generic position of *Benthotrema hoplognathi* Yamaguti, 1938, which has a bipartite seminal vesicle and therefore does not fit our definition of *Benthotrema* Manter, 1934. In fact, it appears close to *Pseudosteringophorus hoplognathi* Yamaguti, 1940, but lacks the 'hairs' lining the cirrus and genital atrium and is described as having an oval seminal receptacle. *B. melanostigmi* Parukhin & Lyadov, 1979 is similar to *Steringophorus melanostigma* (Noble & Orais, 1975) and appears to be wrongly assigned. *B. hilsii* Zaidi & Khan, 1977 is not well enough known for useful comparisons to be made. The description of *B. pyriformis* Wang, 1987 is not available to me: the species was listed by Wang (1989), as from *Triacanthus brevirostris*, but with no reference to a Wang, 1987 paper. *Neobenthotrema pyriformis* Wang, 1991 from *T. brevirostris* is probably the same thing and is placed in the Faustulidae. Manter (1934), in a key on p. 284, used the name *Bathytrema* in what appears to be a *lapsus* for *Benthotrema*.

Choanomyzus Manter & Crowcroft, 1950 was placed tentatively in the family Opistholebetidae by the original authors (Manter & Crowcroft, 1950) and this attribution has been followed, without comment, by such subsequent authors as Manter (1954), Yamaguti (1971) and Sey (1991). Manter & Crowcroft (1950), however, pointed out the many resemblances to the Fellodistomidae, 'including the folded cirrus, lack of seminal receptacle ... posterior extent of uterus ... lobed ovary, a Y-shaped excretory vesicle, and symmetrical testes'. An amphistomatous body plan is not known elsewhere in the Fellodistomidae, but the other characters are strongly suggestive that this is the best family to accommodate this unusual worm. Apart from the amphistomatous condition, it lacks most of the diagnostic characters associated with the Opistholebetidae, such as the pregonadal uterus, the I-shaped excretory vesicle, the canalicular seminal vesicle and the widely distributed follicular vitellarium. *Choanomyzus* is therefore included in the following key as a member of the Fellodistomidae.

In 1983, Bray considered *Complexobursa* Oshmarin & Mamaev, 1963 synonymous with *Proctoeces*, but later work has shown that this genus is distinct and valid (Hall *et al.*, 1999). Yamaguti (1971) erected the new subfamily Complexobursinae in the Lissorchiidae Poche, 1926. The second nominal species, *C. magna* Bilqees, 1980, is here still considered a synonym of *Proctoeces maculatus* (Looss, 1901). My examination of specimens of *C. vjetnamensis* Oshmarin & Mamaev, 1963, from *Pelates quadrilineatus*, in Moreton Bay, indicates that the terminal genitalia are complex and somewhat difficult to interpret, and I interpret them differently from earlier workers. There is a bipartite seminal vesicle, with the distal part much longer than the proximal, which leads into a long, recurved pars prostatica filled with anuclear cell-like bodies. The ejaculatory duct is short, narrow and inconspicuous but leads into a more or less hemispherical chamber, which has been considered the pars prostatica by earlier workers. This organ is prominent and filled with numerous radiating fibres. I interpret this organ as an outgrowth of the genital atrium, the remainder of which has a complexly wrinkled wall.

Dove & Cribb (1995) assigned *Coomera* to the Fellodistominae, but considered that 'its position may change ultimately'. Hall *et al.* (1999) found it closest to *Proctoeces* based on partial 18S rRNA sequences. Members of other recognized subfamilies, e.g. Monascinae, were not included, so the position of this genus is still not definitively resolved, but it can be said that it is not closer to *Fellodistomum* than to *Proctoeces*. Hall *et al.* (1999) interpreted certain features of the *Coomera* terminal genitalia as homologous with those of *Proctoeces*, but have suggested that the 'rank of subfamily is suspended within the Fellodistomidae' (*sensu stricto*).

Thulin (1973) described *Protomonascus nybelini* being unaware of the erection of *Elopsium ghanense* Fischthal & Thomas, 1972 (Fischthal & Thomas, 1972). He quickly recognized, however, that the worms were synonymous (Thulin, 1974). Ahmad (1982) described five new species. Two species, *E. indicus* Ahmad, 1982 and *E. manteri* Ahmad, 1982, are described as having bipartite seminal vesicles and bear some resemblance to *Steringophorus* Odhner, 1905.

Bray & Gibson (1980) considered that the uterus in *Fellodistomum* did not reach significantly into the post-testicular region in this monotypic genus, those species with post-testicular uterine slings being in the related genus *Steringophorus*. Molecular evidence has indicated (Bray *et al.*, 1994b), however, that at least one of the putative *Steringophorus* species, *S. agnotus* (Nicoll, 1909), is distinctly more closely related to *F. fellis* than to *S. furciger*, the type-species of *Steringophorus*. Bray *et al.* (1994b) concluded that the diagnostic feature of *Fellodistomum* is 'bulk of vitellarium anterior to posterior edge of ventral sucker'. In addition to *F. fellis* and *F. agnotus*, *Fellodistomum* now contains the gall-bladder inhabitant *F. sebastodis* Yamaguti & Matumura, 1942.

Gissutrema Machida & Kamegai, 1997 appears rather anomalous and difficult to place, but is placed *pro tempore* in the Acanthocolpidae Lühe, 1906.

The specimens of *Hypertrema ambovatum* Manter, 1960 we have available (Bray & Gibson, 1998) have very fine acicular spines buried in the tegument. The eggs appear collapsed, giving the double-shelled appearance noted by Manter (1960) and Overstreet & Martin (1974). Examination of a paratype (USNPC 38894) has shown it to be very similar to our specimens and, again, it appears that the double-shelled egg is probably an artefact of the thick shell at one end of the egg and collapse of the eggs. An identical shape is seen, but more clearly, in the collapsed larger eggs of some campulids. Spines were not seen on most of the surface, but it appears that much of the tegument is sloughed and there is some evidence of fine spines around the oral sucker. This is most unusual for a fellodistome (*sensu stricto*) and must place a slight doubt on the status of this genus.

Labriferoides Ganapati, Rao & Rao, 1962 is not a valid generic name. It does not satisfy the criteria of availability set out in the International Commission on Zoological Nomenclature (ICZN) Article 13(b), in that no type-species is designated or indicated. The form was described by Ganapati *et al.* (1962) based on an 'allocreadioid' metacercaria from a wood-boring bivalve from the northern Bay of Bengal. The illustration indicates that it is probably a member of the genus *Proctoeces*. As the worm is described, illustrated and discussed, *Labriferoides* cannot be considered a *nomen nudum*, but is what the ICZN refers to as an 'unavailable name'.

The adult morphology of *Lintonium* is similar to that of other members of the subfamily Fellodistomidae, but it is said to differ in its life-cycle and cercarial tail, which is described as having 'setaceous tufts', as in baccigerines (Cable, 1954; Stunkard, 1978; Bray, 1988). Yamaguti (1970) erected a new subfamily for this genus on morphological grounds. For reasons stated above, I do not recognize this subfamily for the purposes of this key. *Gastris* Lühe, 1906 is pre-occupied and is indistinguishable from *Lintonium*, as pointed out by Crowcroft (1950). Madhavi (1975) suggested that *Paradiplangus* Gupta, 1968 and *Paradiplobulbus* Bilqees, 1972 were synonymous with *Lintonium* and there appears no reason to doubt this. As pointed out by Yamaguti (1971), *L. novikovi* Baeva, 1965, with its post-testicular ovary, is not a member of *Lintonium*. It is clearly a hemiuroid and may be a sclerodistomoidid, as suggested by Gibson & Bray (1979).

Manter (1934) interpreted the internal seminal vesicle in *Lissoloma* Manter, 1934 as bipartite, with a short coiled tubular external seminal vesicle surrounded by a 'membrane'. My observations of the type-specimen (USNPC 8680) suggest that the whole of the seminal vesicle is internal, with the 'membrane' being a continuation of the cirrus-sac wall. The eggs are unusual, being thin-shelled, and their shape varies, due, presumably, to surrounding pressures (as in the zoogonines). Manter described and figured short filaments on the eggs, but it could well be that this is just an artefact caused by their collapse. The situation is somewhat like that found in *Hypertrema* Manter, 1960, but with a weakly tanned or untanned capsule. This genus is based on a single specimen, so there is some suspicion that the condition of the eggs may be anomalous.

Manter (1934) formed the concept of *Lomasoma* Manter, 1935 under the pre-occupied name *Lomaphorus* Manter, 1934 and made it clear that the genus was characterized by the 'longitudinal folds' of the body wall. His description of 'two longitudinal ventrolateral folds thrown into lobes' is most readily understood by reference to his figures 17 (of *L. wardi* (Manter, 1934)) and 20 (of *L. monolenei* (Manter, 1934)). The third species he recognized, *L. gracilis* (Manter, 1934), had 'inconspicuous' ventromarginal lobes, and his figure 22 shows this worm to have slightly irregular margins, but no exceptional structures. Two further species have been described: *L. stefanskii* Dollfus, 1960 has 'bords latéraux un peu incurvés ventralement, donnant l'apparence d'un faible bourrelet marginal' (Dollfus, 1960b) and *L. kergeleni* Parukhin & Lyadov, 1979 has no indications of incurved margins or marginal lobes in the figure (A) of Parukhin & Lyadov (1979). The type-species, *L. wardi*, is similar to *Steringophorus*, but I was unable to verify the V-shaped excretory vesicle in the type-specimen (USNPC 8676). The present key is based on the assumption that Manter was correct in describing the shape of the excretory vesicle.

Bray & Gibson (1980) questioned the validity of *Markevitschiella* Skrjabin & Koval, 1957, which is very similar to *Steringotrema* Odhner, 1911, but were reluctant to make a decision on literature evidence. If considered valid, the diagnostic features would be the post-testicular ovary and the divided vitellarium. The genus is based on *Steringotrema nakazawai* Kobayashi, 1921. Yamaguti (1971) stated that Kobayashi's (1921) original specimen is strongly contracted and, if in extended condition, it would be similar to *Steringotrema* in shape and position of organs. After examination of flattened specimens of *M. nakazawai* (Kobayashi, 1921) (NSMT-Pl 141 and 331), I can see no reason to disagree with Yamaguti's (1971) conclusions. The vitellarium tends to break up into separate patches of follicles, but this may well be a result of the flattening, and we know that this can occur ontogenetically in another fellodistomine, *Symmetrovesicula chaetodontis* Yamaguti, 1938 (see Bray *et al.*, 1994a).

Examination of the type-specimen of *Megalomyzon robustus* Manter, 1947 (USNPC 37020) has shown, contrary to Manter's description, that the ovary is trilobed. A distinct uterine seminal receptacle is also visible.

Yamaguti (1953) considered *Megenteron* Manter, 1934 a synonym of *Steringotrema*, but Bray & Gibson (1980) and Bray & Campbell (1995) retained the genus, differentiating it by its long caeca, multilobate (acinous) ovary, relatively small ventral sucker and long hindbody, and from *Steringophorus* by its long caeca and V-shaped excretory vesicle.

Mesolecitha Linton, 1910 was considered close to *Proctoeces* by Manter (1947), but it differs in having spines (or spine-like papillae) on the cirrus and fine spines in the metraterm. It also differs in its saccular (rather than tubular) seminal vesicle. Linton (1910, fig. 172) indicated that this worm may have a muscular papilla beside the male-duct opening, much like that in *Proctoeces*. Examination of syntypes, including sections (USNPC 8471), does not clarify this conclusively, as the sections are damaged at the crucial point. It appears that the ejaculatory duct is strongly muscular and may form some sort of knob, but whether this is equivalent to the structure in *Proctoeces* is not clear. The ejaculatory duct appears to be lined with fine spines, but, considering the condition of the specimens, there must be some suspicion that these are muscles exposed by maceration of the surface. A similar situation arises in relation to the question of tegumental spines. Manter (1947) disagreed with Linton (1910) on the presence of minute spines on the tegument, but Fischthal & Thomas (1968), in describing the second species, *M. ghanense* Fischthal & Thomas, 1968, stated 'tegument finely spined'. All the specimens of *Mesolecitha* I have seen, including two paratypes of *M. ghanensis* (USNPC 63195), are in poor condition, but I can detect no convincing tegumental spines.

The single caecum in *Monascus* Looss, 1907 has been used to validate subfamily and even family rank for this genus. The discovery of the single caecum in *Coomera* by Dove & Cribb (1995) and the studies of Hall *et al.* (1999) indicate that this morphological feature does not serve to distinguish these forms at the family or subfamily level. In *Monascus* a short reduced left caecum is sometimes described, but not as yet in *Coomera*. Dove & Cribb (1995) pointed out, however, that *Cercaria praecox* Walker, 1971 has some characters similar to *Coomera* and has a reduced left caecum. *Haplocladus* Odhner, 1911 has been generally recognized as a synonym of *Monascus*, at least since Dollfus (1947) pointed it out (see Bray & Gibson (1980) for discussion). Re-examination of the holotype of *Karachitrema trilobata* Bilqees, 1973 (BMNH 1982.5.13.13) has confirmed my earlier opinion (Bray, 1983) and that of Hafeezullah (1984) that *Karachitrema* is also a synonym of *Monascus*. Odhner (1911a) described a uroproct in *Monascus* and Dollfus (1947) repeated Odhner's statement but did not see the uroproct personally (footnote, p. 322). Authors who have studied juvenile specimens of *M. filiformis* (e.g. Rees, 1947; Køie, 1979; Girola *et al.*, 1992; Martorelli & Cremonte, 1998) have stated more or less unequivocally that it does not occur, but my observations of a set of serial sections suggest that it may occur, in the adult at least. Careful observation of live worms and further sections are needed to be certain, as the single caecum often overlies the excretory vesicle in whole-mounts, giving the superficial appearance of a uroproct.

When Layman (1930) erected *Plectognathotrema*, he did not place it in a family, but it has usually been considered a cephaloporid. The original, and so far only, description of the type-species, *P. cephalopore* Layman, 1930 omitted several important characters, such as the tegumental armament, female proximal system and excretory system. Kamegai (1970) described another species, *P.* (*Alloplectognathotrema*) *tsushimaense* Kamegai, from the same host as the type-species, the monacanthid *Thamnaconus* (=*Navodon, Cantherhines*) *modestus*, from near the original locality in the Sea of Japan. Two further species are now known, *P. ovata* Parukhin, 1964 and *P. hydrolagi* Olson, Hanson & Pratt, 1970. The excretory system is not reported in either, and *P. ovata*, apparently, lacks tegumental spines. *P. hydrolagi* is reported from a holocephalan, which casts some doubt on its relationship with other *Plectognathotrema* spp., which are reported from monacanthids. *P. ovata* is too poorly described to add useful information. The genus is therefore defined mainly on Kamegai's (1970) description of *P. tsushimaense*. Kamegai's (1970) new subgenus *P.* (*Alloplectognathotrema*) was used at generic rank by Yamaguti (1971), based mainly on the presence of rosette-shaped adhesive organs in the ventral sucker. The description of *P. cephalopore* is inadequate, but it is so similar to *P. tsushimaense* in all other respects that it is presumed here that the species are at least congeneric and probably synonymous. Although *Plectognathotrema* has often been placed in the Cephaloporidae, it is not really amphistomatous. The large ventral sucker is described as 'closer to posterior extremity than to anterior' (Kamegai, 1970). The lack of tegumental spines, the lack of a canalicular seminal vesicle and the possibly V-shaped excretory vesicle indicate that this genus has affinities with the Fellodistomidae, within which it is included, following Price (1937).

Proctoeces Odhner, 1911 has often been considered distinct enough to be the type-genus of a distinct subfamily, for which Bray & Gibson (1980) used the name Xenoperinae Poche, 1926, believing that the genus *Xenopera* Nicoll, 1915 was valid. Bray (1983) examined what was probably type-material of the type-species *X. insolita* Nicoll, 1915, considered it a synonym of *Proctoeces maculatus* (Looss, 1901) and on this basis used the subfamily name Proctoecinae. The synonymy of *Proctoeces* and *Xenopera* had, in fact, been first recognized in 1925 by Dollfus (in a handwritten note on a reprint (see Bray & Gibson, 1980)) and was mentioned in a publication by Dollfus (1952, p. 376, footnote). The studies of Hall *et al.* (1999) indicated that *Proctoeces* is close to some fellodistomine genera and that no purpose is served by sustaining the subfamily.

Prudhoeus Bray & Gibson, 1980 was originally placed in the Fellodistominae, but Bray (1988) pointed out its similarity to *Lintonium*, implying that the Lintoniinae is the best repository. This latter subfamily was, however, recognized almost entirely on its life-cycle peculiarities and is here considered synonymous with the Fellodistominae, so *Prudhoeus*, with no life-cycle elucidated, is best retained in the Fellodistominae. Three species are known: *P. nicholsi* Bray & Gibson, 1980, *P. africanus* Gaevskaya & Aleshkina, 1983 and *P. oligolecithosum* Wang, 1984. *P. africanus* was considered similar to *Symmetrovesicula* Yamaguti, 1938 by Bray *et al.* (1994a), and *P. oligolecithosum* is not described in enough detail to allow for the confirmation of its identification.

Bray (1988) considered it likely that *Pseudantorchis* Wang, 1982 is a synonym of *Steringotrema*. The ovary in the forebody and the testes at ventral sucker level were thought to have been caused by flattening at fixation.

Gomes & de Fábio (1972) placed *Varelacreptotrema* Gomes & de Fábio, 1972 in the allocreadiid subfamily Creptotrematinae Skrjabin & Koval, 1966, but Bray (1988) pointed out its similarity to *Lintonium* and *Prudhoeus*. I have examined specimens of *V. travassosi* kindly lent by the Instituto Oswaldo Cruz (FIOCRUZ 30.593b–f). The specimens are rather poor, heavily flattened, distorted and fragmentary, but they do not alter my earlier opinion that they are fellodistomids. Unfortunately, many features are not visible in these specimens. The proximal female system is not clearly seen. Laurer's canal cannot be detected and the character of the seminal receptacle cannot be seen with any certainty. It appears, however, that a uterine seminal receptacle may be present. The posterior part of the alimentary system cannot be seen. The excretory system is obscured in the hindbody by the eggs and flattening, but it appears that the lateral arms reach to about the level of the pharynx in the forebody, a typical fellodistomid arrangement. Associated with the distal cirrus-sac is a distinct, irregularly oval sac. It is not clear whether it opens into the cirrus-sac or genital atrium, but its wall is similar to that of the genital atrium. This enigmatic organ was described as 'uma formação que lembra um opérculo' by Gomes & de Fábio (1972). It is not possible to make a new whole-worm illustration from these specimens, but a new drawing of the terminal genitalia has been prepared.

There are several features of *Pseudohypertrema* Bilqees, 1976 that indicate that it may not be a fellodistomid. The large continuous vitelline field, the contents of the cirrus and the lack of excretory arms in the forebody are unusual fellodistomid features. The details of the proximal female system are unknown. This is considered a *genus incertae sedis*.

The subfamily Piriforminae Skrjabin & Koval, 1957 was recognized on the basis of the presence of a cyclocoel; the only case in this family. On examination of the holotype and seven other specimens of *Piriforma macrorhamphosi* Yamaguti, 1938 (MPM 22851, 22852), I have failed to detect this structure. In fact, the gut is obscured by eggs in all specimens. Yamaguti (1938a), however, stated that he saw sections. There is some possibility that *Piriforma* is an opecoelid, but it is retained as a fellodistomid *pro tempore*.

Gymnotergestia Nahhas & Cable, 1964 has hitherto been known only from the original description by Nahhas & Cable (1964), but I have seen specimens of an unnamed species from *Arrhamphus sclerolepis* from Moreton Bay, Queensland, Australia, which have been used for the present illustration.

Dollfus (1973) pointed out that most descriptions of the type-species of *Theledera* Linton, 1910, *T. pectinata* (Linton, 1905), including that of Hopkins (1940), in fact refer to a distinct species, which he named *Tergestia manteri* Dollfus, 1973. *Theledera pectinata*, in contrast to *Tergestia* spp., has its intestinal bifurcation in the forebody, a condition Dollfus considered validated *Theledera* at the subgeneric level. Bray & Gibson (1980) considered that this character could be used at the generic level and pointed out that in *Theledera* the uterus tended not to extend posteriorly to the testes. The problem with *Theledera* is the poorly known type-species, but non-type-species exhibit the generic characters clearly. Linton (1905) figured (his figures 200 and 201) specimens from *Bairdiella chrysura* with the intestinal bifurcation distinctly in the forebody, but Hopkins (1940) illustrated '*Tergestia pectinata*' from the same host with the intestinal bifurcation distinctly in the hindbody. Of the worms described in *Tergestia* since Bray & Gibson (1980), *T. atropi* Shen, 1990, *T. bengalensis* Gupta & Singh, 1985 and *T. selenei* Amato, 1982 belong in my concept of *Tergestia*, while *T. atulis* Shen, 1990 and *T. hainanensis* Shen, 1990 belong in my concept of *Theledera*.

Family Fellodistomidae Nicoll, 1909

(Syns Steringophoridae Odhner, 1911; Xenoperidae Poche, 1926; Monascidae Dollfus, 1947)

Diagnosis: Body large to small, globular to elongate. Body surface usually lacking spines, occasionally bearing muscular ornamentation or apparently very fine spines. Oral sucker subterminal, globular. Ventral sucker globular, large to small, equatorial or in anterior half of body. Prepharynx short to apparently absent. Pharynx well developed, globular to elongate. Oesophagus absent to long. Caeca normally two, blind, narrow to wide, reaching to testes or beyond, occasionally with single caecum (? opening into excretory vesicle) or cyclocoel. Testes two, oval to globular, entire or deeply lobed, symmetrical, oblique or tandem, in anterior or posterior hindbody, normally postovarian. Cirrus-sac well developed, oval to claviform, containing seminal vesicle, pars prostatica, prostatic cells and ejaculatory duct. Seminal vesicle usually bipartite, occasionally globular or convoluted and tubular. Pars prostatica usually wide with filamentous lining, occasionally greatly reduced. Spermatophores may be produced. Ejaculatory duct usually wide and convoluted, occasionally small and narrow. Genital atrium short to long. Genital pore ventral, usually sinistral, in middle or posterior half of forebody. Ovary entire to multilobate, in hindbody, usually anterior to testes. Laurer's canal present. Uterine seminal receptacle present. Uterus normally extends posteriorly to testes. Eggs numerous, small, operculate; eggshell occasionally ornamented. Vitellarium follicular, typically in two lateral fields, occasionally confluent medially. Excretory pore terminal; vesicle Y- or V-shaped, with anterior arms reaching about level of pharynx. Parasitic in intestine, pyloric caeca, bile-duct and gall-bladder of marine, and occasionally freshwater, teleosts, occasionally occurring as adults in molluscs (bivalves and gastropods). Type-genus *Fellodistomum* Stafford, 1904.

Key to subfamilies

1a. Cyclocoel present .... Piriforminae Skrjabin & Koval, 1957.
1b. Caeca end blindly (or uroproct reported) .... 2.

2a. Cirrus-sac usually bipartite; oral sucker usually with ring of muscular papillae; muscular flanges (collarettes) usually on margins of anterior forebody; pharynx elongate .... Tergestiinae Skrjabin & Koval, 1957.
2b. Cirrus-sac unipartite; oral sucker lacking ring of muscular papillae; muscular flanges on margins of anterior forebody lacking; pharynx usually oval .... Fellodistominae Nicoll, 1909.

Subfamily Fellodistominae Nicoll, 1909

(Syns Haplocladinae Odhner, 1911; Xenoperinae Poche, 1926; Monascinae Dollfus, 1947; Markevitschiellinae Skrjabin & Koval, 1957; Proctoecinae Skrjabin & Koval, 1957; Lintoniinae Yamaguti, 1970; Complexobursinae Yamaguti, 1971)

Diagnosis: Body large and robust to small. Body surface almost always unarmed. Ventral sucker usually larger than oral sucker, but occasionally smaller, in middle of body or in anterior half. Pharynx globular to elongate. Caeca normally two, narrow to wide, reaching to testes or beyond, occasionally with single caecum or reduced left caecum. Testes in anterior hindbody, postovarian. Cirrus-sac well developed, oval to claviform, containing many prostatic gland cells, muscular papillae may form lateral to opening of male duct. Pars prostatica wide, straight or gently curved. Ejaculatory duct wide or narrow, pocketed or not. Spermatophores occasionally present. Genital atrium short to elongate. Genital pore in mid-forebody, sinistral. Ovary entire to multilobate, usually anterior to right testis. Uterus usually extending posteriorly to testes. Metraterm joins genital atrium from left. Eggs numerous, operculate, small, often variable in size; eggshell may be ornamented. Vitellarium follicular, in two (occasionally four) lateral fields, in fore- and/or hindbody. Parasitic in intestine, pyloric caeca, bile-duct and gall-bladder of marine teleosts, occasionally progenetic in marine molluscs. Type-genus *Fellodistomum* Stafford, 1904.

Key to genera

1a. Single full-length caecum present .... **2.**
1b. Two caeca of similar length present .... **3.**

2a. Testes tandem; pharynx elongate; uroproct reported ....
.... ***Monascus* Looss, 1907. (Figs 34.1, 34.2)**
(Syns *Haplocladus* Odhner, 1911; *Karachitrema* Bilqees, 1973)
Diagnosis: Body elongate, subcylindrical. Oral sucker more or less terminal, slightly larger than ventral sucker. Ventral sucker in anterior third of body. Prepharynx absent. Pharynx elongate. Oesophagus short, with tegumental lining. Caecum single, long, with gastrodermis (rudiment of left caecum sometimes seen). Testes two, tandem, usually separated by slings of uterus, in posterior half of hindbody, usually sinistral. Cirrus-sac broadly oval, normally anterosinistral or partly dorsosinistral to ventral sucker. Seminal vesicle bipartite, internal. Pars prostatica well developed, tubular, wide, curved, with filamentous lining and many external gland-cells. Ejaculatory duct wide, muscular, diverticulate, merges into complex genital atrium. Spermatophores occasionally seen. Genital atrium large, complexly folded. Genital pore sinistrally submedian, ventral in posterior half of forebody. Ovary entire to trilobed, pre-testicular, separated from anterior testis by uterine slings, sinistral in about middle of body. Uterine coils fill much of hindbody, extending to posterior extremity. Vitellarium in two symmetrical, lateral fields, between levels of ventral sucker and testes. Excretory vesicle Y-shaped. In intestine of marine teleosts (predominantly Perciformes). Cosmopolitan. Type-species *M. filiformis* (Rudolphi, 1819).

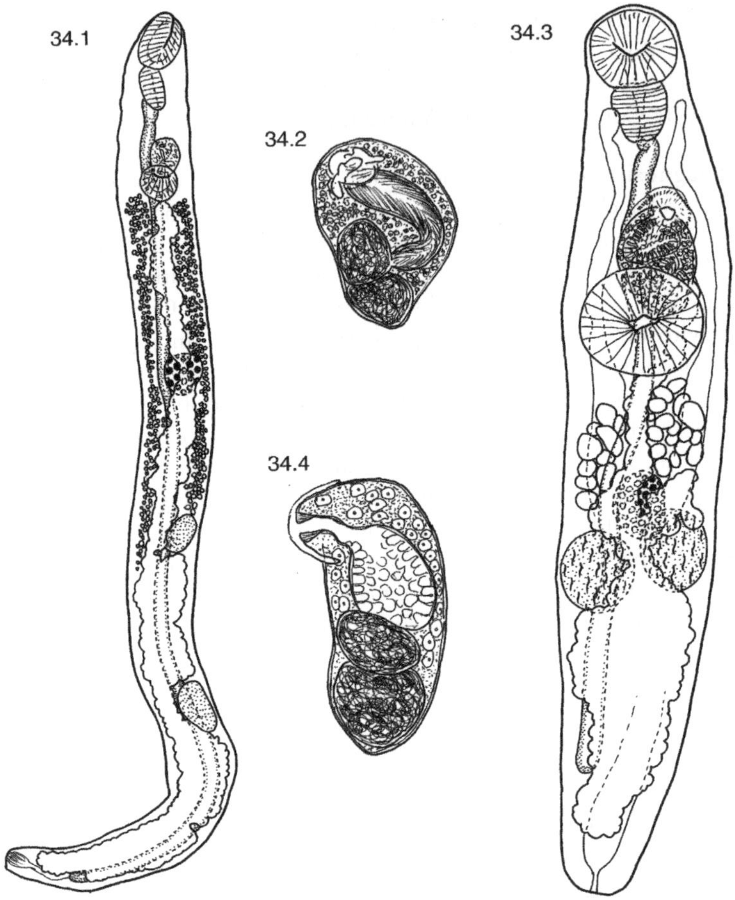

Figs 34.1–34.4.
34.1, 34.2. *Monascus filiformis* (Rudolphi, 1819). 34.1. Ventral view, uterus in outline, specimen from *Liparis liparis* (BMNH 1980.4.15.38). (Original.) 34.2. Terminal genitalia, specimen from *Trachurus trachurus* (BMNH 1988.3.8.194). (Original.)
34.3, 34.4. *Coomera brayi* Dove & Cribb, 1995. 34.3. Ventral view, uterus in outline (BMNH paratype 1995.6.9.6). (Original.) 34.4. Terminal genitalia. (Original.)

2b. Testes obliquely symmetrical; pharynx conical; uroproct absent .... *Coomera* Dove & Cribb, 1995. (Figs 34.3, 34.4)

Diagnosis: Body elongate-oval. Prepharynx short. Pharynx conical. Oesophagus short, with tegumental lining. Caecum single, extends close to posterior extremity, with gastrodermis. Testes two, smooth, oval, obliquely symmetrical, in anterior hindbody, separated by coils of uterus. Cirrus-sac largely anterior to ventral sucker. Seminal vesicle bipartite. Pars prostatica large, lined with anucleate vesicles, with numerous external prostate cells. Ejaculatory duct narrow, smooth. Cirrus large, muscular, fills much of common genital atrium when everted. Genital pore anterosinistral to ventral sucker, surrounded by radiating muscles. Ovary globular to oval, median, pre-testicular. Vitelline follicles in two lateral bands between ventral sucker and testes.

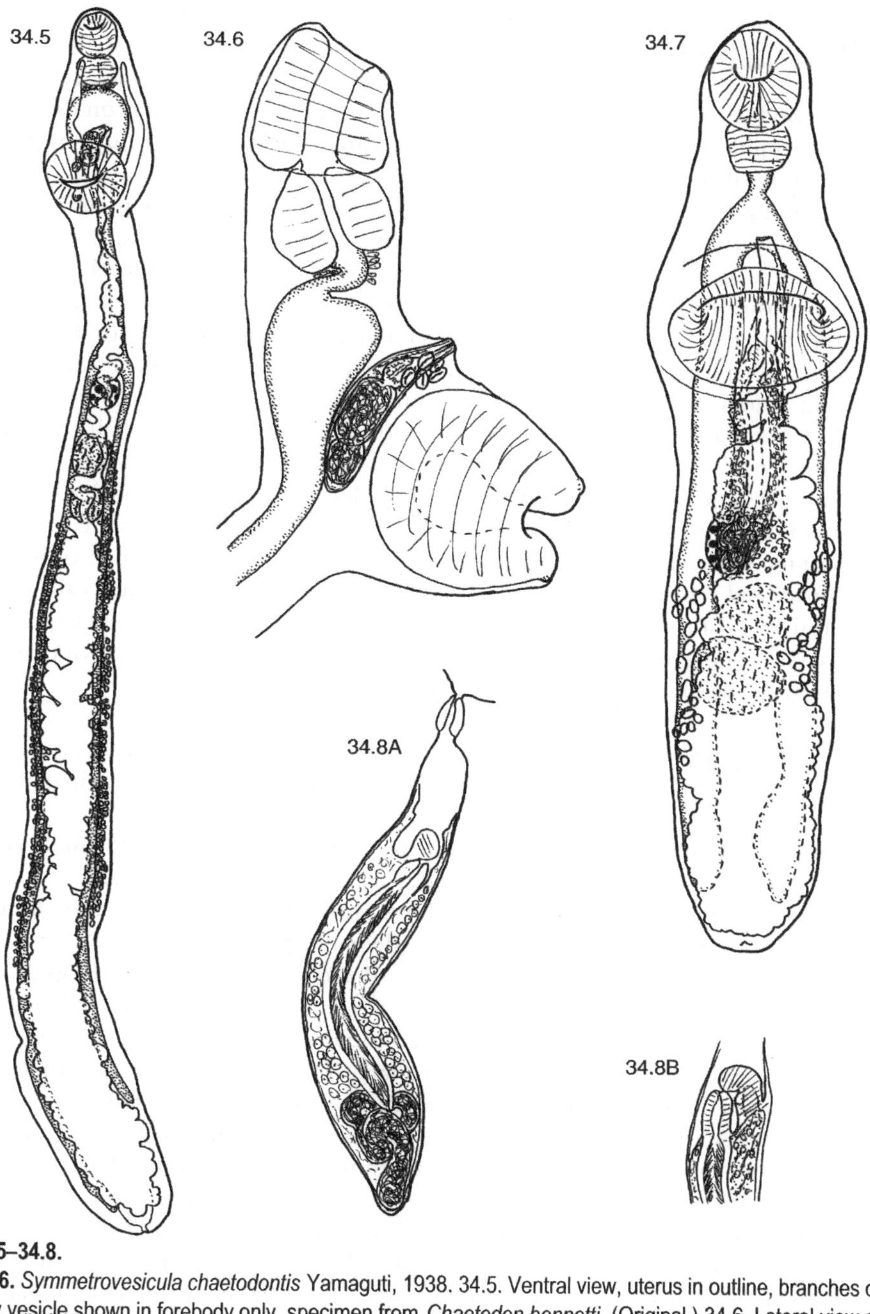

Figs 34.5–34.8.

34.5, 34.6. *Symmetrovesicula chaetodontis* Yamaguti, 1938. 34.5. Ventral view, uterus in outline, branches of excretory vesicle shown in forebody only, specimen from *Chaetodon bennetti*. (Original.) 34.6. Lateral view of anterior extremity showing short oesophagus and anterior end of inflated pseudoesophagus, protuberant ventral sucker and cirrus-sac, specimen from *Chaetodon speculum*. (Original.)

34.7, 34.8. *Proctoeces maculatus* (Looss, 1901). 34.7. Ventral view, uterus in outline, specimen from *Choerodon cyanodus*. (Original.) 34.8. Views of terminal genitalia, specimens from *Choerodon cyanodus*. A. Lateral view of cirrus-sac and genital atrium. B. Ventral view of distal extremity of cirrus-sac, showing muscular ejaculatory duct and papilla. (Original.)

Uterus fills most of hindbody. Excretory pore terminal. Excretory vesicle Y-shaped. In *Monodactylus*. Eastern Australian coast. Type-species *C. brayi* Dove & Cribb, 1995.

3a. Body broadly oval to elongate; ventral sucker lacking papillae on lips; pseudo-esophagus absent; intestinal bifurcation in forebody or just overlapping ventral sucker .... 4.

3b. Body very elongate; ventral sucker with papilla on each lip; pseudoesophagus long; intestinal bifurcation in anterior hindbody .... ***Symmetrovesicula* Yamaguti, 1938. (Figs 34.5, 34.6)**

Diagnosis: Body elongate, narrow. Ventral sucker oval, protuberant, with single muscular papilla on each of anterior and posterior lips, in anterior fifth of body length. Prepharynx short. Pharynx subglobular. Oesophagus short. Pseudoesophagus long. Intestinal bifurcation in anterior hindbody. Caeca long, narrow, reach close to posterior extremity. Testes oblique to almost tandem, in anterior half of hindbody. Cirrus-sac claviform, overlapping ventral sucker. Internal seminal vesicle bipartite. Pars prostatica wide, vesicular. Ejaculatory duct long; cirrus formed. Genital pore submedian, in posterior forebody. Ovary subglobular, pretesticular. Uterus mainly post-testicular, reaches close to posterior extremity, intercaecal; some coils between gonads and ventral sucker. Vitelline fields between level of anterior testis and posterior third of post-testicular region; fields may break up into rosettes in larger worms. Excretory pore terminal. Excretory vesicle V-shaped. In *Chaetodon*. Western Pacific Ocean. Type-species *S. chaetodontis* Yamaguti, 1938.

4a. Internal seminal vesicle coiled, tubular .... 5.

4b. Internal seminal vesicle usually bipartite, occasionally oval or tripartite ... 8.

5a. Genital atrium shallow; muscular papillae on cirrus-sac beside male aperture absent .... 6.

5b. Genital atrium variable, often long and narrow, extending into hindbody; muscular papillae on cirrus-sac beside male aperture .... ***Proctoeces* Odhner, 1911. (Figs 34.7, 34.8)**

(Syn. *Xenopera* Nicoll, 1915)

Diagnosis: Body oval to elongate oval. Ventral sucker oval to rounded, in anterior third of body. Prepharynx short. Pharynx large. Oesophagus short or absent. Intestinal bifurcation in mid-forebody. Caeca extend to middle of post-testicular region or beyond. Testes oval to rounded, entire, oblique to tandem in mid-hindbody. Cirrus-sac claviform, variable in position. Internal seminal vesicle narrow, tubular coiled, may have globular chamber proximally. Pars prostatica long, wide, with muscular wall surrounded by numerous gland-cells. Ejaculatory duct short, muscular. Distinct muscular papilla developed from dorsal wall of cirrus-sac protruding into genital atrium adjacent to ejaculatory duct. Genital atrium highly contractile, usually long, often reaching into hindbody. Genital pore sinistrally submedian, bifurcal or post-bifurcal. Ovary globular to slightly lobate, pretesticular. Uterus extensive in hindbody, fills post-testicular region. Metraterm distinct. Vitelline follicles irregular, in short lateral fields in mid-hindbody, may be weakly developed. Excretory pore terminal. Excretory vesicle Y-shaped. In marine teleosts, often progenetic in invertebrates. Cosmopolitan. Type-species *P. maculatus* (Looss, 1901).

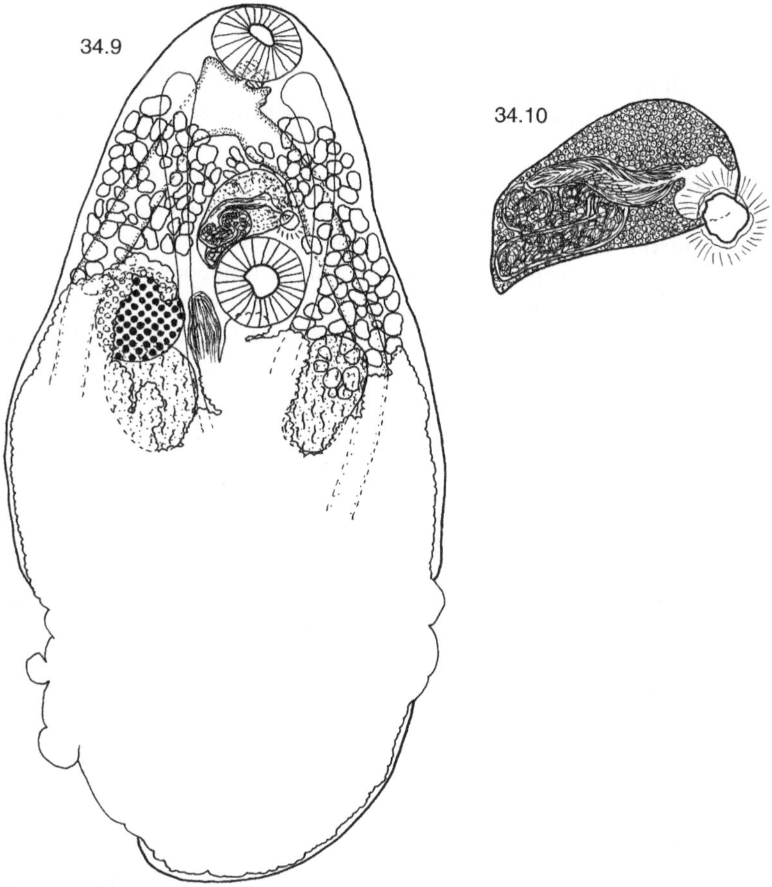

Figs 34.9–34.10. *Benthotrema plenum* Manter, 1934. 34.9. Ventral view, uterus in outline (USNPC 'type' 8681, flattened, damaged in hindbody). (Original.) 34.10. Terminal genitalia. (Original.)

6a. Body oval; vitellarium mainly in forebody .... ***Benthotrema* Manter, 1934. (Figs 34.9, 34.10)**

(Syn. *Bathytrema* Manter, 1934, *lapsus*)

Diagnosis: Body large, plump, robust, oval to fusiform. Ventral sucker rounded, in anterior half of body. Prepharynx short to absent. Pharynx small. Oesophagus short to absent. Intestinal bifurcation in mid-forebody. Caeca short, extend to ventral sucker or to or only just beyond testes. Testes symmetrical or subsymmetrical, in anterior hindbody, oval, entire. Cirrus-sac broadly claviform, in posterior forebody. Internal seminal vesicle coiled, tubular. Pars prostatica well developed. Ejaculatory duct of moderate length. Genital pore sinistrally submedian, in posterior forebody. Ovary pretesticular, dextral, at ventral sucker level, oval, entire. Uterus extensive in post-testicular region, may extend between testes. Eggs thick-shelled with or without tubercles. Vitellarium two lateral fields of follicles between testes and oesophagus, may be confluent. Excretory pore terminal. Excretory vesicle Y-shaped. In marine teleosts. Atlantic, Pacific. Type-species *B. plenum* Manter, 1934.

6b. Body elongate oval or fusiform; vitellarium in hindbody .... **7.**

7a. Body elongate fusiform; ovary three- to five-lobed or entire; eggs with single thick capsule .... ***Elopsium* Fischthal & Thomas, 1972. (Figs 34.11, 34.12)**
(Syn. *Protomonascus* Thulin, 1973)

Diagnosis: Body elongate, subcylindrical. Ventral sucker rounded, small, in anterior third of body. Prepharynx short. Pharynx oval. Oesophagus short. Intestinal bifurcation in anterior forebody. Caeca reach close to posterior extremity. Testes rounded, entire, oblique to symmetrical, in posterior half of hindbody, separated by uterus. Cirrus-sac elongate oval, in posterior forebody and overlapping ventral sucker. Internal seminal vesicle coiled, tubular or bipartite. Pars prostatica long, narrow, surrounded by thick sheath of gland-cells. Ejaculatory duct short, muscular. Genital pore anterosinistral to ventral sucker. Ovary three- to five-lobed to entire, pretesticular, separated from anterior testis by uterus. Uterus from close to posterior extremity to ventral sucker, mainly

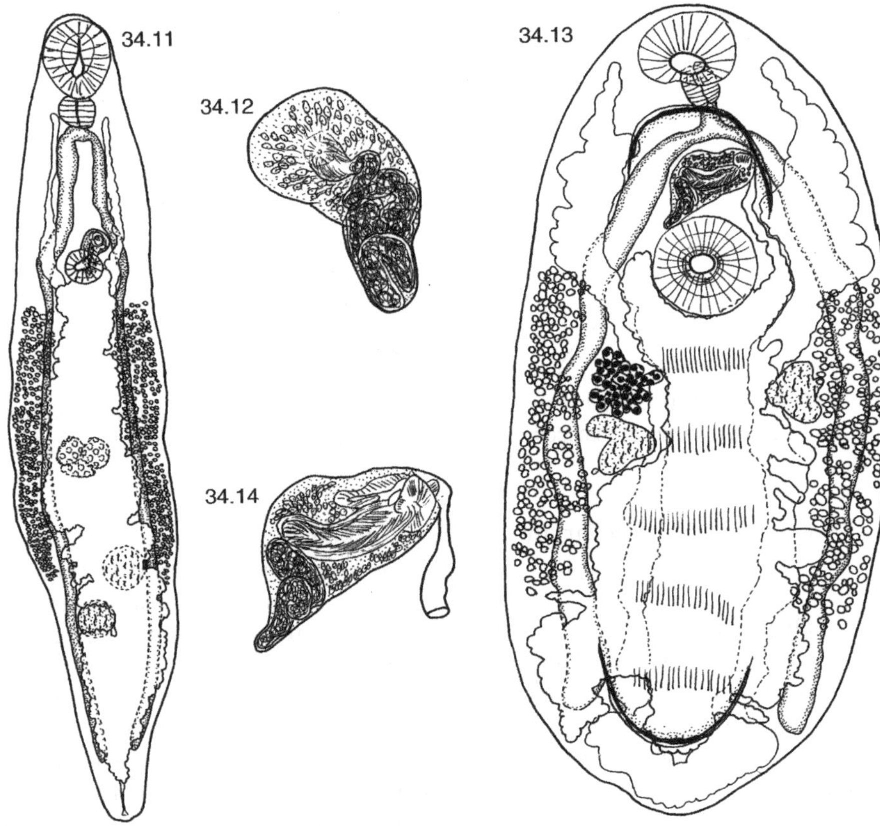

Figs 34.11–34.14.
34.11, 34.12. *Elopsium ghanense* Fischthal & Thomas, 1972. 34.11. Ventral view constructed from two fragments, uterus in outline (USNPC paratype 71742). (Original.) 34.12. Cirrus-sac, dorsal view. (Original.)
34.13, 34.14. *Lissoloma brotulae* Manter, 1934. 34.13. Ventral view, uterus in outline (USNPC 'type' 8680). (Original.) 34.14. Terminal genitalia. (Original.)

intercaecal. Eggs with single thick capsule. Vitelline fields reach from ventral sucker to level of anterior testis or intertesticular zone or into post-testicular zone. Excretory pore terminal. Excretory vesicle Y-shaped. In marine teleosts. Atlantic and Indian Oceans. Type-species *E. ghanense* Fischthal & Thomas, 1972.

7b. Body elongate oval, ends folded ventrally; ovary acinous; eggs with appearance of thin, double capsule .... *Lissoloma* Manter, 1934. (Figs 34.13, 34.14)

Diagnosis: Body oval; edges smooth; anterior and posterior ends incurved ventrally. Ventral sucker rounded, in anterior third of body, of similar size to oral sucker. Prepharynx absent. Pharynx oval. Oesophagus short. Intestinal bifurcation in mid-forebody. Caeca reach close to posterior extremity. Testes with irregular margins, obliquely subsymmetrical in anterior hindbody. Cirrus-sac ovoid, but with narrow pointed extension proximally. Internal seminal vesicle tubular, coiled. Pars prostatica long, wide, rectilinear. Ejaculatory duct deeply pocketed, muscular, forms 'genital lobes' in atrium. Genital pore sinistrally submedian, postbifurcal. Ovary deeply lobate, immediately anterior to right testis. Uterus loosely coiled in hindbody, between gonads, mainly intercaecal. Eggs thin-shelled, oval to pyriform, apparently with two shells; inner shell with short filament (may be artefact of collapse); bipartite embryo often seen. Metraterm thin-walled. Vitelline fields between ventral sucker and about

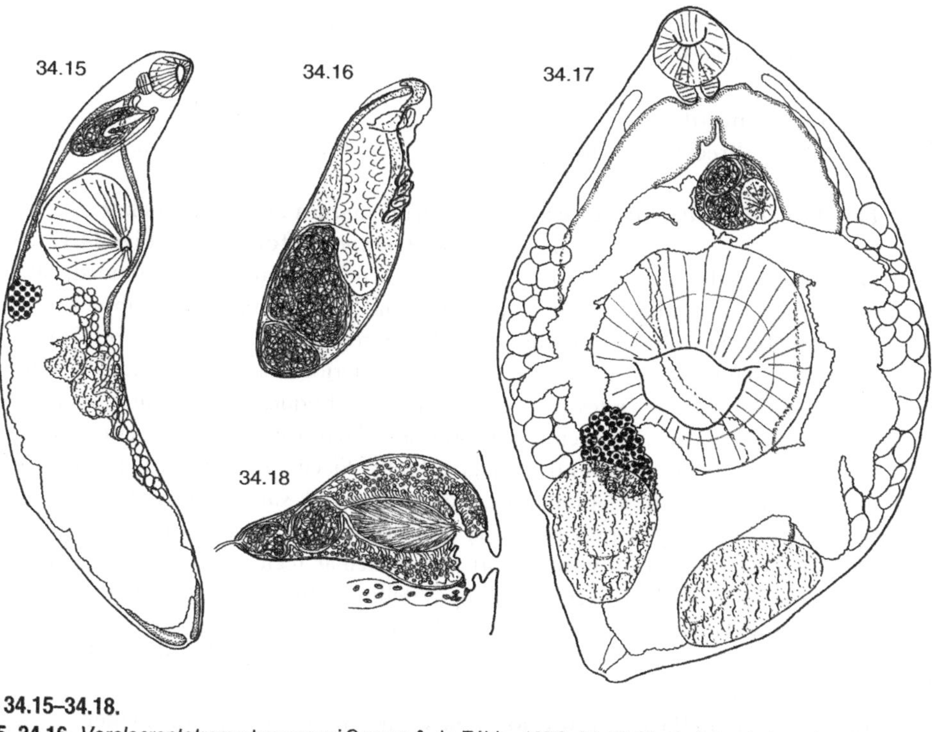

Figs 34.15–34.18.

34.15, 34.16. *Varelacreptotrema travassosi* Gomes & de Fábio, 1972. 34.15. Ventrolateral view of type-specimen, uterus in outline. (After Gomes & de Fábio, 1972.) 34.16. Terminal genitalia, showing enigmatic organ (FIOCRUZ 30.593-b). (Original.)

34.17, 34.18. *Fellodistomum fellis* (Olsson, 1868). 34.17. Ventral view of slightly flattened worm. (After Bray & Gibson, 1980.) 34.18. Terminal genitalia. (After Bray & Gibson, 1980.)

middle of post-testicular zone. Excretory pore terminal. Excretory vesicle V-shaped, wide, moniliform. In deep-water teleosts. Type-species *L. brotulae* Manter, 1934.

8a. Oral sucker oval to subglobular; ovary oval to lobate .... **9.**
8b. Oral sucker with anterior papilla-like extension; ovary reniform ....
.... ***Varelacreptotrema*** **Gomes & de Fábio, 1972. (Figs 34.15, 34.16)**

Diagnosis: Body elongate-oval. Oral sucker with papilliform anterodorsal projection. Ventral sucker large, subglobular, in anterior half of body. Prepharynx short. Pharynx oval. Oesophagus short, distinct. Intestinal bifurcation in mid-forebody. Caeca reach close to posterior extremity. Testes oval, entire, oblique, in anterior hindbody. Cirrus-sac oval, thick-walled, in forebody. Internal seminal vesicle bipartite. Pars prostatica wide, coiled. Ejaculatory duct long, narrower than pars prostatica, usually extruded as cirrus. Small, oval, enigmatic organ associated with distal cirrus-sac. Genital pore submedian, prebifurcal or bifurcal. Ovary reniform, pretesticular, in anterior hindbody. Uterus fills most of hindbody reaching close to posterior extremity. Metraterm not strongly developed. Vitellarium in two symmetrical fields, from level of posterior margin of ventral sucker to about middle of post-testicular zone. Excretory pore terminal. Excretory vesicle shape unknown; arms reach to level of pharynx. In marine teleosts (*Balistes*). Brazil. Type-species *V. travassosi* Gomes & de Fábio, 1972.

9a. Excretory vesicle Y-shaped .... **10.**
9b. Excretory vesicle V-shaped .... **19.**

10a. Ovary multilobed .... **11.**
10b. Ovary three- to four-lobed or entire .... **12.**

11a. Bulk of vitellarium anterior to posterior margin of ventral sucker ....
.... ***Fellodistomum*** **Stafford, 1904. (Figs 34.17, 34.18)**

Diagnosis: Body large, robust, oval or fusiform. Ventral sucker rounded, large or small, at about middle of body or just anterior. Prepharynx short to absent. Pharynx well developed, globular. Oesophagus absent to distinct. Intestinal bifurcation in anterior forebody. Caeca reaching close to posterior margin of testes or well into post-testicular region. Testes two, symmetrical to oblique, oval, in anterior hindbody or close to posterior extremity. Cirrus-sac claviform. Seminal vesicle bipartite. Pars prostatica wide, surrounded by gland-cells. Ejaculatory duct wide, diverticulate. Genital atrium small. Genital pore close to ventral sucker, sinistrally submedian. Spermatophores may be present. Ovary multilobate, just anterior to right testis. Uterus not extending significantly into short post-testicular region or extensive posterior to testes when post-testicular region is long. Vitelline follicles in two lateral fields, lateral to ventral sucker and extending into forebody. Excretory vesicle Y-shaped. Parasitic in bile-duct and gall-bladder of marine teleosts. North Atlantic and Pacific Oceans. Type-species *F. fellis* (Olsson, 1868).

11b. Bulk of vitellarium posterior to posterior margin of ventral sucker ....
.... ***Steringophorus*** **Odhner, 1905. (Figs 34.19, 34.20)**
(Syns *Leioderma* Stafford, 1904 *nec* Willemoes-Suhm, 1873; *Callodistomoides* Yamaguti, 1970; *Abyssotrema* Campbell, 1975; *Occultacetabulum* Reimer, 1985)

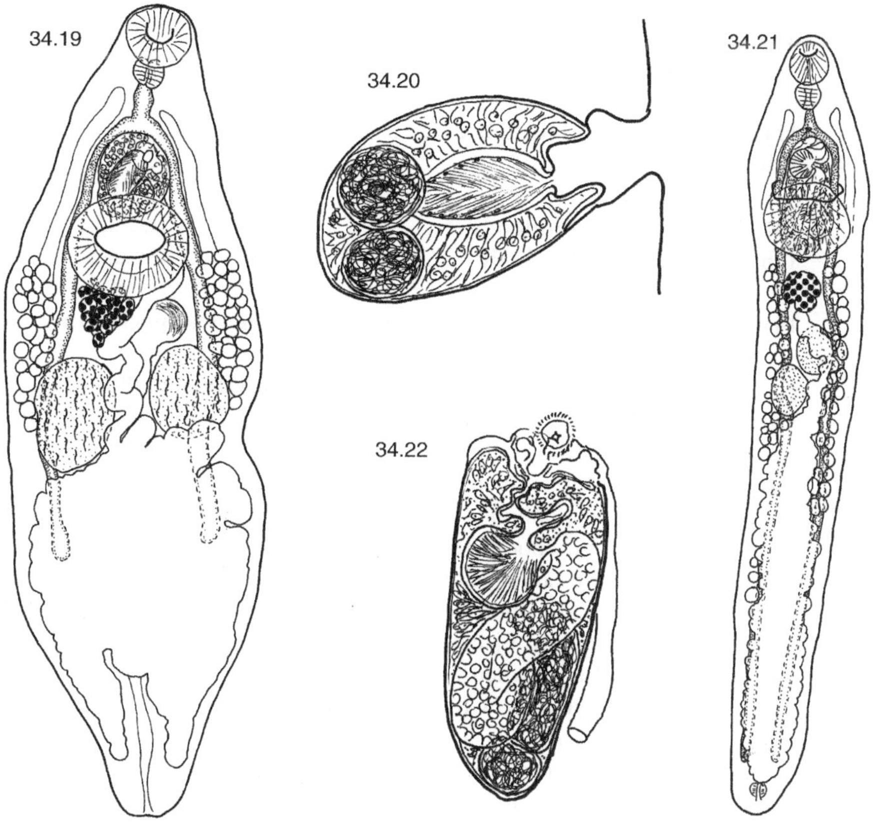

Figs 34.19–34.22.
34.19, 34.20. *Steringophorus furciger* (Olsson, 1868). 34.19. Ventral view, uterus in outline, of specimen from *Glyptocephalus cynoglossus* (BMNH 1990.6.20.42–47). (Original.) 34.20. Cirrus-sac of specimen from *Hippoglossoides platessoides* (BMNH 1978.1.30.1–4). (Original.)
34.21, 34.22. *Complexobursa vjetnamensis* Oshmarin & Mamaev, 1963. 34.21. Ventral view, uterus in outline, specimen from *Pelates quadrilineatus*. (Original.) 34.22. Terminal genitalia, specimen from *Pelates quadrilineatus*. (Original.)

Diagnosis: Body usually large, oval to elongate oval, deep-bodied to dorsoventrally flattened. Ventral sucker usually larger than oral sucker, in anterior half of body. Prepharynx short or absent. Pharynx globular. Oesophagus absent to long. Intestinal bifurcation in mid-forebody. Caeca wide to narrow, extent variable, extending to testes, to about middle of post-testicular region or occasionally beyond. Testes oval, entire, indented or deeply lobed, symmetrical to tandem, in anterior or mid-hindbody. Cirrus-sac oval. Seminal vesicle bipartite. Pars prostatica wide, surrounded by gland-cells. Ejaculatory duct wide, diverticulate. Genital atrium often diverticulate. Spermatophores may be present. Genital pore close to anterior margin of ventral sucker, sinistrally submedian. Ovary multilobate, just pretesticular. Uterus mainly coiled posteriorly to testes. Eggshells occasionally ornamented. Vitelline follicles in two lateral fields between level of ventral sucker and level just posterior to testes.

Excretory pore terminal. Excretory vesicle Y-shaped. Parasitic in intestine of marine teleosts. Widespread, particularly in deep-water teleosts. Type-species *S. furciger* (Olsson, 1868).

12a. Ovary three- to four-lobed .... **13.**
12b. Ovary entire .... **18.**

13a. Vitellarium in hindbody .... **14.**
13b. Vitellarium in forebody .... **15.**

14a. Testes oblique; ovary pretesticular; hemispherical pocket in wall of genital atrium lined with radiating filaments .... ***Complexobursa*** **Oshmarin & Mamaev, 1963. (Figs 34.21, 34.22)**

Diagnosis: Body elongate fusiform, wider anteriorly. Ventral sucker relatively large, in anterior third of body, protuberant. Prepharynx short. Pharynx oval. Oesophagus short. Intestinal bifurcation in mid-forebody. Caeca extend to fairly near posterior extremity. Testes two, oblique, in anterior hindbody, oval, entire. Cirrus-sac elongate oval, just reaches hindbody. Internal seminal vesicle bipartite; proximal part subglobular; distal part larger, elongate, extends to about middle of cirrus-sac. Pars prostatica long, broad, coiled, initially passes posteriorly, then anteriorly, lined with anuclear cell-like bodies. Ejaculatory duct short, narrow, weakly muscular. Genital atrium complexly folded when withdrawn, with distinct subglobular pocket lined with radiating filaments usually withdrawn into anterior part of cirrus-sac. Genital pore mid-ventral in mid-forebody. Ovary trilobed, occasionally with indistinct lobation, pretesticular, in anterior hindbody. Uterus mainly post-testicular. Vitelline fields extend from ovary to well into post-testicular region. Excretory pore dorsally subterminal. Excretory vesicle Y-shaped. In marine teleosts. Indo-Pacific. Type-species *C. vjetnamensis* Oshmarin & Mamaev, 1963.

14b. Testes tandem; ovary post-testicular; genital atrium very deep, extensible, with saccular outgrowth proximally .... ***Allosteringophorus*** **Machida, 1987. (Figs 34.23, 34.24)**

Diagnosis: Body elongate. Ventral sucker in anterior quarter of body. Prepharynx absent. Pharynx oval. Oesophagus short. Intestinal bifurcation in mid-forebody. Caeca extend to posterior extremity. Testes tandem, in anterior third of hindbody, oval, entire. Cirrus-sac elongate, extending well into hindbody. Internal seminal vesicle oval. Pars prostatica long, distinct, narrow. Ejaculatory duct long, with wide wrinkled walls. Genital atrium long, may extend into hindbody, with saccular outgrowth posteriorly, outgrowth surrounded by gland-cells. Genital pore slightly sinistrally submedian, just postbifurcal. Ovary post-testicular, separated from posterior testis, trilobed. Uterus between posterior extremity and anterior testis. Metraterm weakly developed. Vitellarium in two lateral fields of follicles, extending from ventral sucker to posterior testis. Excretory pore terminal. Excretory vesicle Y-shaped. In marine teleosts (*Prionurus*). Japan. Type-species *A. prionuri* Machida, 1987.

15a. Ventral sucker at posterior extremity; seminal vesicle unipartite; excretory pore on dorsal papilla .... ***Choanomyzus*** **Manter & Crowcroft, 1950. (Fig. 34.25)**

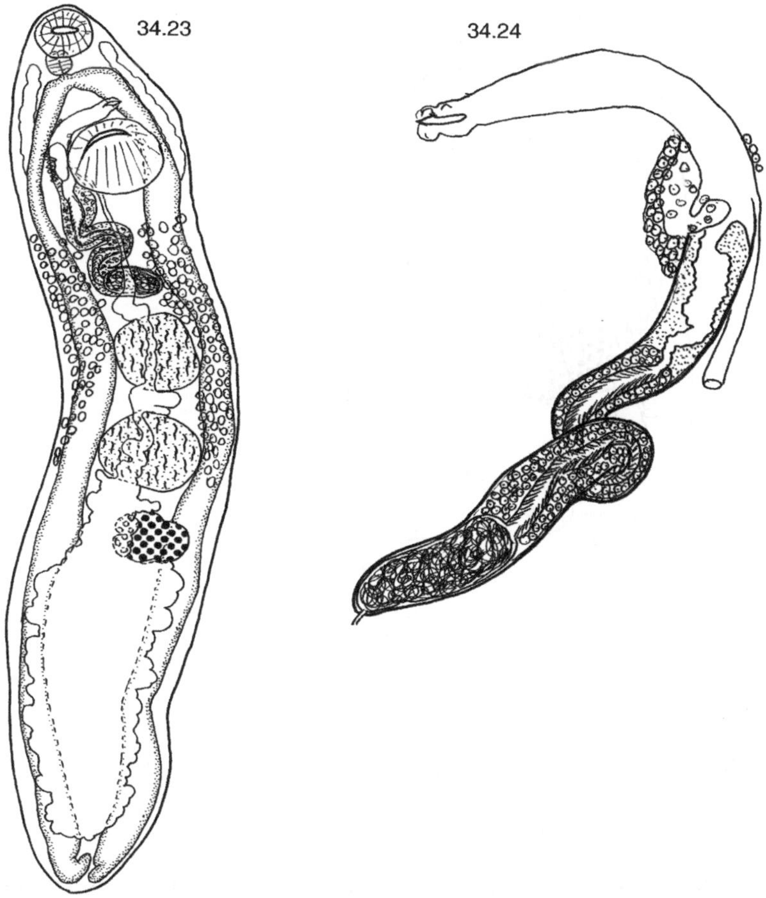

Figs 34.23–34.24. *Allosteringophorus prionuri* Machida, 1987. 34.23. Ventral view, flattened, uterus in outline (NSMT-PI 1638). (Original.) 34.24. Terminal genitalia (NSMT-PI 1638). (Original.)

Diagnosis: Body oval. Tegument unarmed; digitate papillae at posterior extremity. Ventral sucker oval to rounded, close to posterior extremity. Prepharynx not evident. Pharynx oval to rounded. Oesophagus very short or not evident. Intestinal bifurcation at level of pharynx or immediately posterior. Caeca reach close to posterior extremity, lateral to ventral sucker. Testes oval, entire, symmetrical to slightly oblique, just postequatorial. Cirrus-sac oval, in anterior half of body. Internal seminal vesicle subglobular. Pars prostatica wide, oval, vesicular. Ejaculatory duct complex, folded. Genital atrium complex, folded, lined with filamentous structures. Genital pore submedian, sinistral, just posterior to pharynx. Ovary deeply four-lobed, pretesticular, adjacent to or overlapping posterior part of cirrus-sac. Seminal receptacle not seen. Laurer's canal present. Uterus extends from about mid-ventral sucker to genital pore, passes between testes and to left of cirrus-sac. Eggs numerous, tanned, operculate. Vitelline fields lateral, between levels of middle of oral sucker and testes. Excretory pore on papilla dorsal to ventral sucker. Excretory vesicle Y-shaped. In marine teleosts. Off Tasmania and New Zealand. Type-species *C. tasmaniae* Manter & Crowcroft, 1950.

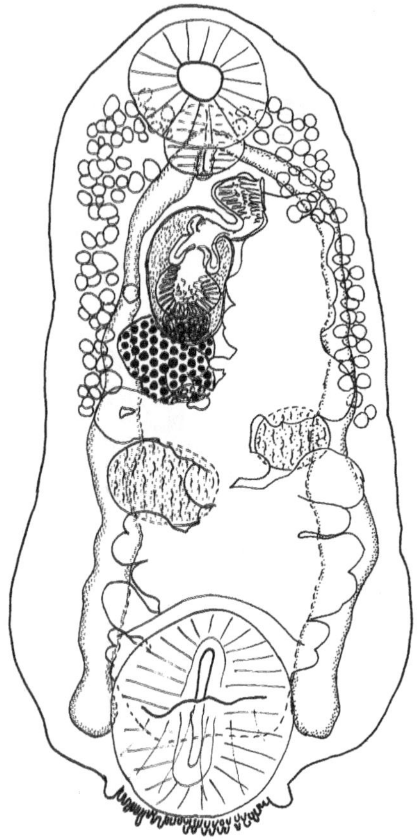

Fig. 34.25. *Choanomyzus tasmaniae* Manter & Crowcroft, 1950. Ventral view, uterus in outline. Specimen from *Aplodactylus arctidens*. (Original.)

15b. Ventral sucker pre-equatorial or subequatorial; seminal vesicle bipartite or tripartite; excretory pore terminal .... 16.

16a. Body almost rounded; ovary intertesticular; vitelline field continuous across forebody; genital pore sinistrolateral to ventral sucker .... *Hypertrema* Manter, 1960. (Figs 34.26, 34.27)

Diagnosis: Body broadly oval to pyriform, plump, large. Tegument with appearance of fine embedded spines. Ventral sucker rounded in anterior half of body, withdrawn into body. Prepharynx short. Pharynx small, oval. Oesophagus short. Intestinal bifurcation in mid-forebody. Caeca extend to level of testes. Testes large, irregular in outline but entire, symmetrical in anterior hindbody, overlapping ventral sucker. Cirrus-sac large, in forebody and overlapping ventral sucker, claviform or constricted near middle. Internal seminal vesicle bipartite or tripartite. Pars prostatica wide, long, lined with filaments and surrounded by dense sheath of gland-cells and connective tissue. Ejaculatory duct wide, merges into wall of genital atrium. Genital atrium large, complex, folded, thick-walled, ensheathed by dense layer of gland-cells, with large muscular papilliform protuberance in posterior or lateral wall. Genital pore sinistral to ventral sucker. Ovary trilobed, with distinct lobes, intertesticular. Uterus fills much of hindbody. Eggshells apparently double (may be artefact of collapse). Metraterm muscular narrow, enters genital atrium close to cirrus-sac. Vitelline follicles in two

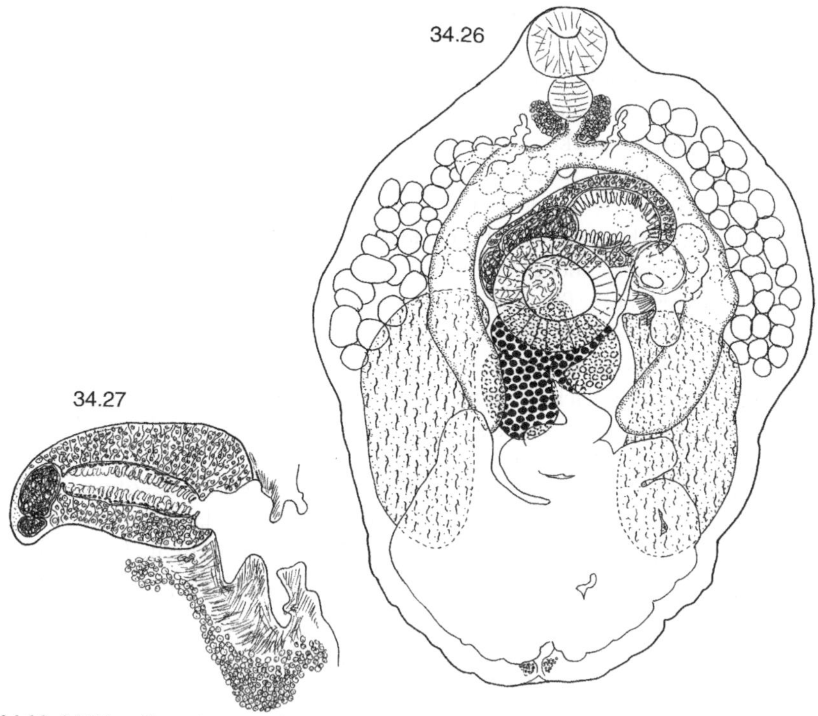

Figs 34.26–34.27. *Hypertrema ambovatum* Manter, 1960. 34.26. Ventral view, uterus in outline. Specimen from *Synaphobranchus kaupi* (BMNH 1998.8.17.3–4). (Original.) 34.27. Terminal genitalia, reconstructed from sections. Specimen from *Synaphobranchus kaupi* (BMNH 1998.8.17.3–4). (Original.)

lateral fields between oral sucker and testes, confluent or nearly so anteriorly. Excretory pore terminal. Excretory vesicle Y-shaped. In deep-water eels. Widespread. Type-species *H. ambovatum* Manter, 1960.

16b. Body oval or fusiform; ovary pretesticular; vitelline fields separate in forebody; genital pore anterior to ventral sucker .... 17.

17a. Body fusiform; oral sucker subterminal; no distinct muscles in lateral forebody; no spines in ejaculatory duct; testes in anterior hindbody .... *Olssonium* Bray & Gibson, 1980. (Figs 34.28, 34.29)

Diagnosis: Body small, fusiform. Ventral sucker pre-equatorial, smaller or similar in size to oral sucker. Prepharynx small but distinct. Pharynx oval, small. Oesophagus short. Intestinal bifurcation in mid-forebody. Caeca narrow, extending to or almost to testes. Testes oval, more or less symmetrical, in anterior hindbody. Cirrus-sac large, overlapping ventral sucker. Seminal vesicle bipartite. Pars prostatica with filamentous lining and external gland-cells. Ejaculatory duct wide, diverticulate. Genital atrium large. Genital pore in posterior forebody, slightly sinistral. Ovary trilobed, anterior to right testis and posterodextral to posterior margin of ventral sucker. Uterus coils in post-testicular region, in intertesticular region and dorsally to ventral sucker. Eggshell smooth, but may have small boss at one or both ends. Vitellarium in two lateral fields in forebody. Excretory pore terminal. Vesicle

Y-shaped. Parasitic in intestine of deep-sea teleosts (*Alepocephalus*). North Atlantic. Type-species *O. turneri* Bray & Gibson, 1980.

17b. Body broadly oval; oral sucker large, withdrawn into body; strong longitudinal muscles in lateral forebody; spines in ejaculatory duct; testes lateral to ventral sucker .... ***Megalomyzon* Manter, 1947. (Figs 34.30, 34.31)**

Diagnosis: Body oval, thick, robust, muscular. Prominent longitudinal muscles run from oral sucker to posterior part of ventral sucker. Oral sucker very large, subglobular, withdrawn into body; mouth aperture small, surrounded by semicircular muscles. Ventral sucker large, weakly muscular, transversely elongate, subequatorial. Prepharynx short. Pharynx oval. Oesophagus absent. Intestinal bifurcation dorsal to anterior edge of ventral sucker. Caeca extend to about level of posterior edge of ventral sucker. Testes oval, entire, symmetrical, lateral to ventral sucker. Cirrus-sac oval, overlaps ventral sucker. Internal seminal vesicle bipartite. Pars prostatica oval, vesicular, lined with filaments, surrounded by gland-cells. Ejaculatory duct short, pocketed, eversible, lined with spines. Genital atrium deep, narrow, with dorsal pocket lined with small spines. Genital pore sinistral, at about posterior edge of oral sucker or pharynx. Ovary trilobed, immediately anterior to right testis, lateral to anterior part of ventral sucker. Uterus fills hindbody. Vitellarium in two small, lateral fields; poral field ventral and lateral to cirrus-sac; aporal field lateral to posterior part of oral sucker.

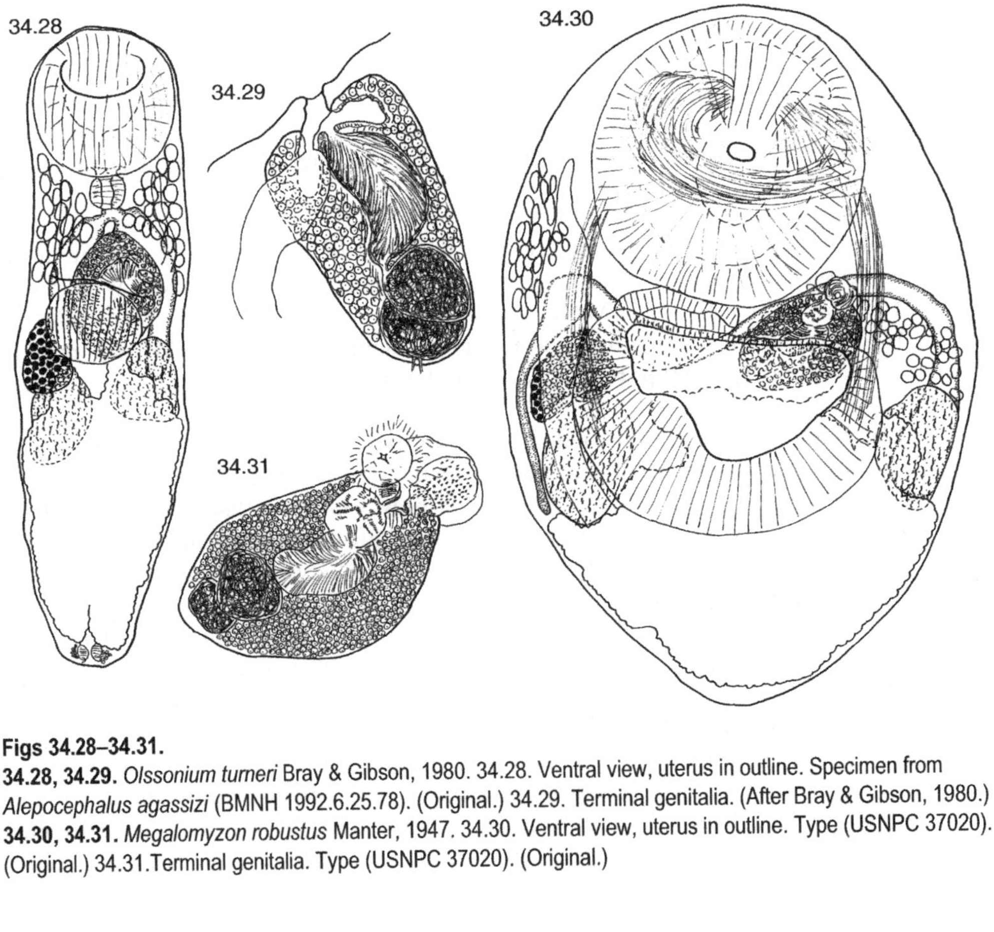

Figs 34.28–34.31.
34.28, 34.29. *Olssonium turneri* Bray & Gibson, 1980. 34.28. Ventral view, uterus in outline. Specimen from *Alepocephalus agassizi* (BMNH 1992.6.25.78). (Original.) 34.29. Terminal genitalia. (After Bray & Gibson, 1980.)
34.30, 34.31. *Megalomyzon robustus* Manter, 1947. 34.30. Ventral view, uterus in outline. Type (USNPC 37020). (Original.) 34.31.Terminal genitalia. Type (USNPC 37020). (Original.)

Excretory pore terminal. Excretory vesicle Y-shaped. In marine teleosts. North-west Atlantic. Type-species *M. robustus* Manter, 1947.

18a. Seminal vesicle bipartite; ejaculatory duct smooth .... ***Pseudosteringophorus* Yamaguti, 1940. (Fig. 34.32)**

Diagnosis: Body fusiform. Oral sucker terminal, with elongate aperture. Ventral sucker bowl-shaped, in mid-body. Prepharynx short. Pharynx oval, small. Oesophagus distinct. Intestinal bifurcation in mid-forebody. Caeca reach ventral sucker. Testes oval, entire, symmetrical, in anterior hindbody. Cirrus-sac recurved, claviform, just reaching ventral sucker. Internal seminal vesicle bipartite. Pars prostatica vesicular, ensheathed by many gland-cells. Ejaculatory duct long, narrow. Cirrus formed. Genital atrium with diverticulum lined with 'hairs', surrounded by gland-cells. Genital pore sinistrally submedian, postbifurcal. Ovary oval, pretesticular, dextrodorsal to ventral sucker. Uterus fills most of hindbody. Eggs numerous, tanned, embryonated. Metraterm not distinct. Vitellarium in form of single field of small follicles between

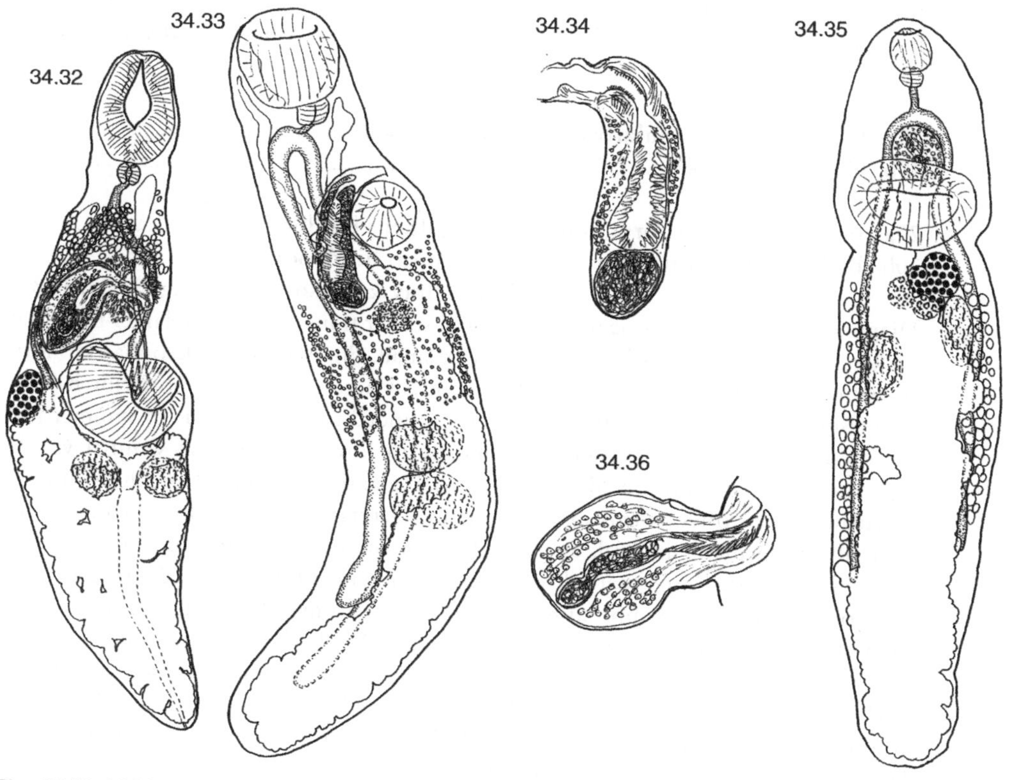

Figs 34.32–34.36.

34.32. *Pseudosteringophorus hoplognathi* Yamaguti, 1940. Ventral view, uterus in outline. (After Yamaguti, 1940.)

34.33, 34.34. *Mesolecitha linearis* Linton, 1910. 34.33. Sublateral view, uterus in outline, flattened. Syntype (USNPC 8471). (Original.) 34.34. Terminal genitalia, reconstructed from damaged sections (USNPC 8471). (Original.)

34.35, 34.36. *Prudhoeus nicholsi* Bray & Gibson, 1980. 34.35. Ventral view, uterus in outline (BMNH 1979.2.14.7–11). (Original.) 34.36. Cirrus-sac (BMNH 1992.3.24.15). (Original.)

intestinal bifurcation and ventral sucker. Excretory pore terminal. Excretory vesicle Y-shaped. In marine teleosts. Indo-Pacific. Type-species *P. hoplognathi* Yamaguti, 1940.

18b. Seminal vesicle oval to subglobular; ejaculatory duct muscular, apparently lined with spines .... ***Mesolecitha* Linton, 1910. (Figs 34.33, 34.34)**

Diagnosis: Body elongate oval. Ventral sucker pre-equatorial. Prepharynx very short. Pharynx oval. Oesophagus short or practically absent. Intestinal bifurcation in mid-posterior forebody. Caeca extend into posterior third of post-testicular region, may be unequal. Testes two, oval, entire, tandem to obliquely tandem, in anterior half of hindbody. Cirrus-sac oval to elongate-oval, overlapping ventral sucker. Internal seminal vesicle oval to subglobular. Pars prostatica rectilinear, wide, lined with anuclear cell-like bodies, surrounded by dense gland-cell mass. Ejaculatory duct muscular, may form cirrus, apparently lined with spines. Genital pore anterosinistral or dextral to ventral sucker. Ovary oval to rounded, entire, pretesticular, separated from anterior testis or not. Uterus fills most of hindbody. Eggs numerous, tanned, operculate. Metraterm lined with fine spines. Vitellarium with relatively few follicles, scattered in region between levels of posterior margin of ventral sucker and anterior part of post-testicular zone, may encircle body or lie in separate lateral fields. Excretory pore terminal. Excretory vesicle Y-shaped. In marine fishes (*Acanthurus*). North Atlantic. Type-species *M. linearis* Linton, 1910.

19a. Ovary oval to irregularly or regularly trilobed .... **20.**
19b. Ovary multilobed .... **23.**

20a. Body broadly pyriform, plump, with bulbous hindbody .... **21.**
20b. Body elongate, with more or less rectilinear hindbody .... ***Prudhoeus* Bray & Gibson, 1980. (Figs 34.35, 34.36)**

Diagnosis: Body small, elongate-oval. Ventral sucker transversely oval, larger than oral sucker, in anterior half of body. Prepharynx small. Pharynx oval. Oesophagus short. Intestinal bifurcation near middle of forebody. Caeca extend into posterior half of post-testicular region. Testes two, irregularly rounded to oval, oblique to symmetrical in anterior hindbody. Cirrus-sac oval, mostly in forebody. Internal seminal vesicle bipartite. Pars prostatica wide with external gland-cells and filamentous lining. Ejaculatory duct often forms extruded cirrus. Genital atrium shallow. Genital pore ventrally submedian, sinistral, at about level of intestinal bifurcation, may be wide when cirrus everted. Spermatophores may be present. Ovary trilobed, pretesticular, anterior to right testis or approximately median. Uterus mainly post-testicular, extends to posterior extremity. Eggs tanned, operculate. Vitellarium in two lateral fields, mainly in ventrolateral plane, between level of posterior margin of ventral sucker and level posterior to middle of hindbody; fields may almost unite ventrally. Excretory pore terminal. Excretory vesicle V-shaped. Parasitic in marine teleosts. Widespread. Type-species *P. nicholsi* Bray & Gibson, 1980.

21a. Vitellarium restricted to hindbody .... ***Lintonium* Stunkard & Nigrelli, 1930. (Figs 34.37, 34.38)**
(Syns *Gastris* Lühe, 1906 *nec* Dallman in Billberg, 1820; *Paradiplangus* Gupta, 1968; *Paradiplobulbus* Bilqees, 1972)

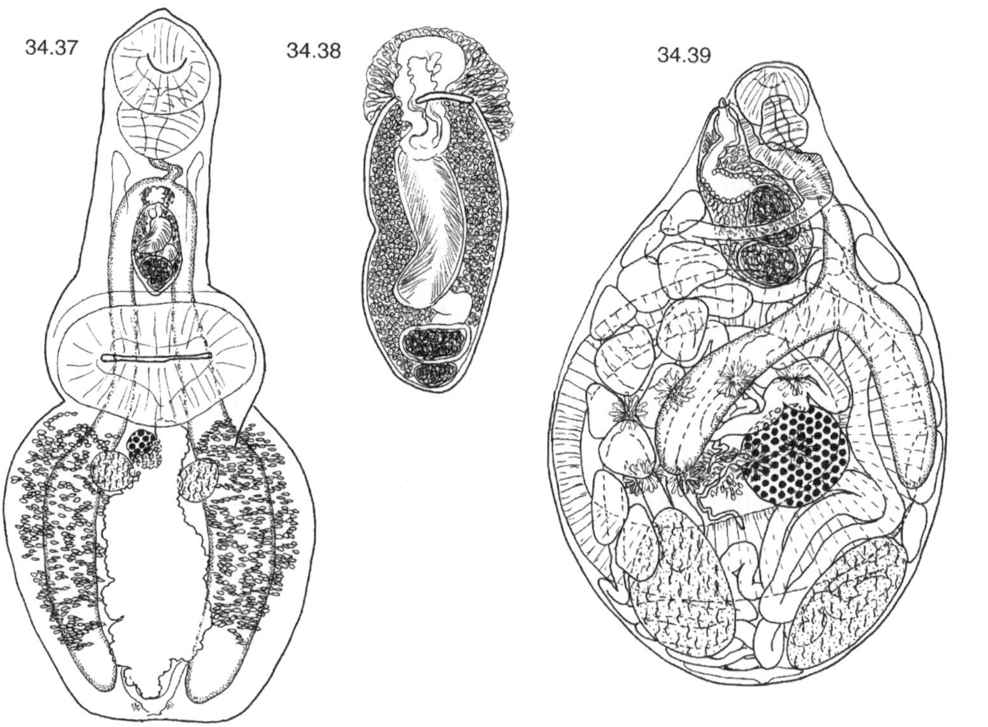

Figs 34.37–34.39.
34.37, 34.38. *Lintonium consors* (Lühe, 1906). 34.37. Ventral view, uterus in outline. Specimen from *Arothron nigropunctatus*. (Original.) 34.38. Terminal genitalia. Specimen from *Arothron nigropunctatus*. (Original.)
34.39. *Plectognathotrema tsushimaense* Kamegai, 1970. Dorsal view. (Modified after Kamegai, 1970.)

Diagnosis: Body plump, pyriform, with bulbous hindbody. Ventral sucker transversely oval, in anterior half of body. Prepharynx absent. Pharynx oval. Oesophagus short. Intestinal bifurcation in mid-forebody. Caeca reach close to posterior extremity, blind. Testes oval to rounded, entire, symmetrical in anterior hindbody. Cirrus-sac oval, mostly in forebody. Internal seminal vesicle bipartite. Pars prostatica rectilinear, fairly wide, lined with oblique filaments, surrounded by numerous gland-cells. Ejaculatory duct short, folded. Genital pore median, more or less bifurcal. Ovary irregularly or regularly trilobed to oval, pretesticular. Uterus mainly post-testicular, intracaecal; distal part extends between testes and into forebody. Vitelline follicles elongate or indistinct; two lateral fields in hindbody, mainly post-testicular; fields may break up into bunches. Excretory pore terminal. Excretory vesicle V-shaped. In tetraodontiform teleosts. Widespread. Type-species *L. vibex* (Linton, 1900).

21b. **Vitellarium in forebody or lateral to ventral sucker, may extend slightly into hindbody .... 22.**

22a. **Transverse row of rosette-shaped adhesive organs in ventral sucker; genital pore sinistro-marginal at level of oral sucker; ovary oval .... *Plectognathotrema* Layman, 1930. (Fig. 34.39)**
(Syn. *Plectognathotrema* (*Alloplectognathotrema*) Kamegai, 1970)

Diagnosis: Body squat-pyriform. Tegument lacking spines. Oral sucker subglobular, subterminal. Ventral sucker enormous, covers most of posterior two-thirds of ventral surface, with transverse rows of rosette-shaped adhesive organs. Prepharynx absent. Pharynx pyriform. Oesophagus long. Intestinal bifurcation at level of anterior margin of ventral sucker. Caeca short, divergent, extend to about mid-ventral sucker. Testes two, oval, entire, symmetrical, at posterior extremity. Cirrus-sac stout claviform, in forebody. Internal seminal vesicle bipartite. Pars prostatica vesicular. Ejaculatory duct long, wide, undulating. Genital pore sinistromarginal at level of oral sucker. Ovary rounded, entire, at about mid-level of ventral sucker. Uterus ascending to bifurcal level, descending to posterior extremity, then re-ascending to mid-oesophagus level. Eggs numerous. Metraterm muscular, well differentiated. Vitellarium follicular; follicles large; fields extending into forebody. Excretory pore terminal. Excretory vesicle V-shaped; arms short, divergent. In marine teleost (*Navodon*). North Pacific. Type-species *P. cephalopore* Layman, 1930.

22b. Ventral sucker lacking internal adhesive organs; genital pore median to sinistrally submedian, bifurcal; ovary trilobed .... *Steringotrema* Odhner, 1911. (Figs 34.40, 34.41)

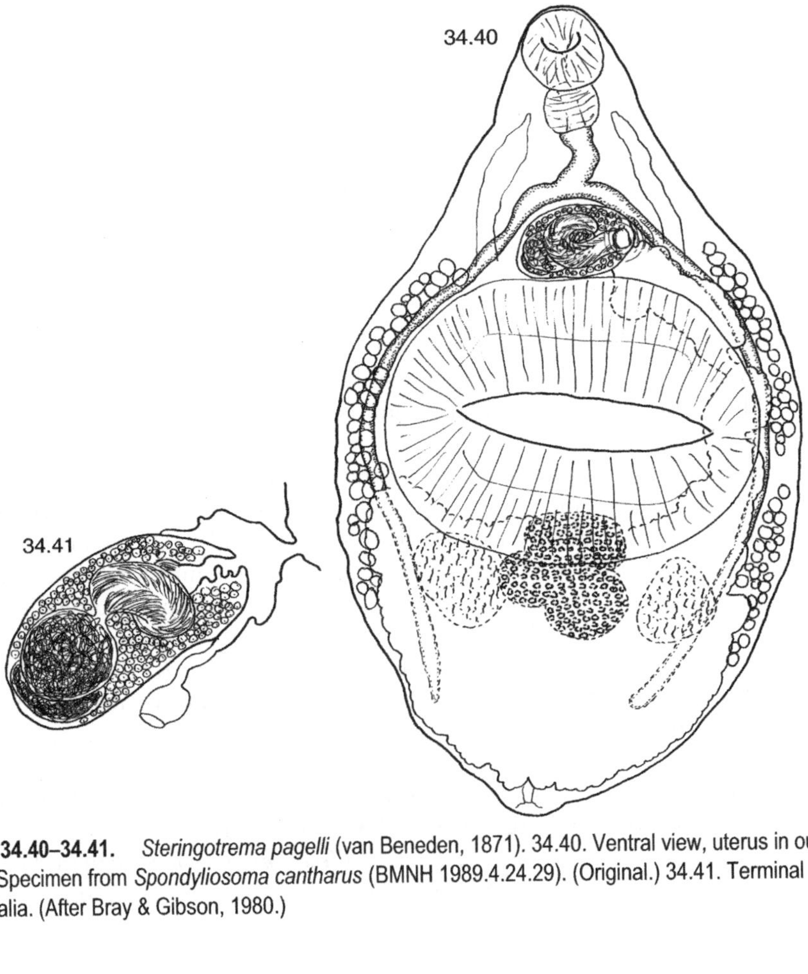

Figs 34.40–34.41. *Steringotrema pagelli* (van Beneden, 1871). 34.40. Ventral view, uterus in outline. Specimen from *Spondyliosoma cantharus* (BMNH 1989.4.24.29). (Original.) 34.41. Terminal genitalia. (After Bray & Gibson, 1980.)

(Syns *Rhodotrema* Odhner, 1911; *Markevitschiella* Skrjabin & Koval, 1957; *Pseudantorchis* Wang, 1982)

Diagnosis: Body small to large, globular to oval. Ventral sucker globular, large, significantly larger than oral sucker, globular, equatorial. Prepharynx short. Pharynx oval. Oesophagus short. Intestinal bifurcation in mid-forebody. Caeca extend back to level of testes or just beyond. Testes two, symmetrical, postovarian, in anterior hindbody. Cirrus-sac oval, in forebody. Internal seminal vesicle bipartite. Pars prostatica wide, with filamentous lining. Ejaculatory duct wide, diverticulate. Genital atrium small. Genital pore median to sinistrally submedian, at about level of caecal bifurcation. Spermatophores may be present. Ovary entire to weakly or strongly trilobate, just anterior, or antero- to posteromedian to right testis. Uterus mainly post-testicular, fills most of hindbody in older specimens. Eggs often with considerable variation in size, occasionally with ornamented shells. Vitellarium in two or four lateral fields between mid-forebody and level of testes, occasionally confluent dorsally, may break up into patches. Excretory pore terminal. Excretory vesicle V-shaped. Parasitic in intestine (occasionally gall-bladder and stomach) of marine teleosts. Widespread. Type-species *S. cluthense* (Nicoll, 1909) (now considered syn. of *S. pagelli* (van Beneden, 1871)).

23a. Lateral body margins with folds thrown into lobes; caeca do not reach posterior extremity .... *Lomasoma* Manter, 1935. (Fig. 34.42)

(Syn. *Lomaphorus* Manter, 1934 *nec* Ameghino, 1889)

Diagnosis: Body oval, margins folded ventrally, with muscular lobes. Ventral sucker rounded, pre-equatorial. Prepharynx very short. Pharynx rounded, small. Oesophagus distinct. Intestinal bifurcation in mid-forebody. Caeca reach middle of post-testicular region or beyond. Testes entire or with irregular margins, obliquely subsymmetrical in anterior hindbody. Cirrus-sac oval, in forebody. Internal seminal vesicle bipartite. Pars prostatica wide, rectilinear, lined with filaments, surrounded by dense gland-cell mass. Ejaculatory duct short. Genital atrium large with thick, pocketed wall and muscular genital lobe. Genital pore sinistrally submedian, postbifurcal. Ovary deeply lobate (acinous), immediately anterior to dextral testis. Uterus fills most of post-testicular zone, passes between testes. Eggs may have filament or short polar process. Lateral vitelline fields extend from ventral sucker to middle of post-testicular region or beyond. Excretory pore terminal. Excretory vesicle V-shaped. In deep-sea teleosts. Atlantic. Type-species *L. wardi* (Manter, 1934).

23b. Lateral body margins relatively smooth; caeca reach close to posterior extremity .... *Megenteron* Manter, 1934. (Figs 34.43, 34.44)

Diagnosis: Body oval to pyriform, tapering anteriorly. Ventral sucker rounded, pre-equatorial. Prepharynx short. Pharynx oval, large. Oesophagus distinct. Intestinal bifurcation in posterior forebody or overlapping ventral sucker. Caeca usually wide, reach close to posterior extremity. Testes lobate, symmetrical in anterior hindbody. Cirrus-sac oval, overlapping ventral sucker. Internal seminal vesicle bipartite. Pars prostatica wide, rectilinear, lined with filaments, ensheathed in dense gland-cells. Ejaculatory duct short. Genital atrium small. Genital pore just anterosinistral to ventral sucker. Ovary deeply lobed (acinous), immediately anterior to right testis. Uterus mainly post-testicular and intercaecal, passes between testes. Metraterm weakly differentiated. Vitelline fields between ventral sucker and testes. Excretory pore

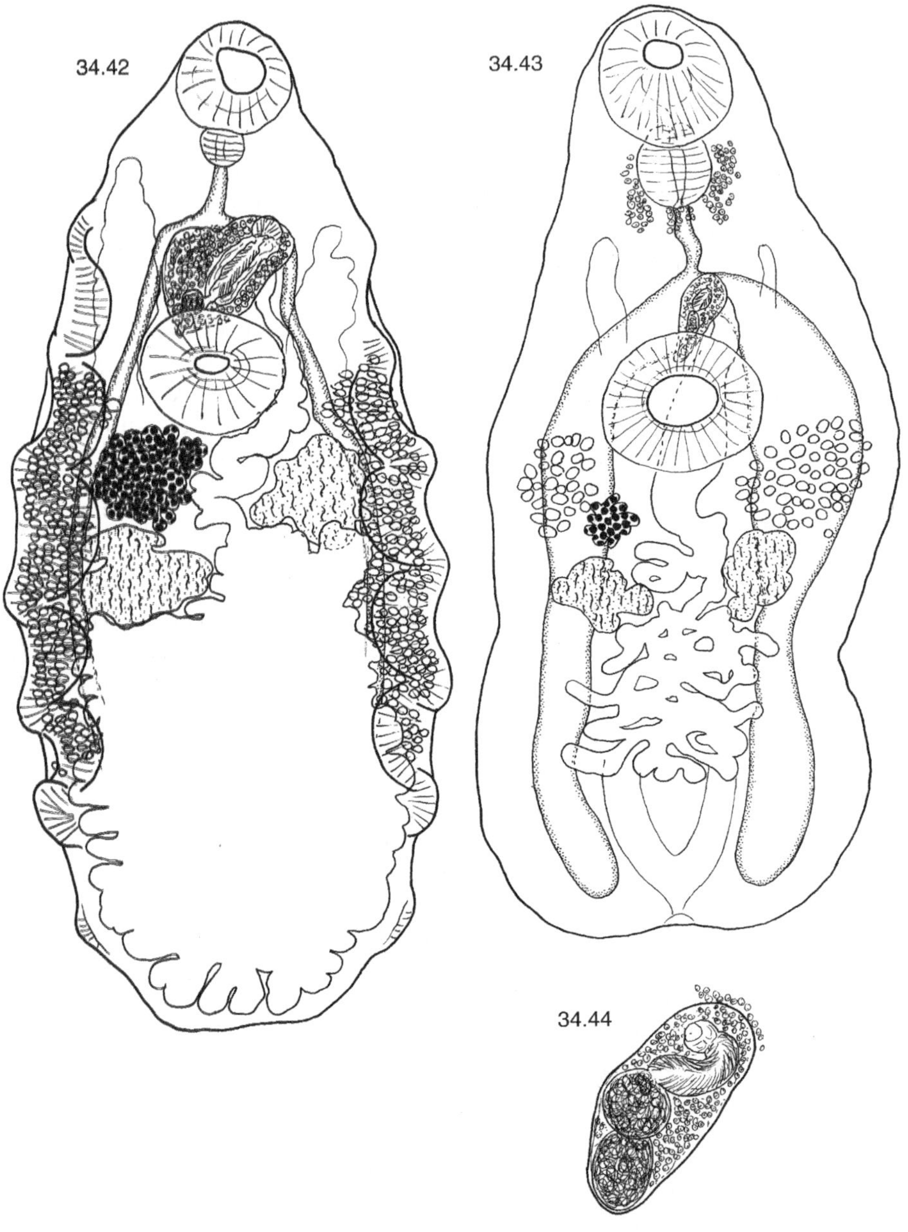

Figs 34.42–34.44.
34.42. *Lomasoma wardi* (Manter, 1934). Ventral view, uterus in outline. Type (USNPC 8676). (Original.)
34.43, 34.44. *Megenteron crassum* Manter, 1934. 34.43. Ventral view. Specimen from *Dicrolene intronigra* (BMNH 1993.11.11.29–31). (Original.) 34.44. Cirrus-sac. Specimen from *Dicrolene intronigra* (BMNH 1993.11.11.29–31). (Original.)

terminal. Excretory vesicle V-shaped. In deep-water Gadiformes. West Atlantic. Type-species *M. crassum* Manter, 1934.

Subfamily Piriforminae Skrjabin & Koval, 1957

(Syn. Piriforminae Yamaguti, 1958)

Diagnosis: Body pyriform. Tegument unarmed. Oral sucker subglobular, subterminal. Ventral sucker large, oval, in posterior half of worm, with transverse muscular lips. Prepharynx absent. Pharynx rounded. Oesophagus distinct. Intestinal bifurcation in anterior forebody. Caeca form cyclocoel in anterior part of short hindbody. Testes oblique, abutted against posterior body wall, rounded to oval, entire. Cirrus-sac claviform. Internal seminal vesicle bipartite. Pars prostatica large, oval, vesicular. Ejaculatory duct short. Genital pore sinistrally submedian distinctly anterior to ventral sucker. Ovary trilobed, anterodextral to right testis. Seminal receptacle absent (?). Laurer's canal present, opens at level of ovary. Uterus pregonadal, mainly at level of ventral sucker or in forebody. Eggs numerous, tanned, operculate, 'elevated punctum' at anopercular ends, embryonated. Vitellarium follicular, with two lateral fields extending from level of pharynx to level of ventral sucker. Excretory pore terminal.

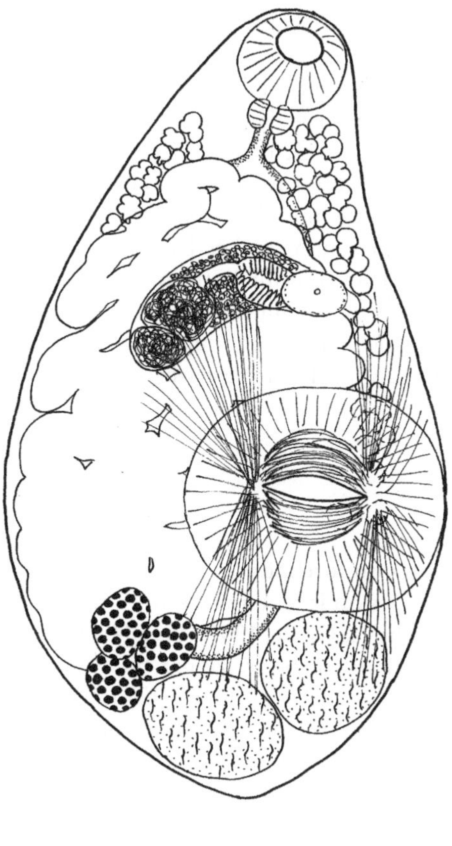

Fig. 34.45. *Piriforma macrorhamphosi* Yamaguti, 1938. Ventral view, uterus in outline. (After Yamaguti, 1938a.)

Excretory vesicle V-shaped. In marine teleosts (*Macrorhamphosus*). Off Japan. Type-genus *Piriforma* Yamaguti, 1938.

Genus *Piriforma* Yamaguti, 1938 (Fig. 34.45)

Diagnosis: With characters of subfamily. Type-species *P. macrorhamphosi* Yamaguti, 1938.

Subfamily Tergestiinae Skrjabin & Koval, 1957

Diagnosis: Body small to medium-sized, elongate, cylindrical. Body surface unarmed, but oral sucker normally surrounded by ring of about 13–20 muscular cephalic lobes, interrupted ventrally and 'neck' region normally with six muscular flanges ('collarettes') on each lateral surface at about level of pharynx. Oral sucker subterminal to terminal, usually larger than ventral sucker, but may be of similar size or smaller. Globular ventral sucker in anterior half of body. Prepharynx absent. Pharynx normally well developed, elongate. Oesophagus well developed, often long. Intestinal bifurcation in forebody or anterior hindbody. Caeca long, reaching close to posterior extremity, blind. Testes two, oval or irregularly lobed, tandem to oblique, in mid- to posterior hindbody. Cirrus-sac normally bipartite, usually extending back dorsally and posteriorly to ventral sucker; posterior part elongate, normally contains seminal vesicle; anterior part globular, normally contains pars prostatica and ejaculatory duct. Seminal vesicle elongate, saccular. Pars prostatica small, vesicular, with indistinct lining. Ejaculatory duct large, muscular, deeply diverticulate, may be completely everted through genital pore to form large, lobed cirrus. Genital atrium deep. Genital pore median to sinistrally submedian, ventral in posterior forebody. Ovary reniform to oval, entire or irregularly lobed, pretesticular, near middle of hindbody. Mehlis' gland dorsal to ovary. Uterine coils extend from ventral sucker back to posterior extremity or to level of testes; uterus opens into base of genital atrium via muscular metraterm. Eggs numerous, small. Vitellarium in two symmetrical, lateral fields of follicles in hindbody; fields may be confluent. Excretory pore terminal. Excretory vesicle Y-shaped. Parasitic in intestine (often rectum) of marine teleosts. Type-genus *Tergestia* Stossich, 1899.

Key to genera

1a. Oral sucker papillate and muscular flanges on forebody .... 2.
1b. Papillae on oral sucker and muscular flanges on forebody lacking .... *Gymnotergestia* Nahhas & Cable, 1964. (Figs 34.46, 34.47)

Diagnosis: Body elongate. Tegument annulated in forebody, giving body margins serrate outline. Oral sucker cup-shaped, terminal, without lobes. Ventral sucker elongate oval; aperture longitudinal. Pharynx elongate conical. Oesophagus short. Intestinal bifurcation in mid-forebody. Caeca extend to posterior extremity. Testes oval, oblique, in mid-hindbody, separated by uterus. Cirrus-sac oval, in forebody and

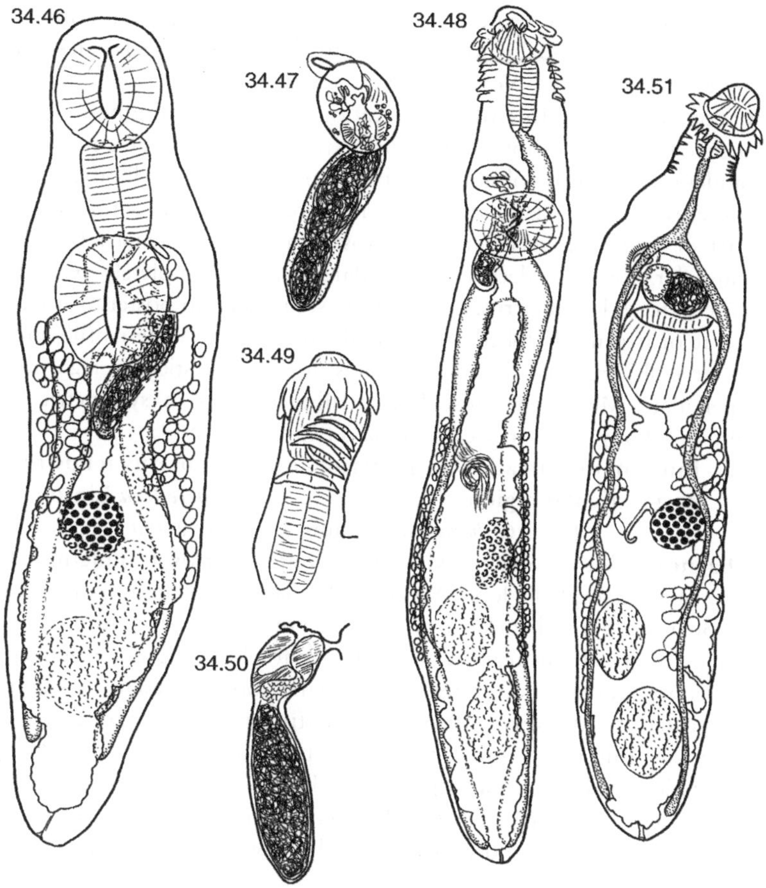

Figs 34.46–34.51.
34.46, 34.47. *Gymnotergestia* sp. 34.46. Ventral view, uterus in outline. Specimen from *Arrhamphus sclerolepis*. (Original.) 34.47. Terminal genitalia. Specimen from *Arrhamphus sclerolepis*. (Original.)
34.48–34.50. *Tergestia laticollis* (Rudolphi, 1819) from *Trachurus trachurus*. 34.48. Ventral view, uterus in outline (BMNH 1990.6.22.33–39). (Original.) 34.49. Lateral view of forebody (BMNH 1977.6.17.3–10). (Original.) 34.50. Terminal genitalia (BMNH 1987.2.27.1–3). (Original.)
34.51. *Theledera kuhliae* (Yamaguti, 1970). Dorsal view, uterus in outline. (After Yamaguti, 1970.)

overlapping ventral sucker. Pars prostatica long, rectilinear, with many gland-cells. Ejaculatory duct folded. Genital pore median in posterior forebody. Ovary entire, separated from anterior testis by uterus. Uterus fills most of postovarian region, extends between ovary and ventral sucker in intercaecal region. Eggs thick-shelled. Metraterm muscular. Vitellarium in two lateral fields between ventral sucker and anterior testis. Type-species *G. chaetodipteri* Nahhas & Cable, 1964.

2a. Intestinal bifurcation in anterior hindbody or dorsal to ventral sucker; uterus tends to extend into post-testicular field .... *Tergestia* Stossich, 1899. (Figs 34.48–34.50)

Diagnosis: Body elongate, cylindrical. Oral sucker oval, subterminal, surrounded by ring of 10–20 cephalic lobes. Six muscular flanges (collarettes) on lateral surface at level of pharynx. Ventral sucker oval, protuberant, in anterior third of body. Oesophagus long. Intestinal bifurcation in anterior hindbody or dorsal to ventral sucker. Caeca reach close to posterior extremity. Testes oval, entire to irregularly lobed, tandem to slightly oblique in posterior half of hindbody. Cirrus-sac bipartite; posterior part elongate; anterior part globular. Internal seminal vesicle in posterior part of cirrus-sac, saccular. Pars prostatica small, vesicular. Ejaculatory duct large, muscular, complex, deeply diverticulate, may be completely everted through genital pore to form lobed cirrus. Genital pore submedian in posterior forebody. Ovary oval, entire or irregularly lobed, near middle of hindbody. Uterus fills much of hindbody, intercaecal, pregonadal, often extends into post-testicular region. Eggs numerous, operculate. Metraterm muscular. Vitelline follicles in symmetrical lateral fields in mid-hindbody. Type-species *T. laticollis* (Rudolphi, 1819).

2b. Intestinal bifurcation in posterior forebody; uterus tends not to extend into post-testicular region .... *Theledera* Linton, 1910. (Fig. 34.51)

(Syns *Cithara* MacCallum, 1917; *Tergestina* Nagaty & Abdel Aal, 1964)

Diagnosis: Body elongate, cylindrical. Oral sucker oval, subterminal, surrounded by ring of 10–20 cephalic lobes. Six muscular flanges (collarettes) on lateral surface at level of pharynx. Ventral sucker oval, protuberant, in anterior third of body. Oesophagus long. Intestinal bifurcation in posterior forebody. Caeca reach close to posterior extremity. Testes oval, entire to irregularly lobed, tandem to slightly oblique close to posterior extremity. Cirrus-sac bipartite; posterior part elongate; anterior part globular. Internal seminal vesicle in posterior part of cirrus-sac, saccular. Pars prostatica small, vesicular. Ejaculatory duct large, muscular, complex, deeply diverticulate, may be completely everted through genital pore to form lobed cirrus. Genital pore submedian in posterior forebody. Ovary oval, entire or irregularly lobed,

Fig. 34.52. *Pseudohypertrema karachiensis* Bilqees,1976. Ventral view, uterus in outline. (After Bilqees, 1976.)

pretesticular, near middle of hindbody. Uterus tends to fill space between anterior testis and ventral sucker, overlaps caeca. Metraterm muscular. Vitellarium in symmetrical lateral fields in mid-hindbody. In marine teleosts. Widespread. Type-species *T. pectinata* (Linton, 1905).

Genus incertae sedis

Genus *Pseudohypertrema* Bilqees, 1976 (Fig. 34.52)

Diagnosis: Body small, plump, oval. Tegument unarmed. Ventral sucker round, in anterior half of body. Prepharynx short or absent. Pharynx oval. Oesophagus indistinct. Intestinal bifurcation in mid-forebody. Caeca reach just past testes, blind. Testes oval, entire, symmetrical, in anterior hindbody. Cirrus-sac claviform, mainly in hindbody. Internal seminal vesicle saccular. Pars prostatica weakly developed. Ejaculatory duct short. Genital atrium small. Genital pore closely anterosinistral to ventral sucker. Ovary oval to irregular, intertesticular, dextrally submedian. Proximal female system not described. Uterus fills post-testicular area. Eggs small, 'double-walled'. Vitelline follicules in continuous field across forebody from level of oral sucker to posterior margins of gonads, absent lateral to testes. Excretory vesicle 'not extending into forebody'. In marine teleosts. Off Pakistan. Type-species *P. karachiensis* Bilqees, 1976.

35 Family Tandanicolidae Johnston, 1927

R.A. Bray

Department of Zoology, The Natural History Museum, London, UK

Introduction

This family has a strong autapomorphy in the distinct accessory copulatory organ associated with the genital atrium. It was erected as a subfamily of the family Brachycoeliidae by Johnston (1927). Later, Dollfus (1937a, b) erected the family Monodhelminthidae. Both of these taxa have subsequently been considered as subfamilies of the Fellodistomidae by many workers, many of whom have been listed by Cribb & Bray (1994) in their review of the taxonomic history of the group. Bray (1988) considered *Tandanicola* Johnston, 1927 as a member of the fellodistome subfamily Monodhelminthinae, but Cribb & Bray (1994) pointed out the priority of the family-group name Tandanicolinae, which they continued to consider a fellodistomid group. Hall *et al.* (1999) presented a molecular and morphological phylogeny, which indicated that the Tandanicolinae and Fellodistominae do not form a clade, and they recognized that the status of family is appropriate for this taxon. Cribb & Bray (1994) reviewed the taxon and this key is based on the one presented in that paper.

Tandanicolids, parasites of the intestine or swim-bladder of marine fishes, have a strong association with siluriform fishes, with over 82% of records from this order and over 47% from the family Ariidae. They are reported mainly from the Indo-Pacific region, with a few records from the Atlantic Ocean.

One tandanicolid species has been described since Cribb & Bray (1994), namely *Mehratrema militaris* Dutta, 1995 from the ariid *Osteogeniosus militaris* in India (Dutta, 1995). It is not clear how this form differs from *Monodhelmis torpedinis* Dollfus, 1937, the type of the largest genus in the family. Cribb & Bray (1994) pointed out that *Parheterorchis* Gupta & Jain, 1991 is likely to be a synonym of *Monodhelmis* and that suggestion has been followed in this key.

Family Tandanicolidae Johnston, 1927

(Syns Tandanicolinae Johnston, 1927; Monodhelminthidae Dollfus, 1937; Monodhelminthinae Dollfus, 1937; Mehratrematinae Srivastava, 1939; Prosogonariinae Mehra, 1963; Buckleytrematinae Yamaguti, 1971)

Diagnosis: Body elongate to oval and plump. Tegument with or without spines. Oral sucker simple, opening subterminally. Ventral sucker larger or smaller than oral sucker, in mid-body, may have embedded specialized muscular lips. Prepharynx present but usually very short. Pharynx present. Oesophagus distinct, short or long. Intestinal bifurcation in forebody. Caeca extend variable distance into hindbody, end blindly. Testes symmetrical or tandem, usually at least partly in forebody. Cirrus-sac present, may be indistinct. Genital atrium large, with thick, irregular, muscular wall. Accessory copulatory organ situated posteriorly in genital atrium; consists of solid, anteriorly directed digitiform process and associated glandular tissue (Fig. 35.3). Genital pore midway between oral and ventral suckers. Ovary (probably always) multilobed; lobes contiguous (may be indistinct) or well separated; pretesticular. Laurer's canal and uterine seminal receptacle probably always present. Uterus usually largely in hindbody. Vitellarium follicular, in two lateral bunches. Excretory vesicle U-shaped; arms extending to near pharynx. Parasites of intestine and swim-bladder of marine and freshwater fishes. Type-genus *Tandanicola* Johnston, 1927.

Key to genera

1a. Ventral sucker with semicircular laminar muscular apparatus embedded in anteroventral part, giving appearance of complex lips .... *Prosogonarium* Yamaguti, 1952. (Fig. 35.1)

Diagnosis: Body oval, plump. Tegument with fine spines. Ventral sucker in mid-body, specialized with semicircular laminar muscular apparatus embedded in anteroventral part giving appearance of complex lips. Caeca extend to mid-hindbody. Testes symmetrical, overlapping ventral sucker posteriorly. Ovary trilobed; lobes distinct. Laurer's canal and uterine seminal receptacle present. Vitelline fields lateral, from ventral sucker to pharynx, almost confluent anteriorly. In marine fishes (Siluriformes). Indo-Pacific. Type-species *P. arii* Yamaguti, 1952.

1b. Ventral sucker not specialized .... 2.

2a. Testes tandem .... *Buckleytrema* Gupta, 1956. (Figs 35.2, 35.3)

(Syn. *Paramonodhelmis* Oshmarin & Mamaev, 1963)

Diagnosis: Body elongate oval. Tegument completely lacking spines. Ventral sucker smaller than oral sucker, not specialized, in mid-body. Caeca extend to mid-hindbody. Testes tandem, contiguous. Ovary trilobed; lobes contiguous, sometimes indistinct. Laurer's canal present. Uterus largely post-testicular, extending to posterior end of body. Vitelline fields lateral, from about level of genital atrium to posterior testis. In marine fishes (Ariidae). Indo-Pacific. Type-species *B. indica* Gupta, 1956.

2b. Testes symmetrical .... 3.

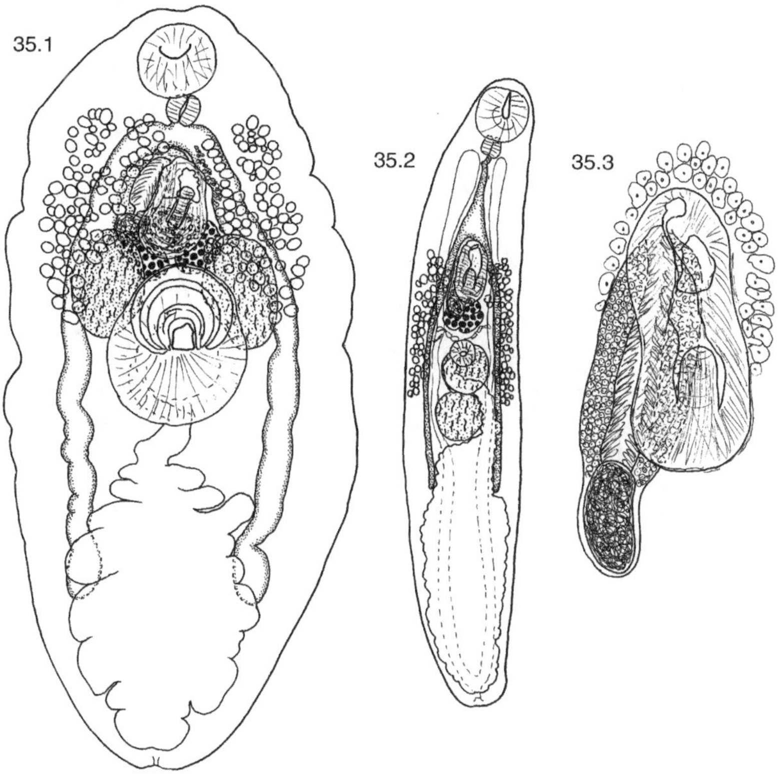

Figs 35.1–35.3.
35.1. *Prosogonarium angelae* Cribb & Bray, 1994. Ventral view of paratype. (Original.)
35.2, 35.3. *Buckleytrema indica* Gupta, 1956. 35.2. Ventral view of specimen from *Arius graeffei*. (Original.) 35.3. Ventral view of terminal genitalia, specimen from *Arius graeffei*. (Original.)

3a. **Body plump, pear-shaped, curved ventrally** .... ***Tandanicola* Johnston, 1927. (Fig. 35.4)**

Diagnosis: Body plump, pear-shaped, curved ventrally. Tegument with fine spines. Ventral sucker larger than oral sucker, not specialized, in mid-body. Caeca extend to mid-hindbody. Testes symmetrical, in forebody. Ovary multilobed; lobes well separated. Laurer's canal present. Canalicular seminal receptacle absent. Uterus largely in hindbody in narrow median band. Vitelline fields lateral, between genital pore and ventral sucker. In freshwater teleosts (*Tandanus*). Australia. Type-species *T. bancrofti* Johnston, 1927.

3b. **Body elongate oval, flattened** .... ***Monodhelmis* Dollfus, 1937. (Fig. 35.5)**
(Syns *Burnellus* Angel, 1971; *Mehratrema* Srivastava, 1939; *Parheterorchis* Gupta & Jain, 1991)

Diagnosis: Body, elongate oval, flattened. Tegument with fine spines. Ventral sucker slightly smaller than oral sucker, not specialized, in mid-body. Caeca extend to near posterior end of body. Testes symmetrical; posterior half of each overlapping ventral sucker. Ovary lobed; lobes distinctly separated or contiguous. Laurer's canal and

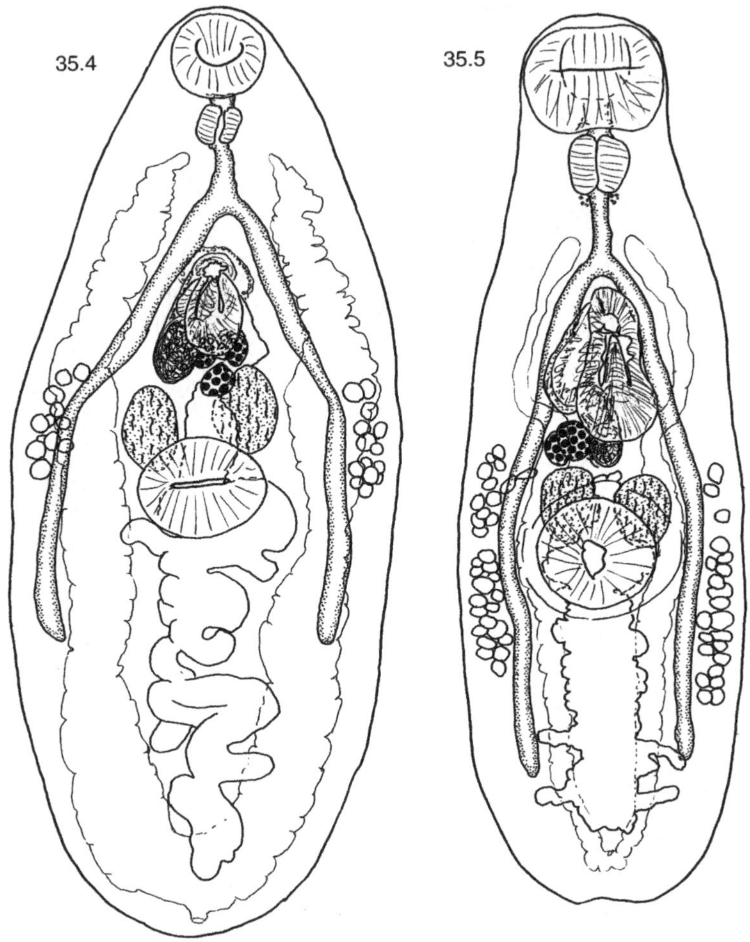

Figs 35.4–35.5.
35.4. *Tandanicola bancrofti* Johnston, 1927. Ventral view. (Original.)
35.5. *Monodhelmis torpedinis* Dollfus, 1937. Ventral view of specimen from *Arius graeffei*. (Original.)

canalicular seminal receptacle not reported. Uterus largely in hindbody between caeca. Vitelline fields lateral, from level of testes to mid-hindbody. In marine fishes. Cosmopolitan. Type-species *M. torpedinis* Dollfus, 1937.

36 Superfamily Hemiuroidea Looss, 1899

D.I. Gibson

Department of Zoology, The Natural History Museum, London, UK

Introduction

Hemiuroids are usually parasitic in the gut, especially the stomach, of fishes. They are found predominantly in marine teleosts, but also occur in freshwater teleosts, elasmobranchs and occasionally in amphibians and reptiles. It is not unusual for progenetic forms to occur in molluscs and other marine, or occasionally freshwater, invertebrates. In addition to the gut, species or even entire groups are known from the tissues, gall-bladder, swim-bladder, body-cavity, mouth, gills, lungs and skin. Overall, the hemiuroids form a very diverse group, not only in habitat, but also in morphology, which has resulted in a good deal of confusion with regard to the validity, composition of and systematic relationships with the group.

This superfamily was first recognized, under the name Hemiurida, by Dollfus (1923) and comprised the families Hemiuridae, Accacoeliidae and Syncoeliidae, although the family-group name is attributed to Looss (1899), who introduced it at the subfamily level. Subsequently, Looss (1907, 1908) carefully redescribed many species of hemiuroids and set a basic pattern on which later authors, notably Odhner (1911b), Poche (1926) and Fuhrmann (1928), were able to build. These early workers based their classifications entirely upon adult morphology and divided the group into a small number of families, although not always indicating their relationships. Odhner (1911b), for example, grouped three families together, the Hemiuridae, Azygiidae and Didymozoidae. Subsequently, the concept of the Hemiuroidea has been subdivided, condensed and subdivided again on numerous occasions. Systematic histories have been compiled by Chauhan (1954), Skrjabin & Guschanskaja (1954, 1956, 1960), Stunkard (1973) and Gibson & Bray (1979). For example, Yamaguti (1971) divided Odhner's concept into the Hemiuroidea, Accacoelioidea, Isoparorchioidea and Didymozooidea, and 19 families, one of which, the Hemiuridae, contained 25 subfamilies. The most recent revision of the superfamily is that of Gibson & Bray (1979), who recognized 14 families within the group, provided diagnoses and keys to generic level and discussed its possible evolution. This work was examined using cladistic methods by Brooks *et al.* (1985). The latter accepted the 'hemiuroids' at the ordinal level as the Hemiuriformes, introducing the Bivesiculidae as the sister group of the remaining eight families which they included; the link with the Bivesiculidae was criticized by Gibson (1987) and Pearson (1992), and this group was separated from the Hemiuroidea but retained within the Hemiuriformes by Brooks *et al.* (1989) and Brooks & McLennan (1993). The conception of the Hemiuridae presented by Brooks *et al.* (1985) was especially broad and included the Isoparorchiidae, Sclerodistomidae, Didymozoidae, Dictysarcidae, Lecithasteridae, Derogenidae and Bunocotylidae. The systematic position of two other 'hemiuroid' groups has long been a matter of contention: these are the Azygiidae and Didymozoidae.

Recent conceptions of the Hemiuroidea, and of the Digenea in general, have been greatly influenced by the work of La Rue (1957), who subdivided the Digenea into two groups, the Epitheliocystidia and the Anepitheliocystidia, depending upon the epithelial or membranous nature of the lining of the cercarial excretory vesicle. The Azygiidae were placed in the latter group, while the remainder of the hemiuroids with 'known' life-histories were placed in the former. This work has resulted in some workers considering the Azygiidae to be distinct from the Hemiuroidea, at least at the superfamily level (see Odening, 1974). Gibson & Bray (1979) suggested that ultrastructure indicates that all cercariae may have a syncytial lining to the excretory vesicle and included the Azygiidae at the base of the hemiuroid tree because of similarities in morphology (adult and cercaria) and life-cycle. However, recent molecular work (Blair *et al.*, 1998) has indicated the exclusion of this family from the clade that forms the Hemiuroidea. In the present work, in which the conception of the Hemiuroidea is otherwise essentially that of Gibson & Bray (1979), the Azygiidae are separated off in their own superfamily to reflect these molecular results. Another group which Gibson & Bray (1979) included close to the base of the hemiuroid tree is the Ptychogonimidae. In relation to this group the molecular data of Blair *et al.* (1998) weakly supported the retention of this family as the sister group of the remainder of the hemiuroids; it is retained within the superfamily in this work.

The Didymozoidae are an enigmatic group, which, because of their atypical morphology and purported atypical digenean life-cycle, have been considered to be a separate subclass of the Trematoda by Baer & Joyeux (1961), a course followed by some authors, even recently (Kearn, 1998). However, links with the hemiuroids have long been recognized (e.g. Odhner, 1911b). In recent years the group has been recognized at the subordinal level by many Russian authors (e.g. Pozdnyakov, 1993) and at the superfamily level by Yamaguti (1971). They were considered to be derived from the Hemiuroidea by Gibson & Bray (1979), but recognized (erroneously in cladistic terms) at the superfamily level; this is corrected in this work, where the group is treated as a family within the Hemiuroidea. The position of the Didymozoidae within the Hemiuroidea was fully supported by the molecular work of Blair *et al.* (1998). Unfortunately, a key to the didymozoids is not included here with the other families of the Hemiuroidea.

The only other major difference between this work and the arrangement proposed by Gibson & Bray (1979) is the subsuming of the family Bunocotylidae, which these authors derived from the Hemiuridae, into the latter family.

The main morphological features of the hemiuroids are a genital pore that is mid-ventral in the forebody, male and female terminal ducts that normally fuse to form a hermaphroditic duct with, often, a hermaphroditic intromittent organ (sinus-organ) and a surrounding muscular sac (sinus-sac), the absence of a prepharynx, a tegument devoid of spines and a Y-shaped excretory vesicle, often with arms uniting dorsally in the forebody.

There are numerous other specialized features of certain families within this group. These include: ecsoma, plications, Juel's organ and Manter's organ. For information, see the Glossary (Volume 3) and, in more detail, Gibson & Bray (1979). One feature of note is the presence of a cirrus and cirrus-sac in one family, the Hirudinellidae; these features are not considered to be homologous with those of other digeneans (Gibson & Bray, 1979) and are likely derived from the sinus-organ and sinus-sac.

Superfamily Hemiuroidea Looss, 1899

(Syns Didymozooidea Poche, 1907; Accacoelioidea Odhner, 1911; Isoparorchioidea Travassos, 1922)

Diagnosis: Body small to large; usually oval to cylindrical, occasionally (Didymozoidae) filamentous, round, bipartite with slender anterior region or variety of other forms. Occasionally encapsulated in tissues in pairs, which exhibit various degrees of hermaphroditism (Didymozoidae). Ecsoma present or absent. Body surface smooth, rugate or plicated (or 'scaly'), never spiny, but occasionally papillate. Oral and ventral suckers well developed, occasionally small or (some Didymozoidae) absent. Ventral sucker normally in middle or anterior half of body, occasionally just inside posterior half of body, occasionally pedunculate, occasionally absent (some Didymozoidae). Prepharynx absent. Pharynx well developed to (in some Didymozoidae) rudimentary or absent; normally oval, occasionally modified. Oesophagus usually short, occasionally long; 'Drüsenmagen' present or absent. Gut-caeca usually end blindly near posterior extremity, occasionally form cyclocoel or uroproct; occasionally atrophied (some Didymozoidae). Testes normally two, rarely one or follicular, normally tandem to symmetrical, pre-ovarian and near middle of body; occasionally in forebody or postovarian, occasionally tubular (Didymozoidae). Seminal vesicle oval to tubular, occasionally constricted into portions, usually thin- but occasionally thick-walled, in fore- or hindbody, normally external to sinus-sac, rarely partly or entirely internal. Pars prostatica tubular to vesicular, long or short, normally external to sinus-sac when present, rarely internal, usually in forebody, occasionally entirely inside or extending into hindbody. Ejaculatory duct usually present, usually short, often within sinus-sac or sinus-organ, normally unites with metraterm to form hermaphroditic duct; rarely (Hirudinellidae) within 'cirrus-sac'. Hermaphroditic duct usually present, usually within sinus-sac and/or sinus-organ. Sinus-sac present or absent, oval to cylindrical, normally enclosing ejaculatory duct and part of metraterm and/or hermaphroditic duct. True cirrus-sac and cirrus absent, but modification of sinus-sac in form of 'cirrus-sac' and associated intromittent organ ('cirrus') occur rarely (Hirudinellidae). Permanent sinus-organ present or absent within genital atrium, conical to tubular, muscular or non-muscular. Temporary sinus-organ sometimes formed from hermaphroditic duct. 'Cirrus-sac', enclosing ejaculatory duct only, and 'cirrus' rarely present (Hirudinellidae). Genital atrium large, small or absent.

Common genital pore midventral in forebody. Ovary usually oval, occasionally lobed, rarely follicular or tubular to (in Didymozoidae) filamentous, usually post-testicular, occasionally pretesticular, rarely intertesticular; normally in hindbody, rarely in forebody. Mehlis' gland usually postovarian, occasionally pre-ovarian. Uterine seminal receptacle plus Laurer's canal and/or Juel's organ or blind seminal receptacle alone normally present. Uterine coils usually fill much of hindbody, occasionally extending well into forebody, rarely entirely in forebody. Eggs normally oval, usually small, numerous, occasionally with spine, filament(s) or threads. Vitellarium tubular, composed of a small number (often seven) oval to tubular lobes or one to three (usually two) entire or lobed masses, filamentous (Didymozoidae) or follicular (Ptychogonimidae), often postovarian, occasionally pre-ovarian, sometimes extending throughout hindbody or into forebody, rarely entirely in forebody. Excretory pore terminal. Excretory vesicle Y-shaped; arms united in forebody or not. Manter's organ (accessory excretory vesicle) rarely present. Parasitic in gut, especially stomach, primarily of marine teleosts, but commonly occur in freshwater teleosts and occasionally in elasmobranchs, amphibians, reptiles and progenetic in invertebrates, also in tissues of marine teleosts, especially scombroids (Didymozoidae), occasionally recorded from gills, skin, body-cavity, swim-bladder and other organs. Type-genus *Hemiurus* Rudolphi, 1809.

Key to families

1a. Parasitic in tissues of marine teleosts (occasionally free in body-cavity); variable in shape from filamentous to globular, non-filamentous forms often bipartite; encysted or free; often in pairs – when in pairs often exhibiting different degrees of hermaphroditism .... Didymozoidae Poche, 1907. [Key not included in this volume]

1b. Otherwise (usually parasitic in gut or associated organs of fishes, occasionally amphibians, rarely reptiles); always completely hermaphroditic .... 2.

2a. Vitellarium composed of numerous widely distributed follicles; in gut of elasmobranchs .... Ptychogonimidae Dollfus, 1937.

2b. Vitellarium otherwise, usually composed of a small number of oval to tubular (occasionally branched) lobes or one to three distinct oval, lobed or unlobed masses .... 3.

3a. Testes two, occasionally one .... 4.

3b. Testes follicular, 11–18 large or many small follicles arranged in rows or irregularly distributed; usually parasitic in buccal or branchial cavities or on skin (? occasionally internally) of elasmobranchs and marine teleosts .... Syncoeliidae Looss, 1899.

4a. Ecsoma absent (take care with this observation as some hemiurids have a very reduced ecsoma) .... 5.

4b. Ecsoma present (sometimes very reduced); body surface often plicated; Juel's organ and uterine seminal receptacle present; vitellarium varies between form

with seven tubular lobes and form with two distinct oval masses; parasitic mainly in gut (especially stomach) of marine teleosts, present rarely in freshwater teleosts and lung of sea-snakes .... Hemiuridae Looss, 1899.

5a. Ovary usually post-testicular, occasionally pretesticular .... 6.
5b. Ovary intertesticular; parasitic on gills (? and in stomach) of marine teleosts .... Bathycotylidae Dollfus, 1932.

6a. Ovary oval or lobed .... 7.
6b. Ovary tubular; parasitic in swim-bladder of freshwater teleosts .... Isoparorchiidae Travassos, 1922.

7a. Ventral sucker anterior to middle of body; parasitic in marine teleosts; seminal vesicle never enclosed in sinus-sac .... 8.
7b. Ventral sucker usually in or near middle of body, occasionally more anterior; significant proportion of uterus usually present in forebody [a small number of marine forms do possess a ventral sucker in the anterior half of the body and uterine coils that do not extend into the forebody, but these forms also possess a seminal vesicle that is enclosed within the sinus-sac]; vitellarium one or two masses, entire or lobed (lobes normally shallow, rarely digitate); seminal vesicle in forebody; ovary and vitellarium pre- or post-testicular; parasitic mainly in gut of freshwater and marine teleosts, occasionally in amphibians, reptiles and freshwater shrimps ...... Derogenidae Nicoll, 1910.

8a. Male intromittent organ ('cirrus') present, enclosed in 'cirrus-sac'; female duct opens into genital atrium independently; large parasites from gut (? or gills) of marine teleosts (immature forms occasionally present in salmonids) .... Hirudinellidae Dollfus, 1932.
8b. 'Cirrus' and 'cirrus-sac' absent; male and female ducts normally unite forming hermaphroditic duct, which is often present within a sinus-organ and enclosed by a sinus-sac .... 9.

9a. Parasitic in gut (occasionally on gills) .... 10.
9b. Parasitic in swim-bladder or gall-bladder .... 13.

10a. Vitellarium one, two or three compact masses (vestige of ecsoma occasionally present) .... Hemiuridae (Bunocotylinae) Dollfus, 1950.
10b. Vitellarium otherwise .... 11.

11a. Vitellarium six to eight (usually seven, occasionally twice this number, occasionally two to seven) oval to digitiform lobes, often arranged in rosette, occasionally branched; Manter's organ absent; pharynx oval .... Lecithasteridae Odhner, 1905.
11b. Vitellarium tubular (filamentous) .... 12.

12a. Manter's organ present; pharynx oval .... Sclerodistomidae Odhner, 1927.

12b. Manter's organ absent; pharynx with narrow anterior extension into base of oral sucker; occasionally present on gills; commonly parasitic in sunfish (Molidae) .... Accacoeliidae Odhner, 1911.

13a. Parasitic in swim-bladder; vitellarium six to eight oval to digitiform lobes, two compact multilobulate masses or two acinous groups of follicles .... Dictysarcidae Skrjabin & Guschanskaja, 1955.

13b. Parasitic in gall-bladder; vitellarium tubular, dendritic, with anteriorly and posteriorly orientated main collecting ducts situated medially .... Sclerodistomoididae Gibson & Bray, 1979.

37 Family Hemiuridae Looss, 1899

D.I. Gibson

Department of Zoology, The Natural History Museum, London, UK

Introduction

The subfamily Hemiurinae was erected by Looss (1899) to include *Hemiurus* Rudolphi, 1809 (also as *Pronopyge* Looss, 1899) and the derogenid *Derogenes* Lühe, 1900 (as *Liopyge* Looss, 1899). Shortly afterwards, Lühe (1901) recognized the group at the family level to include *Hemiurus*, *Lecithocladium* Lühe, 1901, *Lecithaster* Lühe, 1901, *Lecithochirium* Lühe, 1901 and *Accacoelium* Monticelli, 1893, several of which are now attributed to other hemiuroid groups. Throughout the early history of the family Hemiuridae, it was generally deemed to include a wide range of dissimilar forms. For example, Fuhrmann (1928) included the Hemiurinae, Derogeninae, Dinurinae, Lecithasterinae, Sclerodistominae and Lecithochiriinae (as Sterrhurinae), a concept that is almost as broad as the Hemiuroidea as treated in this work. Such a wide conception of the family has, even very recently, been very popular. Workers such as Manter & Pritchard (1960) and Yamaguti (1971) have included all of these forms, with the exception of the sclerodistomines, within the Hemiuridae. Although some of these groups, such as the Derogeninae (= Halipeginae), have at various times been removed from the Hemiuridae, it was not until the works of Skrjabin & Guschanskaja (1954, 1955a, b, 1956, 1957b) that a major change occurred. These authors retained the Hemiuridae for only the Hemiurinae and the Aphanurinae, and they recognized the Dinuridae, Lecithasteridae, Lecithochiriidae, Elytrophallidae and the Halipegidae at family level. This system, although not accepted in its entirety, was given some support by La Rue (1957) and Baer & Joyeux (1961), who recognized the Dinuridae, Halipegidae, Lecithasteridae and Lecithochiriidae, the latter authors including only the Hemiurinae and the Prosorchiinae within the Hemiuridae. Yamaguti (1971) did not accept this narrower view of the group and included many non-ecsomate subfamilies within the family, i.e the Arnolinae, Bunocotylinae, Derogeninae, Gonocercinae, Halipeginae, Hysterolecithinae, Intuscirrinae, Lecithasterinae, Lecithophyllinae, Liopyginae, Macradenininae, Opisthadeninae, Prolecithinae, Prosorchiinae, Quadrifoliovariinae and Trifoliovariinae, among the 26 which he recognized. More recently, in their detailed revision of the Hemiuroidea, Gibson & Bray (1979) have taken an intermediate position and restricted the Hemiuridae to ecsomate forms, thus recognizing the Derogenidae (= Halipegidae), Lecithasteridae and Sclerodistomidae as being distinct families, and transferring the Aphanurinae to the Bunocotylidae. Within the Hemiuridae these authors accepted the following subfamilies: Hemiurinae, Dinurinae, Elytrophallinae, Glomericirrinae, Hypohepaticolinae, Lecithochiriinae, Lethadeninae, Plerurinae and Pulmoverminae. Subsequently, Brooks *et al.* (1985) in their cladistic phylogeny of the Digenea formulated an even broader concept than that of Yamaguti and included the Isoparorchiidae, Sclerodistomidae, Didymozoidae, Dictysarcidae, Lecithasteridae, Derogenidae and Bunocotylidae within the Hemiuridae. The major discrepancy between the concepts of Gibson & Bray (1979) and Brooks *et al.* (1985) is caused by the fact that the latter did not use the ecsoma as the major apomorph of the group. The broad concept of the Hemiuridae of Brooks *et al.* (1985) is not supported by the molecular work of Blair *et al.* (1998), which shows it to be paraphyletic. Gibson & Bray's (1979) narrower concept and classification were followed by Gibson & Bray (1986) in their revision of the Hemiuridae of the north-east Atlantic and by Gibson (1996). It is also followed, in a modified form, in the present work.

The major difference between the present work and that of Gibson & Bray (1979) relates to the position of the Bunocotylidae. The subfamily Bunocotylinae was erected by Dollfus (1950) for *Bunocotyle* Oghner, 1928 and accepted by many authors, some of whom added other genera. It was used at the family level by Gibson & Bray (1979), who included within it the Bunocotylinae, Aphanurinae, Opisthadeninae and Theletrinae. They derived the group from the family Hemiuridae, differentiating it essentially on the secondary loss of an ecsoma. Discussion on the apparent vestige of an ecsoma and the form of the seminal receptacle in this group was given by Gibson & Bray (1979). As members of the group, especially the aphanurines, are clearly hemiurids without an ecsoma, the Bunocotylidae is treated here as a synonym of the Hemiuridae and its component subfamilies are subsumed into the latter family (see also Brooks *et al.*, 1985). Brooks *et al.* (1985) treated the Bunocotylidae *sensu* Gibson & Bray (1979) as a subfamily within their conception of the Hemiuridae. Recently, a cladistic analysis of the Bunocotylinae *sensu* Brooks *et al.* (1985) was presented by León-Règagnon *et al.* (1998). The position of the Opisthadeninae, included in the Bunocotylidae by Gibson & Bray (1979) because of the shape of the vitellarium and the presence of transverse ridges around the body in some genera, is problematical, as it differs from hemiurids and resemble lecithasterids in possessing a blind seminal receptacle. Because the form of the vitellarium is different from that normally encountered in lecithasterids, and a blind seminal receptacle is likely to have evolved on more than one occasion in the hemiuroids, e.g. didymozoids, this group is included with the Hemiuridae in this work. This action is sustained by the work of León-Règagnon *et al.* (1998), whose analysis supported the clades referred to here as the Bunocotylinae and Opisthadeninae, but found that other genera separated independently. Further comments on some of these genera can be found in Bray & Cribb (2000a).

Many generic synonymies and remarks were made by Gibson & Bray (1979); as these will not be commented upon here, the reader is referred to this work. However, the following are observations on some of the more recent changes. These mainly reflect taxa described from fishes of the Northern Indian Ocean by authors unaware of

Gibson & Bray's (1979) revision. Few of the new genera erected since 1979 are considered valid, many being based upon rather poor descriptions, often of only a single worm; furthermore, type-material is usually lacking or unavailable.

The Biporidae (type- and only genus *Bengalotrema* Malhotra, Nanda, Mukherjee, Ghosh, Sukul & Capoor, 1989 from the fish *Hilsa ilisha*; treated below as a synonym of *Lecithocladium*) is clearly hemiurid in nature (presence of ecsoma and plicated tegument) but is claimed by Malhotra *et al.* (1989) to have many features here considered impossible in a hemiurid, e.g. pretesticular ovary, separate male and female genital pores, a follicular vitellarium, a cirrus-sac; these putative features indicate that the original description was somewhat wanting!

Colletostomum Sahai & Srivastava, 1978 and *Clefticolletta* Sahai & Srivastava, 1978 were sunk into *Lecithocladium* by Gibson & Bray (1986); to these can be added *Bengalotrema* and *Magnapharyngium* Bilqees, 1971. *Linguastomachicola* Srivastava & Sahai, 1978, known from only two specimens, is identical to *Stomachicola* Yamaguti, 1934 but for a linguiform protrusion from within the oral sucker; even if confirmed, it is unlikely to warrant acceptance as a valid generic criterion. Similarly, *Cameronia* Bilqees, 1971 (almost unrecognizable from its description), *Segmentatum* Bilqees, 1971 (holotype in The Natural History Museum examined) and *Cestodera* Bilqees, 1971 are also here considered synonyms of *Stomachicola*. *Anaplerurus* Srivastava & Sahai, 1978 and *Umatrema* Srivastava & Sahai, 1978 appear to be synonyms of *Lecithochirium* (and consequently the Umatrematinae Srivastava & Sahai, 1978 becomes a synonym of the Lecithochiriinae). *Daniella* Sahai & Srivastava, 1977 and *Bapatina* Srivastava & Sahai, 1977 at face value appear to be synonyms of *Anahemiurus* Manter, 1947. However, Bray (1990) considered the former to be a synonym of *Parahemiurus* Vaz & Pereira, 1930, although Sahai & Srivastava (1977) claimed that the body is scaled. *Dentiacetabulum* Sahai & Srivastava, 1977 is here listed under *Parahemiurus*, as no crenulate plications ('scales') were mentioned. As the presence of crenulate plications is, in isolation, a feature of questionable generic value, and in view of the fact that it is readily misinterpreted in poorly preserved material, *Anahemiurus* is here considered a synonym of *Parahemiurus*. *Engraulitrema* Gupta & Jain, 1992 is unrecognizable, but has been listed under *Parahemiurus* in view of the forms likely to occur in engraulids. Similarly, the Pseudoprosorchiinae Gupta & Jain, 1992 is listed under the Hemiurinae. *Exoticotyle* Sahai & Srivastava, 1978, *Clupenuroides* Al-Yamani & Nahhas, 1981, *Neoerilepturus* Shalaby & Hassanine, 1996 and *Chrysophrytrema* Gupta & Jain, 1992 are considered synonyms of *Erilepturus* Woolcock, 1935, *Chrysophrytrema* being virtually identical to forms described under *Uterovesiculurus* Skrjabin & Guschanskaja, 1954. *Qasimmetra* Sahai & Srivastava, 1978 appears to be a synonym of *Qadriana* Bilqees, 1971. The latter genus, which had been inadequately described by Bilqees (1971), was included by Gibson & Bray (1979) as a questionable lecithasterine lecithasterid. The examination of the holotype, subsequently deposited in the Collection of the British Museum (Natural History), has highlighted the clear presence of a withdrawn ecsoma. Further study indicated a synonymy with the dinurine *Atheria* Hafeezullah, 1975, which was much better described by Hafeezullah (1975).

Lecithomonoium Shen, 1987 is known from only a single specimen, its diagnostic feature being the presence of only one vitelline mass; however, the loss of one vitelline mass is known to occur in *Parahemiurus* (see Yamaguti, 1934), a genus that occurs in the same host as *Lecithomonoium*, and some illustrations show the masses superimposed. Hence, *Lecithomonoium* is considered a synonym of *Parahemiurus*, which, but for the apparent nature of the vitellarium, it resembles. Similarly, *Trilecithotrema* Gu & Shen, 1981, also known only from a single specimen, appears morphologically identical to *Parahemiurus*, but it is claimed to have three vitelline masses. It is here considered a synonym of the latter genus, in view of the lobation of the vitelline masses in some specimens and their tendency to overlie one another.

Paralecithochirium Zhang, 1994 is an enigmatic taxon, the description of which resembles *Lecithochirium*, except that its eggs are said to be filamented; it is apparently, therefore, somewhat similar to *Hypohepaticola* Yamaguti, 1934. However, it is a parasite of freshwater rather than marine fishes and, apart from its small ecsoma, bears some similarities to certain halipegine genera from Chinese freshwater fishes. The presence of a hemiurid in completely fresh water is surprising, although forms, including *Lecithocladium*, *Parahemiurus*, *Erilepturus* and *Ectenurus* Looss, 1907, do occur in coastal lagoons and estuarine areas, occasionally reaching distances of 100 km from the sea. Through the kindness of Professors Zhang Tongfu and Zhang Jianying I have been able to examine specimens of *Paralecithochirium leptobotiae* Zhang, 1994. I can confirm the presence of an ecsoma and that the terminal genitalia appear to be lecithochiriine in nature; however, I could not find the filaments on the eggs and the lobation of the vitellarium is less distinct than in the published illustrations of Zhang (1994). In many ways the specimens resemble *Prolecithochirium*, which Yamaguti (1970) erected for a Hawaiian marine species. It differs from the latter in the shape of the seminal vesicle and the slight lobation of the vitelline masses.

Another new lecithochiriinae, *Boreascotia*, was erected recently by Bray & Zdzitowiecki (2000) from, unusually, the intestine of a sub-Antarctic marine fish. This genus, which has an uncommonly long seminal vesicle, is accepted here.

The systematic position of several other nominal hemiurid genera which are known to enter coastal brackish or freshwater is problematical. For example, *Clupenurus* Srivastava, 1935 was made a questionable synonym of *Lecithocladium* by Gibson & Bray (1986); however, the same or a similar form has been described by Khan & Bilqees (1990) in freshwater lakes in Pakistan as a species of *Ectenurus*. Based on this description and that of Srivastava (1935), *Clupenurus* is here considered a synonym of *Ectenurus*. It is also worth noting that recently described species attributed to *Clupenurus* from marine fishes are also comparable and similarly confusing. The same is also true of *Prosterrhurus* Fischthal & Kuntz, 1963, which was considered a synonym of *Ectenurus* by Yamaguti (1971), but recognized as a questionable genus close to *Erilepturus* by Gibson & Bray (1979). This species was described from the intestine of freshwater fish in Egypt; however, several typical marine forms of hemiurid were described from the intestine of the same species of host, suggesting that the host, acquired from a fish market, was of brackish-water origin. It is worth noting that Khan & Bilqees's (1990) material was obtained from *Hilsa* and *Labeo*, *Clupenurus* was described from *Hilsa* and *Prosterrhurus* from *Labeo*. Having examined the holotype of *Prosterrhurus*, it is apparent that there are errors in the orginal description relating to the nature of the pars prostatica and that the apparent absence of plications is probably caused by the poor condition of the specimen (which looks as though it has been frozen). In agreement with Yamaguti (1971), *Prosterrhurus* is here considered a synonym of *Ectenurus*. This may suggest doubts concerning the purported identity of the host. Needless to say, it is certain that there has been considerable confusion in the literature regarding the genera *Lecithocladium*, *Ectenurus* and *Erilepturus*. *Lecithocladium*-like forms from *Hilsa ilisha* are in an especially confused state; Gibson & Bray (1986) listed them all under *L. harpodontis* Srivastava, 1937.

The genus *Johniophyllum* and the subfamily Johniophyllinae were erected by Skrjabin & Guschanskaja (1954) for *Lecithocladium johnii* Yamaguti, 1938 on the basis of the absence of a sinus-sac. My observations of the holotype indicate that a sinus-sac is present and that this species is a typical elytrophalline. It has been synonymized below with *Elytrophallus* Manter, 1940 rather than *Lecithocladium* in view of the absence of plications (although the holotype has been heavily flattened) and of the digitate rather than tubular nature of the vitelline lobes.

Gibson & Bray (1979), although recognizing that the Hypohepaticolinae were close to the Lecithochiriinae, accepted the subfamily on the basis of the filamented eggs and the apparently exceptional site of the worms on the liver; however, in their work of 1986, these authors showed that the body-cavity was not an unusual site for certain species of *Lecithochirium*. Two further species other than the type-species have been attributed to *Hypohepaticola*, i.e. one of these is not a hemiurid and the other lacks filamented eggs. In this work, therefore, *Hypohepaticola* is transferred to the Lecithochiriinae.

Voitrema Yamaguti, 1971, known only from a single specimen from a Fijian ribbon-fish, is here treated as a synonym of *Plerurus* Looss, 1907, species of which occur regularly in *Trichiurus* spp. The genus *Merlucciotrema* Yamaguti, 1971, which was synonymized with *Plerurus* by Gibson & Bray (1979), is recognized here, based on the work of Bray (1996).

In the Aphanurinae, *Neoaphanurus* Tang, Shi & Pan in Tang, Shi, Cao, Guan & Pan, 1983 is considered to be a synonym of *Aphanurus* Looss, 1907, as the authors appear to have confused the seminal vesicle with the uterine seminal receptacle. Bray (1990) synonymized *Minutocauda* Srivastava & Sahai, 1977 with *Aphanurus*. *Machidatrema* León-Règagnon, 1998 was erected as a 'bunocotyline', but it is treated here as a hysterolecithine lecithasterid, following Bray & Cribb (2000a).

Family Hemiuridae Looss, 1899

(Syns Lecithochiriidae Lühe, 1901; Dinuridae Looss, 1907; Bunocotylidae Dollfus, 1950; Elytrophallidae Skrjabin & Guschanskaja, 1954; Biporidae Malhotra, Nanda, Mukherjee, Ghosh, Sukul & Capoor, 1989)

Diagnosis: Body usually small, but fusiform to elongate. Ecsoma present, reduced, vestigial or absent. Body surface smooth or with annular plications; latter occasionally serrate, giving scaly appearance. Presomatic pit or ventrocervical groove occasionally present. Ridges around the body in region of suckers may occur in forms lacking ecsoma. Oral and ventral suckers well developed, usually close together. Pharynx well developed. Oesophagus usually short. 'Drüsenmagen' normally present. Gut-caeca normally terminate blindly, usually within ecsoma or near posterior extremity, rarely form cyclocoel. Testes two, tandem, oblique or symmetrical, pre-ovarian, in hindbody. Seminal vesicle tubular, saccular or constricted into portions, muscular or thin-walled, in fore- or hindbody. Pars prostatica of variable length, usually tubular, but occasionally vesicular, may be linked to seminal vesicle by aglandular duct. Ejaculatory duct, if present, usually short, occasionally long. Sinus-sac usually well developed, occasionally reduced or absent. Prostatic or ejaculatory vesicle occasionally present within sinus-sac. Hermaphroditic duct usually enclosed within sinus-sac. Permanent sinus-organ and genital atrium well developed, small or absent; temporary sinus-organ may form hermaphroditic duct in some cases. Genital pore midventral at level of oral sucker or pharynx. Ovary oval, usually entire, post-testicular. Mehlis' gland postovarian. Laurer's canal and canalicular seminal receptacle absent. Juel's organ and uterine seminal receptacle normally present (absent in Opisthadeninae). Blind seminal receptacle normally absent (present in Opisthadeninae). Uterus coiled mainly in pre- and/or postovarian region of hindbody; few or no coils present in forebody; initially descending into or towards ecsoma and then ascending towards forebody. Eggs numerous, small, embryonated, rarely with polar filament. Vitellarium varies between forms with seven tubular branches (three on one side of body, four on other) and forms with two (occasionally one or three) distinct, oval masses, mainly postovarian. Excretory vesicle Y-shaped; arms united in forebody or not. Parasitic mainly in gut, especially stomach, of marine teleosts, occasionally present in gut of freshwater teleosts and lung of sea-snakes. Type-genus *Hemiurus* Rudolphi, 1809.

Key to subfamilies

1a. Ecsoma present .... 2.
1b. Ecsoma absent [take care with this feature, as it is occasionally reduced in some taxa in subfamilies that have this structure, and sometimes appears to be present as a vestige in others that lack this feature] .... 9.

2a. Ejaculatory (or prostatic) vesicle present within sinus-sac, occasionally partly external [this vesicle should not be confused with a pars prostatica, which is also present] .... 3.
2b. Ejaculatory (or prostatic) vesicle absent .... 4.

3a. Long, convoluted hermaphroditic duct and thin-walled permanent sinus-organ present; seminal vesicle bipartite, anterior part muscular; vitellarium two irregularly oval masses .... Glomericirrinae Yamaguti, 1958.
3b. Hermaphroditic duct relatively straight; permanent sinus-organ absent; seminal vesicle tubular or saccular and partitioned, usually thin-walled; vitellarium seven digitiform to oval lobes or with tendency to form two distinct, often lobed, lateral masses; ecsoma sometimes reduced or apparently absent .... Lecithochiriinae Lühe, 1901.

4a. Sinus-sac absent or poorly developed, when present usually of 'open' type; seminal vesicle entirely or mainly thin-walled, usually constricted into portions; ecsoma sometimes poorly developed .... Plerurinae Gibson & Bray, 1979.
4b. Sinus-sac present, usually well developed, occasionally small; seminal vesicle in various forms; ecsoma usually well developed (but may be withdrawn) .... 5.

5a. Vitellarium two symmetrical to slightly oblique, entire or lobed masses .... 6.
5b. Vitellarium seven distinct oval to tubular lobes .... 8.

6a. Seminal vesicle oval or bipartite, in fore- or hindbody; ecsoma well developed; parasitic in gut of marine teleosts .... 7.
6b. Seminal vesicle tubular, extending well into hindbody; ecsoma reduced; parasitic in lung of sea-snakes .... Pulmoverminae Sandars, 1961.

7a. Body surface smooth; seminal vesicle anterodorsal to ventral sucker, oval, with thick muscular wall; pars prostatica vesicular, with muscular wall; sinus-sac very small .... Lethadeninae Yamaguti, 1971.
7b. Body surface with plications or 'scales'; seminal vesicle in hindbody, oval or bipartite, thin- or partly to entirely thick-walled; pars prostatica tubular; sinus-sac well developed, often tubular .... Hemiurinae Looss, 1899.

8a. Seminal vesicle with thick muscular wall, oval; permanent sinus-organ normally delicate and amuscular .... Elytrophallinae Skrjabin & Guschanskaja, 1954.
8b. Seminal vesicle thin-walled and oval, tubular or constricted into portions; permanent sinus-organ large and muscular, reduced to small papilla or apparently absent .... Dinurinae Looss, 1907.

9a. Uterine seminal receptacle present; vitellarium one or two masses .... 10.
9b. Blind seminal receptacle present; vitellarium two or three masses .... Opisthadeninae Yamaguti, 1970.

10a. Parasites up to 6 mm in length; large concentration of uterine coils between ovary and testes .... Theletrinae Gibson & Bray, 1979.
10b. Parasites small, rarely more than 2 mm in length, commonly less than 1 mm; no large concentration of uterine coils present between ovary and testes .. 11.

11a. Body surface smooth; ridges present around body at level of oral sucker and posterior margin of ventral sucker; major part of uterine field pre-ovarian; vitellarium single .... Bunocotylinae Dollfus, 1950.
11b. Body surface usually plicated, occasionally smooth; no ridges around body at level of suckers; large or major part of uterine field postovarian; vitellarium single or double .... Aphanurinae Skrjabin & Guschanskaja, 1954.

Subfamily Hemiurinae Looss, 1899

(Syn. ?Pseudoprosorchiinae Gupta & Jain, 1992)

Diagnosis: Ecsoma well developed. Body surface plicated or 'scaly' (i.e. with crenulate or serrate plications). Presomatic pit absent. Testes tandem to oblique. Seminal vesicle thin-walled, or partly or slightly muscular, bipartite or oval, in hindbody. Pars prostatica tubular, long; gland-cells occasionally delimited by membrane. Sinus-sac present, often tubular, not enclosing prostatic vesicle. Permanent sinus-organ absent, but hermaphroditic duct may be protruded to form temporary sinus-organ. Genital atrium usually small, but variable in length. Ovary oval. Vitellarium composed of two distinct oval masses, but these may show slight tendency towards lobation in 'three and four' arrangement. Excretory arms united in forebody. Parasitic in stomach of marine teleosts. Type-genus *Hemiurus* Rudolphi, 1809.

Key to genera

1a. Seminal vesicle bipartite .... ***Hemiurus*** **Rudolphi, 1809. (Fig. 37.1)**
(Syns *Apoblema* Dujardin, 1845; *Pronopyge* Looss, 1899; *Metahemiurus* Skrjabin & Guschanskaja, 1954)

Diagnosis: Body surface with normal plications. Seminal vesicle constricted into two portions (? occasionally three), one of which may have thick, muscular wall. Type-species *H. appendiculatus* (Rudolphi, 1802).

1b. Seminal vesicle oval .... ***Parahemiurus*** **Vaz & Pereira, 1930. (Fig. 37.2)**
(Syns *Anahemiurus* Manter, 1947; *Daniella* Sahai & Srivastava, 1977; *Bapatina* Srivastava & Sahai, 1977; *Dentiacetabulum* Sahai & Srivastava, 1977; *Trilecithotrema* Gu & Shen, 1981; *Lecithomonoium* Shen, 1987; ?*Engraulitrema* Gupta & Jain, 1992)

Diagnosis: Body surface plicated; plications occasionally crenulate or serrate ('scaly'). Seminal vesicle oval, with muscular wall of variable thickness. Type-species *P. merus* (Linton, 1910).

Subfamily Aphanurinae Skrjabin & Guschanskaja, 1954

(Syn. Ahemiurinae Chauhan, 1954)

Diagnosis: Body normally small. Ecsoma absent, but vestige may be present. Body surface usually with distinct annular plications, occasionally (?) smooth (some species of *Aphanurus*). Ridges around body at level of suckers absent. Ventral sucker normally in anterior half of body. Oesophagus normally short. Gut-caeca end blindly near posterior extremity. Testes tandem to symmetrical, normally well posterior to ventral sucker, not separated from ovary by large concentration of uterine coils. Seminal vesicle tubular in forebody, or saccular (oval, elongate or bipartite) in hindbody (or at least posterior to middle of ventral sucker). Pars prostatica tubular or vesicular; short or long. Ejaculatory duct long, short or apparently absent. Sinus-sac present, enclosing hermaphroditic duct, or (?) absent. Ovary oval, immediately or almost immediately post-testicular. Blind seminal receptacle absent. Uterine seminal receptacle and Juel's

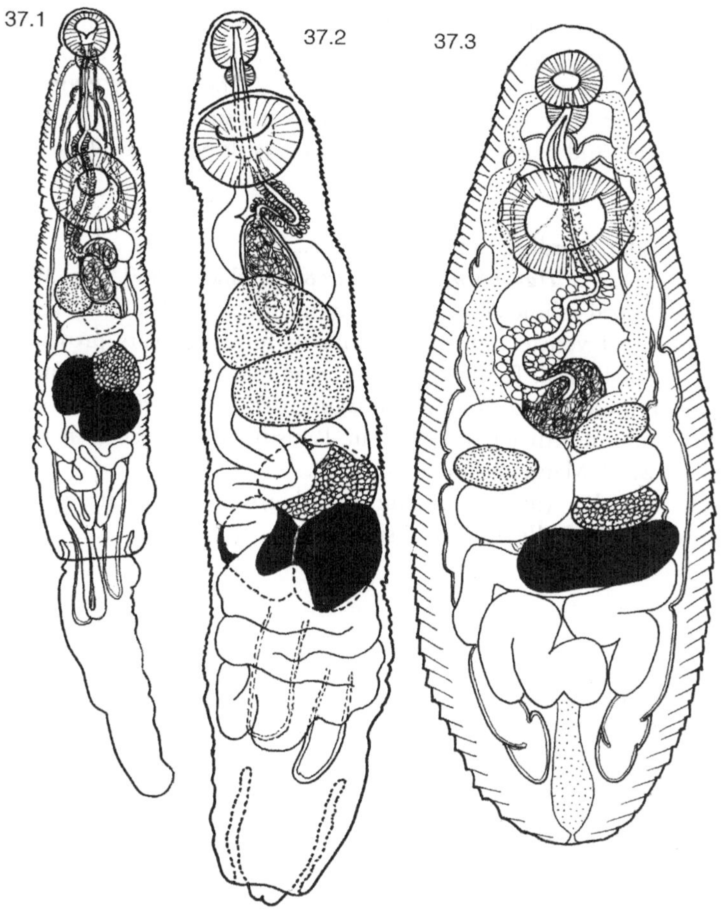

Figs 37.1–37.3.
37.1. *Hemiurus communis* Odhner, 1905. (After Gibson & Bray, 1986.)
37.2. *Parahemiurus merus* (Linton, 1910). (After Bray, 1990.)
37.3. *Aphanurus stossichi* (Monticelli, 1891). (After Looss, 1907.)

organ present. Large or major part of uterine field postovarian. Vitellarium one or two compact masses, usually immediately posterior, occasionally lateral, to ovary. Excretory arms united in forebody; excretory pore often large; actual pore may be withdrawn within vestige of ecsoma. Parasitic mainly in stomach or oesophagus of marine teleosts. Type-genus *Aphanurus* Looss, 1907.

Key to genera

1a. Vitellarium composed of two distinct masses .... **2.**
1b. Vitellarium composed of one distinct mass ....
.... ***Aphanurus* Looss, 1907. (Fig. 37.3)**
(Syns *Chauhanurus* Skrjabin & Guschanskaja, 1954; *Helaphanurus* Slusarski, 1957; *Minutocauda* Srivastava & Sahai, 1977; *Neoaphanurus* Tang, Shi & Pan in Tang, Shi, Cao, Guan & Pan, 1983)

Diagnosis: Body surface normally plicated, occasionally (?) smooth. Small vestige of ecsoma may occur. Testes oblique, occasionally symmetrical or tandem. Seminal vesicle oval to elongate oval, in hindbody; wall may be muscular. Pars prostatica tubular, long. Ejaculatory duct long or short. Sinus-sac present, tubular. Temporary sinus-organ sometimes present as small cone. Vitellarium a single, large, entire or slightly indented, postovarian mass. Parasitic in oesophagus and stomach of essentially marine teleosts from marine and brackish-water environments. Type-species *A. stossichi* (Monticelli, 1891).

2a. Seminal vesicle tubular, winding in forebody .... *Duosphincter* Manter & Pritchard, 1960. (Fig. 37.4)

Diagnosis: Strongly developed sphincter muscles surround apertures of suckers. Testes oblique to tandem. Seminal vesicle tubular; winding in forebody. Pars prostatica tubular, short. Sinus-sac small, oval. Temporary sinus-organ may form. (Small seminal receptacle (? Juel's organ) present, according to Yamaguti, 1970.) Vitellarium two oval masses, oblique to tandem, immediately postovarian. Parasitic in stomach of marine teleosts. Type-species *D. zancli* Manter & Pritchard, 1960.

2b. Seminal vesicle saccular (oval, elongate or bipartite; often attenuated anteriorly), posterior to middle of ventral sucker .... 3.

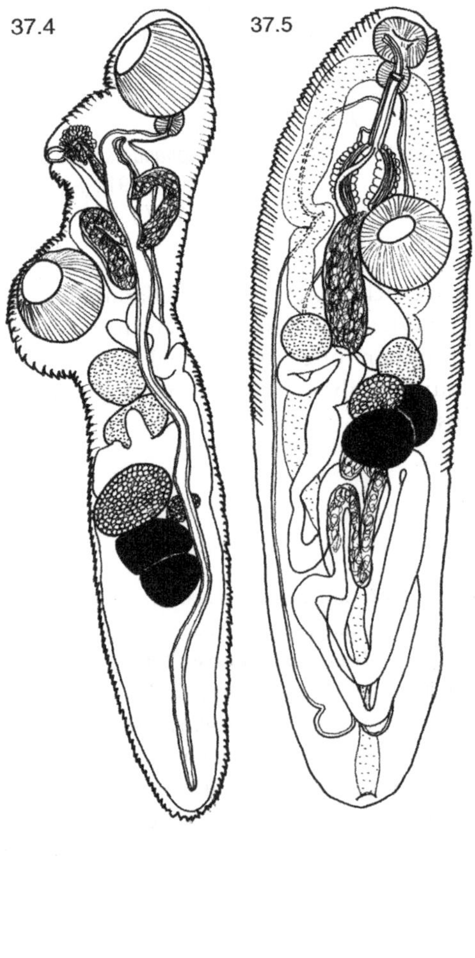

Figs 37.4–37.5.
37.4. *Duosphincter zancli* Manter & Pritchard, 1960. (After Yamaguti, 1970.)
37.5. *Myosaccium ecaude* Montgomery, 1957. (After Yamaguti, 1971.)

3a. Pars prostatica vesicular, with muscular wall .... ***Myosaccium* Montgomery, 1957. (Fig. 37.5)**
(Syn. *Neogenolinea* Siddiqi & Cable, 1960)

Diagnosis: Testes symmetrical to tandem. Seminal vesicle saccular, attenuated anteriorly, at level of posterior margin of ventral sucker. Pars prostatica vesicular, with strong, muscular wall, in forebody. Sinus-sac tubular, enclosing hermaphroditic duct, which may form temporary sinus-organ. Eggs (?) with or without short terminal filament (the structure of collapsed eggs may apparently suggest that a short filament and spine are present). Vitellarium two oval or slightly indented masses, oblique to tandem; one mass usually lateral, other immediately posterolateral or posterior, to ovary. Parasitic in stomach of marine teleosts (Clupeidae). Type-species *M. ecaude* Montgomery, 1957.

3b. Pars prostatica tubular .... **4.**

4a. Seminal vesicle oval; sinus-sac present .... ***Ahemiurus* Chauhan, 1954. (Fig. 37.6)**

Diagnosis: Testes symmetrical to oblique. Seminal vesicle oval, in hindbody. Pars prostatica tubular, long. Ejaculatory duct long. Sinus-sac present, elongate oval. Vitellarium two oval, compact masses, symmetrical, postovarian. Parasitic in stomach of marine teleosts. Type-species *A. karachii* (Srivastava, 1937).

4b. Seminal vesicle apparently bipartite and attenuated anteriorly; sinus-sac apparently absent .... **(?)*Aphanuroides* Nagaty & Abdel-Aal, 1962.[1] (Fig. 37.7)**

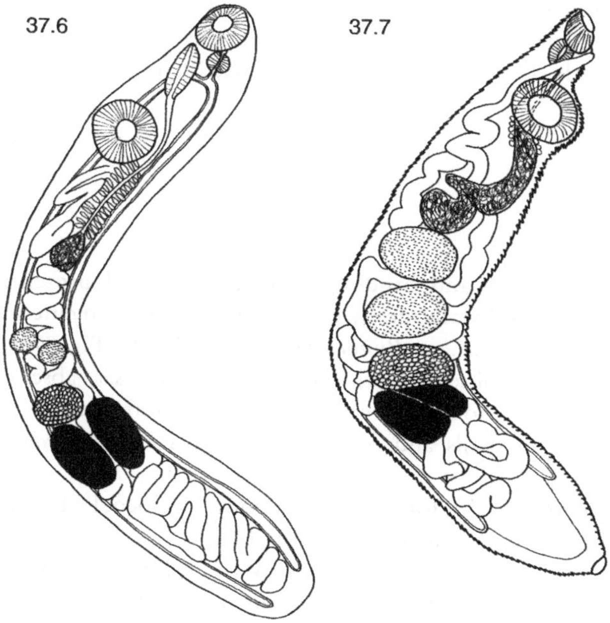

Figs 37.6–37.7.
37.6. *Ahemiurus karachii* (Srivastava, 1937). Reported plications not shown. (Modified after Chauhan, 1954.)
37.7. *Aphanuroides lethrini* Nagaty & Abdel-Aal, 1962. (After Yamaguti, 1971.)

[1] This appears to be a very dubious genus, but the features not described and some of those included prohibit linking it with other taxa.

Diagnosis: [Inadequately described.] Testes tandem. Seminal vesicle saccular, (?) bipartite, attenuated anteriorly, extending between anterior testis and posterior half of ventral sucker. Pars prostatica tubular. [Terminal genitalia not described in detail.] Sinus-sac (?) absent. Short hermaphroditic duct and genital atrium apparently present. [Figures of Nagaty & Abdel-Aal (1962) suggest that temporary sinus-organ may form (?).] Vitellarium two compact masses, symmetrical to oblique, postovarian. Excretory arms (?). Parasitic in gut of marine teleosts. Type-species *A. lethrini* Nagaty & Abdel-Aal, 1962.

Subfamily Bunocotylinae Dollfus, 1950

Diagnosis: Body small. Ecsoma absent, although vestige may be present. Body surface smooth. Ridges (flanges) present around body at level of oral sucker and posterior margin of ventral sucker; additional ridge often present close to posterior extremity. Transverse septate partitions of body may occur. Ventral sucker in anterior half of body. Oesophagus normally short. Gut-caeca end blindly or form cyclocoel. Testes tandem to oblique, not separated from ovary by large concentration of uterine coils. Seminal vesicle saccular, oval to elongate, in forebody or dorsal to ventral sucker. Pars prostatica short, tubular or vesicular. Sinus-sac absent or small and tubular to oval. Short hermaphroditic duct may extend to form temporary sinus-organ. Genital atrium absent or small. Ovary oval. Uterine seminal receptacle present (?). Juel's organ not reported. Blind seminal receptacle absent. Vitellarium a single, unlobed mass, immediately postovarian. Excretory vesicle expanded distally; arms united in forebody; pore wide. Parasitic in gut of freshwater and euryhaline teleosts, occasionally progenetic in snails and copepods. Type-genus *Bunocotyle* Odhner, 1928.

Key to genera

1a. Transverse fibrous septa in fore- and hindbody absent; cyclocoel present; sinus-sac absent; parasitic in freshwater or euryhaline teleosts, or progenetic in snails and copepods .... *Bunocotyle* Odhner, 1928. (Fig. 37.8)

Diagnosis: Transverse fibrous septa absent. Cyclocoel present. Sinus-sac absent. Uterine seminal receptacle (?) presumed to be present. Parasitic in gut of freshwater or euryhaline teleosts, or progenetic in snails and copepods. Type-species *B. cingulata* Odhner, 1928.

1b. Transverse fibrous septa present in fore- and hindbody; cyclocoel absent; sinus-sac present; parasitic in euryhaline teleosts (mugilids) .... *Saturnius* Manter, 1969. (Fig. 37.9)

Diagnosis: Small papillae or corrugations may be associated with suckers. Internal transverse, fibrous septa present in fore- and hindbody. Gut-caeca end blindly. Sinus-sac present, contains eversible hermaphroditic duct. Parasitic under lining of stomach of euryhaline teleosts (mugilids). Type-species *S. segmentatus* Manter, 1969.

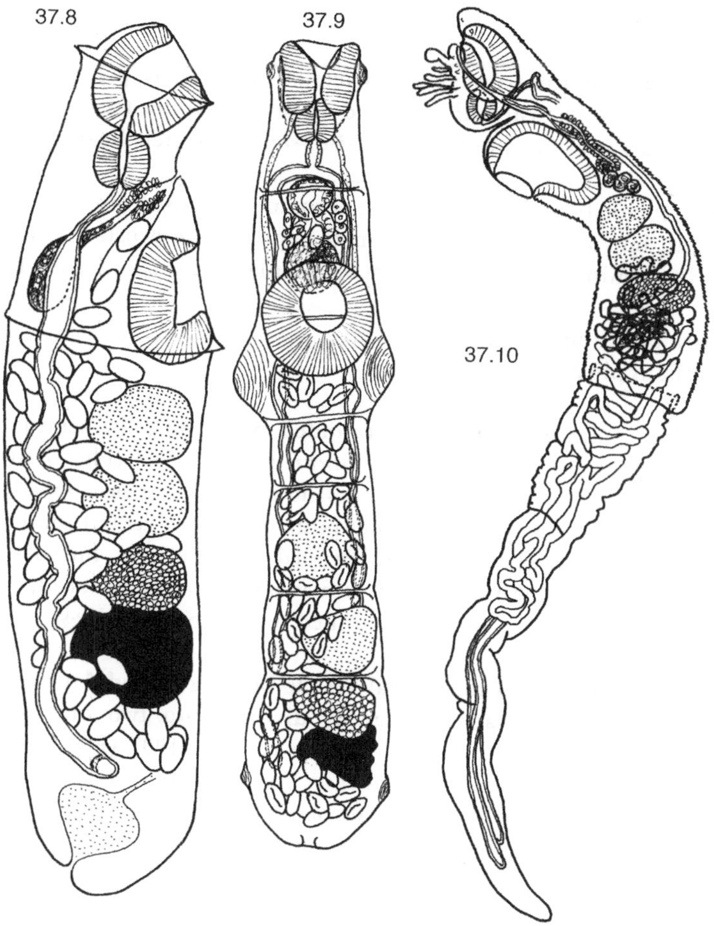

Figs 37.8–37.10.
37.8. *Bunocotyle cingulata* Odhner, 1928. (After Odhner, 1928.)
37.9. *Saturnius papernai* Overstreet, 1977. (After Dimitrov *et al.*, 1998.)
37.10. *Dinurus barbatus* (Cohn, 1902). The digitiform projections on the forebody are not features of other species in the genus. (After Gibson, 1976.)

Subfamily Dinurinae Looss, 1907

(Syn. Stomachicolinae Yamaguti, 1958)

Diagnosis: Ecsoma well developed, occasionally large. Body surface plicated or smooth (apparently occasionally striated). Presomatic pit absent. Testes symmetrical to tandem, usually oblique. Seminal vesicle thin-walled, oval to tubular; may be constricted into two to four portions, in forebody, dorsal to ventral sucker or in hindbody. Pars prostatica tubular or vesicular, short or long, may be linked to seminal vesicle by aglandular duct. Sinus-sac present, small or large, usually oval, not enclosing prostatic vesicle. Permanent sinus-organ large and muscular, reduced to small papilla or apparently absent. Genital atrium usually well developed, deep or shallow (often depending upon contraction). Ovary usually oval, occasionally reniform or lobed. Terminal portion of uterus may or may not form distinct vesicle just outside sinus-sac. Vitellarium normally seven tubular lobes; three on one side, four on other. Excretory arms united or not united in forebody. Normally parasitic in stomach of marine teleosts. Type-genus *Dinurus* Looss, 1907.

Key to genera

1a. **Seminal vesicle usually (but not always) constricted into portions; permanent sinus-organ present, but occasionally reduced to small papilla (sectioning usually required); pars prostatica usually linked to seminal vesicle by distinct aglandular duct .... 2.**

1b. **Seminal vesicle not constricted into portions; permanent sinus-organ usually apparently absent, but may be present as small papilla; pars prostatica not normally linked to seminal vesicle by distinct aglandular duct .... 6.**

2a. **Body surface with plications .... 3.**

2b. **Body surface without plications .... 4.**

3a. **Pars prostatica long, may be sparsely surrounded by gland-cells; seminal vesicle trilocular .... *Dinurus* Looss, 1907. (Fig. 37.10)**

Diagnosis: Body surface with plications. Seminal vesicle trilocular, or occasionally quadrilocular, in anterior hindbody or occasionally posterodorsal to ventral sucker. Pars prostatica long, may be densely or sparsely invested by gland-cells, linked to seminal vesicle by aglandular duct. Sinus-sac and permanent sinus-organ present, of variable size. Ovary oval. Excretory arms not united in forebody. Type-species *D. tornatus* (Rudolphi, 1819).

3b. **Pars prostatica short, connected to seminal vesicle by long aglandular duct; seminal vesicle variable, tubular, saccular or divided into two or three sections .... *Ectenurus* Looss, 1907. (Fig. 37.11)**

(Syns *Magnacetabulum* Yamaguti, 1934; *Clupenurus* Srivastava, 1935; *Parectenurus* Manter, 1947; *Prosterrhurus* Fischthal & Kuntz, 1963)

Diagnosis: Body surface with plications (may be missing in poorly preserved material). Seminal vesicle saccular, tubular or divided into two or three sections, posterodorsal to ventral sucker or in anterior hindbody. Pars prostatica short (? or missing), connected to seminal vesicle by long, aglandular duct. Sinus-sac and permanent sinus-organ present, small. Ovary oval. Excretory arms not united in forebody. Type-species *E. lepidus* Looss, 1907.

4a. **Glandular region of pars prostatica short; excretory arms unite in forebody; distal end of uterus often vesicular .... 5.**

4b. **Glandular region of pars prostatica long; excretory arms not united in forebody; distal end of uterus not vesicular .... *Paradinurus* Vigueras, 1958. (Fig. 37.12)**

Diagnosis: Body surface smooth. Seminal vesicle trilocular, at level of ventral sucker. Pars prostatica tubular, long, densely invested with gland-cells, connected to seminal vesicle by short, aglandular duct. Sinus-sac oval, thick-walled, relatively large. Permanent sinus-organ well developed. Ovary oval. Excretory arms not united in forebody. Type-species *P. manteri* Vigueras, 1958.

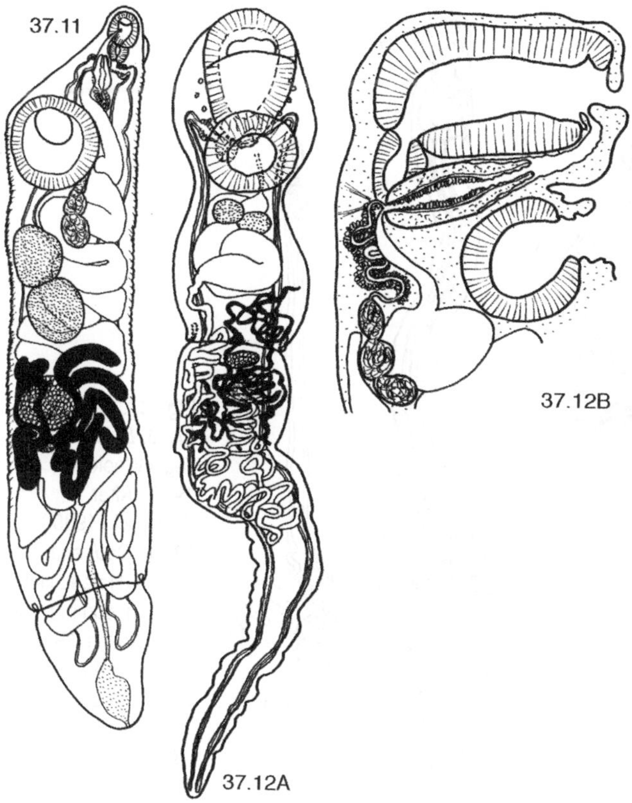

Figs 37.11–37.12.
37.11. *Ectenurus lepidus* Looss, 1907. (After Gibson & Bray, 1986.)
37.12. *Paradinurus manteri* Vigueras, 1958. A. Ventral view. B. Sagittal view showing terminal genitalia. (After Gibson, 1976.)

5a. Aglandular region of pars prostatica long; seminal vesicle dorsal or postero-dorsal to ventral sucker; sinus-sac usually dilated proximally .... *Erilepturus* Woolcock, 1935. (Fig. 37.13)

(Syns *Uterovesiculurus* Skrjabin & Guschanskaja, 1954; *Exoticotyle* Sahai & Srivastava, 1978; *Clupenuroides* Al-Yamani & Nahhas, 1981; *Chrysophrytrema* Gupta & Jain, 1992; *Neoerilepturus* Shalaby & Hassanine, 1996)

Diagnosis: Body surface smooth (or finely transversely striated). Seminal vesicle variable, (?) oval, tubular to trilocular in the same species, dorsal or posterodorsal to ventral sucker. Pars prostatica short, tubular; connected to seminal vesicle by long, aglandular duct. Sinus-sac present, dilated proximally (? or tubular). Permanent sinus-organ present, small. Ovary oval. Distal extremity of uterus (outside sinus-sac) may be vesicular. Excretory arms united in forebody. Type-species *E. tiegsi* Woolcock, 1935.

5b. Aglandular region of pars prostatica short; seminal vesicle in forebody; sinus-sac elongate-oval .... *Qadriana* Bilqees, 1971. (Fig. 37.14)

(Syns *Atheria* Hafeezullah, 1975; *Qasimmetra* Sahai & Srivastava, 1978)

Diagnosis: Body stout. Body surface smooth. Pre-oral lobe large. Seminal vesicle saccular, in forebody. Pars prostatica short, tubular, connected to seminal vesicle by short, aglandular duct. Sinus-sac present, elongate-oval. Permanent sinus-organ present, long, muscular. Ovary oval. Distal extremity of uterus may be vesicular. Excretory arms united in forebody. Type-species *Q. fusiformis* Bilqees, 1971.

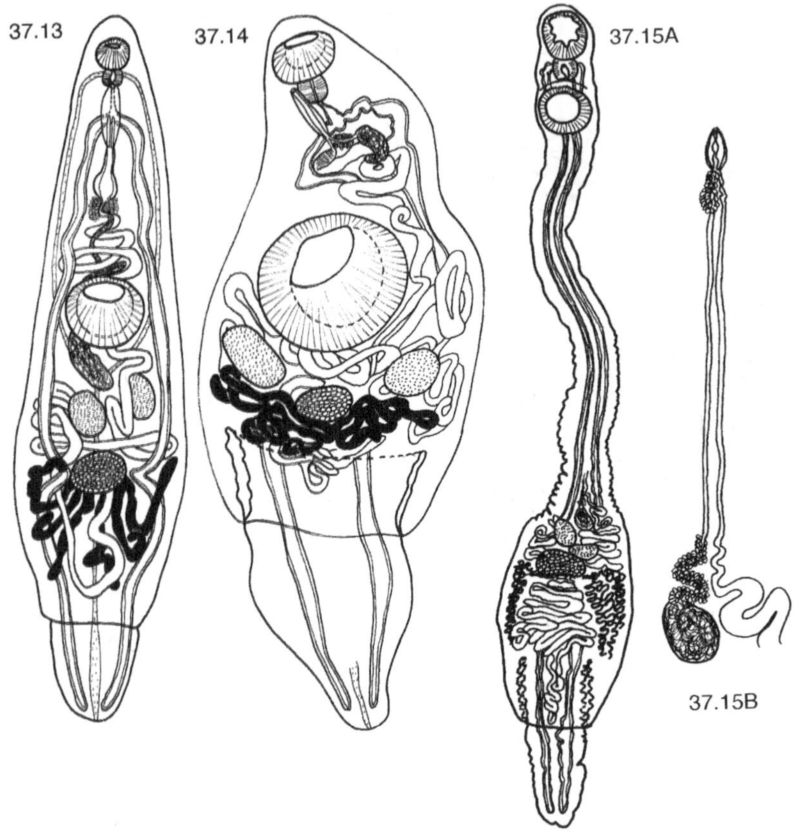

Figs 37.13–37.15. **37.13.** *Erilepturus tiegsi* Woolcock, 1935. (After Woolcock, 1935.) **37.14.** *Qadriana zakiae* (Hafeezullah, 1975) n. comb. (After Hafeezullah, 1975; as *Atheria zakiae*.) **37.15.** *Mecoderus oligoplitis* Manter, 1940. A. Ventral view. B. Terminal genitalia. (After Manter, 1940.)

6a. Anterior part of hindbody greatly attenuated; pars prostatica in two parts separated by long aglandular duct .... *Mecoderus* Manter, 1940. (Fig. 37.15)

Diagnosis: Anterior part of hindbody attenuated. Body surface smooth. Seminal vesicle saccular, well back in hindbody, just anterior to testes. Pars prostatica in two parts; one anterior to and other posterior to attenuated part of body and connected by long aglandular duct, not connected to seminal vesicle by distinct aglandular duct. Sinus-sac present, small, oval. Permanent sinus-organ apparently absent. Ovary oval. Excretory arms united in forebody. Type-species *M. oligoplitis* Manter, 1940.

6b. Anterior part of hindbody with normal configuration; pars prostatica undivided .... 7.

7a. Ecsoma large; seminal vesicle oval or elongate-oval .... 8.

7b. Ecsoma normal; seminal vesicle tubular and sinuous .... *Tubulovesicula* Yamaguti, 1934. (Fig. 37.16)

(Syns *Lecithurus* Pigulewsky, 1938; *Transversolecithus* Sahai & Srivastava, 1978)

Diagnosis: Body surface smooth. Seminal vesicle tubular, sinuous, in hindbody. Pars prostatica with long, wide lumen, sinuous or straight, not connected to seminal vesicle by distinct aglandular duct. Sinus-sac present, oval. Permanent sinus-organ normally

37.16 37.17

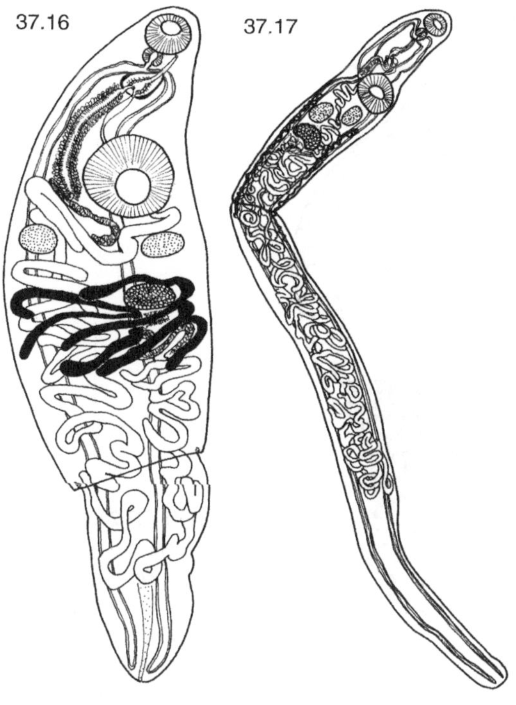

Figs 37.16–37.17.
37.16. *Tubulovesicula spari* Yamaguti (1934). (After Yamaguti, 1934.)
37.17. *Allostomachicola secundus* (Srivastava, 1937). (After Srivastava, 1939.)

absent, but may occur as small papilla. Ovary oval to round. Vitelline lobes tubular, but often stout. Excretory arms united in forebody. Parasitic in stomach, body-cavity and body tissues of marine teleosts (also reported from intestine of sea-snake). Type-species *T. spari* Yamaguti, 1934.

8a. Seminal vesicle in forebody .... ***Allostomachicola* Yamaguti, 1958. (Fig. 37.17)**

Diagnosis: Ecsoma enormous. Body surface smooth. Seminal vesicle elongate-oval, in forebody. Pars prostatica short, vesicular (? also reported as long, tubular), not connected to seminal vesicle by distinct aglandular duct. Sinus-sac present, small, oval. Permanent sinus-organ apparently absent. Ovary reniform, may be indistinctly lobed. Majority of uterus within ecsoma, normally fills more than half of ecsoma. Excretory arms (?) united in forebody. Type-species *A. secundus* (Srivastava, 1937).

8b. Seminal vesicle in hindbody .... ***Stomachicola* Yamaguti, 1934. (Fig. 37.18)**

(Syns *Pseudostomachicola* Skrjabin & Guschanskaja, 1954; *Acerointestinecola* Jahan, 1970; ?*Cameronia* Bilqees, 1971; *Segmentatum* Bilqees, 1971; *Cestodera* Bilqees, 1971; *Indostomachicola Gupta* & Sharma, 1973; *Linguastomachicola* Srivastava & Sahai, 1978)

Diagnosis: Body may be very long. Ecsoma enormous. Body surface smooth. (?Linguiform projection may arise from lumen of oral sucker (rare).) Seminal vesicle oval, in hindbody. Pars prostatica tubular, long, sinuous, not connected to seminal vesicle by distinct aglandular duct; external gland-cells may not be evenly distributed throughout length. Sinus-sac present, small, oval. Permanent sinus-organ absent or

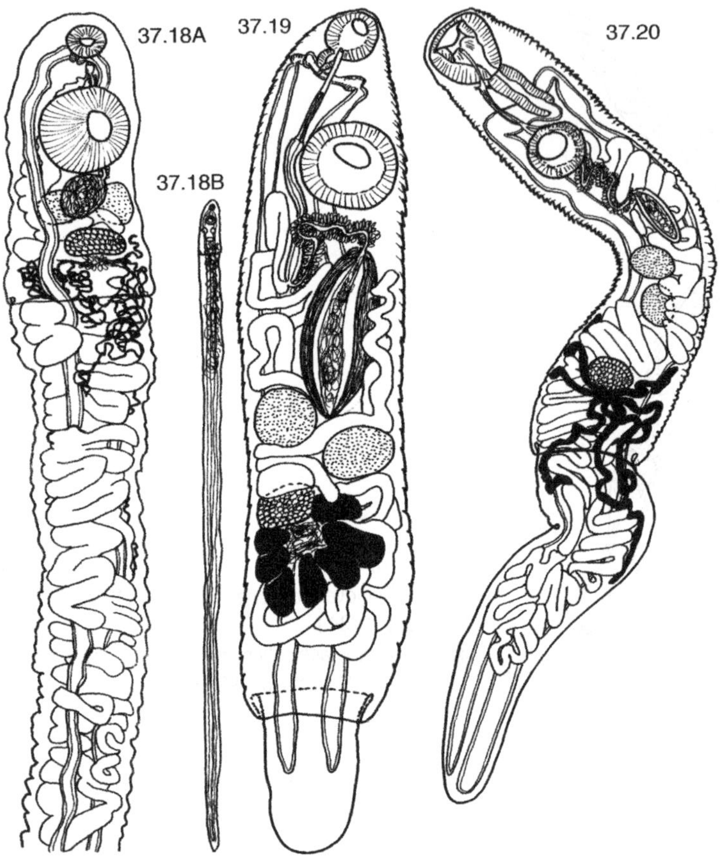

Figs 37.18–37.20.
37.18. *Stomachicola muraenesocis* Yamaguti, 1934. A. Anterior region. B. Entire worm. (After Yamaguti, 1934.)
37.19. *Elytrophalloides oatesi* (Leiper & Atkinson, 1914). (After Gibson, 1976.)
37.20. *Lecithocladium excisum* (Rudolphi, 1819). (After Gibson & Bray, 1986.)

reduced to rudiment. Ovary oval to reniform. Majority of uterine coils within ecsoma, normally fill less than half of ecsoma. Excretory arms united in forebody. Type-species *S. muraenesocis* Yamaguti, 1934.

Subfamily Elytrophallinae Skrjabin & Guschanskaja, 1954

(Syns Musculovesiculinae Skrjabin & Guschanskaja, 1954; Johniophyllinae Skrjabin & Guschanskaja, 1954)

Diagnosis: Ecsoma well developed. Body surface smooth or plicated. Presomatic pit absent, but ventrocervical groove often present. Testes tandem to symmetrical, usually oblique. Seminal vesicle with exceptionally thick, muscular wall, oval, not constricted into portions, in forebody, dorsal to ventral sucker or in hindbody. Pars prostatica tubular, long or short, usually linked to seminal vesicle by short, aglandular duct. Sinus-sac present, commonly tubular, long, not enclosing ejaculatory or prostatic vesicle. Sinus-organ usually well developed, but delicate and amuscular. Genital atrium usually deep (depending upon contraction). Ovary oval. Eggs rarely filamented. Vitellarium seven tubular to tear-shaped lobes; three on one side, four on the other, which may form rosette. Excretory arms united in forebody. Parasitic mainly in stomach of marine teleosts. Type-genus *Elytrophallus* Manter, 1940.

Key to genera

1a. Body surface with plications .... **2.**
1b. Body surface without plications .... **3.**

2a. Sinus-sac long and narrow, reaching to level of seminal vesicle; vitelline lobes tear-shaped .... ***Elytrophalloides* Szidat, 1955. (Fig. 37.19)**
Diagnosis: Body surface with plications. Occasionally with pair of symmetrical muscular pads laterally in forebody. Seminal vesicle large, reaching back to level of testes. Pars prostatica sinuous, in hindbody. Sinus-sac long, normally reaching back to level of seminal vesicle. Vitelline lobes tear-shaped. Parasitic in stomach of marine teleosts (in southern hemisphere). Type-species *E. oatesi* (Leiper & Atkinson, 1914) (syn. *E. merluccii* Szidat, 1955 - type by original designation).
2b. Sinus-sac relatively long, but not reaching to level of seminal vesicle; vitelline lobes tubular .... ***Lecithocladium* Lühe, 1901. (Fig. 37.20)**
(Syns *Magnapharyngium* Bilqees, 1971; *Colletostomum* Sahai & Srivastava, 1978; *Clefticolletta* Sahai & Srivastava, 1978; *Bengalotrema* Malhotra, Nanda, Mukherjee, Ghosh, Sukul & Capoor, 1989)
Diagnosis: Body surface with plications. Oral sucker often funnel-shaped and larger than ventral sucker. Pharynx elongate. Seminal vesicle large, in hindbody. Pars prostatica long and sinuous, mainly or entirely in hindbody. Sinus-sac tubular, narrow, not reaching level of seminal vesicle and usually entirely or mainly in forebody. Vitelline lobes long and tubular. Parasitic in stomach of marine teleosts. Type-species *L. excisum* (Rudolphi, 1819).

3a. Seminal vesicle in forebody; eggs may be filamented .... ***Musculovesicula* Yamaguti, 1940. (Fig. 37.21)**
Diagnosis: Body surface smooth. Seminal vesicle elongate, in forebody or overlapping ventral sucker. Pars prostatica short and indistinct, in forebody. Sinus-sac elliptical to pyriform, short. Vitelline lobes digitiform. Eggs may be filamented. Parasitic in stomach of marine teleosts (eels). Type-species *M. gymnothoracis* Yamaguti, 1940.
3b. Seminal vesicle in hindbody; eggs not filamented .... ***Elytrophallus* Manter, 1940. (Fig. 37.22)**
(Syn. *Johniophyllum* Skrjabin & Guschanskaja, 1954)
Diagnosis: Body surface smooth. Seminal vesicle small to large, in hindbody. Pars prostatica sinuous, mainly or entirely in hindbody. Sinus-sac long, tubular, thick-walled. Vitelline lobes tear-shaped to digitiform. Parasitic in stomach of marine teleosts. Type-species *E. mexicanus* Manter, 1940.

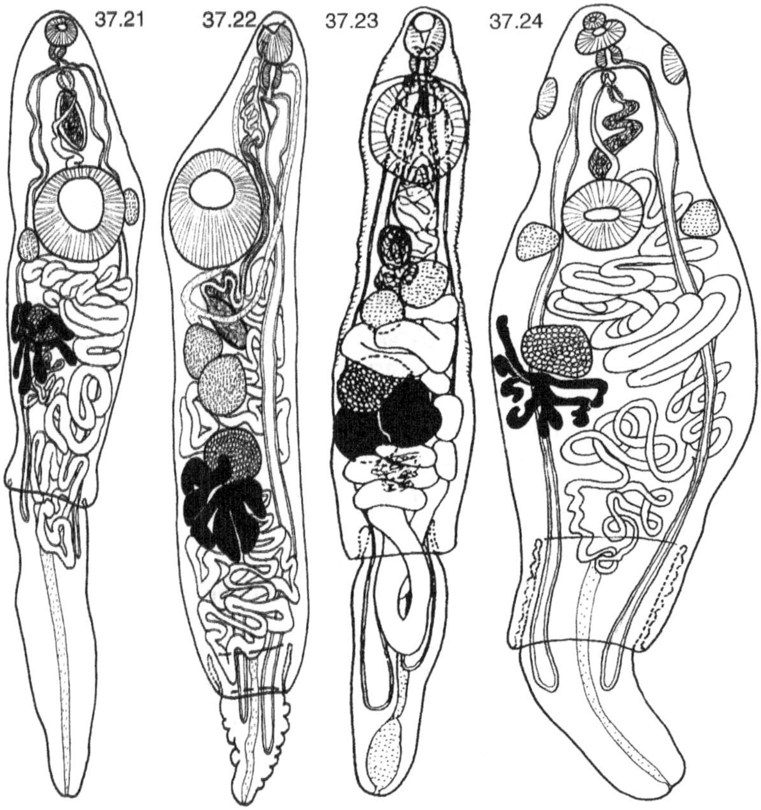

Figs 37.21–37.24. **37.21.** *Musculovesicula gymnothoracis* Yamaguti, 1940. (After Yamaguti, 1940.) **37.22.** *Elytrophallus mexicanus* Manter, 1940. (After Yamaguti, 1971.) **37.23.** *Glomericirrus macrouri* (Gaevskaja, 1975). (After Gibson & Bray, 1986.) **37.24.** *Tricotyledonia genypteri* Fyfe, 1954. (After Fyfe, 1954.)

Subfamily Glomericirrinae Yamaguti, 1958

Diagnosis: Ecsoma well developed. Body surface plicated. Presomatic pit absent. Testes oblique to tandem. Seminal vesicle bipartite, in hindbody or dorsal to ventral sucker; both parts globular to spindle-shaped; anterior part muscular. Pars prostatica tubular, short, linked to seminal vesicle by aglandular duct. Claviform sinus-sac present, in forebody or reaching into hindbody, enclosing prostatic vesicle. Hermaphroditic duct convoluted. Sinus-organ present, amuscular, long, convoluted. Genital atrium well developed. Vitellarium two irregularly oval, symmetrical masses. Excretory arms united in forebody. Parasitic in stomach of marine teleosts. Type-genus *Glomericirrus* Yamaguti, 1937.

Genus *Glomericirrus* Yamaguti, 1937 (Fig. 37.23)

Diagnosis: With characters of subfamily. Type-species *G. amadai* Yamaguti, 1937.

Subfamily Lecithochiriinae Lühe, 1901

(Syns Sterrhurinae Looss, 1907; Subfamily Hypohepaticolinae Skrjabin & Guschanskaja, 1954; Brachyphallinae Skrjabin & Guschanskaja, 1955; Trithelaminae Yeh, 1955; Tricotyledoniinae Skrjabin & Guschanskaja, 1957; Dissosaccinae Yamaguti, 1958; Umatrematinae Srivastava & Sahai, 1978)

Diagnosis: Ecsoma usually well developed, occasionally reduced. Body surface usually smooth, but occasionally plicated or rugate. Muscular 'shoulder-pads' present or absent. Presomatic pit and ventrocervical groove present or absent. Testes tandem to symmetrical, usually oblique. Seminal vesicle usually elongate, occasionally globular; elongate forms constricted into two portions, which are occasionally separated by duct, or tubular and convoluted; in bipartite forms anterior half may have thicker wall; normally in forebody, but may extend into hindbody. Pars prostatica short, vesicular or tubular, may extend slightly into base of sinus-sac, linked to seminal vesicle by short, aglandular duct. Sinus-sac present, rarely of 'open' type, enclosing distinct ejaculatory or prostatic vesicle and metraterm. Permanent sinus-organ absent. Genital atrium usually small or absent, occasionally well developed. Ovary oval. Uterus mainly pre-ovarian or roughly equally distributed in pre- and postovarian fields. Eggs rarely filamented. Vitellarium seven digitiform to oval lobes in lateral groups of three and four, or with tendency to become two distinct lateral masses, which often exhibit three and four lobes. Excretory arms united in forebody. Typically parasitic in gut (normally stomach, also rarely reported from intestine) of marine, rarely freshwater, teleosts, occasionally present as adult in body-cavity. Type-genus *Lecithochirium* Lühe, 1901.

Key to genera

1a. Vitellarium seven distinct oval to digitiform lobes .... 2.
1b. Vitellarium two entire masses which may be indistinctly three- and four-lobed .... 7.

2a. Large, muscular 'shoulder-pads' present .... 3.
2b. Large, muscular 'shoulder-pads' absent .... 4.

3a. Small accessory sucker present anterior to oral sucker .... *Tricotyledonia* Fyfe, 1954. (Fig. 37.24)
(Syn. *Grassitrema* Yeh, 1955)

Diagnosis: Body surface smooth. Muscular 'shoulder-pads' present. Presomatic pit absent. Pre-oral accessory sucker present. Seminal vesicle bipartite; posterior part elongate; anterior to posterior margin of ventral sucker. Pars prostatica vesicular, partly enclosed by sinus-sac, leads into small, aglandular ejaculatory vesicle, connected to seminal vesicle by short, aglandular duct. Vitellarium seven digitiform lobes. Parasitic in stomach of marine teleosts. Type-species *T. genypteri* Fyfe, 1954.

3b. Small accessory sucker absent .... *Cyatholecithochirium* Yamaguti, 1970. (Fig. 37.25)

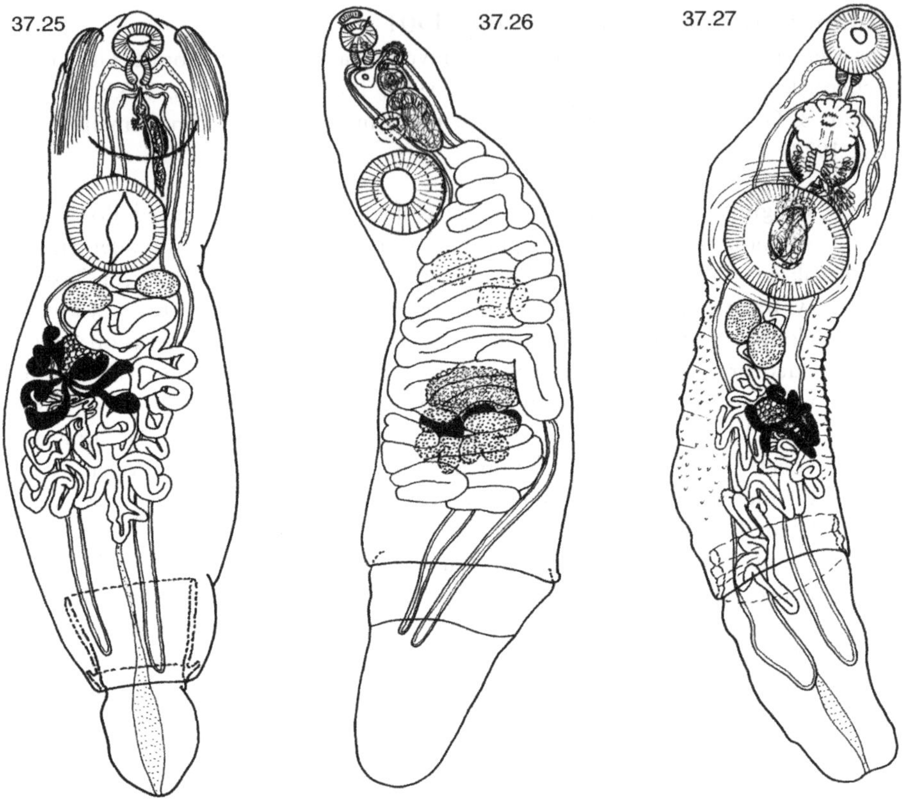

Figs 37.25–37.27.
37.25. *Cyatholecithochirium gymnothoracis* Yamaguti, 1970. (After Yamaguti, 1970.)
37.26. *Catarinatrema verrucosum* Freitas & Santos, 1971. (Modified after Freitas & Santos, 1971.)
37.27. *Plicatrium lycodontis* (Myers & Wolfgang, 1953). (After Yamaguti, 1971.)

Diagnosis: Body surface smooth. Muscular 'shoulder-pads' present. Pre-oral accessory sucker absent. Seminal vesicle bipartite; anterior part with thick wall, in forebody. Pars prostatica may be partly within sinus-sac. Vitellarium seven digitiform lobes in two groups of three and four. Parasitic in stomach of marine teleosts. Type-species *C. gymnothoracis* Yamaguti, 1970.

4a. Small muscular pad present anterior to oral sucker .... ***Catarinatrema*** **Teixeira de Freitas & Santos, 1971. (Fig. 37.26)**

Diagnosis: May bear papillae on ecsoma. Presomatic pit present. Muscular pad present anterior to oral sucker. Seminal vesicle bipartite, in forebody. Pars prostatica tubular. Vitellarium two masses of three and four short, digitiform lobes. Parasitic in stomach and intestine of marine teleosts. Type-species *C. verrucosum* Teixeira de Freitas & Santos, 1971.

4b. Only pre-oral lobe present anterior to oral sucker .... **5.**

5a. Large, eversible genital atrium present .... ***Plicatrium*** **Manter & Pritchard, 1960. (Fig. 37.27)**

Diagnosis: Papillae may occur on body surface. Presomatic pit absent. Seminal vesicle bipartite, anterodorsal to ventral sucker. Pars prostatica tubular. Large, eversible genital atrium present, wrinkled or convoluted when everted. Vitellarium seven digitiform lobes. Parasitic in intestine of marine teleosts. Type-species *P. lycodontis* (Myers & Wolfgang, 1953).

5b. Normal small genital atrium present .... **6.**

6a. Eggs lacking polar filament .... ***Lecithochirium* Lühe, 1901. (Fig. 37.28)**
(Syns *Sterrhurus* Looss, 1907; *Ceratotrema* Jones, 1933; *Jajonetta* Jones, 1933; *Separogermiductus* Skrjabin & Guschanskaja, 1955; *Magniscyphus* Reid, Coil & Kuntz, 1965; *Neohysterolecitha* Ahmad, 1977; *Anaplerurus* Srivastava & Sahai, 1978; *Umatrema* Srivastava & Sahai, 1978)

Diagnosis: Ecsoma well or poorly developed. Body surface smooth. Pre-oral lobe rarely with two lateral knobs. Presomatic pit and/or ventrocervical groove often present. Seminal vesicle bipartite, tripartite or occasionally coiled, in forebody. Pars prostatica tubular, with wide lumen, to vesicular. Short, narrow extension of pars prostatica and/or ejaculatory duct may be present within sinus-sac. Ejaculatory (or prostatic) vesicle linked posteriorly to anterodorsally with pars prostatica or ejaculatory duct. Temporary sinus-organ may form. Vitellarium two lateral masses, usually divided into three and four oval to digitiform lobes. Parasitic in stomach of

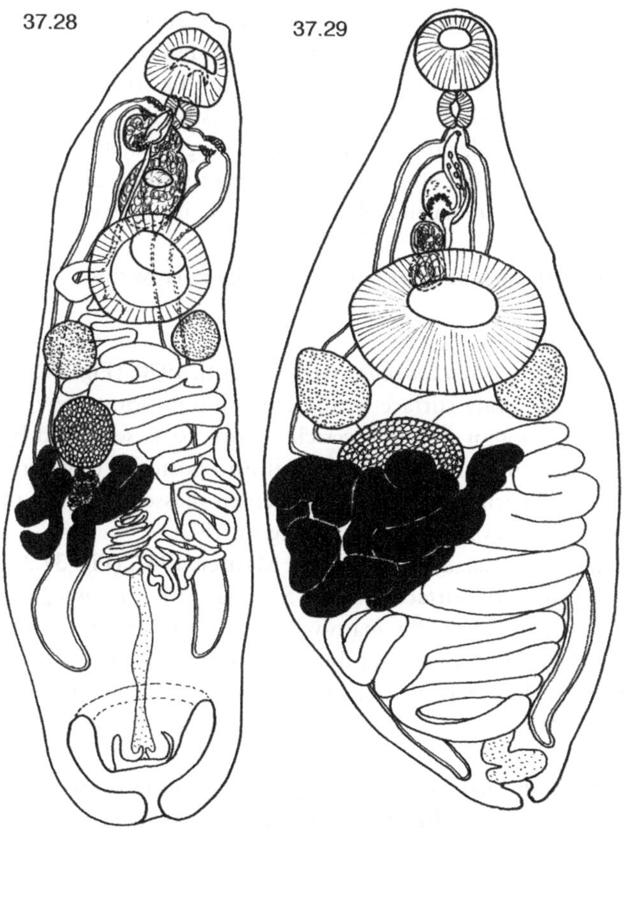

Figs 37.28–37.29.
37.28. *Lecithochirium rufoviride* (Rudolphi, 1819). (After Gibson & Bray, 1986.)
37.29. *Hypohepaticola callionymi* Yamaguti, 1934. Gibson & Bray (1979) reported a withdrawn ecsoma in the type-material. (After Yamaguti, 1934.)

marine teleosts; also recorded from body-cavity, intestine, hepatic ducts and gills of marine teleosts. Type-species *L. rufoviride* (Rudolphi, 1819).

6b. Eggs with polar filament .... *Hypohepaticola* Yamaguti, 1934. (Fig. 37.29)

Diagnosis: Body spindle-shaped. Ecsoma reduced; usually withdrawn. Body surface smooth. Presomatic pit absent. Testes symmetrical at level of middle or posterior margin of ventral sucker. Seminal vesicle anterior or anterodorsal to ventral sucker, constricted into two portions, elongate saccular, may be sinuous. Pars prostatica short, slightly vesicular, may be linked to seminal vesicle by short, aglandular duct. Sinus-sac present, oval, enclosing prostatic vesicle, part of metraterm and hermaphroditic duct. Small sinus-organ may form. Genital atrium present. Ovary oval. Much of uterus postovarian. Eggs with long, polar filament. Vitellarium seven digitiform lobes (three on one side, four on the other), forming postovarian rosette. Excretory arms united in forebody. Parasitic under connective tissue membrane of liver and in gut (stomach) of marine teleosts. Type-species *H. callionymi* Yamaguti, 1934.

7a. Seminal vesicle reaches into hindbody, tubular or composed of two parts separated by narrow duct .... 8.

7b. Seminal vesicle in forebody, tubular, subglobular or bipartite; if latter, parts not separated by duct .... 10.

8a. Seminal vesicle tubular, reaching back to anterior testis .... *Boreascotia* Bray & Zdzitowiecki, 2000. (Fig. 37.30)

Diagnosis: Ecsoma not large (withdrawn in known specimens). Body surface smooth. Presomatic pit absent. Seminal vesicle long, thin-walled, tubular but narrower distally, convoluted, reaching back into hindbody to level of anterior testis. Pars prostatica tubular. Sinus-sac small, oval, thin-walled, not completely enclosing prostatic vesicle. Vitellarium two symmetrical lateral masses, almost entire to slightly three- and four-lobed. In gut (?intestine) of sub-Antarctic marine teleosts. Type-species *B. megavesicula* Bray & Zdzitowiecki, 2000.

8b. Seminal vesicle bipartite, with two saccular parts linked by narrow duct ... 9.

9a. Body surface plicated anteriorly .... *Pseudodinosoma* Yamaguti, 1970. (Fig. 37.31)

Diagnosis: Ecsoma well developed. Body surface with crenulate plications (? giving 'scaly' appearance). Presomatic pit absent. Seminal vesicle in two parts separated by narrow duct; one part (convoluted) anterior and other (claviform) posterior to ventral sucker. Pars prostatica tubular. Vitellarium two slightly indented masses. Parasitic in stomach of marine teleosts. Type-species *P. macrorchis* Yamaguti, 1970.

9b. Body surface smooth .... *Dissosaccus* Manter, 1947. (Fig. 37.32)

Diagnosis: Ecsoma well developed. Body surface smooth. Seminal vesicle in two parts separated by narrow duct; one part normally mainly anterior and other mainly posterior to ventral sucker. Pars prostatica (?) tubular. Vitellarium two slightly indented masses. Parasitic in stomach of marine teleosts. Type-species *D. laevis* (Linton, 1898).

10a. Seminal vesicle bipartite; body surface plicated; deep presomatic pit present; ecsoma not reduced .... *Brachyphallus* Odhner, 1905. (Fig. 37.33)

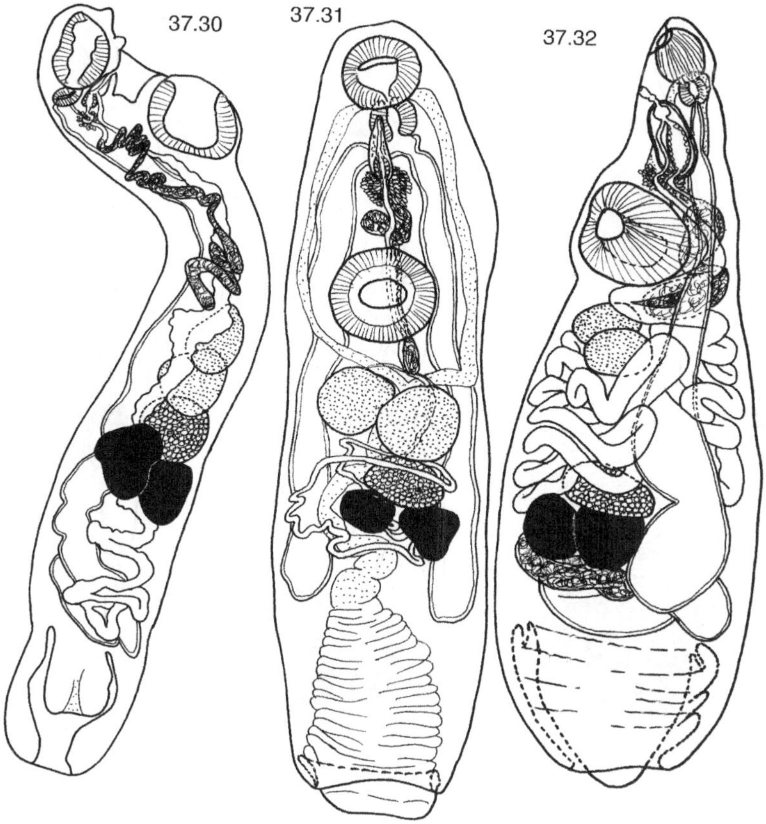

Figs 37.30–37.32. **37.30.** *Boreascotia megavesicula* Bray & Zdzitowiecki, 2000. (After Bray & Zdzitowiecki, 2000.) **37.31.** *Pseudodinosoma macrorchis* Yamaguti, 1970. (After Yamaguti, 1970.) **37.32.** *Dissosaccus laevis* (Linton 1898). (After Yamaguti, 1971.)

Diagnosis: Body surface plicated; plications may be crenulate. Presomatic pit present, circular or oval, deep, glandular. Seminal vesicle bipartite, thin-walled, occurring mostly in forebody; anterior part small; posterior part large. Pars prostatica tubular. Temporary sinus-organ may be seen. Vitellarium two lateral masses, entire, irregularly lobed or indistinctly three- and four-lobed. Parasitic in gut (stomach) of marine and migratory teleosts. Type-species *B. crenatus* (Rudolphi, 1802).

10b. Seminal vesicle not bipartite; body surface smooth; presomatic pit absent; ecsoma reduced .... 10.

11a. Seminal vesicle a wide, convoluted tube; vitelline masses unlobed; in marine fishes .... *Prolecithochirium* Yamaguti, 1970. (Fig. 37.34)

Diagnosis: Ecsoma reduced. Body surface smooth. Presomatic pit absent. Seminal vesicle tubular, convoluted and widening posteriorly, in forebody. Pars prostatica tubular. Vitellarium two compact masses. Parasitic in stomach of marine teleosts. Type-species *P. pterois* Yamaguti, 1970.

11b. Seminal vesicle subglobular; vitelline masses distinctly or indistinctly lobed; in freshwater fishes .... *Paralecithochirium* Zhang, 1994. (Fig. 37.35)

Diagnosis: Ecsoma reduced. Body surface smooth. Presomatic pit absent. Seminal vesicle subglobular, in forebody. Pars prostatica short. Vitellarium two distinctly or

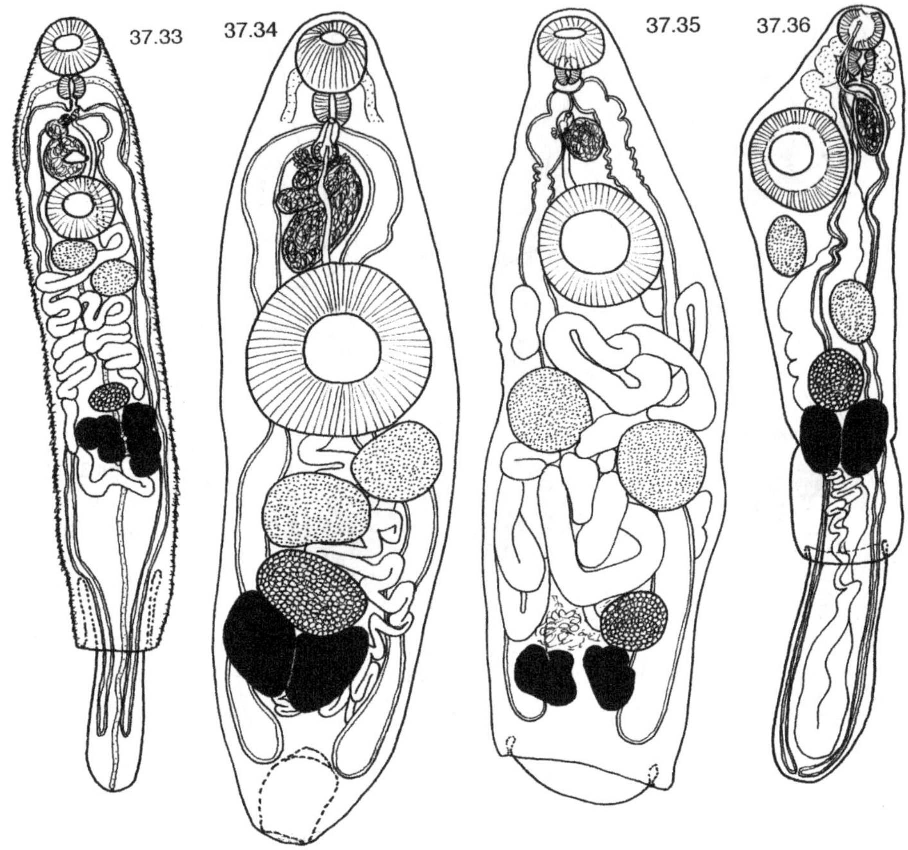

Figs 37.33–37.36.
37.33. *Brachyphallus crenatus* (Rudolphi, 1802). (After Odhner, 1905.)
37.34. *Prolecithochirium pterois* Yamaguti, 1970. (After Yamaguti, 1970.)
37.35. *Paralecithochirium leptobotiae* Zhang, 1994. (Original, from type-material.)
37.36. *Lethadena profunda* (Manter, 1934). (Modified after Manter, 1934.)

indistinctly lobed masses. [Eggs said to be filamented at both poles, but these filaments could not be seen in the specimens examined for this study.] Parasitic in stomach of freshwater fishes (China). Type-species *P. leptobotiae* Zhang, 1994.

Subfamily Lethadeninae Yamaguti, 1971

Diagnosis: Ecsoma well developed. Body surface smooth. Presomatic pit absent. Caeca extend almost to end of ecsoma. Testes oblique to tandem. Seminal vesicle oval, with thick, muscular wall, anterodorsal to ventral sucker. Pars prostatica vesicular, with muscular wall; external gland-cells absent or weakly developed; separated from seminal vesicle by aglandular duct and from sinus-sac by long ejaculatory duct. Sinus-sac small, tubular, enclosing only distal region of ejaculatory duct, without ejaculatory or prostatic vesicle. Sinus-organ present (? temporary), small. Genital atrium short. Vitellarium two symmetrical, unlobed, oval masses, often inside ecsoma. Excretory arms not united in forebody. Parasitic in stomach of marine teleosts. Type-genus *Lethadena* Manter, 1947.

Genus *Lethadena* Manter, 1947 (Fig. 37.36)

Diagnosis: With characters of subfamily. Type-species *Lethadena profunda* (Manter, 1934).

Subfamily Opisthadeninae Yamaguti, 1970
(Syn. Intuscirrinae Skrjabin & Guschanskaja, 1959)

Diagnosis: Body spindle-shaped to elongate. Ecsoma absent. Body surface smooth. Transverse ridges in body-wall present or absent around body at level of oral sucker and/or posterior margin of ventral sucker (these are often not obvious). Ventral sucker normally in anterior half of body. Presomatic pit (?) reported in *Mitrostoma*. Oesophagus normally short. Gut-caeca end blindly near posterior extremity. Testes tandem to oblique, usually well posterior to ventral sucker and near ovary, not separated from ovary by large concentration of uterine coils. Seminal vesicle tubular to saccular (? rarely bipartite), in fore- or hindbody. Pars prostatica long or short, tubular or vesicular. Ejaculatory duct long to short or apparently absent. Sinus-sac present, oval to elongate-oval, enclosing hermaphroditic duct. Sinus-organ (? temporary) occasionally present. Genital atrium usually present, small. Ovary normally oval, occasionally bilobed, normally close to testes. Blind seminal receptacle present, large, usually dorsal or anterodorsal to ovary. Juel's organ and uterine seminal receptacle absent. Uterus varies from mainly pre- to mainly postovarian. Vitellarium forms two, occasionally three, entire or slightly indented masses, posterior or posterolateral to ovary. Excretory arms usually, but not always, united in forebody. Parasitic in stomach, occasionally intestine, of marine teleosts. Type-genus *Opisthadena* Linton, 1910.

Key to genera

1a. Seminal vesicle entirely in hindbody .... **2.**
1b. Seminal vesicle in forebody (occasionally dorsal or posterodorsal to ventral sucker) .... **4.**

2a. Ejaculatory duct short or absent; pars prostatica reaches forward to level of caecal bifurcation; (?) presomatic pit apparently present .... ***Mitrostoma* Manter, 1954. (Fig. 37.37)**

Diagnosis: Body elongate. Thickened projection present on each side of body at level of posterior margin of ventral sucker. Nipple-shaped protuberance (? vestige of ecsoma) may be present at posterior extremity. Weakly muscled pre-oral lobe bears mouth. Structure resembling (?) presomatic pit apparently present anterior to ventral sucker. Ventral sucker with 'sphincter muscles in anterior and posterior halves'. Testes tandem, close to ovary, near middle of hindbody. Seminal vesicle tubular, convoluted,

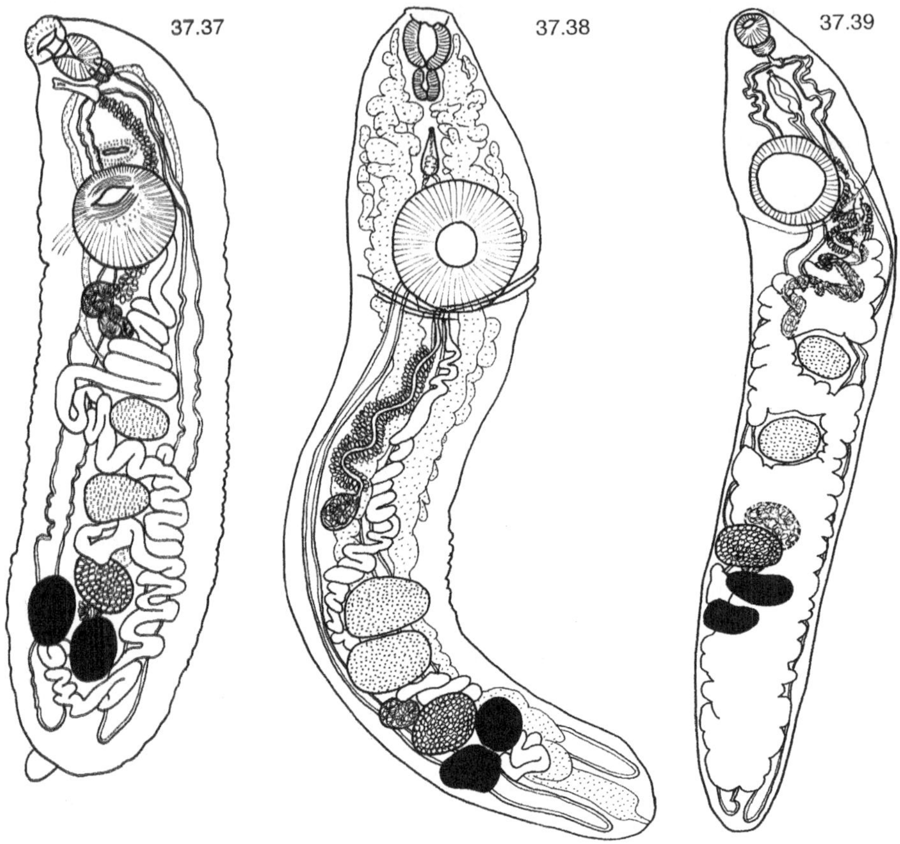

Figs 37.37–37.39.
37.37. *Mitrostoma nototheniae* Manter, 1954. (After Manter, 1954.)
37.38. *Opisthadena dimidia* Linton, 1910. (After Manter, 1947.)
37.39. *Neopisthadena habei* Machida, 1980. (Modified after Machida, 1980.)

entirely in hindbody. Pars prostatica tubular, reaches forward to caecal bifurcation. Ejaculatory duct short or absent. Sinus-sac short, pyriform; (?) protrusible. Ovary in posterior half of hindbody. Uterus mainly pre-ovarian, but does extend into post-vitelline region. Vitellarium two oblique to symmetrical, postovarian masses. Excretory arms united in forebody. Parasitic in intestine of marine teleosts. Type-species *M. nototheniae* Manter, 1954.

2b. Ejaculatory duct long; pars prostatica does not reach further forward than ventral sucker; presomatic pit absent .... **3.**

3a. Seminal vesicle appears globular; testes posterior to middle of body .... ***Opisthadena* Linton, 1910. (Fig. 37.38)**

Diagnosis: Body elongate. Transverse ridge (fold) of body-wall around body immediately posterior to ventral sucker. Testes tandem, posterior to middle of body. Seminal vesicle in hindbody, appears globular, but actual vesicle may be coiled within globular, muscular sac. Pars prostatica tubular, not reaching further forward than posterior margin of ventral sucker. Ejaculatory duct long. Sinus-sac oval. Hermaphroditic duct may be subdivided. Sinus-organ apparently present as small cone, at least temporarily. Ovary close to testes. Uterus mainly pre-ovarian. Vitellarium two symmetrical to tandem masses, postovarian. Excretory arms diverticulate, united in forebody. Parasitic in stomach of marine teleosts (especially *Kyphosus*). Type-species *O. dimidia* Linton, 1910.

3b. Seminal vesicle tubular, convoluted; testes in middle of body .... *Neopisthadena* Machida, 1980. (Fig. 37.39)

Diagnosis: Body elongate. Transverse ridge (fold) of body-wall around body immediately posterior to ventral sucker. Testes tandem, near middle of body. Seminal vesicle in hindbody, clearly tubular, sinuous and not enclosed in muscular sac. Pars prostatica tubular, reaches forward to level of ventral sucker. Ejaculatory duct long. Sinus-sac oval. No sinus-organ reported. Uterus mainly pre-ovarian. Vitellarium two tandem to oblique masses, postovarian. Excretory arms united in forebody. Parasitic in stomach of marine teleosts (*Kyphosus*). Type-species *N. habei* Machida, 1980.

4a. Vitellarium two symmetrical, oblique or tandem masses .... *Genolinea* Manter, 1925. (Fig. 37.40)

(Syns *Parasterrhurus* Manter, 1934; *Intuscirrus* Acena, 1947; *Pseudobunocotyla* Yamaguti, 1965)

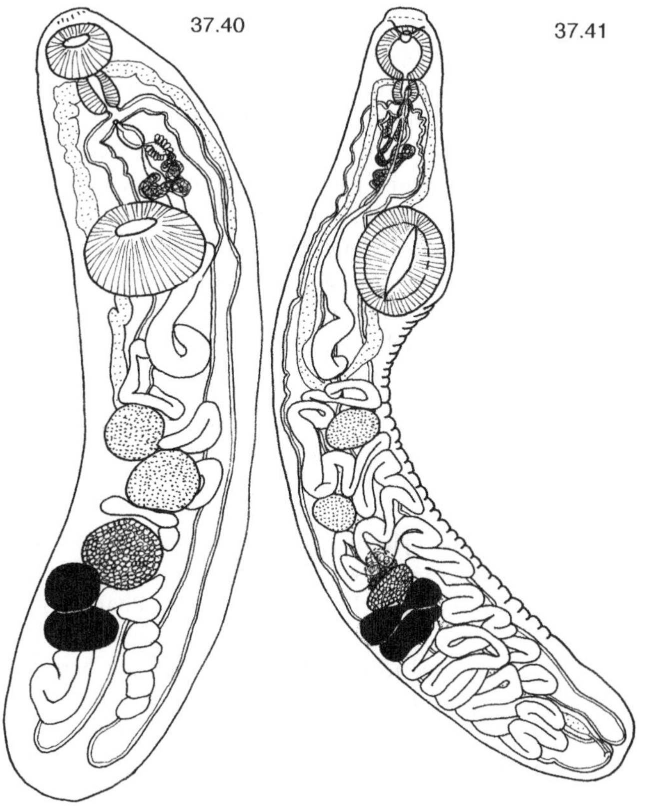

Figs 37.40–37.41.
37.40. *Genolinea laticauda* Manter, 1925. (Modified after Manter, 1925.)
37.41. *Neotheletrum lissosomum* (Manter, 1940). (After Yamaguti, 1971.)

Diagnosis: Body spindle-shaped to slightly elongate. Transverse ridge usually present around body immediately posterior to ventral sucker (often inconspicuous and frequently not reported); similar ridge may surround oral sucker. Large pre-oral lobe may be present. Ventral sucker normally in anterior half of body, (?) occasionally near middle; sphincter muscles sometimes present around aperture. Testes tandem to oblique, close to ovary. Seminal vesicle small, tubular, convoluted in forebody, occasionally dorsal or posterodorsal to ventral sucker. Pars prostatica tubular to vesicular, short. Ejaculatory duct short or absent. Sinus-sac oval to elongate-oval, small. Sinus-organ occasionally present (? temporary). Ovary near middle of hindbody. Uterus usually in both pre- and postovarian fields, occasionally postvitelline distribution is limited. Metraterm sometimes reported as spinous (?). Vitellarium two compact (occasionally lobed), symmetrical, oblique or tandem masses, posterior or posterolateral to ovary. Excretory arms united in forebody. Parasitic mainly in stomach of marine teleosts. Type-species *G. laticauda* Manter, 1925.

4b. Vitellarium three masses; anterior pair of masses symmetrical; posterior mass largest .... *Neotheletrum* Gibson & Bray, 1979. (Fig. 37.41)

Diagnosis: Body small, elongate to spindle-shaped. Tegumental fold around body posterior to ventral sucker apparently absent. Ventral sucker in anterior half of body. Oesophagus short, often with small diverticulum. Testes oval, oblique to symmetrical, usually separated from ventral sucker by loops of uterus, occasionally sandwiched between ventral sucker and ovary. Seminal vesicle small, tubular to saccular (? occasionally bipartite), in forebody. Pars prostatica short, tubular to vesicular. Ejaculatory duct short or absent. Sinus-sac small, oval. Small temporary sinus-organ may form. Hermaphroditic duct short. Genital atrium small. Ovary oval (may occasionally be bilobed), post-testicular, near middle of hindbody. Uterus almost entirely in hindbody, usually with roughly equal amounts in pre- and postovarian fields, occasionally with majority of uterus in postovarian field. Eggs small, numerous, without filaments. Vitellarium forms three compact, entire or slightly indented masses; anterior pair symmetrical, connected by narrow isthmus; posterior mass larger, may be slightly bilobed (vitellarium is essentially two tandem masses, anterior of which is divided into two distinct lobes; vitellarium may appear as two tandem masses in lateral view), postovarian. Excretory arms united in forebody. Parasitic in stomach of marine teleosts. Type-species *N. lissosomum* (Manter, 1940).

Subfamily Plerurinae Gibson & Bray, 1979

Diagnosis: Body small, spindle-shaped to cylindrical. Ecsoma reduced or well developed. Body surface smooth, or occasionally with crenulate plications giving scaly appearance. Presomatic pit present or absent. Prepharynx absent. Pharynx well developed. Oesophagus short. 'Drüsenmagen' present. Gut-caeca terminate blindly inside ecsoma. Testes pre-ovarian, symmetrical to tandem, usually oblique. Seminal vesicle elongate, saccular and constricted into two, three or four sections, thin-walled, although certain sections may have thicker walls, in forebody or partly in hindbody. Pars prostatica vesicular or tubular, may be partly enclosed by muscles of sinus-sac, commonly linked to seminal vesicle by aglandular duct. Sinus-sac apparently absent or poorly developed, when present usually open at base. Permanent sinus-organ absent. Ejaculatory (prostatic) vesicle absent. Hermaphroditic duct commonly vesicular proximally and tubular distally. Genital atrium usually deep, but may be shallow. Genital pore midventral in forebody. Ovary entire or lobed. Laurer's canal absent. Canalicular and blind seminal receptacles absent. Juel's organ present. Uterine seminal receptacle present. Uterus convoluted, passes back from ovary into ecsoma and then forward into forebody. Vitellarium postovarian, composed of two masses, four- and three-lobed; lobes being small to digitiform. Excretory vesicle Y-shaped; arms united in forebody. Parasitic in stomach of marine teleosts. Type-genus *Plerurus* Looss, 1907.

Key to genera

1a. Seminal vesicle in forebody .... **2.**
1b. Seminal vesicle at least partly extended into hindbody .... **4.**

2a. Presomatic pit absent; vitelline lobes digitiform .... **3.**
2b. Presomatic pit present; vitelline lobes short ....
.... ***Synaptobothrium* von Linstow, 1904. (Fig. 37.42)**
Diagnosis: Body surface smooth. Presomatic pit present, circular or oval, deep, glandular. Testes oblique. Seminal vesicle bipartite (? or tripartite), thin-walled, occurring mostly in forebody but may extend dorsal to ventral sucker; anterior part small; posterior part long. Pars prostatica tubular with wide lumen, may extend into base of sinus-sac. Sinus-sac weakly developed, of 'open' type. Ovary oval. Vitellarium two lateral masses with three and four short lobes. Eggs may be reniform. Type-species *S. caudiporum* (Rudolphi, 1819).

3a. Sinus-organ present .... ***Merlucciotrema* Yamaguti, 1971. (Fig. 37.43)**
Diagnosis: Ecsoma well developed. Body surface smooth. Testes symmetrical. Seminal vesicle in forebody, saccular to bipartite, tapers anteriorly to form narrow duct, which links to pars prostatica. Pars prostatica vesicular, large, oval, with dense covering of external gland-cells, which may be delimited from parenchyma. Hermaphroditic duct surrounded by muscle bundles, which do not form complete sinus-sac. Sinus-organ muscular, distinct. Genital atrium deep or may form outer wall of extruded genital papilla. Ovary oval. Vitelline lobes digitiform. Type-species *M. praeclarum* (Manter, 1934).
3b. Sinus-organ absent .... ***Plerurus* Looss, 1907. (Fig. 37.44)**
(Syns *Paraplerurus* Fischthal & Kuntz, 1963; *Voitrema* Yamaguti, 1971)
Diagnosis: Ecsoma reduced or well developed. Body surface smooth. Testes symmetrical to oblique. Seminal vesicle in forebody, elongate, saccular, two-, three- or four-lobed, often sigmoid, thin-walled. Pars prostatica vesicular, may be linked to seminal vesicle by short aglandular duct (some authors maintain that this is a tubular region of the pars prostatica). Sinus-sac apparently absent or poorly developed and of 'open' type. Hermaphroditic duct tubular, deep, possibly eversible. Ovary oval to lobed. Vitelline lobes tubular to digitiform. Type-species *P. digitatus* (Looss, 1899).

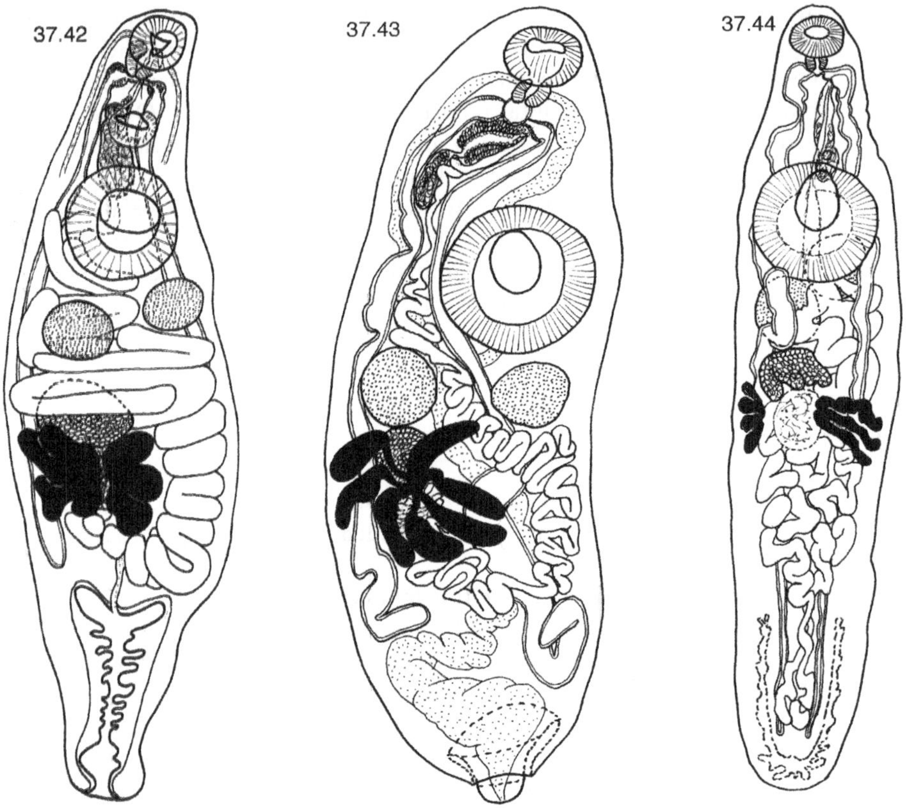

Figs 37.42–37.44.
37.42. *Synaptobothrium caudiporum* (Rudolphi, 1819). (After Gibson & Bray, 1986.)
37.43. *Merlucciotrema praeclarum* (Manter, 1934). (After Yamaguti, 1971.)
37.44. *Plerurus digitatus* (Looss, 1899). (After Looss (1899) and Yamaguti (1971): Looss drew the structure posterior to the ovary as a seminal receptacle, but this is likely to have been either the uterine seminal receptacle or Juel's organ.)

4a. Body surface has scaly appearance .... ***Dinosoma* Manter, 1934. (Fig. 37.45)**
Diagnosis: Body surface with crenulate plications, giving scaly appearance. Testes symmetrical to tandem. Seminal vesicle posterodorsal to ventral sucker, saccular, bipartite or wide, sinuous. Pars prostatica vesicular, may be connected to seminal vesicle by long, aglandular duct. Sinus-sac apparently absent. Hermaphroditic duct long, narrow, with small vesicle proximally. Ovary oval. Vitellarium two indented or lobed masses. Type-species *D. rubrum* Manter, 1934.
4b. Body surface smooth .... ***Adinosoma* Manter, 1947. (Fig. 37.46)**
Diagnosis: Body surface smooth. Testes oblique. Seminal vesicle large, saccular, bipartite, posterodorsal to ventral sucker. Pars prostatica vesicular, but elongate, connected to seminal vesicle by long, aglandular duct. Sinus-sac apparently absent. Hermaphroditic duct long, with poorly developed hermaphroditic vesicle proximally. Ovary unlobed. Vitellarium two indented or lobed masses. Type-species *A. robustum* (Manter, 1934).

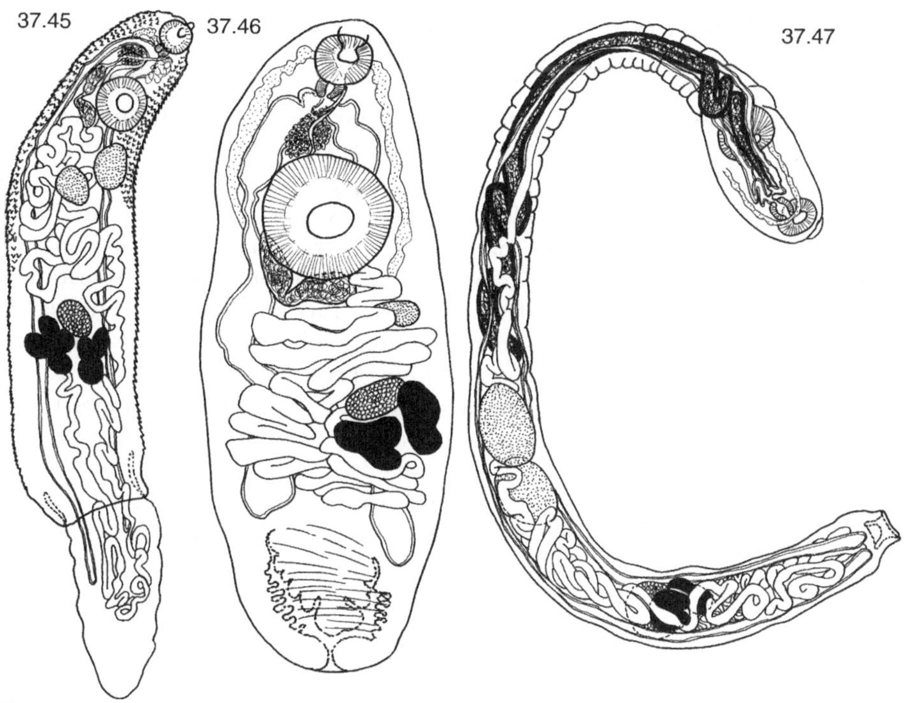

Figs 37.45–37.47.
37.45. *Dinosoma rubrum* Manter, 1934. (Modified after Yamaguti, 1971.)
37.46. *Adinosoma robustum* (Manter, 1934). (After Manter, 1934.)
37.47. *Pulmovermis cyanovitellosus* Coil & Kuntz, 1960, dorsal view. (After Yamaguti, 1971.)

Subfamily Pulmoverminae Sandars, 1961

Diagnosis: Ecsoma reduced. Body surface smooth (? spines reported within suckers). Presomatic pit absent. Testes tandem to oblique. Seminal vesicle tubular, long, thick-walled, extends to or almost to level of testes. Pars prostatica short, vesicular, partly enclosed by sinus-sac. Sinus-sac present, not enclosing ejaculatory or prostatic vesicle. Sinus-organ variable in length (? temporary). Genital atrium small. Ovary occasionally divided into dorsal and ventral lobes. Vitellarium two lateral, closely aligned masses, normally with three and four lobes. Excretory arms united in forebody. Parasitic in lung of sea-snakes. Type-genus *Pulmovermis* Coil & Kuntz, 1960.

Genus *Pulmovermis* Coil & Kuntz, 1960 (Fig. 37.47)
(Syns *Hydrophitrema* Sandars, 1960; *Laticaudatrema* Telford, 1967)

Diagnosis: With characters of subfamily. Type-species *P. cyanovitellosus* Coil & Kuntz, 1960.

Subfamily Theletrinae Gibson & Bray, 1979

Diagnosis: Body elongate. Ecsoma absent. Body surface smooth; papillae may be present ventrally in fore- or hindbody; transverse ridge may be present around body near posterior margin of ventral sucker and possibly around oral sucker. Ventral sucker in anterior half of body. Oesophagus short. Gut-caeca end blindly near posterior extremity, (?) or at level of ovary. Testes two, pre-ovarian, tandem to oblique, near middle of hindbody, separated from ovary by majority or large part of uterine coils. Seminal vesicle tubular or (?) saccular; in forebody, but sometimes reaching back to posterior margin of ventral sucker. Pars prostatica short, tubular or slightly vesicular. Ejaculatory duct short or apparently absent. Sinus-sac usually small, oval or elongate-oval, weakly developed, enclosing hermaphroditic duct, may extrude slightly through genital pore. Permanent sinus-organ absent (?), but temporary sinus-organ may form. Genital atrium small or absent. Ovary oval, near posterior extremity or at least well inside posterior half of hindbody. Laurer's canal and both canalicular and blind seminal receptacle absent (Laurer's canal reported present (?) in *Indoderogenes*). Uterine seminal receptacle present. Juel's organ assumed to be present. Uterus almost entirely or mainly in hindbody (small part of uterus is coiled in forebody of *Monolecithotrema*), mainly pre-ovarian, with large proportion of uterine coils between ovary and testes. Eggs numerous, small, without filaments. Vitellarium one entire or two tandem to oblique, entire or slightly lobed masses, posterior or posterolateral to ovary. Excretory vesicle arms united in forebody. Normally parasitic in stomach of marine teleosts. Type-genus *Theletrum* Linton, 1910.

Key to genera

1a. Vitellarium single (great care should be taken with this observation, as at certain angles the vitellarium of *Theletrum* appears to be single); transverse ridge posterior to ventral sucker absent; some uterine coils in forebody .... *Monolecithotrema* Yamaguti, 1970. (Fig. 37.48)

Diagnosis: Transverse ridges around body absent. Papillae may be present ventrally in forebody. Gut-caeca terminate blindly at posterior extremity. Testes close together, but sometimes separated by uterus. Seminal vesicle tubular, usually extending back dorsally to ventral sucker. Pars prostatica tubular, poorly developed, linked to seminal vesicle by aglandular duct. Sinus-sac small, poorly developed, may be slightly extruded through genital pore. Ovary well inside posterior half of hindbody. Small proportion of uterus may be coiled in forebody; small part of uterus extends posteriorly to vitellarium. Vitellarium one large, entire mass, immediately postovarian. Parasitic in stomach of marine teleosts. Type-species *M. kala* Yamaguti, 1970.

1b. Vitellarium double; transverse ridge posterior to ventral sucker present or absent; uterus not coiled in forebody .... 2.

2a. Seminal vesicle tubular; transverse ridge normally present posterior to ventral sucker; gut-caeca terminate near posterior extremity .... *Theletrum* Linton, 1910. (Fig. 37.49)

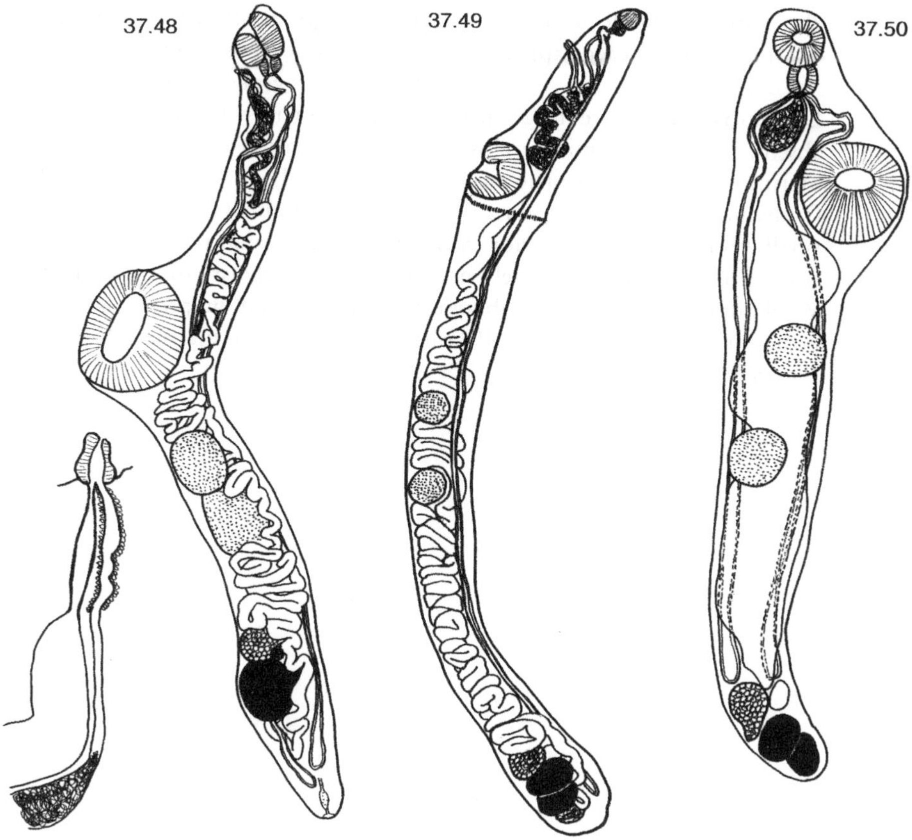

Figs 37.48–37.50.
37.48. *Monolecithotrema kala* Yamaguti, 1970. Entire worm and terminal genitalia. (After Yamaguti, 1970.)
37.49. *Theletrum fustiforme* Linton, 1910. (After Yamaguti, 1971.)
37.50. *Indoderogenes purii* Srivastava, 1937. (Modified after Srivastava, 1941.)

Diagnosis: Transverse tegumental ridge present around body immediately posterior to ventral sucker and possibly around oral sucker. Papillae may be present ventrally in hindbody. Gut-caeca terminate at posterior extremity. Testes fairly close together, but separated by uterus. Seminal vesicle tubular. Pars prostatica slightly vesicular. Sinus-sac elongate-oval, thin-walled, sheath-like, may be partly extruded through genital pore. Ovary close to posterior extremity. Uterine coils entirely or almost entirely in hindbody; few or no coils posterior to vitellarium. Vitellarium two oblique, entire or slightly lobed masses, close together, posterior or posterolateral to ovary, at posterior extremity of body. Parasitic in stomach (occasionally intestine) of marine teleosts. Type-species *T. fustiforme* Linton, 1910.

2b. Seminal vesicle saccular; transverse ridge posterior to ventral sucker not reported; gut-caeca apparently terminate at level of anterior margin of ovary .... (?) *Indoderogenes* Srivastava, 1937.[2] (Fig. 37.50)

Diagnosis: Transverse ridge around body posterior to ventral sucker apparently absent. Gut-caeca terminate close to anterior margin of ovary. Testes separated by uterine coils. Seminal vesicle saccular (flask-shaped). Pars prostatica tubular. Sinus-sac (?) absent. Hermaphroditic duct short. Sinus-organ small (? temporary). Genital atrium small. Ovary close to posterior extremity. Laurer's canal reported present (?). Uterine coils entirely or almost entirely in hindbody, entirely or almost entirely pre-ovarian. Vitellarium two tandem to oblique masses, posterior or posterolateral to ovary, at posterior extremity of body. Parasitic in stomach of marine teleosts. Type-species *I. purii* Srivastava, 1937.

[2] This appears to be a very dubious genus (described in more detail by Srivastava, 1941), which is known from only three specimens. It was discussed by Gibson & Bray (1979). The features not described and some of those included prohibit linking it with other taxa.

38 Family Accacoeliidae Odhner, 1911

D.I. Gibson

Department of Zoology, The Natural History Museum, London, UK

Introduction

Accacoeliids are unusual in that the majority of taxa occur in a single fish species, the sunfish *Mola mola*. This association no doubt relates to the fact that this fish is medusophagous and the metacercariae of this family occur in nektonic organisms, especially cnidarians and ctenophores. The group is readily recognized by the presence of an anterior extension to the pharynx, which penetrates the base of the oral sucker. The family group was first erected by Odhner (1911b) as a subfamily and first used at the full family level, in passing, by Dollfus (1923; see also Dollfus, 1935b), although Looss (1912) had earlier apparently recognized it as such by the use of the vernacular name 'Accacoeliiden' and the change of rank was erroneously attributed to him by Yamaguti (1953, 1958a, 1971) and Skrjabin & Guschanskaja (1959).

Various classifications have been proposed. Dollfus (1935b, 1960a) recognized two subfamilies, the Accacoeliinae and the Tetrochetinae Dollfus, 1935 ('Tetrochetinen' of Looss, 1912). Yamaguti (1958a) accepted four subfamilies, the Accacoeliinae and Tetrochetinae plus two new subfamilies, the Orophocotylinae and the Rhynchopharynginae. Skrjabin & Guschanskaja (1959) agreed with this but added the Guschanskianinae Skrjabin, 1959, an arrangement accepted by Yamaguti (1971). However, Dollfus (1960a) deemed this unsatisfactory. In their revision of the group, Bray & Gibson (1977) preferred the original Dollfus version, but considered that all of the previously known genera should be accommodated in a single subfamily, the Accacoeliinae; however, they erected a new subfamily, the Paraccacladiinae, for a new genus that differed significantly from all other accacoeliids in the form of the intestine, the form of the vitellarium and the postovarian position of Mehlis' gland. Understandably, this classification was accepted by Gibson & Bray (1979) in their revision of the Hemiuroidea and is followed here.

Family Accacoeliidae Odhner, 1911

Diagnosis: Body large or small, commonly elongate. Ecsoma absent. Body surface smooth, but forebody may be papillate. Oral and ventral suckers well developed. Ventral sucker normally in anterior half of body, may be pedunculate. Pharynx well developed, with narrow anterior extension into base of oral sucker, occasionally modified posteriorly. Oesophagus usually long, occasionally short. 'Drüsenmagen' present. Gut-caeca usually H-shaped, terminate blindly or more commonly form uroproct. Testes two, oblique or in tandem, in hindbody, normally close to middle of body, pre-ovarian. Seminal vesicle thin-walled, tubular, sinuous or convoluted, commonly extending into hindbody. Pars prostatica tubular; external gland-cells may be delimited. Short ejaculatory duct commonly present within sinus-sac. Hermaphroditic duct present or absent. Sinus-sac and sinus-organ present or absent. Genital atrium present. Ovary oval, post-testicular. Mehlis' gland pre- or postovarian, linked to anterior or posterior region of ovary by oviduct. Laurer's canal and uterine seminal receptacle present. Juel's organ and canalicular or blind seminal receptacle absent. Uterus extensive; coils entirely or almost entirely in hindbody, usually pass close to posterior extremity before looping forward again. Eggs numerous, small, non-filamented. Vitellarium with one or two main collecting ducts, composed of numerous filamentous tubules (? or occasionally chains of follicles) in various parts of the fore- or hindbody. Excretory vesicle Y-shaped; arms initially in dorsal and ventral fields, united in forebody. Metacercariae usually in nektonic cnidarians or ctenophores. Parasitic in gut or occasionally on gills of marine teleosts. Type-genus *Accacoelium* Monticelli, 1893.

Key to subfamilies

1a. Oesophagus long and narrow; gut-caeca H-shaped; uroproct present; Mehlis' gland pre-ovarian, linked to anterior region of ovary by oviduct; vitellarium with single main collecting duct and system of branching tubules .... Accacoeliinae Odhner, 1911.

1b. Oesophagus short; gut-caeca not distinctly H-shaped and end blindly; Mehlis' gland postovarian, linked to posterior region of ovary by oviduct; vitellarium with symmetrical pair of collecting ducts and system of branching tubules .... Paraccacladiinae Bray & Gibson, 1977.

Subfamily Accacoeliinae Odhner, 1911

(Syns Tetrochetinae Looss, 1912, emend. Dollfus, 1935; Accacladiinae Yamaguti, 1958; Oropho-cotylinae Yamaguti, 1958; Rhynchopharynginae Yamaguti, 1958; Guschanskianinae Skrjabin, 1959)

Diagnosis: Body usually elongate. Body surface smooth, but forebody may be papillate. Lateral flanges occasionally present on ventral sucker. Pharynx occasionally modified to form two muscular bulbs (*Rhynchopharynx*). Oesophagus long and thin. Gut-caeca H-shaped. Uroproct present. Sinus-sac and sinus-organ present or absent;

sinus-sac well or poorly developed when present. Mehlis' gland pre-ovarian, linked by oviduct to anterior region of ovary. Vitellarium filamentous, with single main collecting duct on right and single system of ramifying branches in fore- or hindbody; left-hand system reduced to small process or small ramifying system. Parasitic on gills or in gut of marine teleosts (especially Molidae). Type-genus *Accacoelium* Monticelli, 1893.

Key to genera

1a. Well-defined sinus-sac and sinus-organ present .... 2.
1b. Well-defined sinus-sac and sinus-organ absent .... 5.

2a. Ectoparasitic on gills; long proboscis-like sinus-organ; vitellarium posterior to anterior testis .... *Accacoelium* Monticelli, 1893. (Fig. 38.1)
Diagnosis: Forebody papillate. Oesophagus reaches to ventral sucker. Ventral sucker on short peduncle. Thick muscular layer in ventral hindbody. Sinus-sac present surrounding base of genital atrium. Sinus-organ long and strongly muscular,

38.1
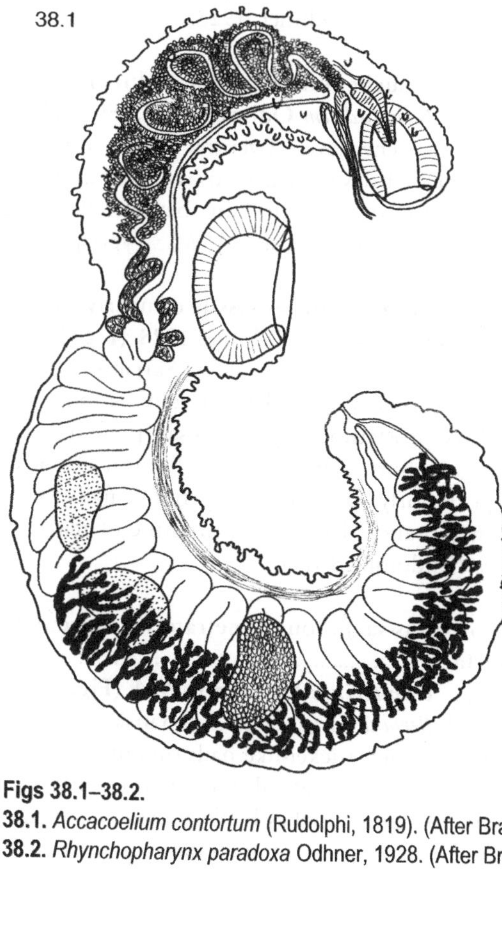

38.2
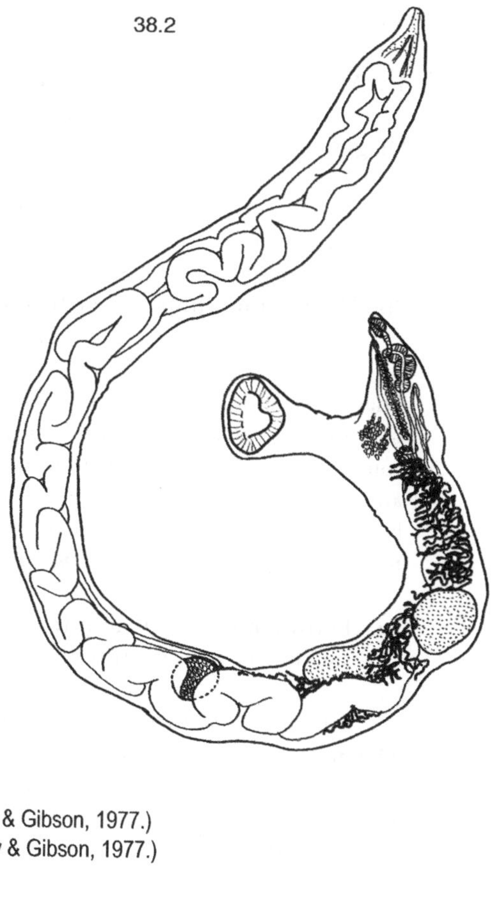

Figs 38.1–38.2.
38.1. *Accacoelium contortum* (Rudolphi, 1819). (After Bray & Gibson, 1977.)
38.2. *Rhynchopharynx paradoxa* Odhner, 1928. (After Bray & Gibson, 1977.)

frequently extended through genital pore. Enormous pars prostatica and associated gland-cells occupy much of forebody. Vitellarium posterior to anterior testis. Parasitic on gills of marine teleosts (*Mola*). Type-species *A. contortum* (Rudolphi, 1819).

2b. Endoparasitic in gut; short cylindrical or dome-shaped sinus-organ; vitellarium not usually extending posteriorly to ovary .... 3.

3a. Pharynx pyriform; no oesophageal bulb; no large glandular organ of unknown function at base of peduncle .... 4.

3b. Pharynx with two bulbs and long anterior elongate portion ensheathed in glandular posterior process of oral sucker; glandular oesophageal bulb immediately posterior to pharynx; large glandular organ of unknown function at base of peduncle .... *Rhynchopharynx* Odhner, 1928. (Fig. 38.2)

Diagnosis: Forebody papillate. Ventral sucker pedunculate. Pharynx consisting of two muscular bulbs (the 'pharynx proper' and the 'snout bladder') and extended anterior 'snout', which may be extended through oral sucker. Oral sucker possesses posterior glandular extension (the 'snout sheath') which envelops 'snout'. Glandular oesophageal bulb present immediately posterior to pharynx. Oesophagus extends to ventral sucker. Large glandular organ of unknown function present at base of peduncle. Pars prostatica long. Sinus-sac surrounds base of genital atrium. Sinus-organ small. Vitellarium extends from anterior region of ventral sucker to ovary. Parasitic in intestine of marine teleosts (*Mola*). Type-species *R. paradoxa* Odhner, 1928.

4a. Vitellarium confined to hindbody; ventral sucker on extensible peduncle .... *Accacladium* Odhner, 1928. (Fig. 38.3)

Diagnosis: Body surface smooth. Oesophagus reaches level of ventral sucker. Ventral sucker on extensible peduncle. Pars prostatica well developed. Sinus-sac surrounding base of genital atrium. Sinus-organ short, cylindrical. Vitellarium between ventral sucker and ovary. Parasitic in intestine of marine teleosts (*Mola*). Type-species *A. serpentulus* Odhner, 1928.

4b. Vitellarium wholly or partly in forebody; ventral sucker sessile or nearly so ... . *Accacladocoelium* Odhner, 1928. (Fig. 38.4)

(Syn. *Guschanskiana* Skrjabin, 1959)

Diagnosis: Body smooth. Lateral flanges on ventral sucker present or absent. Oesophagus extends to ventral sucker. Pars prostatica long. Sinus-sac present surrounding base of genital atrium. Sinus-organ short, cylindrical. Vitellarium extends anteriorly to oral sucker, may extend posteriorly just past ovary but usually not beyond anterior testis; reduced fraction may branch. Parasitic in intestine of marine teleosts (*Mola*). Type-species *A. nigroflavum* (Rudolphi, 1819).

5a. Diffuse muscular region (? vestige of sinus-sac) around metraterm; muscular sucker-like pads on anterodorsal surface .... *Odhnerium* Yamaguti, 1934. (Fig. 38.5)

(Syns *Mneiodhneria* Dollfus, 1935; *Caballeriana* Skrjabin & Guschanskaja, 1959)

Diagnosis: Body surface smooth, but with muscular sucker-like pads on anterodorsal surface. Ventral sucker pedunculate; with flange-like lateral muscular extensions. Oesophagus reaches to ventral sucker. Pars prostatica reaches halfway back to ventral sucker. Diffuse muscular region surrounds distal parts of metraterm and genital

38.3 38.4

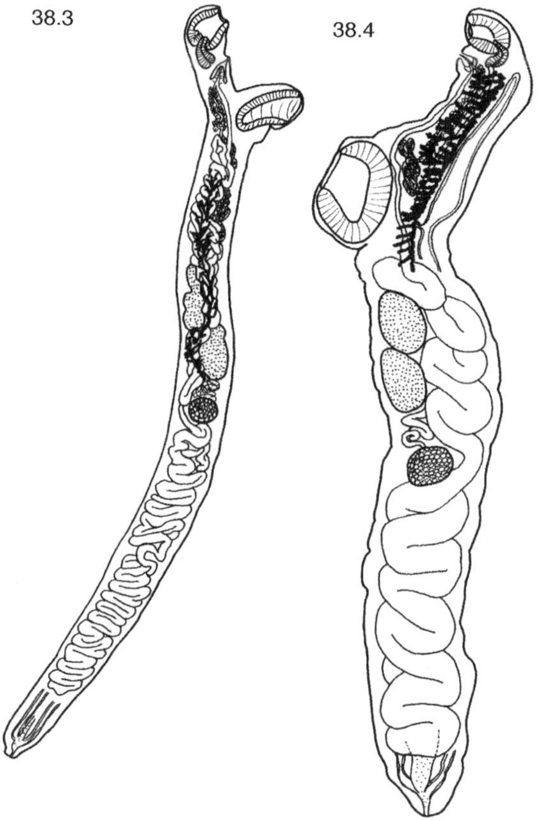

Figs 38.3–38.4.
38.3. *Accacladium serpentulus* Odhner, 1928. (After Bray & Gibson, 1977.)
38.4. *Accacladocoelium nigroflavum* (Rudolphi, 1819). (After Bray & Gibson, 1977.)

atrium (may be vestige of sinus-sac). Male duct enters genital atrium from side through small papilla. Vitellarium tubular, extending from pharynx to ovary. Parasitic in intestine of marine teleosts (*Mola*). Type-species *O. calyptrocotyle* (Monticelli, 1893).

5b. No sinus-sac detectable; no muscular pads present on anterodorsal surface .... 6.

6a. Vitellarium a tubular branching structure on either side of hindbody; no flange on ventral sucker .... *Tetrochetus* Looss, 1912. (Fig. 38.6)
(Syn. *Paratetrochetus* Hanson, 1955)

Diagnosis: Body surface smooth. Ventral sucker pedunculate. Oesophagus long. Diverticula present at intestinal bifurcation. Pars prostatica short, straight, narrow. Sinus-sac and sinus-organ absent. Male and female ducts open together into shallow genital atrium. Vitellarium tubular, in hindbody; reduced half may be branched. Parasitic in intestine of medusophagous and carnivorous marine teleosts. Type-species *T. raynerii* (Nardo, 1833).

6b. Vitellarium comprising (?) few follicles in four rows between anterior testis and base of peduncle; small flange on ventral sucker .... *Orophocotyle* Looss, 1902. (Fig. 38.7)

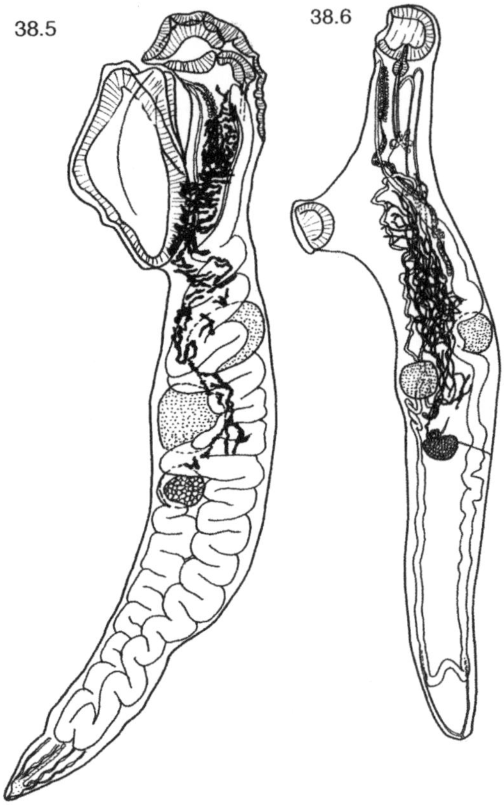

Figs 38.5–38.6.
38.5. *Odhnerium calyptrocotyle* (Monticelli, 1893). (After Bray & Gibson, 1977.)
38.6. *Tetrochetus raynerii* (Nardo, 1833). (After Bray & Gibson, 1977.)

Diagnosis: [Inadequately described.] Body surface smooth. Ventral sucker pedunculate; bears small flange. Oesophagus not reaching ventral sucker. Pars prostatica short. Sinus-sac not reported. Sinus-organ absent. Vitellarium reported to consist of few (?) follicles in four rows between testes and ventral sucker. Parasitic in intestine of marine teleosts (*Ranzania*). Type-species *O. planci* (Stossich, 1899).

Subfamily Paraccacladiinae Bray & Gibson, 1977

Diagnosis: Body elongate. Body surface smooth, but with papillae on outer surface of ventral sucker. Ventral sucker on short peduncle. Pharynx extended into base of oral sucker. Oesophagus short, wide. Anterior caecal shoulders small. Gut-caeca terminate blindly near posterior extremity. Pars prostatica elongate, convoluted. Sinus-sac present surrounding base of genital atrium; musculature diffuse. Sinus-organ short, cylindrical. Mehlis' gland postovarian, linked to posterior region of ovary by oviduct. Vitellarium with symmetrical pair of main collecting ducts and ramifying systems of tubules, posterior to ovary. Mature forms parasitic in rectum of carnivorous marine teleosts; immature forms parasitic in rectum of medusophagous marine teleosts. Type-genus *Paraccacladium* Bray & Gibson, 1977.

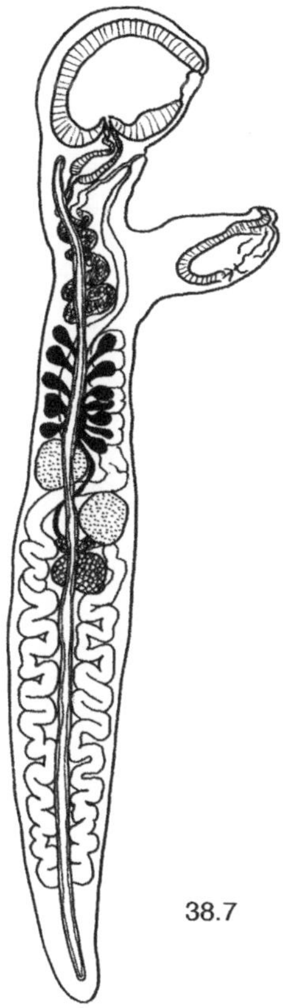

38.7

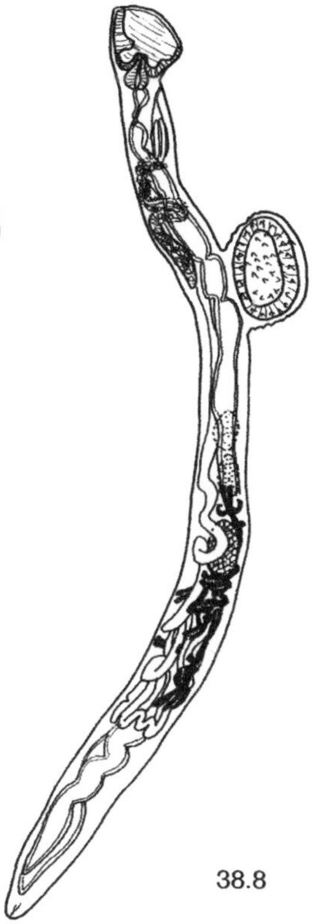

38.8

Figs 38.7–38.8.
38.7. *Orophocotyle planci* (Stossich, 1899). (After Looss, 1902.)
38.8. *Paraccacladium jamiesoni* Bray & Gibson, 1977. (After Bray & Gibson, 1977.)

Genus *Paraccacladium* Bray & Gibson, 1977 (Fig. 38.8)

Diagnosis: With the characters of the subfamily. Type-species *P. jamiesoni* Bray & Gibson, 1977.

39 Family Bathycotylidae Dollfus, 1932

D.I. GIBSON

Department of Zoology, The Natural History Museum, London, UK

Introduction

The family Bathycotylidae was erected by Dollfus (1932b) for the single genus *Bathycotyle* Darr, 1902. Dollfus included the family within the Hemiuroidea, close to the hirudinellids and somewhat distinct from the sclerodistomids. Since that time, its systematic position and monogeneric status has remained essentially unchanged in the various subsequent classifications of the Hemiuroidea. Gibson & Bray (1979) placed the group among the primitive hemiuroids close to the Isoparorchiidae and Accacoeliidae, whereas Brooks *et al.* (1985), using a cladistic analysis of Gibson & Bray's data, linked it, like Dollfus, most closely with the Hirudinellidae. The unusual features of this rare group include an intertesticular ovary, the absence of a sinus-sac and sinus-organ and a site on the gills (usually attached by the ventral sucker to the end of a gill filament of *Coryphaena*). The description of the sclerodistomid *Kenmackenzia* by Gibson (1983), another unusual large form with an intertesticular ovary, again raises the question of a link with the Sclerodistomidae, but Gibson confirmed the absence of Manter's organ in *Bathycotyle*, a genus that he considered to be monotypic.

Family Bathycotylidae Dollfus, 1932

Diagnosis: Body large, elongate, but stout. Ecsoma absent. Body surface smooth, but may be wrinkled. Oral and ventral suckers well developed; latter just in anterior half of body. Pharynx well developed. Oesophagus short. 'Drüsenmagen' present. Gut-caeca end blindly close to posterior extremity. Testes two, tandem, separated by ovary, in mid-hindbody. Seminal vesicle thin-walled, tubular, convoluted, small, well forward in forebody. Pars prostatica tubular, indistinct. Sinus-organ and sinus-sac absent. Genital atrium small, but deep. Genital pore midventral close to posterior margin of oral sucker. Ovary oval to reniform, intertesticular. Mehlis' gland posterior or lateral to ovary. Laurer's canal and uterine seminal receptacle present. Juel's organ and canalicular or blind seminal receptacle absent. Uterus fills much of hind- and forebody. Eggs numerous, small, non-filamented. Vitellarium several filamentous tubules in hindbody. Excretory vesicle Y-shaped; arms united in forebody. Parasitic on gills (? or in stomach) of pelagic marine teleosts (*Coryphaena* and ?scombroids). Type-genus *Bathycotyle* Darr, 1902.

Genus *Bathycotyle* Darr, 1902 (Fig. 39.1)

Diagnosis: With characters of family. Type-species *B. branchialis* Darr, 1902.

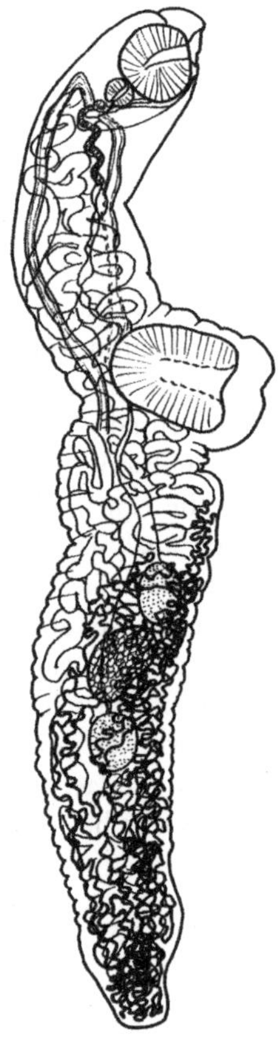

Fig. 39.1. *Bathycotyle branchialis* Darr, 1902 (as *B. coryphaenae* Yamaguti, 1938). (After Yamaguti, 1938b.)

40 Family Derogenidae Nicoll, 1910

D.I. Gibson

Department of Zoology, The Natural History Museum, London, UK

Introduction

Throughout most of its history, this group has been considered within the family Hemiuridae, as used in its broad sense by workers such as Yamaguti (1971) and Brooks *et al.* (1985). The subfamily Derogeninae was erected by Nicoll (1910) for *Derogenes* Lühe, 1900 and in 1926 Poche erected the family Halipegidae for *Halipegus* Looss, 1899, although the latter has subsequently been used mainly at the subfamily level. The Derogenidae was first used at full family level by Dollfus (1950), where he included within it *Halipegus*. Skrjabin & Guschanskaja (1956) recognized the Halipegidae, including within it the Halipeginae, Derogeninae (as the Derogenetinae), Dictysarcinae, Gonocercinae, Genarchinae and Bunocotylinae. The basic classification of the family, as the Derogenidae, used here is that of Gibson & Bray (1979), who included three subfamilies, the Derogeninae, Gonocercinae and Halipeginae, excluding the Dictysarcinae and Bunocotylinae, which they recognized at the family level. All three of the recognized subfamilies have one or more synonyms (see below).

According to Gibson & Bray (1979), this is a family which does not have a constant seminal storage and disposal apparatus in the female system. They considered that the variations of this apparatus probably occur because they are a diverse and successful, but relatively primitive, group, which appear to have evolved at about the time when the first modifications of the primitive arrangement of the seminal storage and disposal apparatus began to occur. The variety of conditions found in this group tend to parallel those which have occurred during the evolution of some of the more advanced hemiuroids, such as the hemiurids and lecithasterids.

The detailed comments and discussion of Gibson & Bray (1979) are not repeated here. The major changes have occurred in the subfamily Halipeginae and are especially associated with those genera included as synonyms of *Halipegus* by Gibson & Bray (1979). The studies of Hamann (1986) on the poorly known type-species of this genus have demonstrated that the 'wide concept', which Gibson & Bray (1979) considered to be the 'most useful' until *Halipegus* was better known, is untenable. Subsequent to this work, Kohn *et al.* (1990) resurrected *Genarchella* Travassos, Artigas & Pereira, 1928 for forms from South and Central American fishes. The same authors also considered *Paravitellotrema* Watson, 1976, *Caballeroiella* Lamothe-Argumedo, 1977 and *Quadripaludus* Jimenez, Guajardo & Briseno, 1981 synonyms of *Genarchella*. *Ghaziatrema* Nasir & Velásquez, 1977 is here also considered a synonym of *Genarchella*, as the filament on the eggs, the only differentiating feature, is difficult to see in whole-mounts. More recently, Jackson & Tinsley (1997) have revived *Dollfuschella* Vercammen-Grandjean, 1960. The position of this group of genera and their relatives has also been aided by recent work on *Genarchopsis* Ozaki, 1925 (by Shimazu, 1995) and *Thometrema* Amato, 1968 (by Kohn *et al*., 1983, 1990). The revision of *Halipegus* (*sensu lato*) is continued in this work; examination of the type-material of *Vitellotrema* Guberlet, 1928, including serial sections, indicates that this genus should be recognized on the basis of the nature of its terminal genitalia, as the sinus-sac encloses the pars prostatica but not the seminal vesicle, and the entire vitelline masses. Similarly, examination of the type-specimen of *Progenarchopsis* Fischthal, 1976 suggests that it should be treated as a synonym of *Dollfuschella*.

The members of the Halipeginae are often difficult to identify, because the critical features are usually parts of the terminal genitalia, which are often very small or poorly developed: in some cases, serial sectioning is almost essential, as whole-mounts do not provide the necessary data or may be misleading.

Chelatrema Gupta & Kumari, 1970, erected in an abstract, was listed by Gibson & Bray (1979) as a *genus inquirendum*. A more detailed description of this genus was published by Gupta & Kumari (1977), who attributed it to the family Hemiuridae and subfamily Arnolinae. However, the latter description clearly indicates that the worms upon which the genus is based are not hemiuroids - in this work they are considered as gorgoderids.

A new genus, *Austrohalipegus*, was added to the family by Cribb (1988b). This Australian freshwater form fits well with other Gondwanan forms of the genus, where the seminal vesicle and pars prostatica are external to the sinus-sac. Gibson & Køie (1991) discuss a possible Tethys origin for the family.

The genus *Gigantogorgoderina* and subfamily Gigantogorgodernininae were erected within the Gorgoderidae by Machida & Kuramochi (1994). It was later realized (Machida & Kamegai, 1997) that the genus is a synonym of *Gonocerca* Manter, 1925.

Family Derogenidae Nicoll, 1910

(Syns Halipegidae Poche, 1926; Liocercidae Ejsmont, 1931; Arnolidae Yamaguti, 1958[1])

Diagnosis: Body normally small, usually spindle-shaped to elongate-oval. Ecsoma absent. Body surface smooth. Oral and ventral suckers well developed; ventral sucker usually near middle of body, occasionally more anterior or posterior. Pharynx well developed. Oesophagus short. 'Drüsenmagen' usually present. Gut-caeca end blindly or form cyclocoel. Testes two, symmetrical to tandem, pre- or postovarian, in hindbody. Seminal vesicle thin-walled, oval, elongate or tubular, not constricted into portions, in forebody, occasionally partly or wholly enclosed within sinus-sac. Pars

[1] Yamaguti (1958a) used this at subfamily level, but it was raised to the family level by Mehra (1962b).

prostatica usually tubular, occasionally vesicular, occasionally enclosed within sinus-sac. Ejaculatory duct short or absent, often within sinus-sac. Sinus-sac normally present, occasionally absent, usually small and oval, often weakly developed, may enclose all or part of pars prostatica and seminal vesicle. Permanent sinus-organ present as small cone or absent. Hermaphroditic duct normally present, occasionally absent, usually short. Genital atrium present or absent, usually small. Genital pore midventral in forebody. Ovary oval, pre- or post-testicular. Mehlis' gland normally postovarian or occasionally at level of ovary. Seminal storage and disposal apparatus variable. Laurer's canal usually present, either opening dorsally to exterior or leading into Juel's organ, often dilated proximally to form small rudimentary seminal receptacle, which is occasionally enlarged to form an apparently functional canalicular seminal receptacle. Juel's organ absent or present in either rudimentary or fully-developed state. Blind seminal receptacle present rarely. Uterine seminal receptacle normally present, rarely absent. Uterus may or may not extend posterior to vitell-arium; significant proportion of uterus usually coiled in forebody. Eggs numerous, with or without filaments or threads, rarely with anopercular spine. Vitellarium one or two masses, entire or lobed (lobes normally shallow, rarely digitate), pre- or post-ovarian, symmetrical, oblique or tandem. Excretory vesicle Y-shaped; arms united in forebody. Parasitic usually in gut (normally stomach) of freshwater and marine teleosts, but occasionally recorded from amphibians, reptiles and freshwater shrimps. Type-genus *Derogenes* Lühe, 1900.

Key to subfamilies

1a. Testes posterior to ovary and vitellarium .... Gonocercinae Skrjabin & Guschanskaja, 1955.
1b. Testes anterior to ovary and vitellarium .... 2.

2a. Parasites primarily of freshwater teleosts, but occasionally present in brackish water or marine teleosts (Caspian, Black and Mediterranean Seas; one species in Danish waters), in amphibians, in reptiles and in freshwater shrimps; ventral sucker occasionally anterior to middle of body; uterus not present posterior to vitellarium; sinus-sac, when present, may enclose part of or entire seminal vesicle and/or pars prostatica .... Halipeginae Poche, 1926.
2b. Parasitic in marine teleosts; ventral sucker not present in anterior half of body; uterus often extends posterior to vitellarium; sinus-sac present, never enclosing any part of pars prostatica or seminal vesicle .... Derogeninae Nicoll, 1910.

Subfamily Derogeninae Nicoll, 1910

(Syns Liopyginae Ejsmont, 1931; Liocercinae Ejsmont, 1931; Genarchinae Skrjabin & Guschanskaja, 1955; Orthoruberinae Nasir & Gomez, 1977)

Diagnosis: Ventral sucker in middle or posterior to middle of body. Gut-caeca end blindly or form cyclocoel. Testes pre-ovarian, symmetrical to oblique. Seminal vesicle small, globular to tubular. Pars prostatica usually tubular, occasionally vesicular, short or long. Sinus-sac present, globular to cylindrical. Permanent sinus-organ present, small, cone-shaped. Hermaphroditic duct normally short. Genital atrium small, often filled by sinus-organ. Ovary closely posterior to testes. Laurer's canal present or absent, opening dorsally or into rudimentary Juel's organ, may dilate proximally forming large rudimentary or functional canalicular seminal receptacle. Blind seminal receptacle present when Laurer's canal and uterine seminal receptacle absent; latter usually present. Uterus coiled throughout hindbody and part of forebody; significant proportion of uterus often present posterior to vitellarium. Eggs without filaments or threads, but may have anopercular spine. Vitellarium two symmetrical to tandem, oval or slightly indented masses, posterior or occasionally lateral and posterolateral to ovary. Parasitic in gut (mainly stomach) of marine teleosts. Type-genus *Derogenes* Lühe, 1900.

Key to genera

1a. Uterus not normally extending posterior to vitellarium .... 4.
1b. Significant proportion of uterus posterior to vitellarium .... 2.

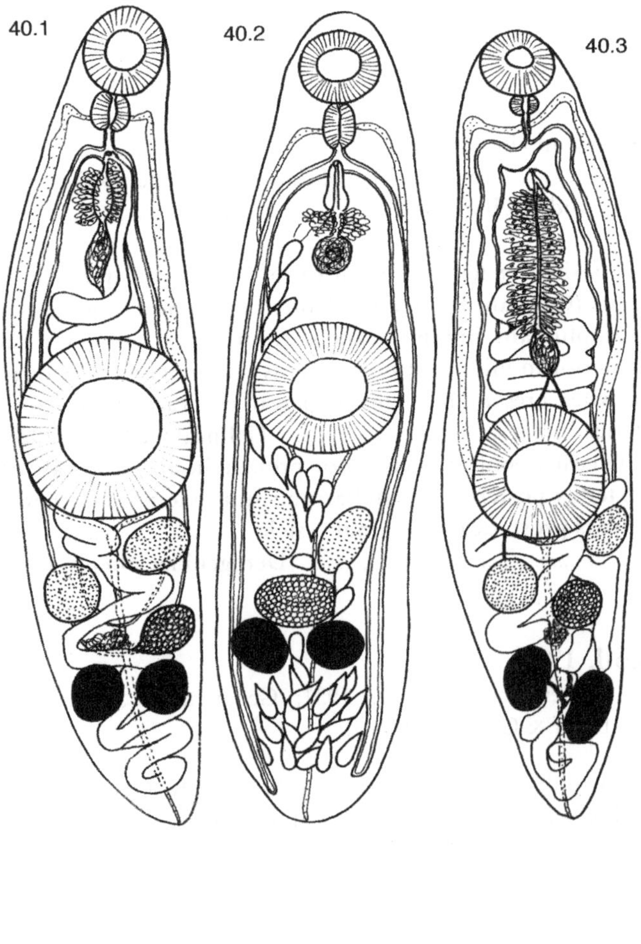

Figs 40.1–40.3.
40.1. *Progonus muelleri* (Levinsen, 1881). (After Odhner, 1905.)
40.2. *Derogenoides ovacutus* Nicoll, 1913. (After Nicoll, 1913.)
40.3. *Derogenes varicus* (Müller, 1784). (After Odhner, 1905.)

2a. Cyclocoel present .... ***Progonus* Looss, 1899. (Fig. 40.1)**
(Syn. *Genarches* Looss, 1902)

Diagnosis: Cyclocoel present. Testes symmetrical. Seminal vesicle elongate, spindle-shaped, elongate-oval or globular. Pars prostatica short; slightly vesicular. Sinus-sac small, globular. Ovary sinistral, halfway between testes and posterior extremity. Canalicular seminal receptacle present. Laurer's canal opens dorsally after passing through cyclocoel. ?Rudimentary Juel's organ present as small dilatation of Laurer's canal at junction with seminal receptacle. Uterine seminal receptacle absent. Uterus extends posteriorly to vitellarium, fills most of hindbody and some of forebody. Eggs without anopercular spine. Vitelline masses entire, symmetrical, postovarian. Parasitic in stomach of marine teleosts. Type-species *P. muelleri* (Levinsen, 1881).

2b. Gut-caeca end blindly .... **3.**

3a. Eggs drawn into sharp point at anopercular pole ....
.... ***Derogenoides* Nicoll, 1913. (Fig. 40.2)**

Diagnosis: Gut-caeca end blindly. Testes symmetrical to oblique. Seminal vesicle small, globular. Pars prostatica short. Sinus-sac somewhat cylindrical with proximal end slightly enlarged. Sinus-organ (?) presumably present. Ovary immediately posterior to testes. Laurer's canal and Juel's organ (?). Seminal receptacle (? rudimentary) reported. Uterine seminal receptacle (?). Much of uterus present posterior to vitellarium. Eggs drawn out to sharp point at anopercular pole. Vitelline masses entire, symmetrical, postovarian. Parasitic in stomach and intestine of marine teleosts. Type-species *D. ovacutus* Nicoll, 1913.

3b. Eggs lacking point at anopercular pole .... ***Derogenes* Lühe, 1900. (Fig. 40.3)**
(Syns *Liopyge* Looss, 1899; *Liocerca* Looss, 1902)

Diagnosis: Gut-caeca end blindly near posterior extremity. Testes symmetrical to oblique. Seminal vesicle globular to tubular and sinuous. Pars prostatica short to long. Sinus-sac globular. Male and female ducts unite within sinus-organ. Ovary usually closely posterior to testes, may be lateral to posterior testis when latter is oblique. In *D. varicus* Laurer's canal opens distally into rudimentary Juel's organ and dilates proximally, forming large rudimentary seminal receptacle. Uterine seminal receptacle present. Uterine field usually extends from posterior extremity to region of genital pore; significant proportion of uterus posterior to vitellarium. Eggs without anopercular spine. Vitelline masses symmetrical to oblique, globular or slightly indented, postovarian. Parasitic in stomach, oesophagus or occasionally gall-bladder of marine (? and freshwater) teleosts. Type-species *D. ruber* Lühe, 1900.

4a. Vitelline masses tandem to oblique, lateral and posterolateral to ovary; blind seminal receptacle present .... ***Leurodera* Linton, 1910. (Fig. 40.4)**
(Syn. *Orthoruberus* Nasir & Gomez, 1977)

Diagnosis: Body oval, stout. Gut-caeca end blindly near posterior extremity. Testes symmetrical, largely or partly extracaecal. Seminal vesicle tubular, slightly sinuous. Pars prostatica short, with few to many external gland-cells, may be partly tubular and vesicular anteriorly. Sinus-sac oval. Sinus-organ small (well developed and conical in *Orthoruberus*). Ovary just posterior to testes, close to posterior extremity. Laurer's canal, Juel's organ and uterine seminal receptacle apparently absent. Blind seminal receptacle present, large, anteroventral, anterolateral or anterodorsal to ovary. Uterus anterior to

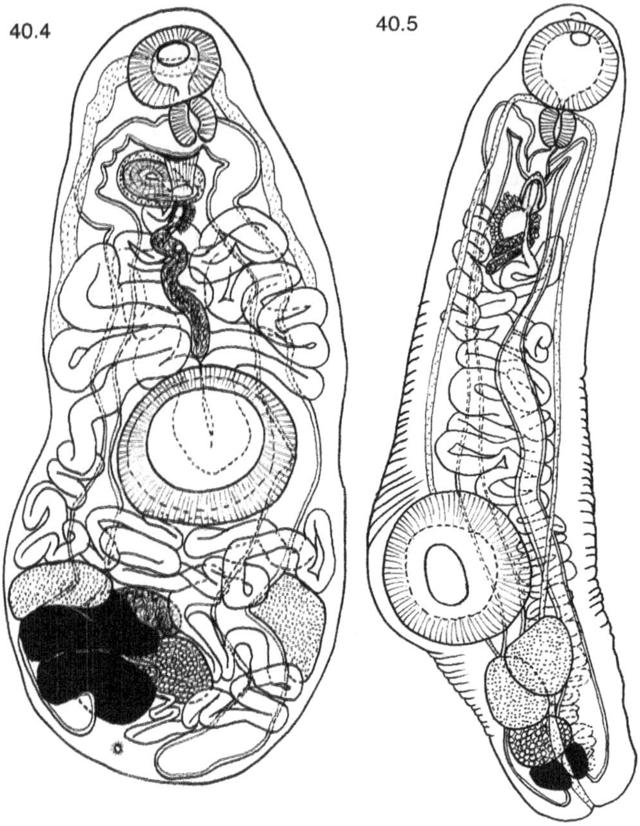

Figs 40.4–40.5.
40.4. *Leurodera decora* Linton, 1910. (After Yamaguti, 1971.)
40.5. *Gonocercella pacifica* Manter, 1940. (After Yamaguti, 1971.)

vitellarium, much of it pretesticular. Eggs without anopercular spine. Vitelline masses entire or slightly indented, tandem to oblique, lateral and posterolateral to ovary. Parasitic in gut (mainly stomach) of marine teleosts (especially Pomadasyidae). Type-species *L. decora* Linton, 1910.

4b. Vitelline masses symmetrical to oblique, postovarian; Laurer's canal and presumably uterine seminal receptacle present .... ***Gonocercella* Manter, 1940. (Fig. 40.5)**

Diagnosis: Ventral sucker in posterior half of body. Gut-caeca end blindly. Testes oblique. Seminal vesicle tubular, coiled. Pars prostatica vesicular. Sinus-sac small. Sinus-organ a muscular cone. Ovary immediately posterior to testes, close to posterior extremity. Laurer's canal opens dorsally (according to MacCallum, 1913). Blind or canalicular seminal receptacle absent. Uterine seminal receptacle presumably present. Juel's organ presumably absent. Uterus mainly coiled in forebody, not reaching posterior to vitellarium. Eggs without anopercular spine. Vitelline masses entire, symmetrical, postovarian, close to posterior extremity. Parasitic in stomach of marine teleosts. Type-species *G. pacifica* Manter, 1940.

Subfamily Gonocercinae Skrjabin & Guschanskaja, 1955

(Syns Hemiperinae Yamaguti, 1958; Gigantogorgoderininae Machida & Kuramochi, 1994)

Diagnosis: Ventral sucker posterior to middle of body. Gut-caeca end blindly. Testes postovarian, tandem to symmetrical, near posterior extremity. Seminal vesicle usually small, oval to tubular, in forebody. Pars prostatica tubular, short, linked to seminal vesicle by short, aglandular duct. Sinus-sac absent or poorly developed. Sinus-organ absent or present as small, blunt cone. Hermaphroditic duct absent or short. Genital atrium small or apparently absent. Ovary between testes and vitellarium. Laurer's canal present, opening dorsally, dilated proximally, forming small rudimentary seminal receptacle. Juel's organ absent. Uterine seminal receptacle present. Uterus entirely pre-ovarian; most of coils usually in forebody. Eggs filamented or not. Vitellarium two entire or indented oval masses, symmetrical, antero- to posterolateral to ovary. Parasitic in stomach or branchial cavity of marine teleosts. Type-genus *Gonocerca* Manter, 1925.

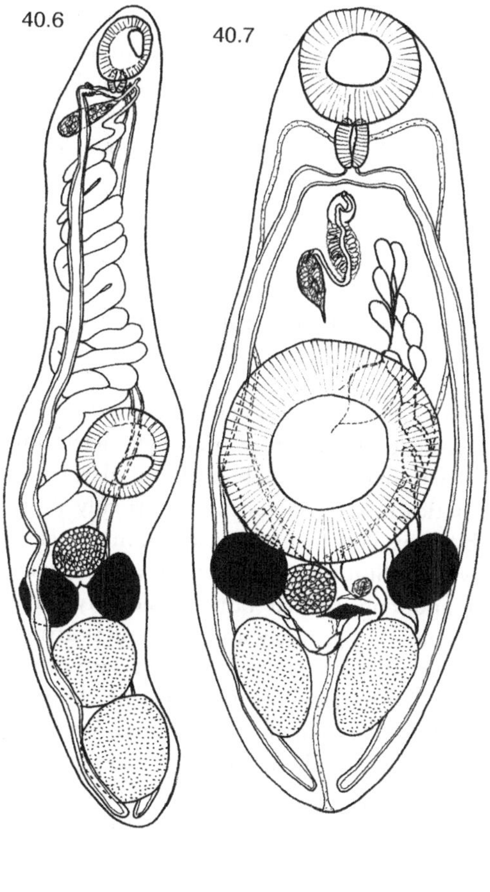

Figs 40.6–40.7.
40.6. *Gonocerca phycidis* Manter, 1925. Lateroventral view. (Modified after Gibson, 1976.)
40.7. *Hemipera ovocaudata* Nicoll, 1913. (After Nicoll, 1913.)

Key to genera

1a. Eggs without filaments; sinus-sac and sinus-organ absent (although reported by some authors) .... *Gonocerca* Manter, 1925. (Fig. 40.6)
(Syn. *Gigantogorgoderina* Machida & Kuramochi, 1994)

Diagnosis: Ventral sucker in posterior half of body. Testes tandem to almost symmetrical, at posterior extremity of body. Seminal vesicle small, thin-walled, oval, closely posterior to genital pore. Pars prostatica short; tubular. Sinus-sac and sinus-organ absent. Hermaphroditic duct absent (assuming that the small cavity into which male and female ducts open is the genital atrium). Genital atrium small or apparently absent. Ovary median. Uterus entirely pre-ovarian, largely in forebody. Eggs without filaments. Vitelline masses lateral or anterolateral to ovary, entire to indented. Parasitic in stomach of marine teleosts (especially in mid-ocean). Type-species *G. phycidis* Manter, 1925.

1b. Eggs filamented; sinus-sac weakly developed; sinus-organ present .... *Hemipera* Nicoll, 1913. (Fig. 40.7)
(Syn. *Hemiperina* Manter, 1934)

Diagnosis: Ventral sucker in posterior half of body. Testes symmetrical to oblique at posterior extremity. Seminal vesicle oval to tubular. Pars prostatica tubular, short. Seminal vesicle and pars prostatica may apparently be enclosed by common, subglobular, parenchymatous capsule. Sinus-sac weakly developed, with diffuse musculature, enclosing base of sinus-organ and proximal region of genital atrium. Sinus-organ a blunt cone. Male and female ducts may open separately on sinus-organ or form short hermaphroditic duct. Ovary median. Uterus entirely pre-ovarian, largely in forebody. Eggs filamented. Vitelline masses antero- to posterolateral to ovary, entire or slightly indented. Parasitic in stomach or branchial cavity of marine teleosts. Type-species *H. ovocaudata* Nicoll, 1913.

Subfamily Halipeginae Poche, 1926

(Syns Arnolinae Yamaguti, 1958; Monovitellinae Ataev, 1970; Ghaziatrematinae Nasir & Velásquez, 1977)

Diagnosis: Ventral sucker usually near middle of body, occasionally more anterior or posterior. Gut-caeca end blindly or form cyclocoel. Testes pre-ovarian, symmetrical to oblique or occasionally tandem to oblique. Seminal vesicle globular to tubular, sometimes entirely or partly internal. Pars prostatica tubular to vesicular, normally short, sometimes internal. Sinus-sac present or absent, usually weakly developed, may enclose pars prostatica and all or part of seminal vesicle. Sinus-organ present or absent, when present usually small, poorly developed and cone-shaped but occasionally large. Hermaphroditic duct usually short, occasionally long, rarely absent. Genital atrium small. Ovary usually close to posterior extremity. Laurer's canal present, with dorsal pore or short and leading into Juel's organ. Rudimentary seminal receptacle often present when Laurer's canal opens dorsally. Blind or canalicular seminal receptacle absent. Uterine seminal receptacle present. Uterus entirely or almost entirely anterior to vitellarium, coils extend into forebody, except in cases with ventral sucker inside anterior half of body. Eggs with or without filaments or threads. Vitellarium one or two masses at posterior extremity of body, usually entire but sometimes with indistinct or digitate lobes. Excretory bifurcation normally in hindbody, occasionally in forebody. Usually parasitic in gut (normally stomach) of freshwater teleosts, also recorded from brackish water and marine teleosts close to the ancient Sarmatic Sea region (Caspian, Black and Mediterranean Seas), amphibians, reptiles and freshwater shrimps (majority of genera occur in Asia). Type-genus *Halipegus* Looss, 1899.

Key to genera

1a. Eggs with filaments or threads .... **2.**
1b. Eggs without filaments or threads .... **11.**

2a. Cyclocoel present .... **3.**
2b. Gut-caeca end blindly .... **4.**

3a. Cyclocoel and excretory bifurcation in forebody .... ***Allotangiopsis* Yamaguti, 1971. (Fig. 40.8)**

Diagnosis: Ventral sucker in middle of body. Cyclocoel present in forebody. Testes symmetrical to oblique. Seminal vesicle tubular, stout, recurved. Pars prostatica short, tubular, not delimited. Sinus-sac forms muscular wall surrounding hermaphroditic duct (?). Sinus-organ absent (?). Laurer's canal and Juel's organ (?). Seminal receptacle reported (presumably either rudimentary type or Juel's organ). Uterine seminal receptacle situated, at least partly, in distal region of uterus. Uterus not reaching

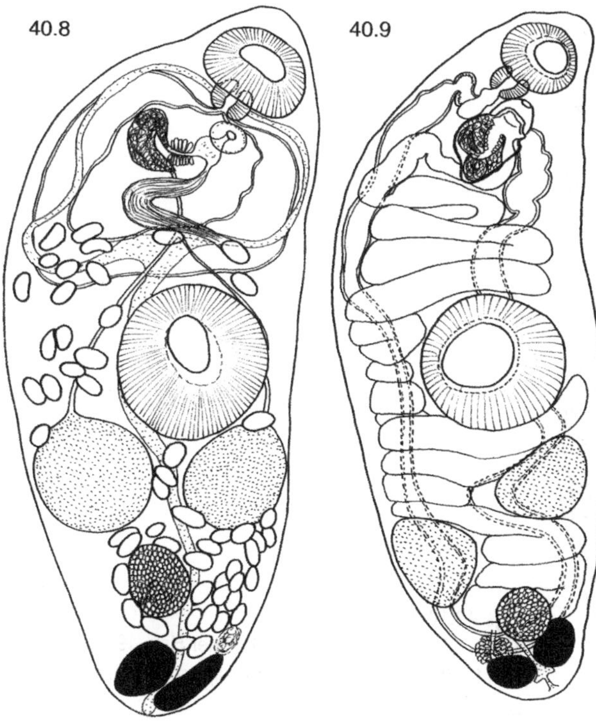

Figs 40.8–40.9.
40.8. *Allotangiopsis shanghaiensis* (Yeh & Wu, 1955). (Modified after Yeh & Wu, 1955.)
40.9. *Genarchopsis goppo* Ozaki, 1925. (Modified after Shimazu, 1995.)

posterior to vitellarium; significant proportion present in forebody. Eggs filamented. Vitellarium two lobed masses, slightly oblique, at posterior extremity. Excretory bifurcation in forebody. Parasitic in gonads of freshwater shrimps as (?) progenetic metacercaria (China). Type-species *A. shanghaiensis* (Yeh & Wu, 1955).

3b. Cyclocoel and excretory bifurcation in hindbody .... ***Genarchopsis* Ozaki, 1925. (Fig. 40.9)**

(Syns *Ophiocorchis* Srivastava, 1933; (?) *Pseudogenarchopsis* Yamaguti, 1971)

Diagnosis: Ventral sucker in posterior half, or occasionally in middle, of body. Cyclocoel present in hindbody; oesophageal pouch often present. Testes usually oblique, occasionally symmetrical. Seminal vesicle tubular to elongate-saccular, coiled. Pars prostatica short, may be slightly vesicular. Sinus-sac present, thin-walled, encloses pars prostatica and seminal vesicle (? sometimes only part of latter). Hermaphroditic duct within sinus-organ. Sinus-organ a strongly muscular, blunt cone. Ovary usually sinistral and well posterior to testes. Laurer's canal opening into well-developed Juel's organ or dorsally. Uterus pre-ovarian, coils extending into forebody. Eggs with long, polar filament. Vitellarium two entire or indented masses at posterior extremity, symmetrical to oblique. Parasitic mainly in stomach of freshwater teleosts, occasionally in amphibians, ? accidentally in snakes (southern Asia and Far East). Type-species *G. goppo* Ozaki, 1925.

4a. Vitellarium a single mass .... ***Chenia* Hsu, 1954. (Fig. 40.10)**

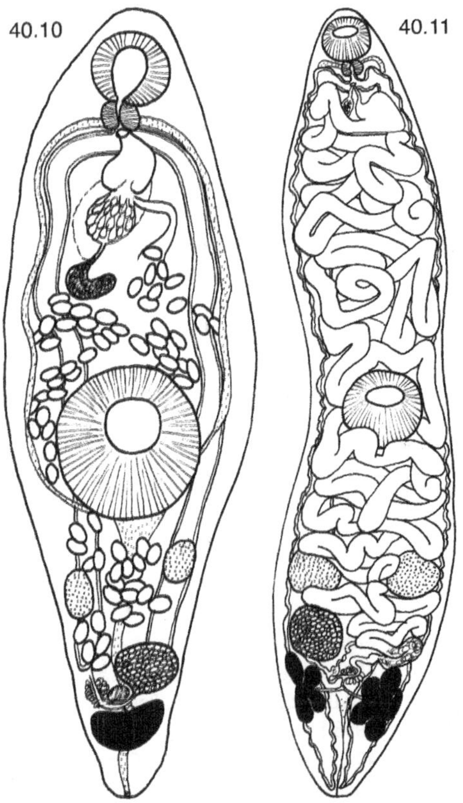

Figs 40.10–40.11.
40.10. *Chenia cheni* Hsu, 1954. (After Hsu, 1954.)
40.11. *Halipegus ovocaudatus* (Vulpian, 1859). (After Looss, 1894.)

Diagnosis: Ventral sucker just posterior to middle of body. Gut-caeca terminate blindly at level of ovary. Testes oblique. Seminal vesicle a curved, elongate sac. Pars prostatica internal (?). Sinus-sac enclosing attenuated anterior portion of seminal vesicle and pars prostatica. Sinus-organ not reported. Seminal receptacle (presumably either rudimentary or Juel's organ) reported. Uterus entirely pre-ovarian, coils extending into forebody. Eggs reniform, with two filaments at one pole. Vitellarium a single compact mass at posterior extremity. Parasitic in gut of freshwater gobiid teleosts (China). Type-species *C. cheni* Hsu, 1954.

4b. Vitellarium two similar masses .... 5.

5a. Sinus-sac absent .... *Halipegus* Looss, 1899. (Fig. 40.11)

Diagnosis: Ventral sucker near middle of body. Gut-caeca end blindly. Testes symmetrical to slightly oblique. Seminal vesicle saccular. Pars prostatica short, may be reduced. Sinus-sac absent. Hermaphroditic duct short to absent. Genital atrium small. Ovary usually separated from testes by uterus. Laurer's canal present. Juel's organ absent. Uterus not extending posteriorly to vitellarium, usually with about equal distribution in fore- and hindbodies. Eggs with long, single filament at anopercular pole. Vitellarium two relatively symmetrical masses, usually clearly four- and five-lobed. Parasitic in upper regions of gut (usually stomach or mouth, occasionally ear) of amphibians, occasionally in fish and reptiles (?accidental). Type-species *H. ovocaudatus* (Vulpian, 1859).

5b. Sinus-sac present but often small or weakly developed .... 6.

6a. Sinus-sac large, enclosing entire seminal vesicle; ventral sucker not in middle of body; eggs with more than single filament; in marine, euryhaline and freshwater fishes (Palaearctic) .... 7.

6b. Sinus-sac small, not enclosing seminal vesicle; ventral sucker near middle of body; eggs with single filament; in Neotropical, southern Nearctic and African freshwater fishes, occasionally amphibians and reptiles .... 8.

7a. Ventral sucker in anterior half of body; vitelline masses entire; in marine or euryhaline fishes .... *Magnibursatus* Naidenova, 1969. (Fig. 40.12)

Diagnosis: Ventral sucker in anterior half of body. Gut-caeca end blindly close to posterior extremity. Testes oblique, in anterior hindbody. Seminal vesicle coiled, tubular, internal. Pars prostatica short, straight, internal. Sinus-sac enclosing seminal vesicle and pars prostatica, plus the metraterm distally, in forebody. Sinus-organ small, conical. Short hermaphroditic duct and small genital atrium present. Laurer's canal ends blindly (? in Juel's organ). Uterine seminal receptacle present. Ovary near posterior extremity, may be separated from testes by uterus. Uterus almost entirely anterior to vitellarium; few coils extending into forebody. Eggs with numerous (? six to eight) filaments (? threads) at each end, may form reticulate pattern. Vitellarium two oblique to symmetrical masses, close together at posterior extremity. Parasitic in stomach of euryhaline or marine teleosts (Black Sea, north-east Atlantic region). Type-species *M. skrjabini* (Vlasenko, 1931).

7b. Ventral sucker in posterior half of body; vitelline masses lobed; in freshwater fishes .... *Anguillotrema* Chin & Ku, 1974. (Fig. 40.13)

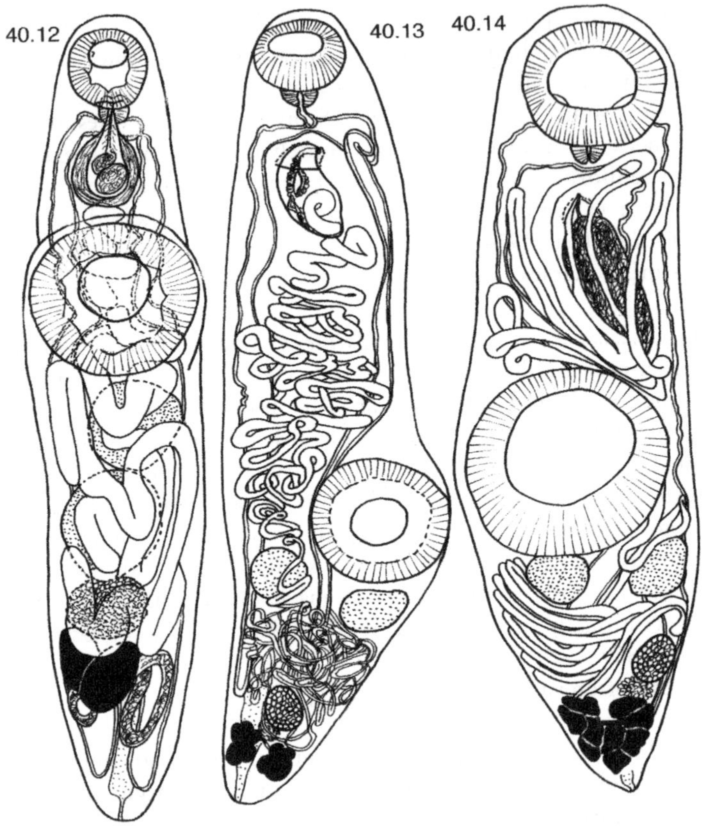

Figs 40.12–40.14.
40.12. *Magnibursatus caudofilamentosa* (Reimer, 1971). (After Gibson & Køie, 1991.)
40.13. *Anguillotrema papillatum* Chin & Ku, 1974. (After Chin & Ku, 1974.)
40.14. *Dollfuschella rodhaini* Vercammen-Grandjean, 1960. (After Jackson & Tinsley, 1997.)

Diagnosis: Ventral sucker in posterior half of body. Gut-caeca end blindly. Testes symmetrical. Seminal vesicle tubular, coiled, internal. Pars prostatica vesicular, internal. Sinus-sac large, weakly muscled, enclosing pars prostatica and seminal vesicle. Sinus-organ muscular, conical. Laurer's canal and Juel's organ (?). Uterus entirely anterior to vitellarium; large proportion in forebody. Eggs with a filament at each end and small threads surrounding base of filament at one end. Vitellarium two masses with four and five lobes, symmetrical at posterior extremity. Parasitic in stomach of eels in freshwater (China). Type-species *A. papillatum* Chin & Ku, 1974.

8a. Sinus-sac enclosing short hermaphroditic duct and pars prostatica .... 9.
8b. Small sinus-sac enclosing only hermaphroditic duct or part of this duct .... 10.

9a. Vitelline masses appear broken into four and five independent follicles; in amphibians and reptiles, Africa .... *Dollfuschella* Vercammen-Grandjean, 1960. (Fig. 40.14)
(Syn. *Progenarchopsis* Fischthal, 1976)

Diagnosis: Ventral sucker near middle of body. Gut-caeca end blindly. Testes symmetrical to slightly oblique. Seminal vesicle saccular, elongate. Sinus-sac present, small, weakly developed, enclosing pars prostatica, ejaculatory duct, distal metraterm and short hermaphroditic duct. Sinus-organ a small, permanent, extensible, conical papilla. Genital atrium small. Ovary separated from testes by uterus. Laurer's canal present, with dorsal pore; proximal region dilated forming rudimentary seminal receptacle. Juel's organ absent. Uterus not extending posteriorly to vitellarium, usually with about equal distribution in fore- and hindbodies. Eggs with single filament at anopercular pole. Vitellarium two relatively symmetrical masses of four and five very distinct lobes with appearance of being isolated follicles. Parasitic in stomach (or oesophagus) of amphibians and reptiles (Africa). Type-species *D. rodhaini* Vercammen-Grandjean, 1960.

9b. Vitelline masses entire; in amphibians and reptiles, Nearctic .... *Vitellotrema* Guberlet, 1928. (Fig. 40.15)

Diagnosis: Ventral sucker near middle of body. Gut-caeca end blindly. Testes symmetrical to slightly oblique. Seminal vesicle saccular, external. Pars prostatica short, internal. Hermaphroditic duct short. Sinus-sac small, enclosing pars prostatica. Sinus-organ apparently not permanent. Genital atrium small. Ovary well separated from testes by uterus. Laurer's canal present, with dorsal pore. Juel's organ absent. Uterus not extending posteriorly to vitellarium, usually with about equal distribution in fore- and hindbodies. Eggs with single filament at anopercular pole. Vitellarium two relatively symmetrical oval masses. Parasitic in stomach (and mouth) of amphibians (urodeles) and snakes (southern Nearctic). Type-species *V. fusipora* Guberlet, 1928.

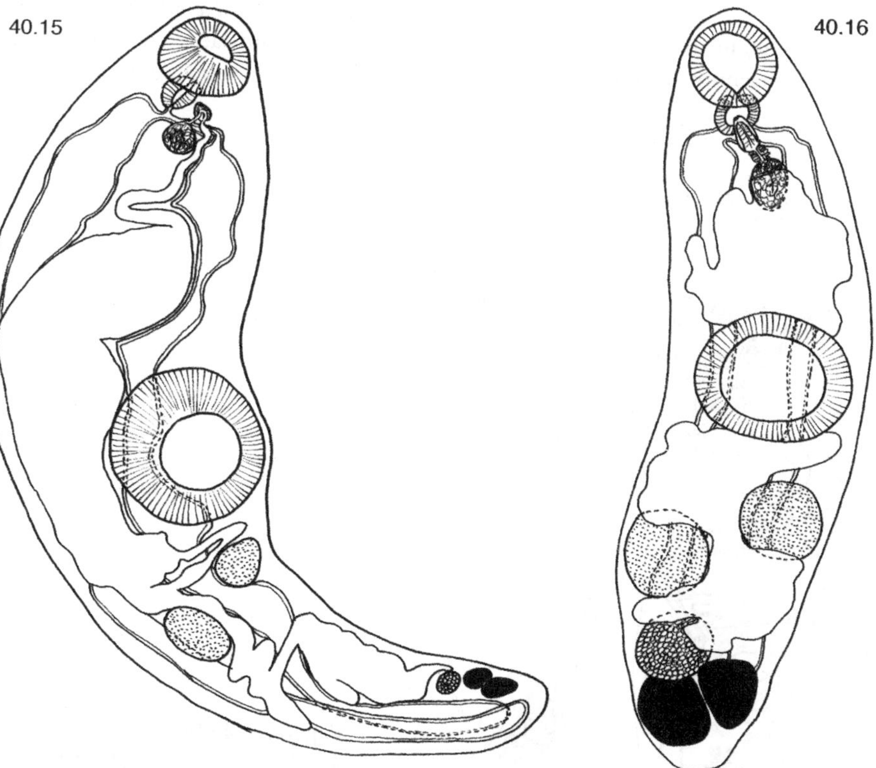

Figs 40.15–40.16.
40.15. *Vitellotrema fusipora* Guberlet, 1928. (After Guberlet, 1928.)
40.16. *Genarchella tropica* (Manter, 1936). (Modified after Scholz *et al.*, 1995.)

10a. Entire hermaphroditic duct within sinus-sac; vitelline masses entire; in Neotropical (including Central American) fishes .... *Genarchella* Travassos, Artigas & Pereira, 1928. (Fig. 40.16)

(Syns *Paravitellotrema* Watson, 1976; *Caballeroiella* Lamothe-Argumedo, 1977; *Ghaziatrema* Nasir & Velásquez, 1977; *Quadripaludus* Jimenez, Guajardo & Briseno, 1981)

Diagnosis: Ventral sucker large, near middle of body. Gut-caeca end blindly. Testes symmetrical to oblique. Seminal vesicle saccular, external. Pars prostatica short, external. Hermaphroditic duct short, within sinus-sac. Sinus-sac present, small, sometimes weakly developed. Sinus-organ present as small, conical or asymmetrical, muscular papilla. Genital atrium normally small. Ovary usually but not always separated from testes by uterus. Laurer's canal present, with dorsal pore; proximal region may be dilated forming rudimentary seminal receptacle. Uterus not extending posteriorly to vitellarium, usually with similar distribution in fore- and hindbodies. Eggs with long, single filament at anopercular pole. Vitellarium two relatively symmetrical masses, usually irregularly oval. Parasitic in upper regions of gut (usually stomach) of freshwater teleosts (South and Central America). Type-species *G. genarchella* Travassos, Artigas & Pereira, 1928.

10b. Hermaphroditic duct extends well outside sinus-sac; vitelline masses with oval to digitate lobes; in Neotropical fishes .... *Thometrema* Amato, 1968. (Fig. 40.17)

Diagnosis: Ventral sucker in middle of body. Gut-caeca end blindly near posterior extremity. Testes oblique to symmetrical. Seminal vesicle elongate, saccular, may be

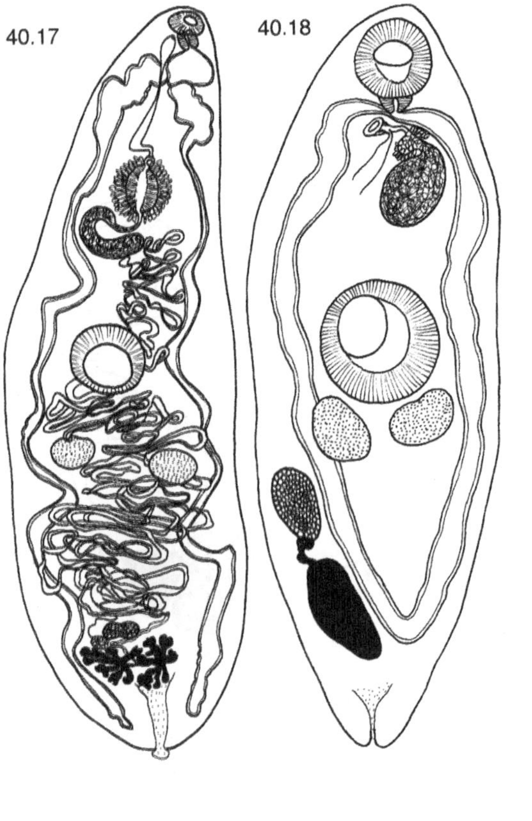

Figs 40.17–40.18.
40.17. *Thometrema magnifica* (Szidat, 1951). (After *T. portoalegrensis* Amato, 1968 of Amato (1968).)
40.18. *Monovitella cyclointestina* Ataev, 1970. (Modified after Ataev, 1970.)

attenuated anteriorly. Pars prostatica vesicular, surrounded by dense, oval mass of gland-cells. Sinus-sac small, enclosing base of sinus-organ. Hermaphroditic duct long, extends outside sinus-sac. Sinus-organ present as small papilla in base of oval genital atrium. Ovary near posterior extremity, separated from testes by loops of uterus. Laurer's canal present. Juel's organ absent. Uterus entirely pre-ovarian; coils extending into forebody. Eggs with one polar filament. Vitellarium two symmetrical masses of about seven to ten oval to short, digitate lobes, which may be irregularly expanded distally. Parasitic in stomach of freshwater teleosts, occasionally in estuarine environments (South America). Type-species *T. magnifica* (Szidat, 1954) (syn. *T. portoalegrensis* Amato, 1968 - type by original designation).

11a. Gut-caeca end blindly .... **14.**
11b. Cyclocoel present .... **12.**

12a. Ventral sucker in anterior half of body; uterine coils retained in hindbody .... **13.**
12b. Ventral sucker near middle of body; uterine coils extend into forebody .... **15.**

13a. Vitellarium a single mass .... ***Monovitella* Ataev, 1970. (Fig. 40.18)**
Diagnosis: Ventral sucker in middle of body. Cyclocoel present in hindbody. Testes symmetrical. Seminal vesicle saccular. Pars prostatica short; vesicular. Sinus-sac reported absent (figure suggests it might be present as weakly muscled, tubular jacket of hermaphroditic duct). Sinus-organ absent (?). Genital atrium short. Ovary lateral, apparently extracaecal, situated between right testis and vitellarium. Laurer's canal and Juel's organ (?). Uterus almost entirely anterior to vitellarium, but descending loop reaches close to posterior extremity; coils extend into forebody. Eggs not filamented. Vitellarium a single, entire mass, lateral, apparently extracaecal between ovary and posterior extremity. Parasitic in intestine of brackish water teleosts (Caspian Sea). Type-species *M. cyclointestina* Ataev, 1970.
13b. Vitellarium two similar masses situated close together .... ***Tangiopsis* Skrjabin & Guschanskaja, 1955. (Fig. 40.19)**
Diagnosis: Ventral sucker in middle of body. Cyclocoel present; caeca unite anterior to testes. Testes symmetrical to oblique. Seminal vesicle tubular, recurved. Pars prostatica small; external gland-cells (?) free in parenchyma. Sinus-sac apparently absent. Temporary sinus-organ may be present (?). Ovary between right testis and vitellarium. Laurer's canal opens dorsally. Juel's organ presumably absent. Rudimentary seminal receptacle apparently present. Uterus not passing posterior to vitellarium, almost entirely pre-ovarian, with coils extending into forebody, apparently filled with spermatozoa throughout most of its length. Eggs without filaments. Vitellarium two slightly indented, symmetrical masses, close together at posterior extremity, united by short duct. Parasitic in stomach of freshwater teleosts (China). Type-species *T. chinensis* (Tang, 1951).

14a. Sinus-sac encloses pars prostatica and seminal vesicle .... ***Arnola* Strand, 1942. (Fig. 40.20)**
(Syn. *Arnoldia* Vlasenko, 1931, pre-occ.)

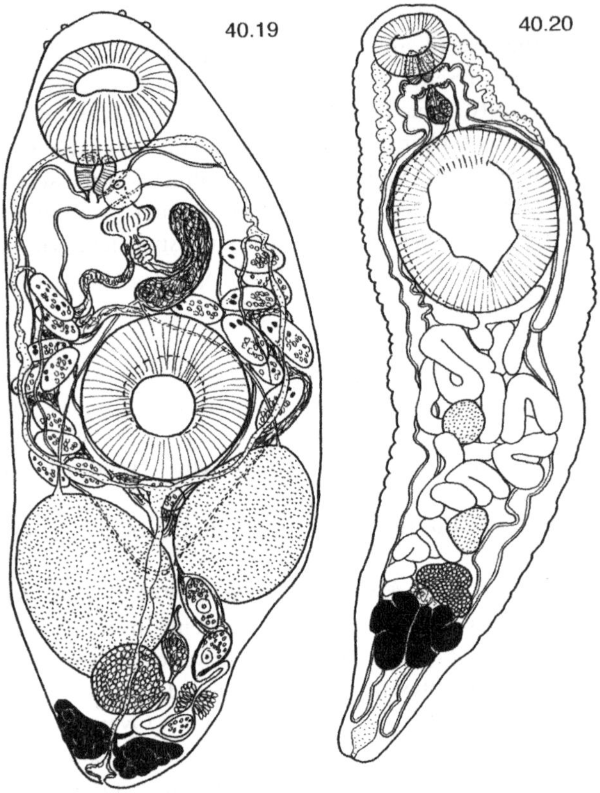

Figs 40.19–40.20.
40.19. *Tangiopsis chinensis* (Tang, 1951). (After Tang, 1951.)
40.20. *Arnola microcirrus* (Vlasenko, 1931). (Original ex *Diplodus annularis*, Adriatic Sea.)

Diagnosis: Ventral sucker in anterior half of body. Gut-caeca end blindly close to posterior extremity. Testes oblique, separated by uterus. Seminal vesicle coiled, tubular, internal. Pars prostatica short, straight; internal. Sinus-sac enclosing seminal vesicle and pars prostatica plus metraterm distally. Permanent sinus-organ absent. Short hermaphroditic duct and genital atrium present. Ovary immediately posterior to hind testis. Short, dilated Laurer's canal, containing sperm, opens into well-developed Juel's organ. Uterine seminal receptacle present. Uterus entirely anterior to vitellarium; coils not extending into forebody. Eggs without filaments. Vitellarium two symmetrical masses, situated close together at posterior extremity, slightly indented, usually indicating three and four lobes. Parasitic in stomach of marine teleosts (*Diplodus*) in Black Sea and Adriatic Sea. Type-species *A. microcirrus* (Vlasenko, 1931).

14b. Sinus-sac small, not enclosing pars prostatica and seminal vesicle .... *Deropegus* McCauley & Pratt, 1961. (Fig. 40.21)

(Syn. *Parahalipegus* Wootton & Powell, 1964)

Diagnosis: Ventral sucker in anterior half of body. Gut-caeca end blindly near posterior extremity. Testes tandem, oblique or symmetrical. Seminal vesicle saccular. Pars prostatica short, tubular to slightly vesicular, surrounded by dense layer of gland-cells. Sinus-sac apparently present, but weakly developed, small. Sinus-organ a muscular cone. Ovary usually separated from testes by loops of uterus. Laurer's canal present, apparently opening dorsally, slightly dilated proximally, forming

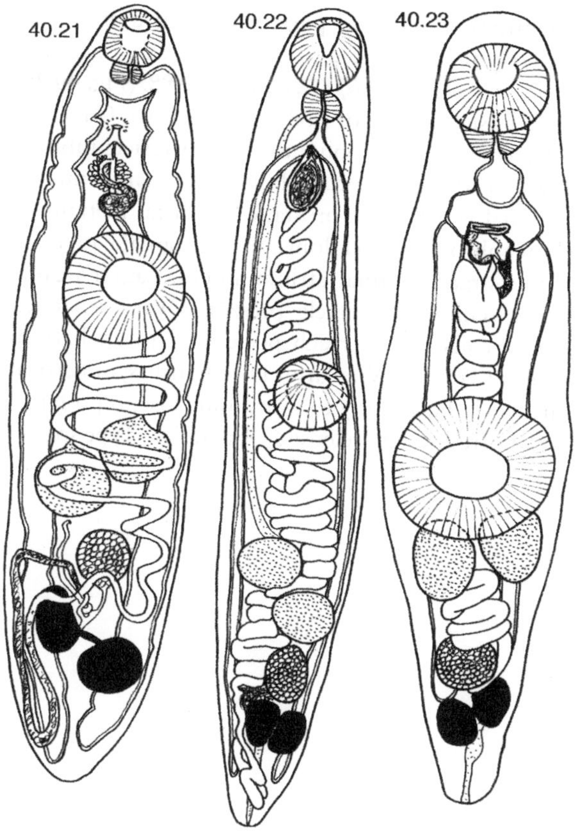

Figs 40.21–40.23.
40.21. *Deropegus aspina* Ingles, 1936. (Modified after McCauley & Pratt, 1961.)
40.22. *Tyrrhenia blennii* Paggi & Orecchia, 1975. (After Paggi & Orecchia, 1975.)
40.23. *Austrohalipegus anguillicola* Cribb, 1988. (After Cribb, 1988b.)

rudimentary seminal receptacle. Juel's organ presumably absent. Uterine coils not extending into forebody; one loop may reach posteriorly to vitellarium. Eggs without filaments. Vitellarium two symmetrical to oblique, entire or slightly lobed masses, close to posterior extremity. Parasitic in stomach of amphibians and teleosts in freshwater (North America). Type-species *D. aspina* (Ingles, 1936).

15a. Sinus-sac encloses seminal vesicle and pars prostatica; uterus extends posteriorly to vitellarium .... ***Tyrrhenia* Paggi & Orecchia, 1975. (Fig. 40.22)**

Diagnosis: Ventral sucker near middle of body. Gut-caeca end blindly near posterior extremity. Testes oblique. Seminal vesicle saccular, but attenuated distally, internal. Pars prostatica (?) short, tubular; internal. Sinus-sac enclosing entire seminal vesicle and pars prostatica, plus metraterm distally. Permanent sinus-organ absent. Hermaphroditic duct short. Genital atrium apparently absent (or small). Laurer's canal present, with small rudimentary seminal receptacle. Juel's organ absent. Uterine seminal receptacle present. Ovary immediately posterior to hind testis. Uterine coils reach posteriorly to vitellarium and extend into forebody. Eggs without filaments. Vitellarium two entire, symmetrical masses, situated close together immediately posteriorly to ovary and close to posterior extremity. Parasitic in stomach and on gills of marine teleosts (*Blennius*) in Mediterranean Sea. Type-species *T. blennii* Paggi & Orecchia, 1975.

15b. Sinus-sac does not enclose seminal vesicle and pars prostatica; uterus does not extend posteriorly to vitellarium .... *Austrobalipegus* Cribb, 1988. (Fig. 40.23)

Diagnosis: Ventral sucker just inside posterior half of body or almost so. Gut-caeca end blindly near posterior extremity. Testes slightly oblique. Seminal vesicle tubular, broad, slightly coiled, external. Pars prostatica short, tubular; external. Sinus-sac weakly developed, not enclosing seminal vesicle, pars prostatica or distal metraterm. Permanent sinus-organ absent. Hermaphroditic duct long, may evaginate to form relatively large, broad temporary sinus-organ. Genital atrium absent. Laurer's canal present. Juel's organ absent. Uterine seminal receptacle present. Ovary separated from testes by coils of uterus. Uterine coils do not reach posteriorly to vitellarium and extend into forebody. Eggs without filaments. Vitellarium two entire, symmetrical masses, situated close together immediately posteriorly to ovary and close to posterior extremity. Parasitic in stomach of freshwater teleosts, Australia. Type-species *A. anguillicola* Cribb, 1988.

41 Family Dictysarcidae Skrjabin & Guschanskaja, 1955

D.I. GIBSON

Department of Zoology, The Natural History Museum, London, UK

Introduction

The Dictysarcidae is not the oldest family-group name linked to this group, but I have considered it necessary to use it for the reasons given below. The genus *Cylindrorchis* was erected by Southwell (1913) associated with a 'provisional description' of *C. tenuicutis* Southwell, 1913, although he stated that it was 'impossible to satisfactorily make out with *certainty*, the precise details of the reproductive system'. This species, recorded from the swim-bladder of *Tetrodon stellatus*, has not been found subsequently. Southwell claimed that this genus had no close relationship with any known genus (it is worth noting that another swim-bladder form, *Isoparorchis* Southwell, 1913, was erected in the same paper). The family Cylindrorchiidae (originally spelled Cylindrorchidae) was founded by Poche (1926) for *Cylindrorchis*. However, Yamaguti (1953) included the genus in the Callodistomidae, but subsequently (Yamaguti, 1958a), apparently unaware of Poche's designation, erected the subfamily Cylindrorchiinae within the Callodistomidae. Skrjabin (1960) and Manter (1962) considered that *Cylindrorchis* could not be a callodistomid. In 1971 Yamaguti accepted the Cylindrorchiidae of Poche and included it in a new superfamily, the Isoparorchioidea, along with the families Isoparorchiidae, Albulatrematidae, Aerobiotrematidae, Tetrasteridae and Dictysarcidae: these families, all erected for individual genera, have in common an almost unique niche, the swim-bladder of physostomatous teleosts. A link with the hemiuroids had been made by Skrjabin & Guschanskaja (1956) when they included the Dictysarcinae within the family Halipegidae (= Derogenidae). The most detailed revision of these swim-bladder forms was carried out by Gibson & Bray (1979), in their revision of the Hemiuroidea, who retained the family Isoparorchiidae as distinct, because it occurs in a freshwater teleost and possesses several primitive features, such as Laurer's canal, a tubular vitellarium and a well-developed, muscular sinus-organ. However, these authors synonymized all of the other families listed above (plus the Dollfustravassosiidae) within the hemiuroid family Dictysarcidae. The appellation Dictysarcidae was used because of the questionable status of *Cylindrorchis*, which they listed as a *genus inquirendum* and considered may prove to be a synonym of *Elongoparorchis* if suggested errors in the interpretation of the reproductive organs of *C. tenuicutis* are substantiated.[1] For this present study, type-material of *C. tenuicutis* was traced to the Collection of the Zoological Survey of India, but the curator considered it too fragile to lend. Gibson & Bray (1979) accepted within the Dictysarcidae the subfamilies Dictysarcinae and Albulatrematinae (syns Pelorohelminthinae and Tetrasterinae) and listed the Cylindrorchiinae as a *subfam. inq.*

Gibson & Bray (1979) went further with their synonymies and considered *Pelorohelmins* Fischthal & Kuntz, 1964, *Tetraster* Oshmarin, 1965 and *Dollfustravassosius* Texeira de Freitas & Kohn, 1967 synonyms of *Elongoparorchis*. These changes were supported by Jones & Khalil (1984).

Family Dictysarcidae Skrjabin & Guschanskaja, 1955

(Syns Cylindrorchiidae Poche, 1926 (?); Aerobiotrematidae Yamaguti, 1958; Pelorohelminthidae Fischthal & Kuntz, 1964; Albulatrematidae Yamaguti, 1965; Tetrasteridae Oshmarin, 1965; Dollfustravassosiidae Teixeira de Freitas & Kohn, 1967)

Diagnosis: Body usually large, oval, stout or flattened. Ecsoma absent. Body surface smooth. Oral and ventral suckers well developed; ventral sucker in anterior half of body. Pharynx well developed. Oesophagus short. 'Drüsenmagen' normally present (?). Gut-caeca end blindly close to posterior extremity. Testes two, large, symmetrical, oval or elongate, pre-ovarian, just posterior to ventral sucker. Seminal vesicle tubular, in forebody. Pars prostatica tubular. Sinus-sac well developed, poorly developed or absent. Permanent sinus-organ absent (? or present as small papilla); temporary sinus-organ may form. Genital atrium small or absent. Hermaphroditic duct well developed or indistinguishable from genital atrium, sometimes appears to be continuation of metraterm with ejaculatory duct entering laterally. Ovary oval or with four (or five) short or elongate lobes, normally separated from testes by loops of uterus. Juel's organ and uterine seminal receptacle present. Laurer's canal and blind or canalicular seminal receptacle absent. Uterus almost entirely retained in hindbody, mainly pre-ovarian or (?) with many loops in postovarian field. Eggs without filaments, may link together and form chains. Vitellarium with six to eight (usually seven, arranged three and four) oval to digitiform lobes, or with two lateral acinous groups of follicles or two compact multilobulate masses, postero- to anterolateral or posterior to ovary. Excretory arms united in forebody. Parasitic in swim-bladder of marine physostomatous teleosts. Type-genus *Dictysarca* Linton, 1910.

Key to subfamilies

1a. Uterus mainly pre-ovarian .... **2.**
1b. Uterine field mainly postovarian .... **Cylindrorchiinae Poche, 1926 *subfam. inq.***

[1] A sectioned specimen from a tetraodontid from off Western Australia has been examined: this appears to have an arrangement of the reproductive system somewhat similar to that of *Elongoparorchis*.

2a. **Ovary oval (entire or irregularly lobed); vitellarium two compact multilobulate masses or two acinous bunches of follicles; hermaphroditic duct indistinguishable from genital atrium .... Dictysarcinae Skrjabin & Guschanskaja, 1955.**
2b. **Ovary four- (or five-) lobed; vitellarium six to eight (usually seven) digitiform to oval lobes; hermaphroditic duct distinguishable from genital atrium .... Albulatrematinae Yamaguti, 1965.**

Subfamily Dictysarcinae Skrjabin & Guschanskaja, 1955

Diagnosis: Body stout. Cuticular ridge may encircle mid-hindbody. Testes oval, entire or irregularly lobed. Sinus-sac and sinus-organ absent. Hermaphroditic duct indistinguishable from genital atrium, tubular, short. Ovary oval, entire or irregularly lobed, in posterior third of hindbody. Uterus mainly pre-ovarian, but some loops present in postovarian field. Vitellarium two compact multilobulate masses or two acinous bunches of follicles, antero- or posterolateral to ovary. Usually parasitic in marine eels. Type-genus *Dictysarca* Linton, 1910.

Key to genera

1a. **Cuticular ridge encircling mid-hindbody absent; ovary irregularly lobed; vitellarium in two, compact, multilobulate masses .... *Dictysarca* Linton, 1910. (Fig. 41.1)**
Diagnosis: No cuticular ridge present encircling mid-hindbody. Testes irregularly lobed. Seminal vesicle sinuous. Pars prostatica well developed. Hermaphroditic duct appears to be continuation of metraterm, with ejaculatory duct entering laterally. Ovary large, irregularly lobed. Vitellarium two compact, multilobulate masses, anterolateral to ovary. Parasitic in moray eels (*Gymnothorax*) and sea-horses (*Hippocampus*). Type-species *D. virens* Linton, 1910.
1b. **Cuticular ridge encircling mid-hindbody present; ovary unlobed; vitellarium two groups of acinous follicles .... *Aerobiotrema* Yamaguti, 1958. (Fig. 41.2)**
Diagnosis: Cuticular ridge present encircling mid-hindbody. Testes entire. Seminal vesicle sigmoid. Pars prostatica straight. Ovary small, entire. Vitellarium two acinous bunches of follicles; one antero- and one posterolateral to ovary. Excretory arms with numerous anastomosing side branches, mostly lying close to caeca. Parasitic in marine eels (*Muraenesox*). Type-species *A. muraenesocis* Yamaguti, 1958.

Subfamily Albulatrematinae Yamaguti, 1965
(Syns Pelorohelminthinae Fischthal & Kuntz, 1964; Tetrasterinae Oshmarin, 1965)

Diagnosis: Body flattened to stout. Testes large, oval or elongate. Sinus-sac present, well or poorly developed. Hermaphroditic duct distinguishable from genital atrium. Ovary four (or five) distinct, oval or elongate lobes, in middle or posterior half of hindbody. Uterus mainly pre-ovarian or extending throughout hindbody. Vitellarium

41.1

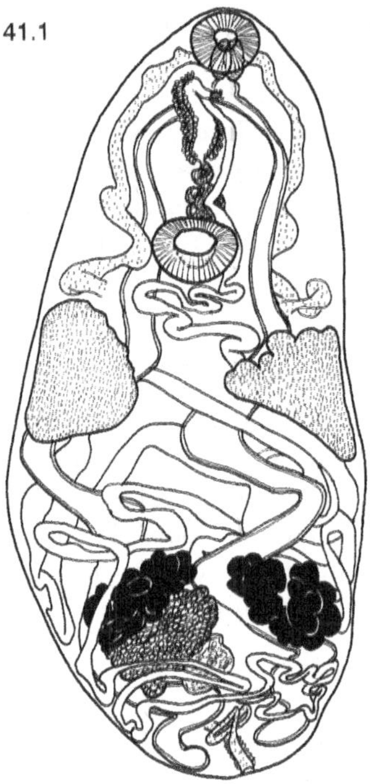

41.2

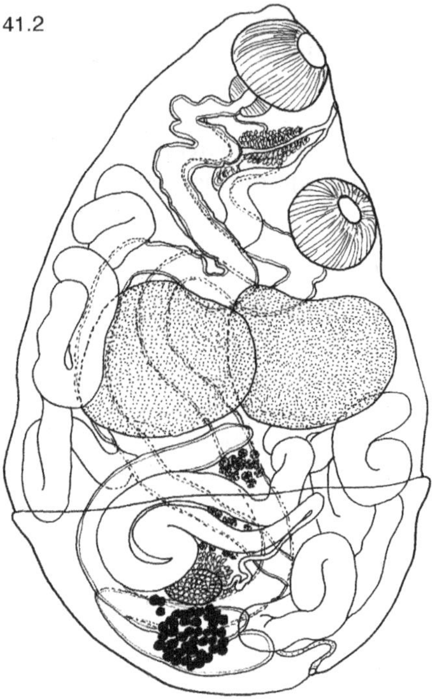

Figs 41.1–41.2.
41.1. *Dictysarca virens* Linton, 1910. (After Yamaguti, 1971.)
41.2. *Aerobiotrema muraenesocis* Yamaguti, 1958. (After Yamaguti, 1958b.)

seven (occasionally six or eight) digitiform to oval lobes, immediately posterior or posterolateral to ovary. Parasitic in marine (? or brackish-water) teleosts. Type-genus *Albulatrema* Yamaguti, 1965.

Key to genera

1a. Vitelline lobes, ovarian lobes and testes oval; ovary in middle of hindbody; significant proportion of uterus postovarian .... *Albulatrema* Yamaguti, 1965. (Fig. 41.3)

Diagnosis: Body stout. Testes oval. Seminal vesicle tubular, narrow, convoluted (according to Yamaguti (1965), the seminal vesicle is replaced by a vas deferens, the distal portion of which is strongly muscular, convoluted and enclosed by an apparently muscular capsule). Pars prostatica sigmoid, delimited. Sinus-sac well developed, bulbous. Temporary sinus-organ may be present. Genital atrium absent or (?) small. Ovary in middle of hindbody; ovarian lobes oval. Uterus extends throughout hindbody, much of it postovarian. Vitelline lobes oval to pyriform. Parasitic in marine (? or brackish-water) teleosts (*Albula*). Type-species *A. ovale* Yamaguti, 1965.

1b. Vitelline lobes and ovarian lobes digitiform; testes elongate; ovary well inside posterior half of hindbody; most of uterus pre-ovarian .... *Elongoparorchis* Rao, 1961. (Fig. 41.4)

(Syns *Pelorohelmins* Fischthal & Kuntz, 1964; *Tetraster* Oshmarin, 1965; *Dollfustravassosius* Teixeira de Freitas & Kohn, 1967)

41.3

41.4

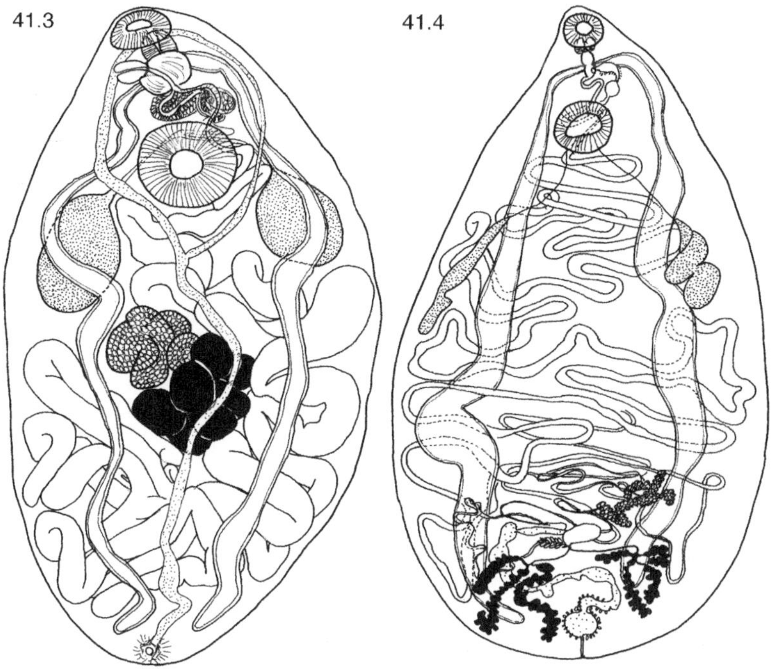

Figs 41.3–41.4.
41.3. *Albulatrema ovale* Yamaguti, 1965. (After Yamaguti, 1970.)
41.4. *Elongoparorchis monilovatus* (Freitas & Kohn, 1967). (After Jones & Khalil, 1984.)

Diagnosis: Body flattened to stout. Testes elongate. Seminal vesicle sinuous, may reach dorsally to ventral sucker. Pars prostatica short. Sinus-sac poorly developed, present only distally as vestige surrounding base of genital atrium. Sinus-organ a small papilla-like structure (? temporary). Hermaphroditic duct relatively long, formed as continuation of metraterm, with ejaculatory duct entering laterally. Genital atrium oval or elongate-oval. Ovary well inside posterior half of hindbody; ovarian lobes digitiform. Uterus mainly pre-ovarian. Eggs may be linked together forming chains. Vitelline lobes digitiform, with condensed follicular appearance. Parasitic especially in catfishes, such as *Arius*. Type-species *E. pneumatis* Rao, 1961.

Subfamily Cylindrorchiinae Poche, 1926 *subfam. inq.*
(Syn. Cylindrorchiinae Yamaguti, 1958)

Diagnosis: [Original description inadequate.] Body elongate-oval. Testes elongate. Terminal genitalia not known. Ovary small, oval, in posterior half of hindbody. Uterus convoluted posterior to ovary, straight anterior to ovary. Vitellarium two clusters of small follicles, anterolateral to ovary. Parasitic in marine (? or brackish-water) teleosts (*Tetraodon*). Type-genus *Cylindrorchis* Southwell, 1913.

Genus *Cylindrorchis* Southwell, 1913 *gen. inq.* (Fig. 41.5)

Diagnosis: With characters of subfamily. Type-species *C. tenuicutis* Southwell, 1913.

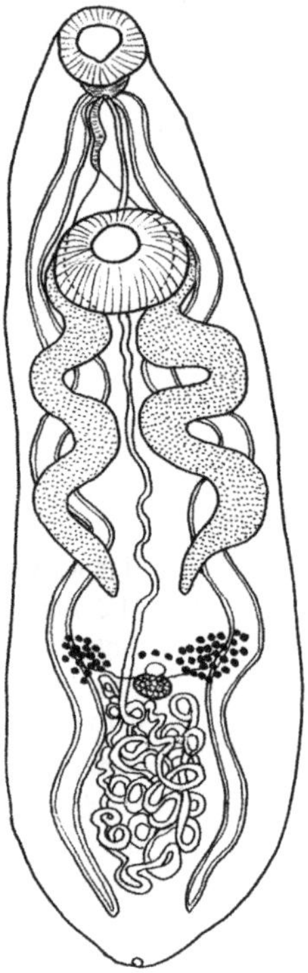

Fig. 41.5. *Cylindrorchis tenuicutis* Southwell, 1913. (After Southwell, 1913.)

42 Family Hirudinellidae Dollfus, 1932

D.I. Gibson

Department of Zoology, The Natural History Museum, London, UK

Introduction

Hirudinellids are very large hemiuroids, usually from the stomach of large, marine teleosts. They are unique among hemiuroids in that they lack a hermaphroditic duct and possess a form of 'cirrus' and 'cirrus-sac'. As these features are not considered homologous with the cirrus and cirrus-sac of other digeneans (see Gibson & Bray, 1979), they are included here in inverted commas. The genus *Hirudinella* was first mentioned by Garcin in 1730, but the first post-Linnaean use of the name was by de Blainville in 1828 (see Gibson, 1976). The genus was linked with the hemiurids by workers such as Nicoll (1915) and Poche (1926), but mentioned in passing by Dollfus (1932b) as representing a family in its own right within the Hemiuroidea, although he did not define the taxon. The family was subsequently defined by Yamaguti (1953, 1958a, 1971), still with a single genus, and by Skrjabin & Guschanskaja (1956, 1957a), who also included *Uroproctinella* (not recognized by Yamaguti, 1971). The family was included in the Azygioidea by Skrjabin & Guschanskaja (1956, 1958). However, the most radical changes to the family were made by Gibson & Bray (1977, 1979), who broadened the concept to include three genera, *Hirudinella, Botulus* Guiart, 1938 and *Lampritrema* Yamaguti, 1940, each of which they considered to be monotypic. At the same time they synonymized several more recently erected genera into the latter three and did not recognize the family-group names, Botulidae, Lampritrematidae, Mediolecithidae and Profundiellinae, erected for these taxa. Gibson & Bray (1977, 1979) found no good features, and no logistic need, for differentiating these genera at the subfamily level.

Family Hirudinellidae Dollfus, 1932

(Syns Botulidae Guiart, 1938; Lampritrematidae Yamaguti, 1940; Mediolecithidae Oshmarin, 1968; Profundiellinae Skrjabin, 1958)

Diagnosis: Body large, stout or elongate, contractile. Ecsoma absent. Body surface smooth, sometimes papillate in forebody or wrinkled. Oral and ventral suckers well developed; latter in anterior half of body. Pharynx well developed. Oesophagus usually short. 'Drüsenmagen' present. Gut-caeca terminate blindly or form uroproct, sometimes fuse subterminally forming cyclocoel, usually diverticulate. Testes two, in tandem, oblique or symmetrical, pre-ovarian, in hindbody. Seminal vesicle tubular, normally thin-walled, occasionally partly thick-walled, convoluted in forebody. Pars prostatica well developed, tubular, usually long. Ejaculatory duct long, muscular, surrounded by muscular 'cirrus-sac', opens into genital atrium through well-developed conical to cylindrical 'cirrus'. Hermaphroditic duct, sinus-sac and sinus-organ absent. Genital atrium large, usually capable of being everted. Genital pore midventral in forebody. Ovary oval, post-testicular, in middle or anterior half of hindbody. Laurer's canal and uterine seminal receptacle normally present. Juel's organ and blind or canalicular seminal receptacle absent. Uterus descending ventrally and ascending more dorsally, coiled mainly at level of vitellarium, but often extending more anteriorly into pre-ovarian region, mainly intercaecal, occasionally reaching extracaecally; metraterm opens into genital atrium directly or through a papilla-like organ situated immediately posteriorly to 'cirrus'. Eggs numerous, small, without filaments. Vitellarium composed of from two to numerous long, straight or convoluted, branched tubules, mainly postovarian, inter- or extracaecal. Excretory vesicle Y-shaped; arms initially dorso-ventrally orientated, convoluted, united in forebody. Parasitic in stomach (occasionally on gills) of large, carnivorous, marine teleosts. Type-genus *Hirudinella* de Blainville, 1828.

Key to genera

1a. Body stout, elongate or keyhole-shaped; uroproct present; 'cirrus-sac' small, globular; seminal vesicle entirely thin-walled .... 2.

1b. Body slender, elongate; uroproct absent; 'cirrus-sac' large, elongate; distal part of seminal vesicle with muscular wall .... *Lampritrema* Yamaguti, 1940. (Fig. 42.1)

(Syn. *Hirudinelloides* Gaevskaja & Kovaleva, 1977)

Diagnosis: Body elongate, slender. Papillae present on forebody. Uroproct absent. Testes in tandem, near middle of hindbody. Most of seminal vesicle thin-walled, but distal portion forms thick-walled, muscular 'pars musculosa'. Thick-walled pars prostatica lies ventral to posterior portion of 'cirrus-sac'. Male duct leads into 'cirrus-sac' some distance from its posterior extremity. 'Cirrus-sac' large, elongate, club-shaped. 'Cirrus' long or short, capable of being extruded some distance through genital pore. Genital atrium deep. Ovary near middle of hindbody. Uterus extends back to near posterior limit of vitellarium, mainly coiled intercaecally in post- and pre-ovarian regions of hindbody. Vitellarium a pair of lateral tubules with short dorsal branches, mainly extracaecal, passing posteriorly from ovary to about halfway to posterior extremity. Excretory arms appear to unite in forebody. Parasitic in stomach (? occasionally on gills) of marine teleosts (*Lampris*, *Brama* and *Thyrsites*). Immature forms recorded from salmonids (stomach, oesophagus or gills). Type-species *L. miescheri* (Zschokke, 1890) (syn. *L. nipponicum* Yamaguti, 1940 – type by original designation).

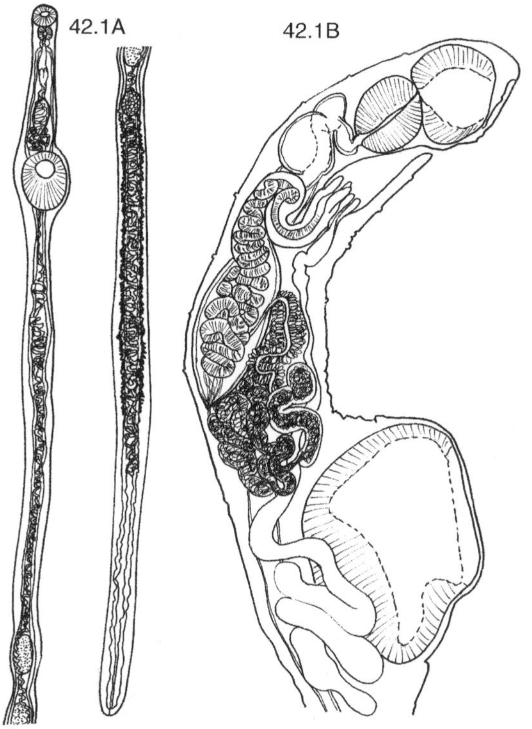

Fig. 42.1. *Lampritrema miescheri* (Zschokke, 1890). A. Entire worm. B. Sagittal view and anterior region showing terminal genitalia. (Modified after Yamaguti (1940), as *L. nipponicum* Yamaguti, 1940.)

2a. Vitellarium in two lateral fields between levels of testes and mid-hindbody; uterus mainly intercaecal, postovarian, at level of vitellarium .... *Hirudinella* de Blainville, 1828. (Fig. 42.2)

(Syns *Hirudinella* Garcin, 1730 (Pre-Linnaean; see Gibson, 1976); *Uroproctinella* Skrjabin & Guschanskaja, 1957)

Diagnosis: Body stout, elongate or keyhole-shaped. Body surface often transversely wrinkled. Uroproct present; gut-caeca may fuse subterminally in older specimens, forming cyclocoel. Testes symmetrical to oblique, in anterior hindbody. Seminal vesicle thin-walled throughout its length. 'Cirrus-sac' relatively small, globular. 'Cirrus' cone-shaped to digitiform. Genital atrium capable of being everted through genital pore. Ovary in anterior hindbody. Uterus mainly intercaecal; coils extending posteriorly from ovary to near posterior limit of vitellarium. Vitellarium in two lateral fields between testes and middle of hindbody. Parasitic in stomach of large carnivorous, marine teleosts (usually scombroids). Type-species *H. ventricosa* (Pallas, 1774) (syn. *H. clavata* (Menzies, 1791) - type by monotypy).

2b. Vitellarium massed close to ventral surface, intercaecal between ovary and posterior extremity; uterus reaching extracaecally, mainly in anterior hindbody .... *Botulus* Guiart, 1938. (Fig. 42.3)

(Syn. *Profundiella* A.S. Skrjabin, 1958; *Mediolecithus* Oshmarin, 1968)

Diagnosis: Body normally stout. Uroproct present. Testes symmetrical to oblique, in anterior hindbody, large. Seminal vesicle thin-walled throughout its length, tubular and convoluted (or ?globular). 'Cirrus-sac' small, globular. 'Cirrus' cone-shaped to

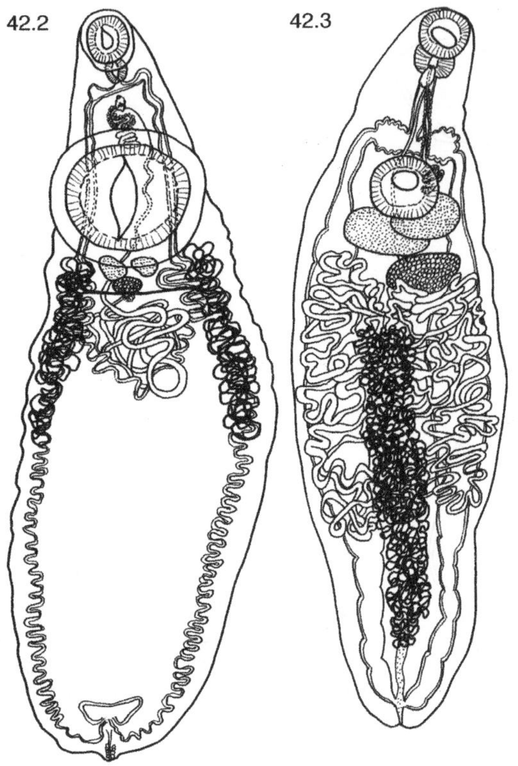

Figs 42.2-42.3.
42.2. *Hirudinella ventricosa* (Pallas, 1774). (After Gibson & Bray, 1977.)
42.3. *Botulus microporus* (Monticelli, 1889). (After Gibson & Bray, 1977.)

digitiform. Genital atrium often everted through genital pore. Ovary in anterior hindbody. Uterus reaching extracaecally in anterior hindbody. Vitellarium a densely tangled mass of tubules in one ventral, intercaecal field between ovary and posterior extremity. Parasitic in stomach of large, carnivorous, marine teleosts (*Alepisaurus*) and (?) accidentally in piscivorous sharks. Type-species *B. microporus* (Monticelli, 1889) (syn. *B. alepidosauri* Guiart, 1938 - type by monotypy).

43 Family Isoparorchiidae Travassos, 1922

D.I. Gibson

Department of Zoology, The Natural History Museum, London, UK

Introduction

This family group was first erected, as a subfamily, by Travassos (1922), for a single genus, *Isoparorchis* Southwell, 1913, from the swim-bladder of freshwater fishes, and raised to full family level by Poche (1926). The status of the family remained the same until the work of Yamaguti (1971), who broadened the concept to include *Elongoparorchis* Rao, 1960 and *Cladolecithotrema* Ichihara, 1970, but this was not accepted by Gibson & Bray (1979), who considered them to be three unrelated hemiuroid genera. Yamaguti (1971) also linked together the Isoparorchiidae with other families of swim-bladder parasites, the Albulatrematidae, Cylindrorchiidae, Aerobiotrematidae, Tetrasteridae, Dictysarcidae and Pelorohelminthidae, in the superfamily Isoparorchioidea. Gibson & Bray (1979) synonymized the latter families with the Dictysarcidae, a marine group which they did not consider to be closely related to *Isoparorchis*. They considered that *Isoparorchis* differs from the other swim-bladder forms in that it occurs in freshwater rather than marine fishes and possesses several quite different (and more 'primitive') features, including a Laurer's canal, a tubular vitellarium and a well-developed, muscular sinus-organ. It also differs from most hemiuroids (apart from didymozoids) in that it has a tubular ovary.

Family Isoparorchiidae Travassos, 1922

Diagnosis: Body large, stout, dorsoventrally flattened. Ecsoma absent. Body surface smooth. Oral and ventral suckers small. Pharynx well developed. Oesophagus short. 'Drüsenmagen' absent. Gut-caeca sinuous, terminate blindly close to posterior extremity. Testes two, symmetrical, pre-ovarian, in anterior hindbody. Seminal vesicle small, thin-walled, tubular, winds in forebody. Pars prostatica tubular. Ejaculatory duct within sinus-sac. Sinus-sac weakly developed, composed of diffuse musculature. Hermaphroditic duct short, opens into genital atrium through stout sinus-organ.

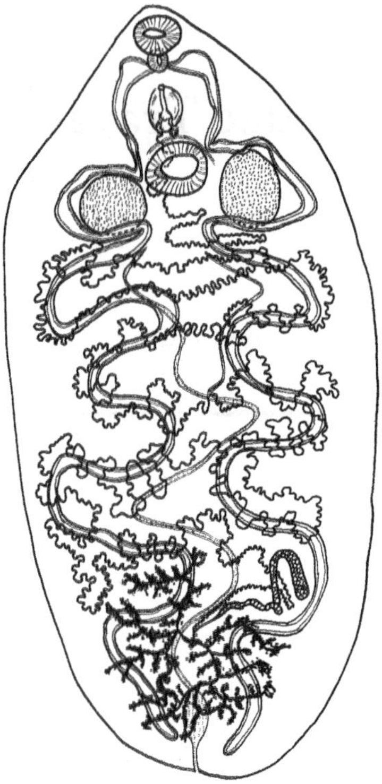

Fig. 43.1. *Isoparorchis hypselobagri* (Billet, 1898). (After Johnston, (1927), as *I. tandani* Johnston, 1927.)

Genital atrium with pair of concentric folds in its wall, may be extruded through genital pore. Genital pore midventral in forebody. Ovary tubular, well posterior to testes. Laurer's canal present, may be slightly dilated proximally, forming small rudimentary seminal receptacle. Uterine seminal receptacle present. Juel's organ and canalicular or blind seminal receptacle absent. Uterus pre-ovarian, mainly coiled in hindbody. Eggs numerous, small, non-filamented. Vitellarium tubular, with dendritic branches arising from about six collecting ducts, postovarian. Excretory vesicle Y-shaped; arms converge dorsal to pharynx but do not unite. Parasitic in swim-bladder of physostomatous freshwater teleosts (Asia and Australasia). Type-genus *Isoparorchis* Southwell, 1913.

Genus *Isoparorchis* Southwell, 1913 (Fig. 43.1)

(Syn. *Leptolecithum* Kobayashi, 1915)

Diagnosis: With characters of family. Type-species *I. hypselobagri* (Billet, 1898) (syn. *L. trisimilitubis* Southwell, 1913 - type by original designation).

44 Family Lecithasteridae Odhner, 1905

D.I. GIBSON

Department of Zoology, The Natural History Museum, London, UK

Introduction

Lecithasterids are primarily parasites of the intestine of marine teleosts. The family group was first used at the subfamily level by Odhner (1905) for *Lecithaster* and *Lecithophyllum* within the family Hemiuridae. This was accepted at this level by most workers, including Manter & Pritchard (1960) and Yamaguti (1971), although the latter's conception was not as broad as the former's. It was first used at full family status by Skrjabin & Guschanskaja (1954), who included within it the Lecithasterinae, Lecithophyllinae, Macradenininae, Hypohepaticolinae, Johniophyllinae and Musculovesiculinae, the latter three of which are ecsomate hemiurids and not recognized here. The concept followed here is that of Gibson & Bray (1979), who, in their revision of the Hemiuroidea, recognized six subfamilies within the Lecithasteridae. Another key to the group was produced by Dutta (1995). Changes subsequent to Gibson & Bray's (1979) work are outlined below; specific comments relating various taxa are included in the latter work. It is worth noting that the Hysterolecithinae are unique within the family in possessing a Juel's organ and a uterine seminal receptacle rather than a blind seminal receptacle. A third arrangement, described by Yamaguti (1940) for *Trifoliovarium* Yamaguti, 1940, i.e. the presence of a canalicular seminal receptacle and Laurer's canal, would appear to be an anatomical misinterpretation of one specimen, which I have examined.

In the Lecithasterinae, *Bilqeesotrema* Gupta & Jain, 1991, based on a single specimen, appears to be a synonym of *Aponurus* Looss, 1907, species of which regularly occur in the same host genus.

The Macradenininae are a subfamily in dire need of revision and their link with the Quadrifoliovariinae needs further investigation. Unfortunately, this is difficult because many of the descriptions are inadequate and some taxa are known only from single specimens. Furthermore, some of the potentially useful features, such as the presence of a cyclocoel, are either difficult to determine or have not been elucidated in detail; hence, the presence of a cyclocoel has not been used as a key feature in relation

 Keys to the Trematoda, Vol. 1 (eds D.I. Gibson, A. Jones and R.A. Bray)

to this subfamily. *Pseudodichadena* Yamaguti, 1971 was considered a synonym of *Dichadena* Linton, 1910 by Nahhas & Carlson (1994). *Paramacradenina* was described by Tang *et al.* (1983) from a single specimen as similar to *Macradenina* Manter, 1947 but with an ecsoma; the presence of a pretesticular ovary is not a hemiurid feature, so I think that this genus is best considered as a questionable synonym of *Macradenina*, a genus with which it has several features in common. *Anadichadena* Dutta, 1995 is difficult to classify on the basis of the information presented by Dutta (1995); this description suggested a close relationship with *Neodichadena* Yamaguti, 1971 (although there are differences indicated by his figure). However, after examining the type-specimens, I can detect a lobed ovary. It is here considered a questionable synonym of *Lecithaster* Lühe, 1901, a genus that Dutta (1995) recorded from the same locality.

In the Prolecithinae, *Cryptodiscus* Srivastava & Sahai, 1978 was synonymized with *Prolecitha* Manter, 1961 by Abdul-Salam & Khalil (1987); the finding of this form in an elasmobranch is likely, as indicated by the latter authors, to be accidental.

Machidatrema, erected as a bunocotyline by León-Règagnon (1998), is transferred to the Hysterolecithinae, following Bray & Cribb (2000a), who discuss in some detail the relationship between the hysterolecithine lecithasterids and the bunocotyline hemiurids. Bray & Cribb (2000a) considered León-Règagnon's (1998) conception of the genus too broad, as it included a species of *Hysterolecithoides* Yamaguti, 1934. *Oligolecithoides* Shen, 1982 was synonymized with *Hysterolecithoides* by Bray & Cribb (2000a).

The same authors (Bray & Cribb, 2000b) have recently provided a key to the species of the Trifoliovariinae. They included a cladistic analysis of the group, which indicated that most of the included taxa, including *Trifoliovarium* Yamaguti, 1940, are poorly resolved and that the current conception of the latter genus is paraphyletic. The synonymy of *Cladolecithotrema* Ichihara, 1970 with *Trifoliovarium* was proposed by Gupta (1979), but not recognized by Gibson & Bray (1979), based essentially on the presence of Laurer's canal in one of Yamaguti's (1940) figures of the type-species of *Trifoliovarium*. However, as indicated above, this was a misinterpretation. In this work *Cladolecithotrema* is treated as a synonym of *Trifoliovarium*. In addition, *Pseudolecithaster* Campbell & Munroe, 1977 had been differentiated from *Trifoliovarium* by Gibson & Bray (1979), based on the absence of Laurer's canal, body shape, a branching or doubling of the vitelline lobes and a more anterior position of the seminal vesicle. Consequent upon the synonymy of *Trifoliovarium* and *Cladolecithotrema*, these differences break down and this genus is also treated here as a synonym of *Trifoliovarium*.

Family Lecithasteridae Odhner, 1905

(Syn. Lobatovitelliovariidae Yamaguti, 1965)

Diagnosis: Body usually small; normally spindle-shaped, occasionally elongate. Ecsoma absent. Body surface smooth. Oral and ventral suckers well developed; ventral sucker normally in anterior half of body. Muscular flange or flanges may be present immediately posterior to ventral sucker. Pharynx well developed. Oesophagus usually short. 'Drüsenmagen' normally present. Gut-caeca usually terminate blindly, but occasionally unite forming cyclocoel. Presomatic pit and ventrocervical groove absent. Testes two, occasionally one, in tandem, oblique or symmetrical, usually, but not always, pre-ovarian, in hindbody. Seminal vesicle generally thin-walled, occasionally muscular; oval, tubular or constricted into portions; in fore- or hindbody. Pars prostatica usually tubular, occasionally vesicular; may be linked to seminal vesicle by aglandular tube. Ejaculatory duct long, short or absent. Hermaphroditic duct present. Ejaculatory (prostatic) vesicle absent. Sinus-sac usually present; well or poorly developed, occasionally absent. Permanent sinus-organ normally absent, but hermaphroditic duct is often protruded to form temporary sinus-organ. Genital atrium large, small or absent. Ovary usually post-testicular, oval or four- (occasionally three-) lobed. Normally blind seminal receptacle present (usually large, thick-walled and situated dorsal or anterodorsal to ovary), and Laurer's canal, Juel's organ and both uterine or canalicular seminal receptacles absent; occasionally (Hysterolecithinae) only Juel's organ and uterine seminal receptacle present. Uterus varies from mainly postovarian to entirely preovarian; main bulk rarely extends into forebody. Eggs numerous, small, rarely filamented. Vitellarium commonly seven-lobed, occasionally two to eight, or four to sixteen (sometimes branched) lobes, often in rosette arrangement, usually immediately postovarian, occasionally pre-ovarian or at level of ovary. Excretory vesicle Y-shaped; arms united in forebody or not. Parasitic in gut, especially intestine, of marine teleosts. Type-genus *Lecithaster* Lühe, 1901.

Key to subfamilies

1a. Uterine seminal receptacle present; blind seminal receptacle absent .... Hysterolecithinae Yamaguti, 1958.
1b. Uterine seminal receptacle absent; blind seminal receptacle present .... 2.

2a. Uterus entirely or almost entirely pre-ovarian; hermaphroditic duct appears to be continuation of uterus .... Trifoliovariinae Yamaguti, 1958.
2b. Uterus partly postovarian; hermaphroditic duct normal .... 3.

3a. Muscular ventrolateral flange or flanges present immediately posterior to ventral sucker; mainly in Acanthuridae .... Quadrifoliovariinae Yamaguti, 1970.
3b. Muscular ventrolateral flange or flanges absent immediately posterior to ventral sucker .... 4.

4a. Seminal vesicle in forebody or dorsolateral to ventral sucker, rarely extending into anterior hindbody; pars prostatica short to medium in length; sinus-sac well developed .... 5.
4b. Seminal vesicle entirely in hindbody; pars prostatica and/or ejaculatory duct long; sinus-sac small or poorly developed; usually parasitic in *Acanthurus* .... Macradenininae Skrjabin & Guschanskaja, 1954.

5a. Sinus-sac relatively well developed, symmetrical; genital atrium not sucker-like; vitellarium postovarian; caeca terminate blindly .... Lecithasterinae Lühe, 1901.

5b. Sinus-sac asymmetrical, with anterior wall modified to form muscular papilla; genital atrium sucker-like; vitellarium pre-ovarian; cyclocoel present; mainly in Belonidae .... Prolecithinae Yamaguti, 1971.

Subfamily Lecithasterinae Odhner, 1905

(Syn. Lecithophyllinae Skrjabin & Guschanskaja, 1954)

Diagnosis: Caeca terminate blindly. Testes two, occasionally one, pre-ovarian. Seminal vesicle in forebody, dorsal to ventral sucker or, occasionally, in anterior hindbody. Pars prostatica short to medium in length. Ejaculatory duct absent or short. Sinus-sac well developed. Permanent sinus-organ absent. Ovary entire or four-lobed. Blind seminal receptacle normally large, usually dorsal to ovary. Uterus reaches postovarian region. Vitellarium seven (rarely six or eight) oval to digitiform lobes in rosette or two linked groups of three and four, immediately postovarian; occasionally antero-posteriorly orientated. Excretory arms united in forebody or not. Normally parasitic in intestine or stomach of marine teleosts. Type-genus *Lecithaster* Lühe, 1901.

Key to genera

1a. Testis single .... *Monorchiaponurus* Fischthal & Nasir, 1974.[1] (Fig. 44.1)
Diagnosis: Testis single. Seminal vesicle saccular; in forebody. Sinus-sac elongate. Hermaphroditic duct tubular. Genital atrium absent. Ovary oval. Seminal receptacle pre-ovarian, posterodextral to testis. Vitellarium seven globular lobes. Parasitic in intestine of marine teleosts. Type-species *M. hemirhamphi* Fischthal & Nasir, 1974.
1b. Testes two .... 2.

2a. Ovary lobed, usually with four lobes; vitelline lobes tear-shaped .... *Lecithaster* Lühe, 1901. (Fig. 44.2)
(Syn. ?*Anadichadena* Dutta, 1995)
Diagnosis: Testes two, obliquely symmetrical, usually oval but occasionally lobed. Seminal vesicle saccular to elongate and sinuous, in forebody, dorsal to ventral sucker or occasionally in anterior hindbody. Sinus-sac oval. Genital atrium short. Ovary normally four-lobed. Seminal receptacle large, globular, dorsal to ovary. Vitellarium a radiating mass of seven tear-shaped lobes. Excretory arms apparently not united in forebody. Parasitic in intestine of marine teleosts. Type-species *L. confusus* Odhner, 1905.
2b. Ovary oval to globular; vitelline lobes globular .... 3.

3a. Genital atrium present, distinct .... *Lecithophyllum* Odhner, 1905. (Fig. 44.3)
Diagnosis: Testes two, obliquely tandem to symmetrical. Seminal vesicle saccular, in forebody or dorsal to ventral sucker. Hermaphroditic duct apparently bipartite. Sinus-sac elongate. Genital atrium present, generally deep. Ovary oval or globular. Seminal receptacle large, dorsal to ovary. Vitellarium seven globular lobes, in lateral

[1] Known from only one specimen. Possibly an anomalous specimen of *Aponurus*, according to Gibson & Bray (1979).

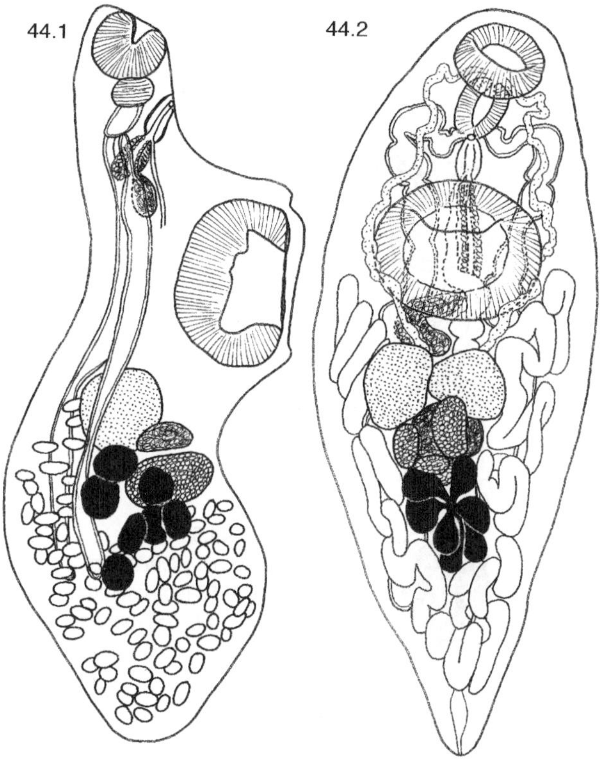

Figs 44.1–44.2.
44.1. *Monorchiaponurus hemirhamphi* Fischthal & Nasir, 1974. (After Fischthal & Nasir, 1974.)
44.2. *Lecithaster confusus* Looss, 1907. (After Looss, 1907.)

groups of three and four. Excretory arms united in forebody. Parasitic in stomach of marine teleosts. Type-species *L. botryophoron* (Olsson, 1868).

3b. Genital atrium small or absent .... ***Aponurus* Looss, 1907. (Fig. 44.4)**
(Syns *Brachadena* Linton, 1910; (?) *Mordvilkoviaster* Pigulewsky, 1938; *Bilqeesotrema* Gupta & Jain, 1991)

Diagnosis: Testes two, tandem to oblique. Seminal vesicle saccular, in forebody or occasionally dorsal to ventral sucker. Hermaphroditic duct usually tubular, occasionally bipartite. Sinus-sac oval to elongate-oval. Genital atrium absent, or occasionally small. Ovary oval or globular. Seminal receptacle small to large, (?) ventral or dorsal to anterior region of ovary. Vitellarium usually seven globular to slightly elongate lobes, in lateral, occasionally anteroposteriorly orientated, groups of three and four. Excretory arms united in forebody. Parasitic in stomach (occasionally intestine) of marine teleosts. Type-species *A. laguncula* Looss, 1907.

Subfamily Hysterolecithinae Yamaguti, 1958

Diagnosis: Body small, elongate to spindle-shaped. Ventral sucker in middle or anterior half of body. Gut-caeca end blindly near posterior extremity. Testes two, oval, symmetrical to obliquely tandem, pre-ovarian, in anterior half of hindbody. Seminal vesicle usually tubular, occasionally elongate saccular, in forebody. Pars prostatica usually tubular, occasionally vesicular, short. Sinus-sac present, often weakly

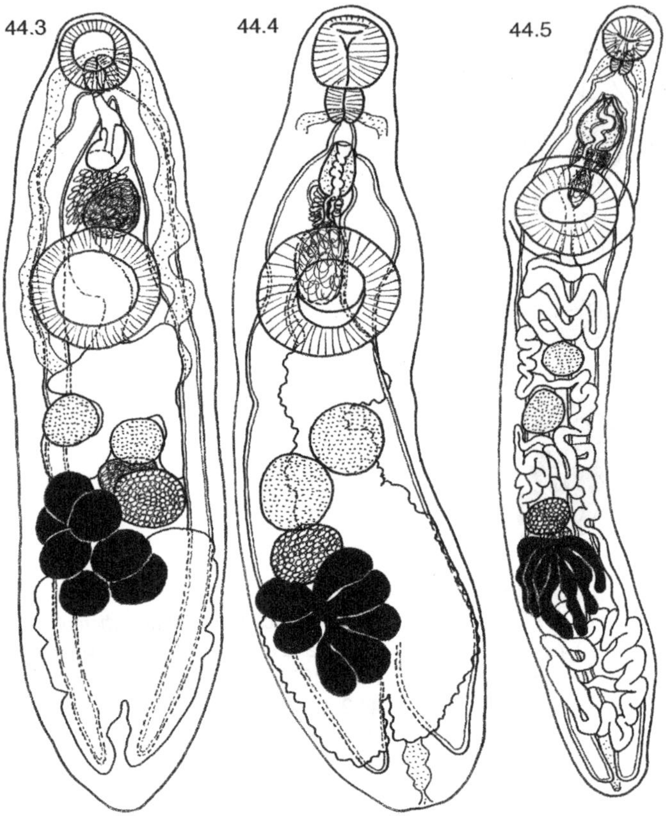

Figs 44.3–44.5.
44.3. *Lecithophyllum anteroporum* Margolis, 1958 (considered by many a synonym of *L. botryophoron* (Olsson, 1868)). (After Margolis, 1958.)
44.4. *Aponurus laguncula* Looss, 1907. (After Bray *et al.*, 1993.)
44.5. *Thulinia microrchis* (Yamaguti, 1934). (After Bray *et al.*, 1993.)

developed, occasionally of open type. Permanent sinus-organ absent; temporary sinus-organ may form. Hermaphroditic duct present within sinus-sac. Genital atrium small or absent. Ovary oval, in anterior or posterior half of hindbody. Laurer's canal and blind or canalicular seminal receptacle absent. Juel's organ and uterine seminal receptacle present. Uterus almost entirely in hindbody, mainly pre- to mainly post-ovarian. Eggs occasionally with filaments (sometimes indistinct and not visible in all specimens). Vitellarium seven (occasionally eight or two to seven) oval to digitiform lobes, postovarian. Excretory arms united in forebody or not. Parasitic mainly in intestine or stomach of marine teleosts (usually perciform genera, especially acanthurids, siganids and pomacentrids). Type-genus *Hysterolecitha* Linton, 1910.

Key to genera

1a. Eggs may be filamented (eggs may need to be teased out of specimen; filaments sometimes not seen, occasionally appear unattached to eggshell); sinus-sac well developed with extrusible sinus-organ; seven digitiform vitelline lobes .... *Thulinia* Gibson & Bray, 1979. (Fig. 44.5)

Diagnosis: Body elongate. Ventral sucker in anterior half of body. Seminal vesicle tubular, in forebody, may reach dorsally to ventral sucker. Pars prostatica tubular, short. Sinus-sac present, well developed; sinus-organ present, extrusible. Genital atrium small. Large extent of uterus in pre- and postovarian fields. Eggs tend to be filamented (one filament at each end). Vitellarium seven (or eight) digitiform lobes. Excretory arms united in forebody. Parasitic in intestine of marine teleosts. Type-species *T. microrchis* (Yamaguti, 1934).

1b. Eggs without filaments; sinus-sac weakly to well developed; sinus-organ temporary if formed; vitellarium two to seven entire or irregular, oval to digitiform lobes .... 2.

2a. Excretory arms united in forebody (visible in good material, but serial sections may be needed); sinus-sac often weakly developed; uterus mainly pre-ovarian .... *Hysterolecitha* Linton, 1910. (Fig. 44.6)

Diagnosis: Ventral sucker usually in anterior half of body, occasionally near middle. Seminal vesicle normally tubular, occasionally elongate saccular. Pars prostatica tubular or vesicular. Sinus-sac present; often weakly developed, may be of 'open' type; sinus-organ not usually formed. Ovary normally in posterior half of hindbody, normally separated from testes by loops of uterus. Uterus usually mainly pre-ovarian, (?) occasionally mainly postovarian. Eggs without filaments. Vitellarium seven oval to digitiform lobes. Excretory arms united in forebody. Type-species *H. rosea* Linton, 1910.

2b. Excretory arms not united in forebody; at least half of uterus postovarian; vitelline masses subglobular; mainly parasitic in *Siganus* .... 3.

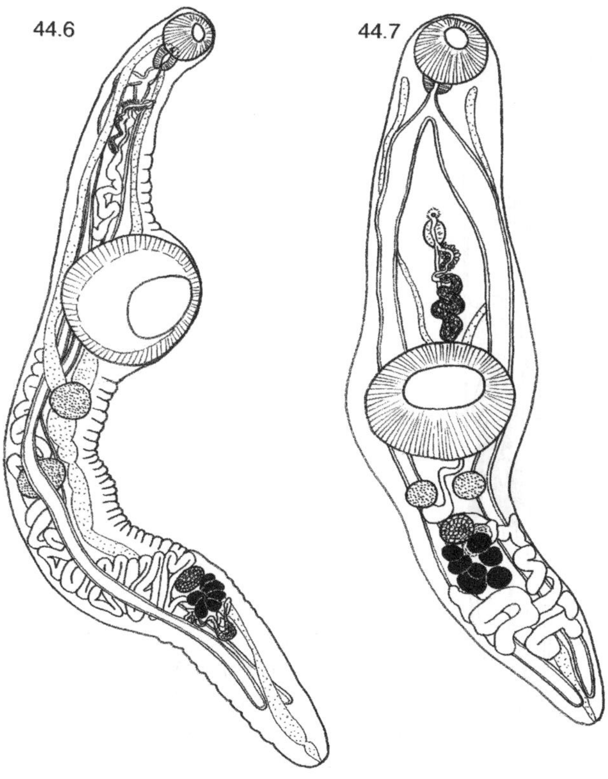

Figs 44.6–44.7.
44.6. *Hysterolecitha rosea* Linton, 1910. (Modified after Yamaguti, 1971.)
44.7. *Hysterolecithoides epinepheli* Yamaguti, 1934. (After Yamaguti, 1934.)

3a. Pars prostatica tubular; two to seven vitelline masses .... *Hysterolecithoides* Yamaguti, 1934. (Fig. 44. 7)
(Syn. *Oligolecithoides* Shen, 1982)

Diagnosis: Ventral sucker in middle of body. Seminal vesicle elongate, tubular. Pars prostatica elongate, tubular. Sinus-sac oval. Genital atrium distinct, with blind anterodorsal sac. Ovary close to testes. Uterus mainly postovarian. Eggs without filaments. Vitellarium two to seven subglobular, entire or irregular masses. Excretory arms not united in forebody. Mainly from stomach or intestine of *Siganus* spp. Type-species *H. epinepheli* Yamaguti, 1934.

3b. Pars prostatica vesicular; two to three vitelline masses .... *Machidatrema* León-Règagnon, 1998. (Fig. 44.8)

Diagnosis: Ventral sucker in middle of body or just anterior. Seminal vesicle elongate, tubular. Pars prostatica short, vesicular. Hermaphroditic duct separated into two regions by constriction; muscular distal region may form temporary sinus-organ. Sinus-sac well developed, oval. Genital atrium distinct. Ovary oval. Large part of uterus postovarian. Eggs without filaments. Vitellarium two to three subglobular masses; entire or irregularly lobed. Excretory arms end blindly in forebody. In stomach of *Siganus* spp. Type-species *M. akeh* (Machida, 1989).

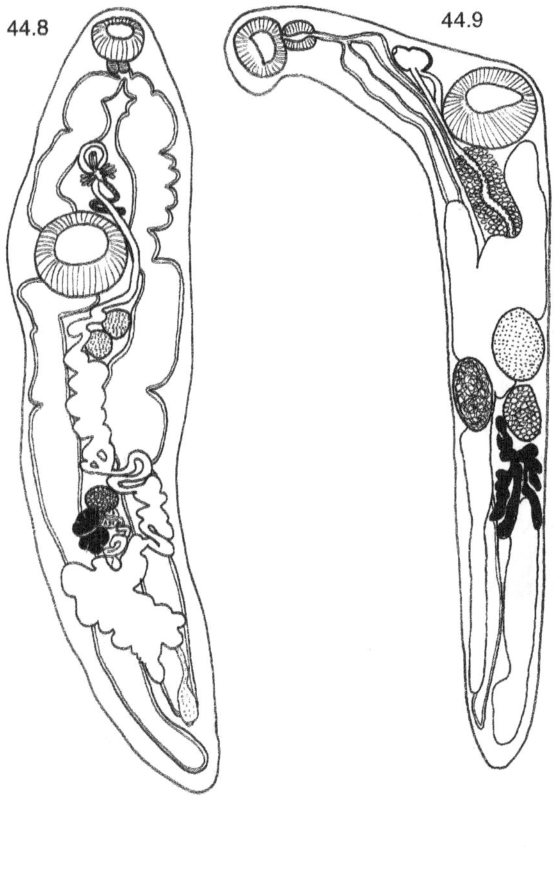

Figs 44.8–44.9.
44.8. *Machidatrema akeh* (Machida, 1989). (After Léon-Règagnon, 1998.)
44.9. *Monorchimacradena acanthuri* Nahhas & Cable, 1964. (After Nahhas & Cable, 1964.)

Subfamily Macradeninae Skrjabin & Guschanskaja, 1954

Diagnosis: Gut-caeca usually end blindly, but cyclocoel sometimes present. Testes two, occasionally one; oval; usually pre-ovarian, but may be at level of ovary or postovarian. Seminal vesicle in hindbody; saccular, tripartite or tubular. Pars prostatica tubular, long. Ejaculatory duct usually long, but may be short or absent. Sinus-sac present, small, may be poorly developed. Permanent sinus-organ absent. Ovary four-lobed or oval. Blind seminal receptacle present. Uterus reaches to postovarian region. Vitellarium immediately anterior, at level of or immediately posterior to ovary; variable, commonly six- to eight-lobed, but may be seven branched lobes, or twelve or fourteen lobes, and group of lobes may be antero-posteriorly orientated. Excretory arms united in forebody or not. Parasitic in intestine or stomach of marine teleosts (normally *Acanthurus* spp.). Type-genus *Macradenina* Manter, 1947.

Key to genera

1a. Vitellarium consisting of six to eight (usually seven) tear-shaped or slightly branched lobes; seminal vesicle saccular (? or tubular) .... 2.
1b. Vitellarium consisting of 12 or more lobes (occasionally seven basic lobes divided into about 40 secondary lobes), which may be tubular or globular; seminal vesicle tubular or tripartite .... 4.

2a. Testis single .... *Monorchimacradena* Nahhas & Cable, 1964. (Fig. 44.9)
Diagnosis: Cyclocoel not reported. Testis single, immediately pre-ovarian. Seminal vesicle sac-like, immediately pretesticular. Pars prostatica and ejaculatory duct about equal in length. Sinus-sac spherical to pyriform, small. Ovary oval. Seminal receptacle as large as or larger than ovary, dorsal to ovary. Vitellarium seven digitiform or slightly branched lobes, united centrally, post-testicular. Excretory arms united in forebody. Type-species *M. acanthuri* Nahhas & Cable, 1964.
2b. Testes two .... 3.

3a. Testes at level of ovary; vitellarium anterolateral to ovary; seminal vesicle saccular (or ?tubular) .... *Dichadena* Linton, 1910. (Fig. 44.10)
(Syn. *Pseudodichadena* Yamaguti, 1971)
Diagnosis: Cyclocoel present. Testes tandem, dorsal to ovary. Seminal vesicle sac-like (tubular according to Yamaguti, 1971), immediately pretesticular. Pars prostatica long, dilated posteriorly. Ejaculatory duct similar in length to pars prostatica. Sinus-sac oval. Ovary oval or three- to four-lobed. Seminal receptacle between posterior margin of ovary and posterior testis. Vitellarium six or seven rounded to tear-shaped lobes, anterolateral to ovary. Excretory arms (?). Type-species *D. acuta* Linton, 1910.
3b. Testes pre-ovarian; vitellarium postovarian; seminal vesicle saccular (? with constriction) .... *Neodichadena* Yamaguti, 1971. (Fig. 44.11)
Diagnosis: Cyclocoel not reported. Testes oblique, immediately pre-ovarian. Seminal vesicle sac-like (may have constriction). Pars prostatica long. Ejaculatory duct about half length of pars prostatica. Sinus-sac small, spherical. Ovary oval. Seminal

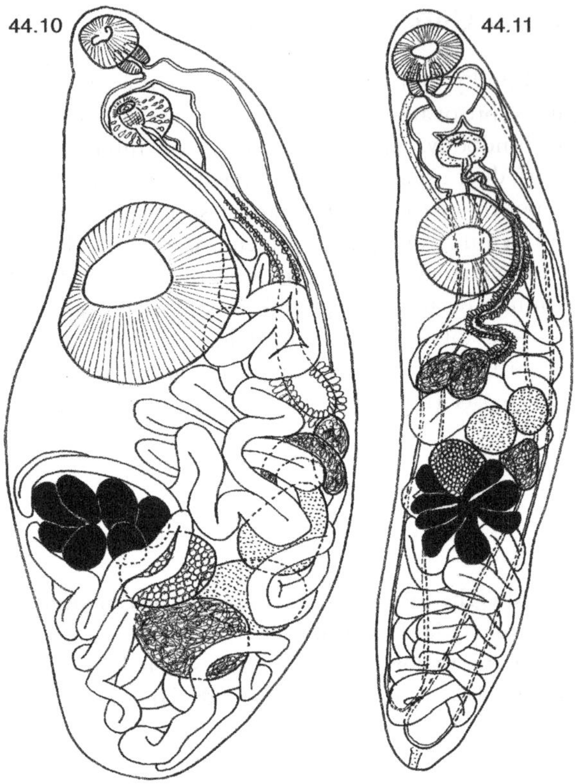

Figs 44.10–44.11.
44.10. *Dichadena acuta* Linton, 1910. (After Yamaguti, 1971.)
44.11. *Neodichadena acanthuri* (Siddiqi & Cable, 1960). (After Yamaguti, 1971.)

receptacle large, lateral to ovary. Vitellarium about eight tear-shaped lobes, postovarian. Excretory arms united in forebody. Type-species *N. acanthuri* (Siddiqi & Cable, 1960).

4a. Seminal vesicle tubular; vitelline lobes elongate, in single group .... 5.
4b. Seminal vesicle tripartite; vitelline lobes globular, in two groups, seven anterior and seven posterior to ovary .... *Acanthuritrema* Yamaguti, 1970. (Fig. 44.12)

Diagnosis: Cyclocoel not reported. Testes tandem to oblique, pre-ovarian. Seminal vesicle tripartite, with posterior part large and spherical, middle part elongate elliptical and anterior part pyriform, connected to pars prostatica by aglandular duct, in anterior hindbody. Pars prostatica short. Ejaculatory duct absent. Sinus-sac thin, membranous, indistinct. Ovary four-lobed. Seminal receptacle large, anterodorsal to ovary. Vitellarium fourteen globular lobes; seven anterior and seven posterior to ovary. Excretory arms not united in forebody. Type-species *A. multivitellosum* Yamaguti, 1970.

5a. Testes postovarian; vitellarium 12 claviform lobes, ventral to ovary .... *Macradenina* Manter, 1947. (Fig. 44.13)
(?Syn. *Paramacradenina* Tang, Shi & Guan in Tang, Shi, Cao, Guan & Pan, 1983)

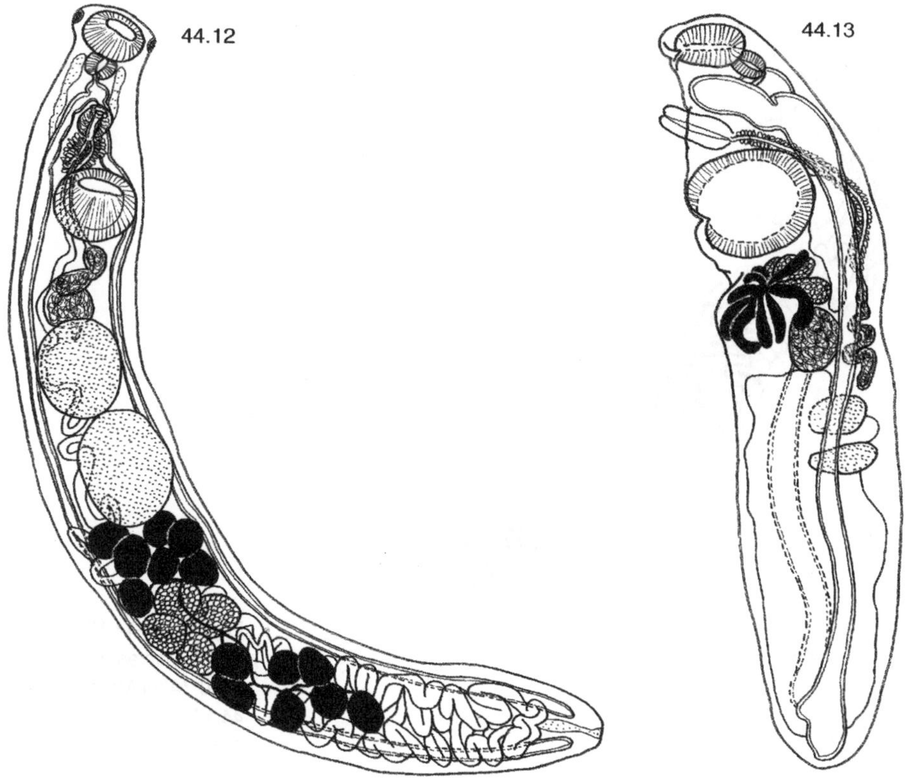

Figs 44.12–44.13.
44.12. *Acanthuritrema multivitellosum* Yamaguti, 1970. (After Yamaguti, 1970.)
44.13. *Macradenina acanthuri* Manter, 1947. (After Manter, 1947.)

Diagnosis: Cyclocoel not reported. Testes tandem, postovarian. Seminal vesicle tubular, loosely coiled at level of ovary. Pars prostatica long. Ejaculatory duct (?) absent. Sinus-sac subcylindrical, may be incomplete posteriorly, appears to be protrusible. Ovary four-lobed, well forward in hindbody. Seminal receptacle globular, large, postovarian. Vitellarium 12 claviform, unbranched tubules, ventral to ovary. Excretory arms (?). Type-species *M. acanthuri* Manter, 1947.

5b. Testes pre-ovarian; vitellarium essentially seven lobes, but divided into about 40 secondary lobes, between ovary and seminal vesicle .... *Macradena* Linton, 1910. (Fig. 44.14)

Diagnosis: Cyclocoel not reported. Testes tandem, pre-ovarian. Seminal vesicle tubular, coiled in anterior hindbody. Pars prostatica long. Ejaculatory duct about one quarter length of pars prostatica. Sinus-sac weakly developed and small. Ovary with four ovoid lobes. Seminal receptacle large, oval, immediately posterior to ovary. Vitellarium originates between ovary and seminal receptacle, consists of seven main radial lobes, which are finely divided to form about 40 fairly short, tubular lobes. Excretory arms united in forebody. Type-species *M. perfecta* Linton, 1910.

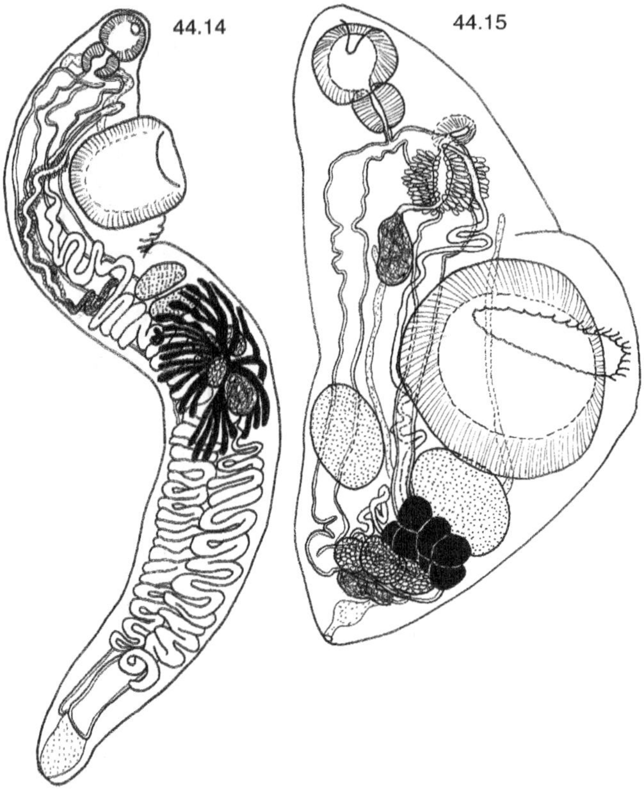

Figs 44.14–44.15.
44.14. *Macradena perfecta* Linton, 1910. (After Manter, 1947.)
44.15. *Prolecitha obesa* Manter, 1961. (Modified after Yamaguti, 1971.)

Subfamily Prolecithinae Yamaguti, 1971

(Syn. Folliovitellotrematinae Gupta & Sharma, 1972 [*sic*])

Diagnosis: Body spindle-shaped. Ventral sucker large, prominent; fluted rim may be apparent. Cyclocoel present. Testes two, diagonal to symmetrical, at level of or slightly posterior to ventral sucker. Seminal vesicle saccular; in forebody. Pars prostatica short, vesicular. Terminal genitalia complex (see Abdul-Salam & Khalil, 1987). Sinus-sac present, muscular, asymmetrical; anterior wall forms muscular sucker- or papilla-like structure (latter may be formed from wall of genital atrium); permanent sinus-organ absent. Genital atrium sucker-like. Ovary four-lobed. Large blind seminal receptacle posterior to ovary, close to posterior extremity of body. Uterus mainly anterior to gonads. Vitellarium seven rounded lobes, immediately pre-ovarian, close to testes. Excretory arms not united in forebody. Parasitic in intestine of marine teleosts (mainly Belonidae). Type-genus *Prolecitha* Manter, 1961.

Genus *Prolecitha* Manter, 1961 (Fig. 44.15)

(Syns *Lobatovitelliovarium* Yamaguti, 1965; *Follicovitellosum* Gupta & Sharma, 1972; *Cryptodiscus* Srivastava & Sahai, 1978)

Diagnosis: With characters of subfamily. Type-species *P. obesa* Manter, 1961.

Subfamily Quadrifoliovariinae Yamaguti, 1970

Diagnosis: Body elongate to spindle-shaped. Muscular ventrolateral flange or flanges present immediately posterior to ventral sucker. Caeca usually terminate blindly, but cyclocoel may be present. Testes two, pre-ovarian. Seminal vesicle in hindbody, saccular, constricted into portions or tubular. Pars prostatica usually short, tubular, in hindbody. Ejaculatory duct long, may be lined with cuticular villi. Sinus-sac oval. Permanent sinus-organ absent. Genital atrium short or apparently absent. Ovary oval or four-lobed. Blind seminal receptacle normally large, anterior or anterodorsal to ovary. Uterus reaches postovarian region. Vitellarium seven postovarian, claviform or oval lobes, or two groups of six to seven digitiform lobes, one pre-ovarian and one postovarian. Excretory arms united in forebody. Parasitic in stomach or pyloric caeca of acanthurid marine teleosts. Type-genus *Quadrifoliovarium* Yamaguti, 1965.

Key to genera

1a. Muscular ventrolateral flange present on one side of body only; cyclocoel present .... *Unilacinia* Manter, 1969. (Fig. 44.16)

Diagnosis: Bilobed muscular flange lateral to ventral sucker, on one side only. Cyclocoel present. Testes oblique. Seminal vesicle saccular, anterodorsal to anterior testis. Ejaculatory duct twice length of pars prostatica. Hermaphroditic duct wide. Sinus-sac broadly ovoid. Ovary oval. Vitellarium seven short, digitiform lobes, postero-ventral to ovary. Parasitic in stomach of marine teleosts (*Naso*). Type-species *U. asymmetrica* Manter, 1969.

1b. Muscular ventrolateral flanges symmetrical; cyclocoel absent .... 2.

2a. Ovary four-lobed; vitelline lobes in two anteroposteriorly orientated groups .... *Quadrifoliovarium* Yamaguti, 1965. (Fig. 44.17)

Diagnosis: Pair of muscular ventrolateral flanges immediately posterior to ventral sucker. Cyclocoel not reported. Testes tandem. Seminal vesicle a wide, convoluted tube, narrowing anteriorly. Pars prostatica short. Ejaculatory duct two to three times length of pars prostatica. Hermaphroditic duct convoluted, may be everted to form temporary sinus-organ. Sinus-sac thin-walled, elliptical to oval. Ovary four-lobed. Vitellarium in two anteroposteriorly orientated rosette-like groups; one pre- and one postovarian, united by collecting duct; each group has six or seven digitiform lobes. Parasitic in stomach and pyloric caeca of marine teleosts (*Naso*). Type-species *Q. pritchardae* Yamaguti, 1965.

2b. Ovary oval; vitelline lobes in one group .... *Bilacinia* Manter, 1969. (Fig. 44.18)
(Syn. *Holacanthitrema* Yamaguti, 1970)

Diagnosis: Pair of bilobed, muscular ventrolateral flanges, symmetrical, immediately posterior to ventral sucker. Cyclocoel not reported. Testes tandem to oblique. Seminal vesicle convoluted, wide and tubular or divided into four to five portions. Pars prostatica sigmoid. Ejaculatory duct one-quarter to nearly equal length of pars prostatica. Hermaphroditic duct wide, straight. Sinus-sac oval, thin-walled. Ovary oval. Vitellarium seven rounded to claviform lobes, immediately postovarian. Parasitic in

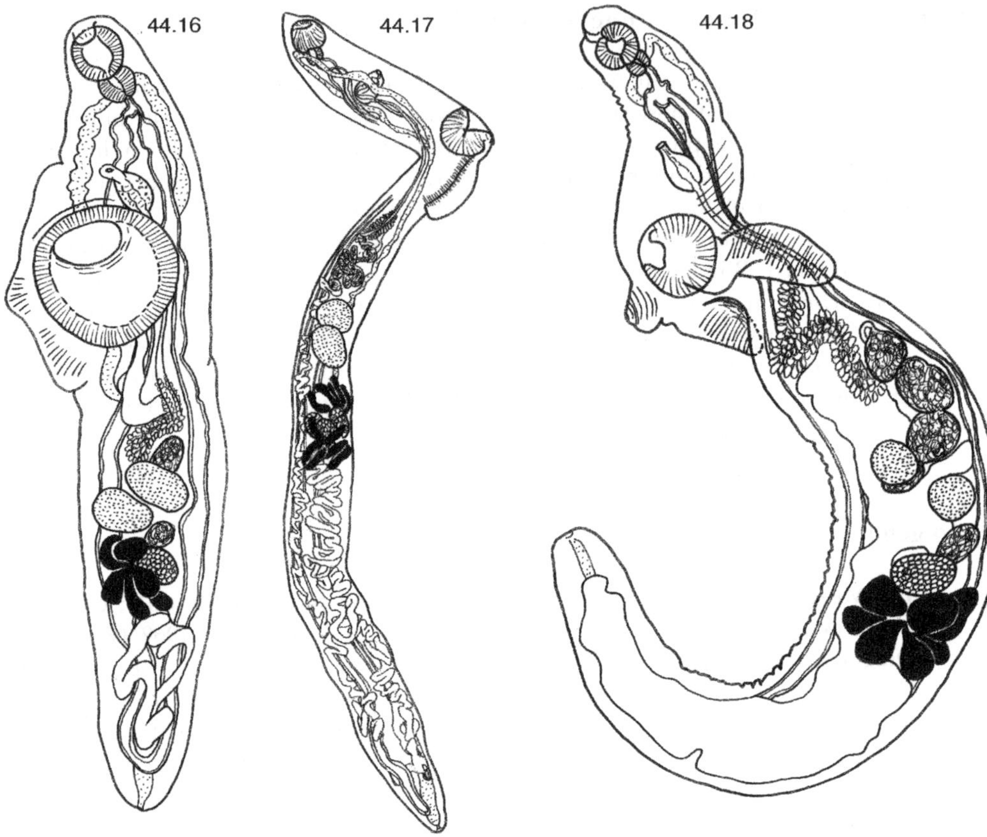

Figs 44.16–44.18.
44.16. *Unilacinia asymmetrica* Manter, 1969. (After Manter, 1969.)
44.17. *Quadrifoliovarium pritchardae* Yamaguti, 1965. (After Yamaguti, 1965.)
44.18. *Bilacinia australis* Manter, 1969. (After Manter, 1969.)

stomach and pyloric caeca of marine teleosts (*Naso* and *Holacanthus*). Type-species *B. australis* Manter, 1969.

Subfamily Trifoliovariinae Yamaguti, 1958

Diagnosis: Body cylindrical, long and thin, fusiform or elongate-oval. Ventral sucker present well inside anterior half of body. Gut-caeca terminate blindly, often wide. Testes two, pre-ovarian in anterior hindbody, separated from ovary by many loops of uterus. Seminal vesicle in forebody, at level of ventral sucker or in hindbody, tubular, moniliform or elongate oval, usually connected to pars prostatica by short aglandular duct. Pars prostatica short, tubular or vesicular. Ejaculatory duct short; enters hermaphroditic duct, which is continuation of uterus, laterally. Sinus-sac present surrounding hermaphroditic duct, weak and membranous, tubular or oval. Permanent sinus-organ absent. Genital atrium absent or small. Female reproductive complex close to posterior extremity. Ovary four-lobed. Blind seminal receptacle large, dorsal or anterodorsal to ovary. Uterus entirely or almost entirely pre-ovarian, coils confined to hindbody. Eggs without filaments. Vitellarium seven or eight claviform or digitiform lobes, which may be bilobed or irregularly branched distally, at level of ovary or immediately postovarian. Excretory arms united in forebody. Parasitic in intestine of marine teleosts. Type-genus *Trifoliovarium* Yamaguti, 1940.

Key to genera

1a. Vitellarium essentially comprises seven claviform or digitiform lobes in rosette arrangement, each of which may be branched (or bilobed); uterus overlaps caeca over much of its length; suckers well separated .... *Trifoliovarium* Yamaguti, 1940. (Fig. 44.19)

(Syns *Cladolecithotrema* Ichihara, 1970; *Pseudolecithaster* Campbell & Munroe, 1977)

Diagnosis: Body elongate-oval, fusiform or long and thin. Oral sucker may be funnel-shaped. Suckers separated. Testes tandem, oblique or symmetrical, widely separated. Seminal vesicle oval, elongate-oval, moniliform or tubular and coiled or sinuous, in hindbody, dorsal to ventral sucker or in forebody. Pars prostatica tubular

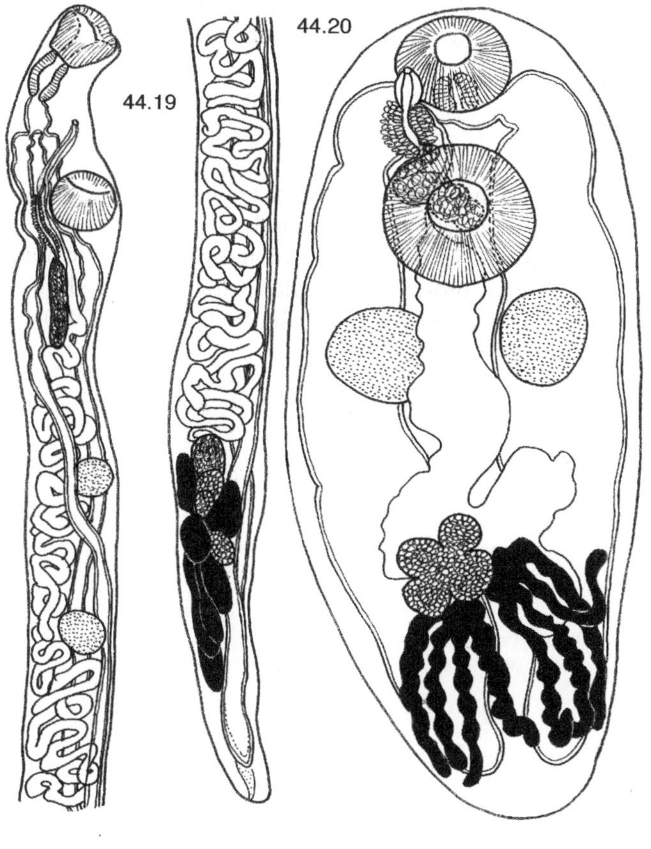

Figs 44.19–44.20.
44.19. *Trifoliovarium acanthocepolae* Yamaguti, 1940. (After Yamaguti, 1940.)
44.20. *Assitrema eichleri* Parukhin, 1976. (Modified after Parukhin, 1976b.)

or vesicular. Sinus-sac oval, elongate-oval or tubular, often small or indistinct. Uterus tends to overlap caeca over much of its length. Vitellarium seven claviform or digitiform lobes, may be branched distally, or rosette of 12 to 16 lobes (may be seven bilobed lobes), at level of ovary or centred just postovarian. Type-species *T. acanthocepolae* Yamaguti, 1940.

1b. Vitellarium comprises eight (two groups of four) tubular, mainly posteriorly orientated, postovarian lobes; uterus tends to be intercaecal over much of its length; suckers close together .... *Assitrema* Parukhin, 1976. (Fig. 44.20)

Diagnosis: Body small, oval to elongate-oval. Gut-caeca wide. Testes symmetrical, separated by coils of uterus. Seminal vesicle short, moniliform or elongate, saccular, winding antero- to posterodorsally to ventral sucker. Pars prostatica tubular. Sinus-sac small, oval. Seminal receptacle present. Uterus almost entirely pre-ovarian, mainly intercaecal. Eggs small. Vitellarium two symmetrical groups of four posterolaterally orientated, digitiform lobes; centre of each group immediately postovarian. Excretory arms (?). Type-species *A. eichleri* Parukhin, 1976.

Notes Added in Proof

Lecithasterinae

The new genus *Weketrema* Bray & Cribb, 2001 is distinguished from other genera in the subfamily by its complete lack of a sinus-sac (Bray & Cribb, 2001).

Macradeninae

The new genus *Zebrasomatrema* Bray & Chambers, 2000 has the following suite of characters which differentiate it from other genera in the subfamily: an anterior flange, 14 subglobular vitelline lobes, a quadripartite seminal vesicle, a long aglandular duct joining the seminal vesicle and the short pars prostatica, a single vitelline field, the lack of a sinus-sac and excretory arms that meet in the forebody (Bray & Chambers, 2000).

45 Family Ptychogonimidae Dollfus, 1937

D.I. Gibson

Department of Zoology, The Natural History Museum, London, UK

Introduction

This family was first erected by Dollfus (1937b) and, presumably in ignorance of this, independently by Yamaguti (1953) for the single genus *Ptychogonimus*, which occurs in the stomach of elasmobranchs. This genus has been discussed previously by Gibson & Bray (1977, 1979); it has an unusual life-history in that it uses a scaphopod as its first intermediate host and transmission to the second intermediate host is effected by means of a motile parthenita. The family was included in the Hemiuroidea by Skrjabin & Guschanskaja (1956), Yamaguti (1971) and Gibson & Bray (1977, 1979). Gibson & Bray (1979) linked it with the Azygiidae as primitive members of the group, a position supported by Brooks *et al.* (1985). Bray *et al.* (1995) added a second genus, *Melogonimus*, and adapted the family diagnosis presented by Gibson & Bray (1979) to accommodate this.

Family Ptychogonimidae Dollfus, 1937

Diagnosis: Body medium-sized, robust, oval to elongate-oval. Ecsoma absent. Body surface smooth or with numerous papillae, without spines or plications. Oral and ventral suckers well developed; ventral sucker in anterior half of body. Pharynx well developed. Oesophagus short. 'Drüsenmagen' absent. Gut-caeca form uroproct or open as ani. Testes two, postovarian, tandem, in middle of hindbody. Seminal vesicle dilated, tubular to saccular, thin- or thick-walled, in forebody or extending posteriorly into anterior hindbody. Pars prostatica tubular or all but absent. Ejaculatory duct long or short. Hermaphroditic duct short. Permanent sinus-organ long or short. Sinus-sac absent; muscles surrounding base of sinus-organ continuous with wall of genital atrium. Wall of genital atrium forms concentric folds around sinus-organ or not. Genital pore midventral in forebody. Ovary oval to suboval, pretesticular in hindbody. Mehlis' gland pre-ovarian. Laurer's canal and uterine seminal receptacle present. Juel's organ and canalicular or blind seminal receptacle absent. Uterine fields in hindbody posterior and/or anterior to testes. Metraterm long. Eggs numerous, without filaments. Vitellarium follicular, in lateral fields of hindbody; fields confluent or not posterior to testes. Excretory vesicle Y-shaped; stem short; arms long, unite twice in forebody forming ring around oral sucker. First intermediate host scaphopod molluscs. Parasitic in stomach of elasmobranchs. Type-genus *Ptychogonimus* Lühe, 1900.

Key to genera

1a. Uroproct present; wall of genital atrium with concentric fold; sinus-organ small; much of uterus post-testicular .... *Ptychogonimus* Lühe, 1900. (Fig. 45.1)

Diagnosis: Tegument smooth. Oral sucker larger than ventral sucker. Gut-caeca form uroproct. Seminal vesicle sinuous, tubular, thin-walled, reaches into anterior hindbody. Pars prostatica distinct. Ejaculatory duct short. Sinus-organ a small cone.

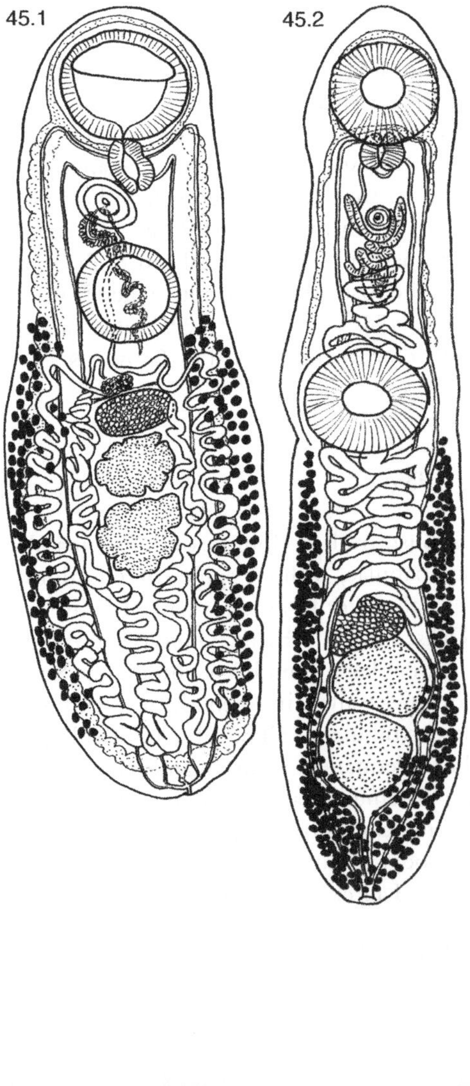

Figs 45.1–45.2.
45.1. *Ptychogonimus megastoma* (Rudolphi, 1819). (After Gibson & Bray, 1977.)
45.2. *Megalogonimus rhodanometra* Bray, Brockerhoff & Cribb, 1995. (After Bray *et al.*, 1995.)

Genital atrium large, broad; wall with three concentric folds that surround sinus-organ. Uterus extends well into post-testicular field. Metraterm not normally convoluted. Lateral vitelline fields not confluent posteriorly. Type-species *P. megastoma* (Rudolphi, 1819).

1b. Two ani present; wall of genital atrium smooth; sinus-organ long, digitiform; uterus pretesticular .... *Melogonimus* Bray, Brockerhoff & Cribb, 1995. (Fig. 45.2)

Diagnosis: Tegument papillate. Suckers of similar size. Caeca open into shallow terminal pit, independently of excretory system, via ani. Seminal vesicle large, oval, thick-walled, in forebody. Pars prostatica indistinct. Ejaculatory duct long. Sinus-organ long, digitiform. Genital atrium deep, smooth-walled. Uterus pretesticular, mainly pre-ovarian. Metraterm convoluted. Lateral vitelline fields confluent in post-testicular region. Type-species *M. rhodanometra* Bray, Brockerhoff & Cribb, 1995.

46 Family Sclerodistomidae Odhner, 1927

D.I. Gibson

Department of Zoology, The Natural History Museum, London, UK

Introduction

The sclerodistomids are normally parasites of the stomach of marine teleosts. The family group was erected by Odhner (1927) as a subfamily of the Hemiuridae, in which he linked *Sclerodistomum* Looss, 1912 with *Hirudinella* de Blainville, 1828 and *Isoparorchis* Southwell, 1913. It was raised to full family status by Dollfus (1932b), who restricted it to *Sclerodistomum* only. The family Prosogonotrematidae and the subfamily Prosorchiinae (within the Hemiuridae) were erected by Vigueras (1940) for *Prosogonotrema* Vigueras, 1940 and Yamaguti (1934) for *Prosorchis* Yamaguti, 1934, respectively. Although accepted as distinct by Skrjabin & Guschanskaja (1956) and Yamaguti (1958a, 1971), these three groups were linked by Gibson & Bray (1977, 1979), who considered them all to represent separate subfamilies within the Sclerodistomidae. A feature of the group is the presence of Manter's organ ('accessory excretory organ'), which occurs, either singly or double, dorsally to the excretory vesicle, into which it opens posteriorly. Most recently, the subfamily Kenmackenziinae was created by Gibson (1983) for *Kenmackenzia* Gibson, 1983, a new genus erected for *Distoma gigas* Nardo, 1827, a huge digenean from the stomach of *Luvarus imperialis*, which Gibson & Bray (1979) had suggested might prove to be a sclerodistomid.

Bhaleraoia Srivastava, 1948 was included as a questionable synonym of *Prosogonotrema* by Gibson & Bray (1979), because, although similar in gross morphology, it is poorly described and purported to possess a uroproct (perhaps a misinterpretation of Manter's organ). The family Bhaleraoiidae, erected by Srivastava (1948), was relegated to synonymy with the Prosogonotrematidae by Yamaguti (1971) and to the Sclerodistomidae by Gibson & Bray (1979). *Mabiarama* Teixeira de Freitas & Kohn, 1967 and the Mabiaramidae, erected by Teixeira de Freitas & Kohn (1967), were considered synonyms of *Sclerodistomum* and the Sclerodistomidae, respectively, by Gibson & Bray (1977, 1979).

Parukhin (1976a) erected a new subfamily, the Pseudosclerodistomoidinae, within the Sclerodistomidae for *Pseudosclerodistomoides kurotchkini*, a new genus and species from the gall-bladder of *Lethrinus miniatus* in the Indian Ocean. Several features of this species, such as the position of the testes and the genital pore and the nature of the seminal receptacle, suggest that it is not a hemiuroid. I tentatively consider these taxa to be lepocreadioids.

There are features of the incomplete and unillustrated description of *Eurycoelum sluiteri* Brock, 1886 which indicate that it may be a sclerodistomid (Skrjabin & Guschanskaja, 1957a; Gibson & Bray, 1979). These may include a sinus-sac, a tubular vitellarium and possibly a Manter's organ. Yet, as recently as 1971, Yamaguti still followed Braun (1892) in listing it as a species of *Hemiurus* Rudolphi, 1809. *E. sluiteri* has been recorded only once from the stomach of a species of *Lutjanus* in Indonesian waters. As suggested by Gibson & Bray (1979), the most likely sclerodistomids to occur in this host-genus are species of *Prosogonotrema*, but Brock's (1886) description gives no information on the position and arrangement of the gonads. In consequence, this taxon must remain a *genus inquirendum*.

Family Sclerodistomidae Odhner, 1927

(Syns Prosogonotrematidae Vigueras, 1940; Bhaleraoiidae Srivastava, 1948; Mabiaramidae Teixeira de Freitas & Kohn, 1967)

Diagnosis: Body usually large, stout or elongate. Ecsoma absent. Body surface smooth, but may be rugate. Oral and ventral suckers well developed; ventral sucker just posterior to middle, in middle or in anterior half of body. Pharynx well developed. Oesophagus short. 'Drüsenmagen' present or absent. Gut-caeca terminate blindly close to posterior extremity. Testes two, symmetrical, oblique or in tandem, pre-ovarian or occasionally separated by ovary, in forebody, dorsal to ventral sucker or in anterior hindbody. Seminal vesicle tubular, convoluted or winding in forebody. Pars prostatica tubular, occasionally with wide lumen, long, convoluted or not, external gland-cells may be delimited. Ejaculatory duct present, unites with metraterm within sinus-organ, forming short hermaphroditic duct. Permanent sinus-organ usually well developed, conical to cylindrical. Sinus-sac well developed, weakly developed or apparently absent. Genital atrium well developed, often almost entirely filled by sinus-organ. Genital pore midventral in forebody. Ovary globular to oval, well posterior to and usually separated from testes by loops of uterus or occasionally intertesticular, in posterior forebody, dorsal to ventral sucker or anterior to mid-hindbody. Mehlis' gland usually posterior or posterolateral, occasionally antero-lateral, to ovary. Laurer's canal present, opens dorsally or into rudimentary Juel's organ. Rudimentary seminal receptacle present or absent. Canalicular or blind seminal receptacle absent. Uterine seminal receptacle present or apparently absent. Uterus mainly in either fore- or hindbody, mainly pre- or postovarian. Eggs small, numerous, non-filamented. Vitellarium composed of convoluted, tubular branches (usually four to seven), which may subdivide, either mainly pre- or postovarian, either mainly in forebody or mainly to entirely in hindbody. Excretory vesicle Y-shaped; stem relatively long; arms united in forebody. Manter's organ (accessory excretory vesicle) present, single or double, dorsal to stem of excretory vesicle. Parasitic in gut (mainly stomach) (?), occasionally in body-cavity, of marine teleosts. Type-genus *Sclerodistomum* Looss, 1912.

Key to subfamilies

1a. **Manter's organ double; testes symmetrical to oblique in hindbody .... Sclerodistominae Odhner, 1927.**
1b. **Manter's organ single .... 2.**

2a. **Testes in forebody; ovary post-testicular .... 3.**
2b. **Testes tandem in hindbody; ovary intertesticular .... Kenmackenziinae Gibson, 1983.**

3a. **Ovary in forebody or at level of ventral sucker; uterus and vitellarium mainly pre-ovarian .... Prosogonotrematinae Vigueras, 1940.**
3b. **Ovary in hindbody; uterus and vitellarium mainly postovarian .... Prosorchiinae Yamaguti, 1934.**

Subfamily Sclerodistominae Odhner, 1927

Diagnosis: Body stout. Ventral sucker near middle of body, occasionally more anterior. 'Drüsenmagen' present. Testes symmetrical to oblique, in anterior hindbody. Pars prostatica long and convoluted. Sinus-sac distinctly present, with diffuse musculature, surrounding base of genital atrium. Sinus-organ a stout cone. Ovary near or just anterior to middle of hindbody. Laurer's canal opens dorsally. Rudimentary seminal receptacle present. Uterus almost entirely in hindbody, mainly posterior to gonads, reaching near to posterior extremity. Vitellarium composed of convoluted, tightly helical, unbranched, filamentous tubules, commonly arranged with three on one side of body, four on other, present laterally in post-testicular region of hindbody. Manter's organ (accessory excretory vesicle) double, symmetrical. Parasitic in stomach (also reported from intestine and body-cavity) of marine teleosts. Type-genus *Sclerodistomum* Looss, 1912.

Genus *Sclerodistomum* Looss, 1912 (Fig. 46.1)
(Syn. *Mabiarama* Teixeira de Freitas & Kohn, 1967)

Diagnosis: With characters of subfamily. Type-species *S. italicum* (Stossich, 1893).

Subfamily Kenmackenziinae Gibson, 1983

Diagnosis: Very large, elongate worms. Oral sucker ventrally subterminal; pre-oral lobe present. Ventral sucker very large, protuberant, well inside anterior half of body, bearing fluted flange within margin. Caeca diverticulate, end blindly near posterior extremity. Testes in tandem, but separated by ovary, in anterior hindbody. Pars prostatica long, convoluted, with delimited layer of surrounding gland-cells. Sinus-sac rudimentary, enclosing base of genital atrium, composed of diffuse musculature. Genital atrium large, muscular, containing large, muscular, campanuliform

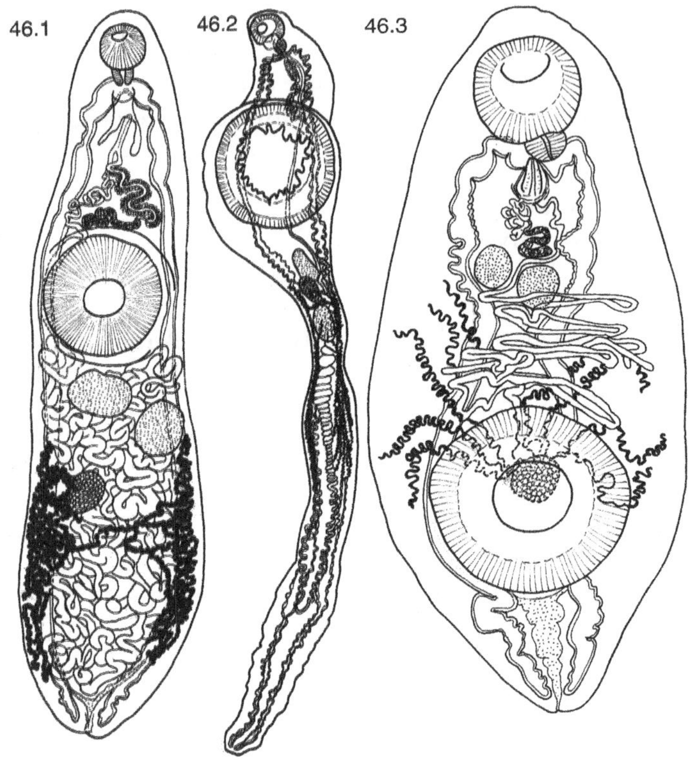

Figs 46.1–46.3.
46.1. *Sclerodistomum diodontis* Yamaguti, 1942. (After Yamaguti, 1942.)
46.2. *Kenmackenzia gigas* (Nardo, 1827). (After Gibson, 1983.)
46.3. *Prosogonotrema subequilatum* Pritchard, 1963. (After Pritchard, 1963.)

sinus-organ, which may extrude through genital pore. Genital pore at level of posterior extremity of oral sucker. Ovary entire, intertesticular. Laurer's canal opens dorsally. Uterine seminal receptacle assumed to occur. Juel's organ absent. Uterus winds into posterior half of hindbody prior to reflexing and winding forward into forebody. Vitellarium filamentous; approximately symmetrical branching system of tightly helical tubules, extending between ovary and middle of hindbody. Terminal pore acts as vent for excretory system and Manter's organ. Excretory vesicle ventral to Manter's organ. Manter's organ single, ends blindly posterior to posterior testis. In mesopelagic teleosts (*Luvarus*). Type-genus *Kenmackenzia* Gibson, 1983.

Genus *Kenmackenzia* Gibson, 1983 (Fig. 46.2)

Diagnosis: With characters of subfamily. Type-species *K. gigas* (Nardo, 1827).

Subfamily Prosogonotrematinae Vigueras, 1940

Diagnosis: Body stout. Ventral sucker in middle or just posterior to middle of body. 'Drüsenmagen' absent. Testes symmetrical, in forebody. Pars prostatica long and sinuous or convoluted. Sinus-sac well developed, but composed of diffuse musculature. Sinus-organ conical to cylindrical. Ovary in forebody or at level of ventral sucker.

Laurer's canal short, opens into large, rudimentary Juel's organ, dilated proximally, forming small rudimentary seminal receptacle. Rudimentary Juel's organ postovarian (distinct in sections). Uterus mainly pre-ovarian, does not extend into hindbody. Vitellarium about three (two to four) convoluted, tubular branches on each side of body, which may subsequently divide, in forebody, often mainly pre-ovarian. Manter's organ (accessory excretory vesicle) single, median. Parasitic mainly in stomach of marine teleosts. Type-genus *Prosogonotrema* Vigueras, 1940.

Genus *Prosogonotrema* Vigueras, 1940 (Fig. 46.3)
(Syn. ?*Bhaleraoia* Srivastava, 1948)

Diagnosis: With characters of subfamily. Type-species *P. bilabiatum* Vigueras, 1940.

Subfamily Prosorchiinae Yamaguti, 1934

Diagnosis: Body elongate-oval to elongate. Ventral sucker normally within anterior half of body, but occasionally in middle. 'Drüsenmagen' absent. Testes tandem to almost symmetrical, in forebody or dorsal to ventral sucker. Pars prostatica broad, usually straight or arcuate; lumen wide; external gland-cells delimited. Sinus-sac apparently absent or (?) weakly developed. Sinus-organ conical to cylindrical. Ovary in hindbody. Laurer's canal long and opening into rudimentary Juel's organ or short and opening to exterior on dorsal surface, may or may not be dilated proximally, forming rudimentary seminal receptacle. Uterus mainly postovarian in hindbody. Vitellarium two to four branching or seven convoluted and filamentous tubules, mainly postovarian in hindbody, reaching near to posterior extremity. Manter's organ (accessory excretory vesicle) single, median. Type-genus *Prosorchis* Yamaguti, 1934.

Key to genera

1a. Laurer's canal long, opening into rudimentary Juel's organ, dilated proximally, forming rudimentary seminal receptacle (feature visible in whole-mounts: seminal receptacle present) .... *Prosorchis* Yamaguti, 1934. (Fig. 46.4)

Diagnosis: Laurer's canal long, opens distally into well-developed rudimentary Juel's organ ('terminal vesicle'), dilated proximally, forming rudimentary seminal receptacle. Type-species *P. psenopsis* Yamaguti, 1934.

1b. Laurer's canal short, opening dorsally to exterior; rudimentary seminal receptacle absent (feature visible in whole-mounts: seminal receptacle absent) .... *Prosorchiopsis* Dollfus, 1947. (Fig. 46.5)

Diagnosis: Laurer's canal short, opening dorsally to exterior; distal region may be slightly dilated and possess glandular wall; rudimentary seminal receptacle absent. Type-species *P. legendrei* Dollfus, 1947.

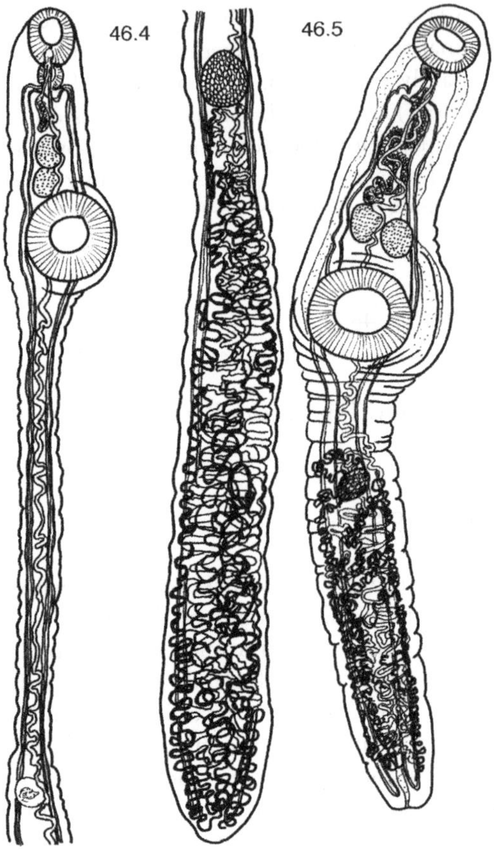

Figs 46.4–46.5.
46.4. *Prosorchis psenopsis* Yamaguti, 1934. (After Yamaguti, 1934.)
46.5. *Prosorchiopsis aluterae* Yamaguti, 1970. (After Yamaguti, 1970.)

47 Family Sclerodistomoididae Gibson & Bray, 1979

D.I. Gibson

Department of Zoology, The Natural History Museum, London, UK

Introduction

The family Sclerodistomoididae was erected by Gibson & Bray (1979) for *Sclerodistomoides* Kamegai, 1971 (see Kamegai, 1971) after examining type-material of *S. pacificus* Kamegai, 1971, because they considered that this genus did not fit into related hemiuroid groups, differing from the accacoeliids in the structure of the pharynx, from the sclerodistomids in the absence of Manter's organ and from both groups in the unique position and orientation of the main collecting ducts of the vitelline system. They considered that its closest relative may be the accacoeliid *Paraccacladium* Bray & Gibson, 1977 and intimated a possible link with '*Lintonium*' *nikovi* Baeva, 1965, which Yamaguti (1971) thought might represent a new sclerodistomid genus. There appear to have been no records of *Sclerodistomoides* since the original description.

Family Sclerodistomoididae Gibson & Bray, 1979

Diagnosis: Body large, stout, elongate-oval and attenuated anteriorly. Ecsoma absent. Body surface smooth. Oral and ventral suckers well developed, small; ventral sucker well inside anterior half of body. Pharynx well developed. Oesophagus short. 'Drüsenmagen' present. Gut-caeca sinuous, end blindly near posterior extremity, very close to stem of excretory vesicle (no union observed). Testes two, slightly lobate, symmetrical to oblique, in hindbody close to ventral sucker, pre-ovarian. Seminal vesicle tubular convoluted in forebody. Pars prostatica short, tubular with wide lumen. Ejaculatory duct short. Hermaphroditic duct short, present within small, cone-shaped permanent sinus-organ. Sinus-sac small and poorly developed or apparently absent. Genital atrium well developed. Genital pore midventral at level of pharynx. Ovary oval, in middle of body, separated from testes by loops of uterus. Mehlis' gland posterolateral to ovary. Laurer's canal and uterine seminal receptacle present.

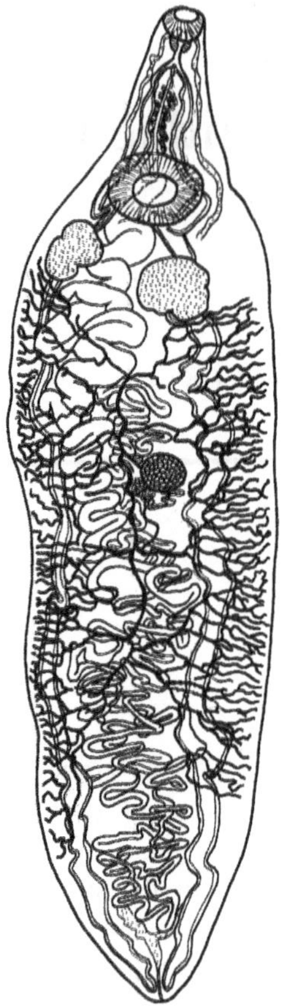

Fig. 47.1. *Sclerodistomoides pacificus* Kamegai, 1971. (Modified after Kamegai, 1971.)

Canalicular or blind seminal receptacle and Juel's organ absent. Uterus intercaecal, coiled almost entirely in hindbody, in both pre- and postovarian fields, with narrow, convoluted descending loop, reaching near to posterior extremity and convoluted ascending loop, which is initially narrow and then much broader. Eggs small, numerous, without filaments. Vitellarium tubular, with two main collecting ducts situated medially, one anteriorly and other posteriorly orientated, laterally directed dendritic branches present between testes and short distance anterior to posterior extremity. Stem of excretory vesicle short; arms initially in dorsal and ventral fields, pass laterally and unite in forebody. Parasitic in gall-bladder of marine teleosts. Type-genus *Sclerodistomoides* Kamegai, 1971.

Genus *Sclerodistomoides* Kamegai, 1971 (Fig. 47.1)

Diagnosis: With characters of family. Type-species *S. pacificus* Kamegai, 1971.

48 Family Syncoeliidae Looss, 1899

D.I. GIBSON

Department of Zoology, The Natural History Museum, London, UK

Introduction

The family Syncoeliidae was originally erected by Looss (1899) as a subfamily that included *Syncoelium* Looss, 1899, *Progonus* Looss, 1899 and *Otiotrema* Setti, 1897. Odhner (1911b) linked both *Derogenes* Lühe, 1900 and *Bathycotyle* Darr, 1902 with this subfamily. It was raised, without comment, to full family status by Dollfus (1923), although he indicated that he included within it the Derogeninae. Poche (1926) kept the Syncoeliinae within the Hemiuridae, but Odhner (1927) retained the family distinct from the Hemiuridae and Derogeninae. Dollfus (1932b) removed *Bathycotyle* from the group, but (Dollfus, 1935b) agreed with Ejsmont (1932) that that the family should include four subfamilies: the Syncoeliinae, Derogeninae, Halipeginae and Liocercinae (= Liopyginae). However, Manter (1938) did not agree and restricted the family to two genera, *Syncoelium* and *Otiotrema*. *Paronatrema* Dollfus, 1937 and *Copiatestes* Crowcroft, 1948 were added to the family by Manter (1940) and Crowcroft (1948), respectively. This is the concept of the family accepted by Yamaguti (1953) and in his subsequent compendia. Yamaguti (1971) recognized three subfamilies, the Syncoeliinae, Otiotrematinae and Paronatrematinae, but the latter was considered a synonym of the Otiotrematinae by Gibson & Bray (1979). The last arrangement is accepted here.

Copiatestes was synonymized with *Syncoelium* by Yamaguti (1953), but Gibson & Bray (1977) indicated that *Syncoelium* can be conveniently split into two distinct genera on the basis of the presence or absence of a sinus-organ. The sinus-organ in *Copiatestes* is a well-developed structure, visible in sections and whole-mounts in both adults and metacercariae (see Gibson, 1976). In *Syncoelium*, which lacks a sinus-organ, the genital atrium appears to be indistinguishable from the hermaphroditic duct, the hind-body tends to be spatulate rather than tubular, the ventral sucker is somewhat less pedunculate and the parasites occur on the gills rather than in the buccal cavity and on the gill-arches.

Family Syncoeliidae Looss, 1899

Diagnosis: Body elongate or dorsoventrally flattened, usually with pedunculate ventral sucker. Ecsoma absent. Body surface smooth, but often papillate on forebody and on peduncle, if present. Oral and ventral suckers well developed, may possess marginal accessory suckers. Glandular cells common in subtegumentary parenchyma and within musculature of suckers. Pharynx well developed. Oesophagus short. Cyclocoel usually present, but gut-caeca may end blindly (?) or form a uroproct. Testes 11–18 distinct, oval follicles (usually arranged in pairs), seven to eight transverse rows of small follicles, or just a large number of irregular follicles, pre-ovarian, in hindbody. Seminal vesicle thin-walled, tubular, winding or sinuous, in forebody. Pars prostatica tubular. Ejaculatory duct short. Hermaphroditic duct and genital atrium present, but indistinguishable when sinus-organ is absent. Permanent sinus-organ and sinus-sac present or absent. Genital pore midventral in anterior forebody. Ovary post-testicular, composed of five large, oval, isolated lobes or numerous irregular follicles. Laurer's canal and uterine seminal receptacle present. Juel's organ and canalicular or blind seminal receptacle absent; rudimentary seminal receptacle may be present. Uterus passes posteriorly but coils mainly in pre-ovarian hindbody. Eggs numerous, small, non-filamented. Vitellarium usually seven (occasionally five or six) isolated, oval lobes, or irregular acinous bunches or rows of follicles, postovarian. Excretory vesicle Y-shaped; arms united in forebody, may initially run in dorsal and ventral fields. Free-floating metacercarial stage present. Parasitic in branchial and buccal cavities, on skin, in (?) oviduct and in (?) gut of elasmobranchs and marine teleosts. Type-genus *Syncoelium* Looss, 1899.

Key to subfamilies

1a. Ovary comprises five large, isolated lobes, and vitellarium five to seven oval, isolated lobes .... Syncoeliinae Looss, 1899.
1b. Ovary comprises numerous irregular follicles arranged in rows, and vitellarium of irregular, acinous bunches or rows of follicles .... Otiotrematinae Skrjabin & Guschanskaja, 1957.

Subfamily Syncoeliinae Looss, 1899

Diagnosis: Small accessory suckers around rim of suckers absent. Peduncle usually present. Cyclocoel present. Testes 11–18 distinct, oval lobes, usually arranged in pairs. Permanent sinus-organ present or absent. Hermaphroditic duct and genital atrium present, but indistinguishable when sinus-organ is absent. Sinus-sac absent or rudimentary. Ovary composed of five large, isolated follicles. Rudimentary seminal receptacle may be present as proximal dilatation of Laurer's canal. Uterus arranged in large, regular loops dorsally and ventrally to gonads. Vitellarium seven (occasionally five or six) small, isolated, oval lobes. Parasitic in branchial and buccal cavities (? with occasional records from gut) of sharks and marine teleosts. Type-genus *Syncoelium* Looss, 1899.

Key to genera

1a. **Permanent sinus-organ absent; hermaphroditic duct indistinguishable from genital atrium; tendency for hindbody to be spatulate; ventral sucker normally sessile or on short peduncle; generally parasitic on gills .... *Syncoelium* Looss, 1899. (Fig. 48.1)**

Diagnosis: Hindbody broad or spatulate. Ventral sucker sessile or surmounted on short peduncle. Permanent sinus-organ absent. Hermaphroditic duct indistinguishable from genital atrium. Parasitic on gills (? and in intestine) of sharks and marine teleosts. Type-species *S. ragazzii* (Setti, 1897).

1b. **Permanent sinus-organ present; hermaphroditic duct readily distinguishable from genital atrium; hindbody usually tubular; ventral sucker usually on well-developed peduncle; usually parasitic in buccal cavity or on gill-arches .... *Copiatestes* Crowcroft, 1948.[1] (Fig. 48.2)**

Diagnosis: Hindbody elongate, tubular. Ventral sucker mounted on well-developed peduncle. Permanent sinus-organ present. Hermaphroditic duct and genital atrium easily distinguishable. Parasitic in branchial (especially gill-arches and gill-rakers) and buccal cavities (? and intestine) of marine teleosts. Type-species *C. thyrsitae* Crowcroft, 1948.

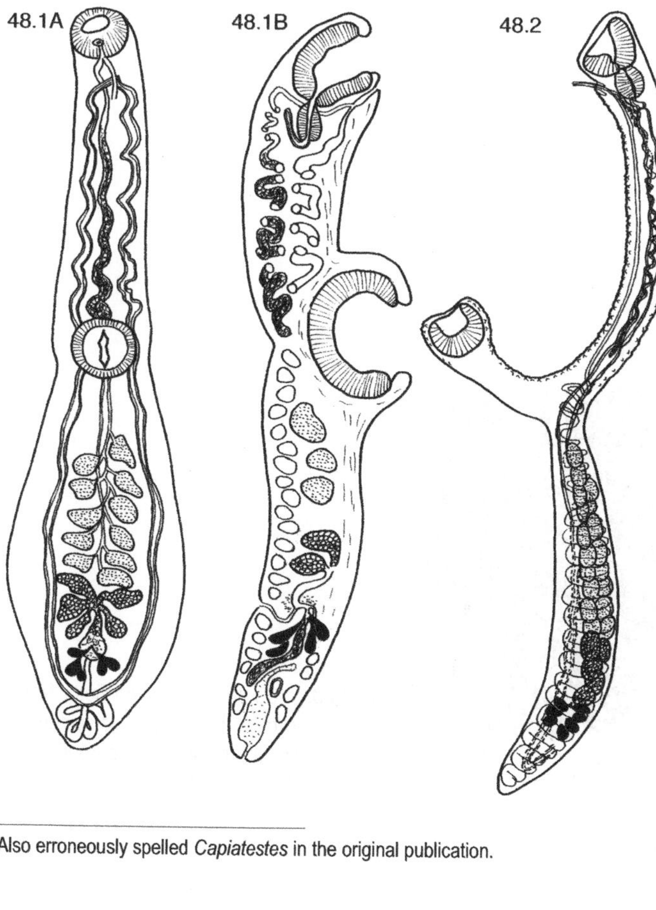

Figs 48.1–48.2.
48.1. *Syncoelium ragazzii* (Setti, 1897). A. Ventral view. B. Sagittal view. (After Looss, 1899.)
48.2. *Copiatestes filiferus* (Leuckart in Sars, 1885). (After Gibson, 1976.)

[1] Also erroneously spelled *Capiatestes* in the original publication.

Subfamily Otiotrematinae Skrjabin & Guschanskaja, 1957

(Syn. Paronatrematinae Skrjabin & Guschanskaja, 1957[2])

Diagnosis: [*Paronatrema* is poorly known.] Accessory suckers may be present around rim of suckers. Cyclocoel present (? or absent; caeca may end blindly or form uroproct). Testes numerous irregular follicles or seven to eight irregular transverse rows of follicles. Hermaphroditic duct present within sinus-sac (? not clearly described in *Paronatrema*). Permanent sinus-organ absent. Ovary composed of numerous irregular follicles, which may be arranged in rows. Vitellarium comprising irregular acinous bunches or rows of follicles. Parasitic in branchial cavity or on skin (? or in oviduct or intestine) of sharks and rays. Type-genus *Otiotrema* Setti, 1897.

Key to genera

1a. Small accessory suckers on suckers absent .... ***Otiotrema*** **Setti, 1897. (Fig. 48.3)**

Diagnosis: Forebody cylindrical; hindbody flattened, alate, recurved. Papillae present within oral sucker; small accessory suckers within suckers absent; ventral sucker

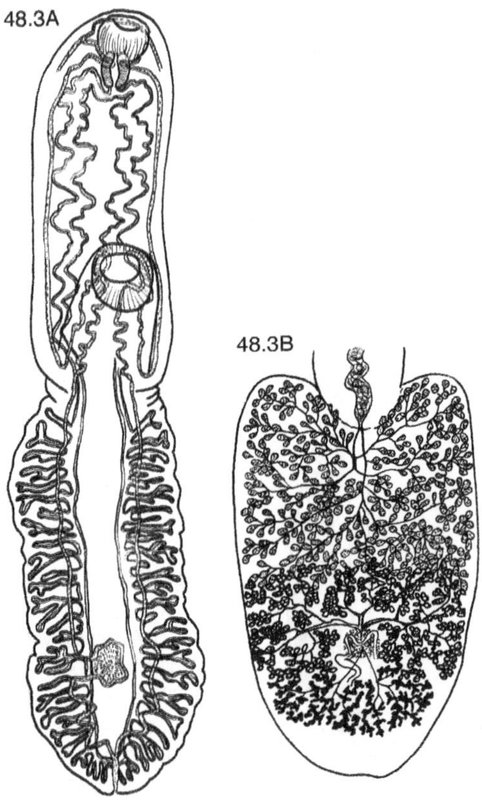

Fig. 48.3. *Otiotrema torosum* Setti, 1897. A. Entire worm, with reproductive system not shown. B. Hindbody, showing reproductive system. (After Looss, 1899.)

2 Proposed, but not named, by Dollfus (1950).

pedunculate. Gut-caeca sinuous in forebody, with numerous diverticulate outgrowths in hindbody, forming cyclocoel. Sinus-sac large, reaching close to dorsal surface. Testes follicular. Ovary and vitellarium consisting of numerous acinous bunches of follicles. Uterus in transverse coils, present in lateral fields of hindbody. Parasitic in branchial cavity (? or intestine) of sharks. Type-species *O. torosum* Setti, 1897.

1b. Small accessory suckers present on oral sucker and/or ventral sucker .... *Paronatrema* Dollfus, 1937. (Fig. 48.4)

Diagnosis: [This genus is poorly known.] Forebody subcylindrical; hindbody flattened, oval. Small accessory suckers present within oral and/or ventral suckers; ventral sucker large, but apparently not pedunculate. Gut-caeca sinuous, apparently end blindly (? or form uroproct or cyclocoel: interpretations uncertain). Testes consist of rows of follicles or segmented tubules. Ovary composed of irregular follicles (interpretations conflict[3]). Vitellarium consists of rows of follicles or segmented tubules. Uterus numerous transverse coils in hindbody. Parasitic on skin or (?) in oviduct or stomach of sharks and rays. Type-species *P. vaginicola* Dollfus, 1937.

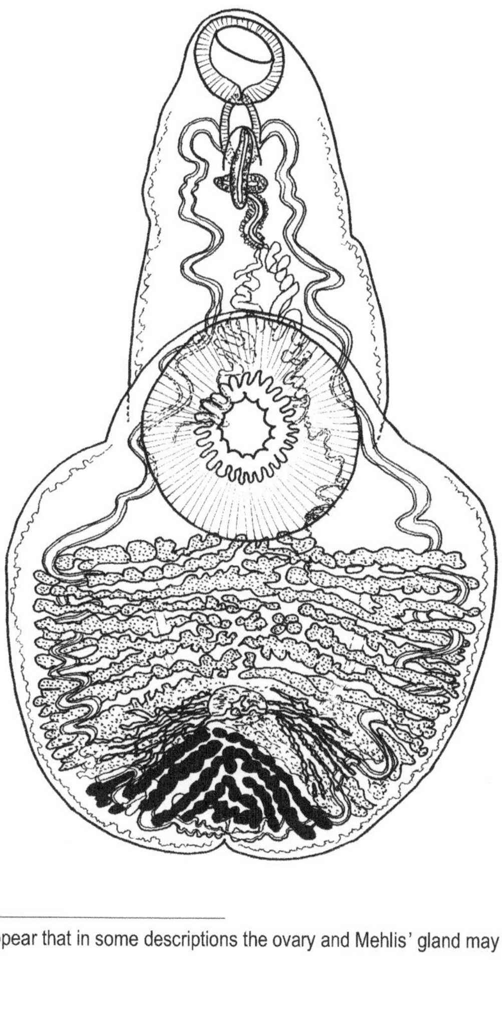

Fig. 48.4. *Paronatrema vaginicola* Dollfus, 1937. (After Dollfus, 1937b.)

[3] It would appear that in some descriptions the ovary and Mehlis' gland may have been confused.

49 Superfamily Schistosomatoidea Stiles & Hassall, 1898

J.W. Smith

29 Donmouth Road, Bridge of Don, Aberdeen, UK

Introduction

The superfamily Schistosomatoidea Stiles & Hassall, 1898 comprises three families: Sanguinicolidae von Graff, 1907 - parasitic in fish; Spirorchiidae Stunkard, 1921 - parasitic in reptiles (turtles); and Schistosomatidae Stiles & Hassall, 1898 - parasitic in reptiles (crocodilians), birds and mammals. Stunkard (1922) wrote that 'probably no other trematode group presents such unusual variety and diversity'.

While the common name 'blood flukes' for these trematodes indicates otherwise, sanguinicolids and spirorchiids are by no means restricted to the circulatory system of their hosts (for further information, see Smith, 1972, 1997a, b). In contrast, the schistosomatids appear to be restricted to the venous system of their hosts, with the exception of *Dendritobilharzia pulverulenta* (Braun, 1901) Skrjabin, 1924, which occurs in the arterial system of anseriform birds (Ulmer & Vande Vusse, 1970).

Life-cycle and morphological data indicate that the three blood fluke families comprise a monophyletic group (Platt & Brooks, 1997). One illustration of their similarity comes from ultrastructural studies on the tegumental outer membrane. In sanguinicolids, spirorchiids and schistosomatids this membrane consists of two lipid bilayers, while it consists of a single lipid bilayer in various other trematodes representing ten families and inhabiting the gut and associated body-cavities of their hosts (McLaren & Hockley, 1977). Further discussion of the evolutionary and phylogenetic relationships of the three schistosomatoid families does not belong here but may be found in Platt & Brooks (1997) and Smith (1972, 1997a, b).

The key to the schistosomatoid families given below would have been simpler had it not been for one spirorchiid and one schistosomatid genus in which the structure of the posterior intestinal caeca is reportedly atypical for the given family; these genera are the spirorchiid *Shobanatrema* Simha & Chattopadhyaya, 1980, from marine turtles, and the schistosomatid *Griphobilharzia* Platt, Blair, Purdie & Melville, 1991, from crocodilians. With regard to the structure of the intestinal caeca in the superfamily as a whole, some sanguinicolid genera have posterior intestinal caeca only, ending blindly and forming an inverted U-shaped intestine; other sanguinicolid genera also possess blind-ending anterior caeca, forming an H- or X-shaped intestine. Spirorchiid genera have posterior caeca only; in all but one genus these caeca end blindly to form an inverted U-shaped intestine - as in some sanguinicolids. The exceptional spirorchiid genus is *Shobanatrema*, for which Simha & Chattopadhyaya (1980) reported that the posterior caeca unite posteriorly to form a common caecum, as in all schistosomatid genera but one. The exceptional schistosomatid genus is *Griphobilharzia*, for which Platt *et al.* (1991) reported the posterior caeca to end blindly, not forming a common caecum.

Blood flukes have considerable economic, veterinary and medical significance. Smith (1972, 1997a, b) reviewed information on the pathological effects of sanguinicolids in fish and of spirorchiids in turtles. With regard to the pathology of schistosomatids in humans, the World Health Organization (WHO, 1990) indicated that schistosomiasis affects over 200 million people in tropical and subtropical regions, causing severe morbidity and mortality.

Two hosts are involved in the life-cycle of blood flukes. As stated above, the definitive (final) hosts are aquatic or terrestrial vertebrates (fish, turtles, crocodilians, birds or mammals). While the intermediate hosts are usually aquatic gastropod molluscs, there are exceptions. Thus, some marine sanguinicolids develop in lamellibranch molluscs, while yet others develop in tube-dwelling polychaete annelids - a feature apparently unique among trematodes (Smith, 1972, 1997a). Cercariae develop in sporocysts in the intermediate host, and directly penetrate the definitive host, there being no metacercarial stage.

Superfamily Schistosomatoidea

Diagnosis: Trematodes, elongate, of small to medium size. Monoecious, or dioecious with sexual dimorphism. Simple or curved tegumental spines, variously disposed, present or absent. Oral sucker, sometimes poorly developed, present or absent. Ventral sucker (acetabulum) present or absent. Prepharynx absent. Pharynx absent. Oesophagus narrow, short or long, straight or sinuous. Anterior intestinal caeca, when present, blind-ending and may be shorter than posterior caeca. Posterior caeca either end blindly or unite to form single caecum. Testis single, or two to numerous testes, varied in shape. Cirrus-sac present or absent. Cirrus present or absent. Ovary single, varied in shape. Seminal receptacle present or absent. Laurer's canal present or absent. Uterus reduced or voluminous, straight or much coiled. Eggs single or numerous, thin- or thick-shelled, operculate or non-operculate; spines or polar filaments present or absent. Genital pores separate or common, dorsal or ventral. Genital sucker present or absent. Vitellarium follicular or comprising tubular acini, variable in distribution. Excretory vesicle V- or Y-shaped. Excretory pore dorsal, ventral, terminal or subterminal. In blood system of marine or freshwater fish or reptiles (turtles and crocodilians), birds or mammals. Type-genus *Schistosoma* Weinland, 1858.

Key to families

1a. Flukes monoecious; anterior intestinal caeca present or absent; posterior caeca blind or reuniting; in marine or freshwater fish or turtles .... 2.

1b. Flukes dioecious, usually sexually dimorphic, males mostly with gynaecophoric canal; anterior intestinal caeca absent; posterior caeca mostly reuniting to form common caecum; in crocodilians, birds or mammals .... Schistosomatidae Stiles & Hassall, 1898.

2a. In marine or freshwater fish; tegumental spines, simple or curved, present; oral sucker absent or poorly developed; ventral sucker absent; anterior intestinal caeca present or absent; posterior caeca blind, terminating various distances from posterior end, not reuniting; eggs thin-shelled; genital sucker absent .... Sanguinicolidae von Graff, 1907.

2b. In marine or freshwater turtles; tegumental spines present or absent; oral and ventral suckers present or absent; anterior intestinal caeca absent; posterior caeca mostly terminating various distances from posterior end, typically not reuniting; eggs thick-shelled; genital sucker present or absent Spirorchiidae Stunkard, 1921.

50 Family Schistosomatidae Stiles & Hassall, 1898

L.F. KHALIL

16 Marshalswick Lane, St Albans, Hertfordshire, UK

Introduction

Flukes of the family Schistosomatidae are peculiar in that they mature in the blood vascular system of their definitive hosts and in that they are generally dioecious. Several species that cause serious diseases to humans and their domestic animals are included in this family. Bloody urine (haematuria) is a disease known since ancient times, but the causative organism was not known. A German scientist, Theodor Bilharz, working in Egypt, discovered the worm that caused the disease after the post-mortem of patients there. He wrote two letters, in May and August, 1851, to his former teacher, von Siebold, telling him of his findings. Shortly thereafter, von Siebold summarized Bilharz's observations in a report published in 1852. In 1856, Bilharz himself published a report on his newly discovered worms, which he named *Distoma haematobium*. Meckel von Hemsbach in 1856 established the genus *Bilharzia* and Weinland in 1858 the genus *Schistosoma* for these worms, as their unusual morphology indicated that they could not be assigned to *Distoma* Retzius, 1786. Although *Bilharzia* has priority over *Schistosoma*, it was suppressed officially by the International Commission for Zoological Nomenclature in favour of the latter. Even though the generic name is now accepted as *Schistosoma*, with *S. haematobium* as its type-species, the name *Bilharzia* has been popular with some medical workers, and in some parts of the world the disease is known as 'bilharzia' rather than 'schistosomiasis'.

Stiles & Hassall (1898) placed all the known schistosomes in a separate subfamily, which they named the Schistosominae, and attached their new subfamily to the family Fasciolidae. Looss (1899) elevated this subfamily to full family status, as the Schistosomidae. Poche (1907) correctly pointed out that Looss's family name is grammatically incorrect and emended the name to Schistosomatidae. With the increased discovery of species and genera of schistosomes, the family, its genera and species received a number of critical studies and revisions; notable among these are the studies of Price (1929), Mehra (1940), Skrjabin (1951), Yamaguti (1958a, 1971), Azimov (1970, 1975) and Farley (1971).

For a long time, the family was divided into four subfamilies, namely, the Schistosomatinae Stiles & Hassall, 1898, Bilharziellinae Price, 1929, Dendritobilharziinae Mehra, 1940 and Gigantobilharziinae Mehra, 1940. Azimov (1970) separated the schistosomes of mammals and birds and placed them in two separate families, the Schistosomatidae, with two subfamilies from mammals, and a new family, the Ornithobilharziidae, with four subfamilies from birds. His monograph (Azimov, 1975) followed his earlier classification. However, his system for separating the two families has been ignored by most workers on the group. Farley (1971) accepted only one family, the Schistosomatidae, and proposed the synonymy of the Dendritobilharziinae with the Gigantobilharziinae. Until recently, species assigned to the Schistosomatidae were reported only from birds and mammals. However, Platt *et al.* (1991) described a new schistosome from crocodiles in Australia, proposing a new genus, *Griphobilharzia*, and assigning it to a new subfamily, the Griphobilharziinae.

Schistosomes of medical and veterinary importance have received a number of studies and revisions. Le Roux (1958), in a very short communication, introduced a new classification for the African schistosomes and proposed dividing the large genus *Schistosoma* into three genera, *Schistosoma* and two new genera, *Afrobilharzia* and *Rhodobilharzia*. He also placed the species *Schistosoma japonicum* Katsurada, 1904 into a new genus, *Sinobilharzia*, and the species *Ornithobilharzia bomfordi* Montgomery, 1906 into another new genus, *Eurobilharzia*. He based his new classification mainly on geographical and host distribution. As he did not attempt to classify the remaining schistosomes and did not provide differential diagnoses, his system of classification has been ignored by other workers. Yamaguti's classification (1971) accepted the division of the family into four subfamilies and recognized 17 genera as valid. Farley (1971) reviewed species and genera of the family, except for *Schistosoma*, recognizing three subfamilies only. He commented on previous synonymies of the various genera and proposed new ones, accepting only 12 genera as valid. Since his review, five new genera (see below) have been assigned to the family, four of which were based on species previously assigned to existing genera.

Chauhan *et al.* (1974) proposed *Baughbilharzia* for *Ornithobilharzia phalacrocoraxi* Baugh, 1963, stating that the main difference between it and *Ornithobilharzia* is that the ventral sucker is pedunculate and areolated in the new genus. This is a minor difference, which can be used for species differentiation but not as a generic criterion. It is proposed here to suppress *Baughbilharzia* and regard it as a synonym of *Ornithobilharzia*.

Khalifa (1974) proposed *Gigantobilharziella* for *Gigantobilharzia gyrauli* Brackett, 1942 and transferred two other species to it. The main difference in the adult species of his new genus is the presence in both sexes of a small oral sucker, which is lacking in some species of *Gigantobilharzia*. Khalifa did not examine the type-species and did not provide any drawings of his new genus, but supported his new proposal by some small differences in the cercaria of this species. On examination of Brackett's (1942) original description and drawings, it was noticed that this oral sucker is very small, measuring 0.03 mm. However, both Yamaguti (1971) and Farley (1971) mentioned the presence or absence of an oral sucker as part of their diagnosis of *Gigantobilharzia*. It is doubtful that the small differences in the cercaria of this species are enough to warrant the creation of a new genus, especially as the adults cannot be differentiated morphologically. Consequently, *Gigantobilharziella* is suppressed and regarded as a synonym of *Gigantobilharzia*.

In his review of the family, Farley (1971) pointed out that, if the description of *Bilharziella yokogawai* Osio, 1927 is correct, a new genus would be required to house it. This species had been previously transferred by McLeod (1937) to *Pseudobilharziella* Ejsmont, 1929, which was later made a synonym of *Trichobilharzia* Skrjabin & Zakharov, 1920. The presence of a gynaecophoric canal starting immediately posterior to the ventral sucker, enclosing the seminal vesicle within its boundary, removed this species from both *Bilharziella* Looss, 1899 and *Trichobilharzia*. When Liu & Bai (1976) found a new species with characters very near to those of *B. yokogawai* in the portal veins of *Anas crecca crecca* in China, they erected a new genus, *Jilinobilharzia*, for their species, naming it *J. crecci* and transferring the species *B. yokogawai* to their new genus as the type-species. This genus is regarded here as valid.

Baugh (1977) proposed *Thailandobilharzia* for *Orientobilharzia harinasutai* (Komiya, 1963) reported from buffaloes in Thailand. He argued that in this species the caecal union is closely posterior to the ventral sucker and the gonads in both sexes are extracaecal. He also pointed out that the cercaria of this species has three penetration glands, while some species of *Orientobilharzia* Dutt & Srivastava, 1955 have five glands. The cercaria is not known for all species of the genus, and it is not possible to base the identification of adult species on the morphology of the larval forms. The extracaecal position of the gonads is not a good enough criterion on its own to distinguish this genus. In view of this, *Thailandobilharzia* is listed here as a synonym of *Orientobilharzia*.

Platt *et al.* (1991) found an unusual schistosome in crocodiles from Australia. They erected *Griphobilharzia* for this species and placed it in a separate subfamily. Although not all the characters of the species were observed because of its small size, it undoubtedly belongs to this family and has characters of taxonomic significance to justify the recognition of both the new genus and the subfamily.

A recent molecular study, by Snyder & Loker (2000), has indicated the presence of 'mammalian' and 'predominantly avian' clades within the family, and that the genus *Schistosoma* arose in Asia and not in Africa.

In this work, four subfamilies are accepted in the family and 14 genera are accepted as valid. In some cases, the diagnoses presented below are incomplete in some details, as relevant data are not available in the literature and museum specimens are unclear (structures are often obscured by the gut contents).

Family Schistosomatidae Stiles & Hassall, 1898

(Syn. Ornithobilharziidae Azimov, 1970)

Diagnosis: Body elongate, dioecious, sometimes with sexual dimorphism; female generally longer and more slender than male. Body of male may be flat with sides rolled ventrally to form gynaecophoric canal or chamber. Suckers well or poorly developed, sometimes absent in one or both sexes; oral sucker variable in development; ventral sucker, when present, postbifurcal in position, generally anterior to genital pore. Mouth leads into short oesophagus; prepharynx and pharynx absent; paired caeca short or long, may reunite at variable distances from bifurcation to form single caecum that extends to near posterior end of body, sometimes with lateral branches. Male with one to numerous testes, anterior or posterior to caecal union or near posterior end of body. Cirrus-sac present or absent. Genital pore posterior to ventral sucker. Female with single compact or elongate ovary, mostly anterior to caecal union. Seminal receptacle usually present. Laurer's canal present or absent. Vitellarium well developed, follicular, posterior to ovary. Uterus short or long, in intercaecal field, generally opens immediately posterior to ventral sucker or at anterior end of body. Eggs non-operculate, with or without lateral or terminal spine. Excretory vesicle tubular. In blood vascular system of birds and mammals, occasionally reptiles; cosmopolitan. Type-genus *Schistosoma* Weinland, 1858.

Key to subfamilies

1a. Male with well-developed gynaecophoric canal or chamber .... 2.
1b. Male with reduced, poorly developed or even without gynaecophoric canal .... 3.

2a. Caeca of both sexes bifurcate, to reunite at various distances posteriorly to form united caecum that terminates near posterior end of body .... Schistosomatinae Stiles & Hassall, 1898.
2b. Caeca do not unite, terminate in anterior quarter of male body, near posterior end of female body .... Griphobilharziinae Platt, Blair, Purdie & Melville, 1991.

3a. Female genital pore immediately posterior to ventral sucker; both suckers present in both sexes .... Bilharziellinae Price, 1929.
3b. Female genital pore at anterior end of body; ventral sucker absent in both sexes .... Gigantobilharziinae Mehra, 1940.

Subfamily Schistosomatinae Stiles & Hassall, 1898

Diagnosis: Both suckers present. Gynaecophoric canal well developed, extending to near posterior end of body. Common caecum usually without lateral branches. Testes usually anterior to caecal union. Genital pores posterior to ventral sucker. Female body usually cylindrical, may be attenuated anteriorly or flattened and slender. Ovary oval or spiral. Small seminal receptacle may be present posterior to ovary. In birds and mammals. Type-genus *Schistosoma* Weinland, 1858.

Key to genera

1a. Gynaecophoric canal commencing some distance posterior to ventral sucker .... *Schistosomatium* Tanabe, 1923. (Fig. 50.1)

Diagnosis: Suckers well developed; ventral sucker smaller than oral sucker. Anterior part of caeca simple; posterior part with lateral diverticula, uniting near posterior end of body. Male: anterior two-fifths of body flattened; remainder forms gynaecophoric canal commencing some distance posterior to ventral sucker. Testes 14–36, irregularly arranged, generally in two rows at anterior end of gynaecophoric canal. Seminal vesicle

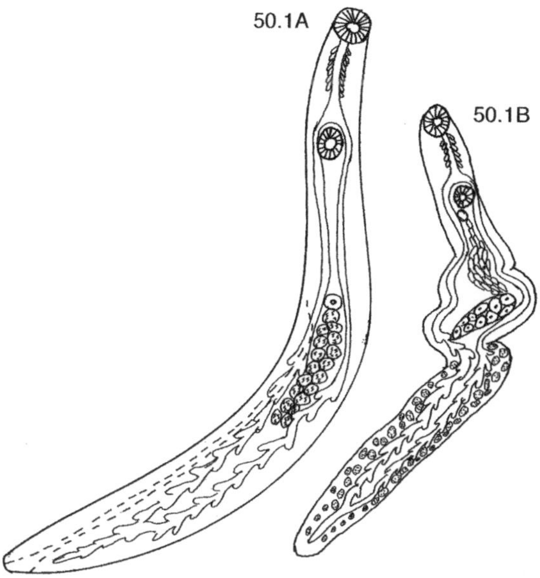

Fig. 50.1. *Schistosomatium pathlocopticum* Tanabe, 1923. A. Male. B. Female. (Modifed after Tanabe, 1923.)

and genital pore immediately pretesticular. Female: body flattened. Ovary oval in anterior half of body. Vitellarium consists of densely packed small follicles, extending from posterior margin of ovary to posterior end of body. Uterus with numerous eggs, opening immediately posterior to ventral sucker. Eggs oval, without spines. In rodents; North America. Type-species *S. pathlocopticum* Tanabe, 1923.

1b. Gynaecophoric canal commencing either anterior or immediately posterior to ventral sucker .... 2.

2a. Testes in posterior third of body .... *Heterobilharzia* Price, 1929. (Fig. 50.2)

Diagnosis: Female longer than male. Both suckers present and well developed. Caeca unite near posterior end of body. Male: forebody short, subcylindrical; hindbody leaflike with lateral sides in-rolled ventrally to form gynaecophoric canal. Tegument covered with small tubercles. Testes numerous, up to 83, in posterior third of body anterior to caecal union. Cirrus-sac present, containing seminal vesicle. Genital pore immediately posterior to ventral sucker. Female: body flattened, slender. Caeca simple anteriorly, with lateral diverticula posteriorly. Ovary tubular, in anterior half of body. Vitellarium occupying all postovarian space. Uterus long, with many eggs. Genital pore immediately posterior to ventral sucker. Eggs oval, without spine. In mammals; North America. Type-species *H. americana* Price, 1929.

2b. Testes in anterior half of body .... 3.

3a. Testes few (< 11) in number, immediately posterior to ventral sucker .... *Schistosoma* Weinland, 1858. (Fig. 50.3)

(Syns *Bilharzia* Meckel von Hemsbach, 1856; *Afrobilharzia* Le Roux, 1958; *Rhodobilharzia* Le Roux, 1958; *Sinobilharzia* Le Roux, 1958 *nec* Dutt & Srivastava, 1955; *Proschistosoma* Gretillat, 1962)

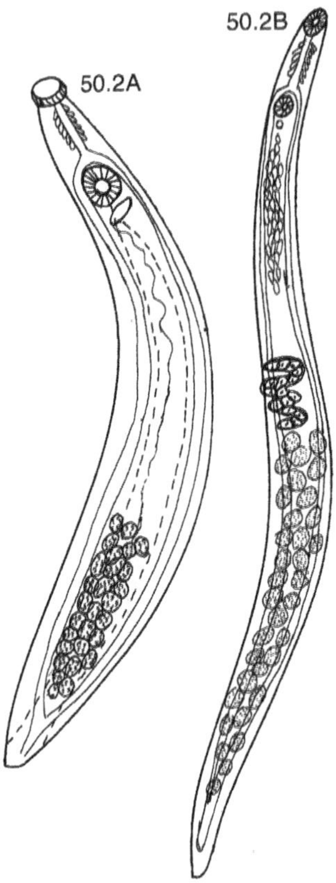

Fig. 50.2. *Heterobilharzia americana* Price, 1929. A. Male. B. Female. (Modified after Price, 1943.)

Diagnosis: Female longer than male. Both suckers well developed. Caeca unite at different positions in different species. Male: forebody short, cylindrical; hindbody long, leaf-like, with lateral sides in-rolled ventrally to form gynaecophoric canal. Tegument with spines and tubercles. Testes not more than ten, in one or two rows posterior to ventral sucker at beginning of gynaecophoric canal. Cirrus-sac absent. Genital pore immediately posterior to ventral sucker. Female: body elongate, filiform, with tegumental spines and tubercles very reduced and confined to ends of body. Ovary elongate, generally in middle third of body. Vitellarium postovarian, extending along common caecum. Uterus short or long. Eggs large, oval or fusiform, with rudimentary tubercle or terminal or lateral spine. In mammals; Africa, Asia and South America. Type-species *S. haematobium* (Bilharz, 1852) Weinland, 1858.

3b. Testes > 10 in number .... **4.**

4a. Vitellarium consists of two complete sets of paired follicles lying on either side of paired caeca ...... ***Bivitellobilharzia*** **Vogel & Minning, 1940. (Fig. 50.4)**

Diagnosis: Male and female of nearly equal length. Suckers present. Caeca unite near posterior end of body. Male: gynaecophoric canal well developed, commencing immediately posterior to ventral sucker, extending to posterior end of body. Tegument with tubercles on dorsal surface. Testes numerous, up to 85, in anterior third of body. Cirrus-sac absent. Genital pore at anterior end of gynaecophoric canal. Female: body

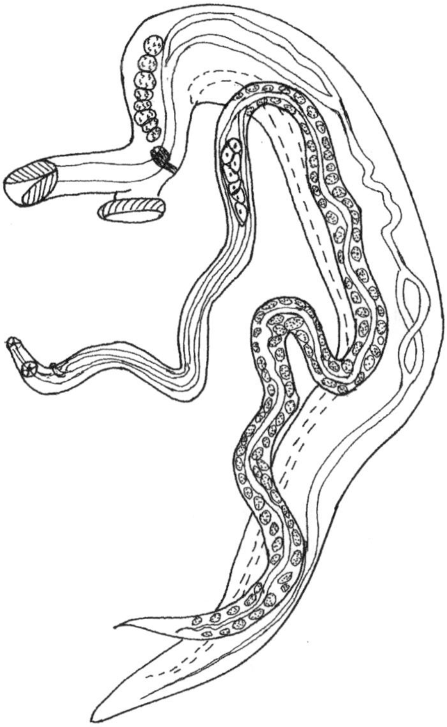

Fig. 50.3. *Schistosoma mansoni* Sambon, 1907. Male and female. (Modified after various sources.)

flattened. Ovary spiral in anterior third of body. Vitellarium consists of two complete sets of paired follicles lying on either side of paired caeca. Uterus usually contains one egg, opens immediately posterior to ventral sucker. Eggs large with polar processes. In elephants; Africa and India. Type-species *B. loxodontae* Vogel & Minning, 1940.

4b. Vitellarium consists of one set of follicles .... 5.

5a. Uterus very short; testes < 66 in number, usually intercaecal .... *Austrobilharzia* Johnston, 1917. (Fig. 50.5)

(Syns *Microbilharzia* Price, 1929; *Sinobilharzia* Dutt & Srivastava, 1961 *nec* LeRoux, 1958)

Diagnosis: Male generally shorter than female. Suckers present. Gynaecophoric canal extends from ventral sucker to posterior end of body. Common caecum generally short. Male: forebody short. Caeca may or may not anastomose before final union. Testes 11–65, commencing either immediately or some distance posterior to ventral sucker. Cirrus-sac present or absent. Female: filiform, slender; posterior end may be flattened. Oral sucker may not be well developed. Ovary spiral in anterior half of body, rarely postequatorial. Vitellarium in postovarian region. Uterus short. Genital pores immediately posterior to ventral sucker. Eggs oval, with or without spines. In birds; Australasia, Asia and North America. Type-species *A. terrigalensis* Johnston, 1917.

5b. Uterus not short; testes usually > 50 in number .... 6.

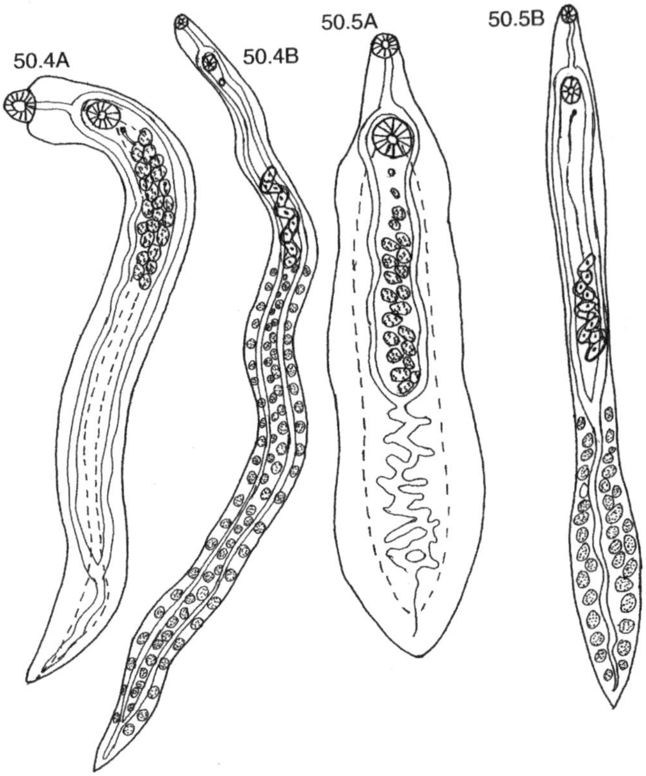

Figs 50.4–50.5.
50.4. *Bivitellobilharzia loxodontae* Vogel & Minning, 1940. A. Male. B. Female. (Modified after Vogel & Minning, 1940.)
50.5. *Austrobilharzia variglandis* (Miller & Northup, 1926). A. Male. B. Female. (Modified after Miller & Northup, 1926.)

6a. Testes up to 80 in number, beginning short distance from ventral sucker; parasites of mammals .... *Orientobilharzia* Dutt & Srivastava, 1955. (Fig. 50.6)
(Syns *Eurobilharzia* Le Roux, 1958; *Thailandobilharzia* Baugh, 1977)

Diagnosis: Male and female of almost equal length. Suckers present. Common caecum may be shorter in male than female. Male: forebody subcylindrical; hindbody flattened, with gynaecophoric canal extending from ventral sucker to end of body. Testes 37–80, situated short distance posterior to ventral sucker. Cirrus-sac absent. Female: body filiform. Ovary oval and spirally twisted, in anterior or middle third of body. Uterus long. Eggs with lateral or terminal spine at one end and sometimes leaf-like appendage at other end. In mammals; Asia. Type-species *O. dattai* (Dutt & Srivastava, 1952).

6b. Testes number > 80; parasites of birds .... **7.**

7a. Testes number > 100; ovary in anterior half of body .... *Ornithobilharzia* Odhner, 1912. (Fig. 50.7)
(Syn. *Baughbilharzia* Chauhan, Srivastava & Chauhan, 1973)

Diagnosis: Males and females of nearly equal length. Suckers present. Common caecum short. Male: gynaecophoric canal extends from ventral sucker to posterior end of body. Caeca tend to anastomose before uniting. Testes numerous, up to 125 in number, usually intercaecal in anterior half of body. Cirrus-sac small or absent.

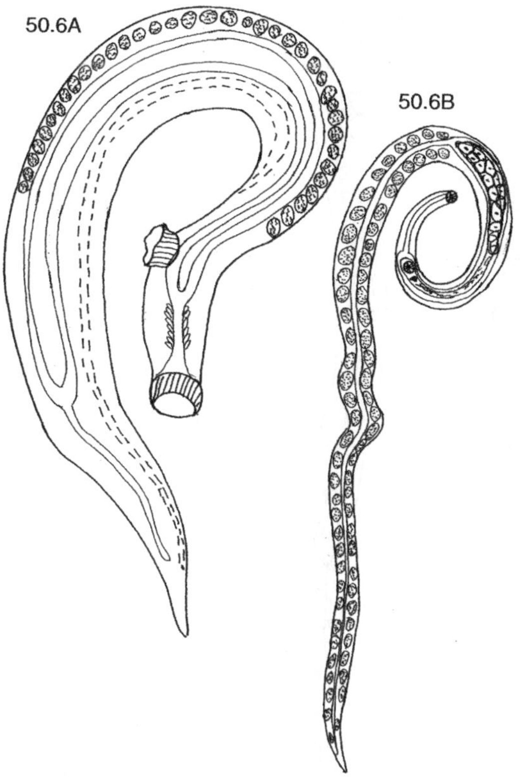

Fig. 50.6. *Orientobilharzia bomfordi* (Montgomery, 1906). A. Male. B. Female. (Modified after Montgomery, 1906.)

Genital pore opens immediately or slightly posterior to ventral sucker. Female: filiform, subcylindrical. Ovary elongate, loosely coiled, in anterior half of body. Vitellarium extensive, occupying nearly two-thirds of body length. Uterus long or short. Genital pore immediately posterior to ventral sucker. Eggs oval, with or without rudimentary spine. In birds; cosmopolitan. Type-species *O. canaliculata* (Rudolphi, 1819) Odhner, 1912.

7b. Testes number > 200; ovary in posterior half of body ....
.... ***Macrobilharzia*** **Travassos, 1922. (Fig. 50.8)**
(Syn. *Paraschistosomatium* Price, 1929)

Diagnosis: Male longer than female. Suckers present. Common caecum short. Male: with well-developed gynaecophoric canal extending from ventral sucker to posterior end of body. Caeca zigzag; common caecum very short. Testes number more than 200, commence short distance posterior to ventral sucker. Female: body flattened, slender, tapering towards extremities. Ovary spiral, intercaecal in posterior half of body. Vitellarium occupying postovarian intercaecal field. Uterus long with numerous eggs. Eggs oval. In birds (*Anhinga*); North and South America. Type-species *M. macrobilharzia* Travassos, 1922.

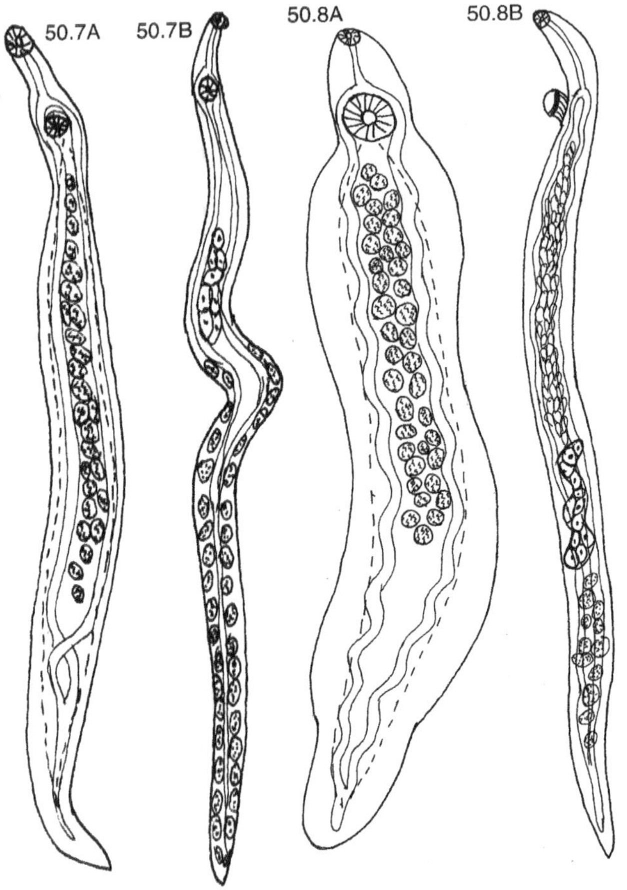

Figs 50.7–50.8.
50.7. *Ornithobilharzia intermedia* Odhner, 1912. A. Male. B. Female. (Modified after Odhner, 1912.)
50.8. *Macrobilharzia macrobilharzia* Travassos, 1922. A. Male. B. Female. (Modified after Kohn, 1964.)

Subfamily Griphobilharziinae Platt, Blair, Purdie & Melville, 1991

Diagnosis: Male much larger than female. Gynaecophoric canal forms chamber where female lies completely enclosed and reversed in relation to male. Both suckers present in male; ventral sucker absent in female. Oesophagus short; caeca do not unite, short in male, long in female. Male: gynaecophoric chamber extends two-thirds of body length from ventral sucker to near posterior end; caeca bifurcate anteriorly to ventral sucker, terminating in anterior quarter of body just posteriorly to beginning of gynaecophoric chamber. Testis single, near posterior end. Terminal genitalia and genital pore not observed. Female: oral sucker terminal; caeca bifurcate near anterior end of body, terminating near posterior end. Vitellarium follicular, dorsal to caeca. Uterus voluminous, extending anteriorly to caecal bifurcation, with numerous eggs. Genital pore not observed. Eggs large, thin-shelled. In crocodiles; Australasia. Type-genus *Griphobilharzia* Platt, Blair, Purdie & Melville, 1991.

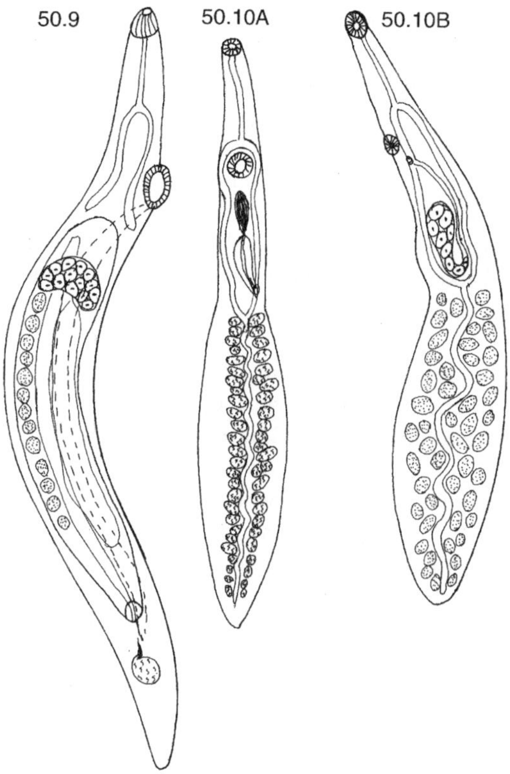

Figs 50.9–50.10.
50.9. *Griphobilharzia amoena* Platt, Blair, Purdie & Melville, 1991. Male and female. (Modified after Platt *et al.*, 1991.)
50.10. *Bilharziella polonica* (Kowalewski, 1895). A. Male. B. Female. (Modified after Kowalewski, 1895.)

Genus *Griphobilharzia* Platt, Blair, Purdie & Melville, 1991 (Fig. 50.9)

Diagnosis: With characters of subfamily. Type-species *G. amoena* Platt, Blair, Purdie & Melville, 1991.

Subfamily Bilharziellinae Price, 1929

Diagnosis: Male and female similar, thread-like or flattened. Gynaecophoric canal absent, poorly developed or reduced, not extending to posterior end. Suckers present. Caeca unite or form common caecum without lateral branches. Male: testes numerous. Cirrus-sac present or absent. Genital pore some distance posterior to ventral sucker. Female: ovary elongate or spherical. Uterus short. Genital pore immediately posterior to ventral sucker. Eggs elongate, with spine. In birds. Type-genus *Bilharziella* Looss, 1899.

Key to genera

1a. **Body of male and female flattened; gynaecophoric canal absent or poorly developed .... *Bilharziella* Looss, 1899. (Fig. 50.10)**
(Syn. *Chinhuta* Lal, 1937)

Diagnosis: Male and female with posterior end of body flattened; male longer than female. Suckers well developed. Common caecum long. Male: gynaecophoric canal absent or poorly developed. Testes up to 110, in posterior region of body, extending along either side of common caecum. Seminal vesicle elongate. Cirrus-sac present. Genital pore opens some distance posterior to ventral sucker. Female: ovary in anterior half of body. Vitelline follicles on either side of common caecum. Uterus short. Genital pore immediately posterior to ventral sucker. Eggs elongate, with small spine. In birds; cosmopolitan. Type-species *B. polonica* (Kowalewski, 1895) Looss, 1899.

1b. Body of male and female thread-like; gynaecophoric canal present, poorly developed or reduced .... 2.

2a. Gynaecophoric canal very small, commences some distance posterior to ventral sucker; male genital pore opens anterior to gynaecophoric canal .... *Trichobilharzia* Skrjabin & Zakharow, 1920. (Fig. 50.11)

(Syn. *Pseudobilharziella* Ejsmont, 1929)

Diagnosis: Posterior end of body of both sexes spatulate. Suckers present. Common caecum long. Male: gynaecophoric canal short. Testes numerous on either side of common caecum. Cirrus-sac small with thin walls; seminal vesicle elongate. Genital pore at anterior end of gynaecophoric canal. Female: ovary elongate, in anterior part of body; seminal receptacle posterior to ovary. Vitellarium extending from seminal receptacle to posterior end. Uterus usually containing single egg. Genital pore immediately posterior to ventral sucker. Eggs elongate, with spine. In anseriform birds; cosmopolitan. Type-species *T. ocellata* (La Valette, 1855) Brumpt, 1931 (syn. *T. kossarewi* Skrjabin & Zakharow, 1920).

2b. Gynaecophoric canal commences immediately posterior to ventral sucker, not extending more than half of body length; male genital pore opens some distance posterior to beginning of gynaecophoric canal .... *Jilinobilharzia* Liu & Bai, 1976. (Fig. 50.12)

Diagnosis: Suckers present. Common caecum long, generally zigzag. Male: testes numerous on either side of common caecum. Seminal vesicle large, within region of gynaecophoric canal, opening some distance posterior to ventral sucker. Female: ovary spiral in anterior part of body. Vitellarium extends from point of caecal union to posterior end of body. Uterus short. Genital pore immediately posterior to ventral sucker. Eggs elongate, with spines. In anseriform birds; Asia. Type-species *J. yokogawai* (Osio,1927) Liu & Bai, 1976.

Subfamily Gigantobilharziinae Mehra, 1940

(Syn. Dendritobilharziinae Mehra, 1940)

Diagnosis: Male and female similar, thread-like or flattened. Gynaecophoric canal absent or poorly developed, not extending to posterior end. Oral sucker present or absent; ventral sucker absent. Common caecum may have lateral diverticula. Male: testes numerous. Cirrus-sac present. Genital pore in anterior part of body. Female: ovary and genital pore in anterior part of body. In birds. Type-genus *Gigantobilharzia* Odhner, 1910.

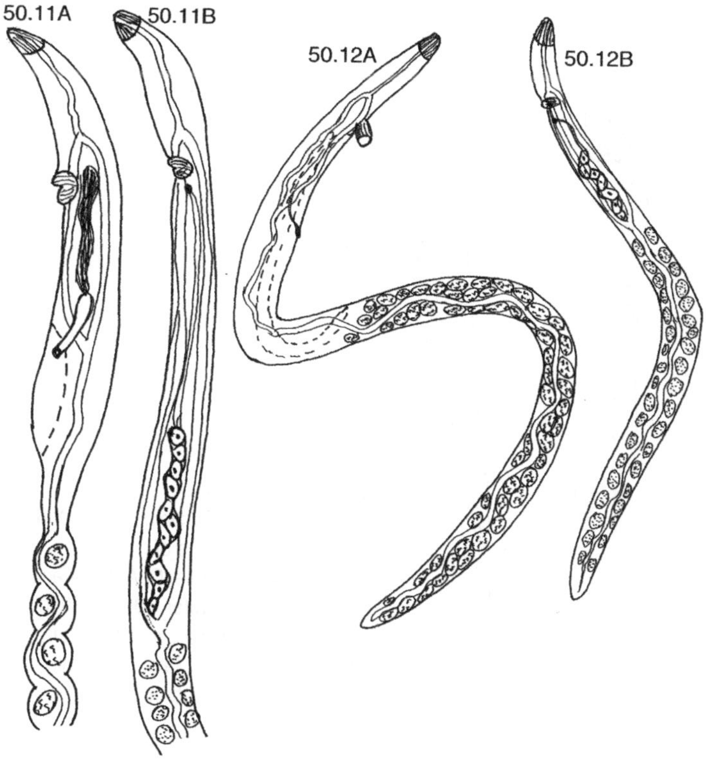

Figs 50.11–50.12.
50.11. *Trichobilharzia waubesensis* (Brackett, 1942). A. Anterior end of male. B. Anterior end of female. (Modified after Brackett, 1942.)
50.12. *Jilinobilharzia crecci* Liu & Bai, 1976. A. Male. B. Female. (Modified after Liu & Bai, 1976.)

Key to genera

1a. Gynaecophoric canal present; body thread-like; oral sucker present or absent; common caeca without lateral diverticula .... *Gigantobilharzia* Odhner, 1910. (Fig. 50.13)
(Syn. *Gigantobilharziella* Khalifa, 1974)

Diagnosis: Body thread-like. Oral sucker absent or weakly developed. Gynaecophoric canal short. Caeca bifurcate near anterior end; common caecum long, without lateral diverticula. Male: testes on either side of common caecum. Cirrus-sac contains part of seminal vesicle. Genital pore at anterior end of gynaecophoric canal. Female: ovary elongate, coiled, anterior to caecal union. Seminal receptacle long, posterior to ovary. Vitellarium extends from caecal union to posterior extremity. Uterus short with few eggs. Eggs oval, with or without terminal spine. In birds; cosmopolitan. Type-species *G. acotylea* Odhner, 1910.

1b. Gynaecophoric canal poorly developed or absent; body flattened; oral sucker absent; common caecum with short lateral diverticula .... *Dendritobilharzia* Skrjabin & Zakharow, 1920. (Fig. 50.14)

Diagnosis: Body of both sexes elongate, flattened and similar. Both suckers absent. Caeca unite in anterior third of body; common caecum long, zigzag, with short lateral diverticula. Male: testes posterior to caecal union, on either side of common caecum. Relationship between seminal vesicle and cirrus-sac not known. Genital pore anterior to caecal union. Female: ovary spiral, intercaecal. Seminal receptacle present.

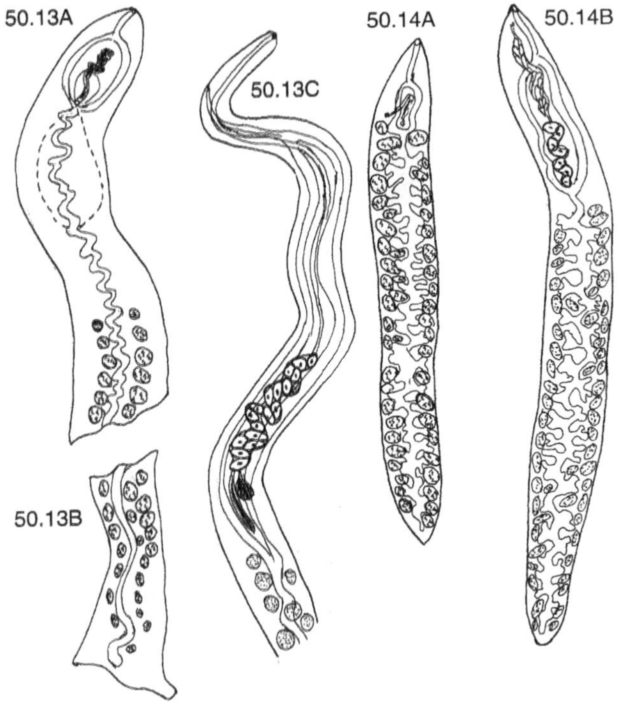

Figs 50.13–50.14.
50.13. *Gigantobilharzia acotylea* Odhner, 1910. A. Anterior end of male. B. Posterior end of male. C. Anterior end of female. (Modified after Odhner, 1910b.)
50.14. *Dendritobilharzia anatinarum* Cheatum, 1941. A. Male. B. Female. (Modified after Cheatum, 1941.)

Vitellarium with numerous follicles along common caecum. Uterus long, usually with numerous eggs. Genital pore at anterior end of body. Eggs spherical with thin shells. In birds, especially Anseriformes; cosmopolitan. Type-species *D. pulverulenta* (Braun, 1901) Skrjabin, 1924 (syn. *D. odhneri* Skrjabin & Zakharow, 1920).

51 Family Sanguinicolidae von Graff, 1907

J.W. Smith

29 Donmouth Road, Bridge of Don, Aberdeen, UK

Introduction

The family Sanguinicolidae was erected by von Graff (1907) for the blood flukes of fish. Twenty genera, 13 of them monospecific, are recognized within the family at the time of writing (June 2000). These genera are characterized by the possession of variously shaped and disposed tegumental spines and oral structures, which enable movement and attachment in a variety of host microhabitats. While the absence of suckers has been considered as diagnostic of the family, Kirk & Lewis (1993) claimed that an oral sucker may be present in a greater number of species than reported previously and may have eluded description because it is poorly developed (see also Overstreet & Køie, 1989; Pearson, 1992). While Yamaguti (1971) considered the absence of an operculum from the eggs to be diagnostic of the Sanguinicolidae, Overstreet & Køie (1989) reported operculate eggs in *Pearsonellum* Overstreet & Køie, 1989, which are 'probably not unique among sanguinicolids'. The literature on this and other aspects of the general biology of sanguinicolids - and of spirorchiid blood flukes in turtles - was reviewed by Smith (1972, 1997a, b).

The early history of taxonomic studies on sanguinicolids is chequered. Odhner (1900a) originally reported *Aporocotyle simplex* Odhner, 1900 to be of uncertain affinities and ectoparasitic on the gills of flatfish. Plehn (1905) originally reported *Sanguinicola armata* Plehn, 1905 and *S. inermis* Plehn, 1905 to be turbellarians endoparasitic in cyprinid fish, erecting for them a new family, the Rhynchostomida; later she regarded them as 'monozoic' cestodes (Plehn, 1908). A further 3 years elapsed before Odhner (1911c) correctly classified both *Aporocotyle* Odhner, 1900 and *Sanguinicola* Plehn, 1905 as trematodes.

Yamaguti (1958a, 1970, 1971) maintained the family Aporocotylidae Odhner, 1912 as distinct from the family Sanguinicolidae von Graff, 1907, and recognized six subfamilies within the latter family. Yamaguti's scheme has met with criticism from various workers. For example, Stunkard (1983) correctly emphasized that the problems encountered in the classification and taxonomy of the schistosomes meet with 'even more difficult situations in the blood flukes of fishes', the diagnostic characters of the two families (Sanguinicolidae and Aporocotylidae) being 'uncertain, indefinite and often inconsistent'. Stunkard (1983) also stated that sanguinicolid blood flukes occur in 'both marine and freshwater hosts, and there is little correlation between hosts and parasites'. More recently, Overstreet & Køie (1989) agreed with Van der Land (1967) and Smith (1972) in rejecting Yamaguti's views, and considered the family Aporocotylidae to be a junior synonym of the family Sanguinicolidae. Furthermore, Overstreet & Køie (1989) rejected Yamaguti's subfamilies, mainly because maintenance of them obscures the true relationships among sanguinicolid genera, and I agree. (It should be noted that Yamaguti (1958a, 1970, 1971) erected a total of six subfamilies within the family Sanguinicolidae, not seven subfamilies as stated in the abstract of Overstreet & Køie's (1989) paper.)

McIntosh (1934a) considered *Janickia* Rašín, 1929 to be a synonym of *Sanguinicola*.

Goto & Ozaki (1930) named *Psettarium* to replace *Plehnia* Goto & Ozaki, 1929, which was pre-occupied. *Plehniella* Szidat, 1951 was considered to be a synonym of *Sanguinicola* by Yamaguti (1958a) and Madhavi & Hanumantha Rao (1970), and is treated so here, although Lebedev & Parukhin (1972) did not agree.

Family Sanguinicolidae von Graff, 1907

(Syns Aporocotylidae Odhner, 1912; Sanguinicolinae of Yamaguti (1958a); Cardicolinae Yamaguti, 1958; Deontacylicinae Yamaguti, 1958; Psettariinae Yamaguti, 1958; Paracardicolinae Yamaguti, 1970; Chimaerohemecinae Yamaguti, 1971)

Diagnosis: Trematodes of small to medium size. Body dorsoventrally flattened, varied in shape. Simple or curved tegumental spines variously disposed. Oral sucker absent or poorly developed. Ventral sucker absent. Pharynx absent. Oesophagus narrow and relatively long, surrounded by gland-cells. Intestine X-, H- or inverted U-shaped. Anterior intestinal caeca (where present) and posterior caeca of approximately equal length, or posterior caeca longer than anterior; all caeca blind. Testis single or two, five, or > 18 testes present, varied in shape, elongate, tubular or ovoid and irregularly lobed, or with deep bilateral lobes, bilateral wings or branches. Cirrus-sac present or absent; if present, contains internal seminal vesicle, prostatic cells and ejaculatory duct. Auxiliary external seminal vesicle present or absent. Ovary single, occasionally lobed, bilobed or multilobulate, or a tubular mass. Seminal receptacle present or absent. Laurer's canal present or absent. Mehlis' gland present or absent. Uterus much coiled, or straight and reduced. Metraterm present or absent. Eggs single or numerous, thin-shelled, reportedly mostly non-operculate. Egg shape varies in different genera from spherical through ovoid to fusiform or triangular; eggs may bear spine or spines. Male and female genital pores separate or common, opening dorsally or ventrally. Vitellarium follicular or consisting of tubular acini, variable in distribution. Excretory vesicle (when reported) V- or Y-shaped, tubular or spherical. Excretory pore (when reported) terminal or subterminal. In blood system[1] of elasmobranch, holocephalan or marine or freshwater teleost fish. Type-genus *Aporocotyle* Odhner, 1900.

[1] Occasionally reported from other sites, but these may be accidental and caused by the rupture of blood-vessels.

Key to genera

1a. Anterior and posterior intestinal caeca form X- or H-shape .... 2.
1b. Posterior intestinal caeca only present, form inverted U-shape .... 12.

2a. Testis single .... 3.
2b. Two or more testes .... 14.

3a. Anterior and posterior intestinal caeca both short and approximately equal in length .... 4.
3b. Posterior intestinal caeca long; anterior caeca usually much shorter than posterior .... 6.

4a. Testis with deep bilateral lobes; male and female genital pores relatively close; parasites of freshwater fish .... *Sanguinicola* Plehn, 1905. (Fig. 51.1)
(Syns *Janickia* Rašín, 1929; *Plehniella* Szidat, 1951)
Diagnosis: Body small, flattened, lanceolate, with marginal spines, or unarmed but with tegumental sheathed, ciliated papillae and club-shaped setae. Protrusible anterior proboscis. Mouth slightly subterminal with small, muscular organ (vestigial oral sucker?). Pharynx absent. Oesophagus extends to anterior third of body. Intestinal caeca relatively short, anterior and posterior, form X- or H-shape; rarely with five short intestinal caeca. Testis single, behind caeca, median, with deep bilateral lobes. Vas deferens median, extending posteriorly to join muscular cirrus-sac. Cirrus-sac and prostatic cells surround ejaculatory duct. Cirrus papilla-like. Auxiliary external seminal vesicle absent. Male genital pore dorsomedial, opens on last quarter of body. Ovary bilobed, post-testicular. Oviduct extends posteriorly and unites with vitelline duct before entering oötype. Seminal receptacle absent. Oötype with solitary egg, thin-shelled, triangular and bearing spine. Uterus reduced, postovarian. Short, uncoiled metraterm. Female genital pore dorsomedial, opens adjacent to male genital pore. Mehlis' gland absent. Laurer's canal absent. Vitellarium follicular, present throughout most of body. Excretory vesicle V-shaped. Excretory pore terminal. In blood vascular system of cyprinids and other families of freshwater teleosts. Europe, north-east Africa, Israel, Asia and North America. Type-species *S. armata* Plehn, 1905.
4b. Testis without deep bilateral lobes; male and female genital pores common or separate; in marine fish .... 5.

5a. Common male and female genital pore at posterior end; C-shaped spines on ventrolateral body margins .... *Selachohemecus* Short, 1954. (Figs 51.2–51.4)
Diagnosis: Body small, flattened, slender, with C-shaped spines on ventrolateral body margins. Suckers and pharynx absent. Oesophagus long. Anterior and posterior intestinal caeca form X- or H-shape; anterior and posterior caeca both relatively short. Testis single, between intestine and ovary in middle third of body. Cirrus and cirrus-sac not reported (absent?). Auxiliary external seminal vesicle absent? Ovary slightly lobed, lying about one-third of body length from posterior end. Seminal receptacle absent. Uterus relatively long, entirely postovarian. Metraterm absent? Eggs thin-shelled, spherical to ovoid. Common male and female genital pore, dorsal, postovarian, near posterior end of body. Laurer's canal absent. Vitellarium follicular,

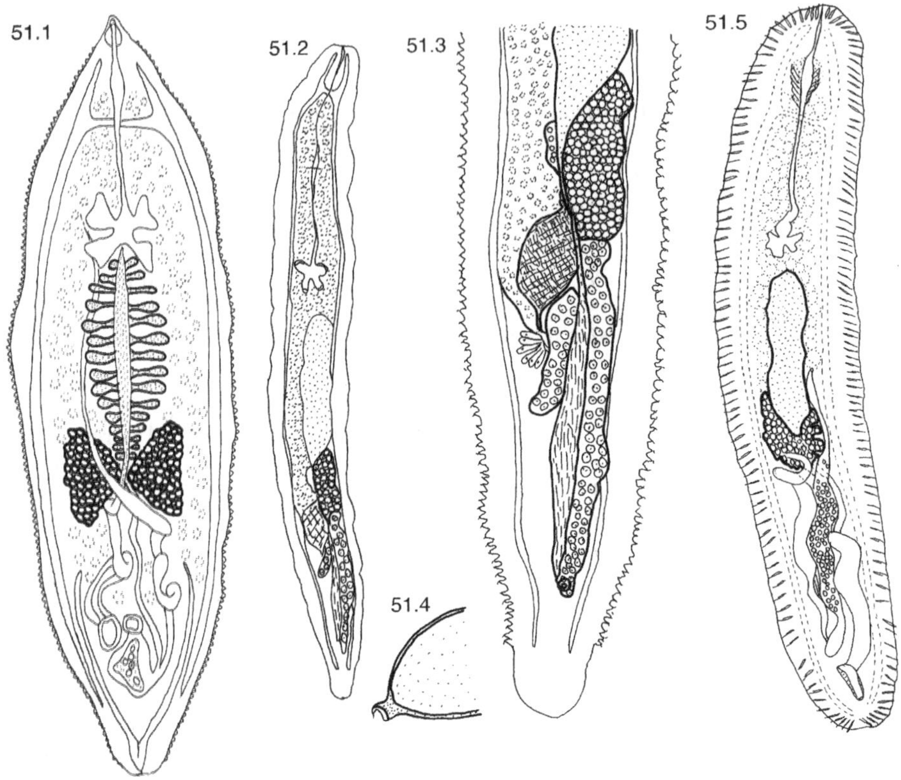

Figs 51.1–51.5.
51.1. *Sanguinicola inermis* Plehn, 1905, whole worm, dorsal view. (After Kirk & Lewis, 1993.)
51.2–51.4. *Selachohemecus olsoni* Short, 1954. 51.2. Whole worm, ventral view. 51.3. Genitalia, ventral view. 51.4. Tegumental spine. (After Short, 1954.)
51.5. *Parasanguinicola vastispina* Herbert & Shaharom, 1995, whole worm, dorsal view. (After Herbert & Shaharom, 1995.)

extensive, mostly anterior to ovary. Excretory vesicle and pore not reported. In heart of sharks (*Rhizoprionodon*). Gulf of Mexico. Type- and only species *S. olsoni* Short, 1954.

5b. Male and female genital pores separate; male pore posterior; female pore equatorial or slightly postequatorial; spines simple, on ventrolateral body margins .... *Parasanguinicola* Herbert & Shaharom, 1995. (Fig. 51.5)

Diagnosis: Body spatulate, flattened, with single row of spines ventrolaterally. Suckers and pharynx absent. Oesophagus relatively long, swollen anteriorly. Anterior and posterior intestinal caeca relatively short, irregular in shape, form X- or H-shape, or with five short caeca. Testis single, posterior to caeca, lying between intestine and ovary. Cirrus-sac present. Cirrus absent. Auxiliary external seminal vesicle absent. Male genital pore dorsal, close to posterior end of body. Ovary single, dendritic, post-testicular, lying posterior to female genital pore. Seminal receptacle absent. Oötype relatively large. Uterus occupies about one-third of body length, with descending limb and ascending limb. Metraterm absent? Eggs small, thin-shelled, triangular. Female genital pore dorsal, dextrally submedian, opens equatorially or slightly postequatorially. Laurer's canal absent. Vitellarium follicular, extending from level of nerve-ring to level of ovary. Excretory vesicle Y-shaped. Excretory pore subterminal. In blood-vessels of *Lates calcarifer*. Malaysian marine waters. Type- and only species *P. vastispina* Herbert & Shaharom, 1995.

6a. Testis between intestinal caeca, extending anteriorly and posteriorly to caeca; rosethorn-shaped spines in transverse rows ventrolaterally .... *Pearsonellum* Overstreet & Køie, 1989. (Figs 51.6, 51.7)

Diagnosis: Body spatulate, flattened, with ventrolateral transverse rows of relatively large rosethorn-shaped spines. Oral disc (oral sucker) poorly developed. Oesophagus relatively long. Anterior and posterior intestinal caeca form X- or H-shape; anterior caeca much shorter than posterior. Testis single, lobed, located mostly between caeca, extending anteriorly and posteriorly to caeca. Cirrus-sac well developed. Auxiliary external seminal vesicle present, sometimes indistinct. Ovary post-testicular, in posterior third of body. Seminal receptacle absent. Uterus extends anterior to ovary.

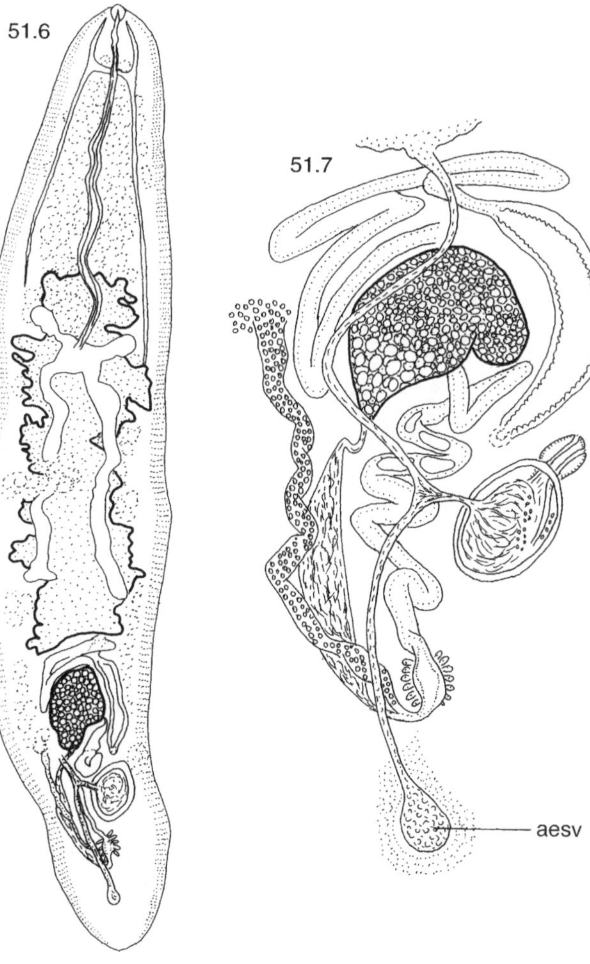

Figs 51.6–51.7. *Pearsonellum corventum* Overstreet & Køie, 1989. 51.6. Whole worm, ventral view. 51.7. Genitalia, ventral view. aesv, auxiliary external seminal vesicle. (After Overstreet & Køie, 1989.)

Metraterm conspicuously muscular. Eggs operculate, thin-shelled, ovoid. Male and female genital pores separate, dorsal, postovarian. Laurer's canal absent. Vitellarium follicular, mostly compacted in large masses, occupying most areas between lateral nerve cords to, or slightly anterior to, dorsal nerve commissure, extending posteriorly to, or occasionally past, mid-level of cirrus-sac. Excretory vesicle nearly spherical, with two distensible, anteriorly directed arms. Excretory pore subterminal. In ventricle and bulbus arteriosus of heart in *Plectropomus leopardus* (type-host) and *Epinephelus* spp. Great Barrier Reef, Australia. Type- and only species *P. corventum* Overstreet & Køie, 1989.

6b. Testis between and lateral to intestinal caeca, but not extending anterior to caeca; spines simple, not rosethorn-shaped .... 7.

7a. Testis with bilateral wings lying between, lateral and posterior to intestinal caeca, pre-uterine; uterus between testis and lobed ovary .... *Deontacylix* Linton, 1910. (Fig. 51.8)

Diagnosis: Body elongate, fusiform, entirely covered with transverse rows of small spines. Mouth subterminal. Suckers and pharynx absent. Oesophagus sinuous. Anterior and posterior intestinal caeca form X- or H-shape; anterior caeca shorter than posterior. Testis single, with bilateral wings, lying between, lateral and posterior to caeca, pre-uterine. Vas deferens leads to seminal vesicle. Cirrus-sac present or absent? Cirrus present. Auxiliary external seminal vesicle absent. Male genital pore opens dorsally, sinistrally submedian, near posterior end of body. Ovary multilobed, median

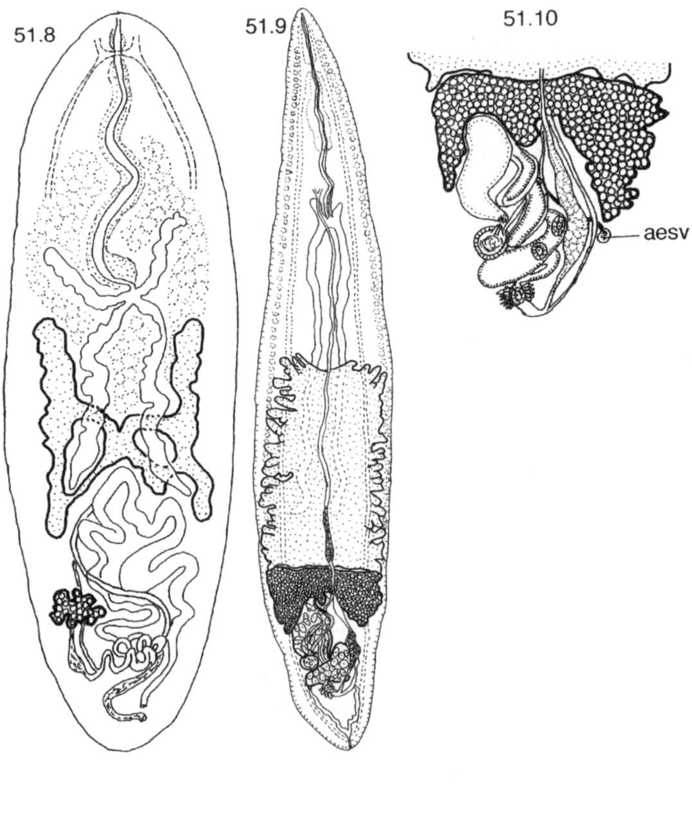

Figs 51.8–51.10.
51.8. *Deontacylix ovalis* Linton, 1910, whole worm, ventral view. (After Linton, 1910.)
51.9–51.10. *Cruoricola lates* Herbert, Shaharom-Harrison & Overstreet, 1994. 51.9. Whole worm, dorsal view. 51.10. Genitalia, dorsal view. aesv, auxiliary external seminal vesicle) (After Herbert *et al.*, 1994.)

or dextral lying near posterior end. Seminal receptacle present, curved. Mehlis' gland present? Uterus closely winding between testis and lobed ovary. Metraterm absent? Laurer's canal absent. Eggs thin-shelled, ovoid. Female genital pore opens dorsally anterior to male genital pore. Vitellarium follicular, distributed laterally to oesophagus and intestinal caeca, and in space between intestinal bifurcation and testis. Excretory vesicle and pore not reported. In coelom (?) and 'gill washing' of *Kyphosus* spp. Florida and Hawaiian marine waters. Type-species *D. ovalis* Linton, 1910.

7b. Testis without bilateral wings, lying between, lateral and posterior to intestinal caeca or between and posterior to caeca; uterus lateral to ovary and/or posterior to ovary .... 8.

8a. Testis large, of uniform width, reaching near lateral body margins; male and female genital pores close together, dorsal and postovarian ....
***Cruoricola* Herbert, Shaharom-Harrison & Overstreet, 1994. (Figs 51.9, 51.10)**

Diagnosis: Body lanceolate, flattened, with single row of submarginal spines except at anterior extremity. Suckers absent. Oesophagus long, about one-quarter of body length. Anterior and posterior intestinal caeca form X- or H-shape; anterior caeca shorter than posterior, which end anterior to ovary. Testis single, pre-ovarian, of uniform breadth, relatively extensive, reaching near lateral body margins. Cirrus-sac contains pear-shaped seminal vesicle, postovarian. Auxiliary external seminal vesicle present, sometimes indistinct. Ovary bilobed, post-testicular, in posterior quarter of body, extending almost to body margins. Seminal receptacle absent. Oötype posterior to uterus. Laurer's canal absent. Uterus postovarian. Metraterm absent? Eggs thin-shelled, ovoid. Male and female genital pores separate but close together, dorsal and post-ovarian. Vitellarium follicular, extensive and pre-ovarian. Excretory vesicle Y-shaped, with short stem and distensible anterolaterally directed arms. Excretory pore not reported. In mesenteric venules of *Lates calcarifer*. Malaysian, Thai and Australian marine waters. Type- and only species *C. lates* Herbert, Shaharom-Harrison & Overstreet, 1994.

8b. Testis not extending laterally to intestinal caeca, but between caeca or more or less posterior to caeca; male and female genital pores not close together .... 9.

9a. Testis relatively small and deeply lobed, mostly posterior to intestinal caeca; cirrus-sac large; uterus postovarian; male genital pore dorsolateral at posterior end; female pore dorsal, opposite mid-level of cirrus-sac ....
.... *Metaplehniella* Lebedev & Parukhin, 1972. (Figs 51.11–51.13)

Diagnosis: Body foliate, flattened, with marginal paired spines. Suckers absent. Mouth subterminal. Oesophagus relatively long. Anterior and posterior intestinal caeca form X- or H-shape, with rounded ends; anterior caeca shorter than posterior. Testis single, deeply lobed, relatively small, mostly posterior to caeca. Cirrus-sac large. Auxiliary external seminal vesicle absent. Male genital pore opens dorsolaterally at posterior end of body. Ovary lobed, medial posterior to testis. Seminal receptacle absent. Laurer's canal absent. Uterus with few coils. Metraterm absent? Eggs ovoid. Female genital pore opens at mid-level of cirrus-sac. Vitellarium follicular, disposed as diffuse mass from anterior end of body to mid-level of cirrus-sac. Excretory vesicle and pore not reported.

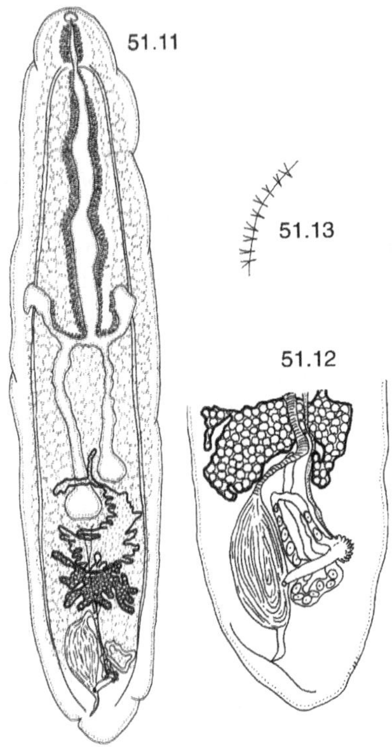

Figs 51.11–51.13. *Metaplehniella lethrini* Lebedev & Parukhin, 1972. 51.11. Whole worm, dorsal view. 51.12. Genitalia, dorsal view. 51.13. Tegumental spines. (After Lebedev & Parukhin, 1972.)

In (? blood-vessels of) stomach wall of *Lethrinus miniatus*. Indian Ocean off India. Type- and only species *M. lethrini* Lebedev & Parukhin, 1972.

9b. Testis relatively large, not deeply lobed, between or partly posterior to intestinal caeca; cirrus-sac absent or pyriform; uterus lateral to ovary and/or postovarian; male genital pore dorsal, near posterior end; female pore dorso-sinistral, anterior to male pore .... 10.

10a. Testis between intestinal caeca; ovary between and/or posterior to caeca; uterus lateral to ovary and/or postovarian; male genital pore dorsal and sinistral near posterior end of body; female pore anterior to male pore .... *Cardicola* Short, 1953. (Figs 51.14, 51.15)

Diagnosis: Body small, flattened, elongate-oval, with ventrally recurved body margins armed with transverse rows of small spines with curved tips. Suckers and pharynx absent. Mouth small, subterminal. Oesophagus long, extending to near middle of body. Anterior and posterior intestinal caeca form X- or H-shape; anterior caeca shorter than, or about same length as, posterior. Testis single, between posterior intestinal caeca. Vas deferens well developed, apparently serving as seminal vesicle. Cirrus-sac absent or pyriform. Cirrus present or absent. Auxiliary external seminal vesicle absent. Male genital pore dorsal, opens on left near posterior end of body. Ovary bilobed, quadrangular or subtriangular, with irregular margins, may overlap or extend beyond caeca. Seminal receptacle absent. Uterus relatively long, lateral to ovary and/or postovarian. Short muscular metraterm. Eggs thin-shelled, pyriform, with spiniform terminal filament. Female genital pore opens dorsosinistrally, anterior to male genital

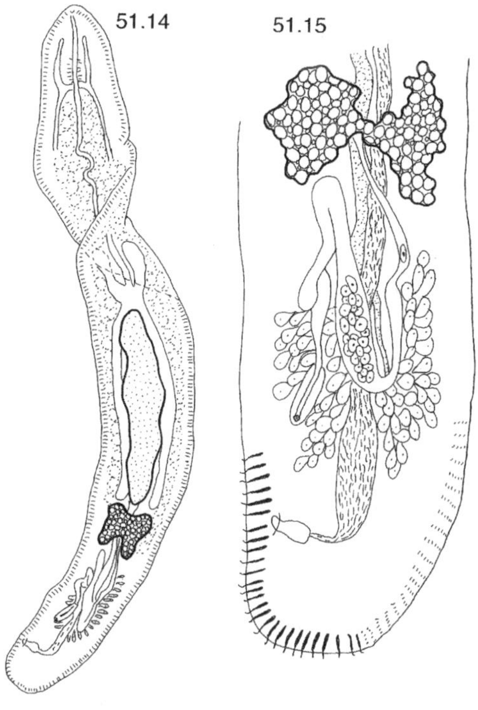

Figs 51.14–51.15. *Cardicola cardiocola* (Manter, 1947) Short, 1953. 51.14. Whole worm, largely dorsal view. 51.15. Genitalia, dorsal view. (After Manter, 1947.)

pore. Laurer's canal absent. Vitellarium follicular, extensive, reaching from nerve-ring to ovary. Extent of excretory vesicle not clear. Excretory pore terminal. In heart and blood-vessels of gills, kidney and liver of teleost fish representing nine families. Atlantic and Pacific Oceans, Hawaiian waters and South China Sea. Type-species *C. cardiocola* (Manter, 1947) Short, 1953.

10b. Testis between and partly posterior to intestinal caeca; uterus lateral and posterior to ovary, or completely postovarian; male genital pore at or near left body margin; female pore opens at or left of dorsal median line; ovary between or posterior to caeca, or between testis and uterus .... 11.

11a. Testis narrower anteriorly than posteriorly; uterus lateral and posterior to ovary; male genital pore at left anterior margin of posterior asymmetrical body lobe; female pore on dorsal mid-line; ovary between testis and uterus ... . *Psettarium* Goto & Ozaki, 1930. (Figs 51.16, 51.17)

(Syn. *Plehnia* Goto & Ozaki, 1929, pre-occupied)

Diagnosis: Body narrow, elongate, flattened, with asymmetrical lobe at posterior end; lateral margins with transverse rows of spines. Mouth subterminal. Suckers and pharynx absent. Oesophagus long. Anterior and posterior intestinal caeca form X- or H-shape; anterior caeca short; right posterior caecum longer than left, terminating about one-quarter of body length from posterior end. Testis single, reticulate, between and posterior to intestinal caeca, narrower anteriorly than posteriorly. Vas deferens sinuous, entering elliptical cirrus-sac with prominent seminal vesicle. Auxiliary external seminal vesicle absent. Male genital pore opens on left anterior margin of posterior lobe. Ovary a tubular mass, occupying space between testis and uterus.

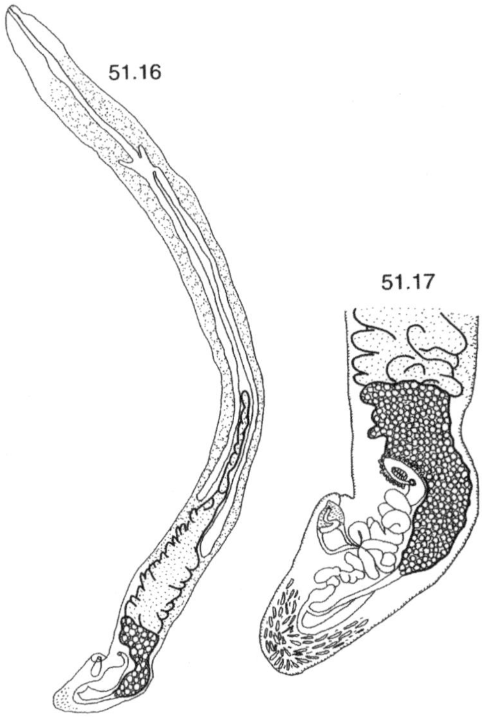

Figs 51.16–51.17. *Psettarium sebastodorum* Holmes, 1971. 51.16. Whole worm, dorsal view. 51.17. Genitalia, dorsal view. (After Holmes, 1971.)

Seminal receptacle absent. Laurer's canal absent. Uterus lateral and posterior to ovary. Uterus loops posteriorly to join metraterm. Eggs thin-shelled, ovoid. Female genital pore opens on dorsal median line slightly anterior to male genital pore. Vitellarium composed of tubular acini, surrounding intestinal caeca, extending from mid-level of oesophagus to anterior margin of ovary. Excretory vesicle and pore not reported. In 'intestine' (?) of *Sphoeroides pardalis*, and in heart, bulbus arteriosus and ventral aorta of *Sebastes* spp. Northern Pacific Ocean. Type-species *P. japonicum* (Goto & Ozaki, 1929).

11b. Testis not narrower anteriorly than posteriorly, between and/or posterior to intestinal caeca; uterus completely postovarian; male genital pore dorsal to left of body mid-line; ovary between or posterior to caeca, lying posterior to testis; female pore sinistrally submarginal .... *Psettarioides* Lebedev & Parukhin, 1972. (Figs 51.18, 51.19)

Diagnosis: Body lanceolate, flattened; lateral margins with spines. Mouth terminal. Suckers absent. Oesophagus long, with buccal capsule. Anterior and posterior intestinal caeca form X- or H-shape; anterior caeca short; posterior caeca long, reaching to posterior quarter of body, with small branches. Testis single, immediately post-bifurcal, reaching to anterior margin of ovary. Vas deferens leads to seminal vesicle within cirrus-sac. Auxiliary external seminal vesicle absent. Male genital pore opens dorsally at or near left body margin. Ovary large, fan-shaped or lobed, between or posterior to caeca and posterior to testis. Uterine seminal receptacle present. Laurer's canal absent. Uterus coiled, completely postovarian. Metraterm absent? Eggs thin-shelled, ovoid. Female genital pore sinistrally submarginal. Vitellarium follicular, diffuse, disposed between intestinal bifurcation and ovary. Excretory vesicle not reported. Excretory pore terminal. In branchial blood-vessels, stomach and intestinal

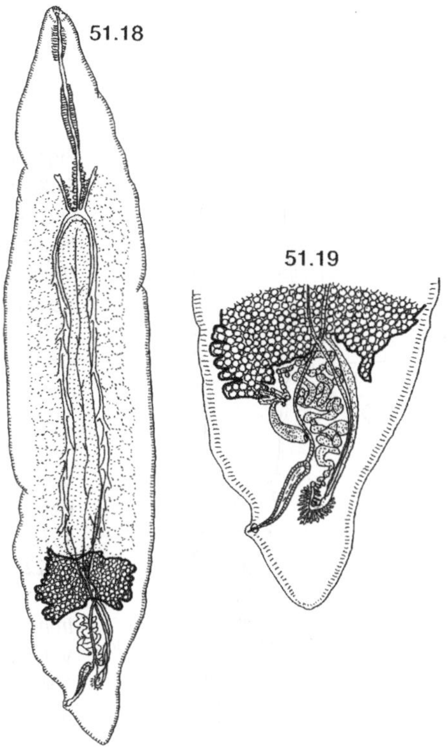

Figs 51.18–51.19. *Psettarioides rachycentri* Lebedev & Parukhin, 1972. 51.18. Whole worm, dorsal view. 51.19. Genitalia, dorsal view. (After Lebedev & Parukhin, 1972.)

wall or kidney of rachycentrid, mullid, istiophorid and tetraodontid marine fish. Cosmopolitan in tropical marine waters. Type-species *P. rachycentri* Lebedev & Parukhin, 1972.

12a. Testis transversely coiled, between and posterior to intestinal caeca; ovary lobed, entirely posterior to caeca; Laurer's canal absent; body with lateral tubercles .... *Orchispirium* Madhavi & Hanumantha Rao, 1970. (Figs 51.20, 51.21)

Diagnosis: Body spatulate, flattened, with tubercles laterally (spines lost?). Suckers and pharynx absent. Oesophagus long, slender, sinuous, surrounded by gland-cells, enlarged posteriorly to form vesicle. Intestine with posterior caeca only, forming inverted U-shape, extending to about two-thirds of body length. Testis single, a transversely coiled tube, between and posterior to caeca. Cirrus-sac poorly developed, enclosing seminal vesicle and cirrus. Auxiliary external seminal vesicle absent. Ovary lobed, entirely posterior to caeca. Seminal receptacle absent. Laurer's canal absent. Uterus post-testicular, relatively long. Metraterm short. Eggs thin-shelled, ovoid. Common genital pore dorsal, sinistrally submedian, postovarian. Vitellarium follicular, with follicles much more extensive on right side than left. Excretory vesicle Y-shaped. Excretory pore not reported. In mesenteric blood-vessels of rays (*Dasyatis*). North-east Indian Ocean. Type- and only species *O. heterovitellatum* Madhavi & Hanumantha Rao, 1970.

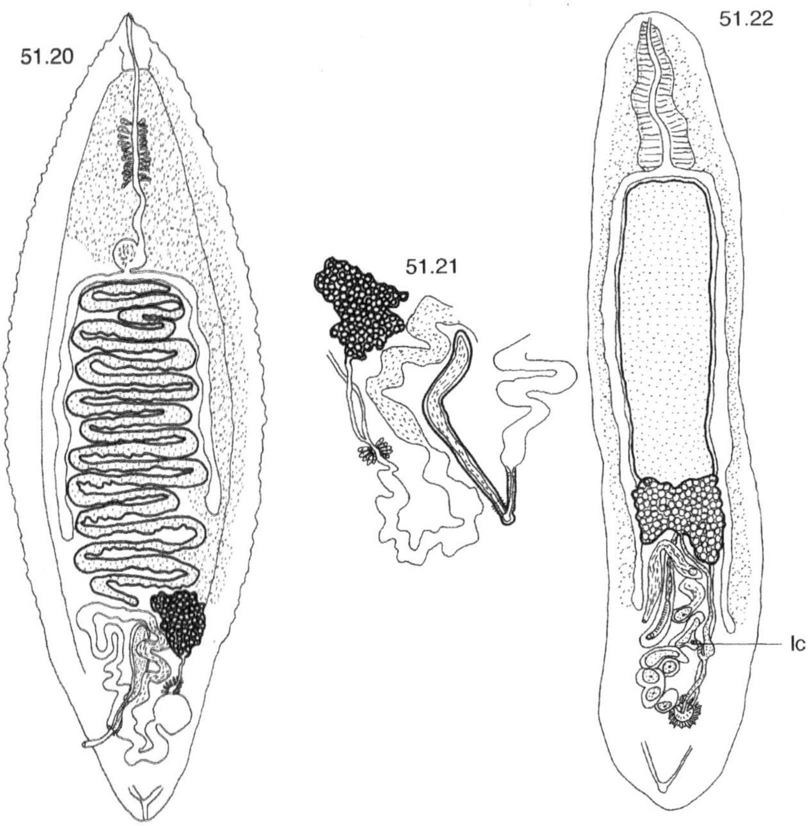

Figs 51.20–51.22. 51.20–51.21. *Orchispirium heterovitellatum* Madhavi & Hanumantha Rao, 1970. 51.20. Whole worm, dorsal view. 51.21. Genitalia, ventral view. (After Madhavi & Hanumantha Rao, 1970.) **51.22.** *Chimaerohemecus trondheimensis* Van der Land, 1967, whole worm, dorsal view. lc, Laurer's canal. (After Van der Land, 1967.)

12b. Testis not transversely coiled; both testis and ovary lateral to intestinal caeca and/or between caeca; ovary bilobed; cirrus-sac well developed; Laurer's canal present; ventrolateral body margins with curved spines .... 13.

13a. Intestinal caeca reach back only to about level of genital pores; ovary bilobed, between caeca .... *Chimaerohemecus* Van der Land, 1967 (Fig. 51.22)

Diagnosis: Body small, flattened, mainly uniform in breadth; anterior and posterior ends slightly tapering. Ventrolateral body margins bear curved spines. Suckers and pharynx absent. Oesophagus relatively short, extending to about one-fifth of body length, surrounded by numerous gland-cells. Intestine with posterior caeca only, forming inverted U-shape; caeca long, about half body length, reaching back to about level of genital pores. Testis single, large, extending from intestinal bifurcation to about two-fifths of body length. Vas deferens short. Cirrus-sac well developed, containing seminal vesicle. Cirrus non-coiled. Auxiliary external seminal vesicle absent. Male genital pore dorsal, in posterior part of body, to left of median line. Ovary bilobed, lying between caeca, immediately post-testicular. Seminal receptacle present. Laurer's canal present. Oötype to right of median line, near posterior extremity. Uterus with winding ascending limb, containing numerous thin-shelled eggs, and straight descending limb. Metraterm absent? Female genital pore opens anterior to male genital pore. Vitellarium follicular, lateral to oesophagus, caeca and gonads. Excretory vesicle

Y-shaped, with very short stem. Excretory pore not reported. In blood system of holocephalans (*Chimaera*). Northern north-east Atlantic Ocean. Type- and only species *C. trondheimensis* Van der Land, 1967.

13b. Intestinal caeca reach to near posterior end of body; ovary bilobed, posterior to testis, between and lateral to caeca .... *Hyperandrotrema* Maillard & Ktari, 1978. (Figs 51.23–51.25)

Diagnosis: Body elongate, flattened, with three to four rows of hook-shaped spines ventrolaterally. Suckers absent. Oesophagus relatively short. Intestine with posterior caeca only, forming inverted U-shape, reaching to near posterior end of body. Testis single, large, between caeca. Cirrus-sac well developed. Auxiliary external seminal vesicle absent. Male genital pore opens dorsolaterally in posterior left quarter of body. Ovary bilobed, posterior to testis, lying between and lateral to caeca. Seminal receptacle absent. Oötype posterior. Laurer's canal present. Uterus completely post-ovarian. Metraterm absent? Eggs thin-shelled, subspherical. Female genital pore opens dorsally close to male pore. Vitellarium follicular, surrounding intestinal caeca. Excretory vesicle Y-shaped. Excretory pore not reported. In heart and blood-vessels of sharks (*Cetorhinus*). Mediterranean Sea and northern north-east Atlantic Ocean. Type- and only species *H. cetorhini* Maillard & Ktari, 1978.

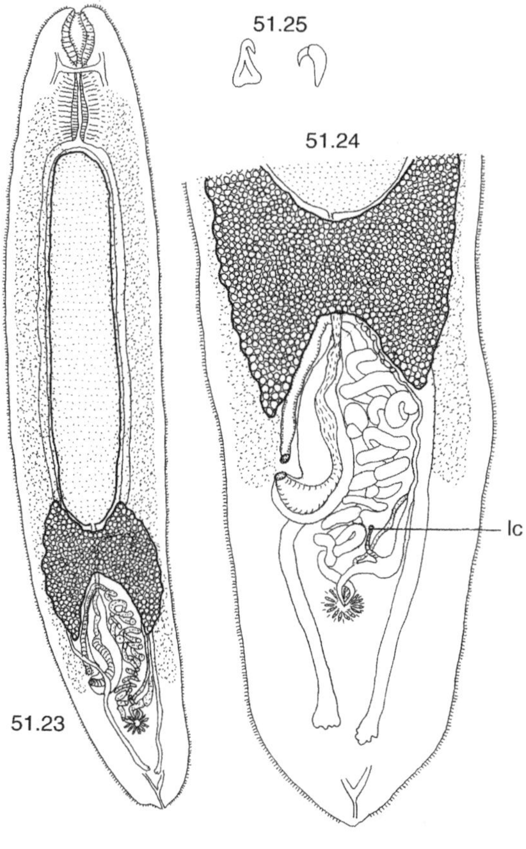

Figs 51.23–51.25. *Hyperandrotrema cetorhini* Maillard & Ktari, 1978. 51.23. Whole worm, dorsal view. 51.24. Genitalia, dorsal view. lc, Laurer's canal. 51.25. Tegumental spines. (Original.)

14a. **Two testes** .... **15.**
14b. **More than two testes** .... **17.**

15a. **Two testes, one pre-ovarian and other lateral to ovary and/or postovarian; anterior testis branched; anterior and posterior intestinal caeca form X- or H-shape** .... ***Neoparacardicola*** **Yamaguti, 1970. (Fig. 51.26)**

Diagnosis: Body elongate, flattened, with deep notch on left posterior margin, with fine spines ventromarginally. Oral sucker weakly developed (according to Overstreet & Køie, 1989). Mouth opens subterminally. Oesophagus narrow, sinuous. Anterior and posterior intestinal caeca of similar length form X- or H-shape; anterior caeca widely divergent; posterior with side branches. Two testes, the anterior one not as dendritic as originally reported (Overstreet & Køie, 1989); posterior testis largely posterior to ovary, elongate and irregularly lobed. Cirrus-sac fusiform, receiving vasa efferentia. Auxiliary external seminal vesicle absent. Ovary with few unequal lobes, extending along right body margin in anterior part of caudal third of body. Seminal receptacle absent. Laurer's canal absent. Uterus winds forward, beginning posterior to ejaculatory duct and then to right of cirrus-sac. Metraterm present. Male and female genital pores separate, open at left marginal notch. Vitellarium follicular, extending in lateral fields from near head end to posterior end of metraterm on left and to anterior end of ovary on right. Excretory system unclear. In blood-vessels of teleosts (*Naso*). Hawaiian marine waters. Type- and only species *N. nasonis* Yamaguti, 1970.

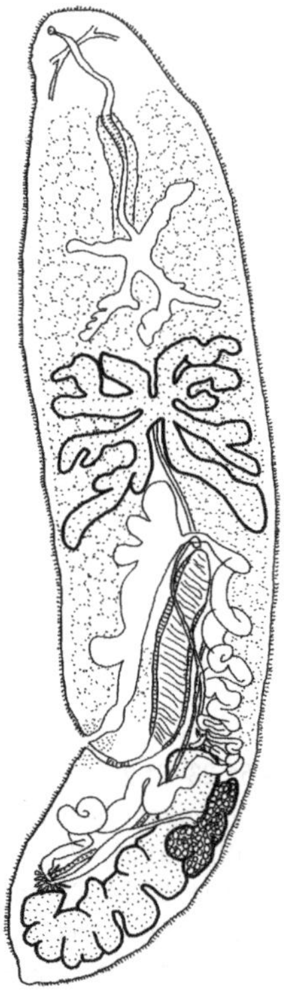

Fig. 51.26. *Neoparacardicola nasonis* Yamaguti, 1970, whole worm, dorsal view. (After Yamaguti, 1970.)

15b. Both testes ovoid, with scalloped margins, situated in posterior half of body; posterior intestinal caeca only, or anterior and posterior caeca form X- or H-shape .... 16.

16a. Intestine with posterior caeca only, forming inverted U-shape; anterior and posterior testis about equal in size; pedunculate oral disc (oral sucker?); ovary deeply indented, lying between testes; uterus pre-ovarian ....
.. *Paracardicoloides* Martin, 1974. (Figs 51.27–51.29)

Diagnosis: Body small, flattened, lanceolate, partly and irregularly spined; each spine sheathed in tegument. Pedunculate oral disc (oral sucker?) raised from rest of body. Oesophagus with bulb near intestinal bifurcation. Intestine with posterior caeca only, forming inverted U-shape, extending halfway along body. Two testes in hindbody. Cirrus-sac well developed, containing seminal vesicle, few prostate cells and narrow cirrus. Auxiliary external seminal vesicle absent. Ovary deeply indented, between testes. Seminal receptacle absent. Laurer's canal absent. Uterus pre-ovarian, with enlarged, sausage-shaped region. Metraterm absent? Eggs thin-shelled. Male and female genital pores open into genital atrium, to right of ovary on dorsal surface. Vitellarium follicular, lateral to most of intestinal tract but also between caeca and anterior to uterus. Excretory vesicle Y-shaped. Excretory pore not reported. In blood-vessels of eels (*Anguilla*). Australia. Type- and only species *P. yamagutii* Martin, 1974.

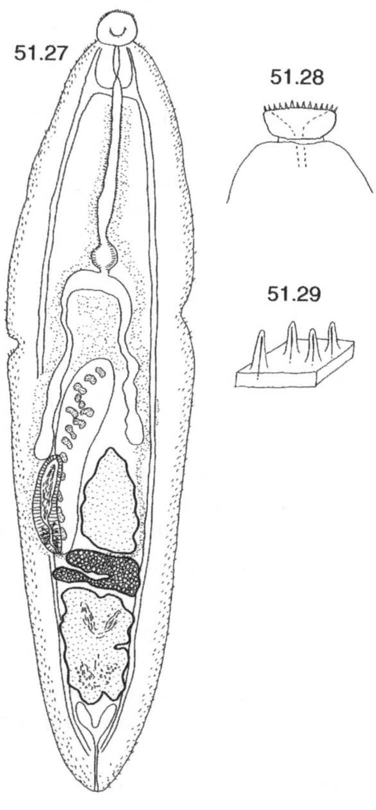

Figs 51.27–51.29. *Paracardicoloides yamagutii* Martin, 1974. 51.27. Whole worm, ventral view. 51.28. Oral disc. 51.29. Tegumental spines. (After Martin, 1974.)

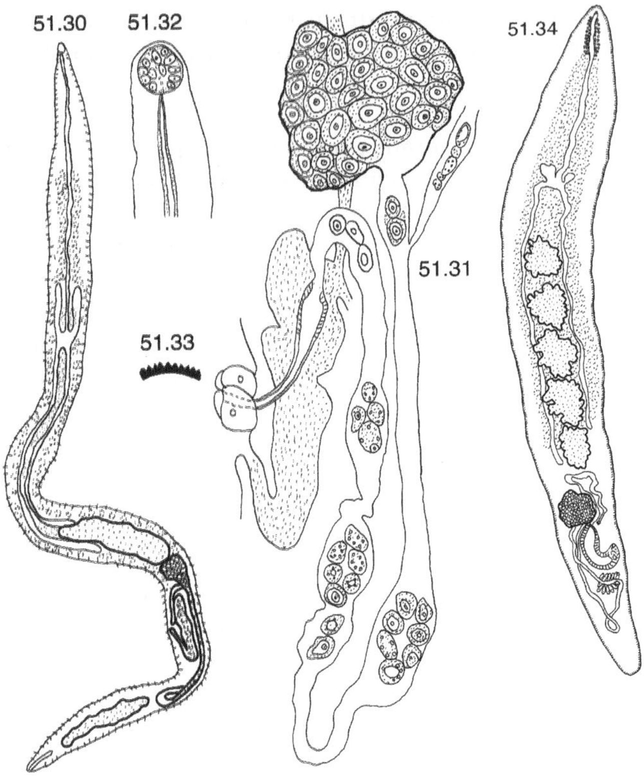

Figs 51.30–51.34.
51.30–51.33. *Paracardicola hawaiensis* Martin, 1960. 51.30. Whole worm, dorsal view. 51.31. Genitalia, dorsal view. 51.32. Rudimentary oral sucker, ventral view. 51.33. Tegumental spines. (After Martin, 1960.)
51.34. *Pseudocardicola emmelichthy* Parukhin, 1985, whole worm, dorsal view. (After Parukhin, 1985.)

16b. Anterior and posterior intestinal caeca form X- or H-shape; posterior testis not as well developed as anterior; pedunculate oral disc absent; oral sucker rudimentary; ovary compact, lying between anterior testis and cirrus-sac; uterus postovarian .... *Paracardicola* Martin, 1960. (Figs 51.30–51.33)

Diagnosis: Body small, slender, flattened, laterally spined along most of length. Oral sucker rudimentary. Oesophagus long, about one-third of body length, slender, slightly sinuous. Anterior and posterior intestinal caeca form X- or H-shape; posterior caeca considerably longer than anterior. Two testes in posterior half of body; one immediately anterior to ovary; the other posterior to uterus and less well developed. Cirrus-sac weak, containing elongate, saccular seminal vesicle. Auxiliary external seminal vesicle absent. Ovary compact, lying between anterior testis and cirrus-sac. Seminal receptacle absent. Laurer's canal absent. Uterus postovarian, relatively long, with few loops. Metraterm absent? Eggs ovoid with terminal spine. Male and female genital pores separate but close together, dorsally near mid-level of seminal vesicle. Vitellarium with small follicles extending from mid-oesophageal level to ovary. Excretory vesicle tubular. Excretory pore terminal. In mesenteric veins of marine fish (*Arothron*). Hawaiian waters. Type- and only species *P. hawaiensis* Martin, 1960.

17a. Five lobed testes, between intestinal caeca; uterus both pre- and postovarian.................................... *Pseudocardicola* Parukhin, 1985. (Fig. 51.34)

Diagnosis: Suckers absent. Oesophagus long, surrounded anteriorly by glands. Anterior and posterior intestinal caeca form X- or H-shape; anterior caeca short; posterior relatively long. Five testes, lobed, between posterior caeca, beginning some distance posterior to intestinal bifurcation and reaching beyond posterior caeca. Well-developed cirrus-sac posterior to ovary; contains seminal vesicle. Auxiliary external seminal vesicle absent. Male genital pore opens dorsally near right body margin. Ovary weakly lobed, lying posterior to hind testis. Seminal receptacle absent. Laurer's canal absent. Uterus extends anterior and posterior to ovary. Metraterm absent? Female genital pore opens on right, posterior to ovary. Vitellarium follicular, beginning posterior to glandular region of oesophagus and extending backwards along posterior caeca. Excretory vesicle and pore not reported. In gill blood-vessels of marine fish (*Emmelichthys*). Indian Ocean. Type- and only species *P. emmelichthy* Parukhin, 1985.

17b. More than five testes; uterus postovarian or largely or wholly pre-ovarian .... 18.

18a. Body spines curved, not in clusters; 19 to 71 testes between intestinal caeca; ovary posterior to caeca and testes; uterus postovarian .... *Paradeontacylix* McIntosh, 1934. (Figs 51.35–51.37)

Diagnosis: Body small, flattened, slender, with ventrolateral transverse rows of rose-thorn-shaped spines. Suckers and pharynx absent. Mouth subterminal. Oesophagus long. Anterior and posterior intestinal caeca form X- or H-shape; anterior caeca shorter than posterior. Testes 19 to 71, arranged in two or three irregular rows between posterior intestinal caeca. Vas deferens leading to seminal vesicle and pars prostatica enclosed in cirrus-sac. Cirrus conical or cylindrical. Auxiliary external seminal vesicle absent. Male genital pore opens dorsally near left body margin. Ovary shield- or

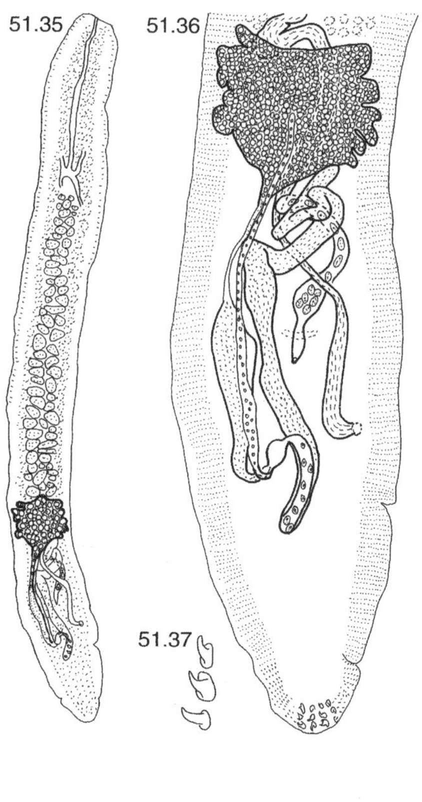

Figs 51.35–51.37. *Paradeontacylix sanguinicoloides* McIntosh, 1934. 51.35. Whole worm, ventral view. 51.36. Genitalia, ventral view. 51.37. Tegumental spines. (After McIntosh, 1934a.)

heart-shaped, lying posterior to caeca and testes. Laurer's canal absent. Uterus coiled, lying posterior to ovary. Metraterm present. Eggs thin-shelled, ovoid. Female genital pore opens dorsally in mid-line, anterior to male pore. Vitellarium follicular, extensive, from near anterior end of body to anterior margin of ovary. Excretory vesicle and pore not reported. In intestinal blood-vessels, heart and gill-arteries of marine fish. Pacific and Atlantic Oceans. Type-species *P. sanguinicoloides* McIntosh, 1934.

18b. Body spines not curved, arranged in clusters; 28–203 testes between intestinal caeca or > 100 testes largely posterior to caeca; ovary near posterior end of body; uterus pre-ovarian or largely so .... 19.

19a. Spines in clusters of 13–50; 28–203 testes between intestinal caeca; uterus pre-ovarian; common genital atrium, pre-ovarian .... *Aporocotyle* Odhner, 1900. (Figs 51.38, 51.39)

Diagnosis: Body spindle-shaped, flattened, with fan-shaped clusters of 13–50 spines on ventrolateral body margins. Armed protusible snout (rudimentary oral sucker?) with whorl of small spines. Mouth terminal, leads to buccal capsule. Oesophagus relatively long. Anterior and posterior intestinal caeca form X- or H-shape; anterior caeca shorter than posterior; latter end near posterior end of body. Testes 28–203, round or irregularly lobed, lying between posterior caeca. Cirrus-sac median, retort-shaped, contains internal seminal vesicle and pars prostatica. Unarmed cirrus opens into common genital atrium, lying in posterior quarter of body, between dorsal mid-line

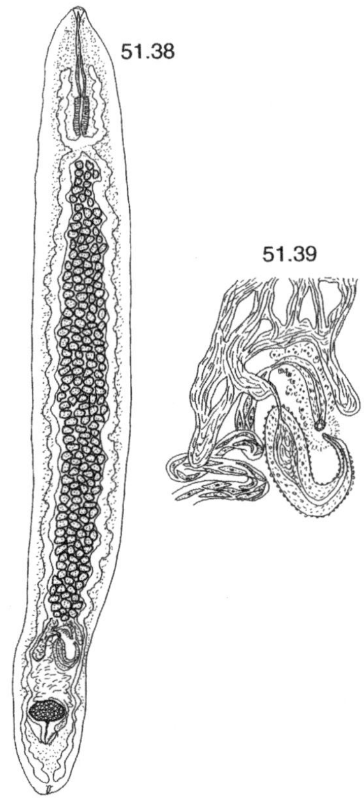

Figs 51.38–51.39. *Aporocotyle simplex* Odhner, 1900. 51.38. Whole worm, ventral view. 51.39. Genitalia, ventral view. (After Thulin, 1980.)

and left lateral body margin, pre-uterine and pre-ovarian. Auxiliary external seminal vesicle absent. Ovary ovoid, median, lying between genital atrium and posterior end. Laurer's canal absent. Uterus pre-ovarian, much coiled, with proximal uterine seminal receptacle. Metraterm present. Eggs fusiform, thin-shelled. Vitellarium follicular, extending from near anterior end to level of ovary, mostly lateral to posterior caeca. Excretory vesicle Y-shaped. Excretory pore subterminal. In gill blood-vessels, heart, bulbus arteriosus and elsewhere of fish of four gadiform families, and of two families each of scorpaeniforms, perciforms and pleuronectiforms. Cooler seas of northern and southern hemispheres. Type-species *A. simplex* Odhner, 1900.

19b. Spines in clusters of three to four; > 100 testes almost entirely posterior to intestinal caeca; uterus largely pre-ovarian; common genital pore, lateral to ovary .... *Plethorchis* Martin, 1975. (Figs 51.40–51.42)

Diagnosis: Body slender, flattened, elongate, with clusters of three to four minute spines laterally. Suckers absent. Mouth terminal or subterminal. Oesophagus slender and elongate, partly glandular. Anterior and posterior intestinal caeca form X- or H-shape; anterior caeca very poorly developed, much shorter than posterior, which terminate anterior to mid-body. Testes numerous, lying between termination of posterior caeca and uterus. Cirrus-sac small, with internal seminal vesicle and very short cirrus. Auxiliary external seminal vesicle absent. Ovary lying dextrolaterally to posterior part of uterus and near posterior end of body. Seminal receptacle absent. Laurer's canal absent. Uterus largely pre-ovarian, lying posterior to caeca and testes.

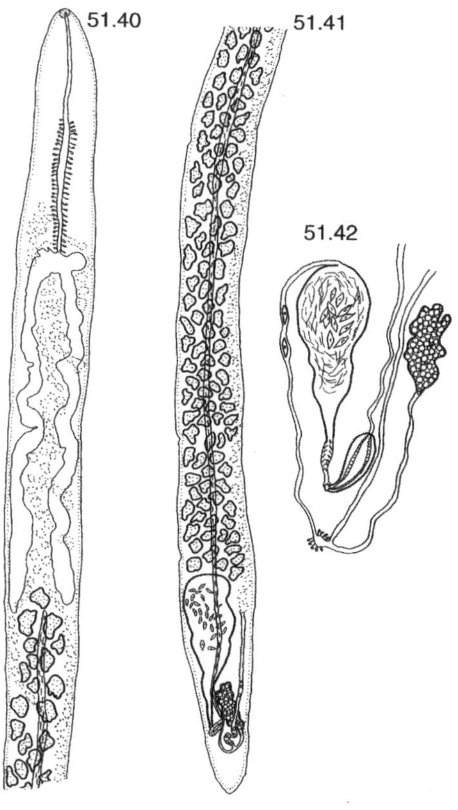

Figs 51.40–51.42. *Plethorchis acanthus* Martin, 1975. 51.40. Anterior half of worm, dorsal view. 51.41. Posterior half of worm, dorsal view. 51.42. Genitalia, dorsal view. (After Martin, 1975.)

Metraterm absent? Eggs thin-shelled, fusiform, with terminal spines. Common genital pore near posterior end of body, ventral to left of mid-line, lateral to ovary. Vitellarium follicular, with more follicles on right side of body. Excretory vesicle V-shaped. Excretory pore not reported. In coelom (?) and blood-vessels of mesenteries, intestine and liver of mugilids. Australia. Type- and only species *P. acanthus* Martin, 1975.

52 Family Spirorchiidae Stunkard, 1921[1]

T.R. Platt

Department of Biology, Saint Mary's College, Notre Dame, Indiana, USA

Introduction

The family Spirorchiidae was erected by Stunkard (1921) for blood flukes from turtles, with *Spirorchis* MacCallum, 1919, as the type-genus, and also subsequently in the same year by MacCallum (1921). The new family name was given precedence over Proparorchiidae Ward, 1921, when its type-genus, *Proparorchis* Ward, 1921, was determined (Stunkard, 1921) to be identical to *Spirorchis*. Yamaguti (1971) recognized nine subfamilies, of which six were monotypic. In the absence of a phylogenetic assessment of the genera, the use of subfamilies is deemed untenable. Smith (1972, 1997a) conducted a thorough review of the literature on the spirorchiids and sanguinicolids, including exhaustive host/parasite and parasite/host lists (Smith, 1997b). While Smith (1997a, b) noted recognized synonymies, he did not attempt to determine the validity of individual taxa.

The current work is based on the examination of type-material whenever possible. In some instances assessments are based solely on information compiled from the literature. No specimens described from the Indian subcontinent were available for examination despite repeated attempts to arrange for loans of this material.

Since the publication of Yamaguti's (1971) synopsis, seven new genera of spirorchiids have been described to add to the 18 valid genera listed and several new

[1] Stunkard (1921) erected the family Spirorchidae [*sic*] to contain several genera of digeneans reported from the circulatory system of turtles. MacCallum (1921), unaware of Stunkard's proposal published earlier in the year, suggested the name Spirorchiidae for the same taxonomic concept. Both spellings have been used in the literature, with Stunkard's original spelling predominant until 1940, when Wall (1940) used the 'ii' form. Platt (1990) reviewed the topic and concluded that Stunkard's original formation was etymologically correct and should be recognized on the basis of priority. Subsequent enquiries to classicists have not altered that opinion. However, within the Digenea, all other family names based on generic names ending in '-orchis' are formed with the 'ii' ending (e.g. Plagiorchiidae, Monorchiidae, etc.). Dr Phillip Tubbs (Executive Secretary, International Commission on Zoological Nomenclature, 1997, personal communication) suggested that 'from the point of view of stability and clarity of nomenclature, which is always the deciding factor, it is important that the "-orchis" family names be treated collectively'. Therefore Stunkard's original spelling is corrected to Spirorchiidae and he is credited with authorship of the family name.

synonymies recognized. This unquestionably represents only a small fraction of spirorchiids that are certainly present in the roughly 75% of turtle genera that have never been examined for parasites or had blood flukes reported as part of their helminth fauna (Platt, 1992).

Mehrotra (1973a, b) described the following new species in abstract form: *Hapalotrema pambanensis*, *Amphiorchis indicus*, *Plasmiorchis stunkardi*, *Coeuritrema sutlejensis*, *C. sheilae* and *Cardiotrema roparensis*. Although the descriptions were meagre, this publication does meet the criteria set forth by the International Commission on Zoological Nomenclature (1999) for a nomenclatural act. Hence all names are considered available from 1973. More complete descriptions, including illustrations of these species, were published as follows: Gupta & Mehrotra (1981) for *H. pambanensis* and *A. indicus*; Gupta & Mehrotra (1975) for *P. stunkardi*; Tandon & Gupta (1982) for *Coeuritrema* spp.; and Tandon & Gupta (1985) for *Caridotrema roparensis*.

Spirorchis MacCallum, 1919, the type-genus, has been treated in either a more or a less restrictive sense over the years (Byrd, 1939; Yamaguti, 1958a, 1971; Platt, 1993). Based on a cladistic analysis of *Spirorchis* and related genera, Platt (1993) concluded that *Aphanospirorchis* Platt, 1990, *Diarmostorchis* Ejsmont, 1927 and *Henotosoma* Stunkard, 1922 should be considered junior synonyms of *Spirorchis*. In addition, *Haematotrema* Stunkard, 1923 and *Proparorchis* Ward, 1921 were considered synonyms of *Spirorchis* by Byrd (1939) and Mehra (1934).

Squaroacetabulum Simha & Chattopadhyaya, 1970 is a synonym of *Amphiorchis* Price, 1934. *S. solus* Simha & Chattopadhyaya 1970 and *A. caborojoensis* Fischthal & Acholonu, 1976 are strikingly similar, including the defining character of *Squaroacetabulum*, a slightly squared ventral sucker (Simha & Chattopadhyaya, 1970; Fischthal & Acholonu, 1976). Unfortunately, specimens of *S. solus* were not available for examination. Both fit well with the original diagnosis of *Amphiorchis*, thus *Squaroacetabulum* must be considered a junior synonym of the former. *S. solus* becomes *A. solus* n. comb.

Carettacola Manter & Larson, 1950 and *Haemoxenicon* Martin & Bamberger, 1952 were described as new taxa from the loggerhead turtle (*Caretta caretta*) and the green turtle (*Chelonia mydas*), respectively (Manter & Larson, 1950; Martin & Bamberger, 1952). The primary differentiating character was the presence of an expanded Laurer's canal in *Carettacola*. Dailey *et al.* (1991) re-examined these taxa and concluded that they were synonymous, with which I concur.

There has been considerable debate in the literature regarding the validity of *Hapalorhynchus* Stunkard, 1922 and *Coeuritrema* Mehra, 1933. Price (1934), Byrd (1939) and Brooks & Mayes (1976) considered *Coeuritrema* a synonym of *Hapalorhynchus*. Mehra (1934, 1939) rejected the synonymy based on the presence of a cirrus-sac in *Coeuritrema* and its absence in *Hapalorhynchus*. Yamaguti (1971) not only recognized the genera as separate, but placed them in separate subfamilies. Separation of these taxa cannot be accomplished based on any suite of characters available either in the specimens available for examination or from the relevant literature. As no specimens of *Coeuritrema* are available for examination, a more thorough assessment of characters determining the validity of these genera must wait. Therefore, *Coeuritrema* is considered a junior synonym of *Hapalorhynchus*, as recommended by several recent workers (Brooks & Mayes, 1976; Platt, 1988; Bourgat, 1990). The following species are recognized as new combinations: *H. mica* (Oshmarin, 1971) n. comb.; *H. ocadiae*

(Takeuti, 1942) n. comb.; *H. oschmarini* (Belous, 1963) n. comb.; *H. sheilae* (Mehrotra, 1973) n. comb.; *H. sutlejensis* (Mehrotra, 1973) n. comb.

Hapalotrema Looss, 1899 was recently reviewed by Cribb & Gordon (1998) and Platt & Blair (1998). Nomenclatural issues based on misidentification of the type-species were submitted to the International Commission on Zoological Nomenclature (see Platt & Blair, 1996) and were resolved by a ruling of the Commission (Anon., 1997). *H. pambanensis* was originally described by Mehrotra (1973a) in an abstract. Nearly a decade later the species was more thoroughly described and illustrated (Gupta & Mehrotra, 1981). Rao (1976) described *H. mehrai* as a new species. Cribb & Gordon (1998) and Platt & Blair (1998) considered the two species identical. As the name *H. pambanenis* is available from 1973, *H. mehrai* is considered a junior subjective synonym of the former instead of the reverse (Platt & Blair, 1998).

Four genera, *Cheloneotrema* Simha & Chattopadhyaya, 1976, *Neocaballerotrema* Simha, 1977, *Satyanarayanotrema* Simha & Chattopadhyaya, 1980 and *Shobanatrema* Simha & Chattopadhyaya, 1980, were described from various marine turtles (Simha, 1977; Simha & Chattopadhyaya, 1980). The descriptions are based on a single or a few specimens, and the illustrations are of limited value in differentiating these taxa from previously described genera of spirorchiids. They are included in the key based on published accounts; however, the validity of these genera is suspect.

Family Spirorchiidae Stunkard, 1921

(Syns Proparorchiidae Ward, 1921; Spirorchiidae MacCallum, 1921)

Diagnosis: Small to medium flukes. Body spines present or absent. Oral and ventral sucker present or absent. Pharynx absent. Oesophagus long, sometimes sinuous; unicellular secretory glands present or absent. Seminal vesicle present. Caeca long, sinuous or straight, reaching to level of anterior testis or terminating near posterior end. Testes single to many (> 50), tubular, irregular to oval, anterior or posterior to genital complex, or in anterior and posterior groups separated by genital complex. Cirrus-sac present or absent; cirrus present. Ovary single, simple to lobed to tubular, location variable. Seminal receptacle variable. Oötype and Laurer's canal present; uterus typically contains single egg, but may be voluminous. Genital pore ventral or dorsal, anterior or posterior to genital complex, median to submedian; genital sucker present or absent. Eggs variable in number and shape, with or without polar filaments; operculum present or absent. Excretory vesicle V- or Y-shaped. Excretory pore ventral, terminal or dorsal. Typically occurring in vascular system of marine and freshwater turtles; occasionally in the intestine. Type-genus *Spirorchis* MacCallum, 1919.

Key to genera

1a. Single intestinal caecum .... ***Unicaecum* Stunkard, 1925. (Fig. 52.1)**
Diagnosis: Body elongate, slender, tapering at both ends, aspinose. Oral sucker small, terminal; ventral sucker absent. Caecum single, terminating near posterior end of body, inflated at oesophageal junction. Testis single, elongate, tubular to lobed. Seminal vesicle elongate, coiled, extending from caecal inflation posteriorly to

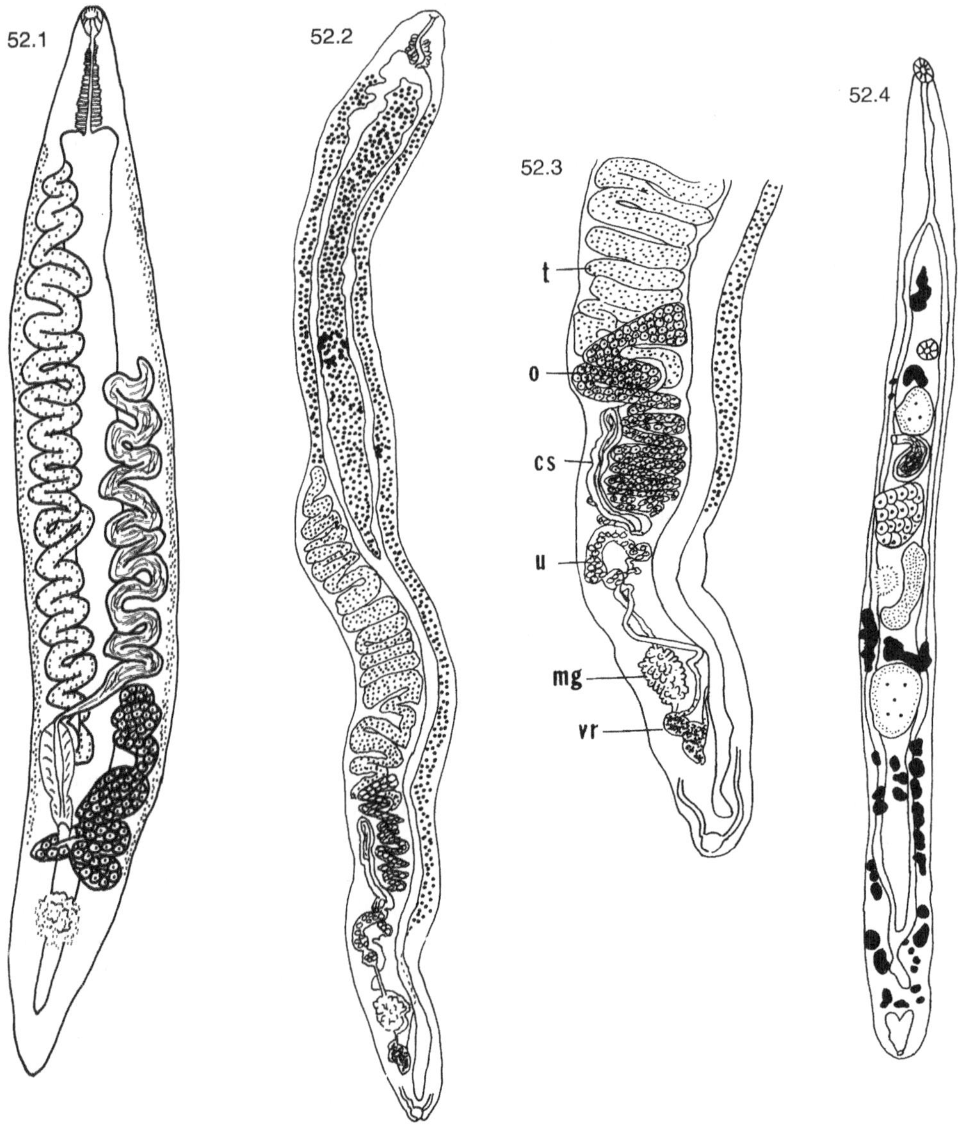

Figs 52.1–52.4.
52.1. *Unicaecum ruszkowskii* Stunkard, 1927. Ventral. (After Schell, 1985.)
52.2, 52.3. *Neospirorchis pricei* Manter & Larson, 1950, 52.2. Ventral. (After Manter & Larson, 1950.) 52.3. Ventral. cs, cirrus-sac; mg, Mehlis' gland; o, ovary; t, testis; u, uterus; vr, vitelline reservoir. (After Manter & Larson, 1950.)
52.4. *Shobanatrema shobanae* Simha & Chattopadhyaya, 1980. Ventral. (After Simha & Chattopadhyaya, 1980.)

cirrus-sac. Cirrus-sac small, muscular, opening ventrally at submedian genital pore in posterior one-fifth of body. Ovary elongate, coiled, lateral and posterior to cirrus-sac. Uterus short containing single, spherical egg. Vitellarium follicular, in lateral margins ranging from level of mid-oesophagus to caecal extremities. Excretory system (?)

I-shaped, saccular to tubular. Parasites of circulatory system of emydid turtles; North America. Type-species *U. ruszkowskii* Stunkard, 1925.

1b. Intestinal caeca two, occasionally fused posteriorly .... 2.

2a. Intestinal caeca fused posteriorly .... 3.

2b. Intestinal caeca not fused .... 4.

3a. Testis single, tubular; ovary tubular; vitellarium extensive, consisting of small follicles .... *Neospirorchis* Price, 1934. (Figs 52.2, 52.3)

Diagnosis: Body elongate, filiform, aspinose. Oral sucker small, terminal; ventral sucker absent. Caeca simple, fused near mid-body, terminating near posterior end. Testis single, tubular, extending posteriorly from near caecal union. Seminal vesicle elongate, coiled. Cirrus-sac small, weakly muscular, located near posterior end of body. Genital pore opening submedian, ventral to lateral, near posterior end. Ovary elongate, coiled, anterior and posterior to cirrus-sac. Uterus elongate, coiled, containing numerous, spherical eggs lacking polar prolongations. Vitellarium follicular, occupying all space from caecal bifurcation to genital pore. Excretory vesicle Y-shaped. Parasites of circulatory system of marine turtles; possibly cosmopolitan. Type-species *N. schistosomatoides* Price, 1934.

3b. Testes two; ovary oval to slightly lobed; vitellarium consisting of large follicles .... *Shobanatrema* Simha & Chattopadhyaya, 1980. (Fig. 52.4)

Diagnosis: Body small, slender, with minute spines. Oral sucker small, subterminal. Ventral sucker membranous, slightly larger than oral sucker, posterior to caecal bifurcation, anterior to anterior testis. Caeca simple, uniting near posterior end to form single caecum. Testes two, oval, located in middle third of body. External seminal vesicle, seminal receptacle and Laurer's canal not described. Cirrus-sac large, retort-shaped, containing internal seminal vesicle and tubular pars prostatica, located between ovary and anterior testis. Genital pore opens over right caecum near posterior edge of ovary. Ovary oval, intertesticular, just posterior to cirrus-sac. Uterus and eggs not described. Vitelline follicles large, primarily in hindbody; several follicles between caecal bifurcation and anterior testis. Excretory vesicle inflated V. Parasites of mesenteric capillaries of marine turtles (*Caretta caretta*); Indian Ocean. Type-species *S. shobanae* Simha & Chattopadhyaya, 1980.

4a. Testis single .... 5.

4b. Testes two or more .... 6.

5a. Testis elongate, oval or lobed; uterus voluminous .... *Uterotrema* Platt & Pichelin, 1994. (Fig. 52.5)

Diagnosis: Body delicate, lanceolate to foliate, with or without constriction at ventral sucker; clusters of spines along lateral margins of hindbody. Posterior end curved ventrally with spines on dorsal terminus. Oral sucker subterminal. Caecal bifurcation at level of ventral sucker; caeca simple, terminating near posterior end. Ventral sucker spined along margins, slightly pedunculate. Testis single, oval or lobed, in posterior quarter of body. External seminal vesicle large, between ventral sucker and cirrus-sac. Cirrus-sac elliptical, containing tubular seminal vesicle and unspined cirrus. Common genital pore opens over left caecum in second quarter of body. Ovary small, oval,

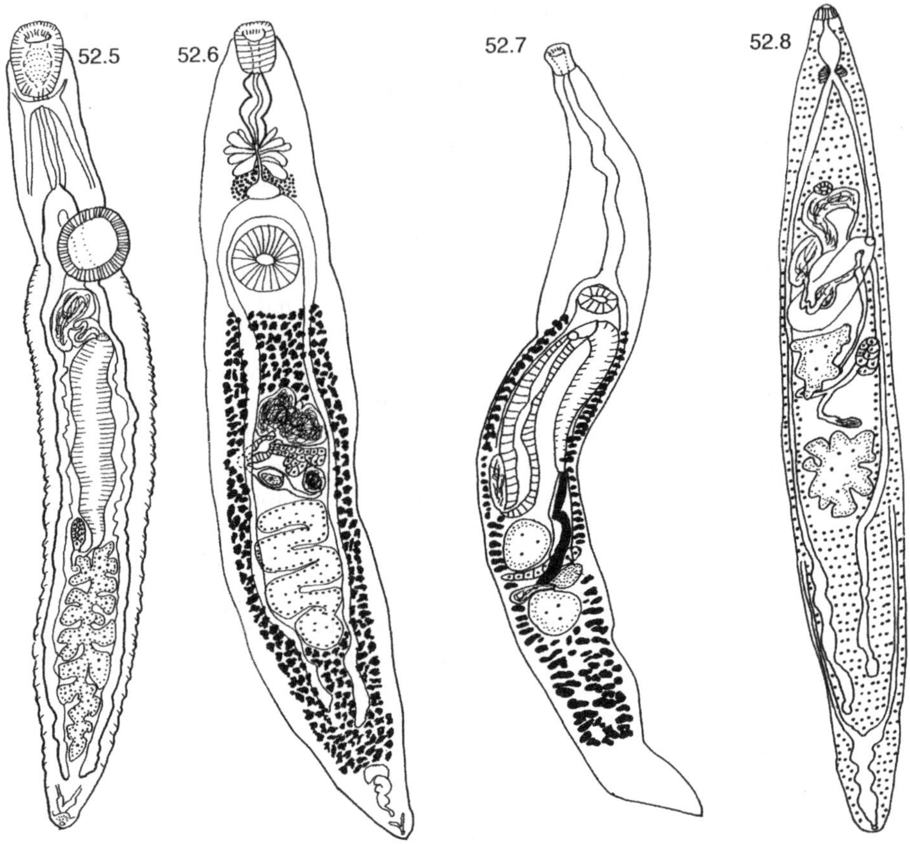

Figs 52.5–52.8.
52.5. *Uterotrema australispinosa* Platt & Pichelin, 1994. Ventral. (After Platt & Pichelin, 1994.)
52.6. *Vasotrema robustum* Stunkard, 1928. Ventral. (Original.)
52.7. *Enterohaematotrema palaeorticum* Mehra, 1940. Ventral. (After Mehra, 1940.)
52.8. *Cardiotrema vaidya* Dwivedi, 1967. Ventral. (After Dwivedi, 1967.)

dextral, at anterior margin of testis. Seminal receptacle saccate. Laurer's canal present, opens mid-dorsally. Uterus voluminous, slightly sinuous, occupies middle third of body, often empty but may contain numerous oval eggs. Vitellarium follicular; follicles small, extracaecal from caecal bifurcation to caecal extremities. Excretory vesicle in form of short-stemmed Y, opens dorsally. Parasites of circulatory system of Australian freshwater turtles (Chelidae). Type-species *U. australispinosa* Platt & Pichelin, 1994.

5b. Testis sinuous or coiled; oesophageal diverticula present .... ***Vasotrema* Stunkard, 1928. (Fig. 52.6)**

(Syn. *Vasatrema* Stunkard, 1926)

Diagnosis: Body delicate, lanceolate, aspinose. Oral sucker well developed, terminal. Ventral sucker well developed, immediately postbifurcal. Oesophagus with numerous diverticula and large number of gland-cells anterior to caecal bifurcation; caeca simple, terminating near posterior end of body. Testis single, coiled to highly folded, in hindbody. Seminal vesicle sinuous to oval, posterior or lateral to cirrus-sac. Cirrus-sac weakly muscular. Common genital pore opens ventrally over left caecum, from just posterior to ventral sucker to near mid-body. Genital pore variable from anterior quarter of body to mid-body. Ovary oval to elliptical, between cirrus-sac and testis. Oviducal seminal receptacle present; Laurer's canal present, opens dorsally. Uterus short, contains single, oval egg. Vitellarium follicular, extracaecal (and occasionally intercaecal), in hindbody reaching to caecal extremities. Excretory vesicle Y-shaped; reserve excretory vesicle (Manter's organ) present; excretory pore terminal. Parasites of the circulatory system of freshwater turtles (Trionychidae); North America. Type-species *V. amydae* (Stunkard, 1926).

6a. Testes two .... 7.
6b. Testes more than two .... 13.

7a. Cirrus-sac present or absent; cirrus-sac or ejaculatory duct anterior to anterior testis .... 8.
7b. Cirrus-sac present, between testes .... 10.

8a. Genital pore median, ventral; cirrus-sac muscular and metraterm well developed .... *Enterohaematotrema* Mehra, 1940. (Fig. 52.7)
(Syn. *Hepatohaematotrema* Simha, 1958)

Diagnosis: Body delicate, lanceolate, aspinose. Oral sucker well developed, terminal. Ventral sucker well developed, immediately posterior to caecal bifurcation. Oesophagus and caeca simple; caeca terminating near posterior end. Testes two, oval to lobed in third quarter of body. Cirrus-sac large, muscular, extending anteriorly from anterior testis to common genital pore just posterior to ventral sucker. Seminal vesicle small, dextral to base of cirrus-sac. Genital pore median, ventral, just posterior to ventral sucker. Ovary transversely elongate, intercaecal. Oviducal seminal receptacle present; Laurer's canal not described. Uterus short, contains single egg; eggs elongate with bipolar processes and/or lateral spine. Vitellarium follicular occupying all available space from posterior margin of ventral sucker to caecal ends. Excretory vesicle Y-shaped; excretory pore terminal. Parasites of the intestine and circulatory system of freshwater turtles (Bataguridae); India. Type-species *E. palaeorticum* Mehra, 1940.

8b. Genital pore sinistral and dorsal .... 9.

9a. Cirrus-sac large, membranous; vitellarium diffuse, extending entire length of body .... *Cardiotrema* Dwivedi, 1967. (Fig. 52.8)

Diagnosis: Body delicate, lanceolate; aspinose. Oral sucker well developed, terminal. Ventral sucker small to rudimentary. Oesophagus short, inflated. Caeca simple with terminal knob-like inflations, terminating near posterior end. Testes two, lobed, in middle third of body. Terminal genitalia anterior to anterior testis. Seminal vesicle large, sinuous, posterior to ventral sucker. Cirrus-sac large, extending anterosinistrally from anterior testis. Genital pore sinistral, dorsal, opening medially to left caecum. Ovary oval sinistral and lateral to anterior testis. Saccate seminal receptacle present; Laurer's canal present, opens dorsally. Uterus short, contains single oval egg with bipolar filaments. Vitellarium follicular, extensive, intracaecal and extracaecal, from oesophagus to posterior end. Excretory vesicle Y-shaped; stem consists of several chambers; excretory pore terminal. Parasites of the circulatory system of freshwater turtles (Bataguridae); India. Type-species *C. vaidya* Dwivedi, 1967.

9b. Cirrus-sac present or absent; vitellarium follicular, varying in extent from mid-oesophagus to caecal extremities .... *Hapalorhynchus* Stunkard, 1922. (Fig. 52.9)

(Syns *Coeuritrema* Mehra, 1933; *Rhynchotrema* Thapar, 1933; *Tremarhynchus* Thapar, 1933)

Diagnosis: Body delicate, lanceolate; aspinose. Oral sucker well developed, terminal. Ventral sucker well developed. Oesophagus simple or with diverticula; caeca terminating near posterior end. Testes two, oval to lobed; in middle third of body. Seminal vesicle large, posterior to ventral sucker. Pars prostatica large, surrounding ejaculatory duct in parenchyma. Cirrus-sac present or absent. Genital pore sinistral, dorsal; opening laterally to left caecum near mid-body. Ovary oval to pyriform; sinistral and intertesticular. Oviducal seminal receptacle present; Laurer's canal present, opens dorsally. Uterus short; contains single egg; eggs oval or tricornuate. Vitellarium follicular, from caecal bifurcation to caecal extremities. Excretory vesicle Y-shaped; excretory pore terminal. Parasites of the circulatory system of freshwater turtles (Pelomedusidae, Chelydridae, Bataguridae); cosmopolitan. Type-species *H. gracilis* Stunkard, 1922.

10a. Genital pore anterior to anterior testis .... *Neocaballerotrema* Simha, 1977. (Fig. 52.10)

Diagnosis: Body elongate, spinose. Oral sucker terminal, oval. Ventral sucker membranous. Oesophagus short; caeca extend past mid-body, but termination not seen. Testes two, oval; anterior testis just posterior to ventral sucker. Seminal vesicle large, posterior to cirrus-sac. Cirrus-sac well developed, transverse, between anterior testis and seminal vesicle. Genital pore submedian, dextral, opening anterior to anterior testis. Ovary slightly lobed, intertesticular; saccate seminal receptacle at posterior edge of ovary. Vitellarium consists of large follicles extending from midway between caecal bifurcation and anterior testis to near posterior extremity. Excretory vesicle Y-shaped. Laurer's canal? Eggs? Parasite of circulatory system of marine turtles (Cheloniidae); India (cosmopolitan?). Type-species *N. caballeroi* Simha, 1977.

10b. Genital pore intertesticular or posterior to posterior testis .... 11.

11a. Genital pore posterior to posterior testis .... *Cheloneotrema*[2] Simha & Chattopadhyaya, 1980. (Fig. 52.11)

Diagnosis: Body elongate, with minute spines. Oral sucker terminal, oval. Ventral sucker large, muscular. Oesophagus slightly sinuous; caeca simple, terminate near posterior end of body. Testes two, lobed, in posterior half of body. Cirrus-sac intertesticular. Seminal vesicle small, anterior to cirrus-sac. Cirrus-sac elongate, well developed, passes sinistrally to ovary. Genital pore ventral, submedian; at posterior margin of posterior testis. Ovary slightly lobed, intertesticular in posterior quarter of body; saccate seminal receptacle at posterior edge of ovary. Vitellarium consists of large follicles extending from midway between caecal bifurcation and anterior testis to near posterior end. Excretory vesicle Y-shaped. Laurer's canal? Eggs? Parasite of

[2] Mis-spelled as *Chelonotrema* and *Chaloneotrema* in original article.

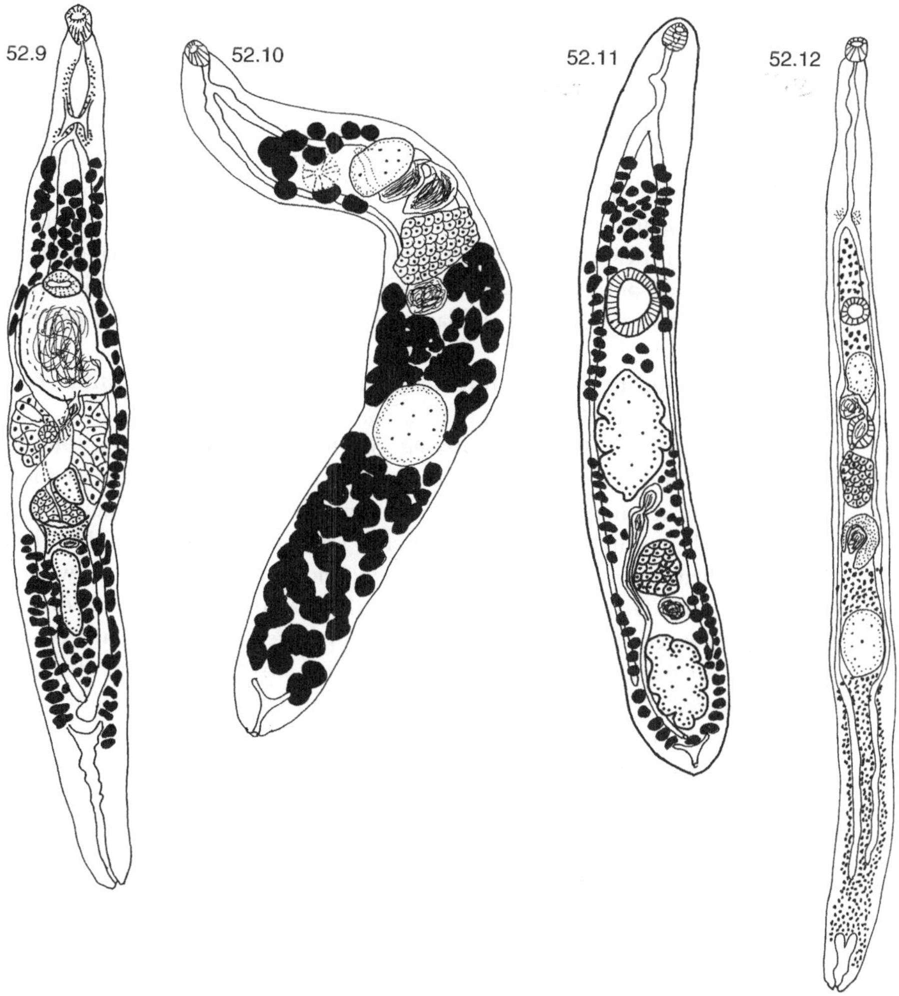

Figs 52.9–52.12.
52.9. *Hapalorhynchus gracilis* Stunkard, 1922. Holotype; dorsal. (After Platt, 1988.)
52.10. *Neocaballerotrema caballeroi* Simha, 1977. Ventral. (After Simha, 1977.)
52.11. *Cheloneotrema testicaudata* Simha & Chattopadhyaya, 1980. Ventral. (After Simha & Chattopadhyaya, 1980.)
52.12. *Amphiorchis amphiorchis* Price, 1934. Holotype; ventral. (Original.)

circulatory system of marine turtles (Cheloniidae); India (cosmopolitan?). Type-species *C. testicaudata* Simha & Chattopadhyaya, 1980.

11b. Genital pore intertesticular .... 12.

12a. Body slender; vitelline follicles small, extending from intestinal bifurcation to caecal extremities .... *Amphiorchis* Price, 1934. (Fig. 52.12)
(Syn. *Squaroacetabulum* Simha & Chattopadhyaya, 1970)

Diagnosis: Body elongate, slender; minute body-spines present or absent. Oral sucker small, oval. Ventral sucker equal to or larger than oral sucker, oval to rectangular in shape. Oesophagus slightly sinuous; caeca simple, terminate near posterior end of body. Testes two, elliptical, lobed or N-shaped, in middle third of body. Seminal vesicle small, anterior to cirrus-sac. Cirrus-sac well developed, between anterior testis and ovary, directed posteriorly. Genital pore ventral, submedian, intertesticular, near mid-body. Ovary slightly lobed, intertesticular; saccate seminal receptacle at posterior edge of ovary; uterus short, containing single egg. Vitellarium follicular, extending from caecal bifurcation to caecal extremities. Excretory vesicle Y-shaped. Laurer's canal present or not observed. Eggs oval with bipolar filaments. Parasite of circulatory system of marine turtle (Cheloniidae); cosmopolitan. Type-species *A. amphiorchis* Price, 1934.

12b. Body robust; vitelline follicles large, extending from seminal receptacle to posterior end of body .... ***Satyanarayanotrema*[3] Simha & Chattopadhyaya, 1980. (Fig. 52.13)**

Diagnosis: Body delicate, aspinose. Oral sucker small, oval. Ventral sucker smaller than oral sucker, immediately anterior to anterior testis. Oesophagus slightly sinuous; caeca simple, terminate near posterior end. Testes two, lobed, in middle third of body. External seminal vesicle absent. Cirrus-sac elongate, transverse, between ovary and anterior testis. Genital pore ventral, sinistral, opening medial to left caecum between ovary and anterior testis. Ovary large, lobed, intertesticular; seminal receptacle saccate, at posterior margin of ovary. Laurer's canal not observed. Uterine structure and eggs not observed. Vitelline follicles extending from seminal receptacle to posterior end, filling most intercaecal space. Excretory vesicle Y-shaped. Parasite of circulatory system of marine turtles (Cheloniidae); India (cosmopolitan?). Type-species *S. satyanarayani* Simha & Chattopadhyaya, 1980 (mis-spelled *S. satyanarayoni* in original article).

13a. Multiple testes anterior and posterior to ovary and terminal genitalia .... ***Hapalotrema* Looss, 1899. (Fig. 52.14)**

Diagnosis: Medium to large worms, robust to slender, with spines anteriorly. Oral sucker well developed, terminal. Ventral sucker located posterior to caecal bifurcation, pedunculate, membranous, with marginal spines. Oesophagus and caeca simple; caeca terminating near posterior end of body. Testes numerous, divided into pre- and postovarian groups. External seminal vesicle present, lateral or anterior to cirrus-sac. Cirrus-sac well developed, containing small internal seminal vesicle, pars prostatica and eversible cirrus, sinistral to ovary. Genital pore ventral, submedian to lateral, sinistral, opening from near mid-body to posterior quarter of body. Genital sucker present or absent; limiting boundary present or absent; genital spines present or absent. Ovary lobed, intertesticular. Oviducal seminal vesicle and Laurer's canal present. Laurer's canal opening dorsally with or without raised muscular ring. Uterus short, contains single egg with bipolar projections. Vitelline follicles circumcaecal, from posterior margin of ventral sucker to caecal extremities, typically intercaecal from ventral sucker to anterior testis. Excretory vesicle Y-shaped. Parasites of circulatory system of marine turtles (Cheloniidae); cosmopolitan. Type-species *H. loossi* Price, 1934.

---

[3] Also mis-spelled as *Satyanorayanotrema* in original article.

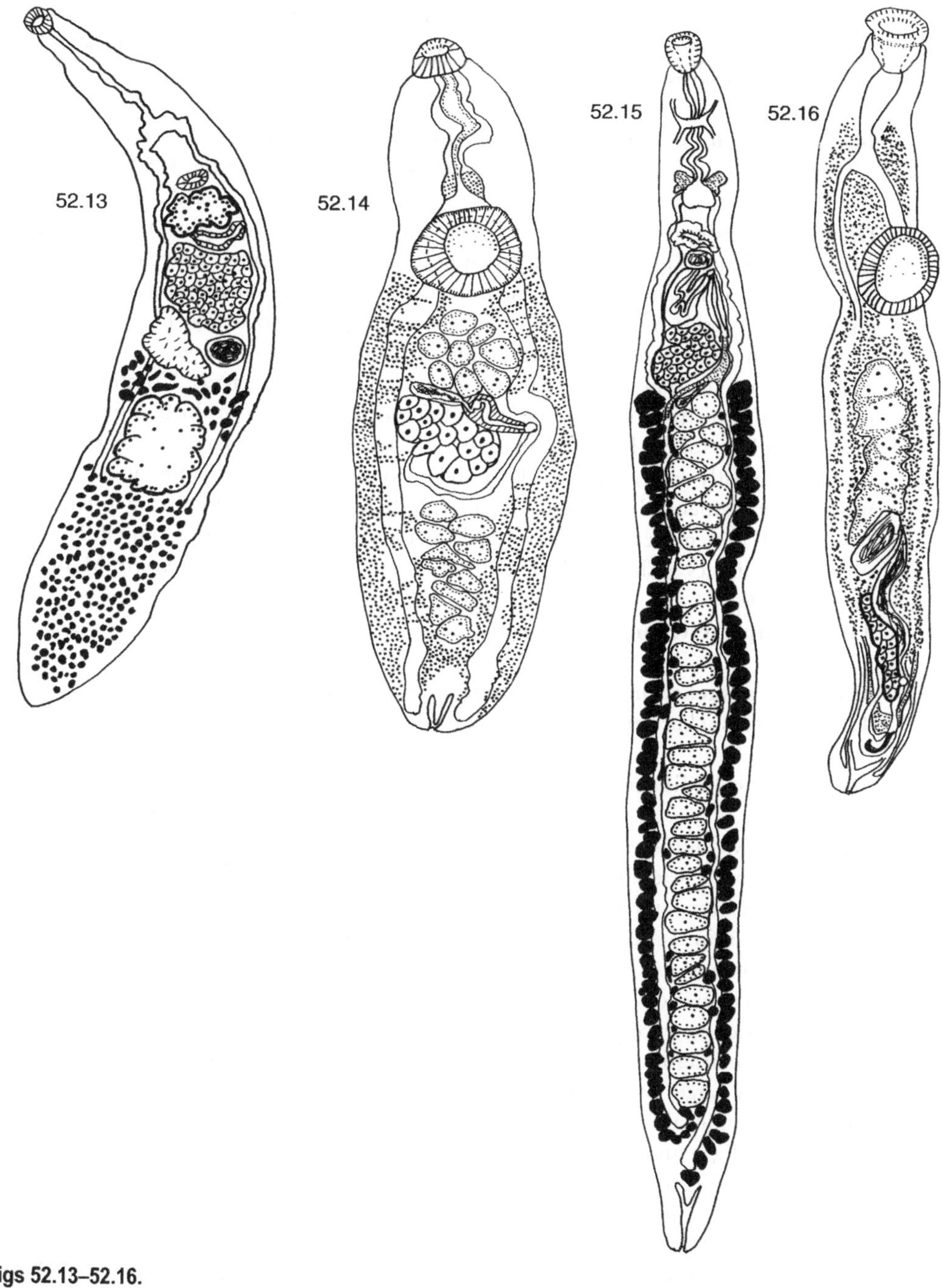

Figs 52.13–52.16.

52.13. *Satyanarayanotrema satyanarayani* Simha & Chattopadhyaya, 1980. Ventral. (After Simha & Chattopadhyaya, 1980.)

52.14. *Hapalotrema mistroides* (Monticelli, 1896). Lectotype; ventral. (After Platt & Blair, 1998.)

52.15. *Carettacola hawaiiensis* Dailey, Fast & Balazs, 1991. Ventral. (Original.)

52.16. *Monticellius similis* (Price, 1934). Holotype; ventral. (Original.)

13b. Multiple testes anterior or posterior to ovary and terminal genitalia; exceptionally multiple testes anterior to ovary and terminal genitalia and one testis posterior .... 14.

14a. Multiple testes posterior to ovary and terminal genitalia .... *Carettacola* Manter & Larson, 1950. (Fig. 52.15)
(Syn. *Haemoxenicon* Martin & Bamberger, 1952.)

Diagnosis: Medium-sized worm, delicate; anterior body-spines present. Oral sucker terminal, well developed. Oesophagus and caeca simple; caeca terminate near posterior end. Ventral sucker immediately postbifurcal, pedunculate, membranous, with marginal spines. Testes many, linear or nearly so, posterior to ovary. External seminal vesicle saccate, between ventral sucker and cirrus-sac. Cirrus-sac large, ovoid, contains small internal seminal vesicle, pars prostatica and eversible cirrus. Common genital pore ventral, sinistral, medial or lateral to left caecum, in anterior third of body. Ovary oval to slightly lobed, between cirrus-sac and testicular field. Laurer's canal present, slightly to greatly enlarged, opens laterally or dorsally. Uterus contains single egg with monopolar filament. Vitellarium follicular; follicles large, extending from anterior testis to caecal extremities. Excretory vesicle Y-shaped; excretory pore terminal. Parasites of marine turtles (Cheloniidae); cosmopolitan. Type-species *C. bipora* Manter & Larson, 1950.

14b. Multiple testes anterior to ovary .... 15.

15a. Ventral sucker present, pedunculate and always spined .... 16.
15b. Ventral sucker present or absent; if present, sessile and unspined .... 17.

16a. Testes < 10, linear; oesophagus < 20% of total body length .... *Monticellius* Mehra, 1939. (Fig. 52.16)

Diagnosis: Medium-sized worm, delicate; anterior body spines present. Oesophagus and caeca simple; caeca terminate near posterior end of body. Ventral sucker between caecal bifurcation and anterior testis, large, pedunculate, membranous, with small spines on margin. Testes multiple, linear or irregular; in middle third of body between ventral sucker and seminal vesicle. External seminal vesicle large, transversely elongate; anterior to cirrus-sac. Cirrus-sac muscular, contains small internal seminal receptacle, pars prostatica and long ejaculatory duct. Ejaculatory duct passes mediodorsally to ovary, terminating in common genital pore opening on ventral surface, posterior to ovary. Ovary large, oval to lobed, in posterior quarter of body. Seminal receptacle, saccate or oviducal; Laurer's canal present, opens dorsally. Uterus short, contains single egg. Egg fusiform with unequal bipolar processes. Vitellarium follicular, extending from caecal bifurcation to caecal extremities, filling intercaecal space where other organs absent. Excretory vesicle Y-shaped; excretory pore terminal. Parasites of marine turtles (Cheloniidae); cosmopolitan. Type-species *M. indicum* Mehra, 1939.

16b. Testis number > 15, irregularly arranged; oesophagus × 40% total body length .... *Learedius* Price, 1934. (Fig. 52.17)

Diagnosis: Medium-sized worms, delicate; anterior body spines present. Oral sucker large, terminal. Oesophagus simple, long; caeca loop anteriorly before turning posteriorly, pass medially near ventral sucker and terminate near posterior end of body. Ventral sucker immediately posterior to caecal bifurcation, large, pedunculate, membranous, with small spines on margin. Testes small, numerous, irregularly distributed in posterior half of body between ventral sucker and seminal vesicle. External seminal vesicle large, transversely elongate, lateral to cirrus-sac. Cirrus-sac weakly developed, contains small internal seminal vesicle, pars prostatica and long ejaculatory duct. Ejaculatory duct passes mediodorsally to ovary, terminating in common genital pore opening on ventral surface posterior to ovary and near posterior end of body. Ovary large, deeply lobed, in posterior quarter of body. Seminal receptacle saccate, blind; Laurer's canal opens dorsally; uterus short, contains single egg. Egg fusiform with unequal bipolar processes. Vitellarium follicular, extends from caecal bifurcation to caecal extremities, fills intercaecal space where other organs absent. Excretory vesicle Y-shaped; excretory pore terminal. Parasites of marine turtles (Cheloniidae); cosmopolitan. Type-species *L. learedi* Price, 1934.

17a. Ventral sucker absent in adults .... *Spirorchis* MacCallum, 1919. (Fig. 52.18)
(Syns *Aphanospirorchis* Platt, 1990; *Diarmostorchis* Ejsmont, 1927; *Haematotrema* Stunkard, 1923; *Henotosoma* Stunkard, 1922; *Proparorchis* Ward, 1921)
Diagnosis: Small to large worms, delicate to robust, aspinose. Oral sucker reduced; ventral sucker absent. Oesophagus simple to sinuous, surrounded by gland-cells; plicate organ and median oesophageal pouch present at junction of oesophagus and intestinal caeca. Caeca simple to sinuous, terminate anterior to testicular field or posterior to terminal genitalia near posterior end of body. Testes four to 11, linear, intercaecal. External seminal vesicle typically large, between posterior testis and cirrus-sac. Cirrus-sac muscular, directed posteriorly, terminates in common genital pore opening medial to left caecum in posterior quarter of body. Ovary oval to lobed, dextral to cirrus-sac, in posterior quarter of body. Oviducal seminal receptacle present; Laurer's canal opens dorsally. Uterus short containing single, oval egg. Vitellarium follicular, typically occupying extra- and intercaecal space from caecal bifurcation to caecal ends. Excretory vesicle Y-shaped; Manter's organ (accessory excretory vesicle) present. Excretory pore terminal. Parasites of freshwater turtles (Chelydridae; Emydidae); North America. Type-species *S. innominatus* Ward, 1921.
17b. Ventral sucker present in adults .... 18.

18a. Intestinal caeca T-shaped at bifurcation; one testis posterior to ovary .... *Spirhapalum* Ejsmont, 1927. (Fig. 52.19)
Diagnosis: Small, delicate, aspinose. Oral sucker small; oesophagus simple to sinuous, surrounded by gland-cells; plicate organ and median oesophageal pouch present at junction of oesophagus and intestinal caeca. Ventral sucker present, sessile, in middle third of body. Caeca simple, terminating posterior to terminal genitalia near posterior end. Pre-ovarian testes four to six, irregular in outline, linear or irregularly spaced, intercaecal; single postovarian testis present. External seminal vesicle typically large, between posterior testis and cirrus-sac. Cirrus-sac small, muscular, directed posteriorly, terminates in common genital pore opening medially to left caecum in posterior quarter of body. Ovary lobed, dextral to cirrus-sac. Oviducal seminal receptacle present; Laurer's canal opens dorsally. Uterus short containing single, oval egg. Vitellarium follicular, typically circumcaecal from caecal bifurcation to caecal ends. Excretory vesicle Y-shaped; Manter's organ present. Excretory pore terminal. Parasites of freshwater turtles

(Bataguridae; Emydidae); Europe, Asia, South-East Asia. Type-species *S. polesianum* Ejsmont, 1927.

18b. Intestinal caeca loop anteriorly near caecal bifurcation; all testes anterior to ovary .... ***Plasmiorchis* Mehra, 1934. (Fig. 52.20)**

(Syns *Gomtiotrema* Sinha, 1934; *Hemiorchis* Mehra, 1939)

Diagnosis: Small to large worms, delicate to robust, aspinose. Oral sucker terminal, muscular; oesophagus simple to sinuous, surrounded by gland-cells; plicate organ and median oesophageal pouch present at junction of oesophagus and intestinal caeca. Ventral sucker present, sessile, in anterior third of body. Caeca loop anteriorly before

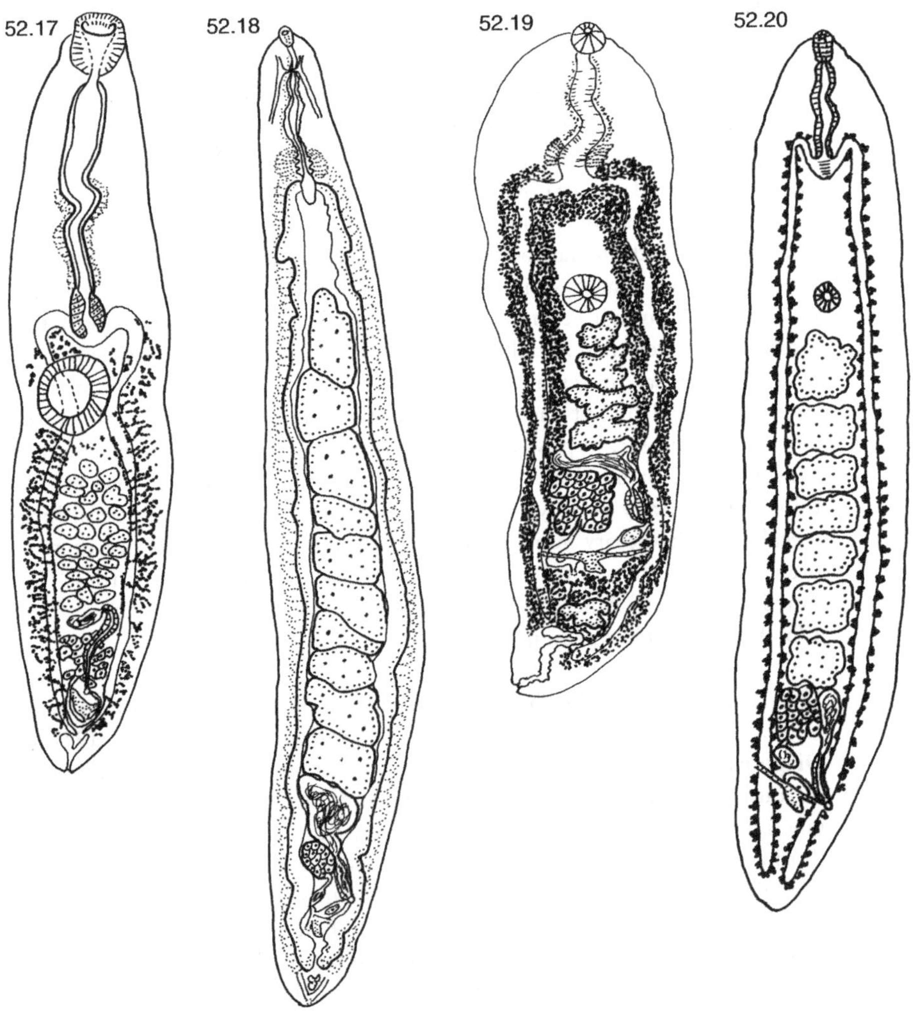

Figs 52.17–52.20.
52.17. *Learedius learedi* Price, 1934. Holotype; ventral. (Original.)
52.18. *Spirorchis innominata* Ward, 1921. Holotype; ventral. (After Platt, 1993.)
52.19. *Spirhapalum polesianum* Ejsmont, 1927. Ventral. (After Ejsmont, 1927.)
52.20. *Plasmiorchis orientalis* Mehra, 1934. Ventral. (After Mehra, 1934.)

turning posteriorly, simple to sinuous, terminate posterior to terminal genitalia near posterior end of body. Testes six to 20+, linear, intercaecal. External seminal vesicle typically large, between posterior testis and cirrus-sac, sinistral to ovary. Cirrus-sac muscular, directed posteriorly, terminates in common genital pore opening near left caecum near posterior end of body. Ovary oval to lobed; dextral to cirrus-sac; oviducal seminal receptacle present. Laurer's canal opens dorsally. Uterus short containing single, oval egg. Vitellarium follicular, circumcaecal or occupying extra- and intercaecal space from caecal bifurcation to caecal extremities. Excretory vesicle Y-shaped; Manter's organ present. Excretory pore terminal. Parasites of freshwater turtles (Bataguridae); India. Type-species *P. orientalis* Mehra, 1934.

53 Superfamily Transversotrematoidea Witenberg, 1944

T.H. Cribb

Department of Microbiology and Parasitology, The University of Queensland, Brisbane, Australia

Introduction

This family group was established by Witenberg (1944) as the subfamily Transversotrematinae (spelled 'Transversotreminae') but not associated with any particular family. It was raised to family status by Yamaguti (1953). The family is one of the most distinctive and also one of the smallest families of digenetic trematodes, having only about ten species. It is distinctive in the transversely elongate shape of the adults, in the ectoparasitism of the adults, which are found beneath the scales of marine and freshwater fishes, and in the cercariae, which (where known) resemble almost completely the developed adult forms and attach directly to the fish, so that the life-cycle involves only two hosts. All confirmed records of transversotrematids are from marine fishes of the Indo-Pacific and from freshwater fishes on land masses bordering these oceans. The family shows no particular association with any other family of trematodes on the basis of morphology or life-cycle and recent molecular studies also suggest an isolated position for the family. It seems appropriate therefore to formally recognize a separate superfamily for this family.

The only real uncertainty in the classification of this superfamily relates to two enigmatic species from freshwater fishes in China. Wang (1981) erected a new family, the Circuitiocoeliidae, for *Circuitiocoelium opsariichthydis* Wang, 1981 from the intestine of *Opsariichthys uncirostris bidens* (Cyprinidae) and later (Wang *et al.*, 1985) from the bagrid *Pseudobagrus fulvidraco*. Pan & Wang (1985) erected the Squamacolidae for *Squamacola parvivitellaria* Pan & Wang, 1985 from under the scales of two species of cyprinids. The description by Wang (1981) was perfunctory but mentioned similarity to the Transversotrematidae. The description by Pan & Wang (1985) was quite detailed and also drew attention to the similarity of the new genus to the Transversotrematidae, but considered that characters such as the pyriform body shape and reduced vitellarium justified the erection of a new family. Pan & Wang (1985) did not refer to *Circuitiocoelium* in their discussion. I have been unable to examine specimens of these two species. Comparison of the two descriptions leads to the conclusion that the two species are synonymous and that the description by Wang interpreted the smallest eggs as vitelline follicles and failed to observe the genital pore in the forebody; it seems highly probable as well that the reported site of infection (the intestine) was made in error. This had been done previously, before the nature of the trematodes was known (e.g. Velasquez, 1958). (In this context, it must be noted, however, that bagrid catfishes are scale-less, which makes the records of *C. opsariichthydis* most surprising and enigmatic.) It is concluded here that only one genus and one species is present. The genus and species combination *Circuitiocoelium opsariichthydis* has seniority, but the description by Pan & Wang (1985) is the more detailed and convincing. The description of another new form, *Crusziella*, by Cribb *et al.* (1992) suggests that the characters exhibited by *Circuitiocoelium*, although striking, can be incorporated within the concept of the family Transversotrematidae. Both the Circuitiocoeliidae and the Squamacolidae are thus here considered synonyms of the Transversotrematidae and four genera are recognized.

The diagnosis of the superfamily given below includes considerable morphological variation for many important organs. For instance, a sinus-sac may be present or absent, the vitellarium may be profuse or reduced, the eggs may or may not embryonate *in utero* and an oral sucker may apparently be present or absent. Despite this, the family appears to be a natural group. Few characters are shared by all genera (spination, eye-spots, cyclocoel, form of excretory vesicle). Perhaps the strongest uniting character for the family will prove to be the nature of the cercaria, which has unique arm processes at its base for those that are known (species of *Transversotrema* Witenberg, 1944 and *Prototransversotrema* Angel, 1969) (see Soparkar, 1924; Crusz *et al.*, 1964; Cribb, 1988a). The use of the sub-scale niche by adult transversotrematids also appears to unite the group.

Superfamily Transversotrematoidea Witenberg, 1944

Diagnosis: Body flattened, transversely elongate or pyriform. Tegument spined throughout. Pair of eye-spots present. True oral sucker present or absent; oral flange present or absent. Ventral sucker median. Prepharynx distinct or very short and inconspicuous. Pharynx present. Oesophagus distinct. Cyclocoel present. Testes symmetrical, deeply lobed. Seminal vesicle tubular or bipartite; if bipartite; two sections separated by narrow duct passing ventral to cyclocoel. Cirrus-sac and pars prostatica absent. Male and female ducts join without specialization at genital pore or at base of distinct sinus-sac. Temporary sinus-organ may form. Genital pore on anterior margin at mid-line or, if mouth present on anterior margin, then postero-dextral to mid-line. Ovary sinistral to left testis or between and anterior to testes, lobed or almost entire. Laurer's canal present, partly dilated by sperm and vitelline remains. Uterine seminal receptacle present. Uterus of varying length, may loop posteriorly to testes. Eggs large; capsules either thick-walled and tanned containing unembryonated zygote or thin-walled, clear and containing embryonating miracidium. Vitellarium consisting of widely distributed follicles, of two small symmetrical fields of follicles inside lateral extremities of cyclocoel, or of two single masses inside lateral extremities of cyclocoel. Excretory pore at small notch in middle of posterior margin; excretory vesicle I-shaped, narrow distally, broad proximally. Life-cycle known for two genera only. Cercariae develop in rediae in prosobranch gastropods, furcocercous, bearing distinctive arm processes at their bases; body in form of nearly fully-developed adult that attaches directly to skin of fish and matures to adult beneath scales without intervening metacercarial form. Adults known from beneath scales of wide variety of families of marine and freshwater fish in Africa, India, Asia, Australia, and Indian and Pacific Oceans. Type-genus *Transversotrema* Witenberg, 1944.

Family Transversotrematidae Witenberg, 1944

(Syns Transversotrematinae Witenberg, 1944; Circuitiocoeliidae Wang, 1981; Squamacolidae Pan & Wang, 1985)

Diagnosis: With characters of superfamily. Type-genus *Transversotrema* Witenberg, 1944.

Key to genera

1a. Vitellarium reduced, restricted within cyclocoel; eggs embryonate *in utero* and contain prominent miracidia with eye-spots .... 2.

1b. Vitellarium follicular, profuse outside cyclocoel; eggs not embryonating *in utero* .... 3.

2a. Body longer than wide; mouth opening on anterior margin of body through oral sucker .... *Circuitiocoelium* Wang, 1981. (Fig. 53.1)

(Syn. *Squamacola* Pan & Wang, 1985)

Diagnosis: Body pyriform. Eye-spots huge, crescent-shaped. Oral sucker present. Seminal vesicle tubular. Male and female ducts join without specialization. Genital pore posterodextral to oral sucker. Ovary between and immediately anterior to testes. Uterus comprising multiple loops anterior and lateral to testes. Egg capsules thin-walled, grow dramatically *in utero*, ultimately contain fully embryonated eye-spotted miracidium. Vitellarium consisting of two single masses inside lateral extremities of cyclocoel. Under scales and (?) in intestine of Asian freshwater fishes. Type-species *C. opsariichthydis* Wang, 1981.

2b. Body much wider than long; mouth opening midventrally .... *Crusziella* Cribb, Bray & Barker, 1992. (Fig. 53.2)

Diagnosis: Body transversely elongate, flattened. Oral sucker absent; mouth opening midventrally, ventral to pharynx. Ventral sucker median, weakly muscular. Seminal vesicle bipartite; parts separated by narrow duct passing ventrally to cyclocoel. Male and female ducts join at base of distinct sinus-sac. Temporary sinus-organ may form. Genital pore on anterior margin at mid-line. Ovary sinistral to left testis. Laurer's canal present. Uterus long, coiled, loops posterior to both testes. Egg capsules thin-walled, grow dramatically *in utero*, ultimately contain fully-embryonated eye-spotted miracidium. Vitellarium two symmetrical small fields of follicles inside lateral extremities of cyclocoel. Under scales of mugilid fishes. Type-species *C. formosa* Cribb, Bray & Barker, 1992.

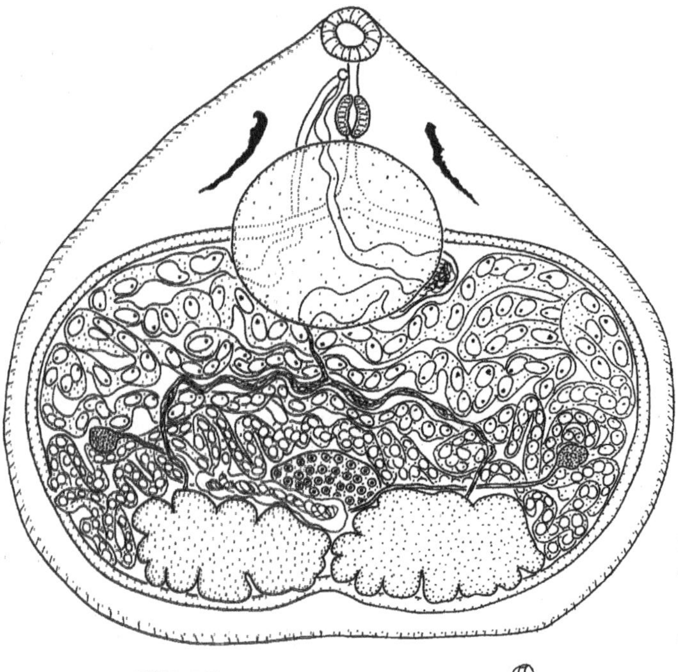

Fig. 53.1. *Circuitiocoelium opsariichthydis* Wang, 1981. (After Pan & Wang (1985), as *Squamacola parvivitellaria* Pan & Wang, 1985.)

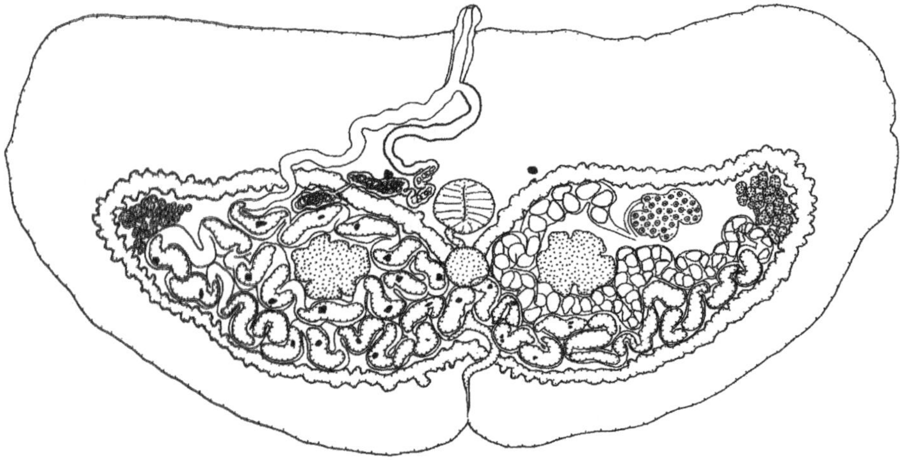

Fig. 53.2. *Crusziella formosa* Cribb, Bray & Barker, 1992. (Original.)

3a. Mouth on anterior margin of body, surrounded by oral flange; seminal vesicle tubular .... ***Prototransversotrema* Angel, 1969. (Fig. 53.3)**

Diagnosis: Body transversely elongate or almost square. Pigment extends from eye-spots in fine network. Mouth opening on anterior margin of body, surrounded by oral flange. Male and female ducts join without specialization at genital pore. Genital pore posterodextral to oral flange. Ovary sinistral to left testis. Uterus with one or two loops. Egg capsules large, thick-walled, tanned, containing unembryonated zygote. Vitellarium consisting of widely distributed follicles. Under scales of many families of marine and freshwater teleosts. Type-species *P. steeri* Angel, 1969.

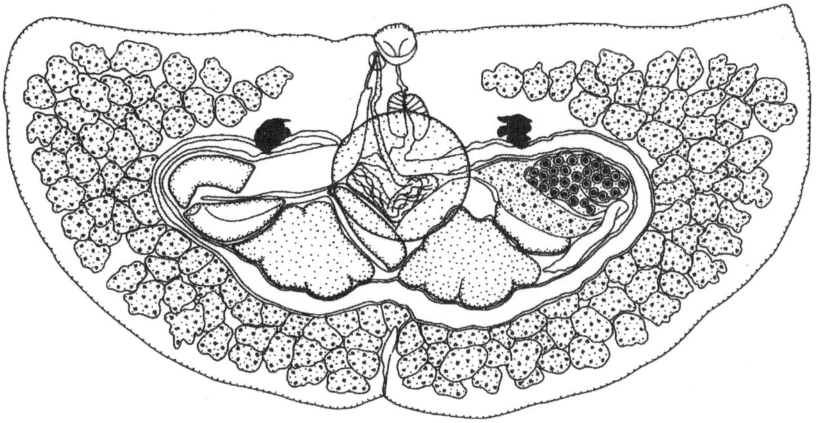

Fig. 53.3. *Prototransversotrema steeri* Angel, 1969. (Original.)

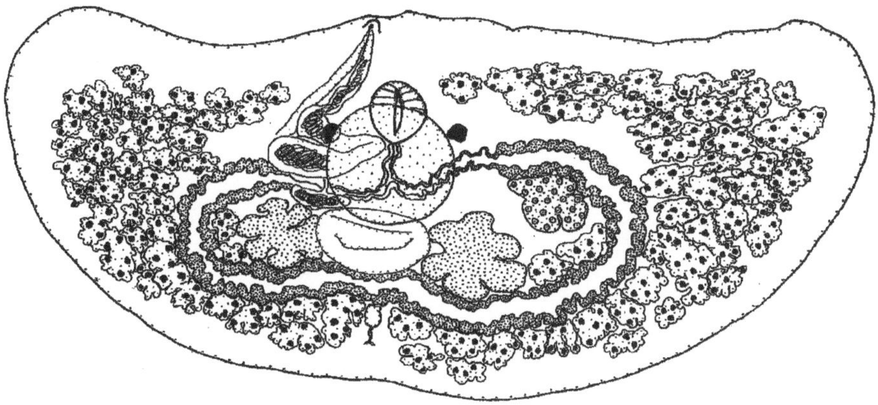

Fig. 53.4. *Transversotrema patialense* (Soparkar, 1924). (Original.)

3b. Mouth midventral; seminal vesicle bipartite .... *Transversotrema* **Witenberg, 1944. (Fig. 53.4)**

Diagnosis: Body transversely elongate. Eye-spots discrete. Mouth opening midventrally without any specialization. Seminal vesicle bipartite, with two parts separated by narrow duct passing ventral to cyclocoel. Male and female ducts join without specialization at genital pore. Genital pore on anterior margin at mid-line. Ovary sinistral to left testis. Uterus with only one loop. Egg capsules large, thick-walled, tanned, containing unembryonated zygote. Vitellarium consisting of widely distributed follicles. Under scales of many families of marine and freshwater teleosts. Type-species *T. haasi* Witenberg, 1944.

Bibliography

Abdul-Salam, J. & Khalil, L.F. (1987) Two digeneans from the needlefish *Ablennes hians* in Kuwait and the description of a new genus and species, *Neohaplosplanchnus ablennis* (Haploplanchnidae). *Systematic Parasitology* 10, 149–158.

Agarwal, G.P. & Agrawal, L.N. (1982) *Bundelatrema orchhaensis* n. sp. (Trematoda: Azygiidae) from the intestine of a fresh water fish *Puntius sarana* (Ham.). In: *Proceedings of the Indian Science Congress*, 69th Session, Part 3, VII, Calcutta, p. 110.

Agrawal, S.C. & Sharma, S.K. (1990) On Paraaspidogasterinae sub.f.nov. *Pseudoaspidogaster betwai* gen. nov., sp. nov. (Trematoda: Aspidogasteridae Poche, 1907) from a fish *Tor tor* (Ham). *Indian Journal of Helminthology* 42, 17–20.

Ahmad, J. (1982) Studies on digenetic trematodes of the families Fellodistomatidae and Gorgoderidae from marine fishes of India. *Rivista di Parassitologia* 43, 301–315.

Allison, L.N. (1943) *Leucochloridiomorpha constantiae* (Mueller) (Brachylaemidae), its life cycle and taxonomic relationships among digenetic trematodes. *Transactions of the American Microscopical Society* 67, 127–168.

Amato, J.F.R. (1968) Contribuição conhecimento da fauna helmintólogia do Rio Grande do Sul. Um nôvo parasito de peixe cascudo "*Plecostomus commersoni*" (Cuv. & Val.) (Trematoda, Hemiuroidea). *Revista Brasileiria de Biologia* 28, 433–438.

Anon. (1993) *Bucephalus* Baer, 1827 and *B. polymorphus* Baer, 1827 (Trematoda): conserved in their accepted usage. *Bulletin of Zoological Nomenclature* 50, 152–154.

Anon. (1997) *Hapalotrema* Looss, 1899 (Digenea): *Hapalotrema loossi* Price, 1934 designated as the type species. *Bulletin of Zoological Nomenclature* 54, 190–191.

Ataev, A.M. (1970) [Description of *Monovitella cyclointestina* gen. et sp. n. and some remarks on the taxonomy of the family Halipegidae.] *Zoologicheskii Zhurnal* 49, 1465–1470. (In Russian.)

Azimov, D.A. (1970) [Revision of the system of trematodes Schistosomatata Skrjabin & Shultz, 1937.] *Zoologicheskii Zhurnal* 49, 1126–1131. (In Russian.)

Azimov, D.A. (1975) [*Schistosomatidae of animals and man (systematics).*] Tashkent: Izdatel'stvo FAN Uzbekskoi SSR, 152 pp. (In Russian.)

Baer, J.G. (1923) Résultats zoologiques du voyage du Dr P.A. Chappuis au Nil supérieur. III. Helminthes. *Revue Suisse de Zoologie* 30, 337–352.

Baer, J.G. (1930) Deux helminthes nouveaux parasites de *Uracotyphlus oxyurus* (Gray), gymnophione de l'Inde méridional (J. Carl et K. Escher. Voyage de recherches zoologiques dans l'Inde méridionale (hiver 1926–27)). *Revue Suisse de Zoologie* 37, 43–52.

Baer, J.G. (1933) Note sur un nouveau trematode *Clinostomum lophophallum* sp. nov. avec quelques considérations générales sur la famille des Clinostomidae. *Revue Suisse de Zoologie* 39, 317–342.

Baer, J.G. (1959) Evolution of host-parasite relationships. *Nature* 184, 760–763.

Baer, J.G. & Joyeux, C. (1961) Classe de Trématodes. In: Grassé, P.-P. (ed.) *Traité de Zoologie*, Vol. IV. Masson, Paris, pp. 561–692.

Bakke, T.A. (1980) A revision of the family Leucochloridiidae Poche (Digenea) and studies on the morphology of *Leucochloridium paradoxum* Carus, 1835. *Systematic Parasitology* 1, 189–202.

Bartoli, P. (1965) Développement expérimental d'une métacercaire nouvelle de la famille des Gymnophallidae Morozov, 1955 (Trematoda: Digenea), parasite de lamellibranches marins. *Bulletin de la Société Zoologique de France* 40, 571–580.

Bartoli, P. (1972) Les cycles biologique de *Gymnophallus nereicola* J. Rebecq et G. Prévot, 1962 et *G. fossarum* P. Bartoli, 1965, espèces jumelles parasites d'oiseaux de rivages marins (Trematoda, Digenea, Gymnophallidae). *Annales de Parasitologie Humaine et Comparée* 47, 193–223.

Bartoli, P. (1974) Recherches sur les Gymnophallidae F.N. Morozov, 1955 (Digenea), parasites d'oiseaux des Côtes de Camargue: systématique, biologie et ecologie. Thèse de Docteur Ès-Sciences, Université de Droit, d'Economie et des Sciences d'Aix-Marseille, 338 pp.

Bartoli, P. (1983) *Gymnophallus rebecqui* n. sp. (syn. *Parvatrema* sp. 1, J. Rebecq, 1964) (Digenea: Gymnophallidae) parasite intestinal d'Anatidés de Camargue (France). *Annales de Parasitologie Humaine et Comparée* 58, 211–225.

Bashkirova, E.I. (1950) [Family Cyclocoelidae Kossack, 1911.] *Osnovy Trematodologii* 4, 329–493. (In Russian.)

Baturo, B. (1992) Case 2251 *Bucephalus* Baer, 1827 and *B. polymorphus* Baer, 1827 (Trematoda): proposed conservation in their accepted usage. *Bulletin of Zoological Nomenclature* 49, 6–11.

Baugh, S.C. (1977) On the systematic position of the Thailand blood fluke *Orientobilharzia harinasutai* Kruatrachue, Bhaibulaya and Harinasuta, 1965. In: *Excerta Parasitologica en Memoria del Doctor Eduardo Caballero y Caballero*. Universidad Nacional Autónoma de México, Mexico City, pp. 121–125.

Bhalerao, G.D. (1935) On two new monostomes from avian host in British India. *Indian Journal of Veterinary Sciences and Animal Husbandry* 5, 49–63.

Bilharz, T. (1856) *Distomum haematobium*, und sein Verhältniss zu gewissen pathologischen Veränderungen der menschlichen Harnorgane. *Wiener Medizinische Wochenschrift* 6(4), 46–52.

Bilqees, F.M. (1971) Marine fish trematodes of West Pakistan. Part IV. Description of three new genera and species. *Pakistan Journal of Scientific and Industrial Research* 14, 254–257.

Bilqees, F.M. (1976) A new trematode genus *Pseudohypertrema* (Fellodistomidae Nicoll, 1909) from fishes of the Karachi coast. *Norwegian Journal of Zoology* 24, 201–203.

Biocca, E. & Ferretti, G. (1958) Un nuovo trematode: *Dollfusinus frontalis* gen. nov., et sp. nov., parasite dei seni naso frontali di *Erinaceus europeus*. *Atti del'Accademia Nazionale dei Lincei, Rendiconti delle Sedute Solenni* 24, 171–175.

Bittner, H. & Sprehn, C.E.W. (1928) Trematodes. Sangwurmer. *Biologie der Tiere Deutschlands. Herausgegeben von Dr. Paul Schulze*, No. 27(5), 1–133.

Blair, D., Bray, R.A. & Barker, S.C. (1998) Molecules and morphology in phylogenetic studies of the Hemiuroidea (Digenea: Trematoda: Platyhelminthes). *Molecular Phylogenetics and Evolution* 9, 15–25.

Blanchard, E. (1847) Recherches sur l'organisation des vers. *Annales des Sciences Naturelles, Zoologie, 3s* 7, 87–108; 8, 119–149, 271–341.

Bolaños, L.L. & Salas, E.M. (1982) *Villarrealina peruanus* gen. et sp. n. (Digenea: Allocreadiidae) en el pez marino *Sciaena gilberti*. *Investigación Ciencia Universidad Federico Villarreal* 1, 51–53.

Bourgat, R. (1990) Extension taxonomique et biogéographique du genre *Hapalorhynchus* (Trematoda, Spirorchiidae). *Bulletin de la Société Française de Parasitologie* 8, 289–294.

Bowers, E.A. & James, B.L. (1967) Studies on the morphology, ecology and life-cycle of *Meiogymnophallus minutus* (Cobbold, 1859) comb. nov. (Trematoda: Gymnophallidae). *Parasitology* 57, 281–300.

Bowers, E.A., Bartoli, P. & James, B.L. (1990) A demonstration of allopatric sibling species within the Gymnophallidae. *Systematic Parasitology* 17, 143–152.

Bowers, E.A., Bartoli, P., Russell-Pinto, F. & James, B.L. (1996) The metacercariae of sibling species of *Meiogymnophallus*, including *M. rebecqui* comb. nov. (Digenea: Gymnophallidae), and their effects on closely related *Cerastoderma* host species (Mollusca: Bivalvia). *Parasitology Research* 82, 505–510.

Brackett, S. (1942) Five new species of avian schistosomes from Wisconsin and Michigan with the life cycle of *Gigantobilharzia gyrauli* (Brackett, 1940). *Journal of Parasitology* 28, 25–42.

Brandes, G.P.H. (1892) Revision der Monostomen. *Centralblatt für Bakteriologie und Parasitenkunde* 12, 504–511.

Braun, M.G.C.C. (1892) Über *Eurycoelum sluiteri* Br. *Zentralblatt für Bakteriologie, Parasitologie, Infektionskrankheiten (und Hygiene)* 11, 727–729.

Braun, M.G.C.C. (1899) Über *Clinostomum* Leidy. *Zoologischer Anzeiger* 22, 484–493.

Braun, M.G.C.C. (1900) Die Arten der Gattung *Clinostomum* Leidy. *Zoologische Jahrbücher, Abteilung für Systematik, Ökologie und Geographie der Tiere* 14, 1–48.

Braun, M.G.C.C. (1902) Fascioliden der Vögel. *Zoologische Jahrbücher, Abteilung für Systematik, Ökologie und Geographie der Tiere* 16, 1–162.

Bray, R.A. (1983) On the fellodistomid genus *Proctoeces* Odhner, 1911 (Digenea), with brief comments on two other fellodistomid genera. *Journal of Natural History* 17, 321–339.

Bray, R.A. (1988) A discussion of the status of the subfamily Baccigerinae Yamaguti, 1958 (Digenea) and the constitution of the family Fellodistomidae Nicoll, 1909. *Systematic Parasitology* 11, 97–112.

Bray, R.A. (1990) A review of the genus *Parahemiurus* Vaz & Pereira, 1930 (Digenea: Hemiuridae). *Systematic Parasitology* 15, 1–21.

Bray, R.A. (1996) *Merlucciotrema praeclarum* (Manter, 1934) (Digenea: Hemiuridae) redescribed from myctophiform and gadiform fishes of the North Atlantic. *Systematic Parasitology* 33, 135–141.

Bray, R.A. & Campbell, R.A. (1995) Fellodistomidae and Zoogonidae (Digenea) of deep-sea fishes of the NW Atlantic Ocean. *Systematic Parasitology* 31, 201–213.

Bray, R.A. & Chambers, C.B. (2000) A new macradenine, *Zebrasomatrema pichelinae* n. g., n. sp. (Digenea: Lecithasteridae), from tangs (Perciformes: Acanthuridae) off the southern Great Barrier Reef, Queensland, Australia. *Systematic Parasitology* 46, 105–109.

Bray, R.A. & Cribb, T.H. (2000a) The status of the genera *Hysterolecithoides* Yamaguti, 1934, *Neotheletrum* Gibson & Bray, 1979 and *Machidatrema* León-Règagnon, 1998 (Digenea: Hemiuroidea). *Systematic Parasitology* 46, 1–22.

Bray, R.A. & Cribb, T.H. (2000b) Species of *Trifoliovarium* Yamaguti, 1940 (Digenea: Lecithasteridae) from Australian waters, with a description of *T. draconis* n. sp. *Systematic Parasitology* 47, 183–192.

Bray, R.A. & Cribb, T.H. (2001) *Weketrema* gen. n., a new genus for *Weketrema hawaiiense* (Yamaguti, 1970) comb. n. (Digenea: Lecithasteridae) recently found in Australian marine fishes. *Folia Parasitologica* 48, 109–114.

Bray, R.A. & Gibson, D.I. (1977) The Accacoeliidae (Digenea) of fishes from the northeast Atlantic. *Bulletin of the British Museum (Natural History) (Zoology)* 31, 51–99.

Bray, R.A. & Gibson, D.I. (1980) The Fellodistomidae (Digenea) of fishes from the northeast Atlantic. *Bulletin of the British Museum (Natural History) (Zoology)* 37, 199–293.

Bray, R.A. & Gibson, D.I. (1998) Further observations on the Digenea (Platyhelminthes) of deep-sea fishes in the northeastern Atlantic: Fellodistomidae and Zoogonidae. *Acta Parasitologica* 43, 194–199.

Bray, R.A. & Zdzitowiecki, K. (2000) *Boreascotia megavesicula* n. g., n. sp. (Digenea: Hemiuridae: Lecithochiriinae) in the nototheniid fish *Lepidonothen macrophthalma* (Norman) from the sub-Antarctic Atlantic. *Systematic Parasitology* 46, 29–32.

Bray, R.A., Cribb, T.H. & Barker, S.C. (1993) The Hemiuroidea (Digenea) of pomacentrid fishes (Perciformes) from Heron Island, Queensland, Australia. *Systematic Parasitology* 24, 159–184.

Bray, R.A., Cribb, T.H. & Barker, S.C. (1994a) Fellodistomidae and Lepocreadiidae (Platyhelminthes: Digenea) from chaetodontid fishes (Perciformes) from Heron Island, southern Great Barrier Reef, Queensland, Australia. *Invertebrate Taxonomy* 8, 545–581.

Bray, R.A., Soto, A. & Rollinson, D. (1994b) The status and composition of the genus *Steringophorus* Odhner, 1905 (Digenea: Fellodistomidae), based on partial small subunit rRNA sequences. *International Journal for Parasitology* 24, 433–435.

Bray, R.A., Brockerhoff, A. & Cribb, T.H. (1995) *Melogonimus rhodanometra* n. g., n. sp. (Digenea: Ptychogonimidae) from the elasmobranch *Rhina ancylostoma* Bloch & Schneider (Rhinobatidae) from the southeastern coastal waters of Queensland, Australia. *Systematic Parasitology* 30, 11–18.

Brinkmann, A., Jr (1957) Fish trematodes from Norwegian waters. IIa. The Norwegian species of the orders Aspidogastrea and Digenea (Gasterostomata). *University of Bergen Årbok (1957), Naturvitenskapelig rekke* 4, 1–32.

Brock, J. (1886) *Eurycoelum sluiteri* n. g., n. sp. Eine vorläufige Mittheilung. *Nachrichten von der Königlichen Gesellschaft der Wissenschaften (und der Georg-Augustus-Universität).* Göttingen 181, 543–547.

Brooks, D.R. & Mayes, M.A. (1976) *Telorchis gutterosi* sp. n. (Trematoda: Telorchiidae) from *Graptemys pseudogeographica* Gray in Nebraska, with reports of additional species of trematodes from Nebraska turtles. *Journal of Parasitology* 62, 901–905.

Brooks, D.R. & McLennan, D.A. (1993) *Parascript. Parasites and the Language of Evolution.* Smithsonian Institution Press, Washington and London, 429 pp.

Brooks, D.R. & Overstreet, R.M. (1978) The family Liolopidae (Digenea) including a new genus and two new species from crocodilians. *International Journal for Parasitology* 8, 267–273.

Brooks, D.R., O'Grady, R.T. & Glen, D.R. (1985) Phylogenetic analysis of the Digenea (Platyhelminthes: Cercomeria) with comments on their adaptive radiation. *Canadian Journal of Zoology* 63, 411–443.

Brooks, D.R., Bandoni, S.M., Macdonald, C.A. & O'Grady, R.T. (1989) Aspects of the phylogeny of the Trematoda Rudolphi, 1808 (Platyhelminthes: Cercomeria). *Canadian Journal of Zoology* 67, 2609–2624.

Brooks, D.R., Catto, J.B. & Amato, J.F.R. (1992) A new phylogenetic classification of the genera of the Proterodiplostomidae Dubois, 1936 (Digenea: Strigeiformes). *Proceedings of the Biological Society of Washington* 105, 143–147.

Bykhovskaya-Pavlovskaya, I.E. (1949) [The parasite fauna of *Perca fluviatilis* L. and the influence of several ecological factors on its mutation.] *Izvestiya Akademii Nauk SSSR, ser. Biologiya* 3, 316–339. (In Russian.)

Bykhovskaya-Pavlovskaya, I.E. (1962) [*Trematodes of birds of the USSR.*] Izdatel'stvo Akademii Nauk SSSR, Moscow, 270 pp. (In Russian.)

Byrd, E.E. (1939) Studies on the blood flukes of the family Spirorchidae. Part II. Revision of the family and description of new species. *Journal of the Tennessee Academy of Science* 14, 116–161.

Byrd, E.E. & Denton, J.F. (1950) The helminth parasites of birds. I. A review of the trematode genus *Tanaisia* Skrjabin, 1924. *American Midland Naturalist* 43, 32–57.

Byrd, E.E. & Reiber, R.J. (1942) Strigeids trematodes of the *Alligator*, with remarks on the prostate gland and terminal portions of the genital ducts. *Journal of Parasitology* 28, 51–73.

Caballero y Caballero, E., Bravo Hollis, M. & Grocott, R.G. (1955) Helmintos de la Republica de Panama. XIV. Tremátodos monogéneos y digéneos de peces marinos del Océano Pacific del norte, con descripcion de nuevas formas. *Anales del Instituto de Biología Universidad Nacional Autónoma de México* 26, 117–147.

Cable, R.M. (1953) The life cycle of *Parvatrema borinqueñae* gen. et. sp. nov. (Trematoda: Digenea) and the systematic position of the subfamily Gymnophallinae. *Journal of Parasitology* 39, 408–421.

Cable, R.M. (1954) A new marine cercaria from the Woods Hole region and its bearing on the interpretation of larval types in the Fellodistomatidae (Trematoda: Digenea). *Biological Bulletin, Marine Biological Laboratory, Woods Hole* 106, 15–20.

Cable, R.M. (1974) Phylogeny and taxonomy of trematodes with reference to marine species. In: Vernberg, W.B. (ed.) *Symbiosis in the Sea*. University of South Carolina Press, Charleston, South Carolina, pp. 173–193.

Cable, R.M. & Nahhas, F.M. (1962) *Bivesicula caribbensis* sp. n. (Trematoda: Digenea) and its life history. *Journal of Parasitology* 48, 536–538.

Canaris, A.G. (1965) A new species of trematode, *Urotocus kenyensis* from the gray woodpecker *Mesopicos goetrae* (Picidae) in Kenya. *Transactions of the American Microscopical Society* 84, 282–284.

Canavan, W.P.N. (1933) A redescription of *Distomum incommodum* Leidy from *Alligator mississippiensis*, and creation of a new genus (*Homoscaphis*) for it. *Parasitology* 25, 501–509.

Carus, C.G. (1835) Beobachtung über einen merkwürdigen schöngefärbten Eingewiedewurm, *Leucochloridium paradoxum* mihi, und dessen parasitische Erzeugung in einer Landschnecke, *Succinea amphibia* Drap., *Helix putris* Linn. *Nova Acta Physico-Medica, Academia Caesarea Leopoldino Carolina* 17, 85–100.

Chatterji, P.N. (1958) On a new avian trematode of the genus *Cyclocoelum* Brandes, 1892. *Journal of Parasitology* 44, 559–565.

Chauhan, A.S., Srivastava, C.B. & Chauhan, B.S. (1974) Studies on the trematode fauna of India. Part 6. Digenea: Schistosomatidae. A monographic aid to the identification of Indian schistosomes. *Journal of the Zoological Society of India* 25 (1973), 83–127.

Chauhan, B.S. (1954) Studies on the trematode fauna of India. Part IV. Subclass Digenea (Protostomata). (A revison of the Hemiuroidea from the Indian Region.) *Records of the Indian Museum* 51, 289–393.

Cheatum, E.K. (1941) *Dendritobilharzia anatinarum* n. sp., a blood fluke from a mallard. *Journal of Parasitology* 27, 165–170.

Chin T.-H. & Ku Y.-M. (1974) The description of *Anguillotrema papillatum* gen. et sp. nov. from China (Trematoda: Halipegidae). *Acta Zoologica Sinica* 20, 35–39. (In Chinese.)

Ching, H.L. (1965) Life cycles of *Lacunovermis conspicuus* n. gen., n. sp. and *Meiogymnophallus multigemmulus* n. gen., n. sp. (Gymnophallidae: Trematoda) from *Macoma inconspicua* and diving ducks from Vancouver, Canada. *Proceedings of the Helminthological Society of Washington* 32, 53–63.

Ching, H.L. (1972) A redescription of *Gymnophalloides tokiensis* Fujita, 1925 (Trematoda: Gymnophallidae). *Canadian Journal of Zoology* 50, 1299–1302.

Ching, H.L. (1973) *Paragymnophallus odhneri* gen. n., sp. n. (Trematoda: Gymnophallidae) for *Gymnophallus somateriae sensu* Odhner (1900, 1905). *Canadian Journal of Zoology* 51, 807–810.

Ching, H.L. (1982) Description of germinal sacs of gymnophallid trematode, *Cercaria margaritensis* sp. n., in the extrapallial fluid of subtidal snails (*Margarites* spp.) in British Columbia. *Canadian Journal of Zoology* 60, 516–520.

Ching, H.L. (1995a) Evaluation of characters of the digenean family Gymnophallidae Morozov, 1955. *Canadian Journal of Fisheries and Aquatic Science* 52 (Suppl. 1), 78–83.

Ching, H.L. (1995b) Four new gymnophallid digeneans from rice rats, willets, and molluscs in Florida. *Journal of Parasitology* 81, 924–928.

Chu, H.J. (1950) *Bucephalopsis kweiyangensis* n.sp. from the giant salamander, *Megalobatrachus japonicus* Temm. in Kweichow, China. *Journal of Parasitology* 36, 120–122.

Cohn, L. (1902) Mitteilungen über Trematoden. *Zoologischer Anzeiger* 25, 712–718.

Cohn, L. (1904) Helminthologische Mitteilungen. 2. *Archiv für Naturgeschichte* 70, 229–252.

Coil, W.H., Reid, W.A. & Kuntz, R.E. (1965) *Paucivitellosus fragilis* gen. et sp. nov. (Bivesiculidae: Digenea), a parasite of *Chelon troscheli* from Formosa. *Transactions of the American Microscopical Society* 84, 365–368.

Corkum, K.C. (1968) Bucephalidae (Trematoda) in fishes of the northern Gulf of Mexico: *Bucephaloides* Hopkins, 1954 and *Rhipidocotyle* Diesing, 1858. *Transactions of the American Microscopical Society* 87, 342–349.

Cremonte, F. (2001) *Bartolus pierrei* n. g., n. sp. (Digenea: Gymnophallidae) from the Península Valdés, Argentina. *Systematic Parasitology* 49, 139–147.

Creplin, F.C.H. (1846) Nächtrage zu Gurlt's Verzaihniss der Thiere, bei welchen Entozoen gefunden worden sind. *Archiv für Naturgeschichte* 12, 129–160.

Cribb, T.H. (1988a) Life cycle and biology of *Prototransversotrema steeri* Angel, 1969 (Digenea: Transversotrematidae). *Australian Journal of Zoology* 36, 111–129.

Cribb, T.H. (1988b) Two new digenetic trematodes from Australian freshwater fishes with notes on previously described species. *Journal of Natural History* 22, 27–43.

Cribb, T.H. (1992) The Brachylaimidae (Trematoda: Digenea) of Australian native mammals and birds, including description of *Dasyurotrema* n. g. and four new species of *Brachylaima*. *Systematic Parasitology* 22, 45–72.

Cribb, T.H. & Barton, P.D. (1991) *Zeylanurotrema spearei* sp. n. (Digenea: Brachylaimidae) from the cane toad, *Bufo marinus*, in Australia. *Zoologica Scripta* 20, 207–213.

Cribb, T.H. & Bray, R.A. (1994) A review of the Tandanicolinae (Monodhelminthinae) (Digenea: Fellodistomidae), with a description of *Prosogonarium angelae* n. sp. *Systematic Parasitology* 27, 187–205.

Cribb, T.H. & Gordon, A.N. (1998) *Hapalotrema* (Digenea: Spirorchidae) in the green turtle (*Chelonia mydas*) in Australia. *Journal of Parasitology* 84, 375–378.

Cribb, T.H. & Pearson, J.C. (1993) *Neodiplostomum spratti* n. sp. (Digenea: Diplostomidae) from *Antechinus* spp. (Marsupialia: Dasyuridae) in Australia, with notes on other diplostomids from Australian mammals. *Systematic Parasitology* 25, 25–35.

Cribb, T.H. & Spratt, D.M. (1991) *Strzeleckia major* n. g., n. sp. and *S. minor* n. sp. (Digenea: Hasstilesiidae) from *Antechinus swainsonii* (Marsupialia: Dasyuridae) in Australia. *Systematic Parasitology* 19, 73–79.

Cribb, T.H., Bray, R.A. & Barker, S.C. (1992) A review of the family Transversotrematidae (Trematoda: Digenea) with the description of a new genus, *Crusziella*. *Invertebrate Taxonomy* 6, 909–935.

Cribb, T.H., Bray, R.A. & Barker, S.C. (1994) Bivesiculidae and Haplosplanchnidae (Digenea) from fishes of the southern Great Barrier Reef, Australia. *Systematic Parasitology* 28, 81–97.

Cribb, T.H., Anderson, G.R., Adlard, R.D. & Bray, R.A. (1998) A DNA-based demonstration of a three-host life-cycle for the Bivesiculidae (Platyhelminthes: Digenea). *International Journal for Parasitology* 28, 1791–1795.

Cribb, T.H., Anderson, G.R. & Bray, R.A. (1999) Faustulid trematodes (Digenea) from fishes of Australia. *Systematic Parasitology* 44, 119–138.

Cribb, T.H., Bray, R.A., Littlewood, D.T.J., Pichelin, S. & Herniou, E.A. (2001) The Digenea. In: Littlewood, D.T.J. and Bray, R.A. (eds) *Interrelationships of the Platyhelminthes.* Taylor and Francis, London, pp. 168–185.

Crowcroft, P.W. (1947) The anatomy of two new digenetic trematodes from Tasmanian food fishes. *Proceedings of the Linnean Society of New South Wales* 71, 108–118.

Crowcroft, P.W. (1948) A new digenetic trematode from the barracouta (Syncoeliidae - Digenea). *Papers and Proceedings of the Royal Society of Tasmania* 1947, 49-57.

Crowcroft, P.W. (1950) Note on *Lintonium vibex* (Linton, 1899) (Digenea - Trematoda). *Parasitology* 40, 316-321.

Crusz, H. & Sanmugasunderam, V. (1973) Parasites of the relict fauna of Ceylon. IV. *Zeylanurotrema lyriocephali* gen. et sp. nov. and other trematodes from mountain lizards and a rodent. *Annales de Parasitologie Humaine et Comparée* 48, 797–810.

Crusz, H., Ratnayake, W.E. & Sathananthan, A.H. (1964) Observations on the structure and life-cycle of the digenetic fish-trematode *Transversotrema patialense* (Soparkar). *Ceylon Journal of Science* 5, 8–17.

Dailey, M.D., Fast, M.L. & Balazs, G.H. (1991) *Carettacola hawaiiensis* n. sp. (Trematoda: Spirorchidae) from the green turtle, *Chelonia mydas,* in Hawaii. *Journal of Parasitology* 77, 906–909.

Dayal, J. (1948) Trematode parasites of Indian fishes. Part 1. New trematodes of the family Bucephalidae Poche, 1907. *Indian Journal of Helminthology* 1, 47–62.

de Blainville, M.H.D. (1820) Fasciole, *Fasciola. Dictionnaire des Sciences Naturelles* 16, 198–204.

de Blainville, M.H.D. (1828) Vers. *Dictionnaire des Sciences Naturelles* 57, 365–625.

Dickerman, E.E. (1954) *Paurorhynchus hiodontis,* a new genus and species of trematode (Bucephalidae: Paurorhynchinae n. subfam.) from the mooneye fish, *Hiodon tergisus. Journal of Parasitology* 40, 311-315.

Diesing, K.M. (1850) *Systema helminthum,* Vol. 1. Braumüller, Vindobonae, 679 pp.

Diesing, K.M. (1855) Neuzehn arten von Trematoden. *Akademie Wissenschaften in Wien Mathematisch-Naturwissenschaftliche Klass* 10, 59–70.

Diesing, K.M. (1858) Revision der Myzhelminthen. Abteilung: Trematoden. *Akademie Wissenschaften in Wien Mathematisch-Naturwissenschaftliche Klass* 32, 307–390.

Dimitrov, G.I., Arapoglou, F.K. & Georgiev, B.B. (1998) First record of *Saturnius papernai* Overstreet, 1977 in the Black Sea, with a review of the genus *Saturnius* Manter, 1969 (Digenea, Bunocotylidae). *Systematic Parasitology* 40, 43–48.

Dollfus, R.-P. (1923) Remarques sur la cycle évolutif des hémiurides. *Annales de Parasitologie Humaine et Comparée* 1, 345-351.

Dollfus, R.-P. (1929) Helmintha I. Trematoda et Acanthocephala. *Faune des Colonies Françaises* 3, 73–114.

Dollfus, R.-P. (1930) Metacercaire de *Nephrocephala. Annales de Parasitologie Humaine et Comparée* 8, 216–217.

Dollfus, R.-P. (1931) Clinostomatoidea comprend seulement les deux familles des Clinostomatidae et des Nephrocephalidae. *Annales de Parasitologie Humaine et Comparée* 9, 492–493.

Dollfus, R.-P. (1932a) Mission Saharienne Augieras Draper, 1927-1928. Trématodes de mammiferes, oiseaux et poissons. *Bulletin du Museum (National) d'Histoire Naturelle, Paris, 2s* 4, 555–563.

Dollfus, R.-P. (1932b) Trématodes. *Résultats scientifiques du voyage aux Indes Orientales Néerlandaises de LL. AA. RR. le Prince et la Princesse Léopold de Belgique (Mémoires de la Musée Royal d'Histoire Nateurelle Belgique, Hors serie)* 2(10), 1–18.

Dollfus, R.-P. (1934) Sur quelques *Brachylaemus* de la faune française récoltés principalement à Richelieu (Indre-et-Loire). *Annales de Parasitologie Humaine et Comparée* 12, 551–575.

Dollfus, R.-P. (1935a) Sur quelques *Brachylaemus* de la faune française récoltés principalement à Richelieu (Indre-et-Loire). *Annales de Parasitologie Humaine et Comparée* 13, 52–79.

Dollfus, R.-P. (1935b) Sur quelques parasites de poissons récolté à Castiglione (Algérie). *Bulletin de la Station d'Aquiculture et de Pêche de Castiglione* (Year 1933), 199–279.

Dollfus, R.-P. (1937a) Trématodes Digenea des sélaciens (Plagiostomes). Catalogue par hôte, distribution géographique. *Annales de Parasitologie Humaine et Comparée* 15, 57–73, 164–176, 259–281.

Dollfus, R.-P. (1937b) Trématodes de Sélaciens et de Chéloniens. Parasitologia mauritanica. Helmintha (III). *Bulletin Comité Études Historiques et Scientifiques d'Afrique Occidentale Français* 19, 397–519.

Dollfus, R.-P. (1939) Distome d'un abcès palpébro-orbitaire chez une panthère. Possibilité d'affinités lointaines entre ce distome et les Paragonimidae. *Annales de Parasitologie Humaine et Comparée* 17, 209–235.

Dollfus, R.-P. (1947) Sur *Monascus filiformis* (Rudolphi, 1819) A. Looss, 1907, trématode de l'intestine de *Cepola rubescens* (L.) en Mediterranée. *Annales de Parasitologie Humaine et Comparée* 22, 319–323.

Dollfus, R.-P. (1948) Sur deux monostomes (Cyclocoelidae) pourvus d'une ventouse ventrale. Observations sur la classification des Cyclocoeloidae Albert Henry, 1923, liste de leur hôtes, répartition géographique. *Annales de Parasitologie Humaine et Comparée* 23, 129–199.

Dollfus, R.-P. (1950) Trématodes récoltés au Congo Belge par le Professeur Paul Brien (mai–août 1937). *Annales de la Musée Royale de Congo Belge (Zoologie), ser. 5* 1(1), 136 pp.

Dollfus, R.-P. (1952) Miscellanea helminthologica maroccana. IV. Affinités naturelles de *Pseudochetosoma salmonicola* R.Ph. Dollfus 1951 (Famille Steganodermatidae nov.) - émendation de la superfamille Haploporoidea W. Nicoll 1935. *Archives de l'Institut Pasteur du Maroc* 4, 369–386.

Dollfus, R.-P. (1958) Trématodes. Sous-classe Aspidogastrea. *Annales de Parasitologie Humaine et Comparée* 33, 305–395.

Dollfus, R.-P. (1960a) Critiques de récentes innovations apportées à la classification des Accacoeliidae (Trematoda - Digenea). Observations sur les métacercaires de cette famille. *Annales de Parasitologie Humaine et Comparée* 35, 648–671.

Dollfus, R.-P. (1960b) Sur un distome de la sole commune, *Solea solea* (L.) en Méditerranée, appartenant à un genre connu jusqu'à présent seulement des Tortugas (Floride). *Vie et Milieu* 11, 188–194.

Dollfus, R.-P. (1973) Du genre *Tergestia* M. Stossich 1899 (Trematoda Digenea). *Annales de Parasitologie Humaine et Comparée* 48, 275–287.

Dove, A.D.M. & Cribb, T.H. (1995) *Coomera*: a new genus of Fellodistomidae (Digenea) from *Monodactylus argenteus* (Monodactylidae) of southern Queensland, Australia. *International Journal for Parasitology* 25, 1159–1162.

Dronen, N.O., Tehrany, M.R. & Wardle, W.J. (1999) Diplostomes from the brown pelican, *Pelecanus occidentalis* (Pelecanidae), from the Galveston Texas Area, including two new species of *Bursacetabulus* gen. n. *Journal of the Helminthological Society of Washington* 66, 21–24.

Dubinina, M. (1937) [The parasitic fauna of *Nycticorax nycticorax* and its change in connection with the host's migration.] *Zoologicheskii Zhurnal* 16, 547–573. (In Russian.)

Dubois, G. (1934) Étude de deux Strigeidae de la collection de l'Institut zoologiques de Naples. *Annuario del Museo Zoologico della R. Universita di Napoli, n.s.* 6, 1–12.

Dubois, G. (1936) Les diplostomes de reptiles (Trematoda: Proterodiplostomidae nov. fam.) du Musée de Vienne. *Bulletin de la Société Neuchâteloise des Sciences Naturelles* 61, 5–80.

Dubois, G. (1938) Monographie des Strigeida (Trematoda). *Mémoires de la Société Neuchâteloise des Sciences Naturelles* 6, 1–535.

Dubois, G. (1951) Nouvelle clé de détermination des groupes systématiques et des genres de Strigeida Poche (Trematoda). *Revue Suisse de Zoologie* 58, 639–691.

Dubois, G. (1953) Systématique des Strigeida. Complément de la monographie. *Mémoires de la Société Neuchâteloise des Sciences Naturelles* 8(2), 1–141.

Dubois, G. (1958) Quelques Strigeida (Trematoda) de la collection R.-Ph. Dollfus. *Bulletin de la Société Neuchâteloise des Sciences Naturelles* 81, 55–67.

Dubois, G. (1959) Revision des Cyclocoelidae Kossack, 1911. *Revue Suisse de Zoologie* 66, 67–147.

Dubois, G. (1968) Synopsis des Strigeidae et des Diplostomatidae (Trematoda). *Mémoires de la Société Neuchâteloise des Sciences Naturelles* 10, 1–258.

Dubois, G. (1970a) Synopsis des Strigeidae et des Diplostomatidae (Trematoda). *Mémoires de la Société Neuchâteloise des Sciences Naturelles* 10, 259–727.

Dubois, G. (1970b) Les fondements de la taxonomie des Strigeata La Rue (Trematoda: Strigeida). *Revue Suisse de Zoologie* 77, 663–685.

Dubois, G. (1970c) Revision des Proterodiplostomatidae Dubois, 1936 (Trematoda: Strigeata). *Anales del Instituto de Biologia Universidad Nacional Autónoma de México (Ser. Zoologia)* 41, 51–60.

Dubois, G. (1976) Description de l'adulte présumé de *Bursotrema tetracotyloides* Szidat, 1960, et d'une nouvelle espèce du genre *Didelphodiplostomum* Dubois, 1944, parasites de *Didelphys azarae* Temminck, 1825 (Trematoda: Alariinae). *Annales de Parasitologie Humaine et Comparée* 51, 341–347.

Dubois, G. (1979) Révision et nouvelle clé de détermination des diplostomes de reptiles (Trematoda: Proterodiplostomidae Dubois, 1936). *Bulletin de la Société Neuchâteloise des Sciences Naturelles* 102, 39–48.

Dubois, G. (1981) Du statut de quelques Strigeata La Rue, 1926 (Trematoda). VII. *Bulletin de la Société Neuchâteloise des Sciences Naturelles* 104, 35–46.

Dubois, G. (1982) Répertoire des synonymes récentes de genres et d'espèces de la superfamille des Strigeoidea Railliet, 1919 (Trematoda). *Bulletin de la Société Neuchâteloise des Sciences Naturelles* 105, 163–183.

Dubois, G. (1983a) Le genre *Holostephanus* Szidat, 1936 (Trematoda: Strigeata: Cyathocotyloidea). Commentaire sur les espèces et érection de *Holostephanoides* gen. nov. *Bulletin de la Société Neuchâteloise des Sciences Naturelles* 106, 87–99.

Dubois, G. (1983b) Les trématodes des genres *Prohemistomum* Orhner, 1913 et *Paracoenogonimus* Katsurada, 1914. *Annales de Parasitologie Humaine et Comparée* 58, 549–556.

Dubois, G. (1987) Systématique des Cyathocotyloidea (Trematoda: Strigeata). *Bulletin de la Société Neuchâteloise des Sciences Naturelles* 110, 41–43.

Dubois, G. (1989) Répertoire des synonymes des Cyathocotyloidea (Trematoda: Strigeata). *Bulletin de la Société Neuchâteloise des Sciences Naturelles* 112, 39–46.

Dubois, G. & Angel, L.M. (1972) Strigeata (Trematoda) of Australian birds and mammals from the helminthological collection of the University of Adelaide. *Transactions of the Royal Society of South Australia* 96, 197–215.

Duggal, C.L. & Toor, J.K. (1986) On *Porphyriotrema indica* n. g., n. sp. (Trematoda: Cyclocoelidae) from the air sac of purple moorhen at Sangur India. *Rivista de Parassitologia* 46, 351–354.

Dujardin, F. (1843) Mémoires sur les helminthes des musaraignes et en particulier sur les trichosomes, les distomes, et les taenias, sur les métamorphoses et leur transmigrations. *Annales des Sciences Naturelles, Zoologie* 20, 329–439.

Dujardin, F. (1845) *Histoire Naturelle des Helminthes ou vers Intestinaux.* Librairie Encyclopédique de Roret, Paris, 652 pp.

Durio, W.O. & Manter, H.W. (1968) Some digenetic trematodes of marine fishes of New Caledonia. Part 1. Bucephalidae, Monorchiidae, and some smaller families. *Proceedings of the Helminthological Society of Washington* 35, 143–153.

Dutta, I.B. (1995) Platyhelminthes: Trematoda: Digenea. In: *Fauna of Chilka Lake*. Wetland Ecosystem Series 1, Zoological Survey of India, Calcutta, pp. 235-277.

Dwivedi, M.P. (1966) On *Capsulodiplostomum crocodilinum* n. g., n. sp. (Trematoda, Digenea), from the Indian crocodile *Crocodilus palustris*. *Revista de Biologia Tropical* 14, 93-97.

Dwivedi, M.P. (1967) Contribution to the family Spirorchiidae Stunkard, 1921 (Digenea: Trematoda). *Indian Journal of Helminthology* 19, 1-14.

Eckmann, F. (1932) Beitrage zur Kenntnis der Trematodenfamilie Bucephalidae. *Zeitschrift für Parasitenkunde* 5, 94-111.

Eckmann, F. (1934) Rectifications de nomenclature. *Annales de Parasitologie Humaine et Comparée* 12, 256.

Ejsmont, L. (1927) *Spirhapalum polesianum* n. g., n. sp. trématode du sang d'*Emys orbicularis*. *Annales de Parasitologie Humaine et Comparée* 5, 220-235.

Ejsmont, L. (1932) Über die Identität von *Proshystera rossittensis* Korkhaus un *Tanaisia fedschenkoi* Skrjabin, nebst einegen Bemerkungen über Trematoden mit verbundenen Darmschenkeln. *Bulletin de l'Académie Polonaise des Sciences et Lettres. Classe des Sciences Mathématiques et Naturelles, Série B: Sciences Naturelles* 2, 531-547.

Erwin, B.E. & Halton, D.W. (1983) Fine structural observations on spermatogenesis in a progenetic trematode, *Bucephaloides gracilescens*. *International Journal for Parasitology* 13, 413-426.

Evans, D.H. & Wares, P.G. (1972) Food habits of striped marlin and sailfish off Mexico and southern California. In: *US Department of the Interior, Bureau of Sport Fisheries and Wildlife Research Report* no. 76, pp. 1-10.

Farley, J. (1971) A review of the family Schistosomatidae: excluding the genus *Schistosoma* from mammals. *Journal of Helminthology* 45, 289-320.

Faust, E.C. & Tang, C.C. (1936) Notes on new aspidogastrid species, with a consideration of the phylogeny of the group. *Parasitology* 28, 487-501.

Feizullaev, N.A. (1980) [*Trematodes of the superfamily Cyclocoeloidea (morphology, biology, phylogeny and systematics).*] ELM, Baku, 212 pp. (In Russian.)

Feizullaev, N.A. (1984) [Taxonomic analysis of trematodes from the family Cyclocoelidae in the light of new data.] In: Sonin, M.D. (ed.) [*Helminths of domesticated and hunted animals.*] Nauka, Moscow, pp. 198-201. (In Russian.)

Feizullaev, N.A. & Mirzoeva, S.S. (1983) [Revision of the superfamily Clinostomoidea (Trematoda) and an analysis of its classification.] *Parazitologiya* 17, 3-11. (In Russian.)

Ferguson, M.A., Cribb, T.H. & Smales, L.R. (1999) Life cycle and biology of *Sychnocotyle kholo* n. g., n. sp. (Trematoda: Aspidogastrea) in *Emydura macquarii* (Pleurodira: Chelidae) from southern Queensland, Australia. *Systematic Parasitology* 43, 41–48.

Fischthal, J.H. & Acholonu, A.D. (1976) Some digenetic trematodes from the Atlantic hawksbill turtle, *Eretmochelys imbricata imbricata* (L.), from Puerto Rico. *Proceedings of the Helminthological Society of Washington* 43, 174–185.

Fischthal, J.H. & Nasir, P. (1974) Some digenetic trematodes from freshwater and marine fishes of Venezuela. *Norwegian Journal of Zoology* 22, 71–80.

Fischthal, J.H. & Thomas, J.D. (1968) Digenetic trematodes of some freshwater and marine fishes from Ghana. *Proceedings of the Helminthological Society of Washington* 35, 126–140.

Fischthal, J.H. & Thomas, J.D. (1972) Additional hemiurid and other trematodes of fishes from Ghana. *Bulletin de l'Institut Fondamental d'Afrique Noir* 34A, 9–25.

Fredericksen, D.W. (1972) Morphology and taxonomy of *Cotylogaster occidentalis* (Trematoda: Aspidogastridae). *Journal of Parasitology* 58, 1110–1116.

Freitas, J.F.T. (1951) Revisão da familia Eucotylidae Skrjabin, 1924. *Memórias do Instituto Oswaldo Cruz* 49, 33–271.

Freitas, J.F.T. (1959) Nota sobre *Tanaisia isopina* Freitas, 1951 (Trematoda: Eucotylidae). *Atas Sociedade Biologia do Rio de Janeiro* 3, 2–4.

Freitas, J.F.T. & Santos, E. (1971) Novo trematodeo parasito de peixe marinho. *Memórias do Instituto Oswaldo Cruz* 69, 79–81.

Fuhrmann, O. (1904) Neue Trematoden. *Centralblatt für Bakteriologie und Parasitenkunde, Erste Abteilung* 37, 58–64.

Fuhrmann, O. (1928) Zweite Klasse des Cladus Plathelminthes. Trematoda. In: Kükenthal, W. & Krumbach, T. (eds) *Handbuch der Zoologie.* Walter de Gruyter, Berlin, Leipzig, 2(2), 140 pp.

Fyfe, M.L. (1954) *Tricotyledonia genypteri* n. g., n. sp., a three suckered trematode from the red ling, *Genypterus blacodes* Bloch & Schn. *Parasitology* 44, 325–328.

Galaktionov, K.V. (1996) [Experimental study of the unusual life cycle of *Parvatrema* sp. (Trematoda: Gymnophallidae).] *Parazitologiya* 30, 487–494. (In Russian.)

Ganapati, P.N., Rao, K.H. & Rao, M.V.L. (1962) A new allocreadioid metacercaria (Trematoda) from the wood-boring pelecypod *Martesia striata* (L.) at Visakhapatnam harbour. *Current Science* 4, 156–157.

Garcin (1730) *Hirudinella marina* or sea-leech. *Philosophical Transactions of the Royal Society* 36, 387–394.

Gibson, C.D. (1998) *The Broadbill Swordfishery of the Northwest Atlantic: the Fishery's Economic and Natural History (Early 1800s through 1995).* Ensign Press, Camden, Maine, 140 pp.

Gibson, D.I. (1976) Monogenea and Digenea from fishes. *Discovery Reports* 36, 179–266.

Gibson, D.I. (1983) *Kenmackenzia* gen. nov. (Sclerodistomidae: Kenmackenziinae subfam. nov.): *Distoma gigas* Nardo revealed. *Journal of Natural History* 17, 189–202.

Gibson, D.I. (1987) Questions in digenean systematics and evolution. *Parasitology* 95, 429–460.

Gibson, D.I. (1996) Trematoda. In: Margolis, L. & Kabata, Z. (eds) *Guide to the Parasites of Fishes of Canada*, Part IV. Canadian Special Publication of Fisheries and Aquatic Sciences No. 124, NRC Press, Ottawa, 373 pp.

Gibson, D.I. & Bray, R.A. (1977) The Azygiidae, Hirudinellidae, Ptychogonimidae, Sclerodistomidae and Syncoeliidae (Digenea) of fishes from the north-east Atlantic. *Bulletin of the British Museum (Natural History) (Zoology)* 32, 167–245.

Gibson, D.I. & Bray, R.A. (1979) The Hemiuroidea: terminology, systematics and evolution. *Bulletin of the British Museum (Natural History) (Zoology)* 36, 35–146.

Gibson, D.I. & Bray, R.A. (1986) The Hemiuridae (Digenea) of fishes from the northeast Atlantic. *Bulletin of the British Museum (Natural History) (Zoology)* 51, 1–125.

Gibson, D.I. & Bray, R.A. (1994) The evolutionary expansion and host–parasite relationships of the Digenea. *International Journal for Parasitology* 24, 1213–1226.

Gibson, D.I. & Chinabut, S. (1984) *Rohdella siamensis* gen. et sp. nov. (Aspidogastridae: Rohdellinae subfam. nov.) from freshwater fishes in Thailand, with a reorganization of the classification of the subclass Aspidogastrea. *Parasitology* 88, 383–393.

Gibson, D.I. & Køie, M. (1991) *Magnibursatus caudofilamentosa* (Reimer, 1971) n. comb. (Digenea: Derogenidae) from the stickleback *Gasterosteus aculeatus* in Danish waters: a zoogeographical anomaly? *Systematic Parasitology* 20, 221–228.

Ginetsinskaya, T.A. & Dobrovolskij, A.A. (1968) [Larval trematode fauna of freshwater molluscs in the Volga delta. 3. Furcocercariae (fam. Cyathocotylidae) and xiphidiocercariae.] *Trudy astrakhanskogo Zapovednika (Sbornik Gel'mintologicheskikh Rabot)* 11, 29–95. (In Russian.)

Girola, C.V., Martorelli, S.R. & Sardella, N.H. (1992) Presencia de metacercarias de *Monascus filiformis* (Digenea, Fellodistomidae) en hidromedusas del Océano Atlántico Sur. *Revista Chilena de Historia Natural* 65, 409–415.

Goldberger, J. (1911) Some known and three new endoparasitic trematodes from American fresh-water fish. *Bulletin of the Hygienic Laboratory, Washington* 71, 7–35.

Gomes, D.C. & de Fábio, S.P. (1972) *Varelacreptotrema travassosi* gen. n. sp. n. parasito de *Balistes vetula* L., 1758 (Trematoda, Creptotrematinae). *Atas da Sociedade de Biologio do Rio de Janeiro* 15, 23–26.

Goodman, J.D. (1990) *Amblosoma reelforti* n. sp. (Trematoda: Brachylaimata: Thapariellidae) from *Viviparus intertextus* in Tennessee and *Thapariella prudhoei* n. sp. from *Lanistes* sp. in Zaire. *Transactions of the American Microscopical Society* 109, 319–324.

Goto, S. & Ozaki, Y. (1930) Brief notes on new trematodes. III. *Japanese Journal of Zoology* 3, 73–82.

Gower, W.C. (1938) Studies on the trematode parasites of ducks in Michigan, with special reference to the mallard. *Memoirs, Michigan State College Agricultural Experimental Station* 3, 1–94.

Gower, W.C. (1939) Host-parasite catalogue of the helminths of ducks. *American Midland Naturalist* 22, 580–628.

Gu C. & Shen J. (1983) [Digenetic trematodes of fishes from the Xisha Islands, Guangdong Province, China. I.] *Studia Marina Sinica* 20, 157–184. (In Chinese.)

Guberlet, J.E. (1928) Two new genera of trematodes from a red-bellied water-snake. *Journal of Helminthology* 6, 205–218.

Gupta, A.N. (1979) Proposed synonymy of *Cladolecithotrema* Ichihara, 1970 with *Trifoliovarium* Yamaguti, 1940 (Hemiuridae - Trematoda). *Anales del Instituto de Biologia, Universidad Nacional Autónomica de México (Serie Zoologia)* 50, 19–23.

Gupta, N.K. & Kumari, A. (1977) *Chelatrema smythi* n. gen., n. sp. (Trematoda: Hemiuridae: Arnolinae) from a fresh-water fish, *Chela bacala*, from Ropar. *Research Bulletin of the Panjab University (new series)* 24(1973), 109–112.

Gupta, N.K. & Kumari, A. (1978) On *Roparhynchus nelsoni* n. g. n. sp. (Bucephalidae: Neoprosorhynchinae) from a freshwater fish, *Xenentoden* [sic] *cancila* (Ham.) from Ropar. *Research Bulletin (NS) of the Punjab University* 25(1974), 123–126.

Gupta, N.K. & Mehrotra, V. (1975) On three blood flukes (Spirorchiidae: Spirorchiinae) from freshwater chelonians and discussion on the synonymy of *Plasmiorchis pellucidus* Mehra, 1934. *Rivista di Parassitologia* 36, 165–170.

Gupta, N.K. & Mehrotra, V. (1981) On two blood flukes (Trematoda) of the family Spirorchidae Stunkard, 1921 from Indian marine turtles. *Acta Parasitologia Polonica* 28, 11–20.

Gupta, R. (1962) *Proeputiodiplostomum indeum* [sic] gen. nov. sp. nov. for a Diplostomum parasite of black-necked stork (Trematoda: Diplostomatidae). *Annual Number. National Academy of Sciences, India* 1961, 88.

Gvozdev, E.V. & Soboleva, T.N. (1972) [On the life cycle of the trematode *Skrjabinotrema ovis* Orloff, Ershoff et Badanin, 1934 (Brachylaemidae).] *Parazitologyia* 6, 435–438. (In Russian.)

Gvozdev, E.V. & Soboleva, T.N. (1983) Biology of trematode *Hasstilesia ovis* (Orloff, Erschoff et Badanin, 1934) - the agent of sheep hasstilesiosis on the mountain pastures of Tien Shan. *Helminthologia* 20, 103–110.

Hafeezullah, M. (1975) Two new trematodes (Digenea: Hemiuridae) from marine fishes from east coast of India. In: Tiwari, K.K. & Srivastava, C.B. (eds) *Dr B.S. Chauhan Commemoration Volume (Presented on his 60th Birthday).* Zoological Society of India, Bhubaneshwar, pp. 203–210.

Hafeezullah, M. (1984) On the status of some digenetic trematodes of marine fishes of India. *Bulletin of the Zoological Survey of India* 6, 209–218.

Hall, K.A., Cribb, T.H. & Barker, S.C. (1999) V4 region of small subunit rDNA indicates polyphyly of the Fellodistomidae (Digenea) which is supported by morphology and life-cycle data. *Systematic Parasitology* 43, 81–92.

Hamann, M.I. (1986) *Halipegus ovocaudatus* (Vulpian, 1859) Looss, 1899 (Hemiuridae, Halipeginae) parasita de *Rana esculenta* Linné de Europa y *Genarchella genarchella* Travassos, Artigas & Pereira, 1928 (Hemiuridae, Halipeginae) parasita *Salminus maxillosus* Valenciennes, 1840 del sudeste de America del Sur. Anatomia y posicion sistematica. *Physis. Buenos Aires, Secc. B* 44(106), 19–24.

Harrah, E.C. (1922) North American monostomes primarily from freshwater hosts. *Illinois Biological Monographs* 7, 225–324.

Heckert, G.A. (1889) *Leucochloridium paradoxum*. Monographische Darstellung der Enwicklungs- und Lebensgeschichte des *Distomum macrostomum*. *Bibliotheca Zoologica* 4, 1–66.

Heider, K. (1900) Über *Braunina*, ein neues Genus aus der Gruppe der Hemistomidae. *Verhandlungen der Deutschen Zoölogischen Gesellschaft, Leipzig* 1900, 19–22.

Hendrix, S.S. & Overstreet, R.M. (1983) Evaluation of the status of *Texanocotyle pogoniae* and *Laterocotyle padreinsulae* (Trematoda: Aspidogastridae). *Journal of Parasitology* 69, 431–432.

Henry, A. (1923) *Helminthiases et helminthes.* Cours professé à l'école national vétérinaire d'Alfort. [Mimeograph; cited from Dollfus, 1948.]

Herbert, B.W. & Shaharom, F.M. (1995) A new blood fluke, *Parasanguinicola vastispina* gen. nov., sp. nov. (Trematoda: Sanguinicolidae), from sea bass *Lates calcarifer* (Centropomidae) cultured in Malaysia. *Parasitology Research* 81, 349–354.

Herbert, B.W., Shaharom-Harrison, F.M. & Overstreet, R.M. (1994) Description of a new blood-fluke, *Cruoricola lates* n. g., n. sp. (Digenea: Sanguinicolidae), from seabass *Lates calcarifer* (Bloch, 1790) (Centropomidae). *Systematic Parasitology* 29, 51–60.

Hoberg, E.P. (1981) *Pseudogymnophallus alcae* gen. n. et sp. n. (Trematoda: Gymnophallidae) from alcids (Charadriiformes) in subarctic seas. *Proceedings of the Helminthological Society of Washington* 48, 190–194.

Holmes, J.C. (1971) Two new sanguinicolid blood flukes (Digenea) from scorpaenid rockfishes (Perciformes) of the Pacific coast of North America. *Journal of Parasitology* 57, 209–216.

Hopkins, S.H. (1940) The excretory systems of *Tergestia* Stossich 1899 and *Cornucopula adunca* (Linton 1905) (Trematoda). *Transactions of the American Microscopical Society* 49, 281–284.

Hopkins, S.H. (1954) The American species of trematode confused with *Bucephalus bucephalopsis haimeanus*. *Parasitology* 44, 353–370.

Horsfall, M.W. (1934) Studies on the life history of the cystocercous cercariae. *Transactions of the Microscopical Society* 53, 311–347.

Hsu, P. (1954) [A new trematode, *Chenia cheni* n. g. n. sp. (Hemiuridae), from a gobiid fish.] *Acta Zoologica Sinica* 6, 33–36. (In Chinese.)

Huehner, M.K. & Etges, F.J. (1972) A new aspidogastrid trematode, *Cotylogasteroides barrowi* sp. n. from freshwater mussels of Ohio. *Journal of Parasitology* 58, 468–470.

International Commission on Zoological Nomenclature (1999) *International Code of Zoological Nomenclature*, 4th edn. International Trust for Zoological Nomenclature, London, 306 pp.

Issaitschikow, I.M. (1928) Zur kenntnis der parasitischen Wurmer einiger Gruppen von Wirbeltieren der russischen Arktis. *Trudy Morskogo Nauchnogo Instituta Moskva* 3(2), 5–79.

Jackson, J.A. & Tinsley, R.C. (1997) The taxonomic status, host range and geographical distribution of *Dollfuschella* Vercammen-Grandjean, 1960 (Digenea: Halipeginae) from *Xenopus* spp. (Anura: Pipidae). *Systematic Parasitology* 36, 1–11.

James, B.L. (1964) The life cycle of *Parvatrema homoeotecnum* sp. nov. (Trematoda: Digenea) and a review of the family Gymnophallidae Morozov, 1955. *Parasitology* 54, 1–41.

Johnston, S.J. (1913) On some Queensland trematodes with anatomical observations and genera. *Quarterly Journal of Microscopical Science* 59, 361–400.

Johnston, S.J. (1915) On *Moreauia mirabilis* gen. et sp. nov., a remarkable trematode parasitic in *Ornithorhynchus*. *Proceedings of the Linnean Society of New South Wales* 40, 278–287.

Johnston, T.H. (1927) New trematodes from an Australian siluroid. *Transactions of the Royal Society of South Australia* 51, 129–136.

Jones, A. & Khalil, L.F. (1984) The relationship between the genera *Elongoparorchis* Rao, 1961, *Pelorohelmins* Fischthal and Kuntz, 1964 and *Dollfustravassosius* Freitas and Kohn, 1967 (Digenea: Hemiuroidea) with comments on the validity of their species. *Journal of Natural History* 18, 95–104.

Jones, D.O. (1943) The anatomy of three digenetic trematodes, *Skrjabiniella aculeatus* (Odhner), *Lecithochirium rufoviride* (Rud.) and *Sterrhurus fusiformis* (Lühe) from *Conger conger* (Linn.). *Parasitology* 35, 40–57.

Joyeux, C.E. & Baer, J. (1927) Notes sur les Cyclocoelidae: Trématodes. *Bulletin de la Société Zoologique de France* 52, 416–434.

Joyeux, C.E. & Foley, H. (1930) Les helminthes des *Meriones shawi shawi* Rozet dans le nord de l'Algérie. *Bulletin de la Société Zoologique de France* 55, 353–374.

Joyeux, C.E. & Houdemer, F.E. (1928) Recherches sur la faune helmintologique de l'Indochine (cestodes et trématodes). *Annales de Parasitologie Humaine et Comparée* 6, 27–58.

Kagan, I.G. (1952) Revision of the subfamily Leucochloridiinae Poche, 1907 (Brachylaimidae). *American Midland Naturalist* 48, 257–301.

Kamegai, S. (1970) *Plectognathotrema* (*Alloplectognathotrema*) *tsushimaense* n. subg., n. sp. (Trematoda: Cephaloporidae, Plectognathotrematinae n. subfam.), from intestine of marine fish, *Navodon modestus*. *Research Bulletin of the Meguro Parasitological Museum* 3, 5–11.

Kamegai, S. (1971) A new digenetic trematode *Sclerodistomoides pacificus* n. g., n. sp. (Sclerodistomidae) from the gall bladder of flying fish from Japan. *Annotationes Zoologicae Japonenses* 44, 105–108.

Kamo, H., Kenichi, O. & Ryou, H. (1962) A unique infection of man with *Clinostomum* sp. a small trematode causing acute laringitis. *Yonago Acta Medica* 6, 12–15.

Kasimov, G.B., Vaidova, S.M. & Feizullaev, N.A. (1962) [Trematodes of birds in the Lenkoran zone of Mugansk and Milsk Steppes of Azerbaidjan.] *Trudy Zoologicheskogo Instituta, Baku* 22, 73–102. (In Azerbaidjani.)

Kasyanov, I.S. (1954) [Elucidation of biological cycle of trematode *Skrjabinotrema ovis* (Brachylaimidae).] *Trudy Gel'mintologicheskoi Laboratorii* 7, 233–257. (In Russian.)

Kearn, G.C. (1998) *Parasitism and Platyhelminths*. Chapman & Hall, London, 544 pp.

Khalifa, R. (1974) Studies on Schistosomatidae Looss, 1899 (Trematoda) of aquatic birds of Poland. II. *Gigantobilharzia mazuriana* sp. n., with a discussion of the subfamily Gigantobilharziinae Mehra, 1940. *Acta Parasitologica Polonica* 22, 265–284.

Khan, A. & Bilqees, F.M. (1990) Four trematodes, including a new species from freshwater fishes of Sind, Pakistan. *Proceedings of Parasitology* 9, 1–44.

Khan, M.H. (1935) On eight new species of the genus *Cyclocoelum* Brandes from North Indian snipes. *Proceedings of the Indian Academy of Sciences* 4, 342–370.

Kirk, R.S. & Lewis, J.W. (1993) The life-cycle and morphology of *Sanguinicola inermis* Plehn, 1905 (Digenea: Sanguinicolidae). *Systematic Parasitology* 25, 125–133.

Kloss, G.R. (1966) Helmintos parasitos de espécies simpátricas de *Astyanax* (Pisces, Characidae). 1. *Papéis Avulsos do Departamento de Zoologia, Secretaria da Agricultura, São Paulo, Brasil* 18, 189–219.

Kniskern, V.B. (1952) Studies of the trematode family Bucephalidae Poche, 1907. Part I. A systematic review of the family Bucephalidae. *Transactions of the American Microscopical Society* 71, 253–266.

Kobayashi, H. (1921) On some digenetic trematodes in Japan. *Parasitology* 12, 380–410.

Koche, B.Y. (1976) Trematode parasites of marine food fishes of India. Part VI. (Digenea: Bucephalidae). Report on a new parasite *Proopisthogyne yamaguti* gen. nov. sp. nov. In: *First National Convention of Indian Helminthologists. Abstracts of Papers* 34, 32.

Kohn, A. (1962) Sôbre um nôvo gênero de trematódeo bucephaliforme parasito de peixe de agua doce. *Revista Brasileira de Biologia* 22, 351–355.

Kohn, A. (1964) Sôbre o gênero *Macrobilharzia* Travassos, 1922 (Schistosomatoidea). *Memórias do Instituto Oswaldo Cruz* 62, 1–6.

Kohn, A. (1967a) Sôbre um nôvo gênero de Prosorhynchinae Nicoll, 1914 e novos dados sôbre *Prosorhynchus bulbosus* Kohn, 1961 e *Rhipidocotyle quadriculatum* Kohn, 1961 (Trematoda, Bucephaliformes). *Memórias do Instituto Oswaldo Cruz* 65, 107–114.

Kohn, A. (1967b) Ocorrência de *Prosorhynchus ozakii* Manter, 1934 (Trematoda, Bucephaliformes) em nôvo hospedador na Baía de Guanabara. *Atas da Sociedade de Biologia do Rio de Janeiro* 11, 69–70.

Kohn, A. (1970a) *Chabaudtrema rarus* gen. n., sp. n. trematódeo bucefaliforme parasito de peixe. *Atas da Sociedade de Biologia do Rio de Janeiro* 13, 147–148.

Kohn, A. (1970b) *Pararhipidocotyle jeffersoni* gen. n., sp. n. trematódeo bucefaliforme parasito de dourado. *Atas da Sociedade de Biologia do Rio de Janeiro* 13, 181–183.

Kohn, A. (1970c) *Bellumcorpus schubarti* (Kohn, 1963) comb. nov. para *Paurorhynchus schubarti* (Kohn, 1963) (Trematoda, Prosorhynchidae). *Atas da Sociedade de Biologia do Rio de Janeiro* 13, 185–186.

Kohn, A. & Fernandes, B.M.M. (1987) Redescription of *Teratotrema dubium* Travassos, Artigas & Pereira, 1928 (Callodistomidae: Digenea). *Memórias do Instituto Oswaldo Cruz* 82, 55–57.

Kohn, A., Fróes, O.M., Fernandes, B.M.M. & Abramson, B. (1983) Redescription of *Thometrema magnifica* (Szidat, 1954) Gibson & Bray, 1979 (Trematoda: Hemiuroidea), with comments on the validity of *T. portoalegrensis* Amato, 1968. *Journal of Helminthology* 57, 143–148.

Kohn, A., Fernandes, B.M.M., Gibson, D.I. & Fróes, O.M. (1990) On the Brazilian species of halipegine genera (Trematoda: Derogenidae) from fishes, with new morphological data, hosts and synonyms. *Systematic Parasitology* 16, 201–211.

Køie, M. (1979) On the morphology and life-history of *Monascus* [= *Haplocladus*] *filiformis* (Rudolphi, 1819) Looss, 1907 and *Steringophorus furciger* (Olsson, 1868) Odhner, 1905 (Trematoda: Fellodistomidae). *Ophelia* 18, 113–132.

Korkhaus, R. (1930) *Proshystera rossittensis*, ein bisher nicht beshreibener Parasit der Vögelniere. *Tierärztliche Rundschau* 36, 293–294.

Kossack, W.F.K. (1911) Über Monostomiden. *Zoologische Jahrbücher. Abteilung für Systematik, Ökologie und Geographie der Tiere* 31, 491–590.

Kowalewski, M. (1895) *Bilharzia polonica* sp. nov. (Studya helmintologiczyne 3.) *Rozprawy Wydzialu Matematyczno-Przyrodniczego Polskiej Akademii Umiejetnosci, Krakow* 31, 41–70.

Kruidenier, J.F. & Gallicchio, V. (1959) The orthography of the Brachylaimidae (Joyeux and Foley, 1930); *Brachylaime microti* n. sp; *B. rauschii* McIntosh, 1950; and addendum to Dollfus' (1935) list of *Brachylaime* (Trematoda, Digenea). *Transactions of the American Microscopical Society* 77, 428–441.

Kurashvili, B.E. (1953) [A new genus and species of trematodes from *Lymnocryptes minima* L.] *Soobshcheniya Akademii Nauk Gruzinskoi SSR* 14, 357–360. (In Russian.)

Kurashvili, B.E. (1957) [*Helminths of commercially important game birds of Georgia from a faunistic and ecological aspect.*] Akademii Nauk, Moscow, 434 pp. (In Russian.)

Kurochkin, Y.V. (1975) *Nuitrema strigeiformis* gen. et sp. nov. A representative of a new family, Nuitrematidae fam. nov. (Trematoda, Bucephalata). *Trudy Biologopochvennogo Instituta, Novaya Seriya* 26, 67–72. (In Russian.)

Lacaze-Duthiers, H. (1854) Mémoire sur le bucéphale haime (*Bucephalus haimeanus*), helminthe parasite des huitres et des bucardes. *Annales des Sciences Naturelles, Zoologie, ser. 4* 1, 294–302.

Lal, M.B. (1936) A new genus of trematodes of the subfamily Typhlocoelinae from the shoveller duck *Spatula clypeata*. *Proceedings of the Indian Academy of Sciences* 4, 45–51.

Lal, M.B. (1939) Studies in helminthology. Trematode parasites of birds. *Proceedings of the Indian Academy of Science, Sect. B.* 10, 111–200.

La Rue, G.R. (1926) Studies on the trematode family Strigeidae (Holostomidae. II. Taxonomy; III. Relationships). *Transactions of the American Microscopical Society* 45, 11–19, 265–280.

La Rue, G.R. (1957) The classification of digenetic Trematoda: a review and a new system. *Experimental Parasitology* 6, 306–344.

Lauckner, G. (1983) Diseases of Mollusca: Bivalvia. In: Kinné, O. (ed.) *Diseases of Marine Animals*, Vol. II. *Introduction, Mollusca: Bivalvia to Scaphopoda*. Biologische Anstalt Helgoland, Hamburg, pp. 477–961.

Layman, E.M. (1930) [Parasitic worms from fishes of Peter the Great Bay.] *Izvestiya Tikhookeanskoi Nauchno-Promyslovi Ostantsii* 3, 1–120. (In Russian and German.)

Lebedev, B.I. (1968) [Trematodes of the family Bucephalidae Poche, 1907 from commercially important marine fishes of New Zealand and Australia.] In: [*Parasites of animals and plants*, Vol. 4.] Akademiya Nauk Moldavskoii SSR, Institut Zoologii, Izdatel'stvo Nauka, Moscow, pp. 156–167. (In Russian.)

Lebedev, B.I. & Parukhin, A.M. (1972) [New sanguinicolids (Trematoda, Sanguinicolidae) of f-ish from the Gulf of Mannar (coast of India).] *Nauchnye Doklady Vysshei Shkoly, Biologicheskie Nauki* 4, 7–14. (In Russian.)

Lebour, M.V. (1908) Fish trematodes of the Northumberland coast. In: *Northumberland Sea Fisheries Committee (1907)*, pp. 23–65.

Lee, S.-H., Chai, J.-Y. & Hong, S.-T. (1993) *Gymnophalloides seoi* n. sp. (Digenea: Gymnophallidae), the first report of human infection by a gymnophallid. *Journal of Parasitology* 79, 677–680.

Leidy, J. (1856) A synopsis of entozoa and some of their ectocongeners observed by the author. *Proceedings of the Academy of Sciences of Philadephia* 8, 42–58.

Leigh, N.H. (1954) Notes on the life-history of *Fibricola texensis* Chandler, 1942, in Florida. *Journal of Parasitology* 40, 45.

Leonov, V.A., Spasskii, A.A. & Kulikov, V.V. (1963) A new parasite of Charadriiformes – *Ovarioptera sobolevi* gen. et sp. nov. (Ovariopteridae). *Helminthologia* 4, 283–289.

León-Règagnon, V. (1998) *Machidatrema* n. gen. (Digenea: Hemiuridae: Bunocotylinae) and phylogenetic analysis of its species. *Journal of Parasitology* 84, 140–146.

León-Règagnon, V., Pérez-Ponce de León, G. & Brooks, D.R. (1998) Phylogenetic analysis of the Bunocotylinae Dollfus, 1950 (Digenea: Hemiuridae). *Journal of Parasitology* 84, 147–152.

Le Roux, P.L. (1958) The validity of *Schistosoma capense* (Harley, 1864) amended as a species. *Transactions of the Royal Society of Tropical Medicine and Hygiene* 52, 12–14.

Lewis, P.D., Jr (1974) Helminths from terrestrial molluscs in Nebraska. III. Life cycle of *Urogonimus certhiae* (McIntosh, 1927) Kagan, 1952 (Digenea: Leucochloridiidae). *Journal of Parasitology* 60, 256–259.

Le Zotte, L.A. (1954) Studies on marine digenetic trematode of Puerto Rico: the family Bivesiculidae, its biology and affinities. *Journal of Parasitology* 40, 148–162.

Linton, E. (1898) Notes on trematode parasites of fishes. *Proceedings of the United States National Museum* 20, 507–548 + 15 pl.

Linton, E. (1900) Fish parasites collected at Woods Hole in 1898. *Bulletin of the United States Fisheries Commission* 19(1899), 267–304.

Linton, E. (1905) Parasites of fishes of Beaufort, North Carolina. *Bulletin of the Bureau of Fisheries* 24, 321–428.

Linton, E. (1910) Helminth fauna of the Dry Tortugas. II. Trematodes. *Papers from the Tortugas Laboratory of the Carnegie Institute of Washington* 4 (Publication 133), 11–98.

Linton, E. (1940) Trematodes from fishes mainly from the Woods Hole region, Massachusetts. *Proceedings of the United States National Museum* 88, 1–172.

Littlewood, D.T.J., Rohde, K. & Clough, K.A. (1999a) The interrelationships of all major groups of Platyhelminthes: Phylogenetic evidence from morphology and molecules. *Biological Journal of the Linnean Society* 66, 75–114.

Littlewood, D.T.J., Rohde, K., Bray, R.A. & Herniou, E.A. (1999b) Phylogeny of the Platyhelminthes and the evolution of parasitism. *Biological Journal of the Linnean Society* 68, 257– 287.

Litvaitis, M.K. & Rohde, K. (1999) A molecular test of platyhelminth phylogeny: inferences from partial 28S rDNA sequences. *Invertebrate Biology* 118, 42–56.

Liu S. (1995) Two new species of trematodes of marine fishes from Fujian, China (Monorchiidae, Bivesiculidae). *Journal of Xiamen University (Natural Science)* 34, 292–295. (In Chinese.)

Liu Z. & Bai G. (1976) [On bird schistosomes from Jilin Province: *Jilinobilharzia crecci* gen. nov., sp. nov. (Schistosomatidae: Bilharziellinae) with a discussion on the taxonomy of the subfamily of Bilharziinae.] *Acta Zoologica Sinica* 22, 385-392. (In Chinese.)

Long S. & Lee W.C. (1964a) [Trematodes from fishes of Tai hu Lake. VI. The description of three new genera and four new species of Bucephalidae Poche, 1907 and the revision of family characteristics.] In: [*Compilation of Abstracts of China Parasitological Society, November 1963.*] Science Press, Beijing, p. 132. (In Chinese.)

Long S. & Lee W.C. (1964b) [Worm parasites from Taihu fishes, digenetic trematodes. VI. Bucephalidae Poche, 1907, with a description of two new genera and four new species and a revision of the family.] *Acta Zoologica Sinica* 16, 567-580. (In Chinese.)

Loos-Frank, B. (1970) Zur Kenntnis der gymnophalliden Trematoden des Nordseeraumes. II. *Lacunovermis macomae* (Lebour, 1908) n. comb. (Syn.: *Gymnophallus macroporus* Jameson and Nicoll, 1913) und seine Metacercariae. *Zeitschrift für Parasitenkunde* 35, 130-139.

Loos-Frank, B. (1971) Zur Kenntnis der gymnophalliden Trematoden des Nordseeraumes. IV. Übersicht über die gymnophalliden Larven aus Mollusken der Gezeitenzone. *Zeitschrift für Parasitenkunde* 36, 206-232.

Looss, A. (1885) Beitrage zür Kenntnis der Trematoden. *Distomum palliatum* nov. spec. und *Distomum reticulatum* nov. spec. *Zeitschrift für Wissenschaftliche Zoologie* 41, 390-446.

Looss, A. (1894) Die Distomen unserer Fische und Frösche. Neue Untersuchungen über Bau und Entwicklung des Distomenkörpers. *Bibliotheca Zoologica, Stuttgart* 16.

Looss, A. (1899) Weitere Beiträge zur Kenntniss der Trematoden-Fauna Aegyptens, zugleich Versuch einer natürlichen Gliederung des Genus *Distomum* Retzius. *Zoologische Jahrbücher (Systematik)* 12, 521-784.

Looss, A. (1901) Natura doceri, eine Erklärung und Begründung einiger Grundsätze, welche mich bei meinem "Versuche einer natürlichen Gliedrung des Genus *Distomum* Retzius" geleitet haben. *Zentralblatt für Bakteriologie, Parasitenkunde, Infektionskrankheiten (und Hygiene)* 29, 191-210.

Looss, A. (1902) Über die Gattung *Orophocotyle* n. g. (Zur Kenntnis der Trematoden fauna des Triester Hafens. 1). *Zentralblatt für Bakteriologie, Parasitologie, Infektionskrankheiten (und Hygiene), I. Orig.* 31, 637-644.

Looss, A. (1907) Zur Kenntnis der Distomenfamilie Hemiuridae. *Zoologischer Anzeiger* 31, 585-620.

Looss, A. (1908) Beiträge zur Systematik der Distomen. Zur Kenntnis der Familie Hemiuridae. *Zoologische Jahrbücher (Systematik)* 26, 63-180.

Looss, A. (1912) Über den Bau einiger anscheinend seltner Trematoden-Arten. *Zoologischer Jahrbücher, Suppl.* 15(1), 324-366 + 3 pl.

Lotz, J.M. & Corkum, K.C. (1975) *Parabrachylaima euglandensis* gen. et sp. n. (Trematoda: Brachylaimidae) from the terrestrial snail, *Euglandina rosea* (Ferussac). *Journal of Parasitology* 61, 870-872.

Lühe, M.F.L. (1901) Über Hemiuriden. (Ein Beitrag zur Systematik der digenetischen Trematoden). *Zoologischer Anzeiger* 24, 394-403, 473-488.

Lühe, M.F.L. (1909) Parasitische Plattwürmer. I. Trematodes. *Die Süsswasserfauna Deutschlands* 17, 1-217.

Lutz, A. (1933) Notas sobre Dicranocercarias brazilieras. *Memórias do Instituto Oswaldo Cruz* 27, 349-402.

Lutz, A. (1935) Beobachtungen und Betrachtungen über Cyathocotylinen und Prohemistominen. *Memórias do Instituto Oswaldo Cruz* 30, 157-182.

MacCallum, G.A. (1913) Notes on four trematode parasites of marine fishes. *Centralblatt für Bakteriologie und Parasitenkunde* 70, 407-416.

MacCallum, G.A. (1921) Studies in helminthology: part 1 - trematodes, part 2 - cestodes, part 3 - nematodes. *Zoopathologica* 1, 77-98.

McCauley, J.E. & Pratt, I. (1961) A new genus *Deropegus* with a redescription of *D. aspina* (Ingles, 1936) nov. comb. *Transactions of the American Microscopical Society* 80, 373–377.

Machalska, J. (1978) The morphological variability and taxonomic status of *Urogonimus macrostomus* (Rudolphi, 1803) (Trematoda, Leucochloridiidae). *Acta Parasitologica Polonica* 26, 1-9.

Machida, M. (1980) Hemiurid trematodes of *Kyphosus* collected around Cape Shionomisaki, Kii Peninsula. *Memoirs of the National Science Museum* 13, 113–120.

Machida, M. & Kamegai, S. (1997) Digenean trematodes from deep-sea fishes of Suruga Bay, central Japan. In: *Deep-sea Fauna and Pollutants in Suruga Bay*. National Science Museum Monographs No. 12, National Science Museum, Tokyo, pp. 19–30.

Machida, M. & Kuramochi, T. (1994) Two new species of trematodes (Gorgoderidae and Lepocreadiidae) from deep-sea fishes of Suruga Bay, Japan. *Bulletin of the National Science Museum, Tokyo* 20, 149–153.

McIntosh, A. (1934a) A new blood trematode, *Paradeontacylix sanguinicoloides* n. g., n. sp., from *Seriola lalandi* with a key to the species of the family Aporocotylidae. *Parasitology* 26, 463–467.

McIntosh, A. (1934b) Two new species of trematodes *Scaphiostomum pancreaticum* n. sp. and *Postharmosomum laruei* n. sp. from chipmunk. *Proceedings of the Helminthological Society of Washington* 1, 2–14.

Macko, J.K. & Busa, V. (1960) Revision der Systematik der Typhlocoeliden. *Helminthologia* 2, 21–34.

Macko, J.K. & Feige, R. (1960) Revision einiger Cyclocoelidengattungen und Arten auf Grund der Variabilitat von *Haematotrephus lanceolatum* (Wedl. 1858). *Helminthologia* 2, 254–265.

McLaren, D.J. & Hockley, D.J. (1977) Blood flukes have a double outer membrane. *Nature* 269, 147–149.

McLeod, J.A. (1937) Two new schistosomid trematodes from water-birds. *Journal of Parasitology* 23, 456–466.

Madhavi, R. (1975) Digenetic trematodes from marine fishes of Waltair coast, Bay of Bengal. Family Fellodistomatidae. *Rivista di Parassitologia* 36, 267–278.

Madhavi, R. & Hanumantha Rao, K. (1970) *Orchispirium heterovitellatum* gen. et sp. n. (Trematoda: Sanguinicolidae) from the ray fish, *Dasyatis imbricatus* Day, from Bay of Bengal. *Journal of Parasitology* 56, 41–43.

Maillard, C. (1975) *Labratrema lamirandi* (Carrère, 1937) (Trematoda: Bucephalidae) parasite de *Dicentrarchus labrax* (L., 1758). Création du genre *Labratrema*. Cycle évolutif. *Bulletin de la Museum National d'Histoire Naturelle, Paris* 283, *Zoologie* 193, 69–80.

Malhotra, S.K., Nanda, S., Mukherjee, M., Ghosh, K., Sukul, N.C. & Capoor, V.N. (1989) *Bengalotrema striatum* gen. et sp. n. (Trematoda, Hemiuroidea) [from] *Hilsa ilisha* of West Bengal with a revised key to the families of superfamily Hemiuroidea (Faust, 1929) emend. Yamaguti (1970). *Indian Journal of Helminthology* 40, 135–140.

Mañé-Garzón, F. & Spector, B.H. (1976) Una nueva especie de Aspidogastrea *Lobatostoma platense* n. sp. del intestino de *Trachinotus glaucus* Bloch. *Revista de Biologia del Uruguay* 4, 67–78.

Manter, H.W. (1925) Some marine fish trematodes of Maine. *Journal of Parasitology* 12, 11–18.

Manter, H.W. (1934) Some digenetic trematodes from deep-water fish of Tortugas, Florida. *Papers from Tortugas Laboratory* 28, 257–345 (Publication 435, Carnegie Institution of Washington).

Manter, H.W. (1938) A collection of trematodes from Florida Amphibia. *Transactions of the American Microscopical Society* 57, 26–37.

Manter, H.W. (1940) Digenetic trematodes of fishes from the Galapagos Islands and the neighboring Pacific. *Allan Hancock Pacific Expedition* 2, 325–497.

Manter, H.W. (1947) The digenetic trematodes of marine fishes of Tortugas, Florida. *American Midland Naturalist* 38, 257–416.

Manter, H.W. (1953) Two new species of Prosorhynchinae (Trematoda: Gasterostomata) from the Fiji Islands. In: Dayal, J. & Singh, K. (eds) *Thapar, G.S., Commemoration Volume.* University of Lucknow, Lucknow, India, pp. 193–200.

Manter, H.W. (1954) Some digenetic trematodes from fishes of New Zealand. *Transactions of the Royal Society of New Zealand* 82, 475–568.

Manter, H.W. (1960) Some additional Digenea (Trematoda) from New Zealand fishes. In: *Libro homenaje al Dr Eduardo Caballero y Caballero.* Secretaria de Educación Publica, Mexico, pp. 197–201.

Manter, H.W. (1961) Studies on digenetic trematodes of fishes of Fiji. I. Families Haplosplanchnidae, Bivesiculidae, and Hemiuridae. *Proceedings of the Helminthological Society of Washington* 28, 67–74.

Manter, H.W. (1962) Notes on the taxonomy of certain digenetic trematodes of South American freshwater fishes. *Proceedings of the Helminthological Society of Washington* 29, 97–102.

Manter, H.W. (1969) Three new genera of hemiurid trematodes from Australian fishes. *Journal of the Fisheries Research Board of Canada* 26, 787–792.

Manter, H.W. & Crowcroft, P.W. (1950) A new genus of amphistome (Trematoda) from a Tasmanian marine fish. *Proceedings of the Helminthological Society of Washington* 17, 122–126.

Manter, H.W. & Larson, M.I. (1950) Two new blood flukes from a marine turtle, *Caretta caretta. Journal of Parasitology* 36, 595–599.

Manter, H.W. & Pritchard, M.H. (1960) Some hemiurid trematodes from Hawaiian fishes. *Proceedings of the Helminthological Society of Washington* 27, 87–102.

Manter, H.W. & Pritchard, M.H. (1961) Studies on digenetic trematodes of Hawaiian fishes: family Bucephalidae. *Journal of Parasitology* 47, 479–582.

Margolis, L. (1958) A new species of *Lecithophyllum* from North Pacific fishes with a consideration of the taxonomy of the genera *Lecithophyllum, Aponurus* and *Brachadena* (Trematoda: Hemiuridae). *Canadian Journal of Zoology* 36, 893–904.

Marshall, W.S. & Gilbert, N.C. (1905) Three new trematodes found principally in black bass. *Zoologischer Jahrbücher, Systematik* 22, 477–488.

Martin, W.E. (1960) Hawaiian helminths. IV. *Paracardicola hawaiensis* n. gen., n. sp. (Trematoda: Sanguinicolidae) from the balloon fish, *Tetraodon hispidus* L. *Journal of Parasitology* 46, 648–650.

Martin, W.E. (1974) *Paracardicoloides yamagutii* gen. et sp. n. from an Australian eel (Trematoda: Sanguinicolidae). *Proceedings of the Helminthological Society of Washington* 41, 22–25.

Martin, W.E. (1975) *Plethorchis acanthus* gen. et sp. n. (Trematoda: Sanguinicolidae) in mullet, *Mugil cephalus* L., from Queensland, Australia. *Proceedings of the Helminthological Society of Washington* 42, 79–82.

Martin, W.E. & Bamberger, J.W. (1952) New blood flukes (Trematoda: Spirorchidae) from the marine turtle, *Chelonia mydas* (L.). *Journal of Parasitology* 38, 105–110.

Martorelli, S.R. & Cremonte, F. (1998) A proposed three-host life history of *Monascus filiformis* (Rudolphi, 1819) (Digenea: Fellodistomidae) in the southwest Atlantic Ocean. *Canadian Journal of Zoology* 76, 1198–1203.

Mas-Coma, S. & Gallego, J. (1975) Algunas consideraciones sistematicas sobre las familias Brachylaemidae Joyeux y Foley, 1930 y Leucochloridiomorphidae Travassos y Kohn, 1966 (Trematoda: Brachylaemoidea). *Revista Iberica de Parasitologia* 5, 339–354.

Matthews, R.A. (1973a) The life cycle of *Prosorhynchus crucibulum* (Rudolphi, 1819) Odhner, 1905, and a comparison of its cercaria with that of *Prosorhynchus squamatus* Odhner, 1905. *Parasitology* 66, 133–164.

Matthews, R.A. (1973b) The life cycle of *Bucephalus haimeanus* Lacaze-Duthiers, 1854 from *Cardium edule* L. *Parasitology* 67, 341–350.

Meckel von Hemsbach, J.H. (1856) *Mikrogeologie. Über die Concrement im thierischen Organismus. Nach des Verfassers Tode herausgegeben und bevorwortet von Dr Theodor von Billroth*. Berlin, 275 pp.

Megnin, J.P. (1890) Un parasite nouveau et dangéreux de l'oie cabouc (*Sarcidiornis melanota*). *Comptes Rendus des Séances de la Société de Biologie* 42, 87–90.

Mehlis, E. (1831) Novae observationes de entozois. In: Creplin, F.C.H. (ed.) *Isis. Encyclopädische Zeitung von Oken*, pp. 68–99, 166–199.

Mehra, H.R. (1934) New blood flukes of the family Spirorchidae Stunkard from Indian fresh-water tortoises with discussion of the synonymy of certain genera and the relationships of the families of blood flukes. Part II. *Bulletin of the Academy of Sciences of the United Provinces of Agra and Oudh* 3, 169–195.

Mehra, H.R. (1939) New blood flukes of the family Spirorchidae Stunkard (Trematoda) from the marine turtle *Chelone* [sic] *mydas* of the Arabian Sea with observations on the synonymity of certain genera and classification of the family. *Proceedings of the National Academy of Sciences of India* 9, 155–167.

Mehra, H.R. (1940) A new distome *Enterohaematotrema* n. g. and a new blood fluke *Hemiorchis bengalensis* n. sp. belonging to the family Spirorchidae Stunkard, and a new species of the genus *Dendritobilharzia* Skrjabin and Zakharow belonging to the family Schistosomatidae Poche, with remarks on the evolution of blood flukes. *Proceedings of the National Academy of Sciences, India* 10, 100–118.

Mehra, H.R. (1950) Evolution of the blood flukes and strigeid trematodes. Recent life-history studies and phylogeny. Phylogeny of supersuperfamily Strigeida Poche (Suborder Strigeata La Rue). *Proceedings of the National Academy of Sciences, India* (1950), 9–38.

Mehra, H.R. (1956) Studies on the family Cyathocotylidae Poche. Part 3. Subfamily Muhlingininae Mehra, 1950, with description of *Muhlingina lutrai* Mehra, 1950. *Records of the Indian Museum* 54, 195–208.

Mehra, H.R. (1962a) Revision of Brachylaemidae Joyeux et Foley, 1930 with new subfamilies Thapariellinae and Urotrematinae and new family Harmotrematidae with its subfamilies Harmotrematinae Yamaguti, 1933 and Helicotrematinae n. subf. *Proceedings of the National Academy of Sciences, India, Sect. Biology* 32, 319–334.

Mehra, H.R. (1962b) On two new families of superfamily Hemiuroidea Faust, 1929; Oesophagicolidae n. fam. from marine snakes and Arnolidae n. fam. from marine fishes. *Proceedings of the National Academy of Sciences, India, Sect. Biology* 32, 93–96.

Mehra, H.R. (1970) Classification of Trematoda Rudolphi. Status of class Trematoda Rudolphi, 1808, class Monogenea van Beneden, 1858 (Monogenoidea Bychowsky, 1957) and subclasses Digenea Beneden, 1858 and Aspidobothrea Burmeister, 1856 of class Trematoda and phylogeny of these groups. In: *Srivastava Commemoration Volume*. Indian Veterinary Research Institute, Izatnagar, UP, pp. 43–54.

Mehrotra, V. (1973a) Digenea from some reptilian hosts in India. Part I. In: *Proceedings of the Sixtieth Indian Science Congress*, Part IV, p. 46.

Mehrotra, V. (1973b) Digenea from some reptilian hosts in India. Part II. In: *Proceedings of the Sixtieth Indian Science Congress*, Part IV, pp. 46–47.

Miller, H.M. & Northup, F.E. (1926) The seasonal infestation of *Nassa obsoleta* (Say) with larval trematodes. *Biological Bulletin of the Marine Biological Laboratory Woods Hole* 50, 490–508.

Miller, J.N. (1936) A study of *Brachylaima* (*Postharmostomum*) *sexconvolutum* n. sp., a trematode parasite of the deer mouse. *Ohio State University Dissertations* 19, 81–86.

Mishchenko, V.F. (1974) [The life-cycle and ontogenesis of *Linstowiella viviparae* (Prohemistimatidae).] *Trudy Gel'mintologicheskoi Laboratorii* 24, 102–112. (In Russian.)

Montgomery, R.E. (1906) Observations on bilharziosis in India. *Journal of Tropical Veterinary Science* 1, 138–174.

Monticelli, F.S. (1888) Saggio di una morphologia dei trematodi. Tesi per ottenere la privata docenza in Zoologia nella R. Università di Napoli, Naples, Stabilmento Tipigrafico Fili, Ferrante, 130 pp.

Monticelli, F.S. (1892a) *Cotylogaster michaelis* n. g., n. sp. e revisione degli Aspidobothridae. In: Taschenberg, E.O.W. (ed.) *Festschrift zum Siebenzigsten Geburtstage R. Leuckarts: dem Verchrten Jubilar Dargebracht von Seinen Dankbaren Schülern &c.* Leipzig, pp. 168–214.

Monticelli, F.S. (1892b) Studii sui trematodi endoparassiti *Monostomum cymbium* Diesing. Contribuzione allo studio dei monostomidi. *Memorias della Reale Accademia delle Scienze di Torino* 42, 683–727.

Morishita, K. (1924a) [A new species of *Cyclocoelum*, monostomous trematode, parasitic in "toratsugumi" (*Turdus dauma aureus*) from Hachijo Island.] *Zoological Magazine, Tokyo* 36, 231–234. (In Japanese.)

Morishita, K. (1924b) On the trematodes of the genus *Cyclocoelum* obtained in Japan, with notes on the phylogeny of the monostomatous trematodes. *Dobutsujaku Zaschi, Tokyo* 424, 189–194.

Morozov, F.N. (1955) [Family Gymnophallidae Morosov, 1955.] In: Skrjabin, K.I. (ed.) [*Trematodes of animals and man.*] *Osnovy Trematodologii* 10, 315–335. (In Russian.)

Mühling, P. (1898) Die Helminthen-Fauna der Wirbeltiere Ostpreussens. *Archiv für Naturgeschichte* 64, 1–118.

Nagaty, H.F. (1937) *Trematodes of Fishes from the Red Sea. Part I. Studies on the Family Bucephalidae Poche, 1907.* Faculty of Medicine Publication No. 12, Egyptian University, Cairo, 172 pp.

Nagaty, H.F. & Abdel Aal, T.M. (1962) Trematodes of fishes from the Red Sea. Part 14. On three new hemiurid species including a new genus. *Journal of the Arab Medical and Veterinary Association* 22, 231–237.

Nahhas, F.M. & Cable, R.M. (1964) Digenetic and aspidogastrid trematodes from marine fishes of Curaçao and Jamaica. *Tulane Studies in Zoology* 11, 167–228.

Nahhas, F.M. & Carlson, K. (1994) *Digenetic Trematodes of Marine Fishes of Jamaica, West Indies.* Hofstra University Marine Laboratory, Ecological Survey of Jamaica, Publication No. 2, Stockton, California, 60 pp.

Nezlobinski, N. (1926) [Helminthological studies in the Ohrida valley. I. On kidney trematodes in birds.] *Glasnik Tsentralnog Khigiyenskog Zavoda, Beograd* 1, 202–217. (In Serb.)

Nickerson, W.S. (1895) On *Stichocotyle nephropis* Cunningham, a parasite of the American lobster. *Zoologische Jahrbücher Anatomie* 8, 447–480.

Nicoll, W. (1909) Studies on the structure and classification of the digenetic trematodes. *Quarterly Journal of Microscopical Science* 53, 391–487.

Nicoll, W. (1910) On the entozoa of fishes of the Firth of Clyde. *Parasitology* 3, 322–359.

Nicoll, W. (1913) New trematode parasites from fishes of the English Channel. *Parasitology* 5, 238–246.

Nicoll, W. (1914a) The trematode parasites of fishes from the English Channel. *Journal of the Marine Biological Association of the United Kingdom* 10, 466–505.

Nicoll, W. (1914b) Trematode parasites from animals dying in the Zoological Society's gardens during 1911–1912. *Proceedings of the Zoological Society of London* 1, 139–154.

Nicoll, W. (1915) A list of the trematode parasites of British marine fishes. *Parasitology* 7, 339–378.

Niewiadomska, K. (1964) The life cycle of *Codonocephalus urnigerus* (Rudolphi, 1819) – Strigeidae. *Acta Parasitologica Polonica* 12, 283–296.

Niewiadomska, K. (1970) An analysis of criteria for generic differentiation within the order Strigeidida (La Rue, 1926). *Acta Parasitologica Polonica* 18, 277–289.

Niewiadomska, K. (1971) On the necessity of dividing the genus *Cotylurus* Szidat, 1928 into two valid genera *Cotylurus* Szidat, 1928 and *Ichthyocotylurus* Odening, 1969 (Strigeidae). *Acta Parasitologica Polonica* 19, 113–120.

Niewiadomska, K. (1973) Some aspects of the biology and evolution of Strigeata La Rue and their importance in systematics in this group of Trematoda. *Acta Parasitologica Polonica* 21, 21–62.

Nitzsch, C.L. (1819) Amphistoma. In: *Allgemeine Encyklopedie der Wissenschaftliche Künste in Alphabetische Folge (Ersch und Gruber)*, Vol. 3. Leipzig, pp. 397–401.

Odening, K. (1961a) Historische und moderne Gesichtspunkte beim Aufbau eines natürlichen Systems der digenetischen Trematoden. *Biologische Beiträge* 1, 73–90.

Odening, K. (1961b) Zur Kenntnis der Trematodenfauna einiger Vögelarten. *Mitteilungen aus dem Zoologischen Museum in Berlin* 37, 125–146.

Odening, K. (1965) Der Lebenszyklus von *Neodiplostomum spathoides* Dubois (Trematoda, Strigeida) im Raum Berlin nebst Beitragen zur Entwicklungsweise verwandter Arten. *Zoologische Jahrbücher (Systematik)* 92, 523–624.

Odening, K. (1974) Verwandtschaft, System und zyklo-ontogenetische Besonderheiten der Trematoden. *Zoologische Jahrbücher (Systematik)* 101, 345-396.

Odhner, T. (1900a) *Aporocotyle simplex* n. g., n. sp., einer neuer Typus von ektoparasitischen Trematoden. *Zentralblatt für Bakteriologie, Parasitenkunde, Infektionskrankheiten (und Hygiene)* 1(27), 62–66.

Odhner, T. (1900b) *Gymnophallus*, eine neue Gattung von Vogeldistomen. *Zentralblatt für Bakteriologie, Parasitenkunde, Infektionskrankeiten (und Hygiene)* 1(28), 12–23.

Odhner, T. (1902) Trematoden aus Reptilien nebst allgemeinen systematischen Bemerkungen. *Öfversigt af Kongl. Vetenskaps-Academiens. Förhandlingar. Stockholm* 59, 19–45.

Odhner, T. (1905) Die Trematoden de arktischen Gebietes. *Fauna Arctica* 4, 289–372.

Odhner, T. (1910a) *Nordostafrikanische Trematoden, grösstenteils vom Weissen Nil (von der schwedischen zoologischen Expedition gesammelt)*. Results of the Swedish Zoological Expedition to Egypt and the White Nile 1901 under the direction of L.A. Jägerskiöld, No. 23A, 170 pp.

Odhner, T. (1910b) *Gigantobilharzia acotylea* n. g., n. sp., ein mit den Bilharzien verwandter Blutparasit von enormer Länge. *Zoologischer Anzeiger* 35, 380–385.

Odhner, T. (1911a) Zum natürlichen System der digenen Trematoden. III. (Ein weiterer Fall von sekundärum Anus). *Zoologischer Anzeiger* 38, 97–117.

Odhner, T. (1911b) Zum natürlichen system der digenen Trematoden. IV. *Zoologischer Anzeiger* 38, 513-531.

Odhner, T. (1911c) *Sanguinicola* M. Plehn - ein digenetischer Trematode! Mit einem Nachtrag über ältere Beobachtungen von Prof. A. Looss, Kairo. *Zoologischer Anzeiger* 38, 33–45.

Odhner, T. (1912) Zum natürlichen System der digenen Trematoden. V. Die phylogenie des *Bilharzia*-typus. *Zoologischer Anzeiger* 41, 54–71.

Odhner, T. (1926) *Protofasciola* n. g., ein Prototypus des grossen Leberegels. *Arkiv för Zoologi* 18(20), 1-7.

Odhner, T. (1927) Über Trematoden aus der Schwimmblase. *Arkiv för Zoologi* 19(15), 1–9.

Odhner, T. (1928) Ein neuer Trematode aus dem Flussbarsch. *Arkiv för Zoologi* 20(1), 1–3.

Orlov, I.V., Ershov, W.S. & Badanin, N.V. (1934) Eine neue Trematodenkrankheit der Schafe, harvorgerufen durch die *Skrjabinotrema ovis* nov. g., n. sp. (Brachylaimidae). *Wiener Tierätzliche Monatsschrift* 21, 321–326.

Overstreet, R.M. (1969) Digenetic trematodes of marine teleost fishes from Biscayne Bay, Florida. *Tulane Studies in Zoology and Botany* 15, 119–176.

Overstreet, R.M. & Køie, M. (1989) *Pearsonellum corventum*, gen. et sp. nov. (Digenea: Sanguinicolidae), in serranid fishes from the Capricornia Section of the Great Barrier Reef. *Australian Journal of Zoology* 37, 71–79.

Overstreet, R.M. & Martin, D.M. (1974) Some digenetic trematodes from synaphobranchid eels. *Journal of Parasitology* 60, 80–84.

Ozaki, Y. (1924) Preliminary notes on Gasterostomata with three new genera. *Zoological Magazine, Zoological Society of Tokyo* 36, 173–201.

Ozaki, Y. (1928) Some gasterostomatous trematodes of Japan. *Japanese Journal of Zoology* 2, 35–60.

Paggi, L. & Orecchia, P. (1975) *Tyrrhenia blennii* g.n. sp.n. (Hemiurata: Halipegidae Poche, 1925) parassita di *Blennius gattorugine* e *Blennius sanguinolentus*. *Parassitologia* 17, 57–64.

Palmieri, J.R., Krishnasamy, M. & Sullivan, J.T. (1979) Strigeoid trematodes of Malaysia with descriptions of a new genus and three new species. *Journal of Helminthology* 53, 51–63.

Pan J.-P. & Wang W.-J. (1985) [An ectoparasitic, digenetic trematode *Squamacola parvivitellaria* gen. et sp. nov. (Squamacolidae fam. nov.), found from freshwater fishes of China.] *Acta Zootaxonomica Sinica* 10, 344–349. (In Chinese.)

Pande, K.C. & Baugh, S.C. (1970) Studies on clinostome metacercariae. IV. An apharyngeal form, *Metaclinostomum srivastavai* n. g. n. sp. In: *H.D. Srivastava Commemoration Volume*. Indian Veterinary Research Institute, Izatnagar, UP, pp. 407–416.

Parukhin, A.M. (1976a) [*Parasitic worms of food fishes of the southern seas*.] Naukova 'Dumka', Kiev. (In Russian.)

Parukhin, A.M. (1976b) *Assitrema eichleri* gen. et sp. nov., eine neue Trematodenart aus dem Indik. *Angewandte Parasitologie* 17, 6–9.

Parukhin, A.M. (1985) [New species of trematodes from commercial fish in the Indian Ocean.] *Nauchnye Doklady Vysshei Shkoly, Biologicheskie Nauki* No. 8, 29–34. [In Russian.]

Parukhin, A.M. & Lyadov, V.N. (1979) New genus and species of trematodes, fish parasites in the subantarctic zone of the Indian Ocean. *Zoologicheskii Zhurnal* 58, 637–642. (In Russian.)

Pearson, J.C. (1956) Studies on the life-cycles and morphology of the larval stages of *Alaria arisaemoides* Augustine and Uribe, 1927 and *Alaria canis* LaRue and Fallis, 1936 (Trematoda: Diplostomidae). *Canadian Journal of Zoology* 34, 295–387.

Pearson, J.C. (1959) *Neodiplostomum intermedium* n. sp. from the allied rat, *Rattus assimilis*, with remarks on the genera *Neodiplostomum* and *Fibricola* (Trematoda: Diplostomatidae). *Parasitology* 49, 111–120.

Pearson, J.C. (1968) Observations on the morphology and life-cycle of *Paucivitellosus fragilis* Coil, Reid & Kuntz, 1965 (Trematoda: Bivesiculidae). *Parasitology* 58, 769–788.

Pearson, J.C. (1972) A phylogeny of life-cycle patterns of the Digenea. *Advances in Parasitology* 10, 153–189.

Pearson, J.C. (1992) On the position of the digenean family Heronimidae: an inquiry into a cladistic classification of the Digenea. *Systematic Parasitology* 21, 81–166.

Pekkarinen, M. & Ching, H.L. (1994) Comparisons of gymnophallid digeneans from North Pacific and Baltic clams, *Macoma balthica* (Bivalvia). *Journal of Parasitology* 80, 630–636.

Pence, D.B. & Bush, A.O. (1973) *Polycyclorchis eudocimi* gen. et sp. n. (Trematoda: Cyclocoelidae) from the trachea of the white ibis, *Eudocimus albus* L. *Journal of Parasitology* 59, 85–89.

Pigulewsky, S.W. (1931) Neue arten von trematoden aus fischen des dnjeprbassins. *Zoologischer Anzeiger* 96, 9–18.

Platt, T.R. (1988) *Hapalorhynchus brooksi* n. sp. (Trematoda: Spirorchiidae) from the snapping turtle (*Chelydra serpentina*), with notes on *H. gracilis* and *H. stunkardi*. *Proceedings of the Helminthological Society of Washington* 55, 317–325.

Platt, T.R. (1990) *Aphanospirorchis kirki* n. gen., n. sp. (Digenea: Spirorchidae), a parasite of the midland painted turtle, *Chrysemys picta marginata*, from northwestern Indiana, with comments on the proper spelling of the family name. *Journal of Parasitology* 76, 650–652.

Platt, T.R. (1992) A phylogenetic and biogeographic analysis of the genera of Spirorchinae (Digenea: Spirorchidae) parasitic in freshwater turtles. *Journal of Parasitology* 78, 616–629.

Platt, T.R. (1993) Taxonomic revision of *Spirorchis* MacCallum, 1919 (Digenea: Spirorchidae). *Journal of Parasitology* 79, 337–346.

Platt, T.R. & Blair, D. (1996) *Hapalotrema* Looss, 1899 (Digenea): proposed designation of *H. loossi* Price, 1934 as the type species. *Bulletin of Zoological Nomenclature* 53, 1–3.

Platt, T.R. & Blair, D. (1998) Redescription of *Hapalotrema mistroides* (Monticell, 1896) and *Hapalotrema synorchis* Luhman, 1935 (Digenea: Spirorchidae), with comments on other species in the genus. *Journal of Parasitology* 84, 594–600.

Platt, T.R. & Brooks, D.R. (1997) Evolution of the schistosomes (Digenea: Schistosomatoidea): the origin of dioecy and colonization of the venous system. *Journal of Parasitology* 83, 1035–1044.

Platt, T.R. & Pichelin, S. (1994) *Uterotrema australispinosa* n. gen., n. sp. (Digenea: Spirorchidae), a parasite of a freshwater turtle *Emydura macquarii* from southern Queensland, Australia, *Journal of Parasitology* 80, 1008–1011.

Platt, T.R., Blair, D., Purdie, J. & Melville, L. (1991) *Griphobilharzia amoena* n. gen., n. sp. (Digenea: Schistosomatidae), a parasite of the freshwater crocodile *Crocodylus johnstoni* (Reptilia: Crocodylia) from Australia, with the erection of a new subfamily, Griphobilharziinae. *Journal of Parasitology* 77, 65–68.

Plehn, M. (1905) *Sanguinicola armata* und *inermis* (n. gen., n. sp.) n. fam. Rhynchostomida. Ein entoparasitisches Turbellar im Blute von Cypriniden. *Zoologischer Anzeiger* 29, 224–252.

Plehn, M. (1908) Ein monozoischer Cestode als Blutparasit (*Sanguinicola armata* u. *inermis* Plehn). *Zoologischer Anzeiger* 33, 427–440.

Poche, F. (1907) Einige Bemerkungen zur Nomenclatur der Trematoden. *Zoologischer Anzeiger* 31, 124–126.

Poche, F. (1926) Das System der Platodaria. *Archiv für Naturgeschichte, Abt. A* 91(1925), 1–459.

Pojmańska, T. (1959) Metacercariae of some Brachylaemidae in land snails of the Białowieża National Park. *Acta Parasitologica Polonica* 7, 343–369.

Pojmańska, T. (1961) Investigations on the occurrence and biology of trematodes of *Sorex araneus araneus* L. in Białowieża National Park. *Acta Parasitologica Polonica* 9, 305–329.

Pojmańska, T. (1967) Variability of *Leucochloridium paradoxum* Carus (= *L. heckerti* Kagan, 1950) in natural and experimental conditions. *Acta Parasitologica Polonica* 14, 381–398.

Pojmańska, T. (1969) *Leucochloridium perturbatum* sp. n. (Trematoda: Brachylaimidae), morphology, individual variability and life cycle. *Acta Parasitologica Polonica* 16, 153–175.

Pojmańska, T. (1971) First record of *Leucochloridiomorpha lutea* (Baer, 1827) in Poland, and a critical review of representatives of the genus *Leucochloridiomorpha* Gower, 1938 (Trematoda, Brachylaimidae). *Acta Parasitologica Polonica* 19, 349–355.

Pojmańska, T. (1972a) *Amblosoma exile* g. n., sp. n. (Trematoda, Brachylaimidae, Leucochloridiomorphinae) - morphology of the adult and metacercaria. *Acta Parasitologica Polonica* 20, 35–44.

Pojmańska, T. (1972b) An attempt to estimate the value of some morphological and biological characters for establishing the taxa of various ranks within the superfamily Brachylaimoidea Allison, 1943 (Trematoda). *Acta Parasitologica Polonica* 20, 249–257.

Pojmańska, T. (1973) *Michajlovia migrata* gen. n., sp. n. (Trematoda, Brachylaimata) - the morphology of the adult. *Acta Parasitologica Polonica* 21, 9–20.

Pojmańska, T. (1975) Life cycle of *Neoleucochloridium holostomum* (Rudolphi, 1819) (Trematoda, Leucochloridiidae). *Acta Parasitologica Polonica* 23, 23–36.

Pojmańska, T. (1978) Life cycle of *Leucochloridium vogtianum* Baudon, 1881 (= *L. phragmitophila* Bychovskaja-Pavlovskaja et Dubinina, 1951 in parte) (Trematoda, Leucochloridiidae). *Acta Parasitologica Polonica* 25, 129–134.

Pozdnyakov, S.E. (1993) The systematics of trematodes of the Suborder Didymozoata. *Parazitologiya* 27, 415–425. (In Russian.)

Pratt, H.S. (1902) Synopses of North American invertebrates. 12. The trematodes. Part 2. The Aspidocotylea and the Malacocotylea, or digenetic forms. *American Naturalist* 36, 887–910.

Price, E.W. (1929) A synopsis of the trematode family Schistosomidae, with descriptions of new genera and species. *Proceedings of the United States National Museum* 75(18), 1–39.

Price, E.W. (1930) *Thaumatocotyle* Odhner, 1910, a preoccupied name. *Journal of Parasitology* 16, 161.

Price, E.W. (1934) New genera and species of blood flukes from a marine turtle, with a key to the genera of the family Spirorchidae. *Journal of the Washington Academy of Sciences* 24, 132–141.

Price, E.W. (1937) Three new genera and species of trematodes from cold-blooded vertebrates. In: Schulz, R.E.S. & Gnyedina, M.P. (eds) *Papers on Helminthology Published in Commemoration of the 30th Year Jubileum of the Scientific, Educational and Social Activities of the Honoured Worker of Science K.I. Skrjabin, M. Ac. Sci. and of the Fifteenth Anniversary of the All-Union Institute of Helminthology*. All-Union Lenin Academy of Agricultural Sciences, Moscow, pp. 483–490.

Price, E.W. (1943) A redescription of *Heterobilharzia americana* Price (Trematoda, Schistosomatidae). *Proceedings of the Helminthological Society of Washington* 10, 85–86.

Pritchard, M.H. (1963) Studies on digenetic trematodes of Hawaiian fishes, primarily families Lepocreadiidae and Zoogonidae. *Journal of Parasitology* 49, 578–587.

Puga, S. & Figueroa, L. (1989) *Pirupalkia queulensis* n. gen., n. sp. (Digenea: Callodistomidae) en el pez *Cilus montii* (Pomadasyidae). *Boletín Chileno de Parasitología* 44, 43–45.

Railliet, A. (1919) Nouveaux trématodes du chien. *Recueil de Médecine Vétérinaire* 95, 5–27.

Ramachandrula, A. & Agarwal, S.M. (1984) Studies on *Lissemysia ocella* n. sp. from *Vivapara bengalensis* at Raipur. *Japanese Journal of Parasitology* 33, 133–139.

Rao, S.L. (1976) On two new blood flukes (Trematoda: Spirorchidae) from the heart of *Chelone mydas* L. *Acta Parasitologica Polonica* 24, 119–124.

Rees, W.J. (1947) A cercaria of the genus *Haplocladus* from *Nucula nucleus* (L.). *Journal of the Marine Biological Association of the United Kingdom* 26, 602–604.

Rivolta, S. & Delparto, P. (1881) *L'ornitojatria o la medicina degli ucceli domestici e semidomestici*. Pisa, 500 pp.

Rohde, K. (1971) Untersuchungen an *Multicotyle purvisi* Dawes, 1941 (Trematoda: Aspidogastrea). 1. Entwicklung und Morphologie. *Zoologische Jahrbücher Anatomie* 88, 138–187.

Rohde, K. (1972) The Aspidogastrea, especially *Multicotyle purvisi* Dawes, 1941. *Advances in Parasitology* 10, 77–151.

Rohde, K. (1973) Structure and development of *Lobatostoma manteri* sp. nov. (Trematoda: Aspidogastrea) from the Great Barrier Reef, Australia. *Parasitology* 66, 63–68.

Rohde, K. (1994) The minor groups of parasitic Platyhelminthes. *Advances in Parasitology* 33, 145–234.

Rohde, K. & Watson, N.A. (1992) Ultrastructure of tegument, ventral sucker and rugae of *Rugogaster hydrolagi* (Trematoda, Aspidogastrea). *International Journal for Parasitology* 22, 967–974.

Rowan, W.B. (1955) A snail intermediate host of the rabbit trematode, *Hasstilesia tricolor* (Stiles and Hassall, 1894) Hall, 1916 (Trematoda: Brachylaemidae). *Transactions of the American Microscopical Society* 74, 1–21.

Rudolphi, C.A. (1803) Neue Beobachtungen über die Eigenwiedewurmer. *Archiv für Zoologie und Zootomie* 3(1), 1–125, 3(2), 1–32.

Rudolphi, C.A. (1809) *Entozoorum sive vermium intestinalium historia naturalis*, Vol. 2. Sumtibus Tabernae, Amstelaedami, 457 pp.

Rudolphi, C.A. (1819) *Entozoorum synopsis cui accedunt mantissa duplex et indices locupletissimi*. August Rücker, Berolini, 811 pp.

Sahai, D. & Srivastava, D.D. (1977) Trematodes of Indian fishes. Part 1. Two new genera of hemiurids (Subfamily Hemiurinae Looss, 1899). *Proceedings of the National Academy of Sciences of India* 47, 7–12.

Santos, V. (1934) Monostomose renal das aves domestica. *Revista do Departamento Nacional da Produccao Animal, Rio de Janeiro* 1, 1–9.

Saxena, V.K. (1952) A new trematode, *Paratylodelphus asiaticus* n. g., n. sp. (Diplostomidae Poirier, 1886) from black necked stork, *Xenorhynchus asiaticus* (Latham). *Business Matters of the National Academy of Sciences, India* 1952, 113–114.

Schell, S.C. (1973) *Rugogaster hydrolagi* gen. et sp. n. (Trematoda: Aspidobothrea: Rugogastridae fam. n.) from the ratfish, *Hydrolagus colliei* (Lay and Bennett, 1839). *Journal of Parasitology* 59, 803–805.

Schell, S.C. (1985) *Trematodes of North America North of Mexico*. University of Idaho Press, Moscow, Idaho, 263 pp.

Schmidt, R. (1965) Zur Kenntnis der Trematodenlarven aus Landmollusken. I. Der Entwicklungszyklus von *Urogonimus macrostomus* (Rudolphi 1803) (Trematoda: Digenea: Leucochloridiidae). *Zeitschrift für Parasitenkunde* 26, 1–17.

Scholz, T., Vargas-Vazquez, J. & Salgado-Maldonado, G. (1995) Revision of *Genarchella* species (Digenea: Derogenidae) parasitizing freshwater fishes in Mexico and Central America. *Journal of Natural History* 29, 1403–1417.

Seo, B.S. (1989) *Fibricola seoulensis* Seo, Rim et Lee, 1964 (Trematoda) and fibricoliasis. *Seoul Journal of Medicine* 31, 61–96.

Sewell, R.B.S. (1922) Cercariae Indicae. *Indian Journal of Medical Research* 10 (Suppl.), 1–370.

Sey, O. (1991) *CRC Handbook of the Zoology of Amphistomes*. CRC Press, Boca Raton, 480 pp.

Shen J. (1995) [Notes on a new genus and species of Treptodemidae (Trematoda: Digenea).] *Studia Marina Sinica* 36, 233–236. (In Chinese.)

Shimazu, T. (1995) Trematodes of the genus *Genarchopsis* (Digenea, Derogenidae, Halipeginae) from freshwater fishes of Japan. *Proceedings of the Japanese Society of Systematic Zoology* 54, 1–18.

Shoop, W.L. (1989) Systematic analysis of the Diplostomidae and Strigeidae (Trematoda). *Journal of Parasitology* 75, 21–32.

Short, R.B. (1954) A new blood fluke, *Selachohemecus olsoni*, n. g., n. sp. (Aporocotylidae) from the sharp-nosed shark, *Scoliodon terra-novae*. *Proceedings of the Helminthological Society of Washington* 21, 78–82.

Simha, S.S. (1977) On a new genus and species of a blood-fluke *Neocaballerotrema caballeroi*, from a marine turtle in India. In: *Excerta Parasitológica en Memoria del Doctor Eduardo Caballero y Caballero*. Universidad Nacional, Mexico, pp. 307–310.

Simha, S.S. & Chattopadhyaya, D.R. (1970) A new genus and species of blood fluke, *Squaroacetabulum solus*, from the ventricle of the heart of a marine turtle, *Chelone* [sic] *mydas*. *Zoologischer Anzeiger* 184, 290–294.

Simha, S.S. & Chattopadhyaya, D.R. (1980) Studies on the trematode parasites of reptiles of India. Contributions to the knowledge of blood flukes from the marine turtles, from the Gulf of Manar [*sic*], South India. *Journal of the Zoological Society of India* 30(1978), 69–82.

Sinitsin, D.F. (1910) Studien über die Phylogenie der Trematoden. II. *Bucephalus* v. Baer und *Cercaria ocellata* De la Vall. *Zeitschrift für Wissenschaftliche Zoologie, Abteilung A* 94, 299–325.

Sinitsin, D.F. (1931) Studien über die Phylogenie der Trematoden. V. Revision of Harmostominae in the light of new facts from their morphology and life history. *Zeitschrift für Parasitenkunde* 3, 786–835.

Skrjabin, K.I. (1913) *Tracheophilus sisowi* n. g., n. sp. Ein Beitrag zur Systematik der Gattung *Typhlocoelum* Stossich und der vermandten Formen. *Zentralblatt für Bakteriologie und Parasitenkunde, Abt. Originale* 69, 90–95.

Skrjabin, K.I. (1920) [Helminthological notices. Information on the helminth fauna of Russian birds.] *Izvestiya Donskogo Veterinarnogo Instituta, Nvocherkassk* 2, 1–7. (In Russian.)

Skrjabin, K.I. (1924) Nierntrematoden der Vögel Russlands. *Zentralblatt für Bakteriologie, Parasitenkunde, Infektionskrankheiten (und Hygiene), II Abteilung* 62, 80–90.

Skrjabin, K.I. (1947a) [Family Clinostomatidae Lühe, 1901.] In: Skrjabin, K.I. (ed.) [*Trematodes of animals and man*.] *Osnovy Trematodologii* 1, 64–97. (In Russian.)

Skrjabin, K.I. (1947b) [Family Eucotylidae Skrjabin, 1924.] In: Skrjabin, K.I. (ed.) [*Trematodes of animals and man.*] *Osnovy Trematodologii* 1, 106–132. (In Russian.)

Skrjabin, K.I. (1948) [Superfamily Fascioloidea Stiles et Goldberger, 1940]. In: Skrjabin, K.I. (ed.) [*Trematodes of animals and man.*] *Osnovy Trematodologii* 2, 5–336. (In Russian.)

Skrjabin, K.I. (1951) [Suborder Schistosomatata Skrjabin et Schulz.] In: Skrjabin, K.I. (ed.) [*Trematodes of animals and man.*] *Osnovy Trematodologii* 5, 624 pp. (In Russian.)

Skrjabin, K.I. (1960) [Family Cylindrorchidae Poche, 1925.] In: Skrjabin, K.I. (ed.) [*Trematodes of animals and man.*] *Osnovy Trematodologii* 18, 5–11. (In Russian.)

Skrjabin, K.I. (1962) [Family Liolopidae Dollfus, 1934.] In: Skrjabin, K.I. (ed.) [*Trematodes of animals and man.*] *Osnovy Trematodologii* 20, 5-26. (In Russian.)

Skrjabin, K.I. & Guschanskaja, L.K. (1954) [Suborder Hemiurata (Markevitsch, 1951) Skrjabin et Guschanskaja, 1954. First part.] In: Skrjabin, K.I. (ed.) [*Trematodes of animals and man.*] *Osnovy Trematodologii* 9, 225-653. (In Russian.)

Skrjabin, K.I. & Guschanskaja, L.K. (1955a) [Suborder Hemiurata (Markevitsch, 1951) Skrjabin et Guschanskaja, 1954. Second part.] In: Skrjabin, K.I. (ed.) [*Trematodes of animals and man.*] *Osnovy Trematodologii* 10, 337–643. (In Russian.)

Skrjabin, K.I. & Guschanskaja, L.K. (1955b) [Suborder Hemiurata (Markevitsch, 1951) Skrjabin et Guschanskaja, 1954. Third part.] In: Skrjabin, K.I. (ed.) [*Trematodes of animals and man.*] *Osnovy Trematodologii* 11, 465–748. (In Russian.)

Skrjabin, K.I. & Guschanskaja, L.K. (1956) [System of trematodes of the suborder Hemiurata (Markevitsch, 1951) Skrjabin et Guschanskaja, 1954.] *Trudy Gel'mintologicheskoi Laboratorii* 8, 144–158. (In Russian.)

Skrjabin, K.I. & Guschanskaja, L.K. (1957a) [Suborder Hemiurata (Markevitsch, 1951) Skrjabin et Guschanskaja, 1954. Fourth part.] In: Skrjabin, K.I. (ed.) [*Trematodes of animals and man.*] *Osnovy Trematodologii* 13, 599–722. (In Russian.)

Skrjabin, K.I. & Guschanskaja, L.K. (1957b) [Supplement to suborder Hemiurata (Markevitsch, 1951) Skrjabin et Guschanskaja, 1954.] In: Skrjabin, K.I. (ed.) [*Trematodes of animals and man.*] *Osnovy Trematodologii* 13, 723–779. (In Russian.)

Skrjabin, K.I. & Guschanskaja, L.K. (1958) [Suborder Azygiata La Rue, 1957.] In: Skrjabin, K.I. (ed.) [*Trematodes of animals and man.*] *Osnovy Trematodologii* 14, 667–819. (In Russian.)

Skrjabin, K.I. & Guschanskaja, L.K. (1959) [Suborder Hemiurata (Markevitsch, 1951) Skrjabin & Guschanskaja, 1954. Fifth part.] In: Skrjabin, K.I. (ed.) [*Trematodes of animals and man.*] *Osnovy Trematodologii* 16, 99–183. (In Russian.)

Skrjabin, K.I. & Guschanskaja, L.K. (1960) Classification and general morphological and biological characterization of trematodes of the suborder Hemiurata. In: Skrjabin, K.I. (ed.) [*Trematodes of animals and man.*] *Osnovy Trematodologii* 17, 533-640. (In Russian: English translation: *IPST, Jerusalem* 1964, 369–444.)

Skrjabin, K.I. & Guschanskaja, L.K. (1962) [Order Bucephalida (Odening, 1960) Skrjabin et Guschanskaja.] In: Skrjabin, K.I. (ed.) [*Trematodes of animals and man.*] *Osnovy Trematodologii* 20, 165-563. (In Russian.)

Smith, J.W. (1972) The blood flukes of cold-blooded vertebrates and some comparison with the schistosomes. *Helminthological Abstracts, Ser. A* 41, 161-204.

Smith, J.W. (1997a) The blood flukes (Digenea: Sanguinicolidae and Spirorchidae) of cold-blooded vertebrates: Part 1. A review of the literature published since 1971, and bibliography. *Helminthological Abstracts* 66, 255–294.

Smith, J.W. (1997b) The blood flukes (Digenea: Sanguinicolidae and Spirorchidae) of cold-blooded vertebrates: Part 2. Appendix I: Comprehensive parasite-host list; Appendix II: Comprehensive host-parasite list. *Helminthological Abstracts* 66, 329–344.

Snyder, S.D. & Loker, E.S. (2000) Evolutionary relationships among the Schistosomatidae (Platyhelminthes: Digenea) and an Asian origin for *Schistosoma*. *Journal of Parasitology* 86, 283-288.

Soparkar, M.B. (1924) A new cercaria from northern India. *Cercaria patialensis* nov. spec. *Indian Journal of Medical Research* 11, 933–942.

Southwell, T. (1913) Notes from the Bengal Fisheries Laboratory, Indian Museum. No. 1. *Records of the Indian Museum* 9, 79–103 + 10 pl.

Sprehn, C.E.W. (1932) *Lehrbuch der Helminthologie.* Gebrüder Bomtraeger, Berlin, 998 pp.

Srivastava, C.B. (1992) Comments on the proposed conservation in their accepted usage of the nominal taxa *Bucephalus* Baer, 1827 and *B. polymorphus* Baer, 1827 (Trematoda). *Bulletin of Zoological Nomenclature* 49, 62–64.

Srivastava, C.B. & Chauhan, B.S. (1973) A review of Indian gasterostomes (Trematoda). *Records of the Zoological Survey of India* 67, 1–13.

Srivastava, H.D. (1935) New hemiurids (Trematoda) from Indian fresh-water fishes. Part II. A rare parasite of the sub-family Dinurinae (Looss, 1907) from *Clupea ilisha. Zeitschrift für Parasitenkunde* 8, 135–138.

Srivastava, H.D. (1939) A new parasite – *Stomachicola secundus* – of the sub-family Dinuridae Looss, 1907. *Indian Journal of Veterinary Science and Animal Husbandry* 9, 77–79.

Srivastava, H.D. (1941) New hemiurids (Trematoda) from Indian marine food fishes. Part VIII – The morphology and systematic relationships of a new parasite – *Indoderogenes purii*, gen. et sp. nov. (sub-family Derogenetinae). *Indian Journal of Veterinary Science and Animal Husbandry* 11, 49–51.

Srivastava, H.D. (1948) A unique digenetic trematode referable to a new family. *Proceedings of the Indian Science Congress* 34, 174–175.

Srivastava, N.N. (1953) On the morphology and systematics of a new avian trematode *Thapariella anastomusa* n. g., n. sp., with a suggestion on the creation of a new family Thapariellidae. *Indian Journal of Helminthology* 5, 101–108.

Stiles, C.W. & Hassall, A. (1898) An inventory of the genera and subgenera of the trematode family Fasciolidae. (Notes on parasites, 48.) *Archives de Parasitologie* 1, 81–99.

Stiles, C.W. & Hassall, A. (1908) Index-catalogue of medical and veterinary zoology. Subjects: Trematoda and trematode diseases. *United States Public Health and Marine Hospital Service Bulletin* 37, 1–401.

Stossich, M. (1902) *Monostomum mutabile* Zeder e le sue forme a affini. *Zoologisches Zentralblatt* 9, 406–407.

Strand, E. (1935) Miscellanea nomenclatorica zoologica et palaeontologica. VIII. *Folia Zoologica et Hydrobiologica, Riga* 8, 176.

Stunkard, H.W. (1921) Notes on North American blood flukes. *American Museum Novitiates* 12, 1–5.

Stunkard, H.W. (1922) The anomalies in the group of blood-infesting trematodes [abstract]. *Anatomical Record* 23, 118–119.

Stunkard, H.W. (1929) The parasitic worms collected by the American Museum of Natural History expedition to the Belgian Congo. Part I. Trematoda. *Bulletin of the American Museum of Natural History* 58, 233–289.

Stunkard, H.W. (1934) The life history of *Typhlocoelum cymbium* (Dies., 1850) Ko., 1911. A contribution to the phylogeny of the Monostome. *Bulletin de la Société Zoologique de France* 59, 447–466.

Stunkard, H.W. (1946) Interrelationships and taxonomy of the digenetic trematodes. *Cambridge Biological Reviews* 21, 148–158.

Stunkard, H.W. (1973) Observations on *Tubulovesicula pinguis* (Linton, 1910) Manter, 1947, and on systematics of the hemiuroid trematodes. *Biological Bulletin of the Marine Biological Laboratory, Woods Hole, Massachusetts* 145, 607–626.

Stunkard, H.W. (1974) The trematode family Bucephalidae – problems of morphology, development, and systematics: description of *Rudolphinus* gen. nov. *Transactions of the New York Academy of Sciences, Series II* 36, 143–170.

Stunkard, H.W. (1975) The excretory system and systematics of the gasterostome trematodes. In: Tiwari, K.K. & Srivastava, C.B. (eds) *Dr B.S. Chauhan Commemoration Volume, 1975.* Zoological Society of India, Orissa, India, pp. 1–11.

Stunkard, H.W. (1976) The life cycles, intermediate hosts, and larval stages of *Rhipidocotyle transversale* Chandler, 1935 and *Rhipidocotyle lintoni* Hopkins, 1954: life cycles and systematics of bucephalid trematodes. *Biological Bulletin of the Marine Biological Laboratory, Woods Hole* 150, 294–317.

Stunkard, H.W. (1978) The life-cycle and taxonomic relations of *Lintonium vibex* (Linton, 1900) Stunkard and Nigrelli, 1930 (Trematoda: Fellodistomidae). *Biological Bulletin of the Marine Biological Laboratory, Woods Hole* 155, 383–394.

Stunkard, H.W. (1983) The marine cercariae of the Woods Hole, Massachusetts region, a review and a revision. *Biological Bulletin of the Marine Biological Laboratory, Woods Hole* 164, 143–162.

Stunkard, H.W. & Uzmann, J.R. (1958) Studies on the digenetic trematodes of the genera *Gymnophallus* and *Parvatrema. Biological Bulletin of the Marine Biological Laboratory, Woods Hole* 115, 276–302.

Sudarikov, V.E. (1959) [Order Strigeidida (La Rue, 1926) Sudarikov, 1959. Part 1. Morphological characteristics of strigeids and superfamily Strigeoidea Railliet, 1919.] In: Skrjabin, K.I. (ed.) [*Trematodes of animals and man.*] *Osnovy Trematodologii* 16, 217–631. (In Russian.)

Sudarikov, V.E. (1960a) [Order Strigeidida (La Rue, 1926) Sudarikov, 1959. Part 2. Superfamily Diplostomatoidea Nicoll, 1937, family Diplostomatidae (Poirier, 1886).] In: Skrjabin, K.I. (ed.) [*Trematodes of animals and man.*] *Osnovy Trematodologii* 17, 155–530. (In Russian.)

Sudarikov, V.E. (1960b) [Suborder Strigeata La Rue, 1926. Part 3. Superfamily Diplostomatoidea Nicoll, 1937. Families Alariidae Tubangui, 1922 and Bolbocephalodidae Strand, 1935. Superfamily Proterodiplostomatoidea Sudarikov, 1960.] In: Skrjabin, K.I. (ed.) [*Trematodes of animals and man.*] *Osnovy Trematodologii* 18, 451–694. (In Russian.)

Sudarikov, V.E. (1961) [Order Strigeidida (La Rue, 1926) Sudarikov, 1959. Part 4. Suborder Cyathocotylata Sudarikov, 1959.] In: Skrjabin, K.I. (ed.) [*Trematodes of animals and man.*] *Osnovy Trematodologii* 19, 267–469. (In Russian.)

Sudarikov, V.E. (1971) [Order Strigeidida (La Rue, 1926) Sudarikov, 1959. Part 5. Metacercariae and mesocercariae. A. Suborder Strigeata La Rue, 1926.] In: Skrjabin, K.I. (ed.) [*Trematodes of animals and man.*] *Osnovy Trematodologii* 24, 69–272. (In Russian.)

Sudarikov, V.E. (1974) The biology of some trematode species from the suborder Cyathocotylata. *Trudy Gel'mintologicheskoi Laboratorii* 24, 182–194. (In Russian.)

Sudarikov, V.E. (1984) [*Trematodes of the fauna of the USSR. Strigeidae.*] Nauka, Moscow, 168 pp. (In Russian.)

Sudarikov, V.E. (1997) On the revision of the trematode system of the order Strigeidida. *Trudy Instituta Parazitologii* 41, 157–167. (In Russian.)

Sudarikov, V.E. & Kurochkin, Y.V. (1968) [Taxonomic position and development of metacercariae of '*Neodiplostomulum scardinii* S. Shulman' – a parasite of the brain of fish.] *Trudy Astrakhanskogo Zapovednika* (*Sbornik Gel'mintologicheskikh Rabot*) 11, 255–273. (In Russian.)

Szidat, L. (1932) Parasiten aus Liberia und Franzosisch-Guinea, Teil II. Trematoden. *Zeitschrift für Parasitenkunde* 5, 506–521.

Szidat, L. (1936a) Parasiten aus Seeschwalben. I. *Zeitschrift für Parasitenkunde* 8, 285–316.

Szidat, L. (1936b) Studien zur Systematik und Enwicklungsgeschichte der Gattung *Leucochloridium* Carus. I. Bemerkungen zur Arbeit von G. Witenberg (1925): Versuch einer Monographie der Trematoden Unterfamilie Harmostominae Braun. *Zeitschrift für Parasitenkunde* 8, 645–653.

Szidat, L. (1962) Über eine ungewohnliche Form parthenogenetischer Vermehrung bei Metacercarien einer *Gymnophallus*-art aus *Mytilus platensis, Gymnophallus australis* n. sp. des Südatlantik. *Zeitschrift für Parasitenkunde* 22, 196–213.

Szidat, L. (1969) Structure, development and behaviour of new strigeatoid metacercariae from subtropical fishes of South America. *Journal of the Fisheries Research Board of Canada* 26, 753–786.

Tanabe, B. (1923) The life history of a new schistosome, *Schistosomatium pathlocopticum* Tanabe, found in experimentally infected mice. *Journal of Parasitology* 9, 183–198.

Tandon, R.S. (1949) A new trematode, *Lissemysia ovata* n. sp. of the family Aspidogastridae Poche, 1907 from freshwater molluscs. *Indian Journal of Helminthology* 1, 85–92.

Tandon, V. & Gupta, N.K. (1982) On some blood flukes (Spirorchiidae: Coeuritrematinae) from freshwater chelonians in India. *Proceedings of the Indian Academy of Science* 91, 275–282.

Tandon, V. & Gupta, N.K. (1985) On the blood fluke, *Cardiotrema roparensis* Mehrotra, 1973 and the validity of *C. longivesticulata* Dwivedi, 1967 (Spirorchiidae: Coeuritrematinae). *Research Bulletin (Science) of the Panjab University* 36, 331–334.

Tang C.-C. (1951) Contribution to the knowledge of the helminth fauna of Fukien. Part 3. Notes on *Genarchopsis chinensis* n. sp., its life history and morphology. *Peking Natural History Bulletin* 19, 217–223. (In Chinese.)

Tang C.-T. & Tang Z.-Z. (1963) Studies on nine species of gasterostomes from Fujian, with observations on the life-cycles of two species. *Acta Zoologica Sinica* 22, 263–278. (In Chinese.)

Tang C.-T., Shi Z.-M., Cao H., Guan J.-Z. & Pan C.-S. (1983) Studies on the trematodes of marine fishes from Fujian. I. Hemiurid trematodes. *Acta Zootaxonomica Sinica* 8, 33–42. (In Chinese.)

Tang Z.-Z. & Tang C.-T. (1989) A study of several cyathocotylid trematodes with description of a new genus and three new species (Strigeidida: Cyathocotylidae). *Acta Zootaxonomica Sinica* 14, 134–144. (In Chinese.)

Taskinin, J., Valtonen, E.T. & Gibson, D.I. (1991) Studies on bucephalid digeneans parasitising molluscs and fishes in Finland. I. Ecological data and experimental studies. *Systematic Parasitology* 19, 81–94.

Tehrany, M.R., Dronen, N.O. & Wardle, W.J. (1999) Revision of *Bursacetabulus* (Diplostomidae: Diplostominae) with the proposal of *Bursatintinnabulus* n. gen., and description of *Bursatintinnabus bassanus* n. sp. and *Bursacetabulus morus* n. sp. from northern gannet, *Morus bassanus* (Aves), from the Texas gulf coast. *Journal of Parasitology* 85, 531–533.

Teixeira de Freitas, J.F. & Kohn, A. (1967) Sobre un nuevo trematodo de peces en el litoral Brasileño. *Anales del Insituto de Biología* 37, 135–142.

Tennent, D.H. (1906) A study of the life history of *Bucephalus haimeanus*; a parasite of the oyster. *Quarterly Journal of Microscopical Science* 49, 635–690.

Thatcher, V.E. (1999) Two new Bucephalidae (Trematoda) from fishes of the Brazilian Amazon. *Acta Amazonica* 29, 331–335.

Thoney, D.A. & Burreson, E.M. (1988) Revision of the Multicalycidae (Aspidocotylea) with comments on postlarval development. *Proceedings of the Helminthological Society of Washington* 55, 62–67.

Thulin, J. (1973) *Protomonascus nybelini* gen. et sp. n. (Trematoda: Fellodistomidae) from *Elops senegalensis* Regan, 1909, on the west coast of Africa. *Journal of Parasitology* 59, 648–649.

Thulin, J. (1974) *Protomonascus nybelini* Thulin, 1973, a synonym for *Elopsium ghanense* Fischthal and Thomas, 1972. *Journal of Parasitology* 60, 433.

Thulin, J. (1980) A redescription of the fish blood-fluke *Aporocotyle simplex* Odhner, 1900 (Digenea: Sanguinicolidae) with comments on its biology. *Sarsia* 65, 35–48.

Timon-David, J. (1955) Cycle évolutif d'un trématode cyclocoelidé: *Pseudohyptiasmus dollfusi* Timon-David, 1950. Recherches expérimentales. *Annales de Parasitologie Humaine et Comparée* 30, 43–61.

Timon-David, J. (1957a) Nouvelles recherches expérimentales sur le cycle évolutif du trématode *Pseudohyptiasmus dollfusi* Timon-David (Digenea, Cyclocoelidae). *Comptes Rendus des Séances de la Société de Biologie, Paris* 151, 592–594.

Timon-David, J. (1957b) Recherches expérimentales sur le cycle evolutif d'un trematode du genre *Urotocus* Looss (Digenea, Leucochloridiidae). *Annales de Parasitologie Humaine et Comparée* 32, 219–242.

Timon-David, J. (1965) Développement expérimental, formes larvaires et cycle vital de *Dollfusinus frontalis* Biocca et Ferretti, 1958 (Trematoda, Digenea, Leucochloridiidae), parasite de sinus frontaux du hérisson. *Annales de Parasitologie Humaine et Comparée* 40, 265–284.

Toman, G. (1992) Digenetic trematodes of marine teleost fishes from the Seychelles, Indian Ocean. III. *Acta Parasitologica* 37, 119–126.

Travassos, L. (1920) Novo tipo de "Telorchinae". *Revista da Sociedade Brasileira de Ciências* [1919] (3), 183–187.

Travassos, L. (1921) Contribuição ao conhecimento dos Cyclocoelidae brasileiros. *Brasil Médico* 35, 121–123.

Travassos, L. (1922) Contribuiçoes para o conhecimento da fauna helmintologica brasileira. XIV. Especies brasileiras da familia Gorgoderidae Looss, 1901. *Brasil-Médico, Rio de Janeiro* 36, I(2), 17–20.

Travassos, L. (1928a) Sur la systémtique de la famille des Clinostomidae Lühe, 1901. *Comptes Rendus des Seánces de la Société de Biologie et de ses Filiales et Associés* 98, 643.

Travassos, L. (1928b) Fauna helminthologica de Matto Grosso (Trematodes, I Parte). *Memórias do Instituto Oswaldo Cruz* 21, 309–341.

Travassos, L. (1929) Notas sobre Cyclocoeliidae. *Memórias do Instituto Oswaldo Cruz* 6, 54.

Travassos, L. & Kohn, A. (1964) Notas sobre Brachylaemoidea (Trematoda). *Atas da Sociedade de Biologia do Rio de Janeiro* 8, 45–48.

Travassos, L. & Kohn, A. (1966) Lista dos gêneros incluídos na superfamília Brachylaemoidea. *Memórias do Instituto Oswaldo Cruz* 64, 11–25.

Travassos, L., Artigas, P. & Pereira, C. (1928) Fauna helmintológica dos peixes de água doce do Brasil. *Archivos do Intituto Biologico de Defesa Agricola e Animal* 1, 5–68.

Travassos, L., de Freitas, J.F.T. & Kohn, A. (1969) Trematódeos do Brasil. *Memórias do Instituto Oswaldo Cruz* 67, 1–886.

Tseng, S. (1930) Sur un gastérostomide immature chez Siniperca. *Annales de Parasitologie Humaine et Comparée* 8, 554–561.

Tubangui, M.A. & Masilungan, V.A. (1935) Trematode parasites of Philippine vertebrates. VII. Additional records on new species. *Philippine Journal of Science* 58, 435–445.

Tubangui, M.A. & Masilungan, V.A. (1936) Trematode parasites of Philippine vertebrates. VIII. Flukes from a cobra and a crocodile. *Philippine Journal of Science* 60, 255–266.

Ulmer, M.J. (1955) Notes on the morphology and host-parasite specificity of *Fibricola cratera* (Barker et Noll, 1915) Dubois, 1932 (Diplostomidae). *Journal of Parasitology* 41, 460–467.

Ulmer, M.J. & Vande Vusse, F.J. (1970) Morphology of *Dendritobilharzia pulverulenta* (Braun, 1901) Skryabin, 1924 (Trematoda: Schistosomatidae) with notes on secondary hermaphroditism in males. *Journal of Parasitology* 56, 67–74.

Vaidova, C.M. (1978) [*Helminths of the birds of Azerbaidzhan.*] Izdatel'stvo Elm, Baku, 235 pp. (In Russian.)

van Beneden, P.J. (1858) [also 1861] Mémoire sur les vers intestinaux. *Comptes Rendus Hebdomadaires des Seances de l'Académie des Sciences, Paris* (*Supplément*) 2, 115 pp.

van Beneden, P.J. (1870) Les poissons des côtes de Belgique, leurs parasites et leurs commensaux. *Memoires de l'Académie des Sciences de Belgique* 38(4 mém.), 100 pp.

van der Kuyp, E. (1953) *Euclinostomum heterostomum* in Celebes. *Documenta de Medicina Geographica et Tropica* 5, 320–332.

Van der Land, J. (1967) A new blood fluke (Trematoda) from *Chimaera monstrosa* L. *Proceedings of the Koninklijke Nederlandse Akademie van Wetenschappen, Section C, Biological and Medical Sciences* 70, 110–120.

Velasquez, C.C. (1958) *Transversotrema laruei*, a new trematode of Philippine fish (Digenea: Transversotrematidae). *Journal of Parasitology* 44, 449–451.

Vigueras, I.P. (1940) Prosogonotrematidae n. fam. y *Prosogonotrema bilabiatum* n. gen., n. sp. (Trematoda, Distoma) parasito de *Ocyurus chrysurus* (Bloch) (Pisces). *Memorias de la Sociedad Cubana de Historia Natural 'Felipe Poey'* 14, 249–252.

Vogel, H. & Minning, W. (1940) Bilharziose bei Elephanten. *Archiv für Schiffs- und Tropenhygiene* 44, 562–574.

von Graff, L. (1907) *Das Schmarotzertum im Tierreich und seine Bedeutung für die Artbildung.* Wissenschaft und Bildung, Einzeldarstellung aus allen Gebieten des Wissens, herausgegeben von Dr P. Herre. Verlag von Quelle and Meyer, Leipzig, 132 pp.

von Siebold, C.T. (1848) *Lehrbuch der vergleichenden Anatomie der wirbellosen Thiere.* Erster theil, *Wirbellose Thiere.* Verlag von Veit, Berlin, 679 pp.

von Siebold, C.T. (1852) Fernere Beobachtungen über das die Pfortader des Menschen behowhnende *Distomum haematobium* ... *Zeitschrift für Wissenschaftliche Zoologie* 4, 72–76.

Wall, L.D. (1940) Life history of *Spirorchis parvus* (Stunkard) Trematoda: Spirorchiidae. *Science* 92, 362–363.

Wang G. (1995) [Systematics of the family Bucephalidae (Digenea).] Thesis, Institute of Hydrobiology, Chinese Academy of Sciences, Wuhan, China, 152 pp. (In Chinese.)

Wang G. & Wang W. (1998a) [Taxonomy of the family Bucephalidae (Digenea).] *Acta Hydrobiologica Sinica* 22 (Suppl.), 89–99. (In Chinese.)

Wang G. & Wang W. (1998b) [Taxonomy and keys to bucephalid species in China, with descriptions of three new species.] *Acta Hydrobiologica Sinica* 22 (Suppl.), 100–110. (In Chinese.)

Wang G. & Wang W. (1998c) [*Rhoporhynchus lini* n.g. n.sp. (Bucephalidae: Bucephalinae) from the swim bladder of yellow catfish, *Pseudobagrus fulvidraco*.] *Chinese Journal of Oceanology and Limnology* 16, 373–376. (In Chinese.)

Wang P.-Q. (1981) Notes on one new family and three new species of trematodes from Vertebrata in Wuyi, Fujian Province. *Wuyi Science Journal* 1, 90–94. (In Chinese.)

Wang P.-Q. (1989) Digenetic trematodes of marine fishes in Pingtan County, Fujian Province, South China. *Wuyi Science Journal* 7, 151–163. (In Chinese.)

Wang P.-Q., Sun Y.-L., Zhao Y.-R., Zhang W.-H. & Wang Y.-L. (1985) Notes on some digenetic trematodes of vertebrates from Wuyishan, Fujian. *Wuyi Science Journal* 5, 129–139. (In Chinese.)

Wang W. & Pan J.-P. (1964) [The description of trematodes (Bucephalidae), including one new genus and five new species that parasitize several freshwater fishes.] In: [*Compilation of Abstracts of China Parasitological Society, November 1963.*] Science Press, Beijing, pp. 131–132. (In Chinese.)

Wedl, C. (1858) Anatomische Beobachtungen über Trematoden. *Sitzungsberichte der Königlich-Bayerischen Academie der Wissenschaften zu München* 26, 241–278.

Weinland, D.F. (1858) *Human Cestoides.* Metcalf and Company, Cambridge, Massachusetts, 93 pp.

WHO (1990) *Health Education and the Control of Schistosomiasis.* WHO, Geneva, 61 pp.

Williams, C.O. (1942) Observations on the life history and taxonomic relationship of the trematode *Aspidogaster conchicola*. *Journal of Parasitology* 28, 467–475.

Witenberg, G.G. (1923) [The trematodes of the family Cyclocoeliidae and a new principle of their systematics.] *Trudy Gosudarstvennogo Instituta Eksperimental'noi Veterinarii* 1, 84–141. (In Russian.)

Witenberg, G.G. (1925) Versuch einer Monographie der Trematoden - unterfamilie Harmostominae Braun. *Zoologische Jahrbücher. Abteilung für Systematic, Ökologie und Geographie der Tiere* 51, 167–254.

Witenberg, G.G. (1926) Die Trematoden der Familie Cyclocoelidae Kossack, 1911. *Beitrag zur Kenntnis der Helminthenfauna Russlands. Zoologische Jahrbücher. Abteilung für Systematik, Ökologie und Geographie der Tiere* 52, 103–186.

Witenberg, G.G. (1928) Notes on Cyclocoelidae. *Annals and Magazine of Natural History, 10th ser.* 2, 410–417.

Witenberg, G.G. (1944) *Transversotrema haasi*, a new fish trematode. *Journal of Parasitology* 30, 179–180.

Wolf, K. (1903) Beitrag zur Kenntnis der Gattung *Braunina* Heider. *Sitzungsberichte der Königlichen Akademie der Wissenschaften* 112, 603–626.

Woodhead, A.E. (1930) Life history studies on the trematode family Bucephalidae. No. II. *Transactions of the American Microscopical Society* 49, 1–17.

Woodhead, A.E. (1931) The germ cell cycle in the trematode family Bucephalidae. *Transactions of the American Microscopical Society* 50, 169–188.

Woolcock, V. (1935) Digenetic trematodes from some Australian fishes. *Parasitology* 27, 309–331.

Wootton, D.M. & Murrell, K.D. (1967) *Ptyalincola ondatrae* gen. et sp. n. (Brachylaimidae), a fluke inhabiting the salivary glands of muskrats. *Journal of Parasitology* 53, 739–742.

Wu J.-Y., Lu J.-Y. & Zhu T.-W. (1997) Notes on digenetic trematodes parasitic in fishes near shallow sea in Guangdong Province. V. Three new species (Digenea: Azygiidae). *Acta Zootaxonomica Sinica* 22, 231–239. (In Chinese, English summary.)

Yamaguti, S. (1933) Studies on the helminth fauna of Japan. Part 1. Trematodes of birds, reptiles and mammals. *Japanese Journal of Zoology* 5, 1–134.

Yamaguti, S. (1934) Studies on the helminth fauna of Japan. Part 2. Trematodes of fishes, I. *Japanese Journal of Zoology* 5, 249–541.

Yamaguti, S. (1938a) *Studies on the Helminth Fauna of Japan. Part 21. Trematodes of Fishes, IV.* Satyû Yamaguti, Kyoto, 139 pp.

Yamaguti, S. (1938b) Studies on the helminth fauna of Japan. Part 24. Trematodes of fishes, V. *Japanese Journal of Zoology* 8, 15–74.

Yamaguti, S. (1939) Studies on the helminth fauna of Japan. Part 26. Trematodes of fishes, VI. *Japanese Journal of Zoology* 8, 211–230.

Yamaguti, S. (1940) Studies on the helminth fauna of Japan. Part 31. Trematodes of fishes, VII. *Japanese Journal of Zoology* 9, 35–108.

Yamaguti, S. (1942) Studies on the helminth fauna of Japan. Part 39. Trematodes of fishes mainly from Naha. *Transactions of the Biogeographical Society of Japan* 3, 329–398.

Yamaguti, S. (1953) *Systema helminthum*. Part I. *Digenetic Trematodes of Fishes*. S. Yamaguti, Tokyo, 405 pp.

Yamaguti, S. (1954) Parasitic worms mainly from Celebes. 4. Trematodes of reptiles and birds. *Acta Medica Okayama* 8, 329–340.

Yamaguti, S. (1958a) *Systema Helminthum*. Vol. 1. *The Digenetic Trematodes of Vertebrates*. Interscience, New York, 1575 pp.

Yamaguti, S. (1958b) Studies on the helminth fauna of Japan. Part 52. Trematodes of fishes, XI. *Publications of Seto Marine Biological Laboratory* 6, 369–384.

Yamaguti, S. (1963). *Systema Helminthum*. IV. *Monogenea and Aspidocotylea*. Interscience, New York, London, pp. 313–325.

Yamaguti, S. (1965) New digenetic trematodes from Hawaiian fishes, I. *Pacific Science* 19, 458–481.

Yamaguti, S. (1970) *The Digenetic Trematodes of Hawaiian Fishes*. Keigaku Publishing Company, Tokyo, 436 pp.

Yamaguti, S. (1971) *Synopsis of the Digenetic Trematodes of Vertebrates*, Keigaku Publishing Company, Tokyo, Vol. 1, 1074 pp.; Vol. II., 349 pl.

Yamaguti, S. (1975) *A Synoptical Review of Life Histories of Digenetic Trematodes of Vertebrates.* Keigaku Publishing Company, Tokyo, 590 pp. + 219 pl.

Yeh Y. & Wu S.C. (1955) Note on a new trematode *Genarchopsis shanghaiensis* n. sp. (Trematoda: Hemiuridae) and its progenesis in freshwater shrimps around Shanghai region. *Acta Zoologica Sinica* 7, 37–42. (In Chinese.)

Yu B.C. (1965) [Two gastrostomid trematodes, *Sinicovothylax chujiangensis*, n. g. and n. sp., and *S. mapaensis*, n. sp., referable to a new family Sinicovothylacidae.] *Acta Zoologica Sinica* 2, 347-354. (In Chinese.)

Zazornova, O.P. & Sysoev, A.V. (1993) Phylogenetic relations between species of the genus *Cotylurus* and its position in the system of the trematode family Strigeidae. *Parazitologiya* 27, 69-76. (In Russian.)

Zeller, E. (1874) Über *Leucochloridium paradoxum* Carus und die weitere Entwicklung seiner Distomenbrut. *Zeitschrift für Wissenschaftliche Zoologie* 24, 564–578.

Zhang J., Qui Z., Ding X., Kuang F.R., Liu L., Xiao Z., Li Q.K., Zhao Z.F., Lee W., Pan J. & Yang T.B. (1999) [*Parasites and parasitic diseases of fishes.*] Science Press, Beijing, 735 pp. (In Chinese.)

Zhang T.-F. (1994) A new genus and species of the family Hemiuridae from fishes in Sichuan Province, China (Trematoda: Digenea). *Acta Zootaxonomica Sinica* 19, 20–24. (In Chinese.)

Index

Printed and bound by CPI Group (UK) Ltd, Croydon, CR0 4YY
16/09/2026
14963436-0006